Statistics

Mean

$$\bar{x} = \frac{1}{N} \sum_{i=1}^{N} x_i$$

where $\bar{x}$ = arithmetic mean, x_i = ith data observation, and N = the number of data.

Variance

$$\text{var}(x) = \frac{1}{N-1} \sum_{i=1}^{N} (x_i - \bar{x})^2$$

Standard Deviation is equal to the square root of variance.

Z-Scores for the Quantiles of the Standard Normal Distribution

1st quantile:	−1.28	6th quantile:	0.25
2nd quantile:	−0.84	7th quantile:	0.53
3rd quantile:	−0.53	8th quantile:	0.84
4th quantile:	−0.25	9th quantile:	1.38
5th quantile:	0.00	10th quantile:	3.50

Useful Conversion Factors

To convert feet to meters, multiply feet by 0.3048.
To convert inches to centimeters, multiply inches by 2.54.
To convert newtons to pounds, multiply newtons by 0.22481.
To convert pounds/square foot to kilograms/square meter, multiply pounds/square foot by 4.88241.
To convert radians to degrees, multiply radians by 57.29578.
1 gram is equal to the weight of 1 cubic centimeter of water.
1 kilogram is equal to 2.20462 pounds.
1 ppm = 1 part per million = 1 milligram per liter.
1 milligram = 0.001 gram.
1 ppb = 1 part per billion = 0.001 ppm.

Numerical Analysis
for the Geological Sciences

JAMES R. CARR

University of Nevada-Reno

Numerical Analysis for the Geological Sciences

PRENTICE HALL, Englewood Cliffs, New Jersey 07632

Library of Congress Cataloging-in-Publication Data

Carr, James R.
    Numerical analysis for the geological sciences / James R. Carr.
        p.   cm.
    Includes index.
    ISBN 0-02-319511-8
    1. Geology—Mathematics.   2. Mathematical analysis.   I. Title.
QE33.2.M3C37   1995
550'.1'5194—dc20   93-46168
                                                    CIP

Editor: Robert McConnin
Production Supervisor: Margaret Comaskey
Production Manager: Nick Sklitsis
Text and Cover Designer: Robert Freese
Illustrations: Rolin Graphics Inc.

 © 1995 by Prentice-Hall, Inc.
A Division of Simon & Schuster, Inc.
Englewood Cliffs, New Jersey 07632

Printed in the United States of America

10 9 8 7 6 5 4 3 2 1

0-02-319511-8

Prentice-Hall International (UK) Limited, *London*
Prentice-Hall of Australia Pty. Limited, *Sydney*
Prentice-Hall Canada Inc., *Toronto*
Prentice-Hall Hispanoamericana, S.A., *Mexico*
Prentice-Hall of India Private Limited, *New Delhi*
Prentice-Hall of Japan, Inc., *Tokyo*
Simon & Schuster Asia Pte. Ltd., *Singapore*
Editora Prentice-Hall do Brasil, Ltda., *Rio de Janeiro*

For Janice, Anna, and Russell
with all my love

Preface

This book is an experiment to assemble between two covers many of the computer applications useful for earth sciences study. In addition, background material is presented to enable students of varying backgrounds to use this book. As an instructor and advisor, I often hear students complain that they don't understand calculus, even though they may have had two or three semesters of instruction in this advanced mathematics subject. It is my hope that Chapter 1 will be of some help to those students, and that the remainder of the text is clear in showing how calculus is applied for earth sciences study. For numerical analyses implemented on digital computers, the mathematical techniques developed by Carl Gauss are vital; hence, a chapter is devoted to him. Matrix algebra is covered in one chapter because students majoring in the earth sciences often do not receive instruction in this area of mathematics. Finally, an overview of statistics is presented as a review for students who have had a course or two in statistics, but more importantly for students who have not had such a course. Chapters following these four introductory chapters present unique applications of computers to earth sciences study.

Many chapters are accompanied by computer software for implementing the methods discussed. Diskettes are supplied with the Instructor's Manual; they may be copied and contain source and executable code for all programs, as well as data sets discussed in the text. All of the computer programs presented herein are written in ANSI Standard FORTRAN-77. Would the language C have been better to use? The answer to this question depends on whom you talk to. Computer scientists will argue affirmatively. I, on the other hand, feel that no language has an advantage over another for numerical calculations, except FORTRAN (FORmula TRANslation), which was developed specifically for this purpose. Further, because I am schooled in FORTRAN, this language is the one I am most comfortable using. Will C replace FORTRAN? I doubt it. In 1980, for instance, computer scientists were touting Pascal as the replacement for FORTRAN; yet Pascal is not now favored as highly by computer scientists as is C, whereas FORTRAN has endured. Before Pascal, computer scientists were advocating PL1, a language virtually unheard of today. Furthermore, a new version of FORTRAN, FORTRAN-90, is about to be released (circa 1994). I believe that FORTRAN will be a standard computer language, especially for science and engineering applications, for some time to come.

Computer graphics are discussed herein, especially as a means for visualizing results from numerical analysis. Labeling of axes, contour lines, tick marks, and so on, is accomplished in a rather elementary manner. Sophisticated labeling of plots requires design of fonts (lettering and numbering style). Such design is well beyond the scope of this book. All graphics programs are herein designed to create a drawing based on results from numerical analysis to visually demonstrate and stimulate the use of each of these methods. Enough axis, tick mark, and legend labels are given to allow interpretation of the graph. Arguably, these graphics are not, in some cases, of journal quality. Commercial graphics drawing packages do a much better job. However, the intent of this book is to demonstrate methods and visualize results, with programs designed to demonstrate the basic elements of each method.

Mathematical geology is an increasingly important subdiscipline of the geological sciences. It saddens me when I hear a student remark that he or she chose geology as a major partly or wholly because of a dislike for mathematics. Nothing disturbs me more than hearing a geological engineering or geophysics major make this remark. I implore all who read this preface to also read the Introduction, paying particular attention to the quotation attributed to Lord Kelvin. May the experiment that is this book be considered a success and may the objectives of mathematical geology be furthered because of it. May failure be a reflection on me.

Acknowledgments

I wish to acknowledge the tremendous help reviewers provided in shaping this text. First, John Mann, University of Illinois-Urbana, and former editor of the journal *Mathematical Geology,* reviewed both the first and second drafts of this text (the present manuscript represents the third draft). It was he who, along with another reviewer, suggested adding finite difference analysis to the text. Moreover, Dr. Mann's knowledge of correct use of grammar is tremendous and his contributions to the correct writing of this text are significant.

No less significant were the reviews rendered by Donald E. Myers, Department of Mathematics, University of Arizona, and Charles (Karl) Glass, Department of Mining and Geological Engineering, University of Arizona. Drs. Myers and Glass are the two most important mentors shaping my academic career. Dr. Myers' review substantially improved Chapters 1 through 4. Dr. Glass' review entirely changes, and substantially improved, the chapter on image processing (Chapter 10).

I thank John Davis, Kansas Geological Survey, Lawrence, Kansas, for his review of the first draft; in particular, his comments improved Chapters 1, 5, and 6. Chapter 11, fractals, was reviewed by Dietmar Saupe, University of Bremen, Germany. Finally, I am indebted to Jeffrey Cawlfield, Department of Geological Engineering, University of Missouri, Rolla. His review resulted in the writing of the contouring program, CTOUR (Chapter 7), and the finite difference program, FDIFF (Chapter 8). His enthusiasm about the entire book was an inspiration. I also wish to thank Mary Dale-Bannister, Washington University, St. Louis, Missouri, for helping me acquire the Magellan and Geologic Remote Sensing Field Experiment CD-ROMs.

In ten years as an academician, I have had the pleasure of meeting and instructing approximately 1500 students. All have had a positive impact on me. All must be recognized, therefore, as contributing to this text.

Finally, many authors acknowledge their spouses in a manner as if to escape divorce. My labor on this text did not interfere with my family life; nor did I take sabbatical leave from the University of Nevada to write this text. Instead, I had so much fun writing and then shaping this book that my stress level actually was reduced. This made me a better husband and father. Furthermore, my wife, Janice, wholeheartedly and steadfastly supported this project. This book is dedicated to the most important people in my life: my wife, Janice, my daughter, Anna, and my son, Russell.

Contents

Introduction   xix
    About This Book   xx
    General Features of the Text   xxi
    A Note for Instructors   xxiii
    A Note on the Software   xxiii
    RIDGES: A Problem Solving Strategy   xxiii
    Chapter Summary   xxiv
    Suggested Readings   xxv

1   A Review of Calculus   1

    1.1   History   1
        1.1.1 The Derivative   3
        1.1.2 The Fundamental Theorem of Integral Calculus   6
    1.2   Applications of Calculus   9
    1.3   An Ironic Aspect of the Twentieth Century   14
    Chapter Summary   15
    Exercises   15
    References and Suggested Readings   16

2   Carl Friedrich Gauss   17

    2.1   A Mathematical Prodigy   17
    2.2   Theory of Errors   18
        2.2.1 Application of Error Theory: Gaussian Quadrature   19
        2.2.2 The General Form for Gaussian Quadrature for $m = 1$ and
            $m = 2$ 24
        2.2.3 A Table of Gaussian Quadrature Locations t and Weights a   26
        2.2.4 Application of Error Theory: Linear Least-Squares Regression   27
    2.3   Gaussian Elimination: Solving Simultaneous Equations   31
        2.3.1 Advanced Topic: Iterative Improvement for Gaussian Elimination   34
    Chapter Summary   35
    Exercises   36
    References and Suggested Readings   37

3   Concepts in Matrix Algebra   44

3.1   The Concept of a Matrix   44
3.2   Linear, Least-Squares Regression Revisited   47
3.3   Eigendecomposition of a Matrix   50
  3.3.1 Roots of an Algebraic Expression   51
  3.3.2 Eigenvectors   52
  3.3.3 Comments   55
  3.3.4 Iterative Solution for Eigendecomposition   56
  3.3.5 Computer Example   56
Chapter Summary   57
Exercises   57
References and Suggested Readings   59

4   An Overview of Probability and Statistics   64

4.1   The Need for Statistics   64
4.2   Statistical Parameters   65
4.3   Probability   67
  4.3.1 Conditional Probability   67
  4.3.2 Bayes' Theorem   68
  4.3.3 Continuous Random Variables   68
  4.3.4 Probability Density Functions for Continuous Random Variables   69
  4.3.5 Expected (Mean) Value of a Continuous Random Variable   69
  4.3.6 The Normal or Gaussian Distribution   70
  4.3.7 Poisson Probability Distribution   71
  4.3.8 Graphical Display of a Sample: The Histogram   71
  4.3.9 Examining Models for Cumulative Distribution Functions   72
4.4   Asymmetrical Data Distributions: Skew   74
4.5   Nonparametric Statistics   76
4.6   Statistical Correlation: Intervariable Relationships   77
4.7   Statistical Correlation: Intravariable Relationships   77
4.8   Additional Numerical Examples   78
  4.8.1 Arithmetic Mean and Median   78
  4.8.2 Variance   78
  4.8.3 Correlation Coefficient   79
  4.8.4 Autocorrelation   79
  4.8.5 Histogram   81
  4.8.6 Bayes' Theorem   83
4.9   Computer Exercise   84
  4.9.1 Histograms   84
Chapter Summary   85
Exercises   85
References and Suggested Readings   87

5   Multivariate Data Analysis   91

5.1   Introduction   92
5.2   An Overview of Multivariate Analytical Methods   92

       5.2.1 Principal Components Analysis   96

       5.2.2 Representing the PCA Results Graphically   98

       5.2.3 Factor Analysis   99

  5.3   Correspondence Analysis   102

       5.3.1 Eckart–Young Theorem   106

       5.3.2 Solutions for Factors in Correspondence Analysis   109

       5.3.3 Example Calculations   109

  5.4   Advanced Topic: Accounting for Missing Data in Correspondence Analysis   111

  5.5   CORSPOND: A Portable FORTRAN-77 Program for Correspondence Analysis   114

  5.6   Applications of CORSPOND   115

       5.6.1 Analysis of Data from Greenacre (1984)   115

       5.6.2 The Stillwater Lakes, Nevada, Environmental Problem Revisited   115

       5.6.3 Advanced Application: Accommodating Missing Data in CORSPOND   125

  5.7   Discriminant Analysis   125

       5.7.1 Application   128

  Chapter Summary   130

  Exercises   131

  References and Suggested Readings   137

6 Geostatistics: The Art of Spatial Analysis   150

  6.1   A Brief History   151

  6.2   The Unique Terminology of Geostatistics   152

  6.3   Kriging: Gauss' Theory of Errors Revisited   152

       6.3.1 Deriving the Solution for Kriging Weights   154

  6.4   Determining Spatial Autocovariance   160

       6.4.1 Obtaining Autocovariance from the Semivariogram   161

  6.5   Calculating the Semivariogram   162

  6.6   FGAM: A Portable FORTRAN-77 Program for Semivariogram Calculation and Plotting   167

       6.6.1 Examples   168

  6.7   Models of Semivariograms Used for Computing Kriging Weights   180

       6.7.1 Equations for the Semivariogram Models   181

       6.7.2 The Negative Semidefiniteness of Semivariogram Models   183

       6.7.3 Nested Semivariogram Models   183

  6.8   Example Calculation to Demonstrate the Use of Kriging   184

       6.8.1 The Exact Interpolation Characteristics of Kriging   187

       6.8.2 A Note on Kriging Variance   187

  6.9   Sample Support   188

       6.9.1 Block Kriging Versus Punctual Kriging   188

  6.10 JCBLOK: A Portable FORTRAN-77 Program for Punctual and Block Kriging   191

       6.10.1 Using JCBLOK with the Lognormal and Indicator Transforms   192

  6.11 Case Studies   193

       6.11.1 An Application of JCBLOK to the NVSIM.DAT Data Set for Gridding   193

   6.11.2 *An Application of JCBLOK to the Block Estimation of Lognormal*
         Data   194
   6.11.3 *Estimation of Indicator Functions Using JCBLOK*   201
Chapter Summary   204
Exercises   205
References and Suggested Readings   206

7   Computer Graphics: Visualizing Quantitative Information   228

   7.1   Computer Graphics and Hardware   229
         7.1.1 *VGA Computer Graphics Environment*   229
   7.2   Viewing and Windowing   232
   7.3   Vector Generation   235
         7.3.1 *Computer-Aided Drawing of Two-Dimensional Graphs*   237
         7.3.2 *GRAPH2D: An Interactive Program for Plotting Two-Dimensional*
               Graphical Images   239
   7.4   Two-Dimensional Coordinate Rotations   245
   7.5   Three-Dimensional Graphical Images   247
         7.5.1 *The Forgetting Transform*   247
         7.5.2. *Three-Dimensional Coordinate Rotations*   248
         7.5.3 *Vanishing Points*   250
         7.5.4 *Plotting Three-Dimensional Grids: Hidden Line Removal and Shading*
               Algorithms   250
   7.6   Mapping Using Color Density Slicing   255
         7.6.1 *Color Density Mapping Using the Program COLORDEN*   256
   7.7   Contour Mapping   258
         7.7.1 *Algorithm*   258
         7.7.2 *CTOUR: A Computer Algorithm for Contour Mapping*   259
Chapter Summary   260
Exercises   262
References and Suggested Readings   263

8   Finite Element Analysis   294

   8.1   Hooke's Law   295
   8.2   Strain and Stress   295
   8.3   Poisson's Ratio   297
   8.4   Truss Systems: Finite Element Analysis in Its Most Simple Form   298
         8.4.1 *Element Stiffness Matrix: Bar Elements*   300
         8.4.2 *Computing the System Stiffness Matrix*   303
   8.5   A Rayleigh–Ritz Derivation of the Element Stiffness Matrix for Bar
         Elements   306
         8.5.1 *An Overview*   306
         8.5.2 *Use of Assumed Displacement Fields in the Rayleigh–Ritz*
               Method   308
   8.6   TRUSS: A FORTRAN-77 Program for Two-Dimensional Finite Element
         Analysis of Truss Systems   311
         8.6.1 *Error Trapping*   313

8.6.2.   *Finite Element Analysis of a Truss System   315*

8.6.3   *A Simple Test Problem   318*

8.7   Two-Dimensional, Solid, Isoparametric, Four-Node Elements: An Advanced Application of Gaussian Quadrature for Deriving the Element Stiffness Matrix   319

8.7.1   *Introduction to Isoparametric, Solid, Four-Node Elements   321*

8.7.2   *Rayleigh–Ritz Derivation of Element Stiffness   321*

8.7.3   *The Material Constitutive Matrix [E].   328*

8.7.4   *An Example of Element Stiffness Matrix Calculation   329*

8.7.5   *Computing Strain and Stress for Quadrilateral Elements   334*

8.8   Nonlinear, Elastic Finite Element Analysis: An Application for the Analysis of Fractured Rock Systems   335

8.8.1   *A Summary of the Continuum Analysis of Nonlinear, Elastic Behavior of Fractured Rock Masses   336*

8.9   QUAD: A FORTRAN-77 Program for Linear, Elastic and Nonlinear, Elastic Finite Element Analysis Using Four-Node, Isoparametric Elements   338

8.10   Applications of QUAD for Finite Element Analysis Using the Two-Dimensional, Solid, Four-Node, Isoparametric Element   339

8.10.1 *Analysis of Displacement Beneath a Spread Footing on Soil   339*

8.10.2 *Analysis of a Rectangular Opening   343*

8.10.3 *Analysis of Stress Around an Elliptical Tunnel   346*

8.10.4 *Analysis of Vertical Fractures for the Tunnel of Example 8.10.3   349*

8.11   A Note on Boundary Conditions   352

8.12   Finite Difference Analysis for Assessing Regional Groundwater Flow   354

8.12.1 *Darcy's Law   354*

8.12.2 *Laplace's Equation   355*

8.12.3 *Finite Difference Approximation of Laplace's Equation for Steady-State Flow   355*

8.12.4 *Example Application   355*

8.13   Finite Element Method in Comparison to the Finite Difference Method   360

Chapter Summary   360

Exercises   361

References and Suggested Readings   363

9   Fourier Analysis   394

9.1   Sinusoidal Waveforms   395

9.1.1 *Cosine Function   395*

9.2   The Periodogram   395

9.3   Fourier Transform   404

9.4   Justifying the Use of the Fourier Transform   407

9.5   The Power Spectrum: Spectral Analysis   408

9.6   Examples of Fourier Analysis   408

9.6.1 *Sunspot Cycles   408*

9.6.2 *Analysis of Seismic Data   411*

9.7   Fast Fourier Transform (FFT)   414

9.8   A Final Example   415

Chapter Summary   416

Exercises   417

References and Suggested Readings   418

10   Numerical Processing of Digital Images   427

10.1   Digital Imagery   427
10.2   Remote Sensing   429
    10.2.1 The Electromagnetic (EM) Spectrum   429
    10.2.2 Satellite Technology   431
    10.2.3 Spatial, Spectral, Temporal, and Radiometric Resolution   433
10.3   Image Processing   434
    10.3.1 The Display of Digital Images Using VGA Hardware   435
    10.3.2 Contrast Enhancement of Digital Images   437
    10.3.3 Using Look-up Tables for Improved Computation Efficiency   445
    10.3.4 CTRAST: A Computer Algorithm for Contrast Enhancement   447
    10.3.5 Filtering Digital Images   448
    10.3.6 Fourier (Frequency) Filtering of Digital Images   453
    10.3.7 Classification of Digital Images   457
10.4   Magellan Radar Images of Venus   461
10.5   Examples of Airborne-Sensor-Acquired Digital Imagery   463
Chapter Summary   464
Exercises   465
References and Suggested Readings   467
Selected Articles Describing Magellan Images   468

11   Fractals   489

11.1   Nineteenth-Century Discoveries   490
    11.1.1 Poincaré   490
    11.1.2 Fricke and Klein   491
    11.1.3 Weierstrass and duBois Reymond   491
    11.1.4 Other Nineteenth-Century Discoveries   491
11.2   Fractals in the Twentieth Century   492
    11.2.1 Fatou and Julia   492
    11.2.2 Richardson   492
    11.2.3 The Sketches of M.C. Escher: A Notion of Infinite Smallness   495
    11.2.4 Hausdorff and Besicovitch   495
    11.2.5 Mandelbrot   496
11.3   Fractals: A Summary of Definitions   498
11.4   A Spectral Notion for Fractal Dimension   500
    11.4.1 The Random Fractal   500
    11.4.2 Background   501
11.5   Use of the Semivariogram for Fractal Studies   509
11.6   Estimating the Fractal Dimension in Practice   511
11.7   Simulation of Fractal Topography and Clouds   514
    11.7.1 Clouds   516
    11.7.2 Why Are These Simulations Useful?   516
11.8   Selected Applications of the Fractal Dimension   518
    11.8.1 Analysis of Rock Surfaces   518
    11.8.2 The Shape and Texture of Concrete Aggregate   519
    11.8.3 Chaos? Characteristics of Rainfall for a Storm   521

11.9 Visualizing Periodicities and Fractal Behavior in One-Dimensional Time
Series: The Wavelet Transform   523
  11.9.1 Specific Information for the Program WAVELET   *527*
Chapter Summary   527
Exercises   528
References and Suggested Readings   529

Appendix A   539
Appendix B   582
Index   585

Introduction

Geology is often stereotyped as a qualitative science. For this reason, many students who choose this science as a major do not anticipate encountering rigorous numerical analysis. Soon after embarking on the journey toward a college degree in geology, however, a student learns that the science is undergoing a rapid metamorphism. The qualitative aspects, such as interpreting rocks, structures, and landscapes, remain dominant, but quantitative aspects play an increasingly important role as geologists and geological engineers struggle to make sense out of vast quantities of numerical data. The science of geology is adapting to the information age, as is all of modern society.

Geology is a relatively young science, and often is seen as a second cousin to the more traditional sciences of physics, chemistry, and biology. Ironically, the study of earth science, in its broadest context, has led to (1) the proposal of a new type of geometry, fractal geometry, proposed by Benoit (pronounced Ben-wah) Mandelbrot, based in part on his studies of coastlines; (2) an elegant extension of Gauss' error theory to spatial analysis via geostatistics, developed by Georges Matheron; and (3) innovative numerical methods for processing satellite-acquired, remotely sensed data to enhance interpretation of earth from space. Fractal geometry alone is sparking a revolution of understanding within the traditional sciences, such as physics. The geologic science is indeed, an inspirational medium for creative thought.

Mathematics is an integral part of geology, as a few examples illustrate: (1) Fourier analysis of seismic and time series data; (2) analysis of stability of soil and rock slopes; (3) statistical and geostatistical analysis of seismic, sedimentologic, paleontologic, geomorphologic, and petrologic data; (4) allied to statistical analysis, probabilistic analysis of earthquakes, volcanic eruptions, floods, and landslides; this aspect of applied mathematics is changing the manner in which engineering design of buildings, excavations, and slopes is approached; and (5) finite element, finite difference, and related techniques to analyze stress and strain in rocks and soils, as well as flow of groundwater. Two journals, *Mathematical Geology* and *Computers and Geosciences,* are devoted to mathematical applications in the earth sciences.

At the risk of belaboring the point, a remarkable quotation from Lord Kelvin (from *Statistics and Data Analysis in Geology* by John C. Davis) is insightful:

> [W]hen you can measure what you are speaking about and express it in numbers, you
> know something about it; but when you cannot express it in numbers, your knowledge is

of a meagre and unsatisfactory kind; it may be the beginning of knowledge, but you have scarcely in your thoughts advanced to the state of science, whatever the matter may be.

This is the banner to wave in the face of the old science, one that has begrudgingly accepted plate tectonics, wherein numbers are thought to have no value for geologic interpretation, and wherein the earth is thought to have an infinite supply of nonrenewable resources, so that exploring neighboring planets and solar systems for such resources will never be necessary. The geological sciences must, and will, change.

About This Book

This book has an intended audience of geologists, geological (and geotechnical) engineers, environmental engineers, civil engineers, mining engineers, and petroleum engineers. Other groups that may find the book useful include climatologists, biologists, and applied mathematicians. What this wide list of disciplines indicates is that numerical analyses described within the covers of this text are broadly useful.

In order, this book covers the following topics:

- calculus
- Carl Friedrich Gauss: error theory and equation solution
- matrix operations
- an overview of statistics
- multivariate data analysis using correspondence analysis
- geostatistics
- graphics: two- and three-dimensional visualization
- finite element analysis (and finite difference analysis)
- Fourier analysis
- digital image processing
- fractal geometry

Each topic is treated at a degree of complexity commensurate with upper level undergraduate courses. Advanced aspects of these topics are presented in many chapters, allowing this book to be used for the education of postbaccalaureate students. Undergraduate students may skip these advanced topics without a loss of continuity in instruction.

As a minimum, students using this book are assumed to have a background (prerequisite knowledge) in integral and differential calculus, physics, statistics, and matrix algebra; introductory chapters are included to review much of this prerequisite material. Chapter 3 should provide sufficient material to allow those students who have not had explicit instruction in matrix algebra to successfully use the rest of the text.

Courses in which this text should be useful include the following (titles offer course suggestions; specific titles at your college or university may vary):

- Geological Engineering (or Geological) Data Analysis
- Numerical Analysis for the Geological Sciences
- Computer Applications in Geological Engineering (or Geology)
- Computer Applications in Civil Engineering

- Computer Applications in Mining Engineering
- Introduction to Geostatistics
- Introduction to Multivariate Statistical Analysis
- Introduction to Remote Sensing
- Computer Processing of Digital Images
- Introduction to Finite Element Analysis
- Introduction to Groundwater Modeling (or Groundwater Hydrology)
- Introduction to Fourier Analysis

In addition, this text may be a useful supplement to courses on geological engineering design, civil engineering design, rock mechanics, computer graphics, fractals, environmental analysis, petroleum reservoir modeling, economic geology, seismology, exploration geophysics, and botanical studies (especially for studying the spatial distribution and density of plant species). In other words, this text describes numerical methods often applied in these courses; hence this text would be a useful supplementary text.

General Features of the Text

Selection and arrangement of the material in this text is based on teaching and research experience. Major features of this book are:

1. *Computer Software.* Many chapters are associated with software useful for implementing algorithms. The software is written in FORTRAN-77. The programs are as follows (listed in order of their appearance in the text):

   LINEAR (Fig. 2.4)—linear, least squares regression
   GAUSS (Fig. 2.7)—Gauss elimination equation solution
   GAUSIT (Fig. 2.10)—adds iterative improvement to Gauss elimination equation solution
   MULMAT (Fig. 3.1)—matrix multiplication: $[A][B] = [C]$
   EIGEND (Fig. 3.6)—eigenvalue/eigenvector computation
   HISTO (Fig. 4.8)—histogram computation and plotting
   CORSPOND (Fig. 5.9)—correspondence analysis
   DISCRIM (Fig. 5.26)—discriminant analysis
   FGAM (Fig. 6.6)—semivariogram computation
   JCBLOK (Fig. 6.23)—punctual and block ordinary kriging, with options for log-normal and indicator transforms
   PALDSGN (Fig. 7.1)—16-color VGA graphics palette design
   BOXDRAW (Fig. 7.5)—demonstrates windowing, viewing, and vector drawing
   GRAPH2D (Fig. 7.7)—two-dimensional graph drawing
   GRID3D (Fig. 7.19)—displays three-dimensional grids with and without shaded relief
   COLORDEN (Fig. 7.21)—displays gridded data using color density slicing
   CTOUR (Fig. 7.26)—computes and displays contour maps
   TRUSS (Fig. 8.14)—finite element analysis of truss systems
   MESHDRAW (Fig. 8.15)—draws finite element meshes before and after analysis; stress level is shown by color code

QUAD (Fig. 8.27)—linear and nonlinear elastic analysis of two-dimensional, solid, isoparametric finite elements

FDIFF (Fig. 8.51)—finite difference analysis of two-dimensional groundwater flow

CSSINE (Fig. 9.2)—creates one-dimensional cosine waveforms

PERIOD (Fig. 9.5)—computes periodograms of one-dimensional time series

FORIER (Fig. 9.11)—computes one-dimensional Fourier transform and associated power spectrum

FORINV (Fig. 9.14)—computes the one-dimensional inverse Fourier transform

FILTER (Fig. 9.15)—filters specified frequencies from a one-dimensional Fourier transform prior to inverse transformation

ADD (Fig. 9.26)—combines two or more cosine waveforms created by the program CSSINE

IMAGIN (Fig. 10.7)—displays digital images using a 16-color palette

IMSTAT (Fig. 10.14)—efficiently computes histograms of digital images

CTRAST (Fig. 10.28)—adjusts the contrast of digital images

CONVOL (Fig. 10.30)—spatial convolution filtering of digital images

IMAG41 (Fig. 10.32)—a utility program to create an 8-bit integer digital image from a 32-bit integer digital image

FORIER2D (Fig. 10.37)—two-dimensional Fourier transform

READRI (Fig. 10.38)—reads output from FORIER2D, then creates a 32-bit integer digital image; IMAG41 is used to convert this image to an 8-bit integer image compatible with IMAGIN

FORFILT (Fig. 10.39)—filters specified frequencies from a two-dimensional Fourier transform

FORINV2 (Fig. 10.40)—two-dimensional inverse Fourier transform

INTFAC (Fig. 10.41)—converts two-dimensional inverse Fourier transform to a digital image

MXLIKE (Fig. 10.44)—maximum likelihood and minimum distance to mean supervised classification of digital images

SPFM1D (Fig. 11.6)—creates a simulated, one-dimensional fractional Brownian motion profile having a fractal dimension $2 - H$

CFRINV (Fig. 11.7)—is used to inverse Fourier transform the spectral simulation produced by SPFM1D

HSCALE (Fig. 11.11)—rescales SPFM1D/CFRINV simulations by r^H

SPFM2D (Fig. 11.15)—creates a simulated, two-dimensional fractional Brownian motion surface having a dimension $3 - H$

INTFC4 (Fig. 11.20)—converts SPFM2D/FORINV2 (Chapter 10) simulations into digital images

WAVELET (Fig. 11.28)—produces a digital image of the convolution of a one-dimensional time series with wavelets of varying size

2. *Literature Citations.* This text includes extensive references to the literature, but presentation of each topic is detailed to avoid requiring students to search for journal articles and other books (many of which are difficult to understand) to obtain an adequate understanding of concepts.
3. *Examples.* Numerous examples are presented in each chapter to illustrate concepts.

4. *Notation.* All notation is modern and consistent with the literature for a particular topic.

A Note for Instructors

If students using this textbook have a background in calculus, matrix operations, and statistics, Chapters 1 through 4 can be skipped; instruction may begin with Chapter 5 or 6, depending on the course. Otherwise, Chapters 1 through 3 (and possibly 4) should be covered in order. Thereafter, chapters may be covered in any order, except that Fourier analysis should be covered prior to fractal geometry. The chapter on graphics is independent of the other chapters, except that examples are used from chapters on multivariate analysis and geostatistics; chapters following the one on graphics often refer to this chapter.

A Note on the Software

Software presented in this text is intended as an aid to understanding concepts, to illustrate how to implement numerical methods on the computer, to help students gain confidence on the computer, and to encourage literacy in a computer language or languages. This software is not intended for commercial applications, especially those on which human safety depends. These programs should not be used in applications for which the user has a formal responsibility.

The author has made considerable effort to ensure that the software presented in this text is error free. Every program has been tested using 80386 and 80486 microprocessors. Errors can enter the programs accidentally between the time of their writing and the printing of the final book. Therefore, the author and the publisher assume no responsibility for errors or omissions. No liability is assumed for damage resulting from the use of information and programs in this book. Nor is it implied or stated that the programs meet any standard of reliability, structure, maintainability, or correctness. The user of the information and programs in this book assumes full risk.

RIDGES: A Problem-Solving Strategy

Solving problems, either by hand or through the design of a computer algorithm, is difficult, if not impossible, unless the paths followed to obtain solutions proceed logically. When faced with the challenge to solve a problem, anxiety can develop, usually because of an insecure feeling about the ability to solve the problem. Such anxiety can be so overwhelming as to inhibit logical thought. How can this tendency for anxiety be overcome?

A recent concept in education is providing a new, refreshing approach to solving (math) problems. This concept has the acronym RIDGES. Although developed for kindergarten through twelfth grade levels, RIDGES is universally applicable—to life in general. It has relevancy, therefore, for college-level instruction and for this text.

RIDGES is described as follows. Given a problem that requires solution— a word problem, for example—the logical sequence to follow in obtaining a solution is:

R—*Read the problem*   Often, a tendency to push the panic button prevents a calm reading of the problem; this is an extremely important first step.

I—*Identify the known quantities*   After reading the problem, identify what is known about the problem; this information is useful for eventually solving the problem.

D—*Diagram*   Draw a diagram of the problem. This is an important step. Simply sketch a picture of the problem. The human brain works much better with a picture, which is emphasized in Chapter 7.

G—*Goal statement*   State the goal or purpose of solving the problem. This, too, is a vital step, because it focuses attention again on important aspects of the problem, requiring the student to think about how the problem needs to be solved.

E—*Estimate the answer*   This is perhaps the most important step. Prior to explicitly solving the problem, develop an educated guess for the solution. This tests a student's fundamental knowledge of the problem. Moreover, such an estimate is useful for assessing the final solution to the problem for accuracy.

S—*Solve the problem*   Prepare necessary equations for solving the problem, obtain an answer, and evaluate the answer against the estimate to assess the adequacy of both solution and understanding.

RIDGES offers a marvelous way to focus attention on a problem as an aid to developing a procedure or procedures for solving the problem. Throughout this text, please use the RIDGES algorithm when solving end of chapter exercises. Further information on RIDGES can be found in Snyder (1988).

Chapter Summary

- Geology is undergoing a metamorphism from a qualitative to a more quantitative science.
- Fractal geometry and geostatistics are mathematical concepts contributed by earth sciences to all other sciences.
- Application of mathematics to earth sciences is extensive, perhaps much more so than many people realize.
- Space technology allows remote scanning of the earth's surface and mathematical methods extract useful information from these scans.
- RIDGES is an excellent tool for aiding the development of procedures for solving math problems:

  READ the problem
  IDENTIFY known quantities
  DIAGRAM the problem
  GOAL statement
  ESTIMATE the result
  SOLVE the problem

- Lord Kelvin was a very wise man.

Suggested Readings

DAVIS, J.C. 1986. *Statistics and Data Analysis in Geology.* New York: Wiley.

MANDELBROT, B.B. 1967. How long is the coast of Great Britain? Statistical self-similarity and fractional dimension. *Science* **155:** 636–638.

MANDELBROT, B.B. 1983. *The Fractal Geometry of Nature.* San Francisco: Freeman.

MATHERON, G. 1963. Principles of geostatistics. *Economic Geology* **58:** 1246–1266.

PETERSON, I. 1988. *The Mathematical Tourist.* New York: Freeman.

SNYDER, K. 1988. RIDGES: a problem-solving math strategy. *Academic Therapy* **23**(3): 261–263.

1

A Review of Calculus

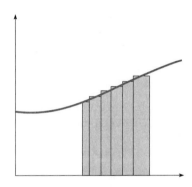

A brief review of calculus is presented to motivate computer applications throughout this text. Some students may feel a need for such a review, especially students of the earth sciences who are often not shown the relevancy of calculus to the discipline. Perhaps this chapter will motivate a student to read again a text on calculus. In particular, this chapter leads to a provocative discussion of an ironic aspect of the twentieth century: the place of calculus in the age of modern computing. Calculus is used as the theoretical premise of many modern numerical methods. But approximate solutions for these methods are implemented in computer programs because digital computers cannot perform calculus directly. Hence, the theme of this text is established.

1.1 History

A simple, continuous function, $y = f(x)$, illustrates a perplexing problem (Figure 1.1). The region beneath $y = f(x)$ and above $y = 0$, and between $x = a$ and $x = b$, is shaded. Suppose the area of the shaded region is to be calculated. How is this done?

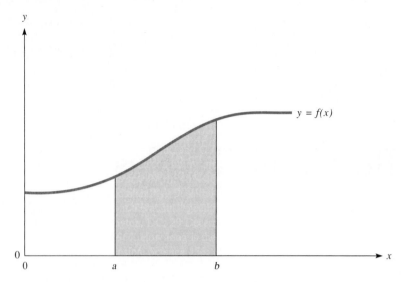

Figure 1.1   Area (hatched) beneath a hypothetical curve $y = f(x)$, bounded by $x = a$, $x = b$, $y = 0$, and $y = f(x)$.

One approach to a solution is to approximate the area as a collection of rectangles (Figure 1.2). The height of each rectangle i is equal to $f(x_i)$, where x_i is the position with respect to the x axis of the midpoint of each rectangle. Clearly, the more rectangles used to approximate the area (that is, the more narrow each rectangle is) the more accurate will be the estimate of total area (Figure 1.3).

Restating this approach to a solution, let each rectangle have the same width, called dx. The height of each rectangle i is equal to $f(x_i)$. From intuitive

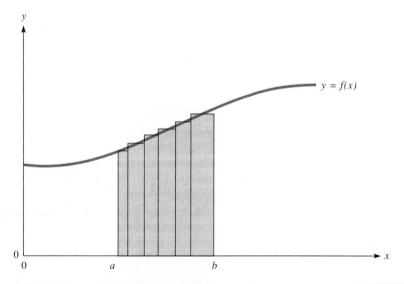

Figure 1.2   An approach to calculating the area shown in Figure 1.1 by subdividing the area into rectangles, calculating the areas of these rectangles, and summing these areas to determine the global area.

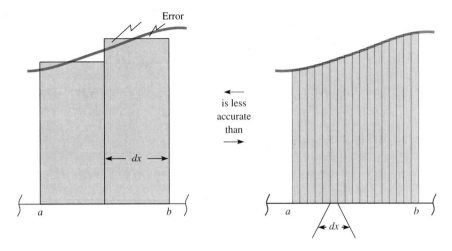

Figure 1.3 A supplement to Figure 1.2. The more rectangles that are used to determine the area of Figure 1.1, the more accurate the determination of area will be.

reasoning, as expressed in the writings of mathematicians of ancient Greece, such as Archimedes, Pythagoras, and Euclid, we know that the area of a rectangle is equal to its width times its height. Thus, approximation of the area of the shaded region (Figure 1.1) is the sum of all rectangular areas:

$$\text{area} \approx \sum_{i=1}^{N} f(x_i)\, dx$$

where N is the number of rectangles used for the approximation.

If dx, the width of each rectangle, is infinitely small (which implies that N is infinitely large), this approximation of area is exact. If this is not intuitively obvious, notice that if dx is very small (Figure 1.3), the error caused by representing the curve $f(x)$ with the edges of the rectangles is reduced because the portion of each rectangle above $f(x)$, as well as that below $f(x)$, is minimized.

The fact that dx is infinitely small is expressed mathematically as a limit:

$$\text{area} = \lim_{dx \to 0; N \to \infty} \sum_{i=1}^{N} f(x_i)\, dx$$

which is the exact area of the shaded region by definition (Figure 1.1).

In 1666, Sir Isaac Newton discussed the calculation of exact areas using an antiderivative (Salas and Hille, 1974):

$$dA = y\, dx$$

Understanding the concept of the antiderivative requires prior understanding of the concept of the derivative.

1.1.1 The Derivative

The derivative was known prior to Newton's writings regarding calculus. The French mathematician Pierre de Fermat noted in the early part of the seventeenth century, for a function $f(A)$ continuous on an interval,

$$a \leq A \leq b$$

that

$$f'(A) = \lim_{E \to 0} \frac{f(A + E) - f(A)}{E}$$

where $f'(A)$ is the derivative of $f(A)$.

A derivative f' of a function f at a point x is a measure of rate of change, or slope, of function f at point x. In Fermat's definition of the derivative, for instance, the numerator, $f(A + E) - f(A)$, is a measure of change of function f over an interval of x between $x = A$ and $x = A + E$. The rate of this change, or slope, is defined as the change in the value of f between $x = A$ and $x = A + E$ divided by the change in x, E. This is, of course, Fermat's definition of the derivative and is illustrated in the following example.

Example 1.1

Let $f(A) = A^2 + 3A - 4$; then
$$f(A + E) = (A + E)^2 + 3(A + E) - 4 = A^2 + 2AE + E^2 + 3A + 3E - 4$$
Moreover, $f(A + E) - f(A)$ is equal to
$$A^2 + 2AE + E^2 + 3A + 3E - 4 - A^2 - 3A + 4$$
which simplifies to $2AE + E^2 + 3E$.

Therefore,

$$\frac{f(A + E) - f(A)}{E} = \frac{2AE + E^2 + 3E}{E} = 2A + E + 3 \qquad E \neq 0$$

Writing the limit as E goes to zero,

$$\lim_{E \to 0} 2A + E + 3 = 2A + 3 = f'(A)$$

yields the derivative.   ∎

1.1.1.1 Finite Difference Approximation of Derivatives

Computing languages such as FORTRAN cannot be used to determine the derivative of a function (that is, given $f(x)$, FORTRAN cannot be used to obtain the functional form of $f'(x)$). Computing languages such as FORTRAN can be used, however, to implement algorithms for approximating the value of a derivative at a point x. One of these algorithms is finite difference approximation (also known as finite divided difference approximation; Chapra and Canale, 1985).

Finite difference approximation is closely related to Fermat's notion of the derivative. For instance, given a function

$$f(x) = 3.6 + 1.9x - 2.7x^4$$

having the derivative

$$f'(x) = 1.9 - 10.8x^3$$

the formal solution for $f'(2.5)$, for example, is -166.85. A derivative is the rate of change of a function f at a point x; in this particular example, the rate of change of f at the point 2.5 is -166.85. Finite difference approximation is based on the notion of rate of change, as seen in the following:

$$f(x) \approx \frac{f(x_{i+d}) - f(x_i)}{d}$$

where the numerator is the rate of change of the function f over a numerical interval d centered about x, the point at which the value of the derivative is desired.

This particular form of the finite difference approximation is called a forward difference, because x_{i+d} is greater than x_i. A backward difference may be computed as an alternative:

$$f'(x) \approx \frac{f(x_i) - f(x_{i-d})}{d}$$

Furthermore, a centered difference may be computed as

$$f'(x) \approx \frac{f(x_{i+d}) - f(x_{i-d})}{2d}$$

where the centered difference is the most accurate of the three approximation approaches (Chapra and Canale, 1985).

We can use finite differences to approximate the solution to the foregoing derivative. Let $d = 0.1$; therefore,

$$x_i = 2.5 \qquad f(2.5) = -97.12$$
$$x_{i+d} = 2.6 \qquad f(2.6) = -114.84$$
$$x_{i-d} = 2.4 \qquad f(2.4) = -81.42$$

The *forward difference approximation* is

$$\frac{-114.84 - (-97.12)}{0.1} = -177.24$$

which has an error of $-166.85 - (-177.24) = 10.39$. The *backward difference approximation* is

$$\frac{-97.12 - (-81.42)}{0.1} = -157.00$$

which has an error of $-166.85 - (-157.00) = -9.85$. The *centered difference approximation* is

$$\frac{-114.84 - (-81.42)}{0.20} = -167.10$$

which has an error of $-166.85 - (-167.10) = 0.25$. The centered difference approximation is the most accurate. The smaller the value of d, the more accurate is the approximation. In the foregoing example, for instance, if d is 0.01 rather than 0.1, the centered difference approximation is equal to -167.00, giving an error of 0.15.

1.1.1.2 Partial Derivatives

Another function involving the concept of the derivative is the partial derivative. Partial derivatives are relevant to computer applications described throughout this text. Given a function of several independent variables, for instance, $f(x, y, z)$, a partial derivative of f is a derivative with respect to only one of these independent variables. Suppose that

$$f(x, y, z) = 15.3 + 10.2x^3 - 5.7y^2 + 45.6z$$

Then the partial derivative of f with respect to x is equal to $30.6x^2$. All terms not associated with x have derivatives equal to zero. On the other hand, the derivative of f with respect to z is 45.6. The term partial refers to the fact that the derivative pertains to only part of the function, any one independent variable.

1.1.2 The Fundamental Theorem of Integral Calculus

We now return to our discussion of area (e.g., Figure 1.1). The sum used to approximate area leads to

$$\text{area} = \int_a^b f(x)\, dx$$

The integration symbol, $\int$, was introduced by Gottfried Wilhelm Leibniz in 1675, and first appeared in one of his publications in 1686. This was not a minor contribution to mathematics. Jacques Solomon Hadamard presents the following debate (Hadamard, 1945, pp. 84–85):

> G Polya's case—I intend to speak only of men who have made quite significant discoveries—is different. He does make an eventual use of words. "I believe," he writes to me, "that the decisive idea which brings the solution of a problem is rather often connected with a well-turned word or sentence. The word or sentence enlightens the situation, gives things, as you say, a physiognomy. It can precede by little the decisive idea or follow on it immediately; perhaps, it arises at the same time as the decisive idea. . . . The right word, the subtly appropriate word, helps us to recall the mathematical idea, perhaps less completely and less objectively than a diagram or a mathematical notation, but in an analogous way. . . . It may contribute to fix it in the mind." Moreover, he finds that a proper notation—that is, a properly chosen letter to denote a mathematical quantity—can give him similar help; and some kind of puns, whether of good or poor quality, may be useful for that purpose. For instance, Polya, teaching in German at a Swiss university, usually made his junior students observe that z and w are the initials of the German words, "Zahl" and "Wert", which precisely denote the respective roles which z and w had to play in the theory which he was explaining.

In other words, a sign or a word, if chosen carefully, can be a significant contribution to mathematics by collecting together many seemingly unrelated ideas under one heading or class of notation. For instance, Fermat introduced the derivative, Newton and Leibniz developed the concept of calculus, and Leibniz further contributed the notation of the integral sign. Later in this book, considerable detail is presented on kriging, contributed by Matheron, and fractals, contributed by Mandelbrot. All are important examples of Hadamard's notion.

In 1677, confronted with the problem of calculating area (cf., Figure 1.1) using rectangles, Leibniz started from

$$\text{area} = \lim_{dx \to 0} \sum_{i=1}^{N} f(x_i)\, dx$$

and noted that

$$\text{area} = \int_a^b f(x)\, dx$$

Letting $F(x)$ be the antiderivative of $f(x)$ gives

$$F(x) = \int f(x)\, dx$$

Leibniz introduced the *Fundamental Theorem of Integral Calculus* for solving this integral. The theorem is as follows: Let f be continuous on $[a, b]$ and F

be the antiderivative of f. Then

$$\int_a^b f(x)\, dx = F(b) - F(a)$$

A continuous function is one such as $y = f(x)$, x in the interval $[a, b]$ (Figure 1.1). A function $f(x)$ is continuous at a point a if three conditions are met: (1) $f(a)$ exists and is finite; (2) $\lim_{x \to a} f(x)$ exists and is finite; and (3) the two values in (1) and (2) are equal. In contrast, a discontinuous function has the characteristics shown in Figure 1.4. This function is not continuous at $x = a$, $x = b$, and $x = c$, because a jump discontinuity exists in $f(x)$ at these points. Therefore, $\lim_{x \to a} f(x)$, $\lim_{x \to b} f(x)$, and $\lim_{x \to c} f(x)$ do not exist.

Derivatives (and antiderivatives) exist only for many continuous functions (Chapter 11 describes one continuous function for which no derivative exists); derivatives and antiderivatives cannot be computed for discontinuous functions at the point(s) of discontinuity.

Before proceeding to applications of calculus, we review the fundamental theorem of integral calculus in greater detail. If $F(x)$ is the antiderivative of $f(x)$ at x, then the derivative of $F(x)$ must be $f(x)$:

$$\lim_{E \to 0} \frac{F(x + E) - F(x)}{E} = f(x)$$

a notion used in the following demonstration of the Fundamental Theorem of Integral Calculus. This demonstration (from Salas and Hille, 1974, pp. 188–192) proceeds as follows. Given $a < c < b$, then

$$F(x) = \int_a^b f(x)\, dx = \int_a^c f(x)\, dx + \int_c^b f(x)\, dx$$

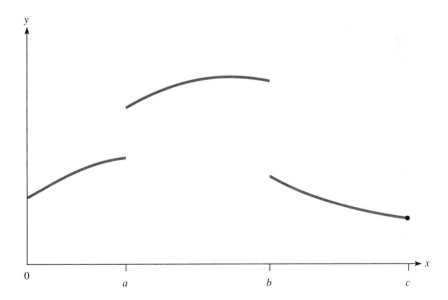

Figure 1.4 An example of a discontinuous function. The function is continuous over the intervals $x = 0$ to $x = a$, $x = a$ to $x = b$, and $x = b$ to $x = c$. But the function is discontinuous at $x = a$ and $x = b$ (at these precise points, no one unique derivative can be found).

Further, if x is in the interval $[a, b]$, then

$$F(x) = \int_a^x f(w)\, dw$$

If $x < x + E < b$, then

$$F(x + E) - F(x) = \int_a^{x+E} f(w)\, dw - \int_a^x f(w)\, dw$$

which simplifies to

$$F(x + E) - F(x) = \int_x^{x+E} f(w)\, dw$$

by acknowledging from the beginning of this demonstration that if $a < x < x + E$, then

$$\int_a^{x+E} f(w)\, dw = \int_a^x f(w)\, dw + \int_x^{x+E} f(w)\, dw$$

Although a nontrivial assumption, let K represent the maximum value of the function f anywhere in the interval $[x, x + E]$. Further, let k represent the minimum value of f in this interval (these assumptions regarding k and K require that $f(x)$ be continuous in the interval $[x, x + E]$). Then, $K[(x + E) - x] = KE$; moreover, $k[(x + E) - x] = kE$, and

$$kE \le \int_x^{x+E} f(w)\, dw \le KE$$

Therefore,

$$k \le \frac{F(x + E) - F(x)}{E} \le K$$

But

$$\lim_{E \to 0} k = f(x) = \lim_{E \to 0} K$$

Thus,

$$\lim_{E \to 0} \frac{F(x + E) - F(x)}{E} = f(x)$$

This shows that $f(x)$ is the derivative of $F(x)$, or that $F(x)$ is the antiderivative of $f(x)$. Given the demonstration to this point, recall the Fundamental Theorem of Integral Calculus:

$$\int_a^b f(x)\, dx = F(b) - F(a)$$

Suppose

$$\int_a^b f(x)\, dx = J(b) - J(a)$$

which supposes J is an antiderivative of f. Then F and J differ only by a constant:

$$J(x) = F(x) + C$$

Because $J(a) = 0$, $F(a) + C$ must also be 0. Therefore,

$$C = -F(a)$$

which means that

$$J(x) = F(x) - F(a)$$

for any x in the interval $[a, b]$, which further means that

$$J(b) = F(b) - F(a)$$

We acknowledge that

$$J(b) = \int_a^b f(w)\, dw$$

Since

$$J(x) = \int_a^x f(w)\, dw$$

for any x in the interval $[a, b]$, because

$$J(b) = F(b) - F(a)$$

Thus,

$$\int_a^b f(w)\, dw = \int_a^b f(x)\, dx = F(b) - F(a)$$

which is the Fundamental Theorem of Integral Calculus.

1.2 Applications of Calculus

Example 1.2   Volume of a Cone

The volume of a cone equals one-third the product of the height of the cone and the area of its circular base. The Greek mathematician Democritus was the first to discover this fact. A demonstration is derived using calculus.

The x and y axes define the cross section of a cone and the z axis parallels the axis of the cone (Figure 1.5). The cross section of the cone at any position z has an area calculated using the formula for the area of a circle. The radius of this circle at any

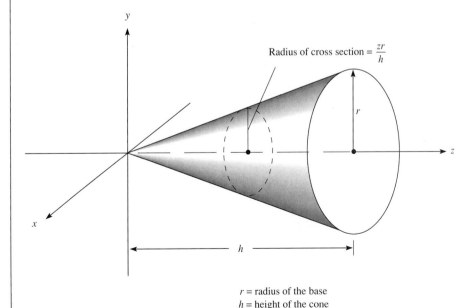

$$\text{Radius of cross section} = \frac{zr}{h}$$

r = radius of the base
h = height of the cone

Figure 1.5   A sketch of a cone in three dimensions, x, y, and z.

point z is equal to zr/h, where r is the radius of the base of the cone and h is the height of the cone. Therefore, the area of the circular cross section of the cone at a point z is equal to

$$\text{cross-sectional area} = \pi \left(\frac{zr}{h}\right)^2$$

The volume of the cone is found by integrating the cross-sectional area over z, as a definite integral from $z = 0$ to $z = h$:

$$\text{volume of cone} = \int_0^h \frac{\pi z^2 r^2}{h^2}\,dz$$

$$= \frac{\pi r^2}{h^2}\int_0^h z^2\,dz$$

$$= \frac{\pi r^2}{h^2}\left(\frac{z^3}{3}\right)\Big|_0^h$$

The volume of the cone is equal to $F(A)$; moreover, $F(A) = F(b) - F(a)$:

$$\text{volume of cone} = \frac{\pi r^2}{3h^2}(h^3) - \frac{\pi r^2}{3h^2}(0^3) = \frac{\pi r^2 h}{3}$$

This is, of course, what Democritus derived.   ∎

Example 1.3   Demonstrating That $z^3/3$ Is the Antiderivative of z^2

In computing the volume of the cone using calculus, the antiderivative of z^2, $z^3/3$, is presented. Fermat's notion of the derivative is used to show that this is the correct antiderivative:

$$f'(z) = \lim_{E \to 0} \frac{f(z + E) - f(z)}{E}$$

In showing that $z^3/3$ is the antiderivative of z^2, the notion that $f(z) = z^3/3$ is used and the derivative of $f(z)$ is determined using Fermat's rule. If the derivative of $z^3/3$ is found to be z^2, then $z^3/3$ is, by definition, the antiderivative of z^2. Letting $f(z) = z^3/3$ yields

$$f(z + E) = \frac{(z + E)^3}{3} = \frac{z^3 + 3z^2 E + 3zE^2 + E^3}{3}$$

Therefore, $f(z + E) - f(z)$ is

$$\frac{z^3 + 3z^2 E + 3zE^2 + E^3}{3} - \frac{z^3}{3} = \frac{3z^2 E + 3zE^2 + E^3}{3}$$

from which we get

$$\frac{f(z + E) - f(z)}{E} = \frac{\tfrac{1}{3}(3z^2 E + 3zE^2 + E^3)}{E} = \tfrac{1}{3}(3z^2 + 3zE + E^2) \qquad E \neq 0$$

The limit of this expression, as E goes to zero, is

$$3\frac{z^2}{3} = z^2$$

thus demonstrating that $z^3/3$ is the antiderivative of z^2.   ∎

Example 1.4   Area of a Circle

In Example 1.2, the formula for the area of a circle is used:

$$\text{area} = \pi r^2$$

Is it possible to derive this relationship using calculus?

SOLUTION

To answer this, a circle (Figure 1.6) is diagrammed that has a center at $x = 0$ and $y = 0$. In Example 1.1, single integration was used to determine the volume of the cone. To derive the formula for the area of a circle, double integration will be used, which has the general form

$$\text{area} = \iint_G 1 \, dA$$

for a region G (in this case, the circle) with an area A.

To determine the area of the circle, only the shaded upper right quadrant of the circle (Figure 1.6) is considered. Focusing only on this quadrant, notice that the range of x is

$$0 \leq x \leq r$$

where r is the radius of the circle. Moreover, y is dependent on x and is written as

$$0 \leq y \leq \sqrt{r^2 - x^2}$$

From the Pythagorean theorem (Figure 1.7), $x^2 + y^2 = r^2$; in fact, this is the equation of the circle. From this relationship, the range of integration for y is obtained. Thus,

$$\text{area} = \int_0^r \int_0^{\sqrt{r^2-x^2}} (1) \, dy \, dx$$

$$= \int_0^r (y)\Big|_0^{\sqrt{r^2-x^2}} \, dx = \int_0^r \sqrt{r^2 - x^2} \, dx$$

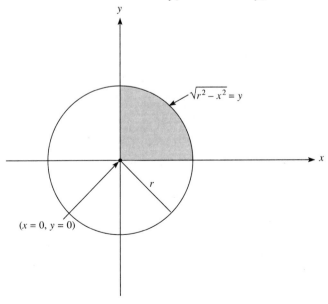

Figure 1.6   A sketch of a circle that is defined by the curve $y = \text{SQRT}(r^2 - x^2)$.

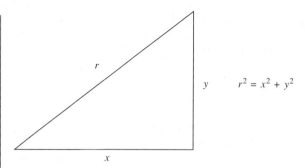

Figure 1.7   The Pythagorean theorem: the distance r is equal to SQRT($x^2 + y^2$).

The last integral is solved as

$$\int_0^r \sqrt{r^2 - x^2}\, dx = \frac{\pi r^2}{4}$$

Because this is the area of only one-quarter of the circle, the total area of the circle is

$$\text{total area} = 4\left(\frac{\pi r^2}{4}\right) = \pi r^2$$

which seems to show that calculus can be used to derive the formula for circular area. Yet, this example has failed to show this. The solution

$$\int_0^r \sqrt{r^2 - x^2}\, dx = \frac{\pi r^2}{4}$$

is obtained from the knowledge that the area of the circle should be equal to pi times the square of the radius. This example shows that the notion of the area of a circle leads to the conclusion that

$$\int_0^r \sqrt{r^2 - x^2}\, dx = \frac{\pi r^2}{4}$$

Alternatively, Gauss quadrature (Chapter 2) may be used as an approximate way to arrive at this integral; or using polar coordinates leads to this solution.   ■

Example 1.5   Volume of a Cone Revisited

Single and double integration have been reviewed and applied. Triple integration is now reviewed for volumetric calculations, using knowledge obtained from the previous two examples. In a generic sense, volume can be determined using triple integration of the form

$$\text{vol} = \iiint_H 1\, dV$$

over a space H that has a volume V. This integration may proceed in any order, provided the limits of integration reflect this order.

In the determination of the volume of the cone, integration is performed in the following order: $dy\, dx\, dz$. Example 1.4 demonstrated how to use double integration to represent area, in that case, circular area. Example 1.3 showed that at any point z along the axis of the cone, the value of the radius of the cross section of the cone is zr/h, where r is the radius of the base. And, as is apparent (Figure 1.5), the range of z

is from 0 to h. Compiling these facts, we find

$$\text{volume of cone} = \int_0^h \int_0^{zr/h} \int_0^{\sqrt{(zr/h)^2 - x^2}} dy\, dx\, dz$$

$$= \int_0^h \int_0^{zr/h} \sqrt{\left(\frac{zr}{h}\right)^2 - x^2}\, dx\, dz$$

$$= \int_0^h \frac{\pi z^2 r^2}{4h^2}\, dz$$

This problem is completed by

$$\text{volume of cone} = \frac{\pi r^2}{4h^2} \int_0^h z^2\, dz$$

$$= \frac{\pi r^2}{4h^2} \left(\frac{z^3}{3}\right)_0^h = \frac{\pi r^2 h}{12}$$

As in Example 1.4, the limits of integration for x and y represent only one-quarter of the cross-sectional area of the cone. Therefore, the total volume of the cone is equal to

$$\text{volume of cone} = 4\left(\frac{\pi r^2 h}{12}\right) = \frac{\pi r^2 h}{3}$$

which coincides precisely with what was determined in Example 1.3. ∎

Example 1.6   Generic Rectangular Area and Volume

A final example is presented in preparation for some of the later chapters. A generic rectangular area (Figure 1.8) and a generic rectangular volume (Figure 1.9) are calculated.

AREA
The generic double-integration formula for area is

$$\text{area} = \iint_G 1\, dA$$

where, as in Example 1.4, G is a region, in this case rectangular, having an area A.

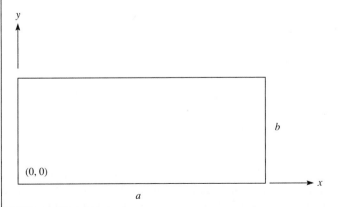

Figure 1.8   A rectangle having dimensions $a \times b$.

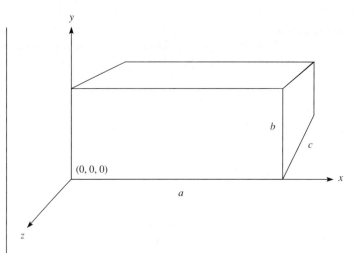

Figure 1.9   A rectangular volume having dimensions $a \times b \times c$.

The rectangle (Figure 1.8) is associated with x from 0 to a, and y from 0 to b. Therefore, the area of the rectangular region is

$$\text{area} = \int_0^a \int_0^b 1 \, dy \, dx = \int_0^a b \, dx = ab$$

That is, the area of a rectangle is equal to its width times its length.

VOLUME

The generic triple-integration formula for volume is

$$\text{volume} = \iiint_H 1 \, dV$$

over a space H having a volume V. The rectangular volume (Figure 1.9) is associated with x from 0 to a, y from 0 to b, and z from 0 to c. Therefore,

$$\text{volume} = \int_0^c \int_0^a \int_0^b 1 \, dy \, dx \, dz$$

$$= \int_0^c \int_0^a b \, dx \, dz = \int_0^c ab \, dz = abc$$

That is, the rectangular volume is equal to the product (width)(length)(height).   ∎

1.3   An Ironic Aspect of the Twentieth Century

This book describes the application of computers to enhance the study of earth sciences. A digital computer cannot perform calculus, except by executing sophisticated programs written specifically to perform symbolic operations; one such program is MAPLE (Char et al., 1991a,b). When using scientific programming languages (such as FORTRAN, Pascal, C, or BASIC), calculus, as described in this chapter, cannot be performed. Instead, integrals are approximated as summations and derivatives are approximated as finite differences.

This is truly ironic. The computer has revolutionized society: Typewriters have been replaced by word processors; personal computers have brought almost supercomputer power into homes and offices. Yet, as useful as computers are, many languages used to implement algorithms on them cannot directly perform calculus. What this has amounted to is that mathematical operations idealized as integrals are approximated as weighted summations; those operations idealized as derivatives are approximated as differences.

In this book, calculus often is used to set up problems, but in writing computer programs to solve these problems, algorithms are implemented that approximate values of derivatives and definite integrals.

Chapter Summary

- Calculus forms the basis for many numerical analysis methods described in this text. Computer algorithms based on these methods, however, use techniques for approximating derivatives and integrals.

Exercises

(An asterisk indicates an advanced or graduate student problem)

1. Plot each of the following functions between $x = 1$ and $x = 10$, and compute the derivative of each. Also find the antiderivative of each of these functions. Show by taking its derivative that each antiderivative is correct.

   (a) $f(x) = 4x - 7$ (b) $f(x) = 9x + 5$
   (c) $f(x) = x^2$ (d) $f(x) = 9x^2 - 7x + 3$
   (e) $f(x) = x^7 + x - 7$ (f) $f(x) = (x + 2)(x + \frac{3}{5})$
   (g) $f(x) = \text{SQRT}(x)$ (h) $f(x) = x^3 + 3$
   (i) $f(x) = e^x$ (j) $f(x) = x^5 + 3x^2 - 2$

2. Use forward, backward, and centered finite differences to approximate the derivative for each function in Problem 1, assuming that $d = 0.01$ and $x = 3.3$. Which difference method is the most accurate for each function?

3. Repeat Problem 2, but use $d = 0.001$. For each function $f(x)$, find the numerical size of approximation error for $d = 0.001$ in comparison to $d = 0.01$.

4. Calculate the definite integrals for each of the functions in Problem 1, assuming that the limits of integration are (1, 5).

5. Using the determination of the volume of a cone as an example, show, using calculus, that the volume of a sphere is equal to $4\pi r^3/3$, where r is the radius of the sphere.

*6. A diagram of a gravel island (Figure 1.10), in general, can be described as a frustum of a cone. Such an island is constructed in the Arctic Ocean, north of Alaska, to serve as a platform used to drill for oil. For much of the year, the Arctic Ocean is covered with ice. This ice is mobile; therefore, it imparts a large, lateral load to the island. A potential failure plane develops as shown. The weight of the island above the failure plane provides the resistance to potential shear failure. Using calculus and dimensions as

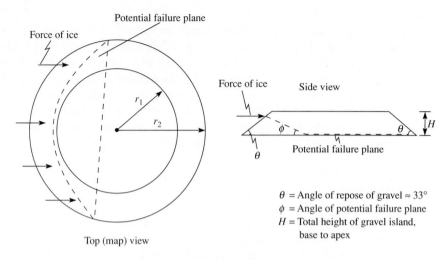

Top (map) view

θ = Angle of repose of gravel $\approx 33°$
ϕ = Angle of potential failure plane
H = Total height of gravel island, base to apex

Figure 1.10   A sketch of the gravel island for Problem 6.

shown, determine a formula for the weight of the island above the failure plane.

References and Suggested Readings

CHAPRA, S.C., and R.P. CANALE. 1985. *Numerical Methods for Engineers With Personal Computer Applications*. New York: McGraw-Hill.

CHAR, B.W., K.O. GEDDES, G.H. GONNET, B.L. LEONG, M.B. MONAGAN, and S.M. WATT. 1991a. *Maple V Language Reference Manual*. New York: Springer.

CHAR, B.W., K.O. GEDDES, G.H. GONNET, B.L. LEONG, M.B. MONAGAN, and S.M. WATT. 1991b. *Maple V Library Reference*. New York: Springer.

EDWARDS, C.H., JR. 1979. *The Historical Development of the Calculus*. New York: Springer.

HADAMARD, J. 1945. *The Psychology of Invention in the Mathematical Field*. Princeton, NJ: Princeton University Press.

PROTTER, M.H., and C.B. MORREY, JR. 1970. *College Calculus with Analytic Geometry*. Reading, MA: Addison Wesley.

SALAS, S.L., and E. HILLE. 1974. *Calculus: One and Several Variables*. New York: Wiley.

2

Carl Friedrich Gauss

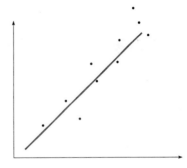

As Newton and Leibniz quarreled over who developed calculus (Newton went so far as to accuse Leibniz of stealing his ideas), the world was enriched by the birth of a mathematical genius. The writings of Carl Friedrich Gauss are vital to modern numerical analysis. Throughout this book, we introduce numerical analysis methods that were contributed by Gauss. Because so much reference is made to Gauss in this book, this chapter is devoted to some of his contributions.

2.1 A Mathematical Prodigy

Johann Friedrich Carl Gauss (Carl Friedrich Gauss) was born on 30 April 1777 in Brunswick, Germany. Gauss clearly was a mathematical prodigy. At the age of three, while watching his father distribute wages to his employees, Carl notified his father that a mistake was made in the distribution. On checking, Carl's father realized the boy was right. A remarkable story for a boy three years of age.

An even more remarkable event occurred several years later. When Gauss was ten years old, attending St. Catherine's elementary school, his teacher presented the class with the problem of writing down the whole numbers from

1 to 100 and determining their sum. The teacher thought this would keep the class busy for quite a while. Carl Gauss, however, completed the problem in only a few seconds and took his slate to the teacher proclaiming, "Ligget se" (there it lies) (Hall, 1970, p. 4).

The teacher eyed Gauss warily while the rest of the class struggled to finish. After all students had finished, the teacher began to grade the slates. Most of the answers were incorrect, with guilty students receiving a crack of the teacher's cane across their knuckles. Gauss sat calmly through this process.

Finally, the teacher arrived at the last slate, the first turned in, that of Carl Gauss. A single number, 5050, appeared on the slate. This is, indeed, the right answer.

"How did you arrive at this answer so quickly?" the teacher asked in stunned amazement. Patiently, Gauss explained that he had discovered the symmetry of arithmetic progressions, that in summing from 1 to 100, $1 + 100 = 101$; $2 + 99 = 101$, $3 + 98 = 101$; . . . ; $50 + 51 = 101$. The progression consists of 50 pairs of numbers, each pair having the sum of 101. Therefore, the total sum is equal to $50(101)$, or 5050.

That Gauss was entirely correct in this notion is seen easily if the progression is written twice, but in opposite order:

$$
\begin{array}{cccccccc}
1 + & 2 + & 3 + & 4 + \cdots + & 98 + & 99 + & 100 \\
\underline{100 +} & \underline{99 +} & \underline{98 +} & \underline{97 + \cdots +} & \underline{3 +} & \underline{2 +} & \underline{1} \\
101 + & 101 + & 101 + & 101 + \cdots + & 101 + & 101 + & 101 = 10{,}100
\end{array}
$$

This is twice what Gauss arrived at, because in writing the series this way, twice the number of pairs are used; hence, the sum is $10{,}100/2 = 5050$. For Gauss to have realized this symmetry at such a young age is astonishing.

At the age of eleven, Gauss enrolled at the Gymnasium Catharineum in Brunswick (the German word gymnasium refers to a school equivalent to a high school in the United States or Canada). After his first written assignment in mathematics was graded, Gauss was told by the instructor that he no longer was required to attend math class because the instructor could teach him nothing more.

Gauss was born into a family of modest income in which hard work was favored over education. He had difficulty convincing his father to permit him to continue his education, but destiny brought Gauss to the attention of Duke Carl Wilhelm Ferdinand of Brunswick, who was fascinated by Gauss' genius. Duke Ferdinand agreed to pay for Gauss' continued education at the Collegium Carolinium. Gauss enrolled in 1792, at the age of fifteen, and graduated in 1795. One year after graduation, Gauss published his first paper, a scientific note about the regular, 17-sided polygon.

Gauss became the foremost representative of the scientific viewpoint in the nineteenth century. Through his research on curved surfaces, he established the mathematical premise for the theory of relativity (formally proposed by Albert Einstein). Moreover, Gauss' research on error theory is of the utmost importance to the remainder of this text.

2.2 Theory of Errors

The method of least squares is one of the most important contributions made by Gauss. Formal credit for this method is often given to Legendre, who first published an article about it in 1806. But from his letters and notes it is clear that Gauss arrived at the method of least squares in 1795 (at the age of 17) (Hall, 1970, p. 74).

Gauss was interested in random errors introduced when making measurements in association with astronomical observations. He defined a function $a(x)$ as the relative error in observation x, which implies that

$$0 \leq a(x) \leq 1$$

Then $a(x)\, dx$ is the probability of an error between x and $x + dx$. The function $a(x)$ is normalized by

$$\int_{-\infty}^{\infty} a(x)\, dx = 1$$

Gauss further wrote that

$$\int_{-\infty}^{\infty} x^2 a(x)\, dx$$

attains a minimum, which establishes the notion that the square of the error is its most suitable weight.

From his astronomical observations, Gauss observed that his measurement errors were well represented by

$$a(x) \propto e^{-x^2/2}$$

This is the equation of the normal distribution, which is well known in probability theory. Although many other probability functions are known today, the normal distribution (also known as Gaussian distribution or the bell-shaped curve) was the only one considered by Gauss.

Legendre, in his publication of 1806, arrived at the method of least squares computationally; hence Gauss surpasses Legendre by noting the theoretical premise. Coincidentally with Gauss and Legendre, Laplace introduced probabilistic considerations into the least-squares approach. Some evidence exists that each of the three mathematicians arrived at his derivation entirely independently of the others (Buhler, 1981, p. 140). After his derivation of the method of least squares, Gauss came to view the method as the most important evidence for the link between mathematics and nature.

2.2.1 Application of Error Theory: Gaussian Quadrature

Calculus is presented as the elegant solution to area and volume calculations, provided that areas and volumes are bounded by integrable curves (Chapter 1). However, computers cannot directly perform calculus. So what is the context of calculus in the modern computing/information age?

Calculus in modern twentieth-century society is as important, if not more so, than it ever was. Certainly, calculus forms the basis or establishes the premise for a large number of computer applications. But, when writing computer programs for solving mathematical equations, algorithms are used to approximate integrals.

One of these algorithms is Gaussian quadrature. This algorithm can be derived using Gauss' method of least squares; Gauss actually used a different procedure to arrive at Gaussian quadrature; but, perhaps approaching this problem from the method of least squares is more understandable.

Let F be the value of the definite integral of a function $f(x)$ over an interval $x = a$ to $x = b$:

$$F = \int_a^b f(x)\, dx$$

Let F be approximated as F^*; the error of this approximation is then $F - F^*$. Let q represent this error. From Gauss' notion of least squares, the value of q^2 is minimized.

In Gaussian quadrature, F^* is approximated as (Dahlquist and Bjorck, 1974, p. 302)

$$\int_a^b f(x)\, dx \approx a_0 f(x_0) + a_1 f(x_1) + \cdots + a_m f(x_m)$$

which is exact for all polynomials $f(x)$ of degree $2m + 1$ or less, when the coefficieints a are computed as shown in the next two examples.

Example 2.1   Two-Point Approximation ($m = 1$)

Gauss was aware of Simpson's and Weddle's formulas for approximating integrals. In these methods, the value of a function f is determined for equally spaced locations x. Gauss reasoned that some other type of spacing might give a better result. He found that the locations x should not be equidistant; instead, they should be located symmetrically about the midpoint of the integration interval. (This entire discussion is taken from Scarborough (1955, p. 145).)

Using Gauss' notion, if $m = 1$, the exact value of the definite integral of f over the region $x = c$ to $x = d$ is determined by finding two values of x, symmetrically located about the midpoint of the integration. Incidentally, the midpoint is equal to $(c + d)/2$. (*Note:* The interval $[c, d]$ is used rather than $[a, b]$ because the symbol a is being used to represent weights in Gaussian quadrature.)

Because $m = 1$, Gaussian quadrature yields the exact solution for a definite integral on $[c, d]$ for up to third-order polynomials. For example, let $f(x) = j_0 + j_1 x + j_2 x^2 + j_3 x^3$. Moreover, let the limits of integration be $c = -1$; $d = 1$. In this case,

$$F = \int_c^d f(x)\, dx = \int_{-1}^1 j_0 + j_1 x + j_2 x^2 + j_3 x^3\, dx$$

approximated for $m = 1$ as

$$F^* = a_0 f(x_0) + a_1 f(x_1)$$

Using the function $q = F - F^*$, the coefficients a_0 and a_1 and the locations x_0 and x_1 are determined to render this approximation exact. But q must be defined explicitly, which requires calculus for determining F. Recalling that

$$F = \int_{-1}^1 j_0 + j_1 x + j_2 x^2 + j_3 x^3\, dx$$

$$= 2j_0 + \frac{2j_2}{3}$$

and that

$$F^* = a_0 f(x_0) + a_1 f(x_1)$$

with

$$a_0 f(x_0) = a_0 j_0 + a_0 j_1 x_0 + a_0 j_2 x_0^2 + a_0 j_3 x_0^3$$
$$a_1 f(x_1) = a_1 j_0 + a_1 j_1 x_1 + a_1 j_2 x_1^2 + a_1 j_3 x_1^3$$

some qualitative reasoning is used to simplify these expressions. Because the interval for this problem is $[-1, 1]$, the midpoint is $(-1 + 1)/2 = 0$. Gauss further reasoned that points x_0 and x_1 should be located symmetrically about the midpoint. In this case, the midpoint is zero, so we let $x_0 = -t$ and $x_1 = t$ (where t is an arbitrarily chosen symbol). Because these points are the same distance (numerical distance) from the midpoint, the weights a are reasoned to be the same. Therefore, let $a_0 = a_1 = a$:

$$F^* = a[f(-t) + f(t)] = 2aj_0 + 2aj_2t^2$$

Recalling the expression for error, we find that

$$q = F - F^* = 2j_0 + \frac{2j_2}{3} - 2aj_0 - 2aj_2t^2$$

Furthermore,

$$q^2 = 4j_0^2 + \frac{8j_0j_2}{3} - 8aj_0^2 - 8aj_0j_2t^2 + \frac{4j_2^2}{9}$$

$$- \frac{8aj_0j_2}{3} - \frac{8aj_2^2t^2}{3} + 4a^2j_0^2 + 8a^2j_0j_2t^2 + 4a^2j_2^2t^4$$

Because a solution is wanted for a and t such that q^2 is a minimum for any value of the coefficients j, the partial derivatives of q^2 with respect to j_0 and j_2 are computed:

$$\frac{\partial q^2}{\partial j_0}: \quad 8j_0 + \frac{8j_2}{3} - 16aj_0 - 8aj_2t^2 - \frac{8aj_2}{3} + 8a^2j_0 + 8a^2j_2t^2 = 0$$

$$\frac{\partial q^2}{\partial j_2}: \quad \frac{8j_0}{3} - 8aj_0t^2 + \frac{8j_2}{9} - \frac{8aj_0}{3} - \frac{16aj_2t^2}{3} + 8a^2j_0t^2 + 8a^2j_2t^4 = 0$$

Because a solution is desired for a and t to yield a minimal value of q^2 for any value of the coefficients j, conveniently consider $j_0 = 1$ and $j_2 = 1$. Substituting these values into the foregoing partial derivatives of q^2 gives

$$\frac{\partial q^2}{\partial j_0}: \quad \frac{4}{3} - 7\frac{a}{3} - at^2 + a^2 + a^2t^2 = 0$$

$$\frac{\partial q^2}{\partial j_2}: \quad \frac{4}{9} - \frac{a}{3} - \frac{5at^2}{3} + a^2t^2 + a^2t^4 = 0$$

One of the roots for the partial derivative of q^2 with respect to j_0 is

$$a = \frac{4}{3(1 + t^2)}$$

One of the roots for the partial derivative of q^2 with respect to j_2 is

$$t = \pm \frac{\sqrt{3}}{3\sqrt{a}}$$

Two simultaneous equations are now obtained for a and t. Substituting the solution for t into the equation for a leads to the solution $a = 1$. Hence,

$$t = \pm \frac{1}{\sqrt{3}}$$

■

Cook (1974, p. 104) simplifies this process by using q rather than q^2:

$$q = 2j_0 + \frac{2j_2}{3} - 2aj_0 - 2aj_2t^2$$

$$\frac{\partial q}{\partial j_0}: \quad 2 - 2a = 0 \qquad \text{hence } a = 1$$

Moreover,

$$\frac{\partial q}{\partial j_2}: \quad \frac{2}{3} - 2at^2 = 0$$

Because the weight a is found to equal 1 from the derivative of q with respect to j_0,

$$\frac{2}{3} - 2at^2 = 0 \qquad \text{with } a = 1 \quad \text{then } t = \pm \frac{1}{\sqrt{3}}$$

which is the same solution found using q^2, but in a simpler fashion. In the next example, q is used instead of q^2, fully recognizing that minimizing q differs substantially from minimizing q^2. The minimum of q could be negative and large, which is not what is wanted for these exercises. In contrast, the minimum of q^2 is always greater than or equal to zero. This problem can be avoided by using the absolute value of q, but differentiating q is then more difficult. For purposes of this text and the derivations of Gaussian quadrature solutions presented herein, using q leads to elegant and correct solutions without having to use its absolute value.

Example 2.2   Three-Point Approximation ($m = 2$)

If $m = 2$, Gaussian quadrature yields the exact answer for definite integrals of polynomials whose order is $2m + 1 = 5$ or less. That is,

$$f(x) = j_0 + j_1 x + j_2 x^2 + j_3 x^3 + j_4 x^4 + j_5 x^5$$

In the interval $[-1, 1]$,

$$F = \int_{-1}^{1} f(x)\, dx$$

$$= 2j_0 + \frac{2j_2}{3} + \frac{2j_4}{5}$$

Using Gauss' approach to approximation, for $m = 2$,

$$F^* = a_0 f(x_0) + a_1 f(x_1) + a_2 f(x_2)$$

Gauss reasoned that the sampling locations should be symmetrically located with respect to the midpoint of the integration interval. In this case, though, the function f is sampled at three locations. How can three points be located symmetrically around the midpoint? In the case of an odd number of sample points, Gauss let one be associated with the exact midpoint of the integration interval, the remaining even number of sampling points are located symmetrically around this point.

In the case of three points, let $x_0 = -t$ and $x_2 = t$, and let $a_0 = a_2 = a$, due to symmetry. If the interval is $[-1, 1]$, then the midpoint is 0 and is associated with point x_1. Hence,

$$F^* = a[f(-t) + f(t)] + a_1 f(0)$$

Substituting $-t$ and t into $f(x)$, and summing the results yields

$$F^* = 2aj_0 + 2aj_2 t^2 + 2aj_4 t^4 + a_1 j_0$$

The error function q is written $q = F - F^*$. That is,

$$q = 2j_0 + \frac{2j_2}{3} + \frac{2j_4}{5} - 2aj_0 - 2aj_2 t^2 - 2aj_4 t^4 - a_1 j_0$$

In this example, partial derivatives of q must be computed separately with respect to j_0, j_2, and j_4:

$$\frac{\partial q}{\partial j_0}: \quad 2 - 2a - a_1 = 0 \qquad \text{which yields } a_1 = 2 - 2a$$

$$\frac{\partial q}{\partial j_2}: \quad \frac{2}{3} - 2at^2 = 0 \qquad \text{which yields } t^2 = \frac{1}{3a}$$

$$\frac{\partial q}{\partial j_4}: \quad \frac{2}{5} - 2at^4 = 0$$

Substituting the solution from the partial derivative of q with respect to j_2 into the solution for the partial derivative of q with respect to j_4 gives

$$\frac{2}{5} - 2at^4 = 0 \qquad \frac{2}{5} = 2a\left(\frac{1}{3a}\right)^2$$

from which a is found to be equal to $\frac{5}{9}$. The solution for t is found using the solution for a:

$$t^2 = \frac{1}{3a} = \frac{3}{5} \qquad t = \pm\frac{\sqrt{3}}{\sqrt{5}}$$

Finally, because $a_1 = 2 - 2a$, $a_1 = \frac{8}{9}$.   ∎

Our understanding of Gaussian quadrature can be enhanced with a visual display of the method (Figure 2.1). Two-point and three-point Gaussian quadrature approximation of a definite integral is demonstrated. Gaussian quadrature is shown as the approximation of the area beneath the curve $f(x)$, bounded by $x = a$ and $x = b$, using rectangles. In other words, Gaussian quadrature is tantamount to a return to area (or volume) calculations as was done prior to the introduction of calculus. Gaussian quadrature simply takes a different approach to defining the rectangles.

Recall that the area of a rectangle is equal to its width times its height. In Gaussian quadrature, the width of each rectangle (Figure 2.1) is equal to a, the Gaussian quadrature weight. The height of each rectangle is equal to $f(-t)$,

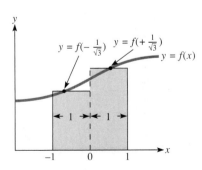
Two-point approximation

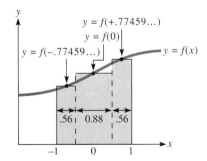
Three-point approximation

Figure 2.1 A visual display of the concept of Gaussian quadrature. Notice that the width of each rectangle is equal to a, the Gaussian quadrature weight, and the height of each rectangle is equal to $f(x)$, where $x = -t$, $+t$, or 0, depending on the number of Gauss points used to approximate the area.

$f(+t)$, or $f(0)$ (for three-point approximation). Therefore, the equation used to approximate integrals in Gaussian quadrature,

$$\text{area} \approx \sum_{i=1}^{N} a_i f(t_i)$$

is exactly analogous to computing the area of rectangles and summing these areas to obtain an estimate of the overall area.

2.2.2 The General Form for Gaussian Quadrature for $m = 1$ and $m = 2$

In Examples 2.1 and 2.2, the integration limits $[-1, 1]$ are used. In the generic sense, for an integration interval of $[c, d]$, the weights a and sampling locations t are determined as follows.

For $m = 1$: For an integration interval of $[-1, 1]$,

$$a = 1 \qquad t = \pm \frac{1}{\sqrt{3}}$$

In general, a is equal to the numerical distance from the midpoint to either end of the integration interval. The generic solution for a is therefore $(d - c)/2$. The sampling locations $-t$ and t are found as

$$t = \frac{c + d}{2} \pm \frac{a}{\sqrt{3}}$$

This leads to the generic solution of $F*$ for two-point Gaussian quadrature approximation:

$$F* = \frac{d - c}{2}\left[f\left(\frac{c + d}{2} - \frac{d - c}{2\sqrt{3}}\right) + f\left(\frac{c + d}{2} + \frac{d - c}{2\sqrt{3}}\right)\right]$$

for an integration interval $[c, d]$ and a polynomial function $f(x)$ of order 3 or less.

For $m = 2$: For the interval $[-1, 1]$, $a = \frac{5}{9}$, $a_1 = \frac{8}{9}$, and

$$t = \pm \frac{\sqrt{3}}{\sqrt{5}}$$

For the interval $[c, d]$,

$$a = \frac{5}{9}\left(\frac{d - c}{2}\right) \qquad a_1 = \frac{8}{9}\left(\frac{d - c}{2}\right) \qquad t = \frac{d + c}{2} \pm \frac{\sqrt{3}(d - c)}{2\sqrt{5}}$$

which leads to the generic Gaussian quadrature solution for $m = 2$:

$$F* = \frac{5(d - c)}{18}\left\{ f\left[\frac{c + d}{2} - \frac{\sqrt{3}(d - c)}{2\sqrt{5}}\right]\right.$$
$$\left. + f\left[\frac{c + d}{2} + \frac{\sqrt{3}(d - c)}{2\sqrt{5}}\right]\right\} + a_1 f\left(\frac{c + d}{2}\right)$$

for an integration interval $[c, d]$ and polynomials of order 5 or less.

Example 2.3

Given the polynomial function,

$$f(x) = 2 - 3x + 4x^2$$

find an approximation for

$$\int_7^{12} f(x)\,dx$$

using Gaussian quadrature.

SOLUTION

The order of this polynomial is two, therefore, two-point Gaussian quadrature approximation is used:

$$F^* = \frac{12-7}{2}\left[f\left(\frac{12+7}{2} - \frac{12-7}{2\sqrt{3}}\right) + f\left(\frac{12+7}{2} + \frac{12-7}{2\sqrt{3}}\right)\right]$$

$$= \frac{5}{2}[f(8.0566) + f(10.9434)] = \frac{5}{2}[237.4654 + 448.2018] = 1714.17$$

A comparison to the exact solution yields

$$F = \int_7^{12} 2 - 3x + 4x^2\,dx$$

$$= 2x - \frac{3x^2}{2} + \frac{4x^3}{3}\Bigg|_7^{12}$$

$$= 2112 - 397.83 = 1714.17$$

Notice how accurate Gaussian quadrature is in this case.   ■

Example 2.4

Given the polynomial function,

$$f(x) = 9 - x + 10x^2 + x^3 + 5x^4$$

solve for

$$\int_4^7 f(x)\,dx$$

using Gaussian quadrature.

SOLUTION

In this case, the polynomial is fourth order, hence three-point Gaussian quadrature is required. Therefore,

$$F^* = a\left\{ f\left[\frac{7+4}{2} - \frac{\sqrt{3}(7-4)}{2\sqrt{5}}\right] + f\left[\frac{7+4}{2} + \frac{\sqrt{3}(7-4)}{2\sqrt{5}}\right]\right\} + a_1 f\left(\frac{7+4}{2}\right)$$

in which

$$a = \frac{5}{9}\left(\frac{7-4}{2}\right) = \frac{5}{6}$$

and

$$a_1 = \frac{8}{9}\left(\frac{7-4}{2}\right) = \frac{4}{3}$$

Therefore,

$$F^* = \tfrac{5}{6}[f(4.3381) + f(6.6619)] + 6730.25 = 17{,}259.77$$

A comparison to the exact solution gives

$$F = \int_4^7 9 - x + 10x^2 + x^3 + 5x^4 \, dx$$

$$= 9x - \frac{x^2}{2} + \frac{10x^3}{3} + \frac{x^4}{4} + x^5 \bigg|_4^7$$

$$= 18{,}589.08 - 1{,}329.33 = 17{,}259.75$$

Hence, a slight error, 0.02, is observed for this approximation.   ▪

2.2.3 A Table of Gaussian Quadrature Locations t and Weights a

Gaussian quadrature locations t and weights a for $m = 1$ and $m = 2$ were derived earlier. Table 2.1 shows Gaussian quadrature locations and weights for up to $m = 9$. These locations and weights assume an integration interval of $[-1, 1]$. To convert to any integration interval $[c, d]$, multiply each weight a by $(d - c)/2$. For situations in which an even number of locations t are used,

Table 2.1 Gaussian Quadrature Locations t and Weights a for $m = 1$ to $m = 9$

	t	a
$m = 1$	0.5773502692	1.0000000000
$m = 2$	0.0000000000	0.8888888889
	0.7745966692	0.5555555556
$m = 3$	0.3399810436	0.6521451548
	0.8611363116	0.3478548452
$m = 4$	0.0000000000	0.5688888889
	0.5384693102	0.4786286704
	0.9061798460	0.2369268850
$m = 5$	0.2386191860	0.4679139346
	0.6612093864	0.3607615730
	0.9324695142	0.1713244924
$m = 6$	0.0000000000	0.4179591836
	0.4058451514	0.3818300506
	0.7415311856	0.2797053914
	0.9491079124	0.1294849662
$m = 7$	0.1834346424	0.3626837834
	0.5255324100	0.3137066458
	0.7966664774	0.2223810344
	0.9602898564	0.1012285363
$m = 8$	0.0000000000	0.3302393550
	0.3242534234	0.3123470770
	0.6133714328	0.2606106964
	0.8360311074	0.1806481607
	0.9681602396	0.0812743884
$m = 9$	0.1488743390	0.2955242248
	0.4333953942	0.2692667194
	0.6794095682	0.2190863626
	0.8650633666	0.1494513492
	0.9739065286	0.0666713443

convert these to

$$t = \frac{c + d}{2} \pm \frac{t(d - c)}{2}$$

where the variable t on the right side of this equation is that listed in Table 2.1.

For an odd number of locations, the middle location (shown as $t = 0.0000000000$ in the table) is always the midpoint of the interval, $(c + d)/2$. The remaining locations are symmetrically distributed about the midpoint, with positions calculated as is done for the even number of integration locations.

2.2.4 Application of Error Theory: Linear Least-Squares Regression

In the preceding sections, Gauss' error theory is used to develop a numerical solution for definite integrals, exact for polynomials of degree $2m + 1$ or less. In this section, error theory is used to fit a straight line of the form $y = mx + b$ to a set of observations (x, y).

Before attempting linear least-squares regression to fit a straight line to observational data (x, y), a graph must be prepared of y versus x to verify that, indeed, y changes with x in a linear fashion. If y does not change with x in a linear fashion, but appears to change in a fashion represented by a polynomial, then the general equation for the polynomial must be used rather than $y = mx + b$ (that is, maybe $y = nx^2 + mx + b$; $y = \exp x$, and so on).

Suppose y varies linearly with x (e.g., Figure 2.2). In this case, $y = mx + b$ is an appropriate model. To define this model exactly for a set of x, y data, the coefficients m and b of the model must be determined. A solution is desired for m and b such that the model $y = mx + b$ is the best fit to the x, y data. Gauss' error theory is used to assure that the model $y = mx + b$ is as accurate as possible for the x, y data. Let y^* represent the estimate of y; that is, $y^* = mx + b$. Further, let y represent the actual value of y measured for a given x.

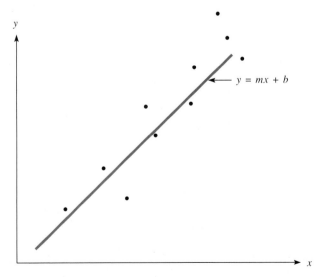

Figure 2.2 An example plot of bivariate data (x, y) for which a model $y = mx + b$ is used to characterize the linear relationship between y and x.

The error q is written as an integral over x:

$$q = \int (y - y^*)^2\, w(x)\, dx$$

Recall from Chapter 1 that integrals are used in this text to set up a problem, but in developing an algorithm for computer solution, integrals are approximated. In this case,

$$q = \sum_{i=1}^{N} (y - y^*)^2$$

where N is the number of pairs of observations (x, y). Moreover, this approximation assumes that the function $w(x) = 1$ for any value of x.

Notice that q is a function of squared error. Laplace, in trying to develop his version of error theory, worked only with the absolute value of the error. Herein, Gauss' notion of squared error is the basis of discussion.

Recalling that $y^* = mx + b$ is the model for the linear least-squares regression, $mx + b$ is substituted for y^* in the expression for q:

$$q = \sum_{i=1}^{N} (y - mx_i - b)^2$$

which after expansion is equal to

$$q = \sum_{i=1}^{N} (y_i^2 - 2mx_iy_i - 2y_ib + 2mx_ib + m^2x_i^2 + b^2)$$

A solution for m and b is sought such that q has a minimal value. Therefore, a partial derivative of q is taken with respect to m:

$$\frac{\partial q}{\partial m}: \quad \sum_{i=1}^{N} (-2x_iy_i + 2x_ib + 2mx_i^2) = 0$$

Let the summation sign be associated with each individual term of this derivative. Move the one negative term to the right-hand side:

$$\frac{\partial q}{\partial m}: \quad m\sum_{i=1}^{N} x_i^2 + b\sum_{i=1}^{N} x_i = \sum_{i=1}^{N} x_iy_i$$

Notice that the multiplier 2 cancels in this process. Taking the partial derivative of q with respect to b gives

$$\frac{\partial q}{\partial b}: \quad \sum_{i=1}^{N} (-2y_i + 2mx_i + 2b) = 0$$

Simplifying this derivative leads to

$$m\sum_{i=1}^{N} x_i + Nb = \sum_{i=1}^{N} y_i$$

because

$$\sum_{i=1}^{N} b = Nb$$

Solving for b using the partial derivative of q with respect to b gives

$$b = \frac{1}{N}\sum_{i=1}^{N} y_i - \frac{m}{N}\sum_{i=1}^{N} x_i$$

Substituting the solution for b into the equation resulting from the partial derivative of q with respect to m yields a solution for m:

$$m \sum_{i=1}^{N} x_i^2 + \left[\frac{1}{N} \sum_{i=1}^{N} y_i \sum_{i=1}^{N} x_i - \frac{m}{N} \left(\sum_{i=1}^{N} x_i \right)^2 \right] = \sum_{i=1}^{N} x_i y_i$$

and with simplification,

$$m = \frac{N \sum_{i=1}^{N} x_i y_i - \sum_{i=1}^{N} y_i \sum_{i=1}^{N} x_i}{N \sum_{i=1}^{N} x_i^2 - (\sum_{i=1}^{N} x_i)^2}$$

This result is overdetermined because only two pairs of observations (x, y) yield sufficient information to solve for m and b. But using all N pairs of observations ensures that the model $y = mx + b$ is the most accurate one for all of the data observations (x, y).

Example 2.5

Given the following (x, y) observations, (a) determine whether the model $y = mx + b$ is appropriate for these data, and (b) find the coefficients m and b for a best linear fit.

x	y
0.699	2.784
0.845	2.619
1.000	2.439
1.176	2.215
1.301	2.087
1.477	1.832
1.699	1.562
1.875	1.327
2.000	1.096

These data were obtained through an experiment to measure the length of the coastline of Great Britain. This type of experiment is described in detail in the discussion

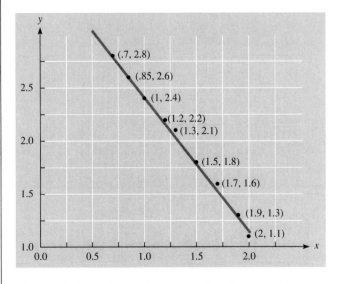

Figure 2.3   Plot of the bivariate data given in Example 2.5.

of fractals in Chapter 11. Results from this experiment are useful for demonstrating linear least-squares regression.

SOLUTION

That a linear model is appropriate for these data is evident (Figure 2.3). The following table can be constructed:

	x	y	xy	x^2
	0.699	2.784	1.946	0.489
	0.845	2.619	2.213	0.714
	1.000	2.439	2.439	1.000
	1.176	2.215	2.605	1.383
	1.301	2.087	2.715	1.693
	1.477	1.832	2.706	2.182
	1.699	1.562	2.654	2.887
	1.875	1.327	2.488	3.516
	2.000	1.096	2.192	4.000
Sum	12.072	17.961	21.958	17.864

The four sums are the ones needed to solve for the coefficients m and b. Using these sums, we find

$$m = \frac{9 * 21.958 - 12.072 * 17.961}{9 * 17.864 - 12.072 * 12.072} = -1.277$$

$$b = \frac{17.961}{9} - \frac{-1.27 * 12.072}{9} = 3.700$$

A computer algorithm, LINEAR (for linear least-squares regression) (Figure 2.4, page 38), and a data file (Figure 2.5) may be used to check the results. Applying the computer algorithm LINEAR to the data (Figure 2.5) gives results (Figure 2.6) that match those calculated by hand.   ∎

```
0.699, 2.784
0.845, 2.619
1.000, 2.439
1.176, 2.215
1.301, 2.087
1.477, 1.832
1.699, 1.562
1.875, 1.327
2.000, 1.096

0;37;44m C:\TEXT=
```

Figure 2.5 A data file for use with the program LINEAR, developed using the data in Example 2.5.

```
            LINEAR REGRESSION SUMMARY

        SLOPE, M              =      -.12782E+01
        INTERCEPT, B          =       .37101E+01
Stop - Program terminated.


0;37;44m C:\TEXT=
```

Figure 2.6 Resultant output from LINEAR for the data file listed in Figure 2.5.

2.3 Gaussian Elimination: Solving Simultaneous Equations

This section is numbered independently of the previous section on error theory because it is somewhat unclear how Gauss arrived at this procedure for solving simultaneous equations. Simultaneous equations are two or more algebraic functions having a common solution, for example, two or more algebraic functions that have the same set of coefficients.

In pursuing the history of Gaussian elimination, disappointment often is encountered. Books on numerical analysis published since 1960 describe Gaussian elimination, but provide no background on its derivation. A reference in Scarborough (1955, p. 507) notes that Gaussian elimination is explained in detail by Encke (*Berliner Astronomisches Jahrbuch,* 1835, pp. 267–272; 1836, pp. 256–259). Encke was one of Gauss' astronomy students, which leads to speculation that Gauss arrived at the elimination theory in his search for a solution to a problem involving astronomical observation.

Hall (1970, p. 64) presents the following observation:

> *Theoria motus* is thus a classical work in theoretical astronomy. Gauss' method for determining an elliptical orbit from three complete observations was later improved with respect to its practical use, primarily by Gauss' student, J.F. Encke, and during the past few years it has been adapted to modern data-processing machines. But the fundamental principles are still those of Gauss, and nothing better has ever been brought forward.

In Gaussian elimination, unknowns associated with simultaneous equations are determined by solving one of the equations for one of the unknowns, then substituting this solution into all remaining equations. This is done repeatedly until only one unknown remains, at which point it is solved. Then, using this solution, other unknowns are solved through back calculation. This is what is done in Section 2.2 for both Gaussian quadrature and linear least-squares regression.

One of the best treatments of Gaussian elimination is that given by Scarborough (1955, pp. 507–512). Given the following set of simultaneous equations,

$$2.63x + 5.21y - 1.694z + 0.938t - 4.23 = 0$$
$$3.16x - 2.95y + 0.813z - 4.21t + 0.716 = 0$$
$$5.36x + 1.88y - 2.15z - 4.95t - 1.28 = 0$$
$$1.34x + 2.98y - 0.432z - 1.768t - 0.419 = 0$$

a common solution is sought. The unknowns x, y, z, and t are solved for in an orderly fashion. We first eliminate x, then y, then z, thereby obtaining a solution for t. We use the solution for t to solve first for z, then for y, and finally for x. This procedure is described as forward elimination followed by back substitution.

Use of the word eliminate requires some explanation. In the context of Gaussian elimination, the term eliminate refers to solving for one of the unknowns in the system by using one of the equations composing the system. The solution for this unknown is substituted into all remaining equations, thereby eliminating its presence. But, the system of equations has not been changed. Once one unknown has been removed, the others are removed in the same way, until only one unknown remains, at which point it is solved.

To eliminate x from the foregoing system of equations, notice that the largest coefficient of x, 5.36, occurs in the third equation. Therefore, this equation is the pivotal equation for x. We can eliminate x in this equation as follows:

$$5.36x + 1.88y - 2.15z - 4.95t - 1.28 = 0$$
$$5.36x = -1.88y + 2.15z + 4.95t + 1.28$$
$$x = \frac{-1.88}{5.36}y + \frac{2.15}{5.36}z + \frac{4.95}{5.36}t + \frac{1.28}{5.36}$$
$$= -0.350746y + 0.401119z + 0.923506t + 0.238806$$

Substituting this solution for x into the remaining three equations yields three simultaneous equations with three unknowns, y, z, and t. For instance, from the first equation,

$$2.63(-0.350746y + 0.401119z + 0.923506t + 0.238806) + 5.21y - 1.694z$$
$$+ 0.938t - 4.23 = 0$$

This simplifies to

$$4.287538y - 0.63906z + 3.36682t - 3.601940 = 0$$

The remaining two equations are found by substituting the solution for x into the second and fourth equations:

$$-4.05836y + 2.08054z - 1.29172t + 1.470627 = 0$$
$$2.510000y + 0.105500z - 0.530500t - 0.0990 = 0$$

Now y is eliminated. The pivotal equation for y is the first because this equation is associated with the largest coefficient, 4.287538, for y. Precision past the decimal point is maintained to minimize round-off error. Such error is an inherent aspect of Gaussian elimination. Using the pivotal equation for y,

$$y = \frac{0.63906}{4.287538}z - \frac{3.36682}{4.287538}t + \frac{3.61940}{4.287538}$$
$$= 0.149051z - 0.785258t + 0.840096$$

Substitute this solution for y into the two remaining equations from the last group of three simultaneous equations. This now gives two simultaneous equations and two unknowns, z and t:

$$1.47564z + 1.89514t - 1.93879 = 0$$
$$0.479618z - 2.50150t + 2.00964 = 0$$

Eliminate z by using its pivotal equation, the one with the $1.47564z$ term:

$$z = -1.28428t + 1.31386.$$

Substituting this into the second equation above gives

$$0.479618(-1.28428t + 1.31386) = 0$$

Now t is solved for: $t = 0.846775$. Using this solution for t, z is solved as

$$z = -1.28428t + 1.31386 = 0.22636$$

Next, we solve for y:

$$y = 0.149051z - 0.785258t + 0.840096 = 0.208898$$

Finally, we solve for x:

$$x = -0.350746y + 0.401119z + 0.923506t + 0.238806 = 1.038335.$$

Solution is now complete.

Example 2.6

Given the following series of simultaneous equations, solve for the unknowns:

$$2x + y - z = -1$$
$$3x - 2y + z = 7$$
$$x + 2y + 2z = 3$$

SOLUTION

The pivotal equation for x is the second. Solving for x yields

$$3x = 2y - z + 7$$

or

$$x = \frac{2y}{3} - \frac{z}{3} + \frac{7}{3}$$

Substituting this solution into the first and third equations yields

$$2\left(\frac{2y}{3} - \frac{z}{3} + \frac{7}{3}\right) + y - z = -1$$

which simplifies to

$$2.333333y - 1.666667z + 5.666667 = 0$$

And

$$\frac{2y}{3} - \frac{z}{3} + \frac{7}{3} + 2y + 2z = 3$$

which simplifies to

$$2.666667y + 1.666667z - 0.666667 = 0$$

Two simultaneous equations are obtained:

$$2.333333y - 1.666667z + 5.666667 = 0$$
$$2.666667y + 1.666667z - 0.666667 = 0$$

Eliminate y using the second equation as the pivotal one:

$$y = -\frac{1.666667z}{2.666667} + \frac{0.666667}{2.666667}$$

Substituting this solution into the first equation gives

$$2.333333\left(-\frac{1.666667z}{2.666667} + \frac{0.666667}{2.666667}\right) - 1.666667z + 5.666667 = 0$$

or

$$-3.125z + 6.25 = 0$$

From this, z is solved for: $z = 2.0$. Using this solution, solve for y:

$$y = -\frac{1.666667z}{2.666667} + \frac{0.666667}{2.666667} = -1.0$$

And, solve for x:

$$x = \frac{2y}{3} - \frac{z}{3} + \frac{7}{3} = 1.0$$

Therefore, the solution is

$$x = 1.0$$
$$y = -1.0$$
$$z = 2.0$$

Gaussian elimination is tedious when performed by hand, but this method is important for the solution of simultaneous equations on a computer. It is used extensively in subsequent chapters, especially in the discussion of kriging and finite element analysis.

A computer algorithm for Gaussian elimination, GAUSS, is presented in Figure 2.7 (page 39) and may be used to verify the solution to this example. A data file (Figure 2.8) is prepared following the user's guide (Figure 2.7). Applying the computer algorithm to this data set yields the solution (Figure 2.9), which matches that obtained by hand: $x = 1$, $y = -1$, and $z = 2$.   ■

```
                                         GAUSS ELIMINATION RESULTS


3                                        X(  1)   =            1.000
2.0,1.0,-1.0                             X(  2)   =           -1.000
3.0,-2.0,1.0                             X(  3)   =            2.000
1.0,2.0,2.0                              Stop - Program terminated.
-1.0,7.0,3.0

0;37;44m C:\TEXT=                        0;37;44m C:\TEXT=
```

Figure 2.8   A data file for use with GAUSS in solving Example 2.6.

Figure 2.9   Resultant output from GAUSS for the data file listed in Figure 2.8.

2.3.1 Advanced Topic: Iterative Improvement for Gaussian Elimination

From our discussion of Gaussian elimination we see that a large number of divisions are involved with this method. Implementing Gaussian elimination on digital computers therefore can result in the introduction of large errors due to rounding of quotients to a decimal precision commensurate with the capability of the computer. These errors are compounded by the Gaussian elimination process. McCarn and Carr (1992), for example, discuss how this error due to numerical precision limitations affects geostatistical analysis.

Two important aspects of equation solution should be considered when attempting to limit round-off error. First is the precision requested for numerical computation by the computer algorithm effecting the solution. In FORTRAN, algorithms for double, or even extended, precision can be used to achieve a greater degree of accuracy. Double-precision calculations use twice the numerical resolution of the computer processor to represent each number. Therefore, more digits to the right of the decimal point are allowed for real numbers and real number calculations are more accurate. The price paid for double precision is a loss of memory space for the program because twice as much memory is required to represent a number as is required for single-precision arithmetic.

A second consideration for achieving an accurate equation solution is the algorithm employed for the solution. Gaussian elimination is one such algorithm. Another is LU decomposition (Press et al., 1986), a method that is related to Gaussian elimination, but that involves fewer calculations, and therefore results in less round-off error. An excellent degree of accuracy, equal to that of LU decomposition, can be achieved with Gaussian elimination and only single-precision arithmetic by using iterative improvement. In this summary of iterative improvement matrix notation is used to represent systems of equations; the notation is described more fully in Chapter 3.

Gaussian elimination is used to solve for $\{x\}$ in the equation system $[A]\{x\} = \{b\}$, where $[A]$ and $\{b\}$ are known and $\{x\}$ is associated with error attributable to rounding off calculations. Therefore, let the solution from Gaussian elimination be written $\{x + e\}$, where e represents the error. With this representation, the error is solved as follows:

$$[A]\{x + e\} - \{b\} = \{d\}$$

where

$$\{d\} = [A]\{e\}$$

Substituting this solution for $\{d\}$ into the previous equation, we find

$$[A]\{e\} = [A]\{x + e\} - \{b\}$$

The term $[A]\{x + e\}$ is easily computed by multiplying $[A]$ by the solution obtained from Gaussian elimination. Once this term is calculated, the vector $\{b\}$, which is also known, is subtracted from it. Calling the result from this subtraction $\{b'\}$, we get

$$[A]\{e\} = \{b'\}$$

and $\{e\}$ is solved using Gaussian elimination.

The solution, $\{x + e\}$, is updated by subtracting $\{e\}$:

$$\{\text{new solution}\} = \{x + e\} - \{e\}$$

But because $\{e\}$ is solved using Gaussian elimination, it, too, is associated with round-off error. The new solution, though more accurate than the first, is not as precise as may be desired. Iteration again proceeds through the following steps until the solution for a current iteration is equal to that for the previous iteration:

1. Calculate $\{e\}$ from $[A]\{e\} = \{b'\}$
2. Calculate $\{x + e\} - \{e\}$
3. Recompute $\{b'\}$ as $[A]\{\{x + e\} - \{e\}\} - \{b\}$
4. Repeat steps 1 through 3 until $\{b'\} = \{0\}$

An expanded version of the Gaussian elimination computer program (Figure 2.10, page 41) allows iterative improvement. Using iterative improvement renders Gaussian elimination as accurate as any other equation solution algorithm, and the accuracy is achieved using only single-precision arithmetic, which conserves computer memory.

Chapter Summary

- Carl Friedrich Gauss is one of the greatest mathematical and scientific geniuses who ever lived.
- The mathematical methods contributed by Gauss surpass in relevancy those contributed by Newton and Leibniz for the solution of problems on digital computers.

- Gauss' error theory is a fundamental contribution to mathematics.
- Gaussian quadrature yields the exact solution to definite integrals for polynomials of degree $2m + 1$ or less.
- Linear least-squares regression is a direct contribution of Gauss and a fundamental aspect of Gauss' error theory.
- Gaussian elimination is an important method for computer solution of simultaneous equations.

Exercises

(An asterisk indicates an advanced or graduate student problem)

1. Use Gaussian quadrature to approximate the definite integrals listed in the exercises for Chapter 1.

*2. Derive the Gaussian quadrature weights and locations for $m = 3$. (*Hint:* Two sets of symmetrically located points will be used. Call one set $-t_1, +t_1$, the other $-t_2, +t_2$. Let the weight for the first pair be designated a_1 and that for the second pair be designated a_2. Now, four unknowns are described and four equations are needed for the solution.)

3. For the following data sets, determine whether $y = mx + b$ is an appropriate model. If so, compute m and b.
Set 1 (from McCuen, 1985, p. 204):

   x: 3 5 6 7 9
   y: 4 8 7 6 10

   Set 2 (from Kreyszig, 1988, p. 1290):

   speed x (mph):   20 30   40   50
   stopping distance (ft): 50 95 150 210

   Set 3 (from Kreyszig, 1988, p. 1290), number of revolutions per min x versus horsepower y of a diesel engine:

   x: 400   500   600   700   750
   y: 580 1030 1420 1880 2100

   Set 4 (from Miller and Freund, 1965, p. 237), tensile force x in thousands of pounds is applied to steel specimens that elongate an amount y in thousandths of an inch:

   x: 1   2   3   4   5   6
   y: 15 35 41 63 77 84

   Set 5 (from Miller and Freund, 1965, p. 237), the amount of potassium bromide y that will dissolve in 100 grams of water at a temperature x is measured:

   x (in degrees Celsius):   0 10 20 30 40 50
   y (in grams):   52 60 64 73 76 81

*4. Suppose the following regression model is more appropriate for a set of data: $y = mx^2 + wx + b$. Use Gauss' error theory to obtain a general solution for m, w, and b, given N pairs of observations (x, y).

5. Solve the following systems of simultaneous equations by hand using Gaussian elimination:

(a) $9x - 9y + 3z = 42$
$-1.5x + 13.5y - 4.5z = -15$
$\dfrac{3x}{8} - 1.5y + 0.5z = \dfrac{5}{2}$

(b) $4x + 4y = 0$
$12x - 16y = 4$

(c) $2x + y + 4z = 15$
$3x + 7y + 2z = 46$
$5x + y + 6z = 26$

(d) $3x + 2y + 7z + 8w = 75$
$x + 4y + 5z + 9w = 82$
$2x + 8y + 2z + w = 37$
$3x + y + z + 6w = 52$

(e) $3x + y = 4$
$x + 7y = 8$

6. Use the computer algorithm GAUSS to verify the answers for Problem 5.

7. In Section 2.3, a solution for a set of four simultaneous equations was determined using Gaussian elimination: $x = 1.038335$, $y = 0.208898$, $z = 0.22636$, and $t = 0.846775$. Substitute these four values into any one of the four simultaneous equations used in Section 2.3. Do you identify any error? If so, how may this error be reduced? Test your answers by trying to reduce the error.

References and Suggested Readings

BUHLER, W.K. 1981. *Gauss: A Biographical Study*. Berlin: Springer.

COOK, R.D. 1974. *Concepts and Applications of Finite Element Analysis*. New York: Wiley.

DAHLQUIST, G., and A. BJORCK. 1974. *Numerical Methods* (translated by N. Anderson). Englewood Cliffs, NJ: Prentice-Hall.

HALL, T. 1970. *Carl Friedrich Gauss* (translated by A. Froderberg). Cambridge, MA: MIT Press.

JAMES, M.L., G.M. SMITH, and J.C. WOLFORD. 1967. *Applied Numerical Methods for Digital Computation with FORTRAN*. Scranton, PA: International Textbook.

KREYSZIG, E. 1988. *Advanced Engineering Mathematics*, 6th ed. New York: Wiley.

McCARN, D.W., and J.R. CARR. 1992. Effect of numerical precision and equation solution algorithm on computation of kriging weights. *Computers and Geosciences* **18**.

McCUEN, R.H. 1985. *Statistical Methods for Engineers*. Englewood Cliffs, NJ: Prentice-Hall.

MILLER, I., and J.E. FREUND. 1965. *Probability and Statistics for Engineers*. Englewood Cliffs, NJ: Prentice-Hall.

PRESS, W.H., B.P. FLANNERY, S.A. TEUKOLSKY, and W.T. VETTERLING. 1986. *Numerical Recipes*. Cambridge, UK: Cambridge University Press.

SCARBOROUGH, J.B. 1955. *Numerical Mathematical Analysis*, 3d ed. Baltimore: The Johns Hopkins Press.

Figure 2.4 A listing for the program LINEAR. Subroutine EIGENJ is borrowed from Davis: *Statistics and Data Analysis in Geology.* Copyright © 1973 John Wiley & Sons. Used with permission.

```fortran
      PROGRAM LINEAR
C
C     AN ANSI STANDARD FORTRAN-77 PROGRAM FOR LINEAR
C     REGRESSION OF Y ON X
C
C*****************************************************
C                  USER'S GUIDE FOR LINEAR
C*****************************************************
C
C     DATA ENTRY TO THIS PROGRAM CONSISTS OF N PAIRS,
C     (X,Y).
C
C     DATA FILE CREATION:
C
C       LINES 1 THROUGH N:
C
C       ENTER THE PAIR, X,Y; ONE PAIR PER LINE
C
C     EXAMPLE:
C
C       GIVEN THE DATA:
C
C             X         Y
C
C             1         5
C             3         10
C             5         12
C
C     THE FILE WOULD BE:
C
C       1,5
C       3,10
C       5,12
C
C*****************************************************
C              END OF INPUT GUIDE
C*****************************************************

      REAL M
C
C     THIS PROGRAM IS NOT LIMITED BY THE SIZE OF THE DATA FILE
C
      OPEN(5, FILE = ' ')
C
C              INITIALIZE THE SUMS:   SUMX, SUMY, SUMXY, AND SUMX2
C
      SUMX = 0.0
      SUMY = 0.0
      SUMXY = 0.0
      SUMX2 = 0.0
      N = 0
C
C     ESTABLISH AN IMPLIED, GO TO LOOP; THIS ALLOWS
C     THE PROGRAM TO BE APPLIED TO DATA SETS OF ANY
C     SIZE, N
C
   10 READ(5,*,END=20) X,Y
      N = N + 1
      SUMX = SUMX + X
      SUMY = SUMY + Y
      SUMXY = SUMXY + X * Y
      SUMX2 = SUMX2 + X * X
      GO TO 10
   20 CONTINUE
C
C     ONCE THE SUMS ARE CALCULATED, COMPUTE THE
C     VALUES FOR M AND B
C
      M = (N*SUMXY - SUMX*SUMY)/(N*SUMX2 - SUMX*SUMX)
      B = (SUMY/N) - M*SUMX/N
      WRITE(*,30) M,B
   30 FORMAT(10X,'LINEAR REGRESSION SUMMARY',//,
     2       10X,'SLOPE, M           = ', 1E15.5,/,
     3       10X,'INTERCEPT, B       = ', 1E15.5)
      STOP
      END
```

38

Figure 2.7 A listing for the program GAUSS. Subroutine EIGENJ is borrowed from Davis: *Statistics and Data Analysis in Geology.* Copyright © 1973 John Wiley & Sons. Used with permission.

```fortran
      PROGRAM GAUSS
C
C     A STANDARD FORTRAN-77 PROGRAM FOR GAUSS ELIMINATION
C     USING A MAXIMUM PIVOT ELEMENT SEARCH FOR EACH
C     UNKNOWN
C
C     THIS PROGRAM IS A FORTRAN-77 UPDATE OF A PROGRAM
C     PRESENTED IN JAMES, M.L., ET. AL., 1967,
C     PP. 192 - 193.
C
C*******************************************************
C                USER'S GUIDE FOR GAUSS
C*******************************************************
C
C     THE INPUT TO THE PROGRAM GAUSS CONSISTS
C     OF A MATRIX, [A], AND A VECTOR OF KNOWN VALUES, {B}.
C     THEN, THE OBJECTIVE OF THIS PROGRAM IS TO SOLVE FOR
C     {X}, SUCH THAT [A]{X} = {B}.
C
C     DATA FILE CREATION:
C
C     LINE 1:  ENTER N
C
C        WHERE N IS THE SIZE OF THE SYSTEM; I.E., [A] IS
C        AN N X N MATRIX; {B} IS AN N X 1 VECTOR AS IS {X}.
C
C     LINES 2 THROUGH N + 1:  ENTER THE MATRIX, [A]:
C
C        USING ONE LINE FOR EACH ROW OF [A]
C
C     LINE N + 2:  ENTER THE VECTOR, {B}, ON ONE LINE.
C
C     EXAMPLE, IF [A] =  1  2    AND {B} = 5
C                        3  4              6
C
C     THEN, THE FILE LOOKS LIKE:
C
C        2
C        1,2
C        3,4
C        5,6
C
C*******************************************************
C                END OF INPUT GUIDE
C*******************************************************
      PARAMETER (NN = 100)
      DIMENSION A(NN,NN+1), B(NN), X(NN)
C
C     CHANGE THE DIMENSION STATEMENT BY SIMPLY MODIFYING
C     THE PARAMETER STATEMENT; PRESENT DIMENSIONING ALLOWS
C     MATRICES UP TO 100 X 100.
C
C     READ THE MATRICES, A AND B INTO CORE
C
      OPEN(5, FILE = ' ')
      READ(5,*) N
C     N IS THE SIZE OF THE MATRIX TO BE SOLVED
      DO I = 1,N
         READ(5,*) (A(I,JK), JK = 1,N)
      END DO
         READ(5,*) (B(JK), JK = 1,N)
C
C     PLACE THE VECTOR, B, INTO THE N+1 COLUMN OF A
C
      DO I = 1,N
         A(I, N+1) = B(I)
      END DO
C
```

Figure 2.7 (continued)

```fortran
C
C       INITIALIZE MATRIX SIZES
C
        M = N + 1
        L = N - 1
C
C       BEGIN THE FORWARD REDUCTION (ELIMINATION)
C
        DO K = 1,L
        JJ = K
        XLARG = ABS(A(K,K))
        K1 = K + 1
C
C       SEARCH FOR LARGEST PIVOT ELEMENT
C
        DO I = K1, N
        CA = ABS(A(I,K))
        DIFF = XLARG - CA
        IF (DIFF .LT. 0) THEN
            XLARG = CA
            JJ = I
        END IF
        END DO
        IDIFF = JJ - K
C
C       SWAP ROWS, IF NECESSARY, DEPENDING ON THE
C       POSITION OF THE MAXIMUM PIVOT ELEMENT.
C
        IF (IDIFF .NE. 0) THEN
        DO J = K,M
            TEMP = A(JJ,J)
            A(JJ,J) = A(K,J)
            A(K,J) = TEMP
        END DO
        END IF
C
C       THIS IS THE MAJOR ASPECT OF GAUSS ELIMINATION:
C       DIVIDE REMAINING ELEMENTS OF A (AND B WHICH
C       IS IN THE LAST COLUMN OF A) BY THE PIVOT
C       ELEMENT, A(K,K)

C
        DO I = K1, N
        QUOT = A(I,K) / A(K,K)
            DO J = K1, M
                A(I,J) = A(I,J) - QUOT * A(K,J)
            END DO
        END DO
        DO I = K1, N
            A(I,K) = 0.0
        END DO
        END DO
C
C       SOLVE, NOW, FOR LAST UNKNOWN
C
        X(N) = A(N,M) / A(N,N)
C
C       USING THIS SOLUTION, BACK SUBSTITUTE TO FIND
C       THE REMAINING UNKNOWNS
C
        DO IN = 1,L
        SUM = 0.0
        I = N - IN
        I1 = I + 1
        DO J = I1, N
            SUM = SUM + A(I,J) * X(J)
        END DO
        X(I) = (A(I,M) - SUM) / A(I,I)
        END DO
C
C       PRINT THE RESULTANT VECTOR, X
C
        WRITE(*, 200)
200     FORMAT(1H0, 'GAUSS ELIMINATION RESULTS', //)
        DO I = 1,N
        WRITE(*,300) I, X(I)
        END DO
300     FORMAT (2X, 'X(',I3,')  = ',F15.3)
        STOP
        END
```

Figure 2.10 A listing for the program GAUSIT. Subroutine EIGENJ is borrowed from Davis: *Statistics and Data Analysis in Geology.* Copyright © 1973 John Wiley & Sons. Used with permission.

```fortran
      PROGRAM GAUSIT
C
C     A STANDARD FORTRAN-77 PROGRAM FOR GAUSS ELIMINATION
C     USING A MAXIMUM PIVOT ELEMENT SEARCH FOR EACH
C     UNKNOWN
C
C     THIS VERSION OF GAUSS ELIMINATION IS USED WITH
C     ITERATIVE IMPROVEMENT TO YIELD CALCULATIONS OF
C     SUPERIOR ACCURACY
C
C     THIS PROGRAM IS A FORTRAN-77 UPDATE OF A PROGRAM
C     PRESENTED IN JAMES, M.L., ET. AL., 1967,
C     PP. 192 - 193.
C
C*********************************************************
C                   USER'S GUIDE
C*********************************************************
C
C     LINE 1:  (FREE FORMAT)
C        READ(5,*) N, NITT
C           WHERE N IS THE SIZE OF THE EQUATION SYSTEM
C                 TO BE SOLVED
C           NITT IS THE NUMBER OF ITERATIVE IMPROVEMENTS
C                 DESIRED; UP TO 10 (MAX.)
C
C     LINES 2 THROUGH N + 1:  (FREE FORMAT)
C        DO I = 1,N
C          READ(5,*) (A(I,J), J = 1,N)
C        END DO
C        IN OTHER WORDS:  ENTER THE MATRIX, [A], ONE
C                         ROW AT A TIME
C
C     LINE N + 2:  (FREE FORMAT)
C        ENTER THE VECTOR, {B}
C
C*********************************************************
C        END OF USER'S GUIDE
C*********************************************************

      PARAMETER (NN = 100)
      DIMENSION A(NN,NN+1), B(NN), X(NN), TEMP(NN,NN+1)
C
C     CHANGE THE DIMENSION STATEMENT BY SIMPLY MODIFYING
C     THE PARAMETER STATEMENT; PRESENT DIMENSIONING ALLOWS
C     MATRICES UP TO 100 X 100.
C
C     READ THE MATRICES, A AND B INTO CORE
C
      OPEN(5, FILE = ' ')
      READ(5,*) N, NITT
C     N IS THE SIZE OF THE MATRIX TO BE SOLVED
      DO I = 1,N
        READ(5,*) (A(I,JK), JK = 1,N)
      END DO
        READ(5,*) (B(JK), JK = 1,N)
C
C     PLACE THE VECTOR, B, INTO THE N+1 COLUMN OF A
C
      DO I = 1,N
        A(I, N+1) = B(I)
      END DO
C
C     INITIALIZE MATRIX SIZES
C
      M = N + 1
      L = N - 1
C
C     PRESERVE THE MATRIX, A, BY PLACING IT IN THE MATRIX, TEMP
C
      DO I = 1,N
      DO J = 1,M
        TEMP(I,J) = A(I,J)
      END DO
      END DO
C
C     CALL THE SUBROUTINE SOLVE FOR A SOLUTION
C
```

Figure 2.10 (continued)

```fortran
C
      CALL SOLVE (TEMP, X, M, L)
C
C     USE ITERATIVE IMPROVEMENT FOR GREATER ACCURACY
C
      DO I = 1, NITT
        CALL GPROVE(A, TEMP, X, M, L)
      END DO
C
C     PRINT THE SOLUTION, X
C
      WRITE(*,100)
      DO I = 1,N
        WRITE(*,200) X(I)
      END DO
100   FORMAT(10X, 'SOLUTION FOR UNKNOWNS', /)
200   FORMAT(10X, 1E15.5)
      STOP
      END
C
C
      SUBROUTINE SOLVE (A, X, M, L)
C
C     SUBROUTINE TO PERFORM GAUSS ELIMINATION
C
      PARAMETER (NN = 100)
      DIMENSION A(NN,NN+1), X(NN)
C
C     BEGIN THE FORWARD REDUCTION (ELIMINATION)
C
      N = L + 1
      DO K = 1,L
      JJ = K
        XLARG = ABS(A(K,K))
        K1 = K + 1
C
C     SEARCH FOR LARGEST PIVOT ELEMENT

C
      DO I = K1, N
        CA = ABS(A(I,K))
        DIFF = XLARG - CA
        IF (DIFF .LT. 0) THEN
          XLARG = CA
          JJ = I
        END IF
      END DO
      IDIFF = JJ - K
C
C     SWAP ROWS, IF NECESSARY, DEPENDING ON THE
C     POSITION OF THE MAXIMUM PIVOT ELEMENT.
C
      IF (IDIFF .NE. 0) THEN
        DO J = K,M
          TEMP = A(JJ,J)
          A(JJ,J) = A(K,J)
          A(K,J) = TEMP
        END DO
      END IF
C
C     THIS IS THE MAJOR ASPECT OF GAUSS ELIMINATION:
C     DIVIDE REMAINING ELEMENTS OF A (AND B WHICH
C     IS IN THE LAST COLUMN OF A) BY THE PIVOT
C     ELEMENT, A(K,K)
C
      DO I = K1, N
        QUOT = A(I,K) / A(K,K)
        DO J = K1, M
          A(I,J) = A(I,J) - QUOT * A(K,J)
        END DO
      END DO
      DO I = K1, N
        A(I,K) = 0.0
      END DO
      END DO
```

42

Figure 2.10 (continued)

```fortran
C
C      SOLVE, NOW, FOR LAST UNKNOWN
C
       X(N) = A(N,M) / A(N,N)
C
C      USING THIS SOLUTION, BACK SUBSTITUTE TO FIND
C      THE REMAINING UNKNOWNS
C
       DO IN = 1, L
         SUM = 0.0
         I = N - IN
         I1 = I + 1
         DO J = I1, N
           SUM = SUM + A(I,J) * X(J)
         END DO
         X(I) = (A(I,M) - SUM) / A(I,I)
       END DO
C
C      RETURN TO MAIN CALLING PROGRAM
C
       RETURN
       END
C
C
       SUBROUTINE GPROVE (A, TEMP, X, M, L)
C
C      PROGRAM TO PROVIDE BETTER ACCURACY IN GAUSS
C      ELIMINATION THROUGH ITERATIVE IMPROVEMENT
C
       PARAMETER (NN = 100)
       DIMENSION A(NN,NN+1), TEMP(NN,NN+1), X(NN)
       DIMENSION R(NN), RR(NN)
C
C      STEP 1:  MULTIPLY THE SOLUTION, X, BY ORIGINAL
C               MATRIX, A
C
       N = M - 1
       DO I = 1,N
       R(I) = 0.0
         DO J = 1,N
           R(I) = R(I) + A(I,J) * X(J)
         END DO
       END DO
C
C      STEP 2:  SUBTRACT THE VECTOR, B, FROM THE VECTOR, R.
C
       DO I = 1, N
         R(I) = R(I) - A(I,M)
       END DO
C
C      STEP 3:  PLACE THE MATRIX, A, INTO THE MATRIX, TEMP
C
       DO I = 1,N
         DO J = 1,N
           TEMP(I,J) = A(I,J)
         END DO
       END DO
C
C      STEP 3A:  PLACE THE VECTOR, R, INTO THE N+1 COLUMN
C                OF THE MATRIX, TEMP.
C
       DO I = 1,N
         TEMP(I,M) = R(I)
       END DO
C
C      STEP 4:  SOLVE FOR THE ERROR, E (PLACED IN THE VECTOR,
C               RR)
C
       CALL SOLVE(TEMP, RR, M, L)
C
C      STEP 5 (FINAL STEP):  UPDATE THE SOLUTION, X.
C
       DO I = 1, N
         X(I) = X(I) - RR(I)
       END DO
C
C      RETURN TO CALLING PROGRAM
C
       RETURN
       END
```

43

3

Concepts in Matrix Algebra

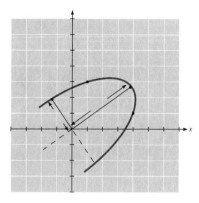

Matrix algebra is used throughout this text. Students should be well versed in the representation of equations in matrix form and in the unique mathematical notation associated with matrix operations.

A matrix is simply a rectangular array of numbers. The numbers may represent known coefficients in a system of equations, or they may represent some sort of physical phenomenon, such as electromagnetic flux, where the matrix is a digital image. Hence matrices play a variety of critical roles in numerical analysis. This chapter presents some basic rules for matrix operations. The remainder of the text demonstrates applications of matrix algebra for earth sciences study.

3.1 The Concept of a Matrix

The explanation of Gaussian elimination in the previous chapter began with the following set of simultaneous equations:

$$2.63x + 5.21y - 1.694z + 0.938t - 4.23 = 0$$
$$3.16x - 2.95y + 0.813z - 4.21t + 0.716 = 0$$

$$5.36x + 1.88y - 2.15z - 4.95t - 1.28 = 0$$
$$1.34x + 2.98y - 0.432z - 1.768t - 0.419 = 0$$

If the constants (those components of the equations not associated with x, y, z, or t) are moved to the right-hand side, then

$$2.63x + 5.21y - 1.694z + 0.938t = 4.23$$
$$3.16x - 2.95y + 0.813z - 4.21t = -0.716$$
$$5.36x + 1.88y - 2.15z - 4.95t = 1.28$$
$$1.34x + 2.98y - 0.432z - 1.768t = 0.419$$

Matrix notation is a convenient way to represent a system of equations. The equations above can be expressed using matrix multiplication protocol:

$$\begin{vmatrix} 2.63 & 5.21 & -1.694 & 0.938 \\ 3.16 & -2.95 & 0.813 & -4.21 \\ 5.36 & 1.88 & -2.15 & -4.95 \\ 1.34 & 2.98 & -0.432 & -1.768 \end{vmatrix} \begin{vmatrix} x \\ y \\ z \\ t \end{vmatrix} = \begin{vmatrix} 4.23 \\ -0.716 \\ 1.28 \\ 0.419 \end{vmatrix}$$

Expressing equations in this form is a rather abrupt leap. Before matrix multiplication can be explained, we must define some basic terms.

A matrix is a rectangular array of numbers:

$$\begin{vmatrix} a_{11} & a_{12} & \cdots & a_{1N} \\ a_{21} & a_{22} & \cdots & a_{2N} \\ \vdots & \vdots & \cdots & \vdots \\ a_{M1} & a_{M2} & \cdots & a_{MN} \end{vmatrix}$$

$$M \times N$$

where the size of a matrix is always specified as $M \times N$, for which M is the number of rows of the matrix and N is the number of columns (e.g., Scarborough (1955), p. 512). A matrix entry is specified using subscripts, with the first number representing the row position of the entry, and the second number representing the column position. For instance, a_{23} represents the entry in matrix a in the second row, third column.

A vector is a special form of a rectangular matrix. Given a matrix of size $M \times N$, if N is equal to 1, a column vector is indicated. If M is equal to 1, a row vector is indicated. A vector is, in other words, a one-dimensional matrix.

Rule for Matrix Multiplication   Two matrices, $[A]$ and $[B]$, may be multiplied in the following order: $[A][B] = [C]$, if, and only if, the number of *columns* N of matrix $[A]$ is equal to the number of rows M of matrix $[B]$. The resultant matrix $[C]$ has the same number of rows as matrix $[A]$ and the same number of columns as matrix $[B]$. Matrix multiplication is not commutative.

In multiplying two matrices, such as

$$[A][B] = [C]$$

each row of matrix $[A]$ is multiplied to each and every column of matrix $[B]$ in a summation process:

$$C_{i,k} = \sum_{j=1}^{N} A_{i,j}B_{j,k}, \qquad i = 1, 2, \ldots, M, \quad k = 1, 2, \ldots, L$$

where matrix $[A]$ has the size $M \times N$, matrix $[B]$ has the size $N \times L$, and matrix $[C]$ has the size $M \times L$.

Example 3.1

The matrix equation

$$
\begin{vmatrix}
2.630 & 5.210 & -1.694 & 0.938 \\
3.160 & -2.950 & 0.813 & -4.210 \\
5.360 & 1.880 & -2.150 & -4.950 \\
1.340 & 2.980 & -0.432 & -1.768
\end{vmatrix}
\begin{vmatrix} x \\ y \\ z \\ t \end{vmatrix}
=
\begin{vmatrix} 4.230 \\ -0.716 \\ 1.280 \\ 0.419 \end{vmatrix}
$$

$$4 \times 4 \qquad\qquad 4 \times 1 \qquad 4 \times 1$$

has the form $[A]\{x\} = \{B\}$. Matrix $[A]$ is square: $M = N = 4$; matrix $\{x\}$ is a column vector, as is the resultant matrix $\{B\}$. Multiplication is performed in the order shown because the number of columns in matrix $[A]$ is four, which equals the number of rows of vector $\{x\}$. Vector $\{B\}$ has the same number of rows as does $[A]$ and the same number of columns as does $\{x\}$. With $M = 4$, $N = 4$, and $L = 1$, we have

$$B_{11} = 2.63x + 5.21y - 1.694z + 0.938t = 4.23$$
$$B_{21} = 3.16x - 2.95y + 0.813z - 4.21t = -0.716$$
$$B_{31} = 5.36x + 1.88y - 2.15z - 4.95t = 1.28$$
$$B_{41} = 1.34x + 2.98y - 0.432z - 1.768t = 0.419$$

This is the system of equations presented at the beginning of Section 3.1. The implied matrix multiplication alluded to earlier is revealed.   ∎

Example 3.2

Given the following matrix system, compute $[C]$:

$$
\begin{vmatrix} 3 & 2 & 9 \\ 7 & 1 & 5 \\ 2 & 4 & 8 \end{vmatrix}
\begin{vmatrix} 2 & 5 \\ 8 & 3 \\ 9 & 1 \end{vmatrix}
= |C|
$$

$$3 \times 3 \qquad 3 \times 2 \quad 3 \times 2$$

SOLUTION

$$C_{11} = 3 \times 2 + 2 \times 8 + 9 \times 9 = 103$$
$$C_{12} = 3 \times 5 + 2 \times 3 + 9 \times 1 = 30$$
$$C_{21} = 7 \times 2 + 1 \times 8 + 5 \times 9 = 67$$
$$C_{22} = 7 \times 5 + 1 \times 3 + 5 \times 1 = 43$$
$$C_{31} = 2 \times 2 + 4 \times 8 + 8 \times 9 = 108$$
$$C_{32} = 2 \times 5 + 4 \times 3 + 8 \times 1 = 30$$

Therefore,

$$
|C| =
\begin{vmatrix}
103 & 30 \\
67 & 43 \\
108 & 30
\end{vmatrix}
$$

∎

Example 3.3

A FORTRAN-77 computer algorithm for matrix multiplication, MULMAT (Figure 3.1, page 60), may be used to solve for matrix $[C]$ in Example 3.2. A data file for solving this problem (Figure 3.2) was created using the user's guide listed at the beginning of the program. Applying the FORTRAN program to this data file yields the output (Figure 3.3), which matches that calculated by hand. •

```
3,3,2
3,2,9
7,1,5
2,4,8
2,5
8,3
9,1

0;37;44m C:\TEXT=
```

Figure 3.2   A listing of a data file for MULMAT to solve Example 3.3.

```
        MATRIX MULTIPLICATION RESULTS


   MATRIX, C, ROW =  1:     103.000     30.000
   MATRIX, C, ROW =  2:      67.000     43.000
   MATRIX, C, ROW =  3:     108.000     30.000
Stop - Program terminated.


0;37;44m C:\TEXT=
```

Figure 3.3   Resultant output from MULMAT for the data file listed in Figure 3.2.

3.2 Linear, Least-Squares Regression Revisited

Gauss' linear, least-squares regression is derived here using matrix algebra. Given N pairs of observations (x, y), the model $y_i = mx_i + b$ is written explicitly for each of the N pairs, yielding N simultaneous equations:

$$y_1 = mx_1 + b$$
$$y_2 = mx_2 + b$$
$$\vdots$$
$$y_N = mx_N + b$$

This system of equations can be written in matrix form:

$$
\begin{vmatrix} y_1 \\ y_2 \\ \vdots \\ y_N \end{vmatrix}
=
\begin{vmatrix} x_1 & 1 \\ x_2 & 1 \\ \vdots & \vdots \\ x_N & 1 \end{vmatrix}
\begin{vmatrix} m \\ b \end{vmatrix}
$$

Let this system be written as $\{y\} = [x]\{m\}$ for simplicity. Then, solving for $\{m\}$, we get

$$\{m\} = [x]^{-1}\{y\}$$

where $[x]^{-1}$ is the inverse of matrix $[x]$.

The inverse of a matrix is defined as

$$[A]^{-1}[A] = [I]$$

where $[I]$ is the multiplicative identity. Every entry in an identity is zero except for the diagonal entries, which are each 1. By this description, the identity $[I]$ is seen to be the matrix equivalent of the number 1; in fact, $[A][I] = [A]$ is another way to define the identity matrix. Matrices $[A]$ and $[I]$ are square matrices; the number of rows in these matrices equals the number of columns. A diagonal entry of a matrix pertains to square matrices. The row number of the entry is equal to the column number for the entry; for instance, a_{11}, a_{22}, or, in general, a_{ii} is a diagonal entry.

The inverse of a matrix is defined for some square matrices. The matrix $[x]$, which is presented for the linear, least-squares regression problem, has a size $N \times 2$, which is square only if $N = 2$. But, a square matrix is formed from $[x]$ for any value of N as follows:

1. Form the transpose of $[x]$: $[x]^T$;
2. Multiply both sides of the system by the transpose:

$$
\begin{array}{ccccc}
[x]^T & \{y\} & = & [x]^T & [x] & \{m\} \\
2 \times N & N \times 1 & & 2 \times N & N \times 2 & 2 \times 1
\end{array}
$$

Notice from matrix multiplication that the product $[x]^T\{y\}$ is a column vector with a size 2×1 (regardless of the size of N). The product $[x]^T[x]$ is a square matrix having a size 2×2 (regardless of the size of N).

The transpose of a matrix is formed as follows. Given a rectangular matrix $[A]$, its transpose $[A]^T$ is formed such that the rows in the transpose $[A]^T$ are the columns of $[A]$. In other words,

$$A_{ij}^T = A_{ji}$$

In the least-squares regression problem, matrix $[x]$ is

$$
\begin{vmatrix}
x_1 & 1 \\
x_2 & 1 \\
\vdots & \vdots \\
x_N & 1
\end{vmatrix}
$$

Therefore, $[x]^T$ is

$$
\begin{vmatrix}
x_1 & x_2 & \cdots & x_N \\
1 & 1 & \cdots & 1
\end{vmatrix}
$$

The product $[x]^T[x]$ is therefore

$$
\begin{vmatrix}
x_1 & x_2 & \cdots & x_N \\
1 & 1 & \cdots & 1
\end{vmatrix}
\begin{vmatrix}
x_1 & 1 \\
x_2 & 1 \\
\vdots & \vdots \\
x_N & 1
\end{vmatrix}
=
\begin{vmatrix}
\sum_{i=1}^{N} x_i^2 & \sum_{i=1}^{N} x_i \\
\sum_{i=1}^{N} x_i & N
\end{vmatrix}
$$

$$
\begin{array}{ccc}
2 \times N & N \times 2 & 2 \times 2
\end{array}
$$

In solving for $\{m\}$, the inverse of the product $[x]^T[x]$ must be computed first. This product is a 2×2 matrix. The inverse of a matrix this size is relatively simple to compute. For a matrix $[C]$ whose size is 2×2,

$$\begin{vmatrix} c_{11} & c_{12} \\ c_{21} & c_{22} \end{vmatrix}$$

the inverse of $[C]$, designated as $[C]^{-1}$, is

$$[C]^{-1} = \frac{1}{\det C} \begin{vmatrix} c_{22} & -c_{12} \\ -c_{21} & c_{11} \end{vmatrix}$$

where det C is the determinant of matrix C and is given by

$$\det C = c_{11}c_{22} - c_{12}c_{21}$$

In summary, the inverse of a 2×2 matrix is computed by reversing the positions of the diagonal entries, multiplying the off-diagonal entries by -1, and dividing the four matrix entries by det C.

Returning to the solution for $\{m\}$, we find that matrix $[C]$ is equal to $[x]^T[x]$ if

$$[C] = [x]^T[x] = \begin{vmatrix} \sum_{i=1}^N x_i^2 & \sum_{i=1}^N x_i \\ \sum_{i=1}^N x_i & N \end{vmatrix}$$

Then

$$[C]^{-1} = \frac{1}{\det C} \begin{vmatrix} N & -\sum_{i=1}^N x_i \\ -\sum_{i=1}^N x_i & \sum_{i=1}^N x_i^2 \end{vmatrix}$$

and

$$\det C = N \sum_{i=1}^N x_i^2 - \left(\sum_{i=1}^N x_i \right)^2$$

Solving for the unknowns m and b yields

$$\begin{vmatrix} m \\ b \end{vmatrix} = [C]^{-1}[x]^T[y]$$

Partially completing the right-hand side gives

$$[x]^T[y] = \begin{vmatrix} x_1 & x_2 & \cdots & x_N \\ 1 & 1 & \cdots & 1 \end{vmatrix} \begin{vmatrix} y_1 \\ y_2 \\ \vdots \\ y_N \end{vmatrix} = \begin{vmatrix} \sum_{i=1}^N x_i y_i \\ \sum_{i=1}^N y_i \end{vmatrix}$$

and substituting for $[C]^{-1}$, we find

$$\begin{vmatrix} m \\ b \end{vmatrix} = \frac{1}{\det C} \begin{vmatrix} N & -\sum_{i=1}^N x_i \\ -\sum_{i=1}^N x_i & \sum_{i=1}^N x_i^2 \end{vmatrix} \begin{vmatrix} \sum_{i=1}^N x_i y_i \\ \sum_{i=1}^N y_i \end{vmatrix}$$

Then

$$m = \frac{N\sum_{i=1}^N x_i y_i - \sum_{i=1}^N x_i \sum_{i=1}^N y_i}{\det C} = \frac{N\sum_{i=1}^N x_i y_i - \sum_{i=1}^N x_i \sum_{i=1}^N y_i}{N\sum_{i=1}^N x_i^2 - (\sum_{i=1}^N x_i)^2}$$

and

$$b = \frac{-\sum_{i=1}^N x_i \sum_{i=1}^N x_i y_i + \sum_{i=1}^N x_i^2 \sum_{i=1}^N y_i}{\det C} = \frac{\sum_{i=1}^N y_i}{N} - \frac{m\sum_{i=1}^N x_i}{N}$$

The same solution for m and b was obtained in Chapter 2. In this example, the transpose of a matrix is introduced along with the concept of matrix inversion, at least for matrices of size 2×2. Computing the inverse of a matrix larger than 2×2 is tedious. Fortunately, Gaussian elimination offers a procedure for solving a series of simultaneous equations without the need to compute an inverse.

3.3 Eigendecomposition of a Matrix

With some types of problems, simultaneous, homogeneous equations of the form

$$a_{11}x_1 + a_{12}x_2 + \cdots + a_{1N}x_N - \lambda x_1 = 0$$
$$a_{21}x_1 + a_{22}x_2 + \cdots + a_{2N}x_N - \lambda x_2 = 0$$
$$\vdots$$
$$a_{N1}x_1 + a_{N2}x_2 + \cdots + a_{NN}x_N - \lambda x_N = 0$$

may arise (see LaFara (1973), pp. 138–140). Homogeneous equations are those for which each right-hand-side value is zero. Viewing the foregoing system of equations as a homogeneous system is demonstrated in the following. In matrix notation, this system is

$$AX = \lambda X$$

which can be written

$$(A - \lambda I)X = 0$$

in which I is the identity matrix, as described in Section 3.2. The right-hand-side value is now shown to be zero, hence the system is a homogeneous one. Many systems of simultaneous equations can be rewritten as a homogeneous system.

In this system of homogeneous equations, X is called an eigenvector, with entries x_i. Lambda is a scalar value known as an eigenvalue. The system of homogeneous equations has a nonzero (nontrivial) solution if, and only if, the determinant of $(A - \lambda I)$ is zero.

To illustrate the computation of eigenvalues and eigenvectors, consider the following matrix (Olea, 1991, p. 22):

$$[A] = \begin{vmatrix} 4 & 1 \\ 1 & 3 \end{vmatrix}$$

Then

$$A - \lambda I = \begin{vmatrix} 4 - \lambda & 1 \\ 1 & 3 - \lambda \end{vmatrix}$$

because

$$\lambda I = \begin{vmatrix} \lambda & 0 \\ 0 & \lambda \end{vmatrix}$$

The determinant of $(A - \lambda I)$ is $(4 - \lambda)(3 - \lambda) - 1$, which is set equal to zero to find a solution for λ:

$$(4 - \lambda)(3 - \lambda) - 1 = \lambda^2 - 7\lambda + 11 = 0$$

This equation has the following roots:

$$\lambda_1 = \frac{7 + \sqrt{5}}{2} = 4.618$$

$$\lambda_2 = \frac{7 - \sqrt{5}}{2} = 2.382$$

which are the eigenvalues for this matrix.

3.3.1 Roots of an Algebraic Expression

How are roots of an equation determined? Many methods exist for finding roots of an algebraic expression. Some of these methods are approximate. One method simply involves graphing the equation to find where the plot crosses the x axis (i.e., finding the value of the expression at $f(x) = 0$).

For example, the plot (Figure 3.4) of the equation

$$f(\lambda) = \lambda^2 - 7\lambda + 11 = 0$$

(used previously to determine the eigenvalues of the 2×2 matrix) may be used to determine the roots of this expression. Notice in Figure 3.4 that this expression has two roots. Moreover, this expression is a parabola, concave upward, with a minimum value determined as

$$\text{min} = f'(\lambda) = 2\lambda - 7 = 0$$

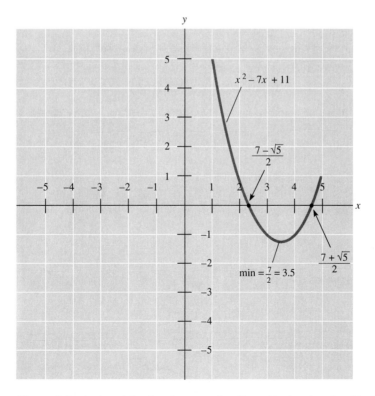

Figure 3.4 A plot of the function $y = x^2 - 7x + 11$, showing that this function has two roots (i.e., two possible values of x when $y = 0$): $x = [7 - \text{SQRT}(5)]/2$ and $x = [7 + \text{SQRT}(5)]/2$.

from which is found the minimum value, $\lambda = \frac{7}{2}$. Notice also that the roots are symmetrical with respect to the minimum point of the plot. If we let a represent the distance of symmetry, then

$$\text{roots} = \tfrac{7}{2} \pm a$$

Solving for a gives

$$\lambda^2 - 7\lambda = -11$$

Substituting $\frac{7}{2} + a$ for λ gives

$$(\tfrac{7}{2} + a)^2 - 7(\tfrac{7}{2} + a) = -11$$

which simplifies to

$$a = \pm \frac{\sqrt{5}}{2}$$

Therefore, the two roots of this equation are

$$\lambda_1 = \frac{7}{2} + \frac{\sqrt{5}}{2} \qquad \lambda_2 = \frac{7}{2} - \frac{\sqrt{5}}{2}$$

These roots are the eigenvalues. In general, the roots of a quadratic expression are found as

$$\frac{-b \pm \sqrt{b^2 - 4ac}}{2a}$$

for a quadratic expression $ax^2 + bx + c = 0$.

3.3.2 Eigenvectors

For each eigenvalue, an eigenvector exists that satisfies

$$(A - \lambda I)X = 0$$

But more than one vector can satisfy this condition because if X is a solution, so is bX, where b is some scalar multiplier. However, the ratio x_i/x_N can be determined uniquely, where these are individual entries in the eigenvector. (See the homogeneous system of equations listed at the start of Section 3.1; the components x are entries in the eigenvector.)

Given the 2×2 matrix $[A]$ having eigenvalues 4.618 and 2.382, the first eigenvector is computed using the first eigenvalue, 4.618, such that

$$(A - \lambda I)X = \begin{vmatrix} 4 - 4.618 & 1 \\ 1 & 3 - 4.618 \end{vmatrix} \begin{vmatrix} x_1 \\ x_2 \end{vmatrix}$$

which leads to two equations:

$$-0.618x_1 + x_2 = 0$$
$$x_1 - 1.618x_2 = 0$$

Following LaFara (1973), let $u_i = x_i/x_N$. In this case, $N = 2$ and so $u_1 = x_1/x_2$ and $u_2 = x_2/x_2 = 1$. Therefore,

$$-0.618u_1 = -1$$

from which we find that $u_1 = 1.618$. Implicitly, $u_2 = 1$; thus, the first eigenvector is $\{1.618 \quad 1\}$ for the first eigenvalue.

To compute the second eigenvector, the second eigenvalue, 2.382, is used such that

$$(A - \lambda I) = \begin{vmatrix} 4 - 2.382 & 1 \\ 1 & 3 - 2.382 \end{vmatrix} \begin{vmatrix} x_1 \\ x_2 \end{vmatrix}$$

which leads to

$$1.618x_1 + x_2 = 0$$
$$x_1 + 0.618x_2 = 0$$

If we let $u_1 = x_1/x_2$ and $u_2 = x_2/x_2 = 1$, then

$$1.618u_1 = -1 \quad \text{or} \quad u_1 = -0.618$$

Implicitly, $u_2 = 1$; therefore, the second eigenvector is $\{-0.618 \quad 1\}$.

VERIFICATION

We know that $(A - \lambda I)X$ must equal zero if $\lambda_1 = 4.618$ and that

$$\begin{vmatrix} 4 - 4.618 & 1 \\ 1 & 3 - 4.618 \end{vmatrix} \begin{vmatrix} 1.618 \\ 1 \end{vmatrix} = 0$$

Therefore, from matrix multiplication

$$-0.618(1.618) + 1 = 0$$
$$-1 + 1 = 0$$
$$0 = 0$$

And for the second equation,

$$1.618(1) - 1.618(1) = 0$$
$$0 = 0$$

For the second eigenvalue,

$$\begin{vmatrix} 4 - 2.382 & 1 \\ 1 & 3 - 2.382 \end{vmatrix} \begin{vmatrix} -0.618 \\ 1 \end{vmatrix} = 0$$

Matrix multiplication yields

$$1.618(-0.618) + 1(1) = 0$$
$$-1 + 1 = 0$$
$$0 = 0$$

and for the second equation,

$$-0.618(1) + 0.618(1) = 0$$
$$0 = 0$$

In Davis (1986) we see that the results of eigendecomposition of this matrix, [A], can be displayed graphically, as shown in Figure 3.5. The eigenvectors define the orientation of the major and minor axes of an ellipse. The eigenvalues define the lengths of each axis from the origin (0, 0) to the circumference of the ellipse; this is also shown in Figure 3.5. Furthermore, the two rows of matrix [A] determine two points, (4, 1) and (1, 3), that plot on the ellipse. This visual display should enhance our understanding of eigendecomposition.

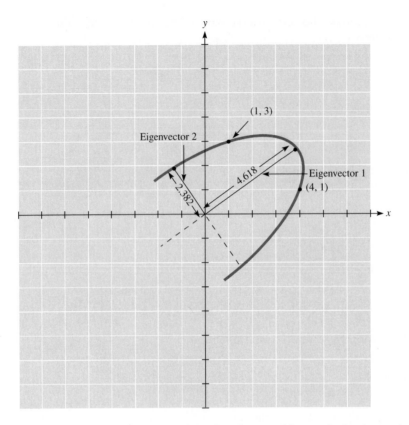

Figure 3.5 A graphical display of the eigendecomposition results for the matrix [*A*].

Example 3.4

Find the eigenvalues and eigenvectors for the following matrix:

$$[A] = \begin{vmatrix} 1.7 & 2.0 & 1.0 \\ 2.0 & -3.5 & -2.0 \\ 1.0 & -2.0 & 1.0 \end{vmatrix}$$

We can write this matrix as a homogeneous system:

$$[A - \lambda I] = \begin{vmatrix} 1.7 - \lambda & 2.0 & 1.0 \\ 2.0 & -3.5 - \lambda & -2.0 \\ 1.0 & -2.0 & 1 - \lambda \end{vmatrix}$$

First, the roots of the equation resulting from setting the determinant of this matrix equal to zero must be computed. But, in this example, the matrix has a size 3 × 3. The determinant of a 3 × 3 matrix is computed as follows:

1. Begin with a matrix [*C*] such that

$$[C] = \begin{vmatrix} c_{11} & c_{12} & c_{13} \\ c_{21} & c_{22} & c_{23} \\ c_{31} & c_{32} & c_{33} \end{vmatrix}$$

2. Compute the determinant of this matrix:

$$\det C = c_{11}[c_{22}c_{33} - c_{23}c_{32}] - c_{12}[c_{21}c_{33} - c_{23}c_{31}] + c_{13}[c_{21}c_{32} - c_{22}c_{31}]$$

The determinant for the 3×3 matrix used in this example is

$$\begin{aligned} \det(A - \lambda I) &= (1.7 - \lambda)\{(-3.5 - \lambda)(1 - \lambda) - [(-2)(-2)]\} \\ &\quad - (2)[2(1 - \lambda) - (-2)(1)] + (1)[2(-2) - (-3.5 - \lambda)(1)] \\ &= -21.25 + 16.75\lambda - 0.8\lambda^2 - \lambda^3 = 0 \end{aligned}$$

The roots of this equation are

$$\lambda_1 = 1.7, \qquad \lambda_2 = -5.0, \qquad \lambda_3 = 2.5$$

The eigenvectors are found using the expression $u_i = x_i/x_N$. For the first eigenvalue, 1.7,

$$\begin{aligned} (1.7 - 1.7)x_1 + 2x_2 + x_3 &= 0 \\ 2x_1 - 5.2x_2 - 2x_3 &= 0 \\ x_1 - 2x_2 - 0.7x_3 &= 0 \end{aligned}$$

Let $u_1 = x_1/x_3$, $u_2 = x_2/x_3$, and $u_3 = x_3/x_3 = 1$. From the first equation, $2u_2 = -1$ or $u_2 = -0.5$. From the second equation, $2u_1 - 5.2u_2 = 2$, because $u_3 = 1$. Since $u_2 = -0.5$, then $u_1 = -0.3$. Therefore, the complete eigenvector is $\{-0.3 \quad -0.5 \quad 1\}$.

The remaining eigenvectors are computed using a similar procedure. For $\lambda_2 = -5.0$ the eigenvector is $\{-0.909 \quad 2.545 \quad 1.000\}$. For $\lambda_3 = 2.5$ the eigenvector is $\{2.500 \quad 0.500 \quad 1.000\}$. ∎

3.3.3 Comments

Comment 1   The sum of eigenvalues for a matrix is equal to the trace of that matrix. The trace of a square matrix is equal to the sum of its diagonal terms:

$$\text{trace of } [A] = \sum_{i=1}^{N} a_{ii}$$

EXAMPLE

For the 2×2 matrix used earlier:

$$\begin{vmatrix} 4 & 1 \\ 1 & 3 \end{vmatrix}$$

the two eigenvalues are 4.618 and 2.382. The sum of these eigenvalues is 7. The sum of the diagonal entries of this matrix, 4 and 3, is also 7.

Comment 2   Each eigenvector is orthogonal to the rest. This implies that the multiplication of the transpose of any one eigenvector with another eigenvector yields a scalar product of zero:

$$\{x_i\}^T\{x_j\} = 0$$

EXAMPLE

For the 2×2 matrix used earlier, the eigenvectors are $\{1.618 \quad 1\}$ and $\{-0.618 \quad 1\}$. Using these vectors gives

$$[x_i]^T[x_j] = |\, 1.618 \quad 1\,| \begin{vmatrix} -0.681 \\ 1 \end{vmatrix} = -1 + 1 = 0$$

which demonstrates the orthogonality of eigenvectors. This characteristic of eigenvectors is important for analyses presented in Chapter 5.

3.3.4 Iterative Solution for Eigendecomposition

Eigenvalues and eigenvectors can be found for symmetrical, positive definite matrices using an iterative technique (LaFara, 1973, pp. 140–141). The eigendecomposition problem is written as follows:

$$\lambda_{k+1} x_{k+1} = AX_k$$

An initial guess at the solution is made:

$$X_0 = \begin{vmatrix} 0 \\ 0 \\ \vdots \\ 0 \\ 1 \end{vmatrix}$$

Iteration proceeds as follows:

STEP 1 $Y_{k+1} = AX_k$

STEP 2 $X_{k+1} = \dfrac{1}{(y_n)_{k+1}} Y_{k+1}$

where y_n is the last component in vector Y. The iteration repeats: step 1 to step 2 to step 1 ... until $X_{k+1} = X_k$; that is, until a solution is reached. With this solution,

$$\lambda = y_n$$

and X is the corresponding eigenvector.

3.3.5 Computer Example

A FORTRAN-77 computer program for eigendecomposition based on the iteration process (Figure 3.6, page 61) can be used to repeat Example 3.4 and compare the results. A data file for use with the computer algorithm to solve this problem is listed in Figure 3.7, developed using the user's guide listed at the beginning of the program. Applying the program to the data file leads to the solution shown in Figure 3.8, which is the same solution that was calculated earlier by hand.

```
3
1.7, 2., 1.
2., -3.5, -2.
1., -2., 1.

0;37;44m C:\TEXT=
```

Figure 3.7   A listing of a data file for use with EIGEND in solving Example 3.4.

```
EIGENDECOMPOSITION RESULTS

THE EIGENVALUES:

E-VALUE    1:              -5.000
E-VALUE    2:               1.700
E-VALUE    3:               2.500
           THE EIGENVECTORS:


VECTOR   1

       -.909
      2.545
      1.000


VECTOR   2

       -.300
       -.500
      1.000


VECTOR   3

      2.500
       .500
      1.000
Stop - Program terminated.

0;37;44m C:\TEXT=
```

Figure 3.8 Resultant output from EIGEND verifying the solution for Example 3.4.

Chapter Summary

- Matrices offer a shorthand way to represent systems of equations.
- Gauss' derivation of the method of least squares is implicitly incorporated when deriving the solution for linear, least-squares regression using matrix algebra. The notions of partial derivatives and the error function are not, however, required for the matrix algebraic derivation. Perhaps this derivation is more understandable.
- Many matrix systems can be considered to be homogeneous systems, for which eigenvalues and eigenvectors are descriptive.

Exercises

(An asterisk indicates an advanced or graduate student problem)

1. Write the following systems of simultaneous equations as matrices:

(a) $2x + 3y + 4z = 9$
$9x + y - z = 9$
$5x + 2y + z = 8$

(b) $8t + 7y = 15$
$t + 3y = 4$

(c) $27u + 5v - 10w = 22$
$3u - 5v - 10w = -12$
$10u - v + 7w = 16$

(d) $3x - 4y + 6z - 7t = -2$
$9x + 3y + z + t = 14$
$-x - y - z - t = -4$
$x + y + z + t = 4$

(e) $7x + y = 8$
$x + y = 2$

2. Use Gaussian elimination to solve for the unknowns in the systems of equations given in Problem 1. Substitute these values into the matrices and verify, using matrix multiplication, that the solutions are correct.

*3. In the exercises in Chapter 2, a problem challenged the development of a solution for a nonlinear, least-squares regression model of the form $y = mx^2 + wx + b$. Derive this solution using matrix algebra. *Hint:* The inverse of a 3×3 matrix is found as follows. Let

$$[C] = \begin{vmatrix} c_{11} & c_{12} & c_{13} \\ c_{21} & c_{22} & c_{23} \\ c_{31} & c_{32} & c_{33} \end{vmatrix}$$

Then

$$[C]^{-1} = \frac{1}{\det C} \begin{vmatrix} cc_{11} & cc_{12} & cc_{13} \\ cc_{21} & cc_{22} & cc_{23} \\ cc_{31} & cc_{32} & cc_{33} \end{vmatrix}$$

where cc_{ij} is the cofactor of c_{ij}. Cofactors are found as follows:

$$cc_{11} = \begin{vmatrix} c_{22} & c_{23} \\ c_{32} & c_{33} \end{vmatrix} = c_{22}c_{33} - c_{23}c_{32}$$

which assumes that

$$cc_{ij} = -1^{i+j} [cc_{ij}]$$

For $i = 1, j = 2,$

$$cc_{12} = - \begin{vmatrix} c_{21} & c_{23} \\ c_{31} & c_{33} \end{vmatrix} = -[c_{21}c_{33} - c_{23}c_{31}]$$

In general, the cofactor of an entry of a 3×3 matrix is found by using rows other than i and columns other than j. Two examples have been given; two more should be enough to allow the completion of the problem:

$$cc_{23} = - \begin{vmatrix} c_{11} & c_{12} \\ c_{31} & c_{32} \end{vmatrix} = -[c_{11}c_{32} - c_{31}c_{12}]$$

and

$$cc_{13} = (+1) \begin{vmatrix} c_{21} & c_{22} \\ c_{31} & c_{32} \end{vmatrix} = c_{21}c_{32} - c_{31}c_{22}$$

The determinant, $\det C$, is found as shown in Example 3.4.

4. Write the transpose of each of the following matrices:
   (a) $\begin{bmatrix} 2 & 3 \\ 4 & 5 \end{bmatrix}$ **(b)** $\begin{bmatrix} 3 & 5 & 6 \\ 1 & 7 & 9 \end{bmatrix}$

(c) $\begin{bmatrix} 2 & 4 & 5 \\ 3 & 6 & 9 \\ 1 & 1 & 1 \end{bmatrix}$ **(d)** $\begin{bmatrix} 1 \\ 2 \\ 3 \end{bmatrix}$

(e) $\begin{bmatrix} 2 & 2 \\ 3 & 3 \end{bmatrix}$ **(f)** $\begin{bmatrix} 1 & 1 & 1 & 1 \\ 2 & 3 & 4 & 5 \end{bmatrix}$

5. For each of the matrices in Problem 4, call each matrix [A], then calculate the following: $[C] = [A]^T[A]$. Verify your answers using the FORTRAN program MULMAT.

6. Determine the trace of each of the following matrices:

   (a) $\begin{bmatrix} 2 & 5 \\ 5 & 7 \end{bmatrix}$ **(b)** $\begin{bmatrix} 3 & 1 & 2 \\ 1 & 5 & 1 \\ 2 & 1 & 9 \end{bmatrix}$

   (c) $\begin{bmatrix} 1 & 0 \\ 0 & 1 \end{bmatrix}$ **(d)** $\begin{bmatrix} 7 & 9 \\ 5 & 4 \end{bmatrix}$

   (e) $\begin{bmatrix} 8 & 1 & 9 \\ 7 & 2 & 4 \\ 5 & 1 & 9 \end{bmatrix}$

7. Calculate the eigenvalues and eigenvectors of the following matrices:

   (a) $\begin{bmatrix} 1 & 0 & 0 \\ 0 & 1 & 0 \\ 0 & 0 & 1 \end{bmatrix}$ **(b)** $\begin{bmatrix} 3 & 2 \\ 2 & 4 \end{bmatrix}$

   (c) $\begin{bmatrix} 3 & 2 \\ 2 & 3 \end{bmatrix}$ **(d)** $\begin{bmatrix} w & 0 & 0 \\ 0 & t & 0 \\ 0 & 0 & z \end{bmatrix}$

   (e) $\begin{bmatrix} 3 & 1 & 2 \\ 1 & -1 & 4 \\ 2 & 4 & 5 \end{bmatrix}$

8. Verify your answers to Problem 6 using computer algorithm EIGEND.

References and Suggested Readings

DAVIS, J.C. 1986. *Statistics and Data Analysis in Geology,* 2d ed. New York: Wiley.

LAFARA, R.L. 1973. *Computer Methods for Science and Engineering.* Rochelle Park, NJ: Hayden.

OLEA, R.A., ed. 1991. *Geostatistical Glossary and Multilingual Dictionary.* New York: Oxford University Press.

SCARBOROUGH, J.B. 1955. *Numerical Mathematical Analysis,* 3d ed. Baltimore: The Johns Hopkins Press.

Figure 3.1 A listing for the program MULMAT. Subroutine EIGENJ is borrowed from Davis: *Statistics and Data Analysis in Geology*. Copyright © 1973 John Wiley & Sons. Used with permission.

```fortran
      PROGRAM MULMAT
C
C     AN ANSI STANDARD FORTRAN-77 PROGRAM FOR
C     MATRIX MULTIPLICATION.
C
C*********************************************************
C          USER'S GUIDE FOR MULMAT
C*********************************************************
C
C     MATRICES [A] AND [B] ARE ENTERED SO THAT THE
C     PRODUCT, [C], CAN BE COMPUTED; THE ORDER OF
C     MULTIPLICATION IS:  [A][B] = [C]
C
C     DATA FILE CREATION:
C
C     LINE 1:  ENTER:  N,M,L  IN THIS ORDER, WHERE
C
C              N = ROW DIMENSION OF [A]
C              M = COLUMN DIMENSION OF [A] = ROW DIM. OF [B]
C              L = COLUMN DIMENSION OF [B]
C
C     LINES:  2 THROUGH N + 1:  ENTER THE MATRIX, [A]
C
C              ENTER ONE LINE FOR EACH ROW OF A:
C
C     LINES:  N + 2 THROUGH N + M + 2:  ENTER THE MATRIX, [B]
C
C              ENTER ONE LINE FOR EACH ROW OF B:
C
C     IF [A] =  1   2    AND [B] =  5   6
C               3   4                7   8
C
C     THEN, THE DATA FILE LOOKS LIKE:
C
C     2,2,2
C     1,2
C     3,4
C     5,6
C     7,8
C
C*********************************************************
C          END OF USER'S GUIDE
C*********************************************************
      PARAMETER (NN = 50)
      DIMENSION A(NN,NN), B(NN,NN), C(NN,NN)
C
C     PROTOCOL:  [A][B] = [C]
C
C     THEREFORE:  ENTER [A] AND [B]; COMPUTE [C]
C
C     FIRST, ENTER THE SIZE OF [A]:  N X M; THEN,
C     ENTER THE COLUMN DIMENSION OF [B]:  L; REMEMBER,
C     THE ROW DIMENSION OF [B] IS M.
C
      OPEN (5, FILE = ' ')
      READ(5,*) N,M,L
      DO I = 1,N
        READ(5,*) (A(I,JK), JK = 1,M)
      END DO
      DO I = 1,M
        READ(5,*) (B(I,JK), JK = 1,L)
      END DO
C
C     PERFORM THE MULTIPLICATION
C
      DO I = 1,N
        DO J = 1,L
          C(I,J) = 0.0
          DO K = 1,M
            C(I,J) = C(I,J) + A(I,K) * B(K,J)
          END DO
        END DO
      END DO
C
C     PRINT THE RESULTS
C
      WRITE (*,200)
200   FORMAT(10X,'MATRIX MULTIPLICATION RESULTS',//)
      DO I = 1,N
        WRITE(*,300) I, (C(I,JK), JK = 1,L)
      END DO
300   FORMAT(2X,'MATRIX, C, ROW =',I3,':',10F10.3)
      STOP
      END
```

Figure 3.6 A listing for the program EIGEND. Subroutine EIGENJ is borrowed from Davis: *Statistics and Data Analysis in Geology.* Copyright © 1973 John Wiley & Sons. Used with permission.

```
      PROGRAM EIGEND
C
C     AN ANSI STANDARD FORTRAN-77 PROGRAM FOR EIGENDECOMPOSITION
C
C*********************************************************
C              USER'S GUIDE FOR EIGEND
C*********************************************************
C
C     THE INPUT TO EIGEND CONSISTS OF A SQUARE MATRIX FOR
C     WHICH EIGENVALUES AND EIGENVECTORS ARE DESIRED.
C
C     FILE CREATION:  DATA ENTRY:
C
C     LINE 1:  ENTER N, THE SIZE OF THE MATRIX
C              WHERE THE MATRIX IS N X N IN SIZE
C
C     LINES 2 THROUGH N + 1:  ENTER THE MATRIX, BY ROW:
C
C     DO I = 1,N
C       READ(5,*) (A(I,J), J = 1,N)
C     END DO
C
C     THAT IS, ENTER THE MATRIX, ONE ROW PER LINE IN THE
C        DATA FILE
C
C     EXAMPLE:  GIVEN THE MATRIX, A:
C
C     [A] =     1    2
C
C               3    4
C
C         THEN, N = 2; THEREFORE,
C
C     THE FILE LOOKS LIKE:
C
C     2
C     1,2
C     3,4
C
C*********************************************************
C              END OF USER'S GUIDE
C*********************************************************
      PARAMETER (NN = 100)
      DIMENSION A(NN,NN), D(NN), B(NN,NN)
C
C     EXPLANATION:  [A] = MATRIX TO BE DECOMPOSED AND
C                         CONTAINS THE EIGENVECTORS
C                   [D] = THE N EIGENVALUES
C
C     ENTER THE MATRIX [A] OF SIZE, N X N:
C
      OPEN (5, FILE = ' ')
      READ(5,*) N
      DO I = 1,N
        READ(5,*) (A(I,JK), JK = 1,N)
      END DO
C
C     CALL THE SUBROUTINE, EIGENJ, FOR THE DECOMP.
C
      CALL EIGENJ(A, D, N, B, NN)
C
C     PRINT THE RESULTS
C
      WRITE (*,200)
200   FORMAT(10X, 'EIGENDECOMPOSITION RESULTS',//)
      WRITE(*,300)
300   FORMAT(10X, 'THE EIGENVALUES:',/)
      DO I = 1,N
        WRITE(*,400) I, D(I)
      END DO
400   FORMAT(2X, 'E-VALUE ',I3,'; ', F15.3)
      WRITE(*,500)
500   FORMAT(10X, 'THE EIGENVECTORS:',/)
      DO I = 1,N
        WRITE(*,600) I
        DO J = 1,N
          WRITE(*,700) A(J,I)
        END DO
      END DO
```

61

Figure 3.6 (continued)

```fortran
600   FORMAT(//, 2X, 'VECTOR', I3, /)
700   FORMAT(2X, F10.3)
      STOP
      END
      SUBROUTINE EIGENJ(A,D,N,B,NN)
      DIMENSION A(NN,NN), B(NN,NN), D(NN)
C
C     A SUBROUTINE FOR THE CALCULATION OF EIGENVALUES AND
C     EIGENVECTORS OF A SQUARE, SYMETRIC MATRIX
C
C     MODIFIED FROM DAVIS, 1973, PP. 166 - 167.
C     [ NOTE THE 1973 VERSION ]
C
C     MATRIX, B, PROVIDES TEMPORARY STORAGE OF EIGENVECTORS
C
C
      ANORM = 0.0
      DO 100 I = 1,N
      DO 100 J = 1,N
         K = I - J
         IF (K .EQ. 0)  THEN
            B(I,J) = 1.0
         ELSE
            B(I,J) = 0.0
            ANORM = ANORM + A(I,J) * A(I,J)
         ENDIF
100   CONTINUE
      XN = 1.0 / FLOAT(N)
      ANORM = SQRT(ANORM)
      FNORM = ANORM * 1.0E-09 * XN
C
C     INITIALIZE INDICATORS AND COMPUTE THRESHOLD
C
123   THR = ANORM
133   THR = THR * XN
      IND = 0
C
C     SCAN DOWN COLUMNS FOR OFF-DIAGONAL ELEMENTS
C     GREATER THAN OR EQUAL TO THE THRESHOLD
C
      DO 200 I = 2,N
      I1 = I - 1
      DO 190 J = 1,I1
         W = ABS(A(J,I)) - THR
         IF (W .GE. 0.0)   THEN
C
C     CALCULATE SINES AND COSINES
C
            IND = 1
            AL = - A(J,I)
            AM = (A(J,J) - A(I,I)) * 0.50
            AO = AL / (SQRT(AL*AL + AM*AM))
            IF (AM .LT. 0.0)  THEN
               AO = - AO
            ENDIF
            SINX = AO / (SQRT(2.0 * (1.0 + SQRT(1.0 - AO*AO))))
            SINX2 = SINX * SINX
            COSX = SQRT(1.0 - SINX2)
            COSX2 = COSX * COSX
C
```

Figure 3.6 (continued)

```fortran
C
C           ROTATE COLUMNS I AND J
C
            DO 150 K = 1, N
              L = K - J
              IF (L .NE. 0)  THEN
                M = K - I
                IF (M .NE. 0)  THEN
                  AT = A(K,J)
                  A(K,J) = AT * COSX - A(K,I) * SINX
                  A(K,I) = AT * SINX + A(K,I) * COSX
                ENDIF
                BT = B(K,J)
                B(K,J) = BT * COSX - B(K,I) * SINX
                B(K,I) = BT * SINX + B(K,I) * COSX
150         CONTINUE
            XT = 2.0 * A(J,I) * SINX * COSX
            AT = A(J,J)
            BT = A(I,I)
            A(J,J) = AT * COSX2 + BT * SINX2 - XT
            A(I,I) = AT * SINX2 + BT * COSX2 + XT
            A(J,I) = (AT - BT) * SINX * COSX +
       2             A(J,I) * (COSX2 - SINX2)
            A(I,J) = A(J,I)
            DO 180 K = 1, N
              A(J,K) = A(K,J)
              A(I,K) = A(K,I)
180         CONTINUE
          ENDIF
190     CONTINUE
200   CONTINUE
      IF (IND .GT. 0)  THEN
        GO TO 133
      ENDIF
      Z = THR - FNORM
      IF (Z .GT. 0.0)  THEN
        GO TO 123
      ENDIF
C
C     SORT EIGENVALUES AND EIGENVECTORS
C

C
      DO 300 I = 2, N
        J = I
229     C = A(J-1,J-1) - A(J,J)
        IF (C .LT. 0.0)  THEN
          AT = A(J-1,J-1)
          A(J-1,J-1) = A(J,J)
          A(J,J) = AT
          DO 250 K = 1, N
            AT = B(K,J-1)
            B(K,J-1) = B(K,J)
            B(K,J) = AT
250       CONTINUE
          J = J - 1
          L = J - 1
          IF (L .GT. 0)  THEN
            GO TO 229
          ENDIF
        ENDIF
300   CONTINUE
C
C     PLACE EIGENVALUES IN VECTOR D IN ASCENDING ORDER
C     PLACE EIGENVECTORS IN MATRIX A IN ASCENDING ORDER
C
      DO 400 I = 1, N
        K = N + 1 - I
        D(I) = A(K,K)
400   CONTINUE
      DO 500 I = 1, N
        K = N + 1 - I
        DO 500 J = 1, N
          A(J,I) = B(J,K)
500   CONTINUE
C
C     NORMALIZE THE EIGENVECTORS
C
      DO 600 I = 1, N
        DO 600 J = 1, N
          A(J,I) = A(J,I) / A(N,I)
600   CONTINUE
      RETURN
      END
```

4

An Overview of Probability and Statistics

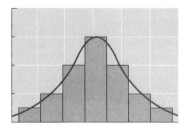

An overview of probability and statistics provides background for multivariate data analysis, geostatistics, Fourier analysis, fractals, and image processing. Some basic concepts are reviewed herein that are applied throughout the book. For those familiar with concepts in statistics and probability theory, this chapter may be bypassed without sacrificing an understanding of later chapters.

4.1 The Need for Statistics

In *Megatrends,* John Naisbitt identifies a major change that occurred in the United States in 1956: For the first time ever, white-collar jobs outnumbered blue-collar jobs. The implications of this are important. White-collar jobs largely involve the handling of information: Accountants, statisticians, teachers, lawyers, doctors, scientists, engineers, and so on, have jobs requiring a good education wherein information processing and dissemination is important. (Blue collar jobs are also important. The distinction drawn here is between jobs wherein a large amount of information is processed and jobs wherein information processing is not as important.)

In 1956, the United States entered the *information age*. How are the exponentially expanding volumes of information generated by this new age processed and interpreted? After all, the mere collection of information, without processing, interpretation, and dissemination, does not foster learning. In 1982, a second revolution began within the information age: the widespread introduction and use of the IBM (International Business Machines) personal computer. The personal computer was not a new invention at this time; Tandy Corporation, Apple Computers, Commodore Computers, and others had models that preceded the IBM PC. But the introduction of such a computer by IBM seemed to spark a revolution. Since 1982, personal computing has changed the educational and work environment radically, placing a computer in almost every office and classroom, and relegating typewriters to the task of occasional label and envelope processing. The ability to manage and process information has expanded in almost immeasurable terms as a result of the personal computer.

Processing the large amounts of data by computer is accomplished by using software for spread sheets, word processing, graphics, statistics, numerical analysis, artificial intelligence, modeling, and so on. Processing, in this context, means converting data to a form more easily interpreted by the human mind than what was possible in its original form. Most of these forms of processing are covered in other chapters. This chapter focuses on statistics for the processing of numerical information.

4.2 Statistical Parameters

Statistics is the science of collecting, processing, and interpreting numerical information. For this science, statistical parameters are a few numbers that, to a degree, characterize a larger amount of numerical information—a population or a sample. A population consists of all individuals or objects of a particular type (Devore, 1991). For students of the geological sciences and engineering, the notion of population is often theoretical. For example, an ore deposit is not actually a collection of individual objects, but it is considered to be a population. A sample is a subset of the population. A few drill holes, for instance, comprise a sample of the entire ore deposit (the population).

A few statistical parameters are useful for the characterization of a sample. A measure of average data value, or magnitude, is the sample mean:

$$\bar{x} = \frac{1}{N}\sum_{i=1}^{N} x_i$$

in which N is the number of values x_i comprising a sample. Often the mean is reported to a numerical accuracy one decimal place beyond the accuracy used for the values x_i; that is, the mean is more precise than the original data (Devore, 1991, p. 15).

Example 4.1

Compute the mean of the following numbers:

1.2   4.6   7.9   3.1   8.4   2.9   5.0   4.3   4.4   9.1

The sum of these 10 values is 50.9. Thus, the mean is equal to 50.9/10, or 5.09. The accuracy of the mean is associated with a greater precision in comparison to the original data.   ∎

Another measure of data magnitude is the median. If sample data are ordered from the most negative value to the largest positive value, then if the number N of sample data is odd, the median value is the single middle data value; if N is even, the median is the average of the two middle values.

Example 4.2

Determine the median for the data given in Example 4.1. First, these data are listed from smallest to largest:

1.2   2.9   3.1   4.3   4.4   4.6   5.0   7.9   8.4   9.1

Because N is 10, an even number, the median is the average of the two middle values, 4.4 and 4.6. The average of these two numbers is 4.5 and is the median for these data.   ∎

In addition to the average numerical magnitude, it is important to know how the data are dispersed (spread out) with respect to this average magnitude, that is, how the numbers vary. This variability, or variance, is measured relative to the arithmetic mean:

$$\text{var}(x) = \frac{1}{N-1} \sum_{i=1}^{N} [x_i - \bar{x}]^2$$

The variance is also known as the second statistical moment of a set of data about the (arithmetic) mean. The notion of moment is invoked by the difference term in the equation for variance. The measure of variability is relative to the mean as is seen by this difference. The arithmetic mean is often called the first statistical moment of a set of data with respect to zero; that is, the mean can be written as the sum of $(x_i - 0)$ divided by N, yielding the same numerical value.

To develop efficient computer algorithms, the expression for variance can be rewritten to allow the components for its computation to be calculated in the same program loop as is used to compute arithmetic mean. Substituting the equation for arithmetic mean into that for variance gives:

$$\text{var}(x) = \frac{1}{N-1} \left[\sum_{i=1}^{N} x_i^2 - \frac{1}{N} \left(\sum_{i=1}^{N} x_i \right)^2 \right]$$

By calculating the sum of x_i and the sum of the squares of x_i in a single program loop, both the arithmetic mean and variance can be calculated simultaneously. This saves N total calculations in comparison to computing the mean and variance in separate loops.

Example 4.3

Use the data of Example 4.1 and calculate their variance. The mean of these data is shown to be equal to 5.09 (Example 4.1). Ten differences are formed:

$(1.2 - 5.09) = -3.89$	$(4.6 - 5.09) = -0.49$
$(7.9 - 5.09) = 2.81$	$(3.1 - 5.09) = -1.99$
$(8.4 - 5.09) = 3.31$	$(2.9 - 5.09) = -2.19$
$(5.0 - 5.09) = -0.09$	$(4.3 - 5.09) = -0.79$
$(4.4 - 5.09) = -0.69$	$(9.1 - 5.09) = 4.01$

The sum of the squares of these differences is equal to 60.17. Thus the variance is equal to 60.17/9, or 6.69. The average squared deviation from the mean value for these data is 6.69.   ∎

Once variance is calculated, the standard deviation can be computed by simply taking the square root of variance:

$$\text{sdev}(x) = \sqrt{\text{var}(x)}$$

The standard deviation of the data used in Examples 4.1 and 4.2 is $6.69^{1/2}$, or 2.59. The significance of the standard deviation is discussed shortly.

4.3 Probability

Uncertainty is an important consideration in scientific study and engineering design. How frequently will earthquakes of a particular magnitude recur? How likely will an engineered slope for a highway fail? The theory of probability encompasses techniques for quantifying the likelihood of particular outcomes.

Rolling a six-sided die and observing the number of dots on the side facing up is a classic example. The six possible outcomes are 1, 2, 3, 4, 5, and 6. The sample space consists of these six possible outcomes. The probability, or likelihood, that any one of these numbers, or events, will occur for a given roll of the die is equal to 1/6, which assumes an equal likelihood that any one of the six outcomes will occur. An event is a subset of the sample space.

Probability is seen, at least in this example, to be a number between 0 and 1. If the probability of A is 1 [$P(A) = 1$], A is called a certain event. If the probability of A is 0.3 [$P(A) = 0.3$], then A has a 30% chance of happening. If the probability of A is 0.3, the probability that A will not occur is equal to $1 - P(A)$, which, in this case, is 0.7, or 70%.

Some problems involving probability are more complex. For example, in the game of five-card draw (poker), each player is dealt five cards from a 52-card deck. How many different 5-card hands are possible using the 52-card deck? In this case, $N = 52$ total cards and a subset of $k = 5$ cards is to be drawn. Thus,

$$\text{number of hands} = \frac{N!}{k!(N - k)!}$$

In this example, the number of possible 5-card hands is equal to $52!/(5!47!)$, or 2,598,960. An assumption is made that the probability of drawing any one of these hands is equal to the reciprocal of this number. Suppose the probability of obtaining a flush (five cards of the same suit) in the first draw is desired. With 13 cards comprising a suit, the number of possible 5-card combinations is equal to $13!/(5!8!)$, or 1287. But, four different suits—clubs, spades, hearts, and diamonds—are possible. Hence, the probability of obtaining a flush in any of the four suits is equal to $(4)(1287)/2,598,960$, or 0.00198 (i.e., approximately 0.20%). This could be interpreted to mean that a flush occurs approximately once every 500 hands.

4.3.1 Conditional Probability

Conditional probability is a notion involving the concept of the intersection between two events, A and B. Given B, the conditional probability that A occurs is written

$$P(A|B) = \frac{P(A' \cap B')}{P(B)}$$

The conditional probability of A given B is seen to be equal to the probability of the intersection of A and B divided by the probability of B. The probability of the intersection between A and B is equal to the total number of outcomes (or area) common to A and B.

Given the formula for computing conditional probability, a multiplication rule for the probability of the intersection between A and B can be described. By multiplying the foregoing formula for conditional probability by $P(B)$, we obtain

$$P(A \cap B) = P(A|B)P(B)$$

This formula is important because the probability of the intersection of A and B is often desired, given available information for $P(B)$ and $P(A|B)$. For instance (Devore, 1991, p. 62, Example 2.28), suppose five individuals are giving blood for the first time. The blood type is not known for any one of the five individuals. Further suppose that blood type A+ is the only blood type desired and only one of the individuals has this blood type. If the five individuals are chosen in random order for blood typing, what is the probability that at least three of the five individuals will have to be typed before finding the person with the right blood type?

If we let event $B = \{$1st type not type A+$\}$ and event $A = \{$2nd type not A+$\}$, then $P(B) = 4/5$. Given B, three of the four remaining individuals do not have A+ blood; therefore, $P(A|B) = 3/4$. In this case, the probability of the intersection is equal to $(4/5)(3/4) = 0.8(0.75) = 0.6$, or a 60% probability that at least three of the five individuals will have to be typed before finding the desired blood type.

4.3.2 Bayes' Theorem

The multiplicative rule for conditional probability leads directly to Bayes' theorem, named for the Reverend Thomas Bayes, an eighteenth-century English clergyman. Bayes' theorem is used later in this text for digital image classification. An understanding is first required of the axiom of total probability. For N mutually exclusive events A,

$$P(B) = \sum_{i=1}^{N} P(B|A_i)P(A_i)$$

Bayes' theorem is developed from this axiom as the following ratio:

$$P(A_j|B) = \frac{P(B|A_j)P(A_j)}{\sum_{i=1}^{N} P(B|A_i)P(A_i)}$$

A useful example for the application of Bayes' theorem to an environmental problem is presented near the end of this chapter.

4.3.3 Continuous Random Variables

A random variable is any rule associating one number with each particular outcome. For example, rolling a die yields six possible outcomes. A random variable X might be defined such that $X = $ the sum of any two outcomes of rolling the die.

A continuous random variable is one whose set of possible values is an entire interval, or intervals, of numbers. In the study of a gold ore deposit in Nevada, for instance, suppose that gold values range from 0 parts per million (ppm) to 10,000 ppm. In this case, X = the gold value at a particular spatial location is a continuous random variable taking on any value between 0 and 10,000 ppm.

4.3.4 Probability Density Functions for Continuous Random Variables

A probability density function $f(x)$ for a continuous random variable X is a function such that for any two numbers a and b,

$$P(a < X \le b) = \int_a^b f(x)\, dx$$

which defines the probability that X assumes a value in the interval a to b. Two properties must hold for the probability density function f:

$$(1) \quad f(x) \ge 0 \qquad \text{for all } x$$
$$(2) \quad \int_{-\infty}^{+\infty} f(x)\, dx = 1$$

A discrete approximation of $f(x)$ is the histogram for relative frequency (Figure 4.1; see also Section 4.3.8).

A cumulative distribution function $F(x)$ for a continuous random variable X is

$$P(X \le x) = \int_{-\infty}^{x} f(y)\, dy$$

A discrete approximation for $F(x)$ is possible using results from a histogram analysis (cf., cumulative relative frequency; see Section 4.3.8).

4.3.5 Expected (Mean) Value of a Continuous Random Variable

The expected value operator is denoted by E. The expected value $E(x)$ of a continuous random variable X is

$$E(X) = \mu_X = \int_{-\infty}^{+\infty} xf(x)\, dx$$

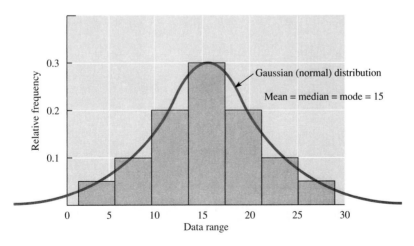

Figure 4.1   An example histogram. Note that this is a graph and records how frequently (frequency) data fall into particular intervals (bins) of the data range. This histogram describes a *normal distribution,* having a bell shape. The curve for such a distribution is superimposed on this histogram.

Example 4.4

Given $f(x)$ such that

$$f(x) = \tfrac{5}{4}(1 - x^4), \qquad 0 \le x \le 1$$
$$f(x) = 0 \qquad \text{for all other } x$$

the expected value of X is found such that

$$E(X) = \int_{-\infty}^{+\infty} xf(x)\, dx = \int_0^1 x \frac{5}{4}(1 - x^4)\, dx$$

$$= \frac{5}{4} \int_0^1 (x - x^5)\, dx = \frac{5}{4} \left(\frac{x^2}{2} - \frac{x^6}{6} \right) \Bigg|_{x=0}^{x=1} = \frac{5}{12}$$

∎

The variance of a continuous random variable is found as an expected value (Devore, 1991, p. 141):

$$\sigma_X^2 = \text{var}(X) = \int_{-\infty}^{+\infty} (x - \mu)^2 f(x)\, dx = E[(X - \mu)^2]$$

4.3.6 The Normal or Gaussian Distribution

A widely used probability density function was introduced by Carl Gauss (Chapter 2):

$$f(x) = \frac{1}{\sqrt{2\pi}\sigma} e^{-(x - \mu)^2/2\sigma^2}, \qquad -\infty < x < \infty$$

where μ is the mean of X and σ is the standard deviation. This function is known as the normal distribution. It is different from the function presented (Chapter 2), which is known as the standard normal distribution and has a mean of 0 and a variance of 1.

Numerical populations are often assumed to have normal distributions. The normal distribution is a symmetrical one. The mean value is the central value in the distribution, hence the median equals the mean. The area beneath the normal curve to the left of the mean equals the area beneath the curve to the right of the mean. Further, 67% of normally distributed values fall within one standard deviation of the mean; 95% fall within two standard deviations of the mean; 99% fall within three standard deviations of the mean (Figure 4.2).

4.3.6.1 Lognormal Distribution

Many types of data encountered in the study of earth sciences, particularly geochemical data, are associated with lognormal distributions. Histograms of the raw data do not show a normal distribution for such data, but natural logarithms of the data are normally distributed. The probability density function for such data is

$$f(x) = \frac{1}{x} \frac{1}{\sqrt{2\pi}\sigma} e^{-[\ln(x) - \mu]^2/2\sigma^2}, \qquad x > 0$$

$$= 0, \qquad x \le 0$$

In this case, the mean μ and the standard deviation σ are for $\ln x$, not for x. The lognormal distribution is an important consideration in geostatistics (Chapter 6).

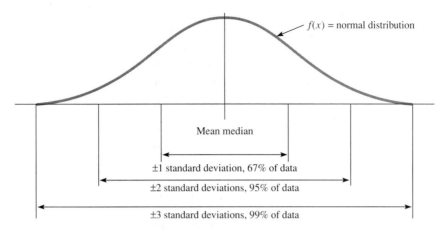

Figure 4.2   A graph of a normal distribution. For such a distribution, 67% of the data fall within one standard deviation of the mean data value, 95% of the data fall within two standard deviations of the mean, and 99% of the data fall within three standard deviations of the mean.

4.3.7 Poisson Probability Distribution

The Poisson distribution is used to describe a frequency or rate at which something occurs. The density for this distribution is (Devore, 1991, p. 119)

$$p(X = x) = \frac{e^{-\lambda}\lambda^x}{x!}, \qquad \lambda > 0, \quad x = 1, 2, 3, \ldots$$

Note that x is an integer greater than or equal to 0 and is a rate or frequency, such as a rate of occurrence in time or over space. Examples would be the number of earthquakes in a particular interval of time or the number of diamonds found over a portion of a diamond mine.

Suppose that x is the number of joints encountered in a 100-m interval on surface exposures of a particular lithologic unit. If x has a Poisson distribution with

$$\lambda = 2$$

the probability of encountering 3 joints in a randomly selected 100-m interval is

$$p(x = 3) = \frac{e^{-2}2^3}{3!} = 0.18 \quad (18\%)$$

Whenever a problem involves the number of occurrences of something over an interval of time or space, a Poisson distribution is often a good model for the process.

4.3.8 Graphical Display of a Sample: The Histogram

How frequently something happens is an inherent aspect of probability. A tool for assessing frequency is the histogram (Figure 4.1). For discrete integer data, a histogram records the number of times (frequency) a data value occurs in a sample. For continuous valued real data, a histogram records the frequency with which numbers fall into particular intervals (bins) of the data range (that is, between maximum and minimum values). A histogram is plotted easily by hand provided the sample is relatively small (no more than 100 data values or so). For samples larger than this, a computer algorithm is helpful.

Histograms are useful for identifying the probability density function for a sample. The effectiveness of this analysis depends on the size of the sample; a probability density function is often more difficult to infer for a smaller sample in comparison to a larger one. The analysis is more effective if frequency is converted to relative frequency by normalizing each frequency by the total number of data N. The following table illustrates these concepts. For the data set

5	14	19	24	26	29	33	37	45	49
7	14	21	24	26	29	33	37	45	50
12	15	21	24	26	29	33	37	46	50
13	15	22	24	26	30	33	38	46	51
13	15	22	25	26	30	35	38	46	51
13	15	22	25	27	30	35	39	46	53
13	16	23	25	27	31	36	39	47	53
14	16	23	25	27	31	36	40	48	57
14	17	24	25	28	32	36	41	48	57
14	17	24	25	28	32	36	41	49	60

the outcome of histogram analysis is

Bin	Frequency	Relative Frequency
0–10	2	0.02
11–20	19	0.19
21–30	35	0.35
31–40	22	0.22
41–50	15	0.15
51–60	7	0.07
Total number of data N:	100	

Another notion, cumulative relative frequency, is also obtainable from the histogram. The cumulative relative frequency for a value or bin K is equal to the sum of the relative frequencies for all values or bins less than and equal to K. Using the previous example, we find

Relative Frequency	Cumulative Relative Frequency
0.02	0.02
0.19	0.21
0.35	0.56
0.22	0.78
0.15	0.93
0.07	1.00

4.3.9 Examining Models for Cumulative Distribution Functions

A histogram is a rough picture of the probability density function. A numerical model (function) can be selected that is thought to represent the cumulative

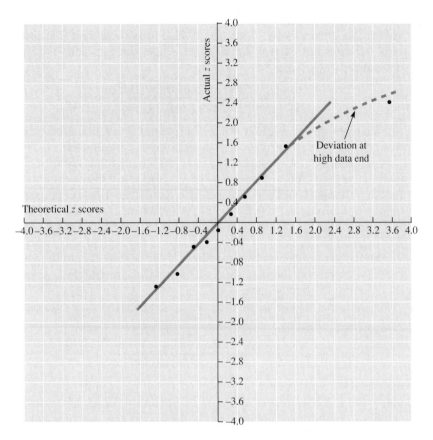

Figure 4.3   A probability plot for Section 4.3.9. This plot shows the relationship between actual data z scores to theoretical z scores. The theoretical values are computed by assuming a normal cumulative distribution function. This plot has a slope of approximately 1.0, hence the normal distribution model fits the data well, except for that region of the data distribution associated with larger values (which may or may not pose a problem for analyses that assume a normal distribution for these data). In other words, the normal distribution is not a good model for the entire data distribution.

distribution function (i.e., the integral of the probability density function). A graphical procedure (Figure 4.3) is used to examine whether data are modeled adequately by a particular cumulative distribution function.

In this procedure, the data are first listed in ascending order, from smallest to largest value. Then each datum is assigned a rank, equal to its relative size in the data set. The smallest data value, for instance, is assigned the rank of 1 and the largest data value is assigned the rank of N, where N is the total number of data values. For example, the 100 data values used to describe histogram analysis in Section 4.3.8 are listed in ascending order.

Data listed in ascending order and ranked are called rank-ordered data. Once ordering is accomplished, the quantiles (percentiles) of the data are determined. All ranks are normalized by N and the quantiles (percentiles) are those data values associated with normalized ranks: 0.1, 0.2, 0.3, 0.4, 0.5, 0.6, 0.7, 0.8, 0.9, and 1.0. If normalized ranks are not exactly equal to these values, linear interpolation is used to approximate a data value for a given quantile. For example, suppose normalized ranks are found: 0.29 and 0.31 and their data values are, respectively, 55.3 and 61.2. Further suppose that the normalized rank 0.3 does not exist in the data set. Then, the estimated data value for the normalized rank 0.3 is

$$55.3 + \frac{(61.2 - 55.3)(0.3 - 0.29)}{0.31 - 0.29} = 58.25$$

For the data set of Section 4.3.8, the quantiles are listed in ascending order from 0.1 to 1.0: 14, 17, 24, 25, 28, 32, 36, 41, 49, and 60.

A probability plot is constructed to assess whether data conform to a particular cumulative distribution model. The object is to select the cumulative distribution function that is thought to represent the data. This function is used to compute values at the quantiles: 0.1, 0.2, ..., 1.0. In other words, a value x is found for each quantile such that $P(x)$ = percentage. For the 0.2 quantile, for instance, x is determined such that $P(x) = 0.2$. These theoretical values of x are plotted versus the actual data at the quantiles. If the assumed cumulative distribution function represents the data well, the plot will be a straight line having a slope of 1 (i.e., a 45° line; Figure 4.3).

Assume that a normal distribution represents the data of Section 4.3.8. If the data quantiles are converted to z scores,

$$z = \frac{\text{data value} - \text{mean}}{\text{standard deviation}}$$

then a table (e.g., Devore (1991), p. 673) can be used to obtain the theoretical values of x. The mean of the data in Section 4.3.8 is approximately 30, the standard deviation is approximately 12.5, and the z scores for the 10 data quantiles are -1.28, -1.04, -0.48, -0.4, -0.16, 0.16, 0.48, 0.88, 1.52, and 2.40. The theoretical values are -1.28, -0.84, -0.53, -0.25, 0.0, 0.25, 0.53, 0.84, 1.38, and 3.5.

Pairs of points are formed as follows. For each quantile, 0.1, 0.2, 0.3, and so on, pair the actual z score with the theoretical value. For the foregoing example, this gives 10 data pairs: $(-1.28, -1.28)$, $(-1.04, -0.84)$, $(-0.48, -0.53)$, $(-0.40, -0.25)$, $(-0.16, 0.0)$, $(0.16, 0.25)$, $(0.48, 0.53)$, $(0.88, 0.84)$, $(1.52, 1.38)$, and $(2.40, 3.5)$. These pairs are plotted to yield a simple x-y plot (Figure 4.3). This plot shows a straight line which has a slope approximately equal to 1; the normal distribution hypothesis appears to be correct for these data.

A probability plot (Figure 4.3) can be developed to assess whether data conform to any cumulative distribution function. All that is required is to select the assumed function, then compute all theoretical values of x such that

$$P(x) = \text{quantile}, \quad \text{quantile} = 0.1, 0.2, \ldots, 1.0$$

Testing for the lognormal distribution is demonstrated in Chapter 6.

4.4 Asymmetrical Data Distributions: Skew

For many data sets, the mean value is not the central value, because it is not equal to the median. An epithermal gold deposit, for instance, may be associated with a large-volume, poor-grade ore zone surrounding a small-volume, rich-grade ore zone. Samples for such deposits are likely to be associated with many more poor-grade than rich-grade assays. Sometimes the way in which a phenomenon is sampled may influence the type of distribution. A large-valued or small-valued portion of the phenomenon may be of interest; consequently, the sample will be associated with predominately large or small values.

These descriptions invoke the notion of a skewed distribution, like that shown in Figure 4.4. Here, the mean and median do not coincide. More important, the majority of data values occur on one side of the mean. A skewed distribution is asymmetrical.

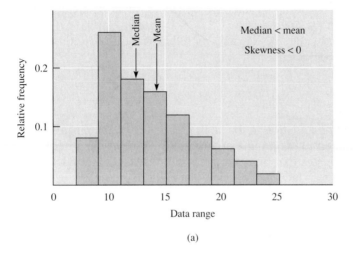

(a)

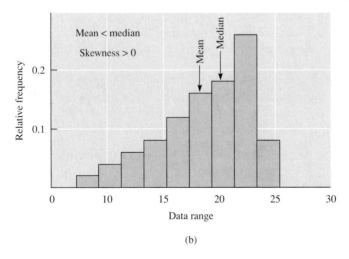

(b)

Figure 4.4 Histograms for samples associated with a skewed or asymmetrical distribution. (a) A skewness to the left. Such a distribution is associated with a positive value of skewness. (b) A skewness to the right. Such a distribution is associated with a negative value of skewness. Notice the relationship between the mean and median for each of these graphs.

For such distributions, the statistic called skewness is important. Skewness is the third statistical moment with respect to the arithmetic mean:

$$\text{skewness}(x) = \frac{1}{N-1} \sum_{i=1}^{N} (x_i - \bar{x})^3$$

Because this expression is cubic, negative differences remain negative. Therefore, for a perfectly symmetrical distribution, such as the normal distribution, the value of skewness is zero because the negative differences (between data values and the arithmetic mean) are equal in absolute value to the positive differences. For samples associated with asymmetrical distributions, however, for which the majority of data are either less than or greater than the mean, skewness is not equal to zero.

Is a nonzero value of skewness significant? To answer this, a coefficient of skewness is computed:

$$\text{coefficient of skewness} = \frac{\text{skewness}(x)^2}{\text{sdev}(x)^6}$$

Using standard deviation to the sixth power allows the magnitude of skewness to be compared to magnitude of variability. A coefficient of skewness greater than one indicates that an asymmetry outweighs in numerical magnitude the variability of the data; hence asymmetry is important.

4.5 Nonparametric Statistics

Our discussion of skewed distributions in Section 4.4 leads to the logical question of how to model such data using a probability density function. Suppose a data set is so sparse that a density function cannot be inferred to represent it. How are such data analyzed from a probabilistic standpoint? To answer this question we must examine nonparametric, or distribution-free, statistical concepts.

A nonparametric random variable, an indicator function, is discussed in detail in Chapter 6; we introduce the subject here. The function is formed as follows. A data value is selected within the total range of data values for a particular data set. Let this selected data value be referred to as a cutoff value C. Then

$$i_j = 1 \quad \text{if } x_j \leq C, \qquad i_j = 0 \quad \text{otherwise}$$

The indicator function i can take one of only two possible values: 0 or 1. The notion of a distribution is not relevant for such a variable. Furthermore, the indicator function has other properties, such as minimization of the influence of data outliers (extreme values), that make it useful for geostatistical analyses (Chapter 6).

Another nonparametric attribute of a number is its rank in a sample. Rank ordering was discussed previously for determining the median of a sample. Distribution assumptions are not relevant for ranks. This fact serves as a useful foundation for nonparametric hypothesis tests, such as the Mann–Whitney test (Davis, 1986).

4.6 Statistical Correlation: Intervariable Relationships

In the study of earth science, the question is often asked: Is the relationship between this phenomenon and that phenomenon significant? For instance, is the relationship between atmospheric carbon dioxide level and mean atmospheric temperature significant? This question is especially relevant to environmental concerns, such as global warming and the greenhouse effect.

Suppose bivariate data, for instance, atmospheric carbon dioxide level and mean atmospheric temperature, are collected over a particular interval of time. The correlation between these data can be quantified statistically using a correlation coefficient r:

$$r = \frac{\text{cov}(x, y)}{\text{sdev}(x)\,\text{sdev}(y)}$$

In this formula, cov is covariance, which has a discrete approximation:

$$\text{cov}(x, y) = \sum_x \sum_y (x - \bar{x})(y - \bar{y})f(x, y)$$

in which

$$\bar{x} = \text{mean of } x = \frac{1}{N}\sum_{i=1}^{N} x_i, \qquad \bar{y} = \text{mean of } y = \frac{1}{N}\sum_{i=1}^{N} y_i$$

Note that the joint probability function $f(x, y)$, appears in this discrete definition of covariance because its theoretical premise is an expected value. A joint probability density function is analogous to $f(x)$, except probability is a function of two variables, x and y. Properties for $f(x)$ must hold also for $f(x, y)$; for instance, the range of $f(x, y)$ is [0, 1], just as it is for $f(x)$. For more on joint probability functions, see Devore (1991).

In practice, covariance for bivariate data (x, y) can be computed as

$$\text{cov}(x, y) = \frac{N\sum_{i=1}^{N} x_i y_i - \sum_{i=1}^{N} x_i \sum_{i=1}^{N} y_i}{N(N - 1)}$$

A topic that is frequently debated is the range of r that is considered to represent strong correlation. If the correlation coefficient r is in the range [0.8, 1.0], then correlation is considered strong and positive; r in the range $[-0.8, -1.0]$ is strong and negative; r in the range $[-0.5, 0.5]$ is weak; other values of r indicate moderate correlation (Devore, 1991, p. 489). Why is a value of $r = 0.5$ (or -0.5) considered a weak correlation? A proper way to evaluate r is to square its value. If $r = 0.5$, then $r^2 = 0.25$. Thus, given a regression equation, $y = f(x)$ if the correlation between x and y is represented by r; further, if $r^2 = 0.25$, then only 25% of the observed y variation is accounted for by the regression equation (Devore, 1991, p. 490).

4.7 Statistical Correlation: Intravariable Relationships

Again using the example of global climate change, rather than being interested in the correlation between carbon dioxide and temperature, suppose the interest, instead, is in the change in temperature with time. A time series of temperatures since the turn of the century is used to illustrate subsequent concepts (Section 4.8.4).

A time series, such as this example of temperature recorded over an interval of time, is a collection of univariate (single-variable) data as a function of time. To evaluate the change in temperature with time, consider the degree of correlation between temperature at one time with that at another. The difference in times is referred to as lag. Because correlation between temperatures is evaluated at two intervals of time, the autocorrelation (correlation with self; e.g., temperature with temperature) for temperature over the lag interval is desired.

Autocorrelation is determined as

$$\text{autocor}(x) = \frac{\text{cov}(x_i, x_{i+L})}{\text{var}(x)}$$

where L is the lag interval. The numerator in this definition is referred to as autocovariance. The formula for autocovariance assumes the data to which it is applied are second-order stationary (i.e., the variance of the data is constant for all portions of the data set).

4.8 Additional Numerical Examples

4.8.1 Arithmetic Mean and Median

Given the following data, calculate the arithmetic mean and median:

12,   17,   17,   12,   17,   24,   15,   16,   16,   8,   13,   21,   30,   14,   17

SOLUTION

This sample contains $N = 15$ data values. The arithmetic mean is

$$\text{mean} = \frac{1}{N}\sum_{i=1}^{N} x_i = \frac{1}{15}(249) = 16.6$$

The median is most easily found by rank ordering these data:

data:	8	12	12	13	14	15	16	16
rank:	1	2.5	2.5	4	5	6	7.5	7.5

data:	17	17	17	17	21	24	30
rank:	10.5	10.5	10.5	10.5	13	14	15

Notice that for ties in data value, the ranks are assigned as the average of the ranks over the data interval. For example, two values of 16 occur in this sample. Their ranks should be 7 and 8, but because the data values are equal, these ranks are averaged and each value of 16 is assigned the rank of (7 + 8)/2, or 7.5. Because N is an odd number, 15, the median for this sample is that number having the rank of 8. In this case, as has just been explained, no unique rank of 8 is found because of the tie for the data value 16. Therefore, the median is 16.

4.8.2 Variance

Given the following data, compute the arithmetic mean, variance, and standard deviation (from Mielke and Johnson (1974), Table 1, data set No. 1, p. 224):

127.96	210.07	203.24	108.91	178.21
285.37	100.85	89.59	185.36	126.94
200.19	66.24	247.11	299.87	109.64
125.86	114.79	109.11	330.33	85.54
117.64	302.74	280.55	145.11	95.36
204.91	311.13	150.58	262.09	477.08
94.33				

SOLUTION

$$\bar{x} = \frac{1}{N} \sum_{i=1}^{N} x_i = \frac{1}{31} (5746.7) = 185.38$$

$$\text{var}(x) = \frac{1}{N-1} \left[\sum_{i=1}^{N} x_i^2 - \frac{1}{N} \left(\sum_{i=1}^{N} x_i \right)^2 \right]$$

from which is found

$$\text{var}(x) = \tfrac{1}{30}[(1, 341, 371.4) - \tfrac{1}{31}(5746.7)^2] = 9202.10$$

The square root of the variance is the standard deviation. For this example, the standard deviation is equal to 95.9.

4.8.3 Correlation Coefficient

Given the following bivariate data (Kreyszig, 1988, p. 1285), compute the correlation coefficient between x and y:

x (temperature, °C)	y (yield of a chemical process, kg/min)
0	0.8
15	1.1
30	0.9
45	1.6
60	1.2
75	1.8

To compute the correlation coefficient between x and y, first determine the covariance between x and y, then normalize the covariance by the product of the standard deviations of x and y.

Covariance between x and y:

$$\text{cov}(x, y) = \frac{N \sum_{i=1}^{N} x_i y_i - \sum_{i=1}^{N} x_i \sum_{i=1}^{N} y_i}{N(N-1)} = \frac{6(322.5) - 225(7.4)}{6(5)} = 9$$

The correlation coefficient is therefore

$$r = \frac{\text{cov}(x, y)}{\text{sdev}(x)\,\text{sdev}(y)} = \frac{9}{(28.06)(0.39)} = 0.82$$

A value of $r = 1$ indicates perfect positive correlation between x and y; a value of $r = -1$ indicates perfect negative correlation between x and y. In this case $r = 0.82$, which indicates a relatively good degree of positive correlation.

4.8.4 Autocorrelation

Given the average temperature in North America since 1900, determine the autocorrelation for lag periods of 1 to 10 years. The following data are used (taken from Hanson et al. (1989), p. 50, Figure 1):

Average Temperature (°C)

Year	Temp	Year	Temp	Year	Temp
1895	10.70	1926	11.50	1957	11.50
1896	11.55	1927	11.60	1958	11.40
1897	11.20	1928	11.50	1959	11.50
1898	11.18	1929	10.90	1960	11.10
1899	11.00	1930	11.55	1961	11.38
1900	11.95	1931	12.25	1962	11.38
1901	11.45	1932	11.30	1963	11.50
1902	11.30	1933	11.95	1964	11.15
1903	10.70	1934	12.60	1965	11.15
1904	11.00	1935	11.40	1966	11.05
1905	10.90	1936	11.55	1967	11.20
1906	11.39	1937	11.25	1968	11.00
1907	11.20	1938	12.00	1969	11.10
1908	11.55	1939	12.05	1970	11.15
1909	11.25	1940	11.35	1971	11.15
1910	11.75	1941	11.80	1972	11.00
1911	11.50	1942	11.35	1973	11.50
1912	10.40	1943	11.50	1974	11.50
1913	11.20	1944	11.40	1975	11.05
1914	11.40	1945	11.30	1976	11.05
1915	11.20	1946	11.90	1977	11.65
1916	10.80	1947	11.40	1978	10.80
1917	10.35	1948	11.20	1979	10.70
1918	11.45	1949	11.50	1980	11.50
1919	11.30	1950	11.15	1981	11.95
1920	10.95	1951	11.00	1982	11.00
1921	12.40	1952	11.70	1983	11.30
1922	11.50	1953	12.15	1984	11.38
1923	11.39	1954	12.15	1985	10.90
1924	10.70	1955	11.30	1986	12.10
1925	11.85	1956	11.60	1987	12.10

SOLUTION

For each lag interval, all pairs of temperatures are found that are separated by the lag interval. For the lag interval of 1 year, for example, 92 such pairs are found. Calling each pair of temperatures (x, y), the correlation between x and y for all pairs is determined as in Example 4.8.3. This correlation is called the autocorrelation for average temperature separated by the different lag intervals of time.

Results are listed for each of the 25 lag intervals:

Lag (years)	Autocorrelation	Lag (years)	Autocorrelation
1.0	0.168	6.0	0.097
2.0	−0.012	7.0	0.166
3.0	0.146	8.0	0.140
4.0	0.132	9.0	−0.025
5.0	0.081	10.0	0.104

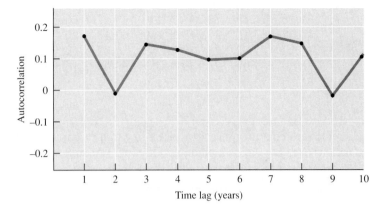

Figure 4.5   A plot of the autocorrelation results from Example 4.8.4. This analysis shows that autocorrelation varies randomly with time (lag), hence the temperature values do not seem to be well correlated over intervals of time.

The results are displayed graphically in Figure 4.5. These data do not display a strong autocorrelation with time. If the autocorrelation were strong, a decreasing behavior from smaller to larger lags would be observed. In the case of the temperature data, the autocorrelation is approximately the same for all lags, which indicates more of a random behavior of temperature with time.

4.8.5 Histogram

For the data listed below, plot a histogram. These are real data. Therefore, first determine the numerical range of this sample and divide the range into 10 equal intervals (bins). Determine the frequency, relative frequency, and cumulative relative frequency for each bin. Also determine the arithmetic mean, median, and standard deviation for these data. Answer the following questions: (1) Does a cumulative distribution function based on the normal distribution represent these data well? (2) Do the mean and median coincide? (3) Do 95% of the data fall within two standard deviations of the mean?

14.59	15.41	26.46	17.11	38.46	20.20
−4.19	19.57	29.95	7.52	23.26	43.73
12.09	15.46	33.58	21.94	31.65	25.92
22.91	33.61	23.06	25.25	26.45	−3.13
14.27	24.43	11.67	27.17	14.90	26.67
24.62	31.76	14.77	16.04	25.58	11.11
35.99	33.40	22.60	18.04	31.93	15.94
32.94	13.22	21.47	10.31	20.37	18.63
10.81	18.26	48.73	9.72	18.60	42.07
44.41	43.28	10.52	46.12	42.23	6.29
29.19	34.93	40.99	29.25	11.13	39.27
20.71	20.35	41.83	23.87	22.03	22.71
23.68	33.79	4.00	27.20	22.16	12.20
38.05	5.45	20.93	9.46	29.13	24.09
31.87	25.33	31.94	21.28	19.27	20.95
37.48	32.55	32.76	22.78	42.96	29.12

(data continue)

26.56	26.25	30.24	11.46	19.97	26.93
15.67	13.10	27.04	15.69	33.17	26.41
33.27	22.72	25.77	11.63	6.82	12.55
33.50	32.94	26.15	18.17	33.98	16.54
51.35	20.06	35.44	26.84		

SOLUTION

A histogram for these data (Figure 4.6) looks like that of a normal distribution. Rank ordering these data yields a median value of 23.68 and the mean is 24.15; hence, the mean and median are similar (though not identical). To determine whether the normal distribution is an appropriate model for these data, a test is designed following the procedure described in Section 4.3.9. The 10 quantiles (percentiles) for these data are determined from the rank order. There are 124 data values, so each rank is normalized by 124. The quantiles are equal to the data values for normalized ranks: $0.1, 0.2, 0.3, \ldots, 1.0$. For these data, the 10 quantiles are close to 10.81, 14.59, 18.26, 20.95, 23.68, 26.41, 29.19, 33.17, 38.05, and 51.35.

As described in Section 4.3.9, these 10 quantiles are converted to z scores. The standard deviation for these data is 10.76; hence the z scores for the 10 quantiles are $-1.24, -0.89, -0.55, -0.30, -0.04, 0.21, 0.47, 0.84, 1.29, 2.53$. Because a normal distribution is being tested for these data, the theoretical values used in Section 4.3.9 may be used for this test. Therefore, the data pairs to be tested are

$(-1.24, -1.28)$	$(-0.89, -0.84)$	$(-0.55, -0.53)$
$(-0.30, -0.25)$	$(-0.04, 0.00)$	$(0.21, 0.25)$
$(0.47, 0.53)$	$(0.84, 0.84)$	$(1.29, 1.38)$
$(2.53, 3.5)$		

These 10 data pairs are plotted in Figure 4.7. The plot is reasonably linear and the slope is nearly 1.0, hence the normal distribution model seems to represent these data.

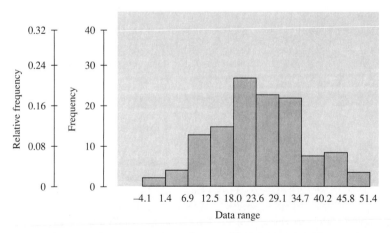

Figure 4.6 Histogram results for Example 4.8.5. The histogram reveals that a normal distribution assumption may be appropriate for these data. This assumption must be tested (Figure 4.7).

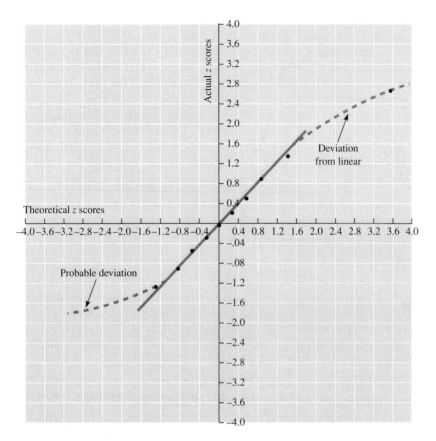

Figure 4.7   A probability plot to test a normal distribution model for the data presented in Section 4.8.5. The slope of this plot is nearly 1.0 and a normal distribution model appears to fit these data well, except that the model is rather poor for that portion of the data distribution associated with larger values (and is possibly poor for that portion of the distribution associated with smaller values). Hence, the normal distribution model is not a good model for the entire data distribution.

Because a normal distribution model is considered appropriate for these data, the third question regarding these data is considered: Do 95% of the data fall within two standard deviations of the mean value? The mean is 24.15 and the standard deviation is 10.76. Thus, to answer this question, we count the number of data values that are less than 2.63 or greater than 45.67. Two data values are less than 2.63 and three data values are greater than 45.67. Thus, five total data values out of 124, or 4% of the total amount of data are outside the two-standard deviation region surrounding the mean data value. Hence, 96% of the data fall within two standard deviations of the mean.

4.8.6 Bayes' Theorem

A simple exercise is designed to illustrate Bayes' theorem. This exercise is adapted from one given in McCuen (1985). A water quality test is conducted to decide whether a domestic well is contaminated by nitrates. Large nitrate concentrations in domestic wells may be caused by agricultural runoff contaminating a groundwater system or by seepage into a groundwater system from septic tanks. Large nitrate concentrations cause birth defects and fatigue.

Suppose that if a well is contaminated by nitrates, the test fails to determine this condition 5% of the time. Further suppose that if the well is not contaminated by nitrates, the test yields a positive result 11% of the time. These are conditional probabilities for test errors.

Let E_1 represent the event of an uncontaminated well and E_2 the event of a contaminated well. Because a well is either contaminated by nitrates or is not, $p(E_1) + p(E_2) = 1$. Let S_1 represent a negative test outcome and S_2 a positive test outcome. The conditional probabilities of test error for either outcome are

$$p(S_2|E_1) = 0.11, \qquad p(S_1|E_2) = 0.05$$

The problem is to determine the probability that a well is contaminated, given that a positive test result is obtained. Bayes' theorem is needed to make this determination. First, the probability that the test is positive, $p(S_2)$, is computed using

$$p(S_2) = p(E_1)p(S_2|E_1) + p(E_2)p(S_2|E_2)$$

No information is yet given for determining the probability of having an uncontaminated well, $p(E_1)$, or having a contaminated well, $p(E_2)$. For the sake of argument, suppose information is available (number of septic tanks, regional geology, agricultural activity, well water test data, and so on) to suggest that $p(E_2)$ is 0.1 (i.e., 10% of the wells tested are expected to be contaminated); then, $p(E_1) = 1 - 0.1 = 0.9$. Therefore,

$$p(S_2) = 0.9(0.11) + 0.1(1 - 0.11) = 0.188$$

In other words, a positive test result is expected to occur (probabilistically) approximately 19 times out of 100 tests. Implicit in this formula is the notion that $p(S_2|E_2) = 1 - p(S_2|E_1)$.

Bayes' theorem is also used to determine the probability of a well being contaminated, given that a positive test result is obtained:

$$p(E_2|S_2) = \frac{p(S_2|E_2)p(E_2)}{p(S_2|E_1)p(E_1) + p(S_2|E_2)p(E_2)} = \frac{(1 - 0.11)(0.1)}{0.188} = 0.473$$

According to this calculation, a 47.3% probability exists that a well is contaminated, given that a positive test result occurs.

4.9 Computer Exercise

4.9.1 Histograms

Use the computer program HISTO (Figure 4.8, page 88) to verify the calculations in Section 4.8.5. A user's guide is provided at the beginning of the program listing.

SOLUTION

The resultant output from HISTO for Example 4.8.5 is shown in Figure 4.9. The arithmetic mean, variance, and skewness are printed below the histogram. These statistics, and the histogram, match what is obtained in Section 4.8.5.

```
        HISTOGRAM RESULTS
          INTERVAL      BIN #      FREQ

 -4.190      1.364        1          2 ****
  1.364      6.918        2          4 ********
  6.918     12.472        3         13 ************************
 12.472     18.026        4         15 ***************************
 18.026     23.580        5         27 *******************************************
 23.580     29.134        6         24 ****************************************
 29.134     34.688        7         21 *****************************************
 34.688     40.242        8          7 *************
 40.242     45.796        9          8 ***************
 45.796     51.350       10          3 ******

         MEAN      =     .24150E+02
         VARIANCE  =     .11588E+03
         SKEW      =     .87055E+02
Stop - Program terminated.


0;37;44m C:\TEXT>
```

Figure 4.9 A listing of the output from HISTO for the data listed in Section 4.8.5. These results from the computer program match those obtained by hand (Section 4.8.5).

Chapter Summary

- In 1956, the United States (as well as other developed nations, such as Canada) entered the information age. Within this age, the age of the personal computer began in 1982.
- Numerical information management is of growing importance in the information age. Statistics help to manage and interpret this information.
- Aspects of statistics useful for subsequent chapters include the arithmetic mean, median, variance, standard deviation, distribution functions, correlation, autocorrelation, and covariance. Bayes' theorem is useful in Chapter 10 on image processing.

Exercises

(An asterisk indicates an advanced or graduate student problem)

1. Compute the mean and median for the following sets of data:
   (a) 4 6 7 7 2 7 9
   (b) 10 1 12 10 3 13 10
   (c) −1 −2 −1 0 1 2 1 1
   (d) 1 0 0 1 0 0 0 1
   (e) 5 5 5 5 4 4 4 3 3

2. Compute the variance and standard deviation for each of the following sets of data (these are data sets 2 through 4, Mielke and Johnson (1974), p. 224):
   (a) 46.65   29.96   25.49   11.85   17.25   50.80   31.93   15.31
       41.01   23.64   30.90   19.51   47.11   75.24   25.39   14.69
        9.06   41.06   57.04   30.93   18.54   45.80   39.64   38.14
       15.94   29.70   37.51   38.78   53.93   39.84   14.40   28.24

(b) | 40.6 | 32.7 | 242.5 | 115.3 | 255.0 | 118.3 | 334.1 | 1697.8 |
|---|---|---|---|---|---|---|---|
| 703.4 | 198.6 | 978.0 | 1656.0 | 7.7 | 274.7 | 274.7 | 200.7 |
| 17.5 | 92.4 | 4.1 | 119.0 | 302.8 | 430.0 | 489.1 | 2745.6 |
| 31.4 | 129.6 | | | | | | |

(c) | 0.030 | 0.020 | 0.015 | 0.045 | 0.100 | 0.100 | 0.125 | 0.190 | 0.390 |
|---|---|---|---|---|---|---|---|---|
| 0.110 | 0.070 | 0.010 | 0.055 | 0.220 | 0.080 | 0.005 | 0.125 | 0.035 |
| 0.085 | 0.060 | 0.010 | 0.065 | 0.020 | 0.260 | 0.030 | 0.015 | 0.025 |
| 0.010 | 0.495 | 0.085 | | | | | | |

3. Use the program HISTO to plot a histogram for the following sets of sediment yield data from Flakman (1972). Verify each outcome from the program by plotting the histogram by hand.

(a) Variable X_1:

0.135	0.135	0.101	0.353	0.353	0.492	0.466	0.466	0.833
0.085	0.085	0.085	0.193	0.167	0.235	0.448	0.329	0.428
0.133	0.149	0.266	0.324	0.133	0.133	0.133	0.356	0.536
0.155	0.168	0.673	0.725	0.275	0.150	0.150	1.140	1.428
1.126								

(b) Variable X_2

4.0	1.6	1.9	17.6	20.6	31.0	70.0	32.2	41.4	14.9
17.2	30.5	6.3	14.0	2.4	19.7	5.6	4.6	9.1	1.6
1.2	4.0	19.1	9.8	18.3	13.5	24.7	2.2	3.5	27.9
31.2	21.0	14.7	24.4	15.2	14.3	24.7			

(c) Variable X_3

40.0	40.0	22.0	2.0	1.0	0.0	23.0	57.0	4.0	22.0
2.0	15.0	1.0	3.0	1.0	3.0	27.0	12.0	58.0	28.0
15.0	48.0	44.0	32.0	19.0	58.0	62.0	27.0	24.0	22.0
10.0	64.0	37.0	29.0	11.0	1.0	44.0			

4. Using the method described in Section 4.3.9, examine the appropriateness of a normal distribution model for each of the data sets listed in Problems 2 and 3. For which of these data sets is the normal distribution model appropriate?

*5. Derive an expression for skewness, as was done in this chapter for variance, allowing parameters for skewness to be computed in the same program loop used to compute parameters for the mean and variance.

6. Use Bayes' theorem to calculate the probability for the situation described in Section 4.8.6, but assume that the test yields a positive test for an uncontaminated well only 6% of the time.

7. Compute the correlation coefficient between x and y for the following data sets and comment on the correlation between x and y:

(a) pair: | 1 | 2 | 3 | 4 | 5 | 6 | 7 | 8 | 9 | 10 |
|---|---|---|---|---|---|---|---|---|---|
| x: | 1 | 3 | 5 | 7 | 9 | 11 | 13 | 15 | 17 | 19 |
| y: | 2 | 3 | 5 | 6 | 6 | 9 | 12 | 13 | 13 | 15 |

(b) pair: | 1 | 2 | 3 | 4 | 5 | 6 | 7 | 8 | 9 | 10 |
|---|---|---|---|---|---|---|---|---|---|
| x: | 2 | 4 | 6 | 8 | 10 | 12 | 14 | 16 | 18 | 20 |
| y: | 30 | 29 | 27 | 23 | 25 | 20 | 17 | 19 | 18 | 10 |

(c) pair: | 1 | 2 | 3 | 4 | 5 | 6 | 7 | 8 | 9 | 10 |
|---|---|---|---|---|---|---|---|---|---|
| x: | 1 | 2 | 3 | 4 | 5 | 6 | 7 | 8 | 9 | 10 |
| y: | 5 | 3 | 1 | 1 | 4 | 2 | 6 | 2 | 3 | 3 |

References and Suggested Readings

DAVIS, J.C. 1986. *Statistics and Data Analysis in Geology,* 2d ed. New York: Wiley.

DEVORE, J.L. 1991. *Probability and Statistics for Engineering and the Sciences,* 3d ed. Monterey, CA: Brooks/Cole.

FLAKMAN, E.M. 1972. Predicting sediment yield in western United States. *Journal of the Hydraulics Division, ASCE* **98**(HY12): 2073–2085.

HANSON, K., G.A. MAUL, and T.R. KARL. 1989. Are atmospheric ''greenhouse'' effects apparent in the climatic record of the contiguous U.S. (1895–1987)? *Geophysical Research Letters* **16**(1): 49–52.

KREYSZIG, E. 1988. *Advanced Engineering Mathematics,* 6th ed. New York: Wiley.

McCUEN, R.H. 1985. *Statistical Methods for Engineers.* Englewood Cliffs, NJ: Prentice-Hall.

MIELKE, P.W., JR., and E.S. JOHNSON. 1974. Some generalized beta distributions of the second kind having desirable application features in hydrology and meteorology. *Water Resources Research,* **10**(2): 223–226.

NAISBITT, J. 1982. *Megatrends: Ten New Directions Transforming Our Lives.* New York: Warner Books.

Figure 4.8 A listing for the program HISTO. A user's guide appears at the beginning of this listing.

```fortran
      PROGRAM HISTO
C*******************************************************
C           USER'S GUIDE FOR HISTO
C*******************************************************
C
C   HISTO IS DESIGNED TO BE FLEXIBLE FOR A VERY WIDE
C   RANGE OF DATA FILES; MOST SIMPLY, HISTO IS USED
C   AS FOLLOWS:
C
C   CREATE A DATA FILE WHEREIN EACH LINE OF THE FILE
C   IS ONE DATA VALUE:
C
C   FOR EXAMPLE:
C
C        4.5, FOR LINE 1;
C        3.6, FOR LINE 2;
C        5.9, FOR LINE 3;
C        AND SO ON; THIS IS JUST AN EXAMPLE
C
C   ONCE THE FILE IS CREATED, RUN HISTO BY SIMPLY
C   TYPING THE WORD, HISTO, AT THE DOS COMMAND
C
C   E.G.:  A>HISTO, IF YOUR DISKETTE IS LOADED IN
C          DISK DRIVE, A.
C
C   THE PROGRAM THEN ASKS YOU SEVERAL QUESTIONS:
C
C   1.  DO YOU WISH TO TRANSFORM YOUR DATA TO
C       NATURAL LOGARITHMS?  1=YES; 0=NO.
C   2.  HOW MANY DATA VALUES ARE THERE PER LINE IN
C       YOUR DATA FILE?  THIS IS WHERE THE FLEXIBILITY
C       OF THE PROGRAM IS ENCOUNTERED; FOR THE EXAMPLE
C       ABOVE, ENTER 1.
C
C   3.  FOR WHICH VALUE ON THE DATA LINE DO YOU WISH
C       THE HISTOGRAM FOR?  AGAIN, THIS IS FOR
C       FLEXIBILITY; FOR THE EXAMPLE ABOVE, ENTER 1
C   4.  HOW MANY BINS IN THE HISTOGRAM?  ENTER THE
C       NUMBER OF SUBDIVISIONS OF THE DATA RANGE YOU
C       DESIRE FOR THE HISTOGRAM.
C
C*******************************************************
C                 END OF INPUT GUIDE
C*******************************************************
C
      COMMON /BIN/ NBIN, IBIN(20), XMIN, XMAX, ASTAT(4)
      DIMENSION Z(1000)
C
C   ACCESS INPUT DATA
C
      OPEN(5, FILE = ' ')
      NDAT = 0
C
C   A PROVISION FOR LOG TRANSFORMATION IS INCLUDED
C   IN CASE ONE WISHES TO TEST DATA FOR LOG-NORMALITY
C
      WRITE(*,*) 'ENTER OPTION FOR LOG-TRANS; 1=YES; 0=NO'
      READ(*,*) ILOG
      WRITE(*,*) 'HOW MANY DATA VALUES APPEAR ON EACH LINE'
      WRITE(*,*) 'IN YOUR DATA FILE?'
      READ(*,*) NUMDAT
      WRITE(*,*) 'WHICH OF THESE VALUES DO YOU DESIRE A'
      WRITE(*,*) 'HISTOGRAM FOR?'
      READ(*,*) IVAR
C
```

Figure 4.8 (continued)

```fortran
C     SCAN THE DATA TWICE, FIRST TO DETERMINE THE RANGE
C     OF DATA VALUES AND THE FIRST FOUR STATISTICAL MOMENTS
C     THEN AGAIN TO CALCULATE THE HISTOGRAM
C
      SUM = 0.0
      SUM2 = 0.0
      SUM3 = 0.0
      SUM4 = 0.0
5     READ(5,*) (Z(JK), JK = 1,NUMDAT)
      IF (ILOG .EQ. 1 .AND. Z(IVAR) .NE. 0.0) THEN
         Z(IVAR) = ALOG(Z(IVAR))
      ENDIF
      IF (ILOG .EQ. 1 .AND. Z(IVAR) .EQ. 0.0) GO TO 5
      NDAT = NDAT + 1
      ZMIN = Z(IVAR)
      ZMAX = ZMIN
      SUM = SUM + Z(IVAR)
      SUM2 = SUM2 + Z(IVAR) ** 2
      SUM3 = SUM3 + Z(IVAR) ** 3
      SUM4 = SUM4 + Z(IVAR) ** 4
10    READ(5,*,END = 20) (Z(JK), JK = 1,NUMDAT)
      NDAT = NDAT + 1
      IF (ILOG .EQ. 1 .AND. Z(IVAR) .NE. 0.0) THEN
         Z(IVAR) = ALOG(Z(IVAR))
      ENDIF
      IF (ILOG .EQ. 1 .AND. Z(IVAR) .EQ. 0.0) GO TO 10
      IF (Z(IVAR) .LT. ZMIN) ZMIN = Z(IVAR)
      IF (Z(IVAR) .GT. ZMAX) ZMAX = Z(IVAR)
      SUM = SUM + Z(IVAR)
      SUM2 = SUM2 + Z(IVAR) * Z(IVAR)
      SUM3 = SUM3 + Z(IVAR) ** 3
      SUM4 = SUM4 + Z(IVAR) ** 4
      GO TO 10
20    CONTINUE
C
C     CALCULATE THE HISTOGRAM
C
      WRITE(*, 21)
21    FORMAT(5X,' ENTER NUMBER OF BINS NOW')
      READ(*,*) NBIN
      REWIND 5
      RANGE = ZMAX - ZMIN
      XMIN = ZMIN
      XMAX = ZMAX
      DO 50 I = 1, NBIN
         IBIN(I) = 0
50    CONTINUE
60    CONTINUE
      READ(5,*, END = 70) (Z(JK), JK = 1,NUMDAT)
      IF (ILOG .EQ. 1 .AND. Z(IVAR) .GT. 0.0) THEN
         Z(IVAR) = ALOG(Z(IVAR))
      ENDIF
      IF (ILOG .EQ. 1 .AND. Z(IVAR) .EQ. 0.0) GO TO 60
      CALL HISTG(RANGE, Z(IVAR))
      GO TO 60
70    CONTINUE
      ASTAT(1) = SUM / FLOAT(NDAT)
      ASTAT(2) = (FLOAT(NDAT)*SUM2-SUM*SUM)/(FLOAT(NDAT)*
     2            FLOAT(NDAT - 1))
      ASTAT(3) = (SUM3/FLOAT(NDAT))-3.*ASTAT(1)*SUM2/FLOAT(NDAT)
     2            + 3.0 * (ASTAT(1) * ASTAT(1)) * SUM/FLOAT(NDAT) -
     3            ASTAT(1) * ASTAT(1) * ASTAT(1)
      ASTAT(4) = (SUM4 / FLOAT(NDAT)) - 4.0 * ASTAT(1) * SUM3
     2           /FLOAT(NDAT)+6.0*ASTAT(1) * ASTAT(1) * SUM2/FLOAT(NDAT)
     3           -4.0 * ASTAT(1) * ASTAT(1) * SUM/FLOAT(NDAT) +
     4            ASTAT(1) * ASTAT(1) * ASTAT(1) * ASTAT(1)
C
```

Figure 4.8 (continued)

```fortran
C       PLOT HISTOGRAM
C
        CALL PLOT
C
C       TERMINATE PROGRAM
C
        STOP
        END
C
C
        SUBROUTINE HISTG(R, Z)
        COMMON /BIN/ NBIN, IBIN(20), XMIN, XMAX, ASTAT(4)
C
C       CALCULATE THE HISTOGRAM
C
        RANGE = R
        AINT  = RANGE / FLOAT(NBIN)
        W = ((Z - XMIN)/AINT) + 1.0
        J = INT(W)
        IF (J.GT.NBIN) J = NBIN
        IF (J .LT. 1) J = 1
        IBIN(J) = IBIN(J) + 1
        RETURN
        END
C


C
        SUBROUTINE PLOT
        COMMON /BIN/ NBIN, IBIN(20), XMIN, XMAX, ASTAT(4)
        CHARACTER*1 ASTER
        ASTER = '*'
        WRITE(*, 10)
10      FORMAT(1H1, 10X, 'HISTOGRAM RESULTS',//,
     2   12X,'INTERVAL',5X,'BIN #',6X,'FREQ',/)
        AINT = (XMAX - XMIN) / FLOAT(NBIN)
        IBINMAX = IBIN(1)
        DO 15 I = 2, NBIN
          IF (IBIN(I) .GT. IBINMAX) IBINMAX = IBIN(I)
15      CONTINUE
        DO 20 I = 1,NBIN
        W = XMIN + AINT * FLOAT(I-1)
        V = W + AINT
        IF (IBIN(I) .EQ. 0) THEN
          WRITE(*,30) W,V,I,IBIN(I)
        ELSE
          KBIN = INT(50 * IBIN(I) / IBINMAX) + 1
          WRITE(*,40) W,V,I,IBIN(I),(ASTER,J=1,KBIN)
        ENDIF
20      CONTINUE
30      FORMAT(2F10.3,I10,I10)
40      FORMAT(2F10.3,I10,I10,1X,92A1)
        WRITE(*,50) (ASTAT(I), I = 1,3)
50      FORMAT(1H0, 9X,'MEAN        = ',1E15.5,//,
     2   10X,'VARIANCE    = ',1E15.5,//,
     3   10X,'SKEW        = ',1E15.5)
        RETURN
        END
```

5

Multivariate Data Analysis

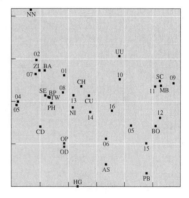

Analysis of multivariate data is merely an extension of bivariate statistical correlation analysis to data derived from more than two variables. This chapter emphasizes the interpretation of multivariate data to identify data intercorrelations. A rather conservative application of multivariate data analysis is presented, fully recognizing that such an analysis is not always scientifically repeatable with respect to interpretation of results.

An emphasis is placed herein on the use of principal components methods, particularly correspondence analysis, for analyzing the similarity among multivariate data. This is but one of a vast array of multivariate analytical methods: Multiple regression analysis, multiple discriminant analysis, multivariate analysis of variance, canonical correlation, linear probability models, conjoint analysis, structural equation modeling, cluster analysis, and multidimensional scaling offer different ways to analyze multivariate data. The purpose of this text is to present a brief discussion of a variety of numerical analysis methods that are applicable to the study of earth science. For more information on these methods, the reader is referred to the excellent books by Hair et al. (1992) and Anderson (1984).

Although this chapter is primarily devoted to principal components analysis and related methods, particularly correspondence analysis, a section is presented on discriminant analysis to demonstrate an analytical technique for predicting the likelihood that a data element (individual) belongs to a particular group of data elements (individuals). This type of analysis differs from principal components and related analyses in that its purpose is to find a linear function that provides the maximum separation between groups of data. This aspect of discriminate analysis is used to verify conclusions drawn from correspondence analysis in application to a water quality data set.

5.1 Introduction

From 1985 through 1986 over 7 million fish and 16,000 waterfowl died at Stillwater Lakes Wildlife Refuge in the State of Nevada (Figure 5.1) (Oleson and Carr, 1990). Stillwater Wildlife Refuge is part of the International Flyway for migratory waterfowl and is protected under international treaty between United States, Canada, and Mexico. An environmental problem at Stillwater Lakes consequently has international ramifications.

What caused these massive deaths at Stillwater Lakes? In an effort to identify the cause or causes of this problem, the United States Geological Survey collected water samples at Stillwater Lakes in June and September 1986 (sites shown in Figure 5.1). Each water sample was analyzed for a variety of chemical attributes, as shown in Tables 5.1 and 5.2. These tables are examples of multivariate data. A number of individuals (in this case, water samples from various sites) are shown, each associated with a number of attributes (in this case, chemical constituents and properties). How can such data be analyzed to facilitate statistical inference (decision making)?

Sites 1 through 4 in Figure 5.1 are background sites and either are disassociated geographically with Stillwater Lakes or are upstream of these lakes. How do water samples from Stillwater Lakes (sites 5 through 16) compare to background sites? An inferred similarity between these groups may indicate that waterfowl deaths are attributable to something other than water chemistry (waterfowl deaths did not occur at sites 1 through 4 in 1985 and 1986). If these groups are different based on chemical attributes of the water samples, then an inorganic cause might be responsible for the deaths. Likewise, how similar are the chemical attributes? Answering this question may lead to an understanding of the source of chemical contamination, if indeed evidence suggests that Stillwater Lakes sites are chemically unique.

These are simply examples to motivate multivariate data analysis. Toward the end of this chapter, results of multivariate analysis (applied to Tables 5.1 and 5.2) are presented, along with inferences based on these results. An assessment of the environmental cause of the waterfowl deaths is also presented. What caused the deaths of the fauna at Stillwater Lakes Wildlife Refuge is discussed later. First some fundamental concepts are discussed to provide insight into multivariate analysis, to describe this type of analysis mathematically, and to demonstrate some simpler examples prior to the analysis of the more complicated data sets in Tables 5.1 and 5.2.

5.2 An Overview of Multivariate Analytical Methods

Multivariate data analysis has been used extensively in the social sciences, especially as an aid in understanding opinion surveys. Opinion surveys are

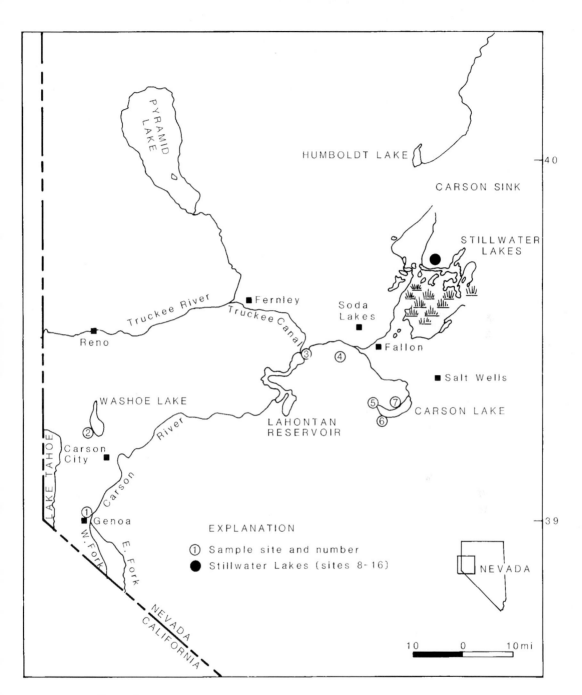

Figure 5.1   A location map of a portion of the state of Nevada (USA) showing Stillwater Lakes Wildlife Refuge.

completed by individuals and the responses are characteristics or attributes. Therefore, in this chapter samples are referred to as individuals and the variables characterizing each sample are referred to as attributes. Use of these terms is commensurate with the historical development of multivariate data analysis in the social sciences. Subsequent discussion focuses on principal components type multivariate analysis methods.

Table 5.1   Water Quality Data Set for Stillwater Lakes, Nevada, and Surrounding Region, June, 1986 (Oleson and Carr, 1990)

IDEN	TW	BP	SC	OD	OP	PH	NN
3	.15E + 02	.66E + 03	.25E + 03	.90E + 01	.10E + 03	.79E + 01	.60E − 01
4	.23E + 02	.66E + 03	.27E + 03	.92E + 01	.13E + 03	.88E + 01	.00E + 00
8	.25E + 02	.66E + 03	.67E + 03	.84E + 01	.12E + 03	.90E + 01	.80E − 01
9	.24E + 02	.66E + 03	.72E + 04	.10E + 02	.15E + 03	.81E + 01	.80E − 01
10	.22E + 02	.66E + 03	.26E + 04	.54E + 01	.72E + 02	.86E + 01	.90E − 01
11	.26E + 02	.66E + 03	.28E + 04	.95E + 01	.14E + 03	.88E + 01	.16E + 00
12	.23E + 02	.66E + 03	.39E + 04	.95E + 01	.13E + 03	.92E + 01	.50E − 01
5	.24E + 02	.66E + 03	.11E + 04	.67E + 01	.93E + 02	.81E + 01	.14E + 00
6	.34E + 02	.66E + 03	.16E + 04	.20E + 02	.30E + 03	.91E + 01	.40E − 01
7	.32E + 02	.65E + 03	.47E + 04	.15E + 02	.25E + 03	.85E + 01	.40E − 01
14	.20E + 02	.66E + 03	.84E + 03	.63E + 01	.79E + 02	.80E + 01	.11E + 00
15	.22E + 02	.66E + 03	.14E + 04	.96E + 01	.13E + 03	.90E + 01	.40E − 01

	AS	BA	BO	CH	CU	PB	MB
3	.70E + 01	.41E + 02	.13E + 03	.10E + 02	.50E + 01	.00E + 00	.30E + 01
4	.14E + 02	.38E + 02	.16E + 03	.50E + 01	.50E + 01	.10E + 01	.40E + 01
8	.43E + 02	.20E + 03	.51E + 03	.50E + 01	.10E + 02	.00E + 00	.10E + 02
9	.80E + 02	.10E + 03	.44E + 04	.50E + 01	.20E + 02	.10E + 01	.11E + 03
10	.56E + 02	.00E + 00	.18E + 04	.50E + 01	.10E + 02	.00E + 00	.45E + 02
11	.51E + 02	.10E + 03	.15E + 04	.50E + 01	.10E + 02	.20E + 01	.38E + 02
12	.88E + 02	.20E + 03	.34E + 04	.50E + 01	.10E + 02	.00E + 00	.74E + 02
5	.10E + 01	.69E + 02	.12E + 04	.50E + 01	.50E + 01	.20E + 01	.28E + 02
6	.10E + 03	.63E + 02	.22E + 04	.50E + 01	.50E + 01	.00E + 00	.20E + 01
7	.74E + 02	.10E + 03	.83E + 04	.50E + 01	.20E + 02	.20E + 01	.40E + 01
14	.23E + 02	.52E + 02	.69E + 03	.50E + 01	.50E + 01	.10E + 01	.14E + 02
15	.80E + 02	.64E + 02	.17E + 04	.50E + 01	.50E + 01	.00E + 00	.26E + 02

	NI	ZI	SE	UU	HG
3	.10E + 01	.12E + 02	.50E + 00	.34E + 01	.30E + 00
4	.30E + 01	.50E + 01	.50E + 00	.46E + 01	.00E + 00
8	.10E + 01	.20E + 02	.50E + 00	.98E + 01	.10E + 00
9	.20E + 01	.30E + 02	.10E + 01	.15E + 02	.10E + 00
10	.10E + 01	.00E + 00	.50E + 00	.30E + 02	.40E + 00
11	.50E + 01	.20E + 02	.50E + 00	.25E + 02	.00E + 00
12	.10E + 01	.20E + 02	.50E + 00	.35E + 02	.30E + 00
5	.13E + 02	.80E + 01	.10E + 01	.36E + 02	.10E + 00
6	.20E + 01	.90E + 01	.50E + 00	.19E + 02	.30E + 00
7	.50E + 01	.20E + 02	.50E + 00	.12E + 02	.10E + 00
14	.10E + 01	.12E + 02	.50E + 00	.89E + 01	.20E + 00
15	.30E + 01	.60E + 01	.50E + 00	.24E + 02	.00E + 00

Note. Collected by the United States Geological Survey, Carson City, Nevada.

Key for attributes:

TW, water temperature (°C)

BP, barometric pressure

SC, specific conductance

OD, oxygen, dissolved (mg/L)

OP, oxygen, dissolved, % sat.

PH, pH, standard units

NN, nitrogen (ammonia), dissolved

AS, arsenic, dissolved

BA, barium, dissolved

BO, boron, dissolved

CH, chromium, dissolved

CU, copper, dissolved

PB, lead, dissolved

MB, molybdenum, dissolved

NI, nickel, dissolved

ZI, zinc, dissolved

SE, selenium, dissolved

UU, uranium, dissolved

HG, mercury, dissolved

Table 5.2 Water Quality Data Set for Stillwater Lakes, Nevada, and Surrounding Region, September, 1986 (Oleson and Carr, 1990)

IDEN	TW	BP	SC	OD	OP	PH	NN
1	.16E + 02	.64E + 03	.43E + 03	.00E + 00	.00E + 00	.72E + 01	.93E + 00
2	.22E + 02	.64E + 03	.39E + 03	.71E + 01	.97E + 02	.87E + 01	.30E − 01
3	.21E + 02	.66E + 03	.20E + 03	.82E + 01	.11E + 03	.78E + 01	.20E − 01
4	.24E + 02	.66E + 03	.22E + 03	.87E + 01	.12E + 03	.86E + 01	.30E − 01
8	.25E + 02	.66E + 03	.49E + 03	.69E + 01	.97E + 02	.88E + 01	.10E + 00
9	.20E + 02	.66E + 03	.12E + 05	.98E + 01	.13E + 03	.82E + 01	.10E + 00
10	.21E + 02	.66E + 03	.18E + 04	.62E + 01	.80E + 02	.82E + 01	.80E − 01
11	.27E + 02	.66E + 03	.40E + 04	.93E + 01	.14E + 03	.86E + 01	.18E + 00
12	.23E + 02	.66E + 03	.40E + 04	.35E + 01	.48E + 02	.91E + 01	.50E − 01
5	.17E + 02	.66E + 03	.14E + 04	.76E + 01	.91E + 02	.80E + 01	.10E + 00
6	.21E + 02	.66E + 03	.84E + 03	.17E + 02	.22E + 03	.90E + 01	.20E − 01
7	.19E + 02	.66E + 03	.32E + 03	.44E + 01	.55E + 02	.78E + 01	.14E + 00
13	.23E + 02	.66E + 03	.57E + 03	.62E + 01	.84E + 02	.81E + 01	.60E − 01
14	.20E + 02	.66E + 03	.66E + 03	.80E + 01	.10E + 03	.82E + 01	.80E − 01
15	.22E + 02	.66E + 03	.23E + 04	.18E + 02	.24E + 03	.10E + 02	.20E − 01
16	.23E + 02	.66E + 03	.11E + 04	.63E + 01	.85E + 02	.81E + 01	.60E − 01

	AS	BA	BO	CD	CH	CU	PB
1	.80E + 01	.41E + 02	.22E + 03	.30E + 01	.50E + 01	.50E + 01	.20E + 01
2	.50E + 01	.12E + 03	.70E + 02	.00E + 00	.50E + 01	.50E + 01	.20E + 01
3	.10E + 02	.47E + 02	.11E + 03	.00E + 00	.50E + 01	.50E + 01	.20E + 01
4	.14E + 02	.90E + 02	.12E + 03	.10E + 02	.50E + 01	.50E + 01	.20E + 01
8	.28E + 02	.23E + 93	.44E + 03	.50E + 01	.50E + 01	.50E + 01	.20E + 01
9	.39E + 02	.10E + 03	.76E + 04	.00E + 00	.20E + 02	.20E + 02	.20E + 01
10	.29E + 02	.26E + 03	.11E + 04	.12E + 02	.50E + 01	.50E + 01	.20E + 01
11	.44E + 02	.10E + 03	.25E + 04	.00E + 00	.50E + 01	.10E + 02	.20E + 01
12	.95E + 02	.10E + 03	.41E + 04	.00E + 00	.10E + 02	.10E + 02	.74E + 02
5	.61E + 02	.60E + 02	.16E + 04	.90E + 01	.50E + 01	.50E + 01	.20E + 01
6	.41E + 02	.56E + 02	.11E + 04	.80E + 01	.50E + 01	.50E + 01	.20E + 01
7	.14E + 02	.21E + 03	.25E + 03	.00E + 00	.50E + 01	.50E + 01	.20E + 01
13	.37E + 02	.24E + 03	.56E + 03	.10E + 01	.50E + 01	.50E + 01	.20E + 01
14	.39E + 02	.11E + 03	.67E + 03	.40E + 01	.50E + 01	.50E + 01	.20E + 01
15	.84E + 02	.10E + 03	.33E + 04	.00E + 00	.50E + 01	.10E + 02	.20E + 01
16	.34E + 02	.21E + 03	.12E + 04	.50E + 01	.50E + 01	.50E + 01	.20E + 01

	MB	NI	ZI	SE	UU	HG
1	.20E + 01	.10E + 01	.30E + 01	.50E + 00	.90E + 00	.00E + 00
2	.40E + 01	.10E + 01	.10E + 02	.50E + 00	.62E + 02	.00E + 00
3	.30E + 01	.10E + 01	.19E + 02	.50E + 00	.23E + 01	.00E + 00
4	.40E + 01	.30E + 01	.21E + 02	.50E + 00	.25E + 01	.00E + 00
8	.16E + 02	.20E + 01	.27E + 02	.50E + 00	.13E + 02	.00E + 00
9	.25E + 03	.30E + 01	.30E + 02	.50E + 00	.14E + 03	.10E + 00
10	.44E + 02	.30E + 01	.54E + 02	.50E + 00	.24E + 02	.00E + 00
11	.10E + 03	.40E + 01	.10E + 02	.50E + 00	.46E + 02	.10E + 00
12	.12E + 03	.30E + 01	.20E + 02	.50E + 00	.48E + 02	.00E + 00
5	.44E + 02	.20E + 01	.12E + 02	.50E + 00	.36E + 02	.20E + 00
6	.50E + 01	.20E + 01	.27E + 02	.50E + 00	.14E + 02	.20E + 00
7	.80E + 01	.20E + 01	.66E + 02	.50E + 00	.36E + 01	.00E + 00
13	.17E + 02	.30E + 01	.28E + 02	.10E + 01	.17E + 02	.10E + 00
14	.21E + 02	.20E + 01	.19E + 02	.50E + 00	.21E + 02	.90E + 00
15	.45E + 02	.30E + 01	.10E + 02	.50E + 00	.11E + 02	.10E + 00
16	.23E + 02	.30E + 01	.20E + 02	.50E + 00	.19E + 02	.00E + 00

Note. Collected by the United States Geological Survey, Carson City, Nevada.

Key for attributes:

TW, water temperature (°C)
BP, barometric pressure
SC, specific conductance
OD, oxygen, dissolved (mg/L)
OP, oxygen, dissolved, % sat.
PH, pH, standard units
NN, nitrogen (ammonia), dissolved
AS, arsenic, dissolved
BA, barium, dissolved
BO, boron, dissolved

CH, chromium, dissolved
CU, copper, dissolved
PB, lead, dissolved
MB, molybdenum, dissolved
NI, nickel, dissolved
ZI, zinc, dissolved
SE, selenium, dissolved
UU, uranium, dissolved
HG, mercury, dissolved
CD, cadmium, dissolved

5.2.1 Principal Components Analysis

Multivariate data (Tables 5.1 and 5.2) are associated inherently with some degree of interdependency (relationship) between individuals and attributes. The challenge is to determine which individuals are similar, which attributes are similar, and the similarity between attributes and individuals.

A multivariate data analysis method is sought that can yield some type of useful plot based on information for all attributes and individuals (Figure 5.2). Once such a plot is obtained, it can be studied to assess data interdependencies. In simple terms, the closer two individuals or attributes are in the plot, the more related they may be.

One multivariate data analysis method that can yield the desired plot is principal components analysis (PCA). This method is perhaps the simplest of the multivariate data analysis techniques. A simple bivariate data set (Table 5.3) from McCuen (1985, p. 242) is used to aid the mathematical understanding of PCA.

In this method, a variance/covariance matrix $[R]$ is formed:

$$[R] = \begin{bmatrix} \text{var}(x) & \text{cov}(x, y) \\ \text{cov}(x, y) & \text{var}(y) \end{bmatrix} = \begin{bmatrix} 2.5 & 4.0 \\ 4.0 & 8.3 \end{bmatrix}$$

These entries are obtained from the information in Table 5.3. Eigendecomposition of $[R]$ is performed:

$$[R - \lambda I] = \begin{vmatrix} (2.5 - \lambda) & 4 \\ 4 & (8.3 - \lambda) \end{vmatrix}$$

which has the following determinant:

$$\det[R - \lambda I] = \lambda^2 - 10.8\lambda + 4.75 = 0$$

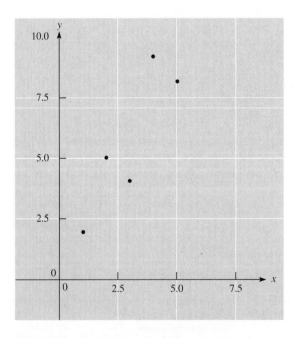

Figure 5.2   An example principal components plot.

Table 5.3 Data Set (Bivariate) from McCuen (1985, p. 242)

	x	y	xy
	1	2	2
	2	5	10
	3	4	12
	4	9	36
	5	8	40
Totals	15	28	100

Note. For these data cov(x, y) = $[5(100) - 15(28)]/[5(4)]$ = 4; r = cov$(x, y)/[\text{sdev}(x)\text{sdev}(y)]$ = 0.88; and sdev(x) = 1.58, sdev(y) = 2.88.

the roots of which are 10.34 (first eigenvalue) and 0.46 (second eigenvalue). Recall from Chapter 3 that the sum of eigenvalues is equal to the trace of the matrix $[R]$, and this trace is equal to the sum of the variances for the two variables var(x) and var(y). Therefore, if each eigenvalue is divided by the trace of $[R]$ and multiplied by 100, the percentage of the total data variance represented by each eigenvalue is obtained.

In the present example (Table 5.3), the trace of $[R]$ is equal to var(x) + var(y), or 10.8. The first eigenvalue, 10.34, consequently represents 100(10.34/10.80) or 95.74% of the total data variance. The second eigenvalue, 0.46, represents 100(0.46/10.8) or 4.26% of the total data variance.

Using the eigenvalues of $[R]$, the two associated eigenvectors are found:

$$|x_1| = \begin{vmatrix} 0.51 \\ 1 \end{vmatrix} \qquad |x_2| = \begin{vmatrix} -1.96 \\ 1 \end{vmatrix}$$

then

$$[x] = \begin{vmatrix} 0.51 & -1.96 \\ 1 & 1 \end{vmatrix}$$

With the two eigenvectors assembled as columns in matrix $[x]$, the original data (Table 5.3) are projected onto the axes defined by these orthogonal (independent) eigenvectors. Let the original data be represented by the matrix $[Y]$ and complete the projection of the data onto the eigenvectors using matrix multiplication:

$$[Y][x] = \begin{vmatrix} 1 & 2 \\ 2 & 5 \\ 3 & 4 \\ 4 & 9 \\ 5 & 8 \end{vmatrix} \begin{vmatrix} 0.51 & -1.96 \\ 1 & 1 \end{vmatrix} = \begin{vmatrix} 2.51 & 0.04 \\ 6.02 & 1.08 \\ 5.53 & -1.88 \\ 11.04 & 1.16 \\ 10.55 & -1.80 \end{vmatrix}$$

These "new" coordinates are used for plotting; for example, the first individual has the coordinates (2.51, 0.04), the second has the coordinates (6.02, 1.08), and so on (Figure 5.3). Spatial proximity in this plot may be interpreted as an

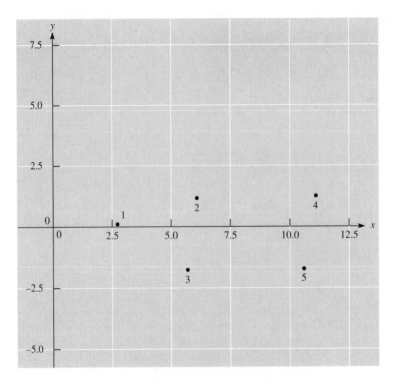

Figure 5.3   A plot of the principal components analysis results for the data of Table 5.3.

indication of the degree of similarity between individuals (samples) (Table 5.3). Individuals 2 and 4, for instance, appear to be more similar than individuals 1 and 4.

5.2.2 Representing the PCA Results Graphically

One way to view eigendecomposition (Figure 5.4) was demonstrated in Chapter 3. From the previous example, the eigenvalues are found to be 10.34 and 0.46, respectively. Eigenvalues are equal to one-half the lengths of the major and minor axes of an ellipse (Figure 5.4). Eigenvectors define the directions of these axes, that is, they show the orientation of the ellipse relative to the original data axes x, y (Figure 5.4). Moreover, each row of the variance/covariance matrix $[R]$ represents coordinates of points that plot on this ellipse.

Variables x and y are well correlated, with a correlation coefficient of 0.88, which explains why the ellipse is highly elongated, with one axis much longer than the other. Notice that data pairs (x, y) (Table 5.3) are plotted to show their positions relative to axes of the ellipse. Observe that the projection of data points onto the axes (eigenvectors) using the multiplication $[Y][x]$ (comparing Figure 5.4 to Figure 5.3) results in a clockwise rotation of the axes defining the ellipse, such that the major axis of the ellipse is parallel to the x axis.

An experiment is suggested by this observation that involves four steps: (1) Draw a pair of perpendicular axes on tracing paper. (2) Lay the perpendicular lines over the axes of the ellipse such that the intersection of the lines on the tracing paper coincides exactly with the intersection of the axes of the ellipse. (3) Plot the five data points on this tracing paper. (4) Lay the axes of the tracing

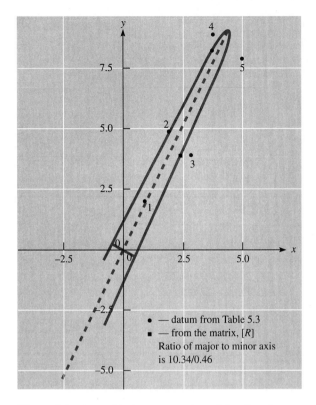

Figure 5.4   A graphical representation of the eigendecomposition results from PCA for the data of Table 5.3. Notice the elongated shape of the ellipse, an indication of good correlation between variables x and y of Table 5.3.

paper over those on Figure 5.3 such that the axis aligned with the long axis of the ellipse in Figure 5.4 is aligned with the horizontal axis of Figure 5.3; make sure the intersection of the lines coincides exactly with the intersection of the axes on Figure 5.3.

What is discovered? In this case, the same geometry of points is found on the tracing paper as is seen in Figure 5.3, with the exception that the points in Figure 5.3 are more widely spaced. This demonstrates the notion that PCA can be viewed as a coordinate transformation.

5.2.3 Factor Analysis

A method of multivariate data analysis that is similar to PCA is factor analysis. A factor is a linear combination of the original variables or individuals (Hair et al., 1992). Factors also describe the inherent dimensionality of a multivariate set of data, that is, the number of factors required to describe the original group of variables or individuals. Factor analysis is a method for analyzing the relationships among a large number of variables or individuals as assessed relative to the factors.

Some controversy exists regarding the application of factor analysis for geologic interpretation (see, for example, Temple (1978)), yet the application of this analytical method for geologic interpretation does seem to have some merit (Klovan, 1966). Moreover, factor analysis is performed by a wide variety of statistical software (a reference on the use of SPSSX (statistical package for the social sciences) is provided by Norusis (1985)).

Factor analysis is performed in one of two ways: R-mode and Q-mode. A distinction is drawn between these modes.

5.2.3.1 R-Mode Factor Analysis

In R-mode analysis, a determination is made regarding how the M attributes characterizing N individuals are interrelated. A procedure is followed that is similar to the previous example used to explain PCA, except that correlation coefficients, rather than variances and covariances, are used to form the matrix $[R]$.

From the data of Table 5.3, matrix $[R]$ (using the same notation as was used for the PCA example) is formed as

$$[R] = \begin{bmatrix} r(x, x) & r(x, y) \\ r(x, y) & r(y, y) \end{bmatrix} = \begin{bmatrix} 1.00 & 0.88 \\ 0.88 & 1.00 \end{bmatrix}$$

because the correlations of x with itself, $r(x, x)$, and that of y with itself, $r(y, y)$, are equal to one (perfect, positive correlation because $\mathrm{cov}(x, x) = \mathrm{var}(x)$; $r(x, x) = \mathrm{cov}(x, x)/\mathrm{sdev}(x)\mathrm{sdev}(x) = \mathrm{var}(x)/\mathrm{var}(x) = 1$; likewise for y).

Eigenvalues of $[R]$ are found as follows:

$$[R - \lambda I] = \begin{vmatrix} (1 - \lambda) & 0.88 \\ 0.88 & (1 - \lambda) \end{vmatrix}$$

which has the determinant

$$\det[R - \lambda I] = \lambda^2 - 2\lambda + 0.2256 = 0$$

the roots of which are 1.88 (first eigenvalue) and 0.12 (second eigenvalue). Moreover, the eigenvectors are $\{1\ 1\}$ for the first eigenvalue and $\{-1\ 1\}$ for the second eigenvalue.

Unlike PCA, these eigenvectors are converted to factors by multiplying each eigenvector by the square root of its eigenvalue. For this example,

$$|\text{ factor } 1| = \sqrt{1.88} \begin{vmatrix} 1 \\ 1 \end{vmatrix} = \begin{vmatrix} 1.37 \\ 1.37 \end{vmatrix}$$

and

$$|\text{ factor } 2| = \sqrt{0.12} \begin{vmatrix} -1 \\ 1 \end{vmatrix} = \begin{vmatrix} -0.35 \\ 0.35 \end{vmatrix}$$

Therefore the coordinates used to plot the original data of Table 5.3 are for x (1.37, -0.35) and for y (1.37, 0.35). An ellipse is developed based on these results (Figure 5.5).

5.2.3.2 Q-Mode Factor Analysis

A Q-mode analysis is used to assess how individual samples (the rows of Table 5.3) are interrelated. The numerical processing required to perform Q-mode factor analysis is more complex than that used for R-mode factor analysis.

In R-mode analysis, the original $N \times M$ data matrix $[Y]$ is converted to an $M \times M$ square matrix of correlation coefficients. In Q-mode analysis, matrix

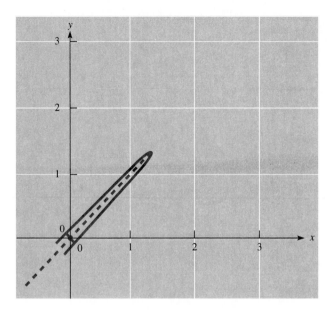

Figure 5.5 A graphical representation of the eigendecomposition results from R-mode factor analysis for the data of Table 5.3. As with the PCA results plotted in Figure 5.4, an elongated ellipse is the result for this analysis.

$[Y]$ is processed to yield an $N \times N$ square matrix using the following expression (Davis, 1986, p. 563):

$$\cos \theta_{ij} = \frac{\sum_{k=1}^{M} Y_{ik} Y_{jk}}{\sqrt{\sum_{k=1}^{M} y_{ik}^2 \sum_{k=1}^{M} y_{kj}^2}}, \qquad i = 1, \ldots, N, \quad j = 1, \ldots, N$$

Each row (individual) in a multivariate data matrix is visualized as a vector defined in M (attribute) space, and the cosine value defined in the foregoing equation is that for the angle between the two vectors (individuals) i and j.

A conversion of the data in Table 5.3 to cosines (Table 5.4) results in a 5×5 matrix. The cosine for an individual with itself is 1 (that is, the angle between two identical vectors is zero). The closer the cosine value is to one, the better related (more similar) are two individuals. A cosine value of zero suggests that two individuals are perfectly independent.

Eigenvalues and eigenvectors for the cosine matrix are calculated (Table 5.5) using the program EIGEND (Chapter 3). Eigenvectors are converted to

Table 5.4 Cosine Matrix for the Data Listed in Table 5.3

	Column 1	*Column 2*	*Column 3*	*Column 4*	*Column 5*
Row 1	1.000	0.997	0.984	0.999	0.995
Row 2	0.997	1.000	0.966	0.999	0.984
Row 3	0.984	0.966	1.000	0.975	0.996
Row 4	0.999	0.999	0.975	1.000	0.990
Row 5	0.995	0.984	0.996	0.990	1.000

Table 5.5   Eigendecomposition
Results for the Cosine Matrix
in Table 5.4

Vector 1	Vector 2
1.002	−0.440
0.996	−1.698
0.991	2.280
0.999	−1.128
1.000	1.000

Note. Eigenvalues: $\lambda_1 = 4.954$, $\lambda_2 = 0.046$, $\lambda_3 = 0.0$, $\lambda_4 = 0.0$, $\lambda_5 = 0.0$. Only two nonzero eigenvalues are found for this matrix, hence the true rank of the matrix is two. Therefore, only two eigenvectors are calculated.

Table 5.6   *Q*-Mode Factors for the Data of Table 5.3

Factor 1	Factor 2
$1.002(4.954^{1/2}) = 2.230$	$-0.440(0.046^{1/2}) = -0.095$
$0.996(4.954^{1/2}) = 2.217$	$-1.698(0.046^{1/2}) = -0.365$
$0.991(4.954^{1/2}) = 2.206$	$2.280(0.046^{1/2}) = 0.490$
$0.999(4.954^{1/2}) = 2.224$	$-1.128(0.046^{1/2}) = -0.243$
$1.000(4.954^{1/2}) = 2.226$	$1.000(0.046^{1/2}) = 0.215$

factors (Table 5.6) using the same procedure used for *R*-mode factor analysis:

$$| \text{factor } i | = \sqrt{\text{eigenvalue}_i} | \text{eigenvector}_i |$$

These factors (5 in this case) are the coordinates used for plotting individuals in Figure 5.6. Comparing this figure to Figure 5.3 developed using PCA reveals that a similar analysis is achieved using factor analysis, except the geometry is different. For example, individuals 2 and 4 are related (that is, in close proximity to one another) and individuals 3 and 5 are related. These same relationships were found using PCA, but, factor analysis has individual 1 plotting between the two groups of individuals 2 and 4 and individuals 3 and 5. Individual 1 plots to the left of the four remaining individuals (Figure 5.3); individuals 2 and 4 and individuals 3 and 5 seem to be related.

5.3 Correspondence Analysis

With factor analysis, as presented in this text, two different computer runs are necessary if both *R*-mode and *Q*-mode analyses are desired (an appropriate objective for a thorough multivariate data analysis). Thus, a method capable of rendering *R*- and *Q*-mode analyses simultaneously would be convenient.

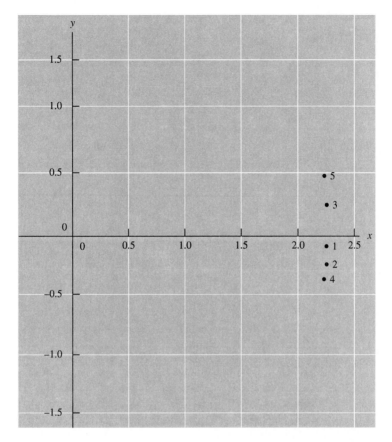

Figure 5.6   A plot of Q-mode factor analysis results for the data of Table 5.3. This figure suggests that individuals 2 and 4 are correlated and that individuals 3 and 5 are correlated, but notice that the geometry of this plot is different from that of PCA (Figure 5.3).

This would also allow a direct analysis of the similarity between individuals and attributes.

Implementing factor analysis in a computer program may be accomplished such that simultaneous R- and Q-mode analyses are obtained. However, in this text we discuss another multivariate data analysis method, correspondence analysis, which also yields simultaneous R-mode and Q-mode analyses. Correspondence analysis has some attractive advantages over factor analysis (Hair et al., 1992, p. 343). In particular, correspondence analysis renders plots showing relative similarity among attributes and individuals, hence simultaneous R- and Q-mode analyses.

Whereas PCA and factor analysis have been known since about 1930, correspondence analysis is a relatively new method. It was first described by Benzecri (1973) for applications in the social sciences. FORTRAN programs for correspondence analysis are given in David et al. (1977) and in Carr (1990). The book by Greenacre (1984) is entirely devoted to the theory and application of correspondence analysis. Davis (1986) and Hair et al. (1992) give clear, straightforward explanations of the method. Our discussion of the theory of correspondence analysis is based on these publications.

Correspondence analysis makes use of the notion of joint and marginal probabilities; the latter is useful for defining the expected value of data entries. For a collection of multivariate data $[Y]$ (e.g., Tables 5.1, 5.2, and 5.3), the first step in correspondence analysis is the computation of the total sum of the data composing $[Y]$:

$$XL = \sum_{i=1}^{N} \sum_{j=1}^{M} y_{ij}$$

where XL is the total sum. Each entry in $[Y]$ is then divided by the total sum:

$$Y_{ij} = \frac{Y_{ij}}{XL}, \quad i = 1, \ldots, N, \quad j = 1, \ldots, M$$

which converts the original entries in $[Y]$ to joint probabilities; that is, after the division by the total sum, each entry Y_{ij} represents the probability of having attribute j in individual i. This is a joint probability.

Next two vectors, $\{W\}$ and $\{T\}$, are formed, which represent the sums of the rows and columns, respectively, of $[Y]$:

$$W_i = \sum_{j=1}^{M} Y_{ij}, \qquad T_j = \sum_{i=1}^{N} Y_{ij}$$

By these definitions, the vector $\{W\}$ represents marginal probabilities for individuals; that is, each value W_i is the probability that individual i occurs in the data set. The vector $\{T\}$ represents marginal probabilities for attributes.

Here our use of the term probability is in reference to the original development of correspondence analysis (Benzecri, 1973) in application to enumeration (count) data. In application to continuous, real-valued data (Tables 5.1 and 5.2), the term probability cannot be correctly applied to describe the entries in $[Y]$, $\{W\}$, and $\{T\}$. Hereafter, rather than using the term probability, the entries in $\{W\}$ and $\{T\}$ are referred to as sums. Entries in $[Y]$ are not referred to as joint probabilities, and no other special term will be used. Entries in $[Y]$ will simply be referred to as Y_{ij}, where i is the row number and j is the column number. Probability is mentioned in the foregoing discussion for its historical context (as used by Benzecri (1973)), and to call attention to why original entries in $[Y]$ are divided by the total sum XL.

Implementing correspondence analysis in a computer program is accomplished easily by forming two matrices, $[S_c]$ and $[S_r]$, as follows:

$$S_{c,ij} = \frac{Y_{ij}}{W_i \sqrt{T_j}}, \qquad S_{r,ij} = \frac{Y_{ij}}{\sqrt{T_j}}$$

Note that both matrices have the dimension $N \times M$, commensurate with $[Y]$. A square $M \times M$ matrix $[S]$ is formed from these two matrices:

$$[S] = [S_c]^T [S_r]$$

where T denotes matrix transposition (Chapter 3).

In correspondence analysis, the matrix $[S]$ is analogous to the matrices $[R]$ used in PCA and factor analysis. But in correspondence analysis, the "similarity measures," that is, the individual entries in the matrix $[S]$, are analogous to the chi-square statistic.

The chi-square statistic is (see McCuen (1985), p. 115)

$$\chi^2 = \sum_{i=1}^{k} \frac{(O_i - E_i)^2}{E_i}$$

where O_i is an observed or measured value, E_i is the expected value of the observation or measurement, and k is the number of observed or measured values. Recall from Chapter 4 that an expected value is based on the notion of probability. An expected value of an observation or measurement is that value having the largest probability of being correct. Such a value may be determined from relative frequencies for actual data or from an assumed distribution function for the observed or measured values.

Strictly speaking, the chi-square statistic analogy is valid only if original entries in [Y] are point-count data; then [Y] is processed to yield joint probabilities. In the context of using continuous, real-valued data, joint probabilities cannot be computed and the entries in the matrix [S] cannot be considered chi-square statistics. But, the formula for the chi-square statistic is useful for explaining why the original data matrix [Y] is processed in the manner just described. In other words, if the original data entries in [Y] are enumeration data, then once [Y] is processed to yield [S], entries in [S] can be considered chi-square statistics. On the other hand, for continuous, real-valued entries in [Y], entries in [S] can be considered to be similarity measures, but are not strictly chi-square statistics.

Certainly, correspondence analysis is applied validly to the analysis of continuous, real-valued data. The foregoing discussion is included merely to draw attention to the strict notion of probability to explain why entries in [Y], {W}, and {T} cannot be considered to be probability values. These matrices are entirely useful for multivariate analysis of continuous, real-valued data, as subsequent examples will show.

Recalling the formula for the chi-square statistic, in correspondence analysis, each entry Y_{ij} in [Y], is analogous to O_i. The expected value $E[Y_{ij}]$ is W_iT_j; i.e., it is the product of the marginal sums for individual i and attribute j.

For a simple example such that $i = j = 1$, the definition of the matrix [S]

$$S_{11} = \frac{Y_{11}^2}{W_1(\sqrt{T_1})^2} = \frac{Y_{11}^2}{W_1T_1}$$

is analogous to the formula for the chi-square statistic as defined above, given the fact that the expected value of Y for any i and j is the product of the marginal sums W and T for i and j.

A better analogy to the chi-square statistic is found by realizing that matrix [S] can be formed as follows (Davis, 1986, p. 581):

$$S_{jk} = \sum_{i=1}^{N} \left(\frac{Y_{ij} - W_iT_j}{\sqrt{W_iT_j}} \right) \left(\frac{Y_{ik} - W_iT_k}{\sqrt{W_iT_k}} \right)$$

For a simple example such that $i = j = k = 1$; we find

$$S_{11} = \frac{(Y_{11} - W_1T_1)^2}{W_1T_1}$$

which shows a direct analogy to the chi-square statistic. Forming the matrix $[S]$ in this manner results in the same eigenvalues and eigenvectors that are obtained from the matrix $[S]$ defined using the matrices $[S_c]$ and $[S_r]$.

5.3.1 Eckart–Young Theorem

This section is similar to a discussion given in Davis (1986). For a real-valued matrix $[A]$ having N rows and M columns, two products can be formed:

$$[C] = [A]^\mathrm{T}[A]$$

and

$$[F] = [A][A]^\mathrm{T}$$

Eigenvalues for $[C]$ and $[F]$ are equal.

DEMONSTRATION

(For two-dimensional matrices with implications for more dimensions.) A matrix $[A]$

$$[A] = \begin{bmatrix} a & b \\ c & d \end{bmatrix}.$$

and its transpose

$$[A]^\mathrm{T} = \begin{bmatrix} a & c \\ b & d \end{bmatrix}$$

can be multiplied to yield

$$[A]^\mathrm{T}[A] = \begin{bmatrix} a & c \\ b & d \end{bmatrix}\begin{bmatrix} a & b \\ c & d \end{bmatrix} = \begin{bmatrix} (a^2 + b^2) & (ab + cd) \\ (ab + cd) & (b^2 + d^2) \end{bmatrix}$$

having eigenvalues equal to the roots of

$$\lambda^2 - (a^2 + c^2 + b^2 + d^2)\lambda + (a^2 + c^2)(b^2 + d^2) - (ab + cd)^2 = 0$$

which in simplified form is

$$\lambda^2 - (a^2 + c^2 + b^2 + d^2)\lambda + a^2d^2 + c^2b^2 - 2abcd = 0$$

Alternatively,

$$[A][A]^\mathrm{T} = \begin{bmatrix} a & b \\ c & d \end{bmatrix}\begin{bmatrix} a & c \\ b & d \end{bmatrix} = \begin{bmatrix} (a^2 + b^2) & (ac + bd) \\ (ac + bd) & (c^2 + d^2) \end{bmatrix}$$

has eigenvalues equal to the roots of the following simplified expression:

$$\lambda^2 - (a^2 + c^2 + b^2 + d^2)\lambda + a^2d^2 + b^2c^2 - 2abcd = 0$$

Because this formula is the same as that obtained for the product $[A]^\mathrm{T}[A]$, the demonstration is complete.

For a matrix $[A]$ with the following values

$$[A] = \begin{bmatrix} 4 & 0 \\ 1 & 3 \end{bmatrix}$$

we find

$$[A]^{\mathrm{T}}[A] = \begin{vmatrix} 4 & 1 \\ 0 & 3 \end{vmatrix}\begin{vmatrix} 4 & 0 \\ 1 & 3 \end{vmatrix} = \begin{vmatrix} 17 & 3 \\ 3 & 9 \end{vmatrix}, \qquad \lambda_1 = 18, \quad \lambda_2 = 8$$

and

$$[A][A]^{\mathrm{T}} = \begin{vmatrix} 4 & 0 \\ 1 & 3 \end{vmatrix}\begin{vmatrix} 4 & 1 \\ 0 & 3 \end{vmatrix} = \begin{vmatrix} 16 & 4 \\ 4 & 10 \end{vmatrix}, \qquad \lambda_1 = 18, \quad \lambda_2 = 8$$

Matrices $[A]$, $[C]$, and $[F]$ are defined above. Placing square roots of the eigenvalues on the diagonal of a matrix $[D]$ whose off-diagonal entries are zero leads to

$$[D] = [V]^{\mathrm{T}}[A][U]$$

which is the Eckart–Young theorem. In this relationship, $[U]$ and $[V]$ are orthogonal matrices such that $[U]^{\mathrm{T}}[U] = I$ and $[V]^{\mathrm{T}}[V] = I$, where I is the identity matrix. These matrices are, respectively, the eigenvectors of $[C]$ and $[F]$. Rewriting this relationship gives a solution for $[A]$:

$$[A] = [V][D][U]^{\mathrm{T}}$$

DEMONSTRATION

Continuing from the demonstration that the eigenvalues of $[C]$ and $[F]$ are equal, eigenvectors of $[C]$ and $[F]$ are computed. We use the procedure described in Chapter 3 for solving eigenvectors. For matrix $[C]$ the first eigenvector is $\{3\ 1\}$, and the second is $\{-\frac{1}{3}\ 1\}$. Rescale the second eigenvector by 3 (recalling from Chapter 3 that $a\{x\}$ is a possible eigenvector):

$$[U] = \begin{vmatrix} 3 & -1 \\ 1 & 3 \end{vmatrix}$$

For matrix $[F]$ the first eigenvector is $\{2\ 1\}$, and the second is $\{-\frac{1}{2}\ 1\}$. Rescaling the second eigenvector by 2 gives

$$[V] = \begin{vmatrix} 2 & -1 \\ 1 & 2 \end{vmatrix}$$

Remember that the following relationships must hold: $[U]^{\mathrm{T}}[U] = I$, and $[V]^{\mathrm{T}}[V] = I$. But, using matrices $[U]$ and $[V]$ as just defined yields

$$[U]^{\mathrm{T}}[U] = \begin{vmatrix} 10 & 0 \\ 0 & 10 \end{vmatrix}, \qquad [V]^{\mathrm{T}}[V] = \begin{vmatrix} 5 & 0 \\ 0 & 5 \end{vmatrix}$$

Neither result is equal to the identity matrix.

Recall from Chapter 3, though, that an infinite number of possible solutions for the eigenvectors exist, because if $\{x\}$ is an eigenvector, then so is $a\{x\}$ (where a is any scalar value). Thus, $[U]$ and $[V]$ are simply rescaled such that their products do equal the identity matrix:

$$[U] = \frac{[U]}{\sqrt{10}} = \begin{vmatrix} \dfrac{3}{\sqrt{10}} & -\dfrac{1}{\sqrt{10}} \\ \dfrac{1}{\sqrt{10}} & \dfrac{3}{\sqrt{10}} \end{vmatrix}, \qquad [V] = \frac{[V]}{\sqrt{5}} = \begin{vmatrix} \dfrac{2}{\sqrt{5}} & -\dfrac{1}{\sqrt{5}} \\ \dfrac{1}{\sqrt{5}} & \dfrac{2}{\sqrt{5}} \end{vmatrix}$$

The matrix $[D]$ is formed such that

$$[D] = \begin{vmatrix} \sqrt{\lambda_1} & 0 \\ 0 & \sqrt{\lambda_2} \end{vmatrix} = \begin{vmatrix} \sqrt{18} & 0 \\ 0 & \sqrt{8} \end{vmatrix} = \begin{vmatrix} 4.24 & 0 \\ 0 & 2.83 \end{vmatrix}$$

Matrix $[A]$ is calculated such that $[A] = [V][D][U]^{\mathsf{T}}$. First the product $[V][D]$ is calculated:

$$[V][D] = \begin{vmatrix} \dfrac{2}{\sqrt{5}} & -\dfrac{1}{\sqrt{5}} \\ \dfrac{1}{\sqrt{5}} & \dfrac{2}{\sqrt{5}} \end{vmatrix} \begin{vmatrix} 4.24 & 0 \\ 0 & 2.83 \end{vmatrix} = \begin{vmatrix} 3.791 & -1.265 \\ 1.895 & 2.530 \end{vmatrix}$$

Multiplying this product with $[U]^{\mathsf{T}}$ gives

$$\begin{vmatrix} 3.791 & -1.265 \\ 1.895 & 2.530 \end{vmatrix} \begin{vmatrix} \dfrac{3}{\sqrt{10}} & \dfrac{1}{\sqrt{10}} \\ -\dfrac{1}{\sqrt{10}} & \dfrac{3}{\sqrt{10}} \end{vmatrix} = \begin{vmatrix} 4 & 0 \\ 1 & 3 \end{vmatrix} = [A]$$

Because $[C]$ and $[F]$ are associated with the same eigenvalues, the Eckart–Young theorem allows for the solution of the eigenvectors of $[F]$ (the matrix $[V]$) from the eigenvectors of $[C]$ (the matrix $[U]$) such that

$$[V] = [A][U][D]^{-1}$$

This has important implications for multivariate data analysis, because the matrix $[C]$ is analogous to an $M \times M$ matrix used to effect an R-mode solution, and the matrix $[F]$ is analogous to an $N \times N$ matrix used to effect a Q-mode solution. The Q-mode solution is obtainable from the R-mode solution using the Eckart–Young theorem.

We recall the manner in which the matrix $[Y]$ is processed in correspondence analysis, and the notion of R-mode and Q-mode multivariate analyses obtained from the discussion of factor analysis. Let a matrix $[X]$ be defined as

$$[X] = [B][S_r]$$

where $[B]$ is a square $N \times N$ matrix such that

$$B_{ii} = (W_i)^{-1/2}$$

and all off-diagonal entries of $[B]$ are zero. Thus, $[X]$ is an $N \times M$ matrix. Using $[X]$, an alternative way to define the matrix $[S]$ is

$$[S] = [X]^{\mathsf{T}}[X]$$

to obtain an R-mode analysis ($[S]$ has the dimension $M \times M$). By defining $[S]$ as

$$[S] = [X][X]^{\mathsf{T}}$$

a Q-mode analysis is obtained ($[S]$ has the dimension $N \times N$).

Previous discussion revealed that eigenvalues are identical for either definition of $[S]$. Because this is true, the Eckart–Young theorem enables the matrices $[U]$ and $[V]$, which are the R-mode and Q-mode eigenvectors of $[S]$, to be written as functions of each other:

$$[U] = [X]^{\mathsf{T}}[V][D]^{-1}$$

and

$$[V] = [X][U][D]^{-1}$$

The relationship between $[U]$ and $[V]$ is important. All that is necessary in correspondence analysis is the R-mode definition of the matrix $[S]$. Computing the eigenvalues and eigenvectors for the R-mode solution enables the derivation of the matrices $[D]$, $[U]$, and $[V]$ using the relationship between $[U]$ and $[V]$. This provides a simultaneous R-mode and Q-mode analysis.

5.3.2 Solutions for Factors in Correspondence Analysis

Once the R-mode solution (eigenvalues and eigenvectors) has been obtained, R-mode and Q-mode factors are calculated as follows (Carr, 1990). Recall that R-mode factors pertain to the attributes:

$$\text{Rfactor}_{jk} = \frac{-(U_{j,k+1}\sqrt{D_{k+1}})}{\sqrt{T_j}}$$

for attribute j; k represents the particular factor. In correspondence analysis, even though matrix $[S]$ is $M \times M$, $M - 1$ nontrivial eigenvalues exist. The first eigenvalue is equal to one and is called the trivial solution, which explains the use of the subscript $k + 1$ in the solution for the R-mode factors.

Once the R-mode solution is obtained, its solution is used to compute the Q-mode factors as follows (Carr, 1990):

$$\text{Qfactor}_{ik} = \sum_{j=1}^{M} \frac{Y_{ij}}{W_i} \frac{\text{Rfactor}_{jk}}{\sqrt{D_{k+1}}}$$

5.3.3 Example Calculations

Using the data of Table 5.3, we find

$$[Y] = \begin{vmatrix} 1 & 2 \\ 2 & 5 \\ 3 & 4 \\ 4 & 9 \\ 5 & 8 \end{vmatrix}, \qquad XL = 43$$

Entries in $[Y]$ are divided by XL and vectors $\{W\}$ and $\{T\}$ are formed:

$$[Y] = \begin{vmatrix} 0.023 & 0.046 \\ 0.046 & 0.116 \\ 0.070 & 0.093 \\ 0.093 & 0.209 \\ 0.116 & 0.186 \end{vmatrix}, \quad |W| = \begin{vmatrix} 0.069 \\ 0.162 \\ 0.163 \\ 0.302 \\ 0.302 \end{vmatrix}, \quad |T| = |0.348 \quad 0.652|$$

In the next step, two matrices, $[S_r]$ and $[X]$, are formed:

$$[S_r] = \begin{vmatrix} 0.039 & 0.057 \\ 0.078 & 0.144 \\ 0.119 & 0.115 \\ 0.158 & 0.259 \\ 0.197 & 0.231 \end{vmatrix}, \quad [X] = [B][S_r] = \begin{vmatrix} 0.148 & 0.217 \\ 0.194 & 0.358 \\ 0.295 & 0.285 \\ 0.288 & 0.471 \\ 0.358 & 0.420 \end{vmatrix}$$

We recall that the matrix $[B]$ is based on the vector $\{W\}$:

$$[B] = \begin{vmatrix} \dfrac{1}{\sqrt{W_1}} & 0 & 0 & 0 & 0 \\[2mm] 0 & \dfrac{1}{\sqrt{W_2}} & 0 & 0 & 0 \\[2mm] 0 & 0 & \dfrac{1}{\sqrt{W_3}} & 0 & 0 \\[2mm] 0 & 0 & 0 & \dfrac{1}{\sqrt{W_4}} & 0 \\[2mm] 0 & 0 & 0 & 0 & \dfrac{1}{\sqrt{W_5}} \end{vmatrix}$$

R-mode formulation of the matrix $[S]$, is

$$[S] = [X]^{\mathrm{T}}[X] = \begin{vmatrix} 0.357 & 0.472 \\ 0.472 & 0.654 \end{vmatrix}$$

Using the concepts of eigendecomposition presented in Chapter 3, the eigenvalues of this matrix are found to be 1.000 and 0.011. Suppose that the Q-mode solution for $[S]$ is formed:

$$[X][X]^{\mathrm{T}} = \begin{vmatrix} 0.069 & 0.106 & 0.106 & 0.145 & 0.144 \\ 0.106 & 0.166 & 0.159 & 0.224 & 0.220 \\ 0.106 & 0.159 & 0.167 & 0.219 & 0.225 \\ 0.145 & 0.224 & 0.219 & 0.305 & 0.301 \\ 0.144 & 0.220 & 0.225 & 0.301 & 0.305 \end{vmatrix}$$

Eigenvalues are found to be 1.000, 0.011, 0.0, 0.0, 0.0, which verifies the conformance of correspondence analysis to the Eckart–Young theorem.

At first glance, this solution for the eigenvalues appears to differ from the R-mode formulation for $[S]$ because from the R-mode formulation two eigenvalues are obtained, whereas for the Q-mode solution five eigenvalues are obtained, the last three of which are zero. Recall from Chapter 3 that as many eigenvalues exist for a matrix as rows. In the foregoing example, R-mode formulation of $[S]$ yields a 2×2 matrix having two eigenvalues; Q-mode formulation of $[S]$ yields a 5×5 matrix having five eigenvalues. But, the number of nonzero eigenvalues is equal to the true rank of the matrix; the rank of a matrix is the total number of linearly independent rows of the matrix. In the foregoing example, although five eigenvalues are computed for the Q-mode formulation of $[S]$, only the first two eigenvalues are nonzero; hence the rank of $[S_{Q\text{-mode}}]$ is equal to the rank of $[S_{R\text{-mode}}]$, which explains why the eigenvalue solution for $[S_{Q\text{-mode}}]$ is determined to be equal to the solution for $[S_{R\text{-mode}}]$.

Notice that the first eigenvalue for both R- and Q-mode formulations of $[S]$ is 1.000. In correspondence analysis, this first eigenvalue is called the trivial solution and is not used for computing factors. Only the $M - 1$ remaining eigenvalues/eigenvectors are considered, where M is the number of attributes.

When using a bivariate data set such as Table 5.3, calculations through the computation of the matrix $[S]$ can be shown, but carrying calculations through

Table 5.7 Complete Data Set from McCuen (1985, p. 242)

Individual	Attribute		
	A	B	C
1	1	2	6
2	2	5	7
3	3	4	5
4	4	9	4
5	5	8	1

the computation of factors is not possible. Table 5.3 is actually a subset of the complete data set of Table 5.7 (McCuen, 1985, p. 242). Using the entire data set leads to the eigenvalue summary and factors shown in Figure 5.7. Notice that only two eigenvalues are listed because the first, equal to 1.00, is not used. Therefore, only $M - 1$, or two, eigenvalues are listed.

A plot of individuals based on the factors for individuals is presented in Figure 5.8 for comparison with PCA (Figure 5.3) and factor analysis (Figure 5.6). As in these previous two figures, the one based on correspondence analysis (Figure 5.8) seems to suggest a correlation between individuals 2 and 4, and possibly between 3 and 5. However, this comparison is not entirely valid, because the results plotted in Figure 5.8 are based on information derived from three attributes per individual, whereas those in Figures 5.3 and 5.6 are based on only two of the three attributes per individual. This example, though, serves simply to demonstrate concepts.

5.4 Advanced Topic: Accounting for Missing Data in Correspondence Analysis

Because correspondence analysis is based on the notion of joint and marginal probabilities (begging the question, for the moment, of continuous, real-valued data) for describing similarity among the data constituents, a straightforward procedure can be implemented to account for missing data values, should this be a necessary consideration for a given set of multivariate data. In this context, a missing data value is the absence of a value representing an attribute for an individual.

The multivariate data set in Table 5.8 (Greenacre, 1984, p. 55) represents the smoking habits of five different groups of employees. Suppose that the light-smoking attribute is not defined for the junior employee group (see Table 5.9). How can the missing data be accounted for?

If a zero is entered for this missing value, the assumption is made that no light smokers are among the junior employees. Since this may not be true, entering a value of zero for missing data is not a satisfactory option. Instead of zero, the mean number of light smokers from the remaining employee groups might be used. Again, this is an assumption and no information is available to support the use of the mean.

```
CORSPOND EXAMPLE.DAT

        NUMBER OF INDIVIDUALS =        5
        NUMBER OF ATTRIBUTES  =        3
        NUMBER OF FACTORS     =        2

        CONTINGENCY TABLE, Y

IDEN

          AA         BB         CC
    01  .10E+01    .20E+01    .60E+01
    02  .20E+01    .50E+01    .70E+01
    03  .30E+01    .40E+01    .50E+01
    04  .40E+01    .90E+01    .40E+01
    05  .50E+01    .80E+01    .10E+01
PLEASE INSERT PAPER AND PRESS ENTER



          EIGENVALUE SUMMARY
              EIGENVALUE           PERCENT VARIATION


              .1727449                96.161
              .0068968                 3.839

          FACTORS FOR ATTRIBUTES

IDEN    FACTORS:   1 TO NUMFAC

    AA    .3428     -.1370
    BB    .2824      .0786
    CC   -.5674     -.0063

          FACTORS FOR INDIVIDUALS

IDEN    FACTORS:   1 TO NUMFAC

    01   -.6675     -.0239
    02   -.3221      .0641
    03   -.1361     -.1287
    04    .2326      .0949
    05    .5854     -.0537
PLEASE INSERT PAPER AND PRESS ENTER
```

Figure 5.7 Eigenvalues and *R*-mode and *Q*-mode factors from a correspondence analysis of the data shown in Table 5.7.

Table 5.8 Data Set from Greenacre (1984, Table 3.1, p. 55)

	Smoking Category			
Employee Group	Non	Light	Medium	Heavy
Senior managers (SM)	4	2	3	2
Junior managers (JM)	4	3	7	4
Senior employees (SE)	25	10	12	4
Junior employees (JE)	18	24	33	13
Secretaries (SC)	10	6	7	2

Note. These are enumeration data; each number represents the number of individuals in each employee group assigned to a particular smoking habit category.

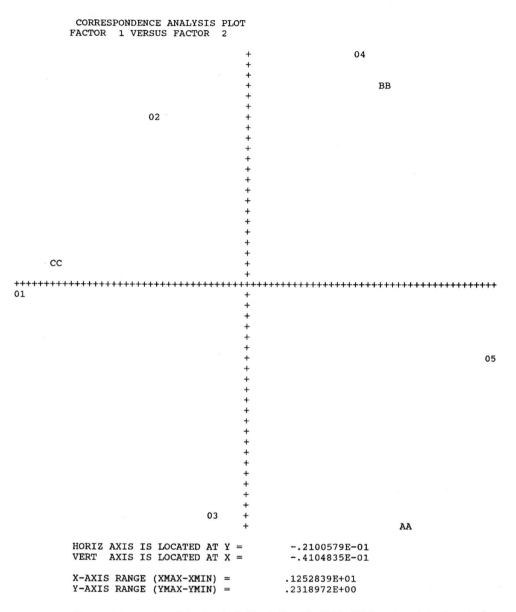

Figure 5.8   A plot of the five individuals listed in Table 5.7 based on their Q-mode factors listed in Figure 5.7.

Instead, the concept of marginal probabilities for individuals and attributes is used. Given a missing value for the jth attribute for the ith individual, the missing value is replaced by the product $W_i T_j$. Assume that the missing value is equal to the product of the marginal probabilities for the ith individual and jth attribute. Recall that for continuous, real-valued data $[Y]$, the notion of marginal probability is not strictly correct. Still, the notion of marginal probability leads to a useful way to accommodate missing data, even if the original data are continuous and real-valued.

Table 5.9   Change in Table 5.8 to Introduce Missing Data

Employee Group	Smoking Category			
	Non	Light	Medium	Heavy
Senior managers (SM)	4	2	3	2
Junior managers (JM)	4	3	7	4
Senior employees (SE)	25	10	12	4
Junior employees (JE)	18	—	33	13
Secretaries (SC)	10	6	7	2

Replacing missing data by the product of marginal probabilities is simple, but it involves iteration. When the product W_iT_j is inserted into the matrix $[Y]$, a global sum XL must be recalculated and all entries in $[Y]$ must again be divided by XL. Then entries in $\{W\}$ and $\{T\}$ must be recomputed. Of course, the entries in $\{W\}$ and $\{T\}$ will now differ from what was obtained for $[Y]$ prior to the replacement of missing values. Therefore, missing data in $[Y]$ must be replaced again using the product W_iT_j; these marginal probabilities are based on the new, revised vectors $\{W\}$ and $\{T\}$. This process is repeated until the vectors $\{W\}$ and $\{T\}$ no longer change.

5.5   CORSPOND: A Portable FORTRAN-77 Program for Correspondence Analysis

A simple, portable FORTRAN-77 computer program is presented in Figure 5.9 (page 138) as a teaching aid for correspondence analysis. A portable computer program is one that will run on a variety of computers with little or no modification. A complete user's guide is included in the beginning of the program listing.

Computer software for correspondence analysis is not readily available. One published program (David et al., 1977) is not portable because it relies on an automatic procedure for assigning plotting identifiers to individuals (rows of the data matrix). This automated procedure does not work well with MS-DOS or Apple MacIntosh systems. Moreover, this program relies on, but does not include a listing for, a subroutine for eigendecomposition, although one, such as EIGENJ (Chapter 3), can be inserted with relative ease. A portable computer program written specifically for personal computers was presented in Carr (1990) to provide a complete listing of a program for correspondence analysis, one that is simple and relatively easy to use.

In this section, a modified version of CORSPOND is presented, which is able to accommodate missing data. (The original version published by Carr (1990) does not have this capability.) The output from this program consists of the following:

1. a summary of eigenvalues and percentage of the original data variance each eigenvalue represents
2. *R*-mode and *Q*-mode factors (Rfactors and Qfactors)

3. two-dimensional line printer plots created by using two factors at a time for plotting. In correspondence analysis of an $N \times M$ data matrix $[Y]$, $M - 1$ nontrivial eigenvalues and eigenvectors are computed, hence $M - 1$ possible factors can be used for plotting. CORSPOND has an option, NUMFAC, that allows program users to select the number of $M - 1$ factors desired for plotting. The use of this option is discussed in following examples.

5.6 Applications of CORSPOND

5.6.1 Analysis of Data from Greenacre (1984)

The data set in Table 5.8, from Greenacre (1984, p. 55), describes a multivariate sample consisting of 5 individuals (groups in this case), each characterized by 4 attributes. No missing data occur in this example. These are enumeration data. Each number in the data table represents the total number of persons in each employee group conforming to one of four smoking habits categories.

A data file (Figure 5.10) is prepared for this example following the user's guide for CORSPOND (Figure 5.9). Application of CORSPOND to this data file yields the results shown in Figure 5.11. The first two eigenvalues are found to account for over 99% of the total data variance. A plot is developed based on factors computed from the first two eigenvalues/eigenvectors (Figure 5.12). This multivariate sample records smoking habits for five different groups of employees working for a hypothetical company. Data similarities are judged on the basis of how closely individuals and attributes plot (Figure 5.12). On the basis of spatial closeness, junior employees (JE) are considered to be light to medium smokers, senior employees (SE) tend not to smoke, and junior managers (JM) tend to smoke heavily. Secretaries (SC) and senior managers (SM) are difficult to categorize based on smoking habits; these two groups are not similar to any smoking category or employee group. These statements are all subjective and are based on a visual inspection of a two-dimensional plot (Figure 5.12). Researchers examining the same plot may come to different conclusions, reinforcing the statement at the beginning of this chapter that this type of multivariate analysis may not be scientifically reproducible.

5.6.2 The Stillwater Lakes, Nevada, Environmental Problem Revisited

Listings of data files for use with CORSPOND for analyzing the Stillwater Lakes, Nevada, data (Tables 5.1 and 5.2) are presented in Figures 5.13 and 5.14. Suppose the following hypothesis is investigated:

An inorganic substance is responsible for the environmental catastrophe at Stillwater Lakes, Nevada, resulting in the deaths of 7 million fish and 16,000 waterfowl.

```
5,4,2,0
0
4.0,2.0,3.0,2.0,'SM'
4.0,3.0,7.0,4.0,'JM'
25.0,10.0,12.0,4.0,'SE'
18.0,24.0,33.0,13.0,'JE'
10.0,6.0,7.0,2.0,'SC'
'NO','LI','ME','HE'
```

Figure 5.10 A listing of a data file for use in analyzing data (Table 5.8).

```
                    NUMBER OF INDIVIDUALS =            5
                    NUMBER OF ATTRIBUTES  =            4
                    NUMBER OF FACTORS     =            2

                    CONTINGENCY TABLE, Y

      IDEN

                   NO          LI          ME          HE
          SM    .40E+01     .20E+01     .30E+01     .20E+01
          JM    .40E+01     .30E+01     .70E+01     .40E+01
          SE    .25E+02     .10E+02     .12E+02     .40E+01
          JE    .18E+02     .24E+02     .33E+02     .13E+02
          SC    .10E+02     .60E+01     .70E+01     .20E+01
      PLEASE INSERT PAPER AND PRESS ENTER



                    EIGENVALUE SUMMARY
                        EIGENVALUE              PERCENT VARIATION


                         .0747592                 87.756
                         .0100172                 11.759
                         .0004136                   .485

                    FACTORS FOR ATTRIBUTES

      IDEN      FACTORS:    1 TO NUMFAC

          NO      .3933      -.0305
          LI     -.0995       .1411
          ME     -.1963       .0074
          HE     -.2938      -.1978

                    FACTORS FOR INDIVIDUALS

      IDEN      FACTORS:    1 TO NUMFAC

          SM      .0658      -.1937
          JM     -.2590      -.2433
          SE      .3806      -.0107
          JE     -.2330       .0577
          SC      .2011       .0789
      PLEASE INSERT PAPER AND PRESS ENTER
```

Figure 5.11   Eigenvalue and factor summary from CORSPOND for the data of Table 5.8.

Correspondence analysis is applied to gain insight into the correctness of this hypothesis. Data (Tables 5.1 and 5.2) are composed of 16 individuals (water sampling locations), each characterized by 19 or 20 attributes. For the June 1986 data set, information for individuals 1 and 2 is missing (Table 5.1); therefore, only 14 individuals are shown. Recall that sample locations 1 through 4 are background sites and locations 5 through 16 are Stillwater Lakes sites. Therefore, the foregoing hypothesis is evaluated by examining the degree of association (similarity) between sample groups 1–4 and groups 5–16. Furthermore, the degree of similarity between groups 5–16 and the chemical attributes will be examined to see if a potentially poisonous attribute is correlated with the Stillwater Lakes sites. This analysis will provide insight to the adequacy of the hypothesis.

Plots from correspondence analysis of the June 1986 data set (Figures 5.15 through 5.17) developed using the first three significant factors (based on the first three significant eigenvectors) and factors and eigenvalue summaries (Figure 5.18) suggest the following:

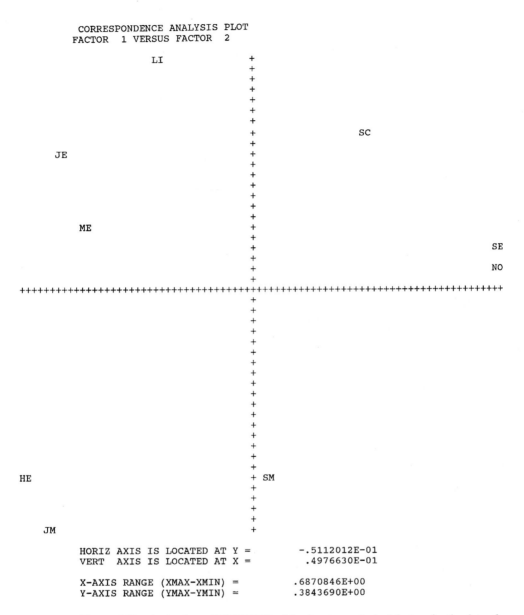

Figure 5.12 A plot from CORSPOND of the first two principal factors for the data of Table 5.8.

1. Background sites 3 and 4 are not associated closely with other sites, except in a plot of the second and third factors (Figure 5.17), which is inferred to represent variability in attributes common to all sites. Sites 3 and 4 are weakly associated with dissolved chromium (CH; this is a two-character computer plotting identifier selected to represent dissolved chromium and is not the symbol for chromium in the periodic table).
2. Some attributes are not consistently associated with any sampling location and are inferred either to be similar at all sites or to vary randomly between

```
12,19,3,0
0
15,659,249,9,104,7.93,.06,7,41,130,10,5, 0,3,1,12,.5,3.4,.3,'03'
23,660,272,9.2,125,8.8,0,14,38,160,5,5,1,4,3,5,.5,4.6,0,'04'
24.5,655,670,8.4,118,9,.08,43,200,510,5,10,0,10,1,20,.5,9.8,.1,'08'
24,655,7230,10.4,148,8.1,.08,80,100,4400,5,20,1,110,2,30,1,15,.1,
'09'
21.5,657,2550,5.4,72,8.6,.009,56,0,1800,5,10,0,45,1,0,.5,30,.4,'10'
26,655,2780,9.5,138,8.8,.16,51,100,1500,5,10,2,38,5,20,.5,25,0,'11'
23,659,3920,9.5,127,9.2,.05,88,200,3400,5,10,0,74,1,20,.5,35,.3,
'12'
24,658,1110,6.7,93,8.1,.14,1,69,1200,5,5,2,28,13,8,1,36,.1,'05'
34,655,1620,20,300,9.1,.04,100,63,2200,5,5,0,2,2,9,.5,19,.3,'06'
31.5,654,4740,15.4,249,8.5,.04,74,100,8300,5,20,2,4,5,20,.5,12,.1,
'07'
19.5,662,843,6.3,79,8,.11,23,52,690,5,5,1,14,1,12,.5,8.9,.2,'14'
22,660,1390,9.6,128,9,.04,80,64,1700,5,5,0,26,3,6,.5,24,0,'15'
'TW' 'BP' 'SC' 'OD' 'OP' 'PH' 'NN' 'AS' 'BA' 'BO' 'CH' 'CU' 'PB'
'MB' 'NI' 'ZI' 'SE' 'UU' 'HG'
```

Figure 5.13   A listing of the input data file to CORSPOND for analysis of the data shown in Table 5.1.

sites. These attributes are nitrogen (NN), zinc (ZN), selenium (SE), barometric pressure (BP), pH (PH), dissolved oxygen (OD and OP), water temperature (TW), barium (BA), mercury (HG), nickel (NI), and lead (PB).

3. Boron (BO) is associated with sites 6 and 7 in all three plots. Specific conductance (SC) is associated with sites 9 and 11 in all three figures.

4. Sites 5, 10, 12, 14, and 15 are weakly associated with sites 6, 7, 9, and 11. Arsenic (AS) is weakly associated with boron (BO); these observations pertain to all three figures.

```
16,20,3,0
0
16,641,431,0,0,7.2,.93,8,41,220,3,5,5,2,2,1,3,.5,.9,0,'01'
21.5,635,385,7.1,97,8.7,.03,5,120,70,0,5,5,2,4,1,10,.5,62,0,
'02'
21,657,197,8.2,107,7.8,.02,10,47,110,0,5,5,2,3,1,19,.5,2.3,0,
'03'
24,657,224,8.7,121,8.6,.03,14,90,120,10,5,5,2,4,3,21,.5,2.5,0,
'04'
25,660,494,6.9,97,8.8,.1,28,230,440,5,5,5,2,16,2,27,.5,13,0,
'08'
20,662,12100,9.8,130,8.2,.1,39,100,7600,0,20,20,2,250,3,30,.5,
140,.1,'09'
20.5,661,1750,6.2,80,8.2,.08,29,260,1100,12,5,5,2,44,3,54,.5,24,
0,'10'
27,660,4030,9.3,137,8.6,.18,44,100,2500,0,5,10,2,100,4,10,.5,46,.1,
'11'
23,662,4020,3.5,48,9.1,.05,95,100,4100,0,10,10,74,120,3,20,.5,48,0,
'12'
17,660,1350,7.6,91,8,.1,61,60,1600,9,5,5,2,44,2,12,.5,36,.2,'05'
21,660,839,17,222,9,.02,41,56,1100,8,5,5,2,27,.5,14,.2,'06'
19,661,319,4.4,55,7.8,.14,14,210,250,0,5,5,2,8,2,66,.5,3.6,0,'07'
23,660,567,6.2,84,8.1,.06,37,240,560,1,5,5,2,17,3,28,1,17,.1,'13'
20,660,656,8,102,8.2,.08,39,110,670,4,5,5,2,21,2,19,.5,21,.9,'14'
22,660,2290,18,241,10,.02,84,100,3300,0,5,10,2,45,3,10,.5,11,.1,
'15'
23,660,1050,6.3,85,8.1,.06,34,210,1200,5,5,5,2,23,3,20,.5,19,0,
'16'
'TW' 'BP' 'SC' 'OD' 'OP' 'PH' 'NN' 'AS' 'BA' 'BO' 'CD' 'CH' 'CU'
'PB' 'MB' 'NI' 'ZI' 'SE' 'UU' 'HG'
```

Figure 5.14.   A listing of the input data file to CORSPOND for analysis of the data shown in Table 5.2.

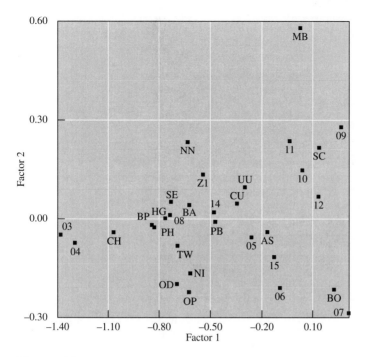

Figure 5.15   Correspondence analysis plot for the June 1986 data set; plot of factors 1 and 2.

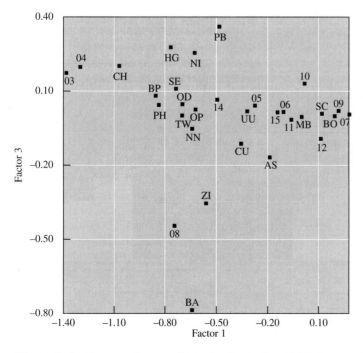

Figure 5.16   Correspondence analysis plot for the June 1986 data set; plot of factors 1 and 3.

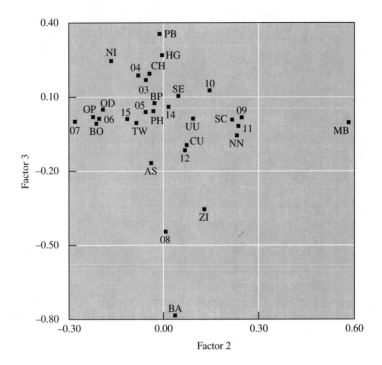

Figure 5.17   Correspondence analysis plot for the June 1986 data set; plot of factors 2 and 3.

5. One site, 8, is an unusual location and is not associated closely with another site or attribute.

These are qualitative assessments made by studying the three plots resulting from correspondence analysis. In summary, for the June 1986 data set, no single attribute is associated with every Stillwater site (sites 5 through 16). Boron (BO) and specific conductance (SC) are related to more Stillwater sites than other attributes. Arsenic (AS) is associated with many Stillwater sites, but is not as well related with these sites as are boron and specific conductance. (Specific conductance, the ability of a water sample to conduct electricity, is a measure of the amount of dissolved solids present in the sample. That specific conductance is associated closely with many Stillwater sites is an indication that water at these sites contained large amounts of dissolved solids.)

A more complete data set (Table 5.2) was collected in September 1986. Completeness, in this context, refers to the fact that attribute information is available for all background sites (sites 1–4), as well as for all Stillwater Lakes sites (5–16). Correspondence analysis of these data (Figures 5.19 through 5.22) suggests that boron (BO) is associated closely with Stillwater site 12 and with sites 5, 9, and 15. Background sites 1–4 are associated with one another, but are not associated closely with the Stillwater group of sites, with the exception of sites 7 and 8. Site 7 is a Carson Lake site and is situated to the south of Stillwater Lakes (Figure 5.1). Site 8 is in an irrigation drain upstream of Stillwater Lakes. The fact that these two sites are similar to background sites suggests, perhaps, an influx of fresh water to these sites since the time the June

Eigenvalue summary	% variation	IDEN	Factors for attributes Factors: 1 to NUMFAC		
.1351050	68.580	TW	−.6920	−.0857	.0005
.0407040	20.662	BP	.8527	−.0269	.0799
.0110530	5.611	SC	.1307	.2163	.0102
.0051254	2.602	OD	−.6981	−.1949	.0517
.0026299	1.335	OP	−.6167	−.2252	.0240
.0016584	.842	PH	−.8345	−.0322	.0436
.0003592	.182	NN	−.6384	.2327	−.0539
.0001785	.091	AS	−.1752	−.0401	−.1674
.0001173	.060	BA	−.6287	.0353	−.7832
.0000557	.028	BO	.2157	−.2131	−.0004
.0000157	.008	CH	−1.0733	−.0454	.1973
.0000000	.000	CU	−.3442	.0653	−.1106
.0000000	.000	PB	−.4786	−.0121	.3592
.0000000	.000	MB	.0115	.5817	−.0049
.0000000	.000	NI	−.6197	−.1656	.2525
.0000000	.000	ZI	−.5512	.1293	−.3561
.0000000	.000	SE	−.7348	.0449	.1075
.0000000	.000	UU	−.3100	.0942	.0180
		HG	−.7670	−.0049	.2738

IDEN	Factors for individuals Factors: 1 to NUMFAC		
3	−1.3788	−.0543	.1719
4	−1.3008	−.0794	.1923
8	−.7366	.0064	−.4479
9	.2335	.2476	.0189
10	.0322	.1488	.1288
11	−.0467	.2368	−.0136
12	.1277	.0726	−.0920
5	−.2651	−.0558	.0416
6	−.0967	−.2050	.0162
7	.2964	−.2842	.0048
14	−.4852	.0162	.0631
15	−.1296	−.1145	.0137

Figure 5.18   Eigenvalue and factor summary for the June 1986 data set; output from CORSPOND for the data of Table 5.1.

1986 data set was collected, because the background sites 1, 3, and 4 are all Carson River sites upstream of Stillwater Lakes.

In summary, for the September 1986 data set, as with the June 1986 data set, no single attribute is associated consistently with all water-sampling sites. Boron and specific conductance are closely associated with some of the Stillwater Lakes locations. Moreover, background sites are different from most of the Stillwater Lakes locations.

Because no single attribute is found to consistently correlate with the Stillwater Lakes sites, it is difficult to associate the environmental problem at Stillwater Lakes with an inorganic toxin. This does not rule out an organic toxin as being responsible for these deaths. Several factors, though, point to the true cause

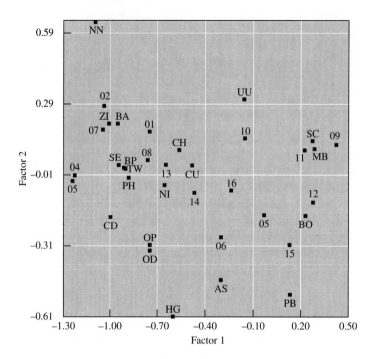

Figure 5.19   Correspondence analysis plot for the September 1986 data set; plot of factors 1 and 2.

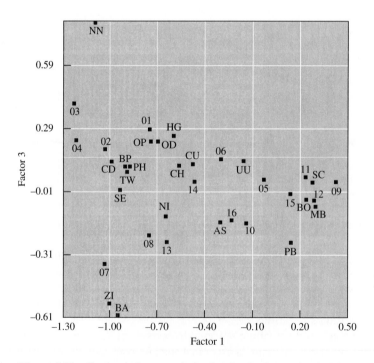

Figure 5.20   Correspondence analysis plot for the September 1986 data set; plot of factors 1 and 3.

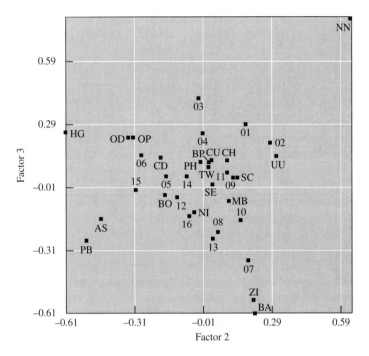

Figure 5.21 Correspondence analysis plot for the September 1986 data set; plot of factors 2 and 3.

of the environmental problem at Stillwater Lakes. One is the avian botulism affecting the waterfowl. This disease was more active in 1985 and 1986 because of a large amount of fish carcasses present at these lakes and the Carson Sink (Figure 5.1). Another factor is the association of specific conductance with many of the Stillwater Lakes sites. This indicates a large amount of total dissolved solids, a sign that evaporation was exceeding inflow during this time.

During the 1980s, a distinct dichotomy in precipitation occurred in western Nevada. The period 1980 through 1984 saw significantly greater than normal snowfall in the Sierra Nevada Mountains to the west of Stillwater Lakes. Two rivers having headwaters in the Sierra Nevada and whose flow in whole or in part reaches Stillwater Lakes are the Truckee and Carson rivers (Figure 5.1). By 1984, Carson Sink, the ultimate base level of the Carson River, rose to its highest level this century. But beginning in 1985, snowfall in the Sierra Nevada abruptly fell to below normal levels. By the end of 1986, Carson Sink was nearly dry. Most of the 7 million fish that died near Stillwater Lakes had lived in Carson Sink in the early part of the 1980s decade and were stranded as the sink became dry beginning in 1985. Therefore, the environmental problem at Stillwater Lakes was most likely due to natural fluctuations in precipitation.

Boron and arsenic are found at many Stillwater Lakes sites, which is likely attributable to geothermal anomalies in this region of Nevada. Boron (specifically, the mineral ulexite) was mined near Stillwater Lakes (e.g., at Salt Wells; Figure 5.1), and active mining for boron still continues in the region. Groundwater near Fallon, Nevada (Figure 5.1), is associated with large arsenic concentations. Boron and arsenic are found in rich concentrations in geothermal

			Factors for attributes		
Eigenvalue summary	% variation	IDEN	Factors: 1 to NUMFAC		
.2210456	77.628	TW	−.9019	.0171	.0846
.0258317	9.072	BP	−.9093	.0189	.1123
.0170258	5.979	SC	.2708	.1413	.0342
.0085897	3.017	OD	−.7428	−.3292	.2283
.0051972	1.825	OP	−.7414	−.3142	.2291
.0030738	1.079	PH	−.8834	−.0162	.1092
.0018747	.658	NN	−1.0966	.6442	.7815
.0011859	.416	AS	−.3042	−.4570	−.1586
.0006121	.215	BA	−.9560	.2123	−.6095
.0001506	.053	BO	.2259	−.1759	−.0458
.0000814	.029	CD	−.9968	−.1947	.1349
.0000369	.013	CH	−.5692	.0982	.1154
.0000267	.009	CU	−.4834	.0317	.1191
.0000128	.004	PB	.1286	−.5196	−.2575
.0000035	.001	MB	.2855	.1052	−.0751
.0000000	.000	NI	−.6566	−.0497	−.1267
.0000000	.000	ZI	−1.0129	.2089	−.5488
.0000000	.000	SE	−.9425	.0358	.0003
.0000000	.000	UU	−.1663	.3171	.1366
		HG	−.6028	−.6097	.2561

	Factors for individuals		
IDEN	Factors: 1 to NUMFAC		
1	−.7536	.1806	.2902
2	−1.0396	.2875	.1982
3	−1.2437	−.0266	.4110
4	−1.2297	−.0083	.2423
8	−.7616	.0559	−.2224
9	.4176	.1247	.0345
10	−.1524	.1562	−.1597
11	.2230	.1006	.0602
12	.2785	−.1224	−.0556
5	−.0354	−.1691	.0433
6	−.3069	−.2787	.1455
7	−1.0467	.1855	−.3579
13	−.6490	.0338	−.2525
14	−.4727	−.0806	.0408
15	.1295	−.3041	−.0193
16	−.2411	−.0713	−.1453

Figure 5.22   Eigenvalue and factor summary for the September 1986 data set; output from CORSPOND for the data of Table 5.2.

fluids near Stillwater Lakes (Oleson and Carr, 1990). The correlation between some Stillwater Lakes sites and these two elements is likely due to the presence of these elements in the natural environment adjacent to Stillwater Lakes and the concentration of these elements in these lakes through leaching of surrounding soils, combined with the great evaporation rates in the interval 1985 through 1986.

```
5,4,2,1
0
-99.0
4.0,2.0,3.0,2.0,'SM'
4.0,3.0,7.0,4.0,'JM'
25.0,10.0,12.0,4.0,'SE'
18.0,-99.,33.0,13.0,'JE'
10.0,6.0,7.0,2.0,'SC'
'NO','LI','ME','HE'
```

Figure 5.23   Input data file to CORSPOND for analysis of artificially created data file having one missing value; the code for this missing value is -99.0.

5.6.3 Advanced Application: Accommodating Missing Data in CORSPOND

To demonstrate the use of CORSPOND for accommodating missing information, data from Greenacre (1984) are artificially altered to create a missing data situation (Table 5.9). In this case, the value for the light smoking attribute for junior employees is missing.

A listing of the data file for this analysis is presented in Figure 5.23. The value -99.0 is chosen to represent a missing datum. Resultant output from CORSPOND for this analysis (Figures 5.24 and 5.25) is similar to that obtained for the full data set (Figures 5.11 and 5.12). Notice how the eigenvalue summary (Figure 5.24) compares to that for the original, complete data set (Figure 5.11). A good degree of similarity exists between eigenvalues and eigenvectors for the two analyses, showing how well missing data values are represented by expected values $(W_i T_j)$. Likewise, the plot in Figure 5.25 compares closely to that for the original data (Figure 5.12).

This is an artificial example because a fully sampled data set (Table 5.8) was altered to have one missing value. Accommodating the presence of this single missing datum yields an analysis similar to that for the fully sampled data set. Correspondence analysis is less precise when a data set is associated with a larger amount of missing information.

5.7 Discriminant Analysis

Foregoing sections describe principal components analysis, factor analysis, and correspondence analysis with applications for analyzing multivariate data interdependence, that is, how similar individuals and attributes are to one another. Occasionally, an objective of multivariate analysis is the segregation of data into different groups, or classes. An example is the Stillwater Lakes data (Tables 5.1 and 5.2). Rather than identifying data intercorrelations, an objective might be to see if water sample sites 1–4 (background) can be separated from sites 5–16 (Stillwater Lakes) on the basis of their chemical attributes.

A method for finding the maximum separability between groups of multivariate data is discriminant analysis. A simple, linear discriminant function is developed,

$$\text{score} = a_1 x_1 + a_2 x_2 + \cdots + a_M x_M$$

```
            NUMBER OF INDIVIDUALS  =        5
            NUMBER OF ATTRIBUTES   =        4
            NUMBER OF FACTORS      =        2

            CONTINGENCY TABLE, Y

   IDEN

              NO         LI         ME         HE
     SM    .40E+01    .20E+01    .30E+01    .20E+01
     JM    .40E+01    .30E+01    .70E+01    .40E+01
     SE    .25E+02    .10E+02    .12E+02    .40E+01
     JE    .18E+02   -.99E+02    .33E+02    .13E+02
     SC    .10E+02    .60E+01    .70E+01    .20E+01
   PLEASE INSERT PAPER AND PRESS ENTER



            EIGENVALUE SUMMARY
               EIGENVALUE              PERCENT VARIATION


               .0744221                   92.578
               .0053357                    6.637
               .0006311                     .785

            FACTORS FOR ATTRIBUTES

   IDEN     FACTORS:   1 TO NUMFAC

     NO       .3534      -.0322
     LI       .0305       .0772
     ME      -.2303       .0464
     HE      -.3363      -.1510

            FACTORS FOR INDIVIDUALS

   IDEN     FACTORS:   1 TO NUMFAC

     SM       .0370      -.1703
     JM      -.2957      -.1337
     SE       .3616      -.0211
     JE      -.2347       .0387
     SC       .2100       .0902
   PLEASE INSERT PAPER AND PRESS ENTER
```

Figure 5.24 Eigenvalue and factor summary for the data of Table 5.9.

and applied to the September 1986 Stillwater Lakes data set (Table 5.2). In this linear discriminant function, the coefficients a_i are determined as described below and the values x_i are the M attributes characterizing multivariate samples. The value, or score, is a sample's discriminant score based on its attributes x. These scores will be used as a measure of separability between data groups.

This treatment of discriminate analysis assumes that two groups of data are analyzed for separability. In the foregoing application of correspondence analysis to the September 1986 Stillwater Lakes data set, a similarity was found among background water sampling locations 1–4; a similarity was found also among water sampling locations 5–16, exclusive of sampling locations 7 and 8. A discriminant analysis is designed to see if water sampling group 1–4 is distinguishable from water sampling group 5–16 (excluding sampling locations 7 and 8).

Before this analysis is attempted, discriminant analysis is developed mathematically. Let the two data groups be referred to as group A and group B. The mathematical objective is to process data groups A and B to yield a solution for the coefficients a_i in the foregoing linear discriminant function.

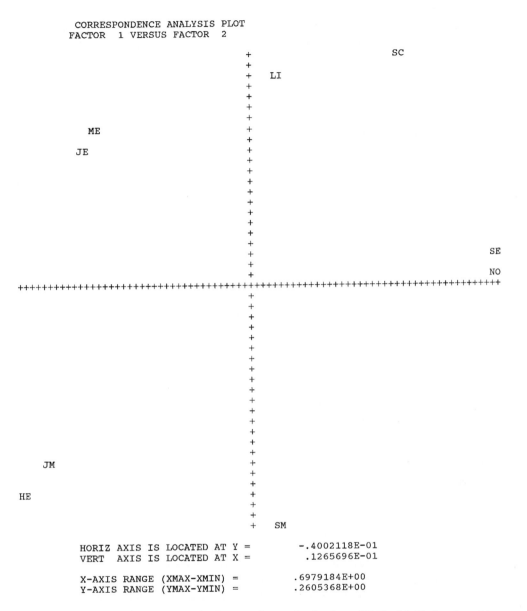

Figure 5.25 A plot of the first two factors for the data of Table 5.9. Notice that with one missing value replaced in the analysis using its expected value, the plot is similar to that of Figure 5.12.

An equation of the form

$$[SP^2]\{a\} = \{W\}$$

is sought for this solution. The matrix $[SP^2]$ is a matrix of pooled variances and covariances between groups A and B (Davis, 1986, p. 480). Entries in the vector $\{W\}$ are differences between the mean values of attributes for each data group:

$$W_i = \text{mean } A_i - \text{mean } B_i, \qquad i = 1, 2, 3, \ldots, M$$

where mean A_i is the mean value of the ith attribute of group A, and mean B_i is the mean value of the ith attribute of group B.

The matrix $[SP^2]$ is obtained in three steps:

$$\text{STEP 1} \quad SP_{A_{jk}} = \sum_{i=1}^{N_A}(A_{ij}A_{ik}) - \frac{\sum_{i=1}^{N_A} A_{ij} \sum_{i=1}^{N_A} A_{ik}}{N_A}$$

$$\text{STEP 2} \quad SP_{B_{jk}} = \sum_{i=1}^{N_B}(B_{ij}B_{ik}) - \frac{\sum_{i=1}^{N_B} B_{ij} \sum_{i=1}^{N_B} B_{ik}}{N_B}$$

$$\text{STEP 3} \quad [SP^2] = \frac{1}{N_A + N_B - 2}([SP_A] + [SP_B])$$

Gaussian elimination is used to solve for coefficients $\{a\}$ in the system: $[SP^2]\{a\} = \{W\}$.

Once vector $\{a\}$ is solved, the discriminant scores for the center of each group, A and B, as well as the score for the midpoint between the two groups, are calculated:

$$\text{score}_{A\text{-center}} = a_1\text{mean } A_1 + a_2\text{mean } A_2 + \cdots + a_M\text{mean } A_M$$
$$\text{score}_{B\text{-center}} = a_1\text{mean } B_1 + a_2\text{mean } B_2 + \cdots + a_M\text{mean } B_M$$
$$\text{score}_{\text{midpoint}} = a_1\text{mean } AB_1 + a_2\text{mean } AB_2 + \cdots + a_M\text{mean } AB_M$$

where

$$\text{mean } AB_i = \frac{\text{mean } A_i + \text{mean } B_i}{2.0}$$

Once these three scores are calculated, scores for each individual in the two data groups are calculated and compared to these three scores to determine membership in group A or group B.

5.7.1 Application

An application of discriminant analysis is made to the September 1986 Stillwater Lakes data set. This data set is separated into two groups: water sampling locations 1–4 and water sampling locations 5–16, excluding sampling locations 7 and 8.

A computer program, DISCRIM, for discriminant analysis is listed (Figure 5.26, page 146). Data entry is as described in the user's guide. Two data sets are developed for the two Stillwater data groups (Figures 5.27 and 5.28). Application of DISCRIM to these data sets yields the following:

$$\text{score}_{A\text{-center}} = 186,726.80$$
$$\text{score}_{B\text{-center}} = 923,993.30$$
$$\text{score}_{\text{midpoint}} = 555,361.00$$

which suggests that the two Stillwater data groups are easily separated. This further supports the conclusion reached by correspondence analysis that Stillwa-

```
16,641,431,0,0,7.2,.93,8,41,220,3,5,5,2,2,1,3,.5,.9,0
21.5,635,385,7.1,97,8.7,.03,5,120,70,0,5,5,2,4,1,10,.5,62,0
21,657,197,8.2,107,7.8,.02,10,47,110,0,5,5,2,3,1,19,.5,2.3,0
24,657,224,8.7,121,8.6,.03,14,90,120,10,5,5,2,4,3,21,.5,2.5,0
```

Figure 5.27 A data file for group A, sites 1–4 for the September 1986 Stillwater Lakes data. This data file is listed exactly as is required for input to DISCRIM.

```
0;37;44m C:\TEXT\CHAPTER5>
20,662,12100,9.8,130,8.2,.1,39,100,7600,0,20,20,2,250,3,30,.5,140,.1
20.5,661,1750,6.2,80,8.2,.08,29,260,1100,12,5,5,2,44,3,54,.5,24,0
27,660,4030,9.3,137,8.6,.18,44,100,2500,0,5,10,2,100,4,10,.5,46,.1
23,662,4020,3.5,48,9.1,.05,95,100,4100,0,10,10,74,120,3,20,.5,48,0
17,660,1350,7.6,91,8,.1,61,60,1600,9,5,5,2,44,2,12,.5,36,.2
21,660,839,17,222,9,.02,41,56,1100,8,5,5,2,5,2,27,.5,14,.2
23,660,567,6.2,84,8.1,.06,37,240,560,1,5,5,2,17,3,28,1,17,.1
20,660,656,8,102,8.2,.08,39,110,670,4,5,5,2,21,2,19,.5,21,.9
22,660,2290,18,241,10,.02,84,100,3300,0,5,10,2,45,3,10,.5,11,.1
23,660,1050,6.3,85,8.1,.06,34,210,1200,5,5,5,2,23,3,20,.5,19,0

0;37;44m C:\TEXT\CHAPTER5>
```

Figure 5.28   A data file for group *B*, sites 5–16, less sites 7 and 8.

ter Lakes sites 5–16 (excluding sites 7 and 8) differ from background sites 1–4.

Data from water sampling sites 7 and 8 was excluded from the data file for Stillwater Lakes sites 5–16 (Figure 5.28). A test is conducted now to use the coefficients calculated by DISCRIM to determine to which data group, *A* or *B*, sites 7 and 8 are closest. The coefficients a_i computed by Discrim are

$a_1 = 657,647.90$ $a_2 = -21,573.89$ $a_3 = 43.35$ $a_4 = 838,810.60$

$a_5 = -63,334.70$ $a_6 = 77,284.47$ $a_7 = 3,007,858.00$ $a_8 = 66,503.95$

$a_9 = 8,271.78$ $a_{10} = -682.49$ $a_{11} = 147,195.90$ $a_{12} = 283,353.20$

$a_{13} = -94,249.40$ $a_{14} = -51,678.19$ $a_{15} = 28,057.29$ $a_{16} = -1,844,882.00$

$a_{17} = -10,208.53$ $a_{18} = -97,825.29$ $a_{19} = -17,428.08$ $a_{20} = 98,724.85$

There are 20 coefficients *a*, one for each of the 20 attributes characterizing each individual.

Scores for all water sampling sites are listed in Table 5.10. An example is presented to show the explicit calculation of the score for water sampling site

Table 5.10   Discriminant Scores for the September 1986 Stillwater Data

Site	Discriminant Score
1	186,728.30
2	186,728.90
3	186,741.20
4	186,709.00
5	923,995.60
6	924,003.90
7	−1,430,979.00
8	4,099,571.00
9	923,993.60
10	923,990.80
11	923,992.90
12	923,991.40
13	923,995.00
14	923,987.90
15	923,986.30
16	923,999.30

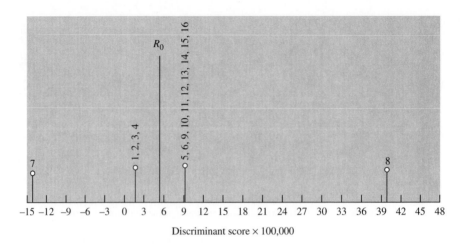

Figure 5.29   A plot of the Stillwater Lakes sites based on their discriminant scores (Table 5.10).

7, using the coefficients just listed and data from Table 5.2:

$$
\begin{aligned}
\text{score}_{\text{site 7}} ={}& 657{,}647.90(19) - 21{,}573.89(660) + 43.35(320) \\
& + 838{,}810.60(4.4) - 63{,}334.70(55) + 77{,}284.47(7.8) \\
& + 3{,}007{,}858.0(0.14) + 66{,}503.95(14) + 8271.78(210) \\
& - 682.49(250) + 147{,}195.9(0) + 283{,}353.2(5) - 94{,}249.4(5) \\
& - 51{,}678.19(2) + 28{,}057.29(8) - 1{,}844{,}882.0(2) - 10{,}208.53(66) \\
& - 97{,}825.29(0.5) - 17{,}428.08(3.6) + 98{,}724.85(0) \\
={}& -1{,}430{,}979.00
\end{aligned}
$$

Notice that discriminate scores for Stillwater Lakes sampling sites 5, 6, and 9–16 are equal to approximately 924,000, scores for background sites 1–4 are equal to approximately 186,700, and sampling sites 7 and 8 are distinctly different from any group. The results of Table 5.10 are plotted in Figure 5.29, along with scores for the centers of groups A and B and the midpoint, to show how well discriminant analysis separated the two data groups.

Sites 7 and 8 are unusual with respect to the majority of the data and to one another. Site 7 is associated with a large negative score. Based on discriminant analysis, this site is closest to group A (background sites 1–4). But because its score is so unusual, classifying site 7 in this group may not be justified. Site 8 is associated with a large positive score. This site is closer to the Stillwater Lakes sites (5–16). But because its score is so much larger than the scores for most of the Stillwater Lakes sites, classifying site 8 in this group may not be justified. In actual scientific or engineering practice, further investigation of sites 7 and 8 would be warranted to determine why these sites are unusual.

Chapter Summary

- Multivariate data analysis is useful for identifying intercorrelations among multiple individuals, each described by multiple attributes.

- Principal components analysis is the simplest of the multivariate analysis methods. Similarity measures used for PCA are variances and covariances.

- Factor analysis is a useful multivariate analysis method. In *R*-mode factor analysis, similarity measures are correlation coefficients *r*; in *Q*-mode factor analysis, similarity measures are cosines of angles between individuals (treated as vectors whose orientations are defined by the attributes).

- Correspondence analysis is a powerful, multivariate analysis method, which yields *R*-mode and *Q*-mode analyses simultaneously. Similarity measures used in correspondence analysis are analogous to the chi-square statistic, and are called a chi-square metric (distance).

- Missing data values are handled in correspondence analysis by replacing missing values with their expected values, the product of the row and column sums on which the missing datum occurs.

- Discriminant analysis is useful for predicting to what group an individual (datum) belongs.

Exercises

(An asterisk indicates an advanced or graduate student problem)

*1. Using the data of Table 5.7, calculate the *R*-mode form of matrix [*S*] for correspondence analysis, then use the program EIGEND from Chapter 3 to verify the results shown in Figure 5.7.

2. Use the correspondence analysis program CORSPOND to analyze each of the following data sets to identify data intercorrelations. Each of these data sets is associated with a reference from which the data are obtained. Although the introduction to this text states that extensive literature citations are avoided when explaining concepts, in this instance, consult the literature source for each data set to see how your analysis compares with conclusions listed there.

   (a) These data are counts of conodont tests recovered from 10-kilogram samples of rock, taken from Davis (1986, p. 583). The columns are conodont species and the rows are stratigraphic units that are members of Missourian Series in Eastern Kansas (USA). Data copyright © 1986 John Wiley & Sons, Inc. Reprinted with permission.

		Conodont Counts									
Individual	*Class*	*A*	*B*	*C*	*D*	*E*	*F*	*G*	*H*	*I*	*J*
1	M	13	10	0	0	37	0	0	0	0	0
2	O	0	0	0	0	11	0	0	0	0	0
3	U	4	2	1	51	26	1	0	0	0	0
4	B	0	7	1	207	350	0	0	34	14	3
5	M	8	28	6	0	60	0	0	0	0	0
6	O	145	20	5	0	10	0	0	0	0	0
7	U	5	134	8	0	353	1	0	4	0	0
8	P	20	60	0	0	920	0	0	0	0	0

(tabulation continues)

Conodont Test Counts (continued)

Individual	Class	Conodont Counts									
		A	B	C	D	E	F	G	H	I	J
9	M	115	255	10	0	1140	0	0	0	0	0
10	S	1	0	0	0	3	0	0	0	0	0
11	S	31	21	7	0	4	1	0	0	0	0
12	?	100	5	0	0	5	0	0	0	0	0
13	U	0	39	1	0	80	0	1	0	0	0
14	P	10	70	0	0	538	0	0	5	0	0
15	M	3	78	5	0	450	0	0	3	0	0
16	O	0	0	0	0	28	0	0	0	0	0
17	U	38	20	3	100	267	3	0	25	0	0
18	B	15	8	0	243	515	0	10	85	55	13
19	M	10	130	10	200	900	0	0	50	0	0
20	O	117	20	0	63	57	0	0	7	0	0

Note. Cyclothem classifications: O, outside shale; S, shoal limestone; U, upper limestone; M, middle limestone; P, "phantom black shale"; and B, black shale. Conodont species: A, *Adetognathus;* B, *Ozarkodina;* C, *Aethotaxis;* D, *Idiognathodus delicatus;* E, *I. elegantulus;* F, *Magnilaterella;* G, *Hindeodella;* H, *Idiprioniodus;* I, *Gondolella;* J, others.

Do you know what a conodont is? If not, please find out. What is the future of the subdiscipline of geology known as paleontology? The science of geology is dependent upon knowledge of past life forms, but fewer and fewer students choose this subdiscipline to specialize in when pursuing postbaccalaureate degrees. Important controversies today involve paleontological studies, such as what caused the extinction of the dinosaur species. Paleontology is at least as vital a science today as it has been in the past.

(b) These data are taken from McCuen (1985, Table B-6, p. 424).

Study of Macroinvertebrate Diversity Data for Altered (or Modified) Streams

Individual	Attributes							
	X1	X2	X3	X4	X5	X6	X7	Y
1	10.2	5.	548.6	1.	3970.	33.	15.	2.584
2	19.4	9.	794.0	6.	10687.	22.	10.	2.486
3	19.2	11.	503.8	30.	6400.	45.	21.	3.456
4	22.9	17.	213.4	30.	3648.	37.	19.	4.093
5	23.9	15.	350.5	2.	1298.	35.	19.	3.720
6	10.7	9.	182.9	1.	622.	48.	35.	3.871
7	24.8	25.	476.6	17.	2816.	47.	34.	3.658
8	11.4	6.	362.7	1.	2703.	24.	9.	2.125
9	8.9	4.	164.6	20.	448.	18.	10.	2.702
10	41.7	43.	432.8	20.	2048.	17.	10.	3.180
11	19.6	19.	377.9	17.	4032.	40.	18.	2.996
12	4.8	4.	195.1	30.	2270.	34.	18.	2.976
13	27.9	28.	1067.0	5.	28080.	28.	11.	3.697
14	4.8	4.	155.2	30.	2112.	24.	11.	2.714

Study of Macroinvertebrate Diversity Data (continued)

Individual	Attributes							
	X1	*X2*	*X3*	*X4*	*X5*	*X6*	*X7*	*Y*
15	16.2	14.	231.6	30.	1664.	30.	24.	3.914
16	15.4	13.	240.8	30.	1856.	36.	30.	3.058
17	38.1	14.	96.5	12.	2359.	36.	22.	3.609
18	26.7	16.	420.6	7.	4570.	29.	14.	3.596
19	11.2	9.	224.3	30.	3000.	24.	13.	3.304
20	13.6	9.	302.0	24.	800.	16.	11.	3.183
21	13.9	14.	232.8	1.	735.	26.	12.	3.604
22	5.4	5.	466.3	17.	2249.	16.	5.	1.790
23	17.8	9.	269.2	17.	109.	14.	6.	2.742
24	9.5	4.	237.7	15.	753.	30.	6.	0.982
25	23.2	18.	223.5	30.	2600.	36.	19.	3.553
26	59.2	30.	537.6	6.	1540.	21.	9.	2.673

Note. Explanation of attributes: X1, mean stream depth (cm); X2, stream depth range (cm); X3, mean stream width (cm); X4, years since channel work (yr); X5, drainage area (acres); X6, total taxa; X7, benthic taxa; X8, Shannon–Weaver diversity index.

What are taxa? What is the Shannon–Weaver diversity index? What does your analysis reveal about these streams?

(c) These data are taken from Imbrie (1963), and are also given in Joreskog et al. (1976, p. 105).

Artificial Data

Individual	Attributes									
	X1	*X2*	*X3*	*X4*	*X5*	*X6*	*X7*	*X8*	*X9*	*X10*
1	5.0	25.0	15.0	5.0	5.0	20.0	10.0	5.0	5.0	5.0
2	10.0	30.0	17.0	17.0	8.0	8.0	5.0	4.0	1.0	0.0
3	3.0	6.0	10.0	13.0	25.0	15.0	13.0	8.0	5.0	2.0
4	7.5	27.9	16.0	11.0	6.5	14.0	7.5	4.5	3.0	2.5
5	4.6	21.2	14.0	6.6	9.0	19.0	10.6	5.6	5.0	4.4
6	3.8	13.6	12.0	9.8	17.0	17.0	11.8	6.8	5.0	3.2
7	8.3	26.6	15.9	14.2	9.1	11.1	6.8	4.6	2.2	1.2
8	6.1	22.7	14.6	10.2	9.9	15.4	9.1	5.3	3.8	2.9
9	7.6	24.2	15.2	13.8	10.8	11.8	7.6	5.0	2.6	1.4
10	3.9	10.3	11.2	12.6	21.3	14.8	11.9	7.3	4.6	2.1

Note. X1, copper (ppm); X2, gold (ppm); X3, silver (ppm); X4, antimony (ppm); X5, arsenic (ppm); X6, lead (ppm); X7, zinc (ppm); X8, molybdenum (ppm); X9, thalium (ppm); X10, mercury (ppm).

Because these are artificial data, a scenario is devised whereby these data may be inferred to have geologic relevancy. Let each individual be a rock chip sample, and let each attribute be as described in the table footnote. How are these attributes associated with the individual rock chip samples?

(d) These data are taken from Griffith and Amrhein (1991, p. 455).

Ontario Drainage Basins Runoff Data

Individual	AREA	Q	RUN	LAT	LONG	SNOW	PRECIP	TEMP
1	3240.	132.2	215.	53.5	89.3	76.0	22.0	25.0
2	865.	50.6	308.	43.1	81.7	88.0	34.0	46.0
3	440.	39.3	471.	45.4	75.2	87.0	34.5	42.5
4	342.	18.0	277.	47.5	80.4	60.0	35.5	46.0
5	1520.	111.5	387.	44.7	79.4	92.0	36.0	40.0
6	1980.	144.6	385.	48.7	86.2	110.0	31.0	33.0
7	650.	38.3	311.	48.9	88.4	96.0	28.2	33.5
8	1780.	134.1	397.	47.9	79.9	90.0	32.0	33.5
9	1170.	31.1	140.	49.6	90.6	92.0	28.5	33.0
10	199.	9.5	252.	43.5	79.8	48.0	30.0	46.0
11	4350.	80.3	97.	47.4	79.6	88.0	31.8	36.5
12	262.	13.7	275.	44.0	78.3	68.0	32.0	44.0
13	1160.	103.8	471.	46.9	84.0	132.0	37.0	37.5
14	303.	10.4	181.	43.8	79.6	60.0	29.5	46.0
15	443.	31.4	374.	45.8	77.1	80.0	28.0	41.0
16	6480.	190.5	155.	48.5	89.6	88.0	28.5	34.0
17	4250.	220.0	273.	51.2	80.8	112.0	30.5	30.0
18	648.	51.5	419.	43.8	81.3	108.0	35.0	45.0
19	1030.	56.1	287.	43.2	80.4	80.0	37.0	44.0
20	1180.	48.7	218.	44.2	79.8	72.0	31.0	44.5
21	4900.	251.9	271.	52.3	88.8	88.0	25.0	28.0
22	404.	34.9	455.	45.2	74.6	91.0	36.0	43.5
23	3960.	278.0	370.	44.5	81.3	120.0	36.0	44.0
24	1350.	105.1	410.	46.2	82.5	85.0	32.5	40.0
25	712.	46.0	341.	44.5	77.3	80.0	34.0	42.0
26	4450.	142.4	169.	50.2	91.6	90.0	26.5	33.5
27	730.	33.6	243.	42.8	81.8	64.0	35.0	47.0
28	4870.	144.8	157.	48.8	92.7	70.0	26.5	36.0

Note. Q, discharge (m³/s); AREA (km²); RUN, Qt/AREA, where t is time, a 2-month interval; LAT, latitude; LONG, longitude; SNOW, snowfall; PRECIP, precipitation amount; TEMP (°C). These data were taken from Environment Canada, Water Survey of Canadian Surface Water Data, for drainage area and spring discharge; from topographic maps for latitude and longitude; and from Environment Canada, Atmospheric Environment Services Climate Summaries for snowfall, temperature, and precipitation.

(e) These data are taken from Harris (1966).

Hypothetical Sandstone Data

Sample	%Quartz	%Feldspar	%Carbonates	%Heavy Min.	%Clay	%Silt
1	2.0	0.0	42.0	0.4	54.0	1.4
2	3.0	0.0	60.0	0.7	34.0	2.3
3	5.0	0.6	80.0	1.1	10.0	3.3
4	8.0	1.2	80.0	1.8	5.0	4.0

Hypothetical Sandstone Data (continued)

Sample	%Quartz	%Feldspar	%Carbonates	%Heavy Min.	%Clay	%Silt
			Attributes			
5	6.0	0.2	35.0	0.6	56.0	2.2
6	8.0	0.8	50.0	0.9	39.0	3.2
7	10.0	1.0	74.0	1.5	10.0	3.5
8	15.0	1.6	68.0	2.2	8.2	5.0
9	20.0	1.0	32.0	0.8	43.0	3.3
10	33.0	1.3	38.0	1.2	22.0	4.5
11	44.0	1.7	38.0	1.8	10.5	4.5
12	55.0	3.0	20.0	2.7	12.3	7.0
13	20.0	1.5	24.0	1.2	48.7	4.6
14	30.0	2.0	18.0	1.8	40.2	8.2
15	47.0	3.0	8.0	2.3	20.7	9.0
16	65.0	5.0	4.0	2.9	3.1	20.0
17	12.0	2.0	18.0	1.9	60.1	6.0
18	22.0	4.0	12.0	2.4	40.6	7.0
19	35.0	8.0	6.0	3.0	20.2	27.8
20	47.0	22.0	3.0	3.5	4.5	20.0
21	7.0	5.0	15.0	2.2	65.8	5.0
22	15.0	8.0	10.0	3.0	50.0	14.0
23	28.0	16.0	5.0	3.8	34.2	13.0
24	44.0	30.0	3.0	4.5	6.5	12.0
25	6.0	5.0	12.0	2.2	70.0	4.8
26	12.0	9.0	9.0	2.9	52.0	15.1
27	30.0	16.0	5.0	3.8	30.2	15.0
28	45.0	22.0	2.0	4.5	4.5	12.0

Note. The attribute labels are volume percentages.

How do these 28 samples compare with one another? How do your eigenvalues and factors compare with factor analysis as presented by Harris (1966)? Are any of the attributes correlated?

3. Use the program DISCRIM to perform a discriminate analysis on the data in Tables 5.11, 5.12, and 5.13. Treat Table 5.11 as group *A* and Table 5.12 as group *B*, then use the coefficients computed by DISCRIM to classify each individual in Table 5.13 to one of the data groups, *A* or *B*. To perform the classification, determine the discriminate score of each individual in Table 5.13. Assign the individual to group *A* if the score is less than score$_{midpoint}$, and to group *B* if the score is greater than this middle score.

Table 5.11 Data from Davis (1986, p. 488) from a Stream Draining an Old Mining District. Copyright © 1986 John Wiley & Sons, Inc. Reprinted with permission.

Ti^a	Mn^b	Ag^c	Ba^a	Co^d	Cr^a	Cu^c	Ni^a	Pb^a	Sr^a	V^a	Zn^a	Au^c
7,280	1300	30.0	720	30	150	73	50	70	60	70	190	0.02
10,300	1200	0.7	1280	20	160	25	50	70	90	50	50	0.02
6,500	700	1.0	1070	20	200	48	70	100	210	50	170	0.01
7,000	1500	0.7	760	30	160	70	40	110	240	40	250	0.01
5,100	1000	0.5	740	20	140	39	50	80	50	60	130	0.02
10,600	2100	0.3	980	30	50	25	30	70	150	160	110	0.01
14,200	2000	0.2	690	30	70	25	50	60	160	70	180	0.01
9,700	900	0.2	680	35	70	38	30	70	80	110	250	0.01
2,300	1500	0.2	710	5	110	50	20	70	80	30	120	0.01
12,100	6300	0.1	1520	30	30	24	30	80	320	160	190	0.02
3,000	1100	0.2	510	5	30	15	30	30	240	30	50	0.02
7,500	2400	0.7	690	30	30	31	10	100	210	40	280	0.03
7,800	1800	4.0	730	55	40	24	30	20	90	320	90	0.01
6,900	1500	1.0	326	30	50	25	10	90	70	200	70	0.04
11,200	3100	1.5	660	50	40	20	40	50	140	280	90	0.01
5,200	1400	0.8	680	35	50	42	20	50	30	150	150	0.01
5,100	1500	0.9	700	25	60	67	40	80	40	190	90	0.01
10,600	2900	0.4	1640	25	20	21	30	30	320	90	200	0.01
11,500	3200	0.7	710	30	30	15	20	20	260	270	180	0.01
7,100	1800	0.9	490	75	50	8	10	30	80	180	100	0.02

[a] Measured to nearest 10 ppm; [b] measured to nearest 100 ppm; [c] measured to nearest fraction of ppm; [d] measured to nearest 5 ppm.

Table 5.12 Data from Davis (1986, p. 489) from a Stream Draining an Area Thought to Be Barren of Ore. Copyright © 1986 John Wiley & Sons, Inc. Reprinted with permission.

Ti^a	Mn^b	Ag^c	Ba^a	Co^d	Cr^a	Cu^c	Ni^a	Pb^a	Sr^a	V^a	Zn^a	Au^c
4820	500	0.1	160	20	70	30	10	0	720	140	200	0.01
3040	500	0.2	150	20	30	82	10	20	1580	160	70	0.01
890	600	0.1	50	10	10	61	10	0	340	40	50	0.02
2100	500	0.1	100	15	30	77	10	0	650	90	80	0.02
5060	700	0.3	140	20	50	154	20	0	1240	140	80	0.01
1980	700	0.1	80	15	20	63	20	0	720	80	110	0.00
3220	600	0.2	160	20	30	45	20	10	1100	120	60	0.01
3280	800	0.2	90	15	10	40	30	20	1480	70	40	0.00
2020	700	0.1	80	15	20	104	20	0	420	80	70	0.00
4600	700	0.3	160	20	60	48	10	20	780	150	50	0.02
3100	500	0.2	100	15	30	65	10	20	710	100	40	0.01
3020	600	0.2	90	15	10	69	0	30	1310	110	30	0.02
1860	500	0.1	70	10	20	63	0	10	480	80	50	0.00
2800	700	0.1	110	15	20	58	10	20	730	120	80	0.01
1040	1600	0.1	20	5	10	37	0	10	140	30	80	0.01
4640	800	0.3	220	15	20	121	20	20	1200	210	160	0.00
4990	900	0.3	190	20	40	59	20	30	480	230	120	0.02
2830	800	0.2	120	15	20	40	10	20	690	140	60	0.00
4500	700	0.2	140	20	30	82	20	10	710	170	70	0.00
2900	600	0.1	80	15	10	99	0	0	760	80	90	0.01

[a] Measured to nearest 10 ppm; [b] measured to nearest 100 ppm; [c] measured to fraction of ppm; [d] measured to nearest 5 ppm.

Table 5.13   Data from Davis (1986, p. 489) from a Stream That May or May Not Drain an Area Associated with an Ore Deposit. Copyright © 1986 John Wiley & Sons, Inc. Reprinted with permission.

Ti^a	Mn^b	Ag^c	Ba^a	Co^d	Cr^a	Cu^c	Ni^a	Pb^a	Sr^a	V^a	Zn^a	Au^c
4,260	800	0.3	180	20	60	128	30	30	460	110	80	0.02
6,500	1200	0.5	380	30	40	72	50	20	320	90	160	0.01
12,200	5200	1.5	630	25	80	39	40	90	210	200	180	0.01
1,080	1600	0.2	80	5	10	102	0	10	160	30	80	0.00
3,820	500	0.2	170	25	40	60	20	10	1100	160	40	0.02
1,020	2400	0.1	20	0	10	28	0	0	1320	20	60	0.00

Note. These data serve as a classification experiment.
[a]Measured to nearest 10 ppm; [b]measured to nearest 100 ppm; [c]measured to nearest fraction of ppm; [d]measured to nearest 5 ppm.

References and Suggested Readings

ANDERSON, T.W. 1984. *An Introduction to Multivariate Statistical Analysis,* 2d ed. New York: Wiley.

BENZECRI, J.P. 1973. *L'Analyse des donnees,* Tome II, *l'Analyse des correspondances.* Paris: Dunod.

CARR, J.R. 1990. CORSPOND: a portable FORTRAN-77 program for correspondence analysis. *Computers and Geosciences* **16**(3): 289–307.

DAVID, M., M. DAGBERT, and Y. BEAUCHEMIN. 1977. Correspondence analysis. *Quarterly of the Colorado School of Mines* **72**(1): 11–57.

DAVIS, J.C. 1973. *Statistics and Data Analysis in Geology.* New York: Wiley.

DAVIS, J.C. 1986. *Statistics and Data Analysis in Geology,* 2d ed. New York: Wiley.

ECKART, C., and B. YOUNG. 1936. The approximation of one matrix by another of lower rank. *Psychometrika* **1**(8): 211–218.

GREENACRE, M.J. 1984. *Theory and Applications of Correspondence Analysis.* London: Academic.

GRIFFITH, D.A., and C.G. AMRHEIN. 1991. *Statistical Analysis for Geographers.* Englewood Cliffs, NJ: Prentice-Hall.

HAIR, J.F., R.E. ANDERSON, R.L. TATHAM, and W.C. BLACK. 1992. *Multivariate Data Analysis with Readings.* New York: Macmillan.

HARRIS, D.P. 1966. Factor analysis: a tool for quantitative studies in mineral exploration. *Proceedings of the Symposium and Short Course on Computers and Operations Research in Mineral Industries,* Vol. 2, Mineral Industries Experiment Station, Special Publication 2-65, Pennsylvania State University, p. GG-37.

IMBRIE, J. 1963. *Factor and Vector Analysis Programs for Analyzing Geologic Data.* Office of Naval Research, Geographic Branch, Tech. Report 6.

JORESKOG, K.G., J.E. KLOVAN, and R.A. REYMENT. 1976. *Geological Factor Analysis.* Amsterdam: Elsevier Scientific.

KLOVAN, J.E. 1966. The use of factor analysis in determining depositional environments from grain size distributions. *Journal of Sedimentary Petrology* **36**: 115–125.

LONG, D.T., and Z.A. SALEEM. 1974. Hydrogeochemistry of carbonate groundwaters of an urban area. *Water Resources Research* **10**(6): 1229.

MCCUEN, R.H. 1985. *Statistical Methods for Engineers.* Englewood Cliffs, NJ: Prentice-Hall.

NORUSIS, M.J. 1985. *SPSS-X Advanced Statistics Guide.* New York: McGraw-Hill.

OLESON, S.G., and J.R. CARR. 1990. Correspondence analysis of water quality data: implications for fauna deaths at Stillwater Lakes, Nevada. *Mathematical Geology* **22**(6): 665–698.

TEMPLE, J.T. 1978. Use of factor analysis in geology. *International Journal for Mathematical Geology* **10**: 379.

Figure 5.9 A listing for the program CORSPOND. Subroutine EIGENJ is borrowed from Davis: *Statistics and Data Analysis in Geology.* Copyright © 1973 John Wiley & Sons. Used with permission.

```
C     PROGRAM CORSPOND
C
C     THIS IS A GENERAL FORTRAN-77 PROGRAM FOR
C     CORRESPONDENCE ANALYSIS.
C
C     PROGRAM CAPABILITIES:
C
C     1.  EIGENVECTOR AND EIGENVALUE COMPUTATION OF [S];
C     2.  COMPUTATION OF FACTORS USING EIGENVECTORS;
C     3.  LINE PRINTER PLOTS FOR SELECTED NUMBER OF FACTORS
C     4.  OUTPUT FILE CREATION FOR FACTORS.
C     5.  ITERATION TO ACCOMMODATE MISSING DATA
C
C     PROGRAM LIMITATIONS:
C
C     A MAXIMUM OF 250 INDIVIDUALS;
C     A MAXIMUM OF 50 ATTRIBUTES PER INDIVIDUAL;
C     A MAXIMUM OF 10 FACTORS FOR PLOTTING.
C
C ***************************************************
C
C                    USER'S GUIDE
C
C ***************************************************
C
C     ALL DATA ENTRY FOLLOWS BY FREE FORMAT:
C
C     DATA FILE, LINE 1:  OPTION INITIALIZATION:
C
C     READ(5,*) N, M, NUMFAC, IMISS
C
C     WHERE:  N = THE NUMBER OF INDIVIDUALS (ROWS IN THE
C                 CONTINGENCY TABLE);
C
C             M = THE NUMBER OF ATTRIBUTES PER INDIVIDUAL
C        NUMFAC = THE NUMBER OF FACTORS TO BE PLOTTED;
C                 EG, IF NUMFAC = 3, THE FIRST THREE
C                 SIGNIFICANT FACTORS WILL BE USED.
C         IMISS = 0, NO MISSING DATA VALUES
C               = 1, FOR MISSING DATA VALUES
C
C     DATA FILE, LINE 2:    CONTINGENCY TABLE OPTION:
C
C     READ (5,*) IDEN
C
C     WHERE:  IDEN = 0 IF INDIVIDUAL IDENTIFIER IS LAST
C                      ENTRY PER DATA LINE; OR
C             IDEN = 1 IF INDIVIDUAL IDENTIFIER IS FIRST
C                      ENTRY PER DATA LINE.
C
C     DATA FILE, LINE 3 (ONLY IF IMISS = 1):
C     FLAG FOR MISSING DATA:
C
C     READ(5,*) XFLAG
C
C     WHERE XFLAG = DATA VALUE INDICATING A MISSING DATUM
C
C     DATA FILE, LINES 4 TO N + 1:
C
C     READ(5,*) (Y(I,J), J = 1,M), QFLAG(I)
C
C     WHERE:    Y(I,J) IS ONE COMPLETE ROW OF ATTRIBUTES FOR
C                THE ITH INDIVIDUAL;
C               QFLAG(I) = A TWO-LETTER CHARACTER SYMBOL FOR
C                THE ITH INDIVIDUAL.
C
C     LAST DATA LINE:    ATTRIBUTE SYMBOLS
C
C     READ(5,*) (RFLAG(JK), JK = 1,M)
C
C     WHERE:  RFLAG(JK) = A TWO-LETTER SYMBOL FOR THE
C                JK-TH ATTRIBUTE.
C
C ***************************************************
```

Figure 5.9 continued

```fortran
      PARAMETER (NSS=250, KN=50, KF=10)
      DIMENSION S(KN,KN), Y(KN,KN), W(NSS), T(KN)
      DIMENSION SC(NSS,KN), SR(NSS,KN), SCT(KN,NSS), B(KN,KN)
      DIMENSION E(KN), D(KN), RFACTOR(KN,KF), QFACTOR(NSS,KF)
      DIMENSION IFLAG(NSS,KN),QMIN(KF),QMAX(KF),RMIN(KF),RMAX(KF)
      CHARACTER*2, QFLAG(NSS), RFLAG(KN), L(40), OVER(NSS+KN,2)
      MM = KN
      KITER = 1
      IPASS = 1
C
C     ACCESS DATA; STORE AS [Y].
C
      OPEN (5, FILE = ' ')
      READ(5,*) N, M, NUMFAC, IMISS
      READ(5,*) IDEN
      IF (IMISS .EQ. 1) READ(5,*) XFLAG
      IF (IDEN .EQ. 0) THEN
         DO 10 I = 1,N
            READ(5,*) (Y(I,J), J = 1,M), QFLAG(I)
10       CONTINUE
      ELSEIF (IDEN .EQ. 1) THEN
         DO 15 I = 1,N
            READ(5,*) QFLAG(I), (Y(I,J), J = 1,M)
15       CONTINUE
      ENDIF
      READ(5,*) (RFLAG(JK), JK = 1,M)
      IF (NUMFAC .GE. M) NUMFAC = M - 1
      CALL ECHO(Y, QFLAG, RFLAG, N, M, NUMFAC, NSS, KN, KF)
C
C     IF DATA ARE MISSING, FLAG THE MISSING LOCATIONS
C
C     INITIALIZE THE FLAGGING ARRAY, IFLAG
C
      IF (IMISS .EQ. 1) THEN
         DO 16 I = 1,N
         DO 16 J = 1,M
         IFLAG (I,J) = 0
16       CONTINUE
      ENDIF
C
C     STEP 1.   COMPUTE THE SUM OF [Y]; CALL IT XL
C
19    CONTINUE
      XL = 0.0
      DO 20 I = 1,N
      DO 20 J = 1,M
         IF (IMISS .EQ. 1 .AND. Y(I,J) .EQ. XFLAG) THEN
            IFLAG(I,J) = 1
            Y(I,J) = 0.0
            GO TO 20
         ENDIF
         XL = XL + Y(I,J)
20    CONTINUE
      IF (KITER .EQ. 1) XSAVE = XL
C
C     STEP 2.   NORMALIZE [Y] BY XL
C
      DO 30 I = 1,N
      DO 30 J = 1,M
         Y(I,J) = Y(I,J) / XL
30    CONTINUE
C
C     STEP 3.   CALCULATE THE VECTORS {W} AND {T}.
C
      DO 40 I = 1,N
      W(I) = 0.0
      DO 40 J = 1,M
         W(I) = W(I) + Y(I,J)
40    CONTINUE
      DO 50 J = 1,M
      T(J) = 0.0
      DO 50 I = 1,N
         T(J) = T(J) + Y(I,J)
50    CONTINUE
C
C     STEP 4.   FORM THE MATRICES, [SC] AND [SR].
C
      DO 60 I = 1,N
      DO 60 J = 1,M
         SC(I,J) = Y(I,J) / (W(I) * SQRT(T(J)))
         SR(I,J) = Y(I,J) / SQRT(T(J))
60    CONTINUE
C
```

Figure 5.9 continued

```
C    STEP 5.    CALCULATE THE MATRIX, [S].
C    STEP 5A:   TRANSPOSE THE MATRIX, SC.
C
     DO 65 I = 1,M
     DO 65 J = 1,N
     SCT(I,J) = SC(J,I)
65   CONTINUE
     DO 70 I = 1,M
     DO 70 J = 1,M
     S(I,J) = 0.0
     DO 70 K = 1,N
     S(I,J) = S(I,J) + SCT(I,K) * SR(K,J)
70   CONTINUE
C
C    STEP 6.   COMPUTE THE EIGENVALUES AND EIGENVECTORS OF [S].
C
     CALL EIGENJ(S,D,M,KN,B)
     SUMEIGVL = 0.0
     DO 80 I = M-1,1,-1
     SUMEIGVL = SUMEIGVL + D(I)
80   CONTINUE
     IF(IMISS .EQ. 1) THEN
     GO TO 620
     ENDIF
81   CONTINUE
     WRITE(*,85)
85   FORMAT(1H0,10X,'EIGENVALUE SUMMARY',/,
    2 15X,'EIGENVALUE',10X,'PERCENT VARIATION',//)
     DO 90 I = M-1,1,-1
     THISVAR = D(I) * 100.0 / SUMEIGVL
     WRITE(*,95) D(I),THISVAR
90   CONTINUE
95   FORMAT(10X,1F15.7,10X,1F10.3)
C

C    STEP 7:    CALCULATE FACTORS
C
     DO 300 I = 1,NUMFAC
     DO 300 J = 1,M
     RFACTOR(J,I) = - (S(J,M-I)/SQRT(T(J)))*SQRT(D(M-I))
300  CONTINUE
     WRITE(*,310)
310  FORMAT(1H0,10X,'FACTORS FOR ATTRIBUTES',//,
    2 1X,'IDEN',5X,'FACTORS:  1 TO NUMFAC',/)
     DO 400 I = 1,M
     WRITE(*,410) RFLAG(I),(RFACTOR(I,JK), JK = 1,NUMFAC)
400  CONTINUE
410  FORMAT(3X,A2,7F10.4)
     DO 500 J = 1,N
     QFACTOR(I,J) = 0.0
     DO 500 K = 1,M
     QFACTOR(I,J) = QFACTOR(I,J) + (Y(I,K)/W(I)) *
    2                RFACTOR(K,J)/SQRT(D(M-J))
500  CONTINUE
     WRITE(*,510)
510  FORMAT(1H0,10X,'FACTORS FOR INDIVIDUALS',//,
    2 1X,'IDEN',5X,'FACTORS:  1 TO NUMFAC',/)
     DO 600 I = 1,N
     WRITE(*,610) QFLAG(I),(QFACTOR(I,JK), JK = 1,NUMFAC)
600  CONTINUE
610  FORMAT(3X,A2,7F10.4)
     IF (IPASS .EQ. 1) THEN
     GO TO 850
     ENDIF
C
```

Figure 5.9 continued

```fortran
C
C      ESTIMATE VALUES FOR MISSING DATA LOCATIONS IF IMISS = 1
C
620    CONTINUE
       IF (IMISS .EQ. 1) THEN
         ERROR = 0.0
         DO 800 I = 1,N
           DO 800 J = 1,M
             IF (IFLAG(I,J) .EQ. 1) THEN
               RATIO = W(I) * T(J)
               XTEMP = RATIO
               ERROR = ERROR + ABS(Y(I,J) - XTEMP)
               Y(I,J) = XTEMP
             ENDIF
800        CONTINUE
         IF (ERROR .GT. 0.0000001) THEN
           KITER = KITER + 1
           GO TO 19
         ENDIF
         IPASS = 1
         GO TO 81
       ENDIF
850    CONTINUE
C
C      STEP 8:  PLOT THE RESULTS
C
       CALL PLOT(RFACTOR,QFACTOR,N,M,NUMFAC,QFLAG,RFLAG,NSS,KN,KF,
      2          QMAX, QMIN, L, OVER, RMIN, RMAX)
C
C      STEP 9:  CREATE OUTPUT FILE FOR FACTORS
C
C
C      CREATE AN OUTPUT FILE FOR FACTORS IF DESIRED
C
       WRITE(*,900)
900    FORMAT(/////,5X,'DO YOU WISH TO SAVE THE FACTORS IN AN',/
      2       ,5X,' OUTPUT FILE? ENTER 1 FOR YES; 2 FOR NO'//)
       READ(*,*) ISAVE
       IF (ISAVE .EQ. 1) THEN
         OPEN(1, FILE = ' ', STATUS = 'NEW')
         DO 1000 I = 1,N
           WRITE(1,*) QFLAG(I),(QFACTOR(I,JK), JK = 1,NUMFAC)
1000     CONTINUE
         DO 1010 I = 1,M
           WRITE(1,*) RFLAG(I),(RFACTOR(I,JK), JK = 1,NUMFAC)
1010     CONTINUE
       ENDIF
C
C      STEP 10:  TERMINATE THE PROGRAM
C
       STOP
       END
       SUBROUTINE PLOT(R,Q,N,M,NUMFAC,QFLAG,RFLAG,NSS,KN,KF,
      2                QMAX, QMIN, L, OVER, RMIN, RMAX)
C
C      SUBROUTINE TO GENERATE LINE PRINTER PLOTS OF
C      CORRESPONDENCE ANALYSIS RESULTS
C
       DIMENSION R(KN,KF), Q(NSS,KF), RMAX(KF), RMIN(KF)
       DIMENSION QMAX(KF), QMIN(KF)
       CHARACTER*2, L(40), QFLAG(NSS), RFLAG(KN), OVER(NSS+KN,2)
C
C      STEP 1:  FIND MAX AND MIN VALUES FOR R AND Q
C
       DO 100 I = 1,NUMFAC
         RMIN(I) = R(1,I)
         RMAX(I) = RMIN(I)
         QMIN(I) = Q(1,I)
         QMAX(I) = QMIN(I)
         DO 30 J = 2,M
           IF(R(J,I) .LT. RMIN(I)) RMIN(I) = R(J,I)
           IF(R(J,I) .GT. RMAX(I)) RMAX(I) = R(J,I)
30       CONTINUE
         DO 40 J = 2,N
           IF(Q(J,I) .LT. QMIN(I)) QMIN(I) = Q(J,I)
           IF(Q(J,I) .GT. QMAX(I)) QMAX(I) = Q(J,I)
40       CONTINUE
100    CONTINUE
C
```

Figure 5.9 continued

```fortran
C
C          STEP 2:    GENERATE PLOTS
C
        DO 200 I = 1,NUMFAC
          K = I + 1
          DO 200 J = K,NUMFAC
          PAUSE 'PLEASE INSERT PAPER AND PRESS ENTER'
          WRITE(*,105) I,J
105       FORMAT(1H0,10X,'CORRESPONDENCE ANALYSIS PLOT',//,
     2          10X,'FACTOR ',I2,' VERUS FACTOR ',I2,/)
C
C          PLOT FACTOR J (Y-AXIS) VERUS FACTOR I (X-AXIS)
C
          XMAX = RMAX(I)
          IF(QMAX(I).GT. XMAX) XMAX = QMAX(I)
          XMIN = RMIN(I)
          IF(QMIN(I).LT. XMIN) XMIN = QMIN(I)
          XRANGE = XMAX - XMIN
          YMAX = RMAX(J)
          IF(QMAX(J).GT. YMAX) YMAX = QMAX(J)
          YMIN = RMIN(J)
          IF(QMIN(J).LT. YMIN) YMIN = QMIN(J)
          YRANGE = YMAX - YMIN
          ITEM = 0
          DO 190 II = 1,46
            DO 110 JJ = 1,40
            IF (II .EQ. 23) THEN
              L(JJ) = '++'
            ELSE
              L(JJ) = ' '
            ENDIF
110         CONTINUE
            IF (II .NE. 23) L(20) = '+ '
C
C
C          SEARCH ON Q AND R FOR LINE II
C
          DO 120 KK = 1,N
            LL = ((YMAX-Q(KK,J))/YRANGE)*46 + 1
            IF (LL .LT. 1) LL = 1
            IF (LL .GT. 46) LL = 46
            IF (LL .EQ. II) THEN
              LK = ((Q(KK,I)-XMIN)/XRANGE)*40 + 1
              IF (LK .LT. 1) LK = 1
              IF (LK .GT. 40) LK = 40
              IF (L(LK) .EQ. ' ' .OR. L(LK) .EQ. '++'
     2          .OR. L(LK) .EQ. '+ ') THEN
                L(LK) = QFLAG(KK)
              ELSE
                ITEM = ITEM + 1
                OVER(ITEM,1) = L(LK)
                OVER(ITEM,2) = QFLAG(KK)
              ENDIF
            ENDIF
120       CONTINUE
          DO 130 KK = 1,M
            LL = ((YMAX-R(KK,J))/YRANGE)*46 + 1
            IF (LL .LT. 1) LL = 1
            IF (LL .GT. 46) LL = 46
            IF (LL .EQ. II) THEN
              LK = ((R(KK,I)-XMIN)/XRANGE)*40 + 1
              IF (LK .LT. 1) LK = 1
              IF (LK .GT. 40) LK = 40
              IF (L(LK) .EQ. ' ' .OR. L(LK) .EQ. '+ ') THEN
                L(LK) = RFLAG(KK)
              ELSE
                ITEM = ITEM + 1
                OVER(ITEM,1) = L(LK)
                OVER(ITEM,2) = RFLAG(KK)
              ENDIF
            ENDIF
130       CONTINUE
C
```

Figure 5.9 continued

```fortran
C
C          PRINT THE BUFFER, L

           WRITE(*,140) (L(JK), JK = 1,40)
140        FORMAT(1X,40A2)
190        CONTINUE
C
C          PRINT THE SUMMARY INFORMATION
C
           YMID = (YRANGE / 2.0) + YMIN
           XMID = (XRANGE / 2.0) + XMIN
           WRITE(*,195) YMID, XMID
195        FORMAT(1H0,10X,'HORIZ AXIS IS LOCATED AT Y = ',
     2     1E20.7,//,10X,1X,
     3     'VERT AXIS IS LOCATED AT X = ',1E20.7)
           WRITE(*,196) XRANGE, YRANGE
196        FORMAT(1H0,10X,'X-AXIS RANGE (XMAX-XMIN) = ',
     2     1E20.7,//,10X,1X,
     3     'Y-AXIS RANGE (YMAX-YMIN) = ',1E20.7)
           IF (ITEM .EQ. 0) GO TO 200
           PAUSE 'PLEASE INSERT PAPER AND PRESS ENTER'
           WRITE(*,197)
197        FORMAT( 10X, 'OVERPRINT SUMMARY')
           DO 198 IO = 1,ITEM
           WRITE(*,199) (OVER(IO,JK), JK = 1,2)
198        CONTINUE
199        FORMAT(10X,A10,' PRINTS OVER', A10)
200        CONTINUE
           RETURN
           END
           SUBROUTINE ECHO(Y, QFLAG, RFLAG, N, M, NUMFAC, NSS, KN, KF)
C

C
C          SUBROUTINE WHICH PRINTS INPUT INFORMATION FOR CHECK
C
           DIMENSION Y(NSS,KN)
           CHARACTER*2, QFLAG(NSS), RFLAG(KN)
           WRITE(*,10) N, M, NUMFAC
10         FORMAT(1H0,   9X, 'NUMBER OF INDIVIDUALS = ', I10,//,
     2        10X, 'NUMBER OF ATTRIBUTES  = ', I10,//,
     3        10X, 'NUMBER OF FACTORS     = ', I10)
           WRITE(*,20)
20         FORMAT(1H0, 9X, 'CONTINGENCY TABLE, Y', //, 1X, 'IDEN',/)
           IF (M .LE. 7) THEN
           WRITE(*,30) (RFLAG(JK), JK = 1,M)
30         FORMAT(5X,7A10)
           DO 40 I = 1,N
           WRITE(*,50) QFLAG(I), (Y(I,J), J = 1,M)
40         CONTINUE
50         FORMAT(1A5, 7E10.2)
           PAUSE 'PLEASE INSERT PAPER AND PRESS ENTER'
           ELSEIF (M .GT. 7 .AND. M .LE. 14)  THEN
           WRITE(*,30) (RFLAG(JK), JK = 1,7)
           DO 60 I = 1,N
           WRITE(*,50) QFLAG(I), (Y(I,J), J = 1,7)
60         CONTINUE
           PAUSE 'PLEASE INSERT PAPER AND PRESS ENTER'
           WRITE(*,70)
70         FORMAT( 10X, 'PAGE TWO OF TWO',/)
           WRITE(*,30) (RFLAG(JK), JK = 8,M)
           DO 80 I = 1,N
           WRITE(*,50) QFLAG(I), (Y(I,J), J = 8,M)
80         CONTINUE
```

Figure 5.9 continued

```
        PAUSE 'PLEASE INSERT PAPER AND PRESS ENTER'
      ELSEIF (M. GT. 14) THEN
        WRITE(*,30) (RFLAG(JK), JK = 1,7)
        DO 90 I = 1,N
          WRITE(*,50) QFLAG(I), (Y(I,J), J = 1,7)
90      CONTINUE
        PAUSE 'PLEASE INSERT PAPER AND PRESS ENTER'
        WRITE(*,95)
95      FORMAT( 10X, 'PAGE TWO OF THREE',/)
        WRITE(*,30) (RFLAG(JK), JK = 8,14)
        DO 100 I = 1,N
          WRITE(*,50) QFLAG(I), (Y(I,J), J = 8,14)
100     CONTINUE
        PAUSE 'PLEASE INSERT PAPER AND PRESS ENTER'
        WRITE(*,105)
        KMM = M
        IF (M .GT. 20) KMM = 20
105     FORMAT( 10X, 'PAGE THREE OF THREE',/)
        WRITE(*,30) (RFLAG(JK), JK = 15,KMM)
        DO 110 I = 1,N
          WRITE(*,50) QFLAG(I), (Y(I,J), J = 15,M)
110     CONTINUE
        PAUSE 'PLEASE INSERT PAPER AND PRESS ENTER'
      ENDIF
      RETURN
      END
C
C     SUBROUTINE EIGENJ(A,D,N,KN,B)
C     DIMENSION A(KN,KN), B(KN,KN), D(KN)
C
C     A SUBROUTINE FOR THE CALCULATION OF EIGENVALUES AND
C     EIGENVECTORS OF A SQUARE, SYMETRIC MATRIX
C
C     MODIFIED FROM DAVIS, 1973, PP. 166 - 167.
C     [ NOTE THE 1973 VERSION ]
C
C     MATRIX, B, PROVIDES TEMPORARY STORAGE OF EIGENVECTORS


C
      ANORM = 0.0
      DO 100 I = 1,N
        DO 100 J = 1,N
          K = I - J
          IF (K .EQ. 0)  THEN
            B(I,J) = 1.0
          ELSE
            B(I,J) = 0.0
            ANORM = ANORM + A(I,J) * A(I,J)
          ENDIF
100   CONTINUE
      XN = 1.0 / FLOAT(N)
      ANORM = SQRT(ANORM)
      FNORM = ANORM * 1.0E-09 * XN
C
C     INITIALIZE INDICATORS AND COMPUTE THRESHOLD
C
      THR = ANORM
123   THR = THR * XN
133   IND = 0
C
C     SCAN DOWN COLUMNS FOR OFF-DIAGONAL ELEMENTS
C     GREATER THAN OR EQUAL TO THE THRESHOLD
C
      DO 200 I = 2,N
        I1 = I - 1
        DO 190 J = 1,I1
          W = ABS(A(J,I)) - THR
          IF (W .GE. 0.0)  THEN
C
C     CALCULATE SINES AND COSINES
C
            IND = 1
            AL = - A(J,I)
            AM = (A(J,J) - A(I,I)) * 0.50
            AO = AL / (SQRT(AL*AL + AM*AM))
            IF (AM .LT. 0.0)  THEN
              AO = - AO
            ENDIF
            SINX = AO / (SQRT(2.0 * (1.0 + SQRT(1.0 - AO*AO))))
            SINX2 = SINX * SINX
            COSX = SQRT(1.0 - SINX2)
            COSX2 = COSX * COSX
C
```

Figure 5.9 continued

```fortran
C
C     ROTATE COLUMNS I AND J
C
      DO 150 K = 1,N
        L = K - J
        IF (L .NE. 0)  THEN
          M = K - I
          IF (M .NE. 0)  THEN
            AT = A(K,J)
            A(K,J) = AT * COSX - A(K,I) * SINX
            A(K,I) = AT * SINX + A(K,I) * COSX
          ENDIF
          BT = B(K,J)
          B(K,J) = BT * COSX - B(K,I) * SINX
          B(K,I) = BT * SINX + B(K,I) * COSX
        ENDIF
150   CONTINUE
      XT = 2.0 * A(J,I) * SINX * COSX
      AT = A(J,J)
      BT = A(I,I)
      A(J,J) = AT * COSX2 + BT * SINX2 - XT
      A(I,I) = AT * SINX2 + BT * COSX2 + XT
      A(J,I) = (AT - BT) * SINX * COSX +
     2          A(J,I) * (COSX2 - SINX2)
      A(I,J) = A(J,I)
      DO 180 K = 1,N
        A(J,K) = A(K,J)
        A(I,K) = A(K,I)
180   CONTINUE
      ENDIF
190   CONTINUE
200   CONTINUE
      IF (IND .GT. 0)  THEN
        GO TO 133
      ENDIF
      Z = THR - FNORM
      IF (Z .GT. 0.0)  THEN
        GO TO 123
      ENDIF
C
C
C     SORT EIGENVALUES AND EIGENVECTORS
C
      DO 300 I = 2,N
        J = I
229     C = A(J-1,J-1) - A(J,J)
        IF (C .LT. 0.0)  THEN
          AT = A(J-1,J-1)
          A(J-1,J-1) = A(J,J)
          A(J,J) = AT
          DO 250 K = 1,N
            AT = B(K,J-1)
            B(K,J-1) = B(K,J)
            B(K,J) = AT
250       CONTINUE
          J = J - 1
          L = J - 1
          IF (L .GT. 0)  THEN
            GO TO 229
          ENDIF
        ENDIF
300   CONTINUE
C
C     PLACE EIGENVALUES IN VECTOR D IN ASCENDING ORDER
C     PLACE EIGENVECTORS IN MATRIX A IN ASCENDING ORDER
C
      DO 400 I = 1,N
        K = N + 1 - I
        D(I) = A(K,K)
400   CONTINUE
      DO 500 I = 1,N
        K = N + 1 - I
        DO 500 J = 1,N
          A(J,I) = B(J,K)
500   CONTINUE
      RETURN
      END
```

145

Figure 5.26 A listing for the program DISCRM.

```fortran
      PROGRAM DISCRM

C     A PROGRAM TO PERFORM DISCRIMINANT ANALYSIS ON TWO
C     GROUPS OF DATA, A AND B.
C
C     ******************************************************
C                  USER'S GUIDE:  DISCRM
C     ******************************************************
C
C     THIS PROGRAM COMPUTES A DISCRIMINANT FUNCTION FOR COMPARING
C     TWO GROUPS OF DATA, CALLED A AND B.
C
C     FIRST, FORM TWO, SEPARATE DATA FILES, ONE FOR GROUP A DATA,
C     THE OTHER FOR GROUP B DATA; INDIVIDUALS IN EACH GROUP ARE
C     ASSUMED TO BE DESCRIBED BY M ATTRIBUTES.  IN THIS CASE
C
C     FOR GROUP A:
C        CREATE A FREE-FORMAT DATA FILE, EACH ROW = AN INDIVIDUAL:
C
C        ROW 1:  A(1,J), J = 1, 2, 3, ..., M
C        ROW 2:  A(2,J), J = 1, 2, 3, ..., M
C                .
C                .
C        ROW N:  A(N,J), J = 1, 2, 3, ..., M
C
C        THIS CONCLUDES THE FILE FOR GROUP A
C
C     FOR GROUP B:  CREATE A FILE IN THE SAME MANNER AS IS USED
C        FOR GROUP A.
C
C     SECOND, ONCE THESE TWO FILES ARE CREATED, RUN THE PROGRAM
C        BY TYPING:  DISCRM.  THE PROGRAM THEN WILL PROMPT THE
C        USER TO ENTER TWO FILE NAMES, ONE FOR GROUP A THE OTHER
C        FOR GROUP B; THE PROGRAM ALSO PROMPTS THE USER TO ENTER
C        A VALUE FOR THE NUMBER OF ATTRIBUTES, M.

C     AN OPTION IS ADDED TO DISCRM TO CLASSIFY A THIRD FILE
C     USING THE DISCRIMINANT COEFFIECIENTS DETERMINED FROM
C     DATA GROUPS A AND B.  TO USE THIS OPTION:
C
C     A QUESTION IS ASKED:  WILL A THIRD FILE BE CLASSIFIED?
C        ENTER 1 FOR YES, 2 FOR NO.
C
C     IF YES IS CHOSEN, THEN ENTER THE NAME OF THE FILE WHEN
C        PROMPTED
C
C     ******************************************************
C              END OF USER'S GUIDE:  DISCRM
C     ******************************************************
C
      PARAMETER (MM=30, NN=200)
      DIMENSION A(NN,MM), B(NN,MM), AMEAN(MM), BMEAN(MM), X(MM)
      DIMENSION D(MM), SPA(MM,MM), SPB(MM,MM), SP2(MM,MM+1)
C
C     ACCESS THE DATA, A AND B
C
      WRITE(*,*) 'ENTER THE NUMBER OF VARIABLES, M, TO BE CONSIDERED'
      READ(*,*) M
      OPEN(5, FILE = ' ')
C     MATRIX A
      NA = 1
10    READ(5,*,END=20)  (A(NA,JK), JK=1,M)
      NA = NA + 1
      GO TO 10
20    NA = NA - 1
      CLOSE(5)
      OPEN(5, FILE = ' ')
C     MATRIX B
      NB = 1
30    READ(5,*,END=40)  (B(NB,JK), JK = 1,M)
      NB = NB + 1
      GO TO 30
40    NB = NB - 1
C
```

146

Figure 5.26 continued

```fortran
C
C       COMPUTE THE VARIABLE MEAN VALUES
C
        DO J = 1,M
        SUM = 0
        DO I = 1,NA
          SUM = SUM + A(I,J)
        END DO
        AMEAN(J) = SUM / FLOAT(NA)
        END DO
        DO J = 1,M
        SUM = 0
        DO I = 1,NB
          SUM = SUM + B(I,J)
        END DO
        BMEAN(J) = SUM / FLOAT(NB)
        END DO
C
C       FORM THE VECTOR, D
C
        DO I = 1,M
        D(I) = AMEAN(I) - BMEAN(I)
        END DO
C
C       FORM THE MATRIX, SP2
C
        DO J = 1,M
        SUMA1 = 0.0
        DO I = 1,NA
          SUMA1 = SUMA1 + A(I,J)
        END DO
        DO K = 1,M
        SPA(J,K) = 0.0
        SUMAA = 0.0
        SUMA2 = 0.0
        DO I = 1,NA
          SUMAA = SUMAA + A(I,J)*A(I,K)
          SUMA2 = SUMA2 + A(I,K)
        END DO
        SPA(J,K) = SUMAA - (SUMA1*SUMA2)/(FLOAT(NA))
        END DO
        END DO

        DO J = 1,M
        SUMB1 = 0.0
        DO I = 1,NB
          SUMB1 = SUMB1 + B(I,J)
        END DO
        DO K = 1,M
        SPB(J,K) = 0.0
        SUMBB = 0.0
        SUMB2 = 0.0
        DO I = 1,NB
          SUMBB = SUMBB + B(I,J) * B(I,K)
          SUMB2 = SUMB2 + B(I,K)
        END DO
        SPB(J,K) = SUMBB - (SUMB1*SUMB2)/(FLOAT(NB))
        END DO
        END DO
        DO I = 1,M
        DO J = 1,M
        SP2(I,J) = (SPA(I,J) + SPB(I,J))/(FLOAT(NA + NB - 2))
        END DO
        END DO
C
C       USE GAUSS ELIM TO SOLVE:   [SP2]{X} = {D}; I.E., SOLVE FOR {X}
C
        CALL GAUSS(SP2, D, X, MM, M)
C
C       PRINT X
C
        CLOSE(5)
        WRITE(*,1000)
1000    FORMAT(//,5X,'DISCRIMINANT ANALYSIS COEFFICIENTS:',//)
        DO I = 1,M
        WRITE(*,*) '        COEFFICIENT: ',I,' EQUALS: ',X(I)
        END DO
C
```

Figure 5.26 continued

```fortran
C     COMPUTE RA, RB, AND R0
C
      RA = 0.0
      RB = 0.0
      R0 = 0.0
      DO I = 1, M
        RA = RA + AMEAN(I) * X(I)
        RB = RB + BMEAN(I) * X(I)
        R0 = R0 + X(I) * (AMEAN(I) + BMEAN(I)) / 2.0
      END DO
      WRITE(*,*) 'RA= ',RA,'  RB= ',RB,'  R0= ',R0
C
C     CALCULATE DISCRIMINANT SCORES FOR SAMPLES IN GROUPS A AND B
C
      WRITE(*,1050)
1050  FORMAT(//,5X,'DISCRIMINANT SCORES, GROUPS A AND B: ',///)
      DO I = 1, NA
        SUM = 0.0
        DO J = 1, M
          SUM = SUM + A(I,J) * X(J)
        END DO
        WRITE(*,*) 'GROUP A:  SAMPLE: ',I,'  SCORE = ', SUM
      END DO
      WRITE(*,*)
      DO I = 1, NB
        SUM = 0.0
        DO J = 1, M
          SUM = SUM + B(I,J) * X(J)
        END DO
        WRITE(*,*) 'GROUP B:  SAMPLE: ',I,'  SCORE = ', SUM
      END DO
C
C     OPTION TO CLASSIFY SAMPLES IN A THIRD FILE
C
      WRITE(*,*) 'DO YOU WISH TO CLASSIFY A THIRD FILE? 1=YES;2=NO'
      READ(*,*) ICLASS
      IF (ICLASS .EQ. 1) THEN
        OPEN(5, FILE = ' ')
        NC = 1
1100    READ(5,*,END=1200) (A(NC,JK), JK = 1,M)
        NC = NC + 1
        GO TO 1100
1200    NC = NC - 1
        WRITE(*,1250)
1250    FORMAT(//,5X,'DISCRIMINANT SCORES FOR THIRD FILE: ',//)
        DO I = 1, NC
          SUM = 0.0
          DO J = 1, M
            SUM = SUM + A(I,J) * X(J)
          END DO
          WRITE(*,*) 'GROUP C:  SAMPLE: ',I,'  SCORE = ', SUM
        END DO
      END IF
      STOP
      END
C
C
      SUBROUTINE GAUSS (A, B, X, NN, N)
      DIMENSION A(NN,NN+1), B(NN), X(NN)
C
C
C     PLACE THE VECTOR, B, INTO THE N+1 COLUMN OF A
C
      DO I = 1,N
        A(I, N+1) = B(I)
      END DO
C
```

Figure 5.26 continued

```fortran
C     INITIALIZE MATRIX SIZES
C
      M = N + 1
      L = N - 1
C
C     BEGIN THE FORWARD REDUCTION (ELIMINATION)
C
      DO K = 1,L
      JJ = K
      XLARG = ABS(A(K,K))
      K1 = K + 1
C
C     SEARCH FOR LARGEST PIVOT ELEMENT
C
      DO I = K1, N
      CA = ABS(A(I,K))
      DIFF = XLARG - CA
      IF (DIFF .LT. 0) THEN
      XLARG = CA
      JJ = I
      END IF
      END DO
      IDIFF = JJ - K
C
C     SWAP ROWS, IF NECESSARY, DEPENDING ON THE
C     POSITION OF THE MAXIMUM PIVOT ELEMENT.
C
      IF (IDIFF .NE. 0) THEN
      DO J = K,M
      TEMP = A(JJ,J)
      A(JJ,J) = A(K,J)
      A(K,J) = TEMP
      END DO
      END IF
C
C     THIS IS THE MAJOR ASPECT OF GAUSS ELIMINATION:
C     DIVIDE REMAINING ELEMENTS OF A (AND B WHICH
C     IS IN THE LAST COLUMN OF A) BY THE PIVOT
C     ELEMENT, A(K,K)
C
      DO I = K1, N
      QUOT = A(I,K) / A(K,K)
      DO J = K1, M
      A(I,J) = A(I,J) - QUOT * A(K,J)
      END DO
      END DO
      DO I = K1, N
      A(I,K) = 0.0
      END DO
      END DO
C
C     SOLVE, NOW, FOR LAST UNKNOWN
C
      X(N) = A(N,M) / A(N,N)
C
C     USING THIS SOLUTION, BACK SUBSTITUTE TO FIND
C     THE REMAINING UNKNOWNS
C
      DO IN = 1,L
      SUM = 0.0
      I = N - IN
      I1 = I + 1
      DO J = I1, N
      SUM = SUM + A(I,J) * X(J)
      END DO
      X(I) = (A(I,M) - SUM) / A(I,I)
      END DO
      RETURN
      END
```

6

Geostatistics: The Art of Spatial Analysis

The word geostatistics may be somewhat misleading. This term is composed of a word, statistics, and a prefix, geo. A literal translation of geostatistics consequently implies that the objective of this chapter is the review of statistical applications relevant to the geological science. In its most frequent usage, however, the term geostatistics refers to spatial estimation, especially using a method known as kriging. This method is an outgrowth of Gauss' Theory of Errors; in fact, kriging is a least-squares procedure.

Contributing to the unfortunate misnomer is the prefix geo, which may imply certain limitations regarding applications. True, geologists, geological engineers, and mining engineers are eager to use geostatistics to map the spatial aspects of an ore deposit. But, botanists are likewise eager to map the spatial distribution of a species of flora, and zoologists are interested in mapping the spatial density of a species of fauna. Environmental management specialists want to map the spatial distribution of sulfur pollutants originating from a coal-fired power plant. Astronomers may want to analyze the spatial distribution of matter in a galaxy. These examples show that the prefix geo is too limiting and that geostatistics is a discipline encompassing powerful generic methods, applicable for spatial analysis, whatever may occupy the space.

6.1 A Brief History

In the context of this book, geostatistics is a discipline encompassing a variety of kriging methods for optimal, spatial interpolation. Optimal, as used here, refers to a minimization of the variance of estimation error. The term geostatistics is used in this text to describe the spatial autocorrelation (autocovariance; Chapter 4) inherent to a spatial phenomenon. This spatial autocorrelation is used as the basis for optimal interpolation (estimation is a synonym).

A brief review of the history of geostatistics will aid in the understanding of terms. Geostatistics was developed within the minerals industry to improve the calculation of mineral (e.g., gold, silver, platinum) reserves. D. G. Krige, a South African mining engineer, approached this problem from a probabilistic viewpoint and his work culminated in a benchmark paper (Krige, 1951). When Georges Matheron, an engineer with Ecoles des Mines, Fontainebleau, France, became aware of Krige's work, he lent his expertise in the theory of probability and statistics to formalize Krige's approach to ore reserves calculation. Out of Matheron's research came a spatial interpolation method which he called *kriging* (pronounced by most practitioners as ''kree-ging'' or ''kree-jing''), in honor of Dr. Krige. As will be shown, kriging is an outgrowth of Gauss' Theory of Errors.

Matheron's first English paper on this matter appeared in 1963 (*Economic Geology* **58:** 1246–1266). The fact that the journal *Economic Geology* was chosen by Matheron for this publication is significant, and shows that as of the early 1960s kriging was known primarily within the minerals industry.

By the middle of the 1970s, knowledge of kriging (and geostatistics) had begun to spread beyond the minerals industry. By this time, Matheron had graduated several classes of students. Some of these students published landmark books on the subject: Michel David (1977), Andre Journel (1978; with C. J. Huijbregts), and Isobel Clark (1979). Andre Journel's book, while primarily documenting theory and applications of geostatistics for the minerals industry, also shows maps of simulated rainfall, suggesting an application of geostatistics for the study of meteorology, and includes an application to piezometric data, which motivates applications of geostatistics for the study of hydrogeology.

During the 1980s, a virtual explosion of applications of geostatistics occurred outside the minerals industry, documented in such journals as *Mathematical Geology* (prior to 1986, *The International Journal for Mathematical Geology*), *Computers and Geosciences,* and *Water Resources Research.* The latter journal especially documents the broad range of geostatistical applications to the study of surface and subsurface hydrology, beginning with a paper by Delhomme (1979), a former student of Matheron's.

Today, geostatistics is widely used for groundwater studies. Spatial analysis is of vital importance for environmental studies of, for example, air and water pollution and aerial distribution of acidic rain (rainfall having a pH less than 5.5, the pH of uncontaminated rainwater); actually, geostatistics is useful for assessing the aerial distribution of any contaminant. By the end of the 1980s, knowledge of geostatistics was fully entrenched in disciplines outside the minerals industry.

Four proceedings volumes offer the best documentation of the historical development of geostatistics and its growth into a wide range of disciplines: Guarascio et al. (1976), Verly et al. (1984), Armstrong (1989), and Soares

(1993). These volumes document the development of different kriging methods, numerous applications in the minerals industry, and applications beyond the minerals industry, such as earthquake ground motion studies, seismic refraction and reflection studies, air pollution monitoring, groundwater studies, remote sensing and image processing, and geotechnical parametric studies.

Geostatistics encompasses generic methods for spatial mapping and correlation studies, which are applicable to many disciplines. The goals of this chapter are (1) to describe a few of these generic methods, (2) to document some applications, and (3) to motivate experimentation to expand the range of geostatistical applications.

6.2 The Unique Terminology of Geostatistics

As with many disciplines, geostatistics has its own unique suite of terms. One such term, kriging, was introduced earlier. Kriging is a linear, unbiased, least-squares spatial interpolation method; a weighted-average, or weighted-mean, estimator whose weights are functions of spatial covariance; and an advanced application of Gauss' Theory of Errors.

Knowledge of the spatial covariance inherent to a spatial phenomenon is required for kriging. A notion of spatial covariance for a particular spatially distributed phenomenon, such as an ore deposit, hazardous waste contamination spill, or earthquake, is developed from a function known in geostatistics as the variogram. As will be shown, the variogram is a representation of variance as a function of spatial lag (distance). Use of this terminology is commensurate with the notion that the variogram is a spatial autocorrelation, or autocovariance, function.

Kriging is used to estimate the value of a regionalized variable at a spatial location. A regionalized variable (Matheron, 1963) is a function, usually called Z, distributed in space (whatever the space may be: one, two, or three dimensions or multiple dimensions beyond three). A regionalized variable has properties intermediate between a purely deterministic process and a purely random one. Such a variable has spatial continuity, so it cannot be purely random, but a deterministic function cannot be used to model the variable (or spatial process). Kriging was developed as a model for such variables.

The terms kriging and variogram are unique to geostatistics and are, indeed, unusual. Hence, their introduction in a separate section of this chapter is warranted. All terms unique to or used in geostatistics are defined in Olea (1991).

6.3 Kriging: Gauss' Theory of Errors Revisited

We begin with a visual stimulus, a two-dimensional region (Figure 6.1). Within this space, data are collected at locations shown by small circles. Each circle is associated with a number, the data value observed at that location. The objective of kriging is to use a number of closest data observations to an estimation location and multiply each of these data observations by a weight, the sum of these products being the best estimate. For example, kriging may be used to make an estimate at location x_0, shown by the square symbol in Figure 6.1, using the data at closest surrounding locations.

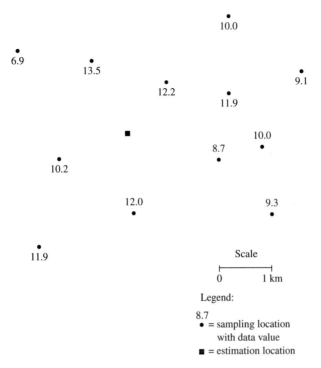

Figure 6.1 An example of a spatial region showing locations at which a phenomenon Z is observed (circles), and an example of an estimation location (square).

In geostatistics, the regionalized variable under study is represented by an uppercase letter, Z. A discrete observation of Z (Figure 6.1, by a small circle) is universally referred to as $Z(x_i)$, where x_i represents the spatial location of the observation. The estimated (interpolated) value of Z at the estimation location (referred to as x_0) is $Z^*(x_0)$.

Matheron defined the kriging estimator over an infinite domain Z as

$$Z^*(x_0) = \frac{1}{\text{space}} \int_{\text{space}} f(Z_{\text{space}}) \, d(\text{space})$$

where space is 1, 2, 3, ... , dimensions.

In geostatistics, using this integral to obtain the estimate requires complete (total) knowledge of Z, which is rarely available. Usually, all that is available are a few discrete observations of Z (cf., Figure 6.1). The integral for $Z^*(x_0)$ is approximated using a procedure exactly analogous to Gaussian quadrature (Chapter 2):

$$Z^*(x_0) = \sum_{i=1}^{N} \lambda_i Z(x_i)$$

where λ_i are weights.

To apply this equation, N (an arbitrarily chosen number, the choice of which is discussed later in this chapter) closest spatial locations are found at which data observations are recorded; $Z(x_i)$ represents the data value at one of these N locations, x_i. Because these values are known, what is required is a procedure for determining weights to be multiplied by each of the N data values. The sum of these products comprises the kriging estimate.

Recalling Gauss' Theory of Errors, the square of the (estimation) error is the most natural function to use when attempting to optimize estimation accuracy. An error function q is defined. This error function (Chapter 2) is a deterministic one:

$$q = \sum_{i=1}^{N} (y - y^*)^2$$

where the objective is to find the equation of a line, $y = mx + b$, offering the best accuracy in predicting y given x.

With kriging, the notion of the error function q is used but developed in a probabilistic context:

$$q = E[Z(x_0) - Z^*(x_0)]^2 = \text{minimum}$$

Moreover,

$$E[Z(x_0) - Z^*(x_0)] = 0$$

Further, $E[Z(x_0)] = $ mean; therefore, $E[Z^*(x_0)] = $ mean. That $E[Z(x)]$ is equal to the mean data value is an assumption known as the *intrinsic hypothesis*. This hypothesis further holds that $E[Z(x)] = \text{mean}(x) = $ mean; that is, the expected value is not dependent on spatial position.

Commensurate with the intrinsic hypothesis, a sufficient condition for unbiasedness is (Journel and Huijbregts, 1978, p. 305)

$$\sum_{i=1}^{N} \lambda_i = 1$$

Thus,

$$E[Z^*(x_0)] = (\text{mean}) \sum_{i=1}^{N} \lambda_i = \text{mean} = E[Z(x_0)]$$

from which is obtained $E[Z(x_0) - Z^*(x_0)] = 0$.

This constraint placed on the kriging weights is a nonbias condition. Nonbiased, or unbiased, estimation is that for which the mean value of all estimates is equal to the original; that is, estimation does not result in a shift in mean data value.

6.3.1 Deriving the Solution for Kriging Weights

The error function q is used to derive a solution for kriging weights. Recall the error function used for kriging:

$$q = E[Z(x_0) - Z^*(x_0)]^2$$

and further restate

$$\text{ERR} = e = Z(x_0) - Z^*(x_0)$$

where e is the error, whose mean value is zero (for unbiased estimation).

The concept of statistical variance was given in Chapter 4. Writing the notion of statistical variance in a probabilistic sense yields

$$\text{var}(x) = E[x]^2 - \bar{x}^2$$

Thus, variance of kriging error may be written

$$\text{var}(e) = E[e]^2 - \bar{e}^2$$

If kriging errors have a mean value of zero, variance of kriging error becomes

$$\text{var}(e) = E[e]^2 = E[Z(x_0) - Z^*(x_0)]^2 = q$$

In other words, the error function q is the expression for the variance of the kriging error if the mean of kriging errors is zero.

To use the function q to derive a solution for the kriging weights, q must be written as a function of these weights. The solution for $Z^*(x_0)$ is introduced into q (as in Chapter 2 for derivation of a least-squares linear regression solution) such that

$$\text{var}(e) = q = E[Z(x_0) - \sum_{i=1}^{N} \lambda_i Z(x_i)]^2$$

which upon expansion is

$$\text{var}(e) = q = E[Z(x_0)]^2 - 2E\left[Z(x_0) \sum_{i=1}^{N} \lambda_i Z(x_i)\right] + E\left[\sum_{i=1}^{N} \lambda_i Z(x_i)\right]^2$$

Notice that the mathematical expectation operator E is associated with each of the three individual terms now seen to compose the error function q.

For clarity, each of the three components of q is discussed separately. The first term,

$$E[Z(x_0)]^2$$

can be written as

$$\text{var}[Z(x_0)] + [\overline{Z}(x_0)]^2$$

where $\overline{Z}(x_0)$ is the local mean of Z at x_0. Recall the intrinsic hypothesis, which holds that the mean and variance of Z are spatially invariant. This implies that

$$\overline{Z}(x_0) = \overline{Z}(x_i) = \overline{Z}$$

and

$$\text{var}[Z(x_0)] = \text{var}[Z(x_i)] = \text{var}[Z]$$

and allows the following:

$$E[Z(x_0)]^2 = \text{var}[Z] + \overline{Z}^2$$

For the moment, this first term comprising q is left as stated, and the second term of q is discussed. This term is rewritten as

$$2 \sum_{i=1}^{N} \lambda_i E[Z(x_0)Z(x_i)]$$

to associate the expected value operator with the product of $Z(x_0)$, a constant, and $Z(x_i)$, a variable. Covariance (Chapter 4) is expressed in terms of an expected value:

$$\text{cov}(x, y) = E[(x - \overline{x})(y - \overline{y})]$$

which for autocovariance is written as

$$\text{cov}(x_i, x_j) = E[(x_i - \overline{x})(x_j - \overline{x})]$$

and may also be written as (Devore, 1991)

$$\text{cov}(x_i, x_j) = E[x_i x_j] - \overline{x}^2$$

The notion of spatial autocovariance is used to expand the second term of q by noticing that

$$E[Z(x_0)Z(x_i)] = \text{cov}[Z(x_0)Z(x_i)] + \overline{Z}^2$$

Therefore, the second term becomes

$$2 \sum_{i=1}^{N} \lambda_i \{ \text{cov}[Z(x_0)Z(x_i)] + \bar{Z}^2 \}$$

which can be rewritten as

$$2 \left\{ \sum_{i=1}^{N} \lambda_i \, \text{cov}[Z(x_0)Z(x_i)] \right\} + 2 \left(\sum_{i=1}^{N} \lambda_i \bar{Z}^2 \right)$$

Because the kriging weights sum to one (for unbiased estimation), this expression further simplifies to

$$2 \left\{ \sum_{i=1}^{N} \lambda_i \, \text{cov}[Z(x_0)Z(x_i)] \right\} + 2\bar{Z}^2$$

As with the first term, this second term is left as stated, and the third term comprising the error function q is discussed. This term,

$$E \left[\sum_{i=1}^{N} \lambda_i Z(x_i) \right]^2$$

is written as

$$E \left[\sum_{i=1}^{N} \sum_{j=1}^{N} \lambda_i \lambda_j Z(x_i)Z(x_j) \right]$$

DEMONSTRATION

(For N = 2, with implications for larger systems.) Given that $N = 2$,

$$\left[\sum_{i=1}^{2} \lambda_i Z(x_i) \right]^2 = [\lambda_1 Z(x_1) + \lambda_2 Z(x_2)]^2$$

$$= \lambda_1^2 Z(x_1)^2 + 2\lambda_1 \lambda_2 Z(x_1)Z(x_2) + \lambda_2^2 Z(x_2)^2$$

In comparison,

$$\sum_{i=1}^{2} \sum_{j=1}^{2} \lambda_i \lambda_j Z(x_i)Z(x_j) = \lambda_1^2 Z(x_1)^2 + \lambda_1 \lambda_2 Z(x_1)Z(x_2)$$
$$+ \lambda_2 \lambda_1 Z(x_2)Z(x_1) + \lambda_2^2 Z(x_2)^2$$

$$= \lambda_1^2 Z(x_1)^2 + 2\lambda_1 \lambda_2 Z(x_1)Z(x_2) + \lambda_2^2 Z(x_2)^2$$

We recall

$$\sum_{i=1}^{N} \sum_{j=1}^{N} \lambda_i \lambda_j E[Z(x_i)Z(x_j)]$$

and use the notion of spatial autocovariance to write this term as

$$\sum_{i=1}^{N} \sum_{j=1}^{N} \lambda_i \lambda_j \{ \text{cov}[Z(x_i)Z(x_j)] + \bar{Z}^2 \}$$

which expands to

$$\left\{ \sum_{i=1}^{N} \sum_{j=1}^{N} \lambda_i \lambda_j \, \text{cov}[Z(x_i)Z(x_j)] \right\} + \left(\sum_{i=1}^{N} \sum_{j=1}^{N} \lambda_i \lambda_j \bar{Z}^2 \right)$$

Because the kriging weights are constrained such that their sum is one, this expression simplifies to

$$\left\{ \sum_{i=1}^{N} \sum_{j=1}^{N} \lambda_i \lambda_j \, \text{cov}[Z(x_i)Z(x_j)] \right\} + \bar{Z}^2$$

With all three terms of the error variance function q expanded and rewritten, this function is written in its entirety:

$$q = \text{var}(Z) + \bar{Z}^2 - 2\left\{ \sum_{i=1}^{N} \lambda_i \, \text{cov}[Z(x_0)Z(x_i)] \right\} - 2\bar{Z}^2$$

$$+ \left\{ \sum_{i=1}^{N} \sum_{j=1}^{N} \lambda_i \lambda_j \, \text{cov}[Z(x_i)Z(x_j)] \right\} + \bar{Z}^2$$

which simplifies to

$$q = \text{var}(Z) - 2 \sum_{i=1}^{N} \lambda_i \, \text{cov}[Z(x_0)Z(x_i)] + \sum_{i=1}^{N} \sum_{j=1}^{N} \lambda_i \lambda_j \, \text{cov}[Z(x_i)Z(x_j)]$$

At this point, the expression for the error variance is not yet complete because the constraint that kriging weights sum to one is not built into this expression (albeit the constraint is implicitly included to derive two of the three components of q). To explicitly include the unbiased constraint on the kriging weights, a Lagrangian function is derived using the error variance function q. A Lagrangian function is one developed to optimize an initial function subject to a constraint.

DEMONSTRATION

Given the situation of a rectangular area circumscribed by a circle (Figure 6.2), what is the maximum rectangular area that can fit inside this circle?

This problem involves optimizing a function area$(x, y) = 4xy$, subject to the constraint that $x^2 + y^2 = r^2$. A Lagrangian function is developed from

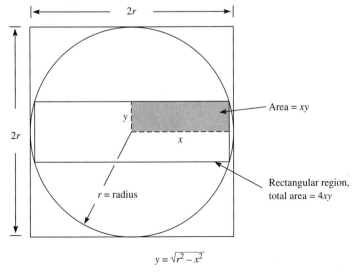

$$y = \sqrt{r^2 - x^2}$$

Figure 6.2 A circular region bounding a rectangular area $4xy$ and bounded by a square region of area $4r^2$.

the function area(x, y), such that

$$L(x, y, \mu) = 4xy - \mu(x^2 + y^2 - r^2)$$

where μ is the Lagrangian multiplier. Notice that the constraint is provided by the term

$$\mu(x^2 + y^2 - r^2)$$

which does not change the value of the function area(x, y) for any x, y, because the term $(x^2 + y^2 - r^2)$ is zero for any x, y. This term, however, has an interesting influence on the optimization of the function $L(x, y, \mu)$.

Recall from Chapter 1 that a derivative of a function $y(x)$ at x is the slope of the function at this point. The problem now is to find the maximum area, $4xy$, subject to the constraint $x^2 + y^2 = r^2$. Therefore, because the function $L(x, y, \mu)$ is a quadratic expression, its shape can be imagined to be that of a parabolic dish, in this case concave downward. The point x, y, μ at which the slopes of the derivatives of L with respect to x, y, and μ are zero is the maximum value of the function $L(x, y, \mu)$. This maximum point is found by taking the partial derivative (Chapter 1) of L with respect to each of the three terms, then setting each derivative equal to zero, yielding three simultaneous equations which are used to solve for x, y, and μ:

$$\frac{\partial L}{\partial x}: \quad 4y - 2\mu x = 0$$

$$\frac{\partial L}{\partial y}: \quad 4x - 2\mu y = 0$$

$$\frac{\partial L}{\partial \mu}: \quad -x^2 - y^2 + r^2 = 0$$

Taking the partial derivative of L with respect to x,

$$y = \mu \frac{x}{2}$$

and substituting this solution into the partial derivative of L with respect to y gives the solution for the Lagrangian multiplier:

$$4x - 2\mu \left(\mu \frac{x}{2} \right) = 0, \qquad \therefore \mu = 2$$

Using this solution,

$$\because 4y = 2\mu x, \quad \therefore y = x$$

Substituting this solution into the partial derivative of L with respect to the Lagrangian multiplier gives

$$2x^2 = r^2, \qquad \therefore x = y = \frac{r}{\sqrt{2}}$$

The optimal area is therefore $4xy = 2r^2$, showing that the maximum rectangular area able to fit inside a circle is a square area equal to one-half the area of a square whose dimensions are equal to 2r.

The significance of the Lagrangian multiplier, equal to 2 in this case, is seen by restating the Lagrangian function L for the foregoing problem as

$$L(x, y, \mu) = area(x, y) - B(x, y)$$

Hence

$$\mu = \frac{\partial \text{ area}(x, y)/\partial x}{\partial B(x, y)/\partial x} = \frac{\partial \text{ area}(x, y)/\partial y}{\partial B(x, y)/\partial y}$$

Using this relationship for the problem in Figure 6.2 gives

$$\frac{\partial \text{ area}(x, y)/\partial x}{\partial B(x, y)/\partial x} = \frac{4y}{2x}, \quad \because y = x, \quad \text{then} \frac{4y}{2x} = 2 = \mu$$

As is done in the foregoing demonstration, expanding the kriging error variance function q as a Lagrangian function to include the constraint on the kriging weights yields

$$q = \text{var}(Z) - 2 \sum_{i=1}^{N} \lambda_i \text{cov}[Z(x_0)Z(x_i)] + \sum_{i=1}^{N} \sum_{j=1}^{N} \lambda_i \lambda_j \text{cov}[Z(x_i)Z(x_j)]$$
$$- 2\mu \left[\left(\sum_{i=1}^{N} \lambda_i \right) - 1 \right]$$

for which the final term

$$2\mu \left[\left(\sum_{i=1}^{N} \lambda_i \right) - 1 \right]$$

provides the constraint. Notice that, unlike in the demonstration, twice the Lagrangian multiplier is used in this expansion. Why this is done will become clear when we compute partial derivatives of q. Because the part of the Lagrangian function that is multiplied by the Lagrangian multiplier is equal to zero, any multiple of the Lagrangian multiplier may be used without changing the function.

Partial derivatives of the error function q are taken with respect to kriging weights and Lagrangian multiplier, yielding in this case $N + 1$ simultaneous equations with which to solve for N kriging weights plus the one Lagrangian multiplier:

$$\frac{\partial q}{\partial \lambda_i} = -2 \text{cov}[Z(x_0)Z(x_i)]$$
$$+ 2 \sum_{j=1}^{N} \lambda_j \text{cov}[Z(x_i)Z(x_j)] - 2\mu = 0$$

and

$$\frac{\partial q}{\partial \mu} = -2 \left[\left(\sum_{i=1}^{N} \lambda_i \right) - 1 \right] = 0$$

from which is determined

$$\sum_{i=1}^{N} \lambda_i = 1$$

Notice that by using twice the Lagrangian multiplier, the coefficient 2 cancels in the partial derivatives. This explains why a multiple of the Lagrangian multiplier is used to develop the Lagrangian function for q.

In taking the partial derivative of q with respect to kriging weights, the following is assumed:

$$\frac{\partial}{\partial \lambda_i} \text{ of } \sum_{i=1}^{N} \sum_{j=1}^{N} \lambda_i \lambda_j \, \text{cov}[Z(x_i) Z(x_j)] = 2 \sum_{j=1}^{N} \lambda_j \, \text{cov}[Z(x_i) Z(x_j)]$$

for each unique λ_i.

DEMONSTRATION

(For N = 2, with implications for larger systems.) Given that $N = 2$,

$$\sum_{i=1}^{2} \sum_{j=1}^{2} \lambda_i \lambda_j \, \text{cov}[Z(x_i) Z(x_j)] = \lambda_1^2 \, \text{cov}[Z(x_1) Z(x_1)]$$

$$+ 2\lambda_1 \lambda_2 \, \text{cov}[Z(x_1) Z(x_2)] + \lambda_2^2 \, \text{cov}[Z(x_2) Z(x_2)]$$

Taking the partial derivatives of this expression with respect to the two kriging weights gives

$$\frac{\partial}{\partial \lambda_1} = 2\lambda_1 \, \text{cov}[Z(x_1) Z(x_1)] + 2\lambda_2 \, \text{cov}[Z(x_1) Z(x_2)] = 2 \sum_{j=1}^{2} \lambda_j \, \text{cov}[Z(x_1) Z(x_j)]$$

$$\frac{\partial}{\partial \lambda_2} = 2\lambda_1 \, \text{cov}[Z(x_1) Z(x_2)] + 2\lambda_2 \, \text{cov}[Z(x_2) Z(x_2)] = 2 \sum_{j=1}^{2} \lambda_j \, \text{cov}[Z(x_j) Z(x_2)]$$

A matrix solution for the kriging weights in the form $[A]\{x\} = \{b\}$ is written by grouping the $N + 1$ equations obtained from the partial derivatives of the Lagrangian expansion of q:

$$\begin{vmatrix} \text{cov}(h_{11}) & \cdots & \text{cov}(h_{1N}) & 1 \\ \text{cov}(h_{21}) & \cdots & \text{cov}(h_{2N}) & 1 \\ \vdots & & \vdots & \vdots \\ \text{cov}(h_{N1}) & \cdots & \text{cov}(h_{NN}) & 1 \\ 1 & \cdots & 1 & 0 \end{vmatrix} \begin{vmatrix} \lambda_1 \\ \lambda_2 \\ \vdots \\ \lambda_N \\ -\mu \end{vmatrix} = \begin{vmatrix} \text{cov}(h_{01}) \\ \text{cov}(h_{02}) \\ \vdots \\ \text{cov}(h_{0N}) \\ 1 \end{vmatrix}$$

which can be expressed simply as

$$[\text{cov}(h_{ij})][\lambda, \mu] = [\text{cov}(h_{0i})]$$

6.4 Determining Spatial Autocovariance

In the matrix system

$$\text{cov}[Z(x_i), Z(x_j)]\lambda_i - \mu = \text{cov}[Z(x_0), Z(x_i)]$$

cov is the spatial autocovariance function for the random function Z. How the spatial autocovariance function for Z is obtained remains to be explained.

Spatial autocovariance (likewise, spatial autocorrelation) is a measure of similarity between Z sampled at two different spatial locations, x_i and x_j. As David (1977) points out, a natural way to measure similarity between two numerical values is simply to compute their difference. Therefore,

$$\text{similarity} \propto [Z(x_i) - Z(x_j)]^{-1}$$

which is an inverse proportionality, suggesting that similarity becomes infinitely large as the difference $Z(x_i) - Z(x_j)$ becomes infinitely small.

Similarity is related to the error function q:

$$(\text{similarity})^{-1} \propto q = E[Z(x_i) - Z(x_j)]^2$$

which equates similarity as used herein with error variance.

Let the lag (Chapter 4) be a distance h separating two locations x_i and x_j. Using h, x_j is written x_{i+h} (i.e., x_j is a distance h away from x_i), which acknowledges the concept of lag distance and will relate spatial autocovariance to h. This also implies the use of the intrinsic hypothesis. Recall that this hypothesis holds that the parameters of Z, in this case, mean and variance, are spatially invariant. By using the intrinsic hypothesis, assume that similarity is a function only of h and consequently is not a function of x (position). Therefore,

$$[\text{similarity}(h)]^{-1} \approx \frac{1}{N} \sum_{i=1}^{N} [Z(x_i) - Z(x_{i+h})]^2$$

Rather than using the name similarity for this function, the variogram is reintroduced to represent the similarity function as follows:

$$2\gamma(h) = \frac{1}{N} \sum_{i=1}^{N} [Z(x_i) - Z(x_{i+h})]^2$$

which is the formal definition of the variogram. Writing this equation instead as

$$\gamma(h) = \frac{1}{2N} \sum_{i=1}^{N} [Z(x_i) - Z(x_{i+h})]^2$$

is known as the semivariogram. Occasionally in the literature, the terms variogram and semivariogram are used interchangeably. This is perhaps tolerable for practitioners familiar with geostatistics who understand the two notions well. For beginning practitioners, however, this careless use of terms can be confusing. Throughout the remainder of this text, the correct term, semivariogram, is used to accurately suggest its calculation method.

6.4.1 Obtaining Autocovariance from the Semivariogram

Restating the expression of the semivariogram as an expected value gives

$$\gamma(h) = \tfrac{1}{2} E[Z(x) - Z(x + h)]^2$$

and expanding gives

$$\gamma(h) = \tfrac{1}{2} [E[Z(x + h)]^2 - 2E[Z(x + h)Z(x)] + E[Z(x)]^2]$$

We use the expressions for variance and covariance in terms of mathematical expectation to write the semivariogram as

$$\gamma(h) = \tfrac{1}{2} \{ \text{var}[Z(x + h)] + \bar{Z}^2 - 2\,\text{cov}[Z(x + h)Z(x)] - 2\bar{Z}^2 \\ + \text{var}[Z(x)] + \bar{Z}^2 \}$$

Only the mean of Z is written by invoking the intrinsic hypothesis, and the mean value of Z cancels in this expression. By further invoking the intrinsic hypothesis to assume that the variance of Z is constant, the expression for the semivariogram becomes

$$\gamma(h) = \tfrac{1}{2} \{ 2\,\text{var}(Z) - 2\,\text{cov}[Z(x + h)Z(x)] \}$$

As was done earlier for the semivariogram, the intrinsic hypothesis is invoked such that $\text{cov}[Z(x + h)Z(x)] = \text{cov}(h)$. The expression for the semivariogram is consequently seen to be a function of autocovariance:

$$\gamma(h) = \text{var}(Z) - \text{cov}(h)$$

Restating this relationship for the covariance,

$$\text{cov}(h) = \text{var}(Z) - \gamma(h)$$

shows how information from the semivariogram is used to derive the matrix entries for the solution of kriging weights.

6.5 Calculating the Semivariogram

Before kriging can be applied for estimation of the value of a regionalized variable Z at unsampled locations, the spatial correlation (spatial autocovariance) of Z must be known. As shown in the previous section, spatial autocovariance is a function of the semivariogram. Furthermore, the semivariogram is a function only of lag distance h (by invoking the intrinsic hypothesis and assuming that this hypothesis is reasonably correct for a given regionalized variable Z).

Because the semivariogram is a function only of lag distance, knowledge of how the semivariogram changes with h is sought. This knowledge is used to evaluate, for a particular regionalized variable Z, the fundamental assumption implicit with kriging: that observations of Z separated by small lags are more likely to be similar (are associated with a greater probability of being similar), hence only the N closest sample locations to the estimation location are used to obtain an estimate. The semivariogram proves whether this is a correct assumption for a regionalized variable Z. If the assumption is found to be correct, the semivariogram defines how rapidly observations of Z change in value as distance of separation between observations (lag distance h) increases.

A graphical display of the semivariogram as a function of h demonstrates the spatial correlation (covariance) for a regionalized phenomenon (Figure 6.3).

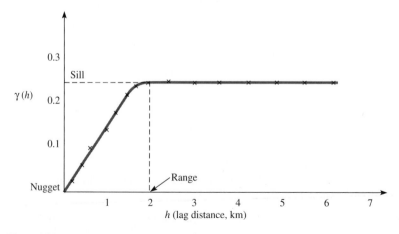

Figure 6.3   A spherical semivariogram.

The shape of this plot is an ideal representation of the semivariogram observed for many regionalized phenomena and is what is known as a spherical semivariogram shape. This is a rather bold statement. For someone newly introduced to geostatistics, this statement seems to be opposed to scientific intuition: How can this shape be described as typical when spatial phenomena never before analyzed using geostatistics are yet to be encountered? As a person's experience with geostatistics grows, a conclusion is reached that this shape is, indeed, observed for many regionalized phenomena. Why this is so is explained with the formal introduction of the spherical, semivariogram model (Section 6.7.1.1).

The plot in Figure 6.3 is the ultimate goal of a semivariogram analysis of sample data representing observations of a spatially distributed phenomenon Z. Perhaps visual inspection of the semivariogram equation

$$\gamma(h) = \frac{1}{2N} \sum_{i=1}^{N} [Z(x_i) - Z(x_{i+h})]^2$$

belies the relationship between the semivariogram and the graphical notion. Therefore, an example calculation is warranted. A notion of two-dimensional space with bounds is shown in Figure 6.4. Within this space, a spatial phenomenon Z occurs. Locations at which Z is observed discretely are shown by small circles. Beneath each circle is a number representing the value of Z at that

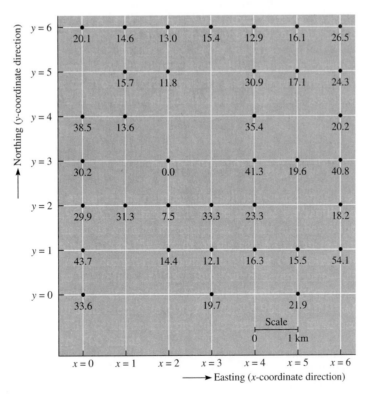

Figure 6.4 A two-dimensional spatial region within which a regionalized variable Z is observed discretely at locations shown by circles. East (x direction) and north (y direction) coordinates are as shown. Each observation location is associated with a data value.

location. Moreover, a scale is shown to define the lag distance between pairs of observations.

Minimum sampling resolution is 1 km (Figure 6.4). When solving the semivariogram equation, a small value of lag distance h is used initially and all pairs of observations $Z(x_i)$ and $Z(x_{i+h})$ physically separated by distances approximately equal to h are identified. Once this is done, the semivariogram equation is solved for that value of h. Lag h is then incremented to become $2h$ and the process is repeated *without using pairs of observations from the previous value of h.*

In this process, an additional constraint may be placed on the search for pairs $Z(x_i)$ and $Z(x_i + h)$ by treating the semivariogram function as a vector; that is, its calculation may be constrained to one spatial direction (north–south, east–west, and so on). To effect such a calculation, not only are pairs of observations of Z found that are separated by distances approximately equal to h, but orientation of the pairs must match, with some allowable and finite amount of variation, a particular desired direction. Directional semivariogram calculations are used to investigate whether or not a spatial phenomenon Z is associated with anisotropic spatial autocovariance; if so, kriging is able to accommodate this anisotropy. How plots of the directional semivariograms are interpreted to infer isotropic or anisotropic autocovariance is described subsequently.

First, we demonstrate how a semivariogram is calculated. An east–west directional semivariogram is computed using the data configuration of Figure 6.4. Thus, only pairs of data locations that are oriented parallel to the easting coordinate axis are used. Calculation begins by choosing an initial value for h, which is also the amount h that is incremented to develop the graph. This initial value of h is called the class size. In Figure 6.4, the class size is selected equal to the minimum sampling resolution, 1 km. Selection of a class size is somewhat arbitrary and subjective. A class size that is too large forces data pairs of widely varying separation distances into a single lag increment; a class size that is too small often results in a relatively few pairs contributing to any one lag increment, further resulting in a rather "noisy" plot of the semivariogram. A class size is selected by experimenting with the class size (increasing and decreasing its value) and noting its influence on the resultant plot of the semivariogram, as is done later (Section 6.6.1.3).

Once the class size of 1 is selected, all pairs are isolated such that

$$0 < h \leq 1$$

Allowing pairs of data observations to be chosen such that their lag distance falls into this range does not seem to make sense because these data are sampled at spatially regular locations (Figure 6.4), hence pairs can be found such that their lag distances exactly equal 1 km. This is true for this example, and regular sampling makes this demonstration easier to see and follow. In practice, however, most data are sampled at irregular locations (Figure 6.1), and a tolerance must be allowed for each increment of h to accommodate spatially irregular sampling patterns.

Returning to the calculation example of Figure 6.4, a search is conducted for pairs of observations $Z(x_i)$ and $Z(x_{i+h})$ satisfying (1) an east–west (horizontal) orientation and (2) lag distances falling into the previously stated interval.

Beginning in the upper left-hand corner we find and list acceptable pairs of data observations for this lag interval. Each pair of data locations is represented by a difference in data values (shown in Figure 6.4) enclosed in parentheses.

DEMONSTRATION

Class size = 1 km
For $h = 1$, such that $h =]0, 1]$:
Pairs:

$(20.1 - 14.6)$	$(14.6 - 13.0)$	$(13.0 - 15.4)$	$(15.4 - 12.9)$
$(12.9 - 16.1)$	$(16.1 - 26.5)$	$(15.7 - 11.8)$	$(30.9 - 17.1)$
$(17.1 - 24.3)$	$(38.5 - 13.6)$	$(41.3 - 19.6)$	$(19.6 - 40.8)$
$(29.9 - 31.3)$	$(31.3 - 7.5)$	$(7.5 - 33.3)$	$(33.3 - 23.3)$
$(14.4 - 12.1)$	$(12.1 - 16.3)$	$(16.3 - 15.5)$	$(15.5 - 54.1)$

Twenty ($N = 20$) pairs are found for lag $h = 1$.
Squared differences:

$$30.25 + \quad 2.56 + \quad 5.76 + \quad 6.25 +$$
$$10.24 + 108.16 + \quad 15.21 + \quad 190.44 +$$
$$51.84 + 620.01 + 470.89 + \quad 449.44 +$$
$$1.96 + 566.44 + 665.64 + \quad 100.00 +$$
$$5.29 + \quad 17.64 + \quad 0.64 + 1489.96$$

$$\sum \text{diff}^2 = 4808.62, \qquad \therefore \gamma(1) = \frac{4808.62}{2(20)} = 120.22$$

Lag distance h is incremented such that $h = h + h = 2h = 2$ km. A search is conducted for pairs oriented in the east–west direction. This time, however, pairs are identified whose lag distances fall in to the following interval:

$$1 < h \le 2$$

Notice that all pairs used for the $h = 1$ km calculation and search are not included in the new lag interval.

Pairs of observations oriented parallel to the east–west direction and with lag distances of approximately 2 km are now listed. A difference in data values within parentheses represents a pair.

DEMONSTRATION

$h = 2$ km such that $h =]1, 2]$:
Pairs:

$(20.1 - 13.0)$	$(14.6 - 15.4)$	$(13.0 - 12.9)$	$(15.4 - 16.1)$
$(12.9 - 26.5)$	$(11.8 - 30.9)$	$(30.9 - 24.3)$	$(35.4 - 20.2)$
$(30.2 - 0.0)$	$(0.0 - 41.3)$	$(41.3 - 40.8)$	$(29.9 - 7.5)$
$(31.3 - 33.3)$	$(7.5 - 23.3)$	$(23.3 - 18.2)$	$(43.7 - 14.4)$
$(14.4 - 16.3)$	$(12.1 - 15.5)$	$(16.3 - 54.1)$	$(19.7 - 21.9)$

For this increment of h, $N = 20$ pairs are found.
Squared differences:

$$
\begin{array}{rrrr}
50.41 + & 0.64 + & 0.01 + & 0.49 + \\
184.96 + & 364.81 + & 43.56 + & 231.04 + \\
912.04 + & 1705.69 + & 0.25 + & 501.76 + \\
4.00 + & 249.64 + & 26.01 + & 858.49 + \\
3.61 + & 11.56 + & 1428.84 + & 4.84
\end{array}
$$

$$
\sum \text{diff}^2 = 6582.65, \qquad \therefore \gamma(2) = \frac{6582.65}{2(20)} = 164.57
$$

Table 6.1 Summary of the East–West Directional Semivariogram Calculation for the Data of Figure 6.4

Interval of h	Number of Pairs	$\gamma(h)$
1.00–2.00	20	120.22
2.01–3.00	20	164.57
3.01–4.00	15	151.23
4.01–5.00	12	194.87
5.01–6.00	8	93.20
6.01–7.00	14	106.31

Calculating a semivariogram by hand is a tedious exercise. But in doing the calculation by hand, even for a few lag intervals, the semivariogram equation becomes easier to understand. Results for larger lag distances are shown in Table 6.1, and in Figure 6.5 a plot is developed based on these results. This plot shows a semivariogram displaying increasing behavior from smaller lags to larger lags, until a lag is reached at which the semivariogram becomes constant or begins to decrease. The lag distance at which this occurs (i.e., the semivariogram stops increasing) is known as the range. The value of the

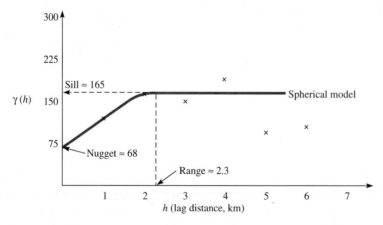

Figure 6.5 A plot of results for an east–west semivariogram calculation for data (Figure 6.4).

semivariogram at the range is known as the sill, and often approximately equals the statistical variance of Z. By extrapolating the plot of the semivariogram back to the origin (i.e., for zero lag), the nugget value is found, which is that value of the semivariogram associated with a zero lag distance. The nugget value is (1) a reflection of white (random) noise, perhaps caused by sampling error, or (2) a function of microspatial autocovariance at a scale below the minimum sampling resolution.

6.6 FGAM: A Portable FORTRAN-77 Program for Semivariogram Calculation and Plotting

A relatively simple, yet powerful semivariogram calculation program is presented as an aid to experimenting with geostatistical data analysis. This program, called FGAM (Figure 6.6, page 207), is a modification of one presented in Carr (1987). The following algorithm is employed:

1. Select the class size h.
2. Evaluate up to 20 increments of h.
3. For all increments simultaneously, define an outer loop on i, $i = 1$ to total number of data locations.
4. Define an inner loop on j, $j = i + 1$ to total number of data locations. (*Note:* The loop on j enables the comparison of distances between all other locations and the ith location. As i increases, j progresses through successively smaller intervals. The reason for this is not obvious. When i is 1, for instance, j ranges from 2 to the total number of data locations. When i increments to 2, j need not equal 1, because this pair was evaluated when i was 1 and j was 2. Furthermore, j need never equal i because the distance between a point and itself is zero, and this does not yield useful information for the calculation.)
5. Compute the distance between locations i and j using
$$dist_{ij} = \sqrt{(x_i - x_j)^2 + (y_i - y_j)^2}$$
6. Determine to which of the 20 increments of class size this pair belongs: increment = (distance/class size) rounded to an integer value.
7. If the increment (an integer) is greater than 20 (the largest allowable increment of h allowed by the algorithm), discard the data pair.
8. If directional semivariograms are calculated, determine to which direction this pair belongs:
   a. Compute difx = $x_i - x_j$.
   b. Compute dify = $y_i - y_j$.
   c. Compute ratio = difx/dify
   d. Compute direction angle = arctan(ratio). If difx < 0 and dify > 0, then angle = angle + 180. Compute angle as k = integer(angle/y) + 1, where y is 45 for a four-directional calculation, or y is 22.5 for an eight-directional calculation.
9. Place the squared difference $[Z(x_i) - Z(x_j)]^2$ into the proper array location:
   a. $\gamma(k, \text{increment}) = \gamma(k, \text{increment}) + $ squared difference.
   b. ntotal(k, increment) = ntotal(k, increment) + 1.
10. End loop j; end loop i.

11. Complete the calculation for each of the 20 increments of h using

$$\gamma(k, \text{increment}) = \gamma(k, \text{increment})/[2 * \text{ntotal}(k, \text{increment})]$$

The program FGAM employs this algorithm. A short user's guide is provided at the beginning of the program listing (Figure 6.6). The selection criteria for the various options allowed by this program are demonstrated in subsequent examples throughout this chapter.

6.6.1 Examples

An application of the program FGAM is made to the dataset NVSIM.DAT (Appendix A). The histogram for these data (Figure 6.7) is computed using the program HISTO (Chapter 4). These data conform closely to a normal distribution (Figure 6.8).

A reason for using the program HISTO prior to geostatistical analysis is to identify the presence of unusual data values in the data set. In some cases measurement error, recording error, accidents, and so on, can result in unusual data values. Of course, unusual data values may be an inherent attribute of the phenomenon under study. Unusual data values have a detrimental influence on geostatistical analysis, especially on the computation of the semivariogram. For this reason, unusual data values must be identified and a decision must be made about how to treat them. More on this topic is given later in the section on data transformation.

Another reason for computing a histogram for spatial data, prior to geostatistical analysis, is to identify the most likely distribution of the data. For normally distributed data, kriging is the best linear, unbiased estimator for the data. Kriging errors are normally distributed, and kriging has minimum variance of estimation error. For distinctly nonnormal data sets, kriging may still be used, but it is not necessarily the best linear, unbiased estimator. A data transformation, two of which are discussed later, may be considered for distinctly nonnormal data sets if the deviation of the data set from a normal distribution is determined to be problematic for estimation.

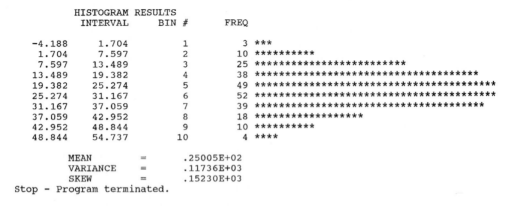

```
        HISTOGRAM RESULTS
          INTERVAL      BIN #        FREQ

 -4.188      1.704        1           3 ***
  1.704      7.597        2          10 **********
  7.597     13.489        3          25 *************************
 13.489     19.382        4          38 **************************************
 19.382     25.274        5          49 *************************************************
 25.274     31.167        6          52 ****************************************************
 31.167     37.059        7          39 ***************************************
 37.059     42.952        8          18 ******************
 42.952     48.844        9          10 **********
 48.844     54.737       10           4 ****

          MEAN       =     .25005E+02
          VARIANCE   =     .11736E+03
          SKEW       =     .15230E+03
Stop - Program terminated.


0;37;44m C:\TEXT=
```

Figure 6.7 A histogram of the NVSIM.DAT data set. Notice that these data appear to be normally distributed (a subjective inference).

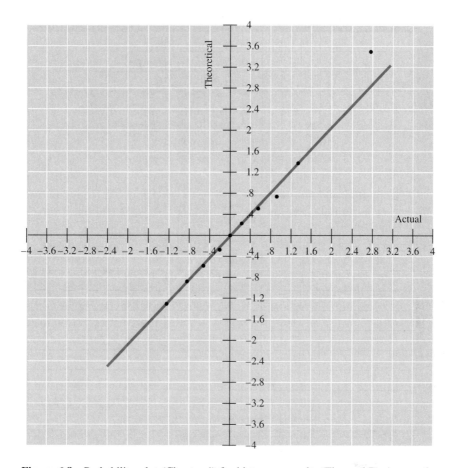

Figure 6.8   Probability plot (Chapter 4) for histogram results (Figure 6.7). A normal
distribution model represents these data well.

For now, though, we are interested only in obtaining the semivariogram,
and histogram analysis to identify the presence of extreme, or unusual, data
values is critical for this objective. No unusual, outlier values are identified for
the NVSIM.DAT data set, and these data conform well to a normal distribution.
Semivariogram analysis is attempted without considering a data transform.
Directional calculations are attempted first.

6.6.1.1 Directional Calculations

One purpose of the semivariogram calculation is to identify anisotropic spatial
correlation behavior. Earlier in this chapter, directional semivariogram calculation was introduced for hand calculation. More detail is given here to show
how to interpret directional semivariogram calculations to identify isotropic or
anisotropic spatial correlation.

Directional calculations are made in four or eight spatial directions. For four
directions

$$0.0 = \text{east–west}$$
$$45.0 = \text{northeast–southwest}$$
$$90.0 = \text{north–south}$$
$$135.0 = \text{northwest–southeast}$$

For eight directions

$$0.0 = \text{east–west}$$
$$22.5 = \text{east–northeast}$$
$$45.0 = \text{northeast}$$
$$67.5 = \text{north–northeast}$$
$$90.0 = \text{north–south}$$
$$112.5 = \text{north–northwest}$$
$$135.0 = \text{northwest}$$
$$157.5 = \text{west–northwest}$$

What is the difference between analyzing four or eight directions? None, really, except that eight directional analysis yields more directional resolution. But rather large data sets are usually required to yield enough information for the eight directional calculations.

Once these semivariograms are computed, the range for each is plotted as a vector whose orientation is the same as that for the semivariogram; the length of the vector is equal to the range. If the ranges are approximately the same for the different directions, spatial isotropy is identified (Figure 6.9a); if the ranges are different, a spatial, geometric aniostropy is identified (Figure 6.9b). Spatial isotropy is identified for the NVSIM.DAT data set (Figures 6.10 through

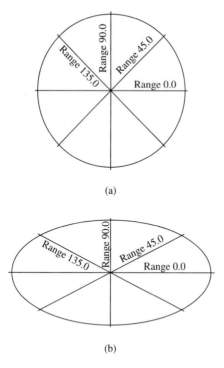

(a)

(b)

Figure 6.9 A two-part figure showing (a) an isotropic spatial structure because the ranges of the semivariograms for the different directions are approximately the same, and (b) an anisotropic spatial structure associated with semivariograms having differing ranges for the different spatial directions. Anisotropy is inferred in this case, because the spatial correlation extends over a greater distance in the east–west (0.0) direction than in the north–south (90.0) direction.

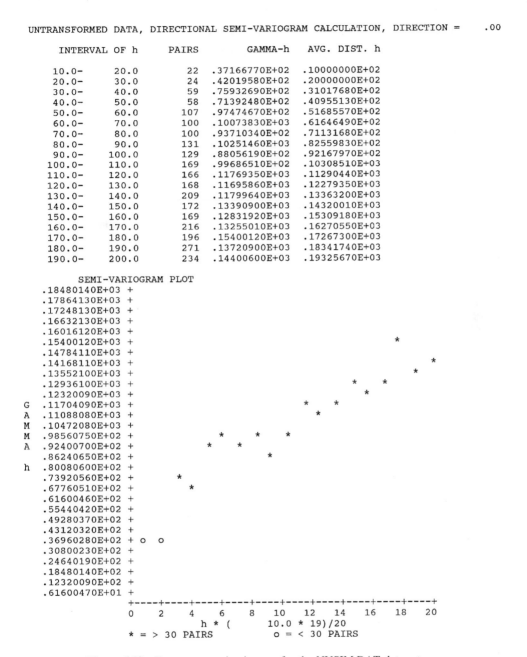

UNTRANSFORMED DATA, DIRECTIONAL SEMI-VARIOGRAM CALCULATION, DIRECTION =   .00

INTERVAL OF h	PAIRS	GAMMA-h	AVG. DIST. h
10.0- 20.0	22	.37166770E+02	.10000000E+02
20.0- 30.0	24	.42019580E+02	.20000000E+02
30.0- 40.0	59	.75932690E+02	.31017680E+02
40.0- 50.0	58	.71392480E+02	.40955130E+02
50.0- 60.0	107	.97474670E+02	.51685570E+02
60.0- 70.0	100	.10073830E+03	.61646490E+02
70.0- 80.0	100	.93710340E+02	.71131680E+02
80.0- 90.0	131	.10251460E+03	.82559830E+02
90.0- 100.0	129	.88056190E+02	.92167970E+02
100.0- 110.0	169	.99686510E+02	.10308510E+03
110.0- 120.0	166	.11769350E+03	.11290440E+03
120.0- 130.0	168	.11695860E+03	.12279350E+03
130.0- 140.0	209	.11799640E+03	.13363200E+03
140.0- 150.0	172	.13390900E+03	.14320010E+03
150.0- 160.0	169	.12831920E+03	.15309180E+03
160.0- 170.0	216	.13255010E+03	.16270550E+03
170.0- 180.0	196	.15400120E+03	.17267300E+03
180.0- 190.0	271	.13720900E+03	.18341740E+03
190.0- 200.0	234	.14400600E+03	.19325670E+03

```
          SEMI-VARIOGRAM PLOT
   .18480140E+03 +
   .17864130E+03 +
   .17248130E+03 +
   .16632130E+03 +
   .16016120E+03 +
   .15400120E+03 +                                              *
   .14784110E+03 +
   .14168110E+03 +                                                 *
   .13552100E+03 +                                              *
   .12936100E+03 +                                    *     *
   .12320090E+03 +                                       *
 G .11704090E+03 +                              *     *
 A .11088080E+03 +                                 *
 M .10472080E+03 +
 M .98560750E+02 +                   *     *     *
 A .92400700E+02 +                *     *
   .86240650E+02 +                         *
 h .80080600E+02 +
   .73920560E+02 +          *
   .67760510E+02 +             *
   .61600460E+02 +
   .55440420E+02 +
   .49280370E+02 +
   .43120320E+02 +
   .36960280E+02 + o   o
   .30800230E+02 +
   .24640190E+02 +
   .18480140E+02 +
   .12320090E+02 +
   .61600470E+01 +
                 +----+----+----+----+----+----+----+----+----+----+
                 0    2    4    6    8   10   12   14   16   18   20
                        h * (        10.0 * 19)/20
                 * = > 30 PAIRS         o = < 30 PAIRS
```

Figure 6.10   East–west semivariogram for the NVSIM.DAT data set.

6.13) because the ranges for the four directional calculations are about the same.

Another type of anisotropy is zonal anisotropy, in which the ranges of semivariograms over all spatial directions are approximately the same, but the sill values are different. Zonal anisotropy is discussed in some detail in Journel and Huijbregts (1978). Programs for kriging in this chapter consider only geometric anisotropy.

UNTRANSFORMED DATA, DIRECTIONAL SEMI-VARIOGRAM CALCULATION, DIRECTION = 45.00

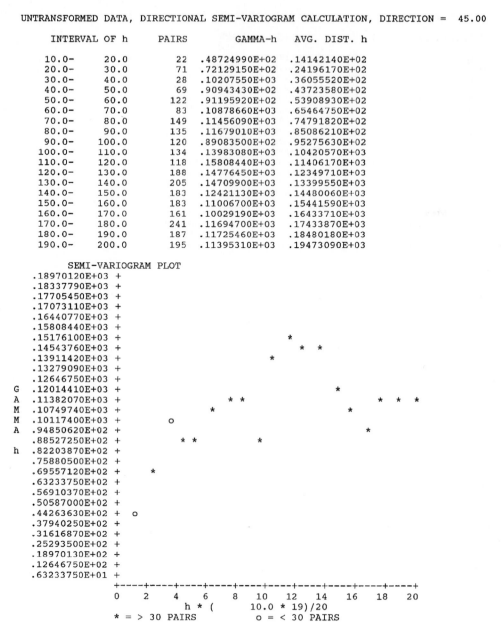

INTERVAL OF h		PAIRS	GAMMA-h	AVG. DIST. h
10.0-	20.0	22	.48724990E+02	.14142140E+02
20.0-	30.0	71	.72129150E+02	.24196170E+02
30.0-	40.0	28	.10207550E+03	.36055520E+02
40.0-	50.0	69	.90943430E+02	.43723580E+02
50.0-	60.0	122	.91195920E+02	.53908930E+02
60.0-	70.0	83	.10878660E+03	.65464750E+02
70.0-	80.0	149	.11456090E+03	.74791820E+02
80.0-	90.0	135	.11679010E+03	.85086210E+02
90.0-	100.0	120	.89083500E+02	.95275630E+02
100.0-	110.0	134	.13983080E+03	.10420570E+03
110.0-	120.0	118	.15808440E+03	.11406170E+03
120.0-	130.0	188	.14776450E+03	.12349710E+03
130.0-	140.0	205	.14709900E+03	.13399550E+03
140.0-	150.0	183	.12421130E+03	.14480060E+03
150.0-	160.0	183	.11006700E+03	.15441590E+03
160.0-	170.0	161	.10029190E+03	.16433710E+03
170.0-	180.0	241	.11694700E+03	.17433870E+03
180.0-	190.0	187	.11725460E+03	.18480180E+03
190.0-	200.0	195	.11395310E+03	.19473090E+03

SEMI-VARIOGRAM PLOT

Figure 6.11 Northeast–southwest (45 degrees) semivariogram for the NVSIM.DAT data set.

6.6.1.2 Omnidirectional Calculation

In the previous section, a spatial isotropy is identified for the NVSIM.DAT data set. For data associated with isotropic spatial behavior, an omnidirectional, or average, semivariogram calculation is appropriate. Such a semivariogram is computed by using the algorithm listed earlier, but skipping the step associated with determining the direction to which each pair belongs. Instead, all directions are combined for any given increment.

```
UNTRANSFORMED DATA, DIRECTIONAL SEMI-VARIOGRAM CALCULATION, DIRECTION = 90.00

   INTERVAL OF h       PAIRS        GAMMA-h      AVG. DIST. h

    10.0-    20.0        18      .26621970E+02   .10000000E+02
    20.0-    30.0        20      .48959720E+02   .20000000E+02
    30.0-    40.0        74      .69006320E+02   .30986840E+02
    40.0-    50.0        55      .71874220E+02   .40917700E+02
    50.0-    60.0       104      .13149810E+03   .51806170E+02
    60.0-    70.0       103      .12537970E+03   .61699840E+02
    70.0-    80.0        95      .14361560E+03   .71493160E+02
    80.0-    90.0       132      .14333730E+03   .82496700E+02
    90.0-   100.0       127      .16781230E+03   .92248850E+02
   100.0-   110.0       180      .16036000E+03   .10338970E+03
   110.0-   120.0       163      .16453010E+03   .11262490E+03
   120.0-   130.0       179      .16213760E+03   .12261300E+03
   130.0-   140.0       188      .13953010E+03   .13404900E+03
   140.0-   150.0       173      .15004540E+03   .14333520E+03
   150.0-   160.0       169      .12683280E+03   .15304980E+03
   160.0-   170.0       181      .11930590E+03   .16264260E+03
   170.0-   180.0       195      .12978330E+03   .17243210E+03
   180.0-   190.0       226      .13875990E+03   .18324060E+03
   190.0-   200.0       185      .10944400E+03   .19357170E+03

        SEMI-VARIOGRAM PLOT
  .20137480E+03 +
  .19466230E+03 +
  .18794980E+03 +
  .18123730E+03 +
  .17452480E+03 +
  .16781230E+03 +
  .16109980E+03 +                         *      * *
  .15438730E+03 +                       *
  .14767490E+03 +                              *
  .14096240E+03 +              *  *
  .13424990E+03 +                          *        *
G .12753740E+03 +        *                              *
A .12082490E+03 +          *
M .11411240E+03 +                              *
M .10739990E+03 +                            *
A .10068740E+03 +                                 *
  .93974910E+02 +
h .87262410E+02 +
  .80549920E+02 +
  .73837430E+02 +
  .67124930E+02 +     * *
  .60412440E+02 +
  .53699950E+02 +
  .46987450E+02 +  o
  .40274960E+02 +
  .33562470E+02 +
  .26849970E+02 +
  .20137480E+02 + o
  .13424990E+02 +
  .67124930E+01 +
                +----+----+----+----+----+----+----+----+----+----+
                0    2    4    6    8   10   12   14   16   18   20
                     h * (        10.0 * 19)/20
                * = > 30 PAIRS          o = < 30 PAIRS
```

Figure 6.12 North–south semivariogram for the NVSIM.DAT data set.

An omnidirectional semivariogram for the NVSIM.DAT data set in Figure 6.14 reveals that more pairs are associated with each increment of lag in comparison to the directional calculations shown in Figures 6.10 through 6.13. This is an expected outcome, because for each lag increment, the omnidirectional calculation is based on pairs of data locations for all spatial directions.

```
UNTRANSFORMED DATA, DIRECTIONAL SEMI-VARIOGRAM CALCULATION, DIRECTION = 135.00

      INTERVAL OF h        PAIRS         GAMMA-h      AVG. DIST. h

       10.0-     20.0        20        .19277480E+02   .14142140E+02
       20.0-     30.0        69        .57492280E+02   .24592770E+02
       30.0-     40.0        47        .63551630E+02   .36055530E+02
       40.0-     50.0        70        .86260590E+02   .44032890E+02
       50.0-     60.0        74        .11777320E+03   .53760170E+02
       60.0-     70.0        78        .10471570E+03   .65361080E+02
       70.0-     80.0       142        .10303150E+03   .75160160E+02
       80.0-     90.0       137        .13456030E+03   .84764880E+02
       90.0-    100.0       132        .10541940E+03   .95384670E+02
      100.0-    110.0       164        .10793140E+03   .10446240E+03
      110.0-    120.0       129        .96618490E+02   .11400580E+03
      120.0-    130.0       183        .87653220E+02   .12353540E+03
      130.0-    140.0       201        .83751080E+02   .13388980E+03
      140.0-    150.0       172        .98545270E+02   .14483460E+03
      150.0-    160.0       203        .10061750E+03   .15446450E+03
      160.0-    170.0       152        .10225240E+03   .16452200E+03
      170.0-    180.0       233        .12463940E+03   .17474570E+03
      180.0-    190.0       203        .12850650E+03   .18507100E+03
      190.0-    200.0       201        .12695570E+03   .19471700E+03

            SEMI-VARIOGRAM PLOT
    .16147240E+03  +
    .15609000E+03  +
    .15070760E+03  +
    .14532510E+03  +
    .13994270E+03  +
    .13456030E+03  +                          *
    .12917790E+03  +
    .12379550E+03  +                                       *  *  *
    .11841310E+03  +
    .11303070E+03  +              *
    .10764830E+03  +                              *
  G .10226580E+03  +                  *  *    *
  A .96883430E+02  +                                   *  *  *
  M .91501010E+02  +                            *
  M .86118610E+02  +           *                    *
  A .80736190E+02  +                                  *
    .75353780E+02  +
  h .69971370E+02  +
    .64588950E+02  +
    .59206540E+02  +        *
    .53824130E+02  +      *
    .48441720E+02  +
    .43059300E+02  +
    .37676890E+02  +
    .32294480E+02  +
    .26912060E+02  +
    .21529650E+02  +
    .16147240E+02  +  o
    .10764830E+02  +
    .53824130E+01  +
                   +----+----+----+----+----+----+----+----+----+----+
                   0    2    4    6    8   10   12   14   16   18   20
                             h * (      10.0 * 19)/20
               * = > 30 PAIRS              o = < 30 PAIRS
Stop - Program terminated.
```

Figure 6.13 Northwest–southeast (135 degrees) semivariogram for the NVSIM.DAT data set.

6.6.1.3 Influence of Class Size on the Semivariogram Calculation

Class size is important when using the program FGAM (in this program the option for selecting class size is called XCLASS, Figure 6.6). How is the class size determined? One possibility, as mentioned earlier in this chapter, is to use the minimum sampling resolution, but this may result in too few data pairs for

UNTRANSFORMED DATA, OMNIDIRECTIONAL SEMI-VARIOGRAM CALCULATION

INTERVAL OF h		PAIRS	GAMMA-h	AVG. DIST. h
10.0-	20.0	82	.33589800E+02	.12121590E+02
20.0-	30.0	184	.60194560E+02	.23341430E+02
30.0-	40.0	208	.74190080E+02	.32823200E+02
40.0-	50.0	252	.80980930E+02	.42559940E+02
50.0-	60.0	407	.10797720E+03	.52760010E+02
60.0-	70.0	364	.11039850E+03	.63328210E+02
70.0-	80.0	486	.11258140E+03	.73501500E+02
80.0-	90.0	535	.12439510E+03	.83746350E+02
90.0-	100.0	508	.11274960E+03	.93758190E+02
100.0-	110.0	647	.12697050E+03	.10375110E+03
110.0-	120.0	576	.13450210E+03	.11330880E+03
120.0-	130.0	718	.12881890E+03	.12312160E+03
130.0-	140.0	803	.12189570E+03	.13388690E+03
140.0-	150.0	700	.12667230E+03	.14405360E+03
150.0-	160.0	724	.11559150E+03	.15380140E+03
160.0-	170.0	710	.11537250E+03	.16344830E+03
170.0-	180.0	865	.13030890E+03	.17364130E+03
180.0-	190.0	887	.13140550E+03	.18404320E+03
190.0-	200.0	815	.12476510E+03	.19404140E+03

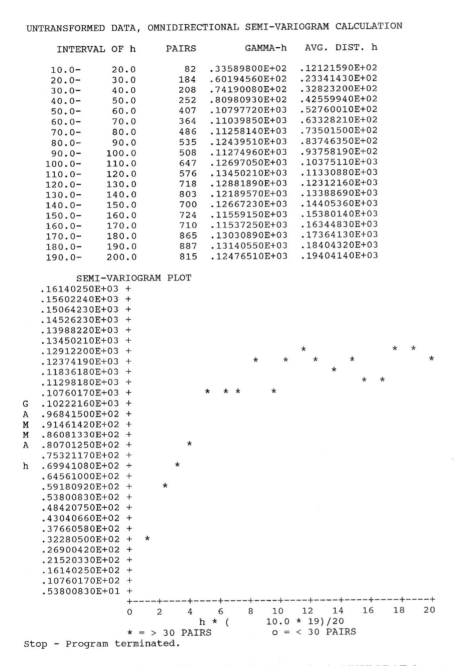

Figure 6.14   Omnidirectional semivariogram for the NVSIM.DAT data set.

smaller lags. Another possibility is to determine the maximum separation distance between any two data locations in a data set and divide this quantity by
20 because FGAM computes up to 20 increments of the class size. Of course,
a guess can be made for the class size. A good selection of class size is known
only by interpreting the calculation results from FGAM.

An omnidirectional semivariogram calculation was shown in the previous section. For the first increment of h, in the interval 10 to 20 km, 82 pairs are found, a sizeable value. Notice, however, that more than 800 pairs are found for the final increment.

Selection of a proper class size is an empirical exercise. A class size of 10 km is used to develop Figure 6.14. Because the number of pairs found for the first increment of lag is 1/10 of that found for the final increment, a class size larger than 10 km may be warranted. Even so, the plot in Figure 6.14 shows a definite spherical shape and is not erratic (i.e., $\gamma(h)$ does not vary widely with h), so the class size of 10 km is acceptable. This is, of course, a subjective assessment.

An experiment is presented to demonstrate the empirical selection of class size. In this experiment, FGAM is applied to the data set LONGB.DAT, an earthquake ground motion data set from the 1933 Long Beach, California (USA), earthquake (Appendix A). An omnidirectional calculation is selected.

Histogram analysis shows that extreme data values, known as outliers, are not a problem with these data (Figure 6.15). Inspection of the LONGB.DAT data set shows that some locations are less than 1 km apart. Therefore, the initial guess at the class size is 1 km. The omnidirectional calculation based on this class size is shown in Figure 6.16. A definite spherical shape is noted in the semivariogram, having a range of 8 km, a sill of 0.7, and a nugget of 0.06. But relatively few pairs are found for each lag increment.

On further inspection of the data set, it appears that some data locations are more than 200 km apart. Therefore, a second guess at class size is the value of 200/20, or 10 km. The omnidirectional calculation for this class size is shown in Figure 6.17. Many more pairs for each lag increment are obtained for this class size. But the semivariogram has a shape similar to that shown in Figure 6.16, albeit the sill, nugget, and range are larger. In fact, the sill value (Figure 6.17) is more similar to statistical variance (Figure 6.15).

In this application, an important discovery is made by accident: the notion of self-similar (fractal; Chapter 11) scales of spatial correlation. A class size of 1 km yields a spherical semivariogram with a shape identical to that obtained

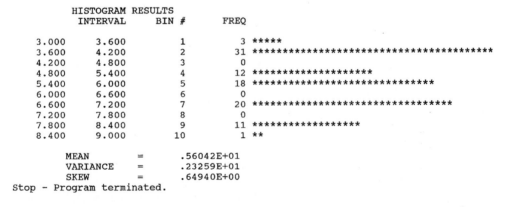

Figure 6.15 Histogram for the LONGB.DAT data set. These data are Mercalli intensity values for the 1933 Long Beach, California (USA), earthquake (Richter magnitude 6.3).

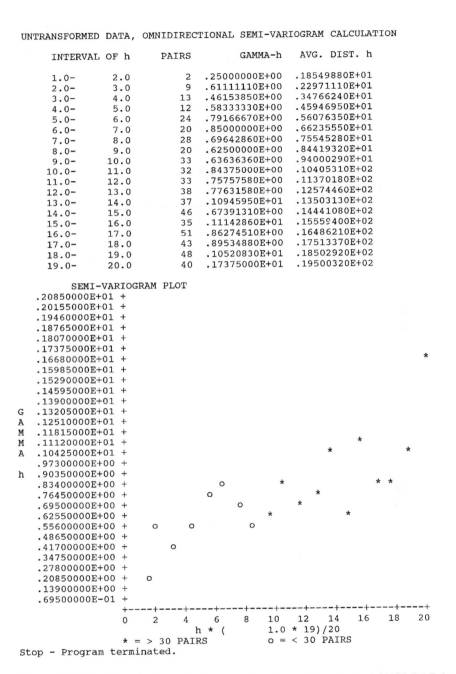

UNTRANSFORMED DATA, OMNIDIRECTIONAL SEMI-VARIOGRAM CALCULATION

INTERVAL OF h		PAIRS	GAMMA-h	AVG. DIST. h
1.0-	2.0	2	.25000000E+00	.18549880E+01
2.0-	3.0	9	.61111110E+00	.22971110E+01
3.0-	4.0	13	.46153850E+00	.34766240E+01
4.0-	5.0	12	.58333330E+00	.45946950E+01
5.0-	6.0	24	.79166670E+00	.56076350E+01
6.0-	7.0	20	.85000000E+00	.66235550E+01
7.0-	8.0	28	.69642860E+00	.75545280E+01
8.0-	9.0	20	.62500000E+00	.84419320E+01
9.0-	10.0	33	.63636360E+00	.94000290E+01
10.0-	11.0	32	.84375000E+00	.10405310E+02
11.0-	12.0	33	.75757580E+00	.11370180E+02
12.0-	13.0	38	.77631580E+00	.12574460E+02
13.0-	14.0	37	.10945950E+01	.13503130E+02
14.0-	15.0	46	.67391310E+00	.14441080E+02
15.0-	16.0	35	.11142860E+01	.15559400E+02
16.0-	17.0	51	.86274510E+00	.16486210E+02
17.0-	18.0	43	.89534880E+00	.17513370E+02
18.0-	19.0	48	.10520830E+01	.18502920E+02
19.0-	20.0	40	.17375000E+01	.19500320E+02

```
                    SEMI-VARIOGRAM PLOT
    .20850000E+01 +
    .20155000E+01 +
    .19460000E+01 +
    .18765000E+01 +
    .18070000E+01 +
    .17375000E+01 +
    .16680000E+01 +                                        *
    .15985000E+01 +
    .15290000E+01 +
    .14595000E+01 +
    .13900000E+01 +
  G .13205000E+01 +
  A .12510000E+01 +
  M .11815000E+01 +
  M .11120000E+01 +                            *
  A .10425000E+01 +                         *        *
    .97300000E+00 +
  h .90350000E+00 +
    .83400000E+00 +               o        *           * *
    .76450000E+00 +            o                *
    .69500000E+00 +                o        *
    .62550000E+00 +                    *          *
    .55600000E+00 +    o     o         o
    .48650000E+00 +
    .41700000E+00 +       o
    .34750000E+00 +
    .27800000E+00 +
    .20850000E+00 +    o
    .13900000E+00 +
    .69500000E-01 +
                  +----+----+----+----+----+----+----+----+----+----+
                  0    2    4    6    8   10   12   14   16   18   20
                            h * (      1.0 * 19)/20
                  * = > 30 PAIRS          o = < 30 PAIRS
Stop - Program terminated.
```

Figure 6.16   Omnidirectional semivariogram, class size = 1 km, for the LONGB.DAT data set.

using a class size of 10 km, except the range, nugget, and sill are smaller. Earthquake ground motion is composed of a suite of waveforms (Chapter 9). A smaller class size in the semivariogram calculation is able to capture smaller scale (higher frequency) ground motion, whereas the larger class size provides more averaging over all wavelengths. If a regionalized phenomenon Z is known or suspected to be composed of several components, each having a different

```
          UNTRANSFORMED DATA, OMNIDIRECTIONAL SEMI-VARIOGRAM CALCULATION

            INTERVAL OF h        PAIRS          GAMMA-h        AVG. DIST. h

                .0-     10.0       161       .67080750E+00     .67225890E+01
              10.0-     20.0       403       .97890820E+00     .15276740E+02
              20.0-     30.0       577       .15545930E+01     .25204120E+02
              30.0-     40.0       616       .20657470E+01     .34824840E+02
              40.0-     50.0       542       .24750920E+01     .44829710E+02
              50.0-     60.0       465       .27666670E+01     .54763900E+02
              60.0-     70.0       384       .27278650E+01     .65085870E+02
              70.0-     80.0       293       .29027300E+01     .74682280E+02
              80.0-     90.0       265       .30622640E+01     .84881390E+02
              90.0-    100.0       225       .29488890E+01     .94826060E+02
             100.0-    110.0       171       .27222220E+01     .10498510E+03
             110.0-    120.0       130       .25846150E+01     .11495900E+03
             120.0-    130.0       114       .29078950E+01     .12457190E+03
             130.0-    140.0        90       .35055560E+01     .13487200E+03
             140.0-    150.0        58       .33706900E+01     .14468180E+03
             150.0-    160.0        34       .40882350E+01     .15471680E+03
             160.0-    170.0        19       .36052630E+01     .16439130E+03
             170.0-    180.0        11       .55454550E+01     .17409220E+03
             180.0-    190.0         2       .10250000E+02     .18251740E+03

                SEMI-VARIOGRAM PLOT
       .12300000E+02 +
       .11890000E+02 +
       .11480000E+02 +
       .11070000E+02 +
       .10660000E+02 +
       .10250000E+02 +
       .98400000E+01 +                                                     o
       .94300000E+01 +
       .90200000E+01 +
       .86100010E+01 +
       .82000000E+01 +
   G   .77900000E+01 +
   A   .73800000E+01 +
   M   .69700000E+01 +
   M   .65600000E+01 +
   A   .61500000E+01 +
       .57400000E+01 +
   h   .53300000E+01 +                                               o
       .49200000E+01 +
       .45100000E+01 +
       .41000000E+01 +
       .36900000E+01 +                                         *
       .32800000E+01 +                                   *   *     o
       .28700000E+01 +            *    *  *         *
       .24600000E+01 +      *   *  *          *   *
       .20500000E+01 +    *
       .16400000E+01 +
       .12300000E+01 +  *
       .82000000E+00 + *
       .41000000E+00 +*
                      +----+----+----+----+----+----+----+----+----+----+
                      0    2    4    6    8   10   12   14   16   18   20
                                     h * (      10.0 * 19)/20
                      * = > 30 PAIRS             o = < 30 PAIRS
   Stop - Program terminated.
```

Figure 6.17   Omnidirectional semivariogram, class size = 10 km, for the LONGB.DAT data set.

range (in the semivariogram sense), this example shows how, by using differing values for class size, to resolve the spatial correlation of each component.

In summary, the calculation of a semivariogram is an experimental and subjective process. Class size is an important influence on semivariogram calculation. Optimally, a class size should be found whereby approximately the same number of pairs result for each lag increment. For some data sets, such a class size cannot be found, most likely due to sampling geometry. For these data sets a class size should be sought yielding an interpretable semivariogram (i.e., showing a clear sill value and range).

Sometimes practitioners use a class size equal to the minimum grid spacing to be used in subsequent kriging. This is an alternative to the three possible choices for class size noted earlier. No definitive procedure for determining the class size exists, so one practitioner is just as right as another when choosing the class size. This does demonstrate, though, why this chapter is subtitled "The Art of Spatial Analysis" because semivariogram calculation and interpretation is partly an art.

6.6.1.4 Data Transformations

If the data distribution is not examined, vis-à-vis a histogram, prior to geostatistical analysis, severe errors may occur. For example, the application of kriging to a strongly skewed data set will result in estimates whose distribution is more normal. In other words, the distribution of the estimates will not match that of the original data. This is a poor result and is an example of analysis-distorting results, a situation to avoid.

For data associated with decidedly nonnormal distributions, a data transformation should be considered. In some instances, the data may conform to a lognormal distribution. If so, a probability plot (Chapter 4) developed using z scores for natural logarithms of the data plotted versus theoretical z scores should yield a straight line. In computing z scores of the natural logarithms of the data, the mean log value is simply subtracted from each data log value, and this difference is divided by the standard deviation of the log values. Theoretical z scores are those for a normal distribution (Chapter 4).

For data conforming to a lognormal distribution, changing each data value to its natural logarithm yields a normally distributed data set to which kriging may be applied. Once estimates of the logarithms are obtained from kriging, these estimates are reverse-transformed back to values in the original data range. This back transform for lognormal data is tricky and is discussed subsequently. Notice that the program FGAM has an option for log transformation that allows the program to be adapted to lognormally distributed data.

In addition to parametric transforms, such as the lognormal transform, where the transformation yields a desired distribution, a nonparametric transformation may be useful. A nonparametric transformation is one that is independent of data distribution assumptions.

One nonparametric data transform is the indicator transform (Journel, 1983). This is a simple transform, given by

$$i(x) = 1 \quad \text{if } Z(x) \leq C, \quad \text{else } i(x) = 0$$

where C is some cutoff or threshold value intermediate between minimum and maximum data values. If the semivariogram for Z appears to be noisy (erratic, difficult to interpret), an indicator transform produced using C equal to the median data value often yields a spherical semivariogram. Carr and others (1985) show how useful the indicator transform can be for such a purpose and for data filtering in general.

Each transformed value $i(x)$ is the probability $p(x)$ in the range $[0, 1]$ that Z is less than or equal to C at location x. If kriging is applied to the transformed data, each estimate of i is likewise interpreted to be the probability that Z is less than or equal to C at the estimation location.

This type of kriging has some interesting applications. For example, in earthquake hazard analysis, if C is selected to be equal to the "damage threshold" ($C = 6.0$ for intensity data or $C = 0.10g$ for acceleration data), a map of $1 - i*$ developed through the application of kriging shows the probability of being greater than or equal to the damage threshold; larger probability values indicate greater hazard (see Carr and Bailey (1985) for examples of such maps).

In another example, an operating mine may choose to estimate i rather than Z by developing i using C equal to the economic cutoff for the mineralization. Estimates of i are obtained to indicate the probability of being less than or equal to the economic cutoff. Large-valued estimates of i approaching 1 actually indicate a small probability of having economic ore at the estimation location.

Many potential applications are yet to be explored. An additional advantage of the indicator transform involves minimizing the influence of outliers, those data that lie outside the distribution of the majority. Recall that one benefit of histogram analysis is identification of the presence of unusual data values in a data set. If present, how are these unusual values dealt with? Simply deleting them is one approach, but some practitioners wince at this suggestion because data are thrown away. Alternatively an indicator transform may be applied to the data, because no matter how unusual the data values are, they are transformed to either 0 or 1.

6.7 Models of Semivariograms Used for Computing Kriging Weights

An explanation is presented here of how the calculation and plotting of the semivariogram contributes to the matrix system used to solve for kriging weights. Recall the following matrix system:

$$[\text{cov}(h_{ij})]\{\lambda \; \mu\} = \{\text{cov}(h_{0i})\}$$

Entries to the left-hand square matrix, $[\text{cov}(h_{ij})]$, are spatial autocovariance values; likewise, information for spatial autocovariance is required for the right-hand vector, $\{\text{cov}(h_{0i})\}$. The semivariogram calculation for Z is used to infer the parameters, range, sill, and nugget (Figure 6.5).

A model function is chosen whose shape approximates that of the semivariogram plot for Z. Graphs of standard semivariogram models are shown in Figures 6.18 and 6.19. Notice that the model whose shape is most similar to Figure 6.5 is the spherical model. The shape of the exponential model is similar to the spherical model, except that the exponential model approaches a sill value asymptotically. Both spherical and exponential models show a linear behavior in the semivariogram plot near the origin. The shape of the Gaussian model (so named because of the similarity of the equation for the model to that of the normal, or Gaussian, distribution) is concave upward at the origin, becoming linear prior to asymptotically approaching the sill value. The linear model does not attain a sill, whereas the spherical, exponential, and Gaussian models do.

What reasons can be given for the variety in semivariogram models? Many regionalized phenomenon are spatially correlated according to the spherical spatial law. Yet, for some spatial phenomena, especially those showing a regular or more deterministic spatial variation, a Gaussian-shaped semivariogram may

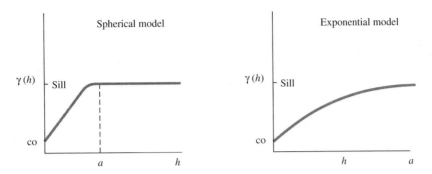

Figure 6.18 Plots of two idealized semivariogram shapes: spherical and exponential.

be obtained. An exponential spatial law suggests that the spatial phenomenon is associated with a finite correlation, even for infinitely large lags (i.e., a sill is never reached). Finally, a linear spatial law indicates a finite correlation for lags as large as can be assessed for a given phenomenon; this correlation varies linearly with lag.

When performing kriging, the computer software assembles the matrix system to be solved for the kriging weights. To enable the computer algorithm to do this, input is required to (1) select the model function that is most similar to the semivariogram plot for Z; (2) define the nugget, range, and sill values; and (3) define any anisotropic shape characteristics revealed by directional variogram calculations (Section 6.6.1.2). This data entry is described in detail in a later section that presents the kriging algorithm.

Once the semivariogram model parameters are specified, the model is used to develop values for autocovariance using the relationship

$$\text{cov}(h) = \text{var}(Z) - \gamma(h) \approx \text{sill} - \gamma(h)$$

6.7.1 Equations for the Semivariogram Models

6.7.1.1 The Spherical Model
From David (1977; p. 102), we have

$$\gamma(h) = CO + C\left(\frac{3}{2}\frac{h}{a} - \frac{1}{2}\frac{h^3}{a^3}\right), \qquad 0 < h \le a$$
$$= CO + C = \text{sill}, \qquad h > a$$
$$\gamma(0) = 0$$

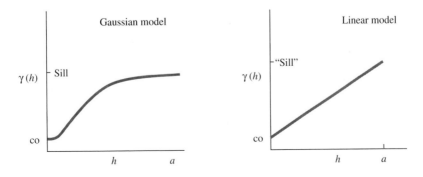

Figure 6.19 Plots of two idealized semivariogram shapes: Gaussian and linear.

where a is the range, CO is the nugget value, and C is equal to the sill minus the nugget value ($C = $ sill $-$ CO).

The origin of the name *spherical* for this semivariogram model explains a great deal about spatial correlation inherent to regionalized phenomena. David (1977, p. 106) provides a clear discussion of this origin by using the notion of a process having a Poisson distribution, where this process is realized as a distribution of masses in space. Each realization, or observation, of this regionalized variable Z is equal to the number of masses within a radius a of the observation location. In other words, each observation is equal to the number of masses within a spherical region surrounding the observation location. Furthermore, the volume of intersection between two adjacent spherical regions, each of volume V, radius a, and a distance h apart, is equal to

$$V\left(1 - \frac{3}{2}\frac{h}{a} + \frac{1}{2}\frac{h^3}{a^3}\right)$$

provided that $h < a$.

The size of the volume V of this intersection describes the amount of correlation between the two spheres. Therefore, the semivariogram is inversely proportional to this volume, such that

$$\gamma(h) \propto V\left[1 - \left(1 - \frac{3h}{2a} + \frac{h^3}{2a^3}\right)\right] = V\left(\frac{3h}{2a} - \frac{h^3}{2a^3}\right)$$

which is directly comparable to the spherical variogram model.

6.7.1.2 The Exponential Model
This equation is taken from David (1977, pp. 108–110):

$$\gamma(h) = CO + C\left[1 - \exp\left(-\frac{|h|}{a}\right)\right], \qquad h > 0$$

$$\gamma(0) = 0$$

where CO, C, and a are as defined for the spherical model. The application of this model is more difficult than the equation implies. As Journel and Huijbregts (1978, p. 164) acknowledge, the range a in the equation for the exponential model is determined using the practical range a' as follows: $a = a'/3$, where a' is such that

$$\gamma(a') \approx 0.95(\text{sill})$$

6.7.1.3 The Gaussian Model
This model is written using the same notation used by David (1977) and the functional form given in Journel and Huijbregts (1978, p. 165):

$$\gamma(h) = CO + C\left[1 - \exp\left(-\frac{h^2}{a^2}\right)\right], \qquad h > 0$$

$$\gamma(0) = 0$$

As with the exponential model, the Gaussian model approaches a sill asymptotically. The range a is determined using the practical range a', such that

$$a = \frac{a'}{\sqrt{3}}, \qquad \text{where } \gamma(a') = 0.95(\text{sill})$$

6.7.1.4 The Linear Model (Without a Sill)

For some spatial data sets, the plot of the semivariogram never attains a sill, perhaps due to a large range (relative to the scale at which the samples were taken) or for some other reason. As long as the semivariogram increases with h, but where this increase is less rapid than h^2, a general power model can be used (Journel and Huijbregts, 1978, p. 165). Herein, this general power model is written to yield a linear model:

$$\gamma(h) = CO + Ch, \qquad h > 0$$
$$\gamma(0) = 0$$

where C is the slope of the linear semivariogram.

What does it mean if the semivariogram increases faster than h^2? Often, this is an indication of a spatial trend inherent to the data, a situation wherein the intrinsic hypothesis does not hold. For such data, universal kriging is necessary (Carr (1990) presents a computer algorithm).

6.7.2 The Negative Semidefiniteness of Semivariogram Models

The matrix system used to solve for kriging weights must be invertible. To achieve this matrix property, all covariance entries in the kriging matrix system must be obtained from positive definite functions. A function f is positive definite if the following is true:

$$\sum_{i=1}^{N} \sum_{j=1}^{N} \lambda_i \lambda_j f(x_{ij}) \geq 0 \qquad \text{for } \lambda \in [\Re]$$

Because covariance is a function of the semivariogram such that

$$\text{cov}(h) = \text{sill} - \gamma(h)$$

to assure that cov is a positive definite function, the semivariogram function must be negative definite, such that

$$\sum_{i=1}^{N} \sum_{j=1}^{N} \lambda_i \lambda_j \gamma(h_{ij}) \leq 0, \qquad \lambda \in [\Re]$$

The task of proving a function to be negative definite or positive definite is an arduous one, but proving that a function does not conform to either notion can be relatively easy (Carr and Glass (1985) present some examples). The semivariogram models defined in this chapter are negative definite. Only a few other models, not defined herein, meet this requirement.

6.7.3 Nested Semivariogram Models

Any positive, linear combination of negative definite functions is also negative definite. Sometimes semivariograms result for a spatial situation that cannot be modeled easily by any one of the four models defined in this chapter. Recall the application of semivariogram analysis to the LONGB.DAT data set (Figures 6.16 and 6.17) where different semivariogram models exist for different spatial scales. Multiple spatial correlation is represented using a linear combination of semivariogram models. Such a linear combination is shown in Figure 6.20 and is referred to as a nested combination of semivariogram models.

When is a nested combination of semivariogram models necessary for the representation of the spatial structure for a set of spatial data? And, if a nested combination of models is necessary, how are sill, range, nugget, model type,

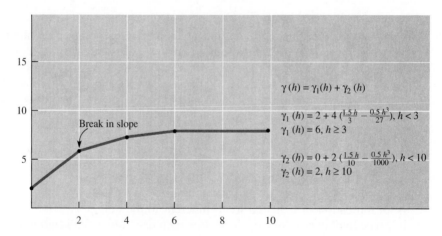

Figure 6.20   An example of a nested combination of semivariogram models and the consequential shape of the model.

and so on determined? If a semivariogram plot is obtained (Figure 6.20), nested semivariogram structure is often manifested by a break in slope of the semivariogram near the origin, such that two different linear portions are revealed. Experimentation is required to figure out how to choose two separate semivariogram models such that their sum, for any lag distance, equals the semivariogram for the actual data (Figure 6.20).

6.8 Example Calculation to Demonstrate the Use of Kriging

Now that the modeling of semivariograms has been discussed, example calculations are presented to complete the summary of kriging. Assume, for the sake of brevity, that the data in Figure 6.4 are associated with isotropic spatial correlation. Therefore, the directional semivariogram of Figure 6.5 is representative of all spatial directions. This semivariogram is modeled using a spherical model (Table 6.2).

Using these model parameters, an estimate is attempted at the unsampled location having the coordinates $x = 2.0$, $y = 4.0$ (Figure 6.4). Furthermore, a decision is arbitrarily made to use the six closest data locations to this estimation

Table 6.2   Semivariogram Model Parameters for Figure 6.5

Model type: spherical
Nugget (CO): 20.0
Sill: 140.0
Range: 3.0
Explicit model:

$$\gamma(h) = 20.0 + 120.0 \left(\frac{1.5h}{3} - \frac{0.5h^3}{27} \right), \quad 0 < h < 3$$

$$\gamma(h) = 140.0, \quad h \geq 3$$
$$\gamma(0) = 0$$

Table 6.3 Data Locations Used for Estimation at $x = 2$, $y = 4$ for Calculation Example Using Figure 6.4

i	(x_i, y_i)	$Z(x_i)$	Distance from $x = 2$, $y = 4$
1	(1, 5)	15.7	1.414
2	(2, 5)	11.8	1.000
3	(0, 4)	28.5	2.000
4	(1, 4)	13.6	1.000
5	(4, 4)	35.4	2.000
6	(2, 3)	0.0	1.000

location, arbitrarily numbered 1 through 6 in Table 6.3. These data locations are explicitly represented by their data values, also shown in this table.

The first step in this demonstration is calculation of the intersample covariance matrix, cov_{ij}. This matrix is symmetrical, that is, $cov_{ij}(I, J) = cov_{ij}(J, I)$, which follows from the observation that distance between locations x_i and x_j is the same as between locations x_j and x_i. The matrix cov_{ij} is also diagonalized; that is, the largest entries in this matrix are on the diagonal. Each diagonal entry $cov_{ij}(I, I)$, is the covariance for a zero lag distance. Because $\gamma(0) = 0$, then

$$cov(0) = sill - \gamma(0) = sill$$

Therefore, each diagonal entry of the matrix cov_{ij}, is equal to the sill value of the semivariogram.

Because the matrix cov_{ij} is symmetric, only the entries for one side of the diagonal must be calculated explicitly; entries for the other side are filled by the symmetrical relationship. In this case where six closest data locations to the estimation location are used, only $6(6 - 1)/2$, or 15 explicit calculations are required (Table 6.4). Once these calculations are complete, the point–sample

Table 6.4 Computation of Covariances for Calculation Example to Obtain an Estimate at $x = 2$, $y = 4$, Figure 6.4

i	j	h_{ij}: Distance, x_i to x_j	$\gamma(h_{ij})$	$cov(h_{ij})$
1	2	1.000	77.78	62.22
1	3	1.414	98.56	41.44
1	4	1.000	77.78	62.22
1	5	3.162	140.00	0.00
1	6	2.236	129.32	10.68
2	3	2.236	129.32	10.68
2	4	1.414	98.56	41.44
2	5	2.236	129.32	10.68
2	6	2.000	122.22	17.78
3	4	1.000	77.78	62.22
3	5	4.000	140.00	0.00
3	6	2.236	129.32	10.68
4	5	3.000	140.00	0.00
4	6	1.414	98.56	41.44
5	6	2.236	129.32	10.68

Note. $cov(h) = sill - \gamma(h)$.

Table 6.5 Covariances for Point–Sample
Distances Listed in Table 6.3

Sample	Distance to Estimation Point	Covariance
1	1.414	41.44
2	1.000	62.22
3	2.000	17.78
4	1.000	62.22
5	2.000	17.78
6	1.000	62.22

covariances are computed. These covariances (Table 6.5) are a function of distance between the six closest data locations and the estimation location (Table 6.3).

With all calculations required to assemble the matrix system to solve for the kriging weights completed, this matrix system is as follows:

$$
\begin{vmatrix}
140.00 & 62.22 & 41.44 & 62.22 & 0.00 & 10.68 & 1 \\
62.22 & 140.00 & 10.68 & 41.44 & 10.68 & 17.78 & 1 \\
41.44 & 10.68 & 140.00 & 62.22 & 0.00 & 10.68 & 1 \\
62.22 & 41.44 & 62.22 & 140.00 & 0.00 & 41.44 & 1 \\
0.00 & 10.68 & 0.00 & 0.00 & 140.00 & 10.68 & 1 \\
10.68 & 17.78 & 10.68 & 41.44 & 10.68 & 140.00 & 1 \\
1 & 1 & 1 & 1 & 1 & 1 & 0
\end{vmatrix}
\begin{vmatrix}
\lambda_1 \\ \lambda_2 \\ \lambda_3 \\ \lambda_4 \\ \lambda_5 \\ \lambda_6 \\ \mu
\end{vmatrix}
=
\begin{vmatrix}
41.44 \\ 62.22 \\ 17.78 \\ 62.22 \\ 17.78 \\ 62.22 \\ 1
\end{vmatrix}
$$

Notice than an extra row, the last row, wherein each entry except for the last is equal to 1, and an extra column, the last column, wherein each entry is equal to 1 except for the last, are added. The additional row assures that kriging weights sum to 1; the additional column maps intersample covariances cov_{ij} to point–sample covariances cov_{0i}, vis-à-vis the Lagrangian multiplier.

GAUSS, a program for Gaussian elimination (Chapter 2), is used to solve for kriging weights:

$$
\begin{vmatrix}
\lambda_1 \\ \lambda_2 \\ \lambda_3 \\ \lambda_4 \\ \lambda_5 \\ \lambda_6 \\ \mu
\end{vmatrix}
=
\begin{vmatrix}
0.036 \\ 0.316 \\ -0.039 \\ 0.267 \\ 0.090 \\ 0.331 \\ -1.713
\end{vmatrix}
$$

An estimate is computed using this solution:

est $= \lambda_1(15.7) + \lambda_2(11.8) + \lambda_3(38.5) + \lambda_4(13.6) + \lambda_5(35.4) + \lambda_6(0.0) = 9.6$

Kriging variance, or variance of the error, for this estimate can be computed as follows:

$$\text{krig var} = \text{sill} - \left(\sum_{i=1}^{N} \lambda_i \, \text{cov}_{0i} \right) - \mu$$

The kriging variance for this calculation demonstration is

$$\text{krig var} = 140.0 - 59.3 + 1.7 = 82.4$$

6.8.1 The Exact Interpolation Characteristics of Kriging

Kriging is an exact interpolator. Such an interpolator is one that, when applied for estimation wherein the estimation location coincides with a sample location, the estimation method returns, as the estimate, the data value for the sample location.

For example, suppose an estimate is made at the location $x = 1.0$, $y = 5.0$ (Figure 6.4), which coincides with a sample location associated with the data value 15.7. In this case, using the 7 closest data locations for estimation, *including the location $x = 1.0$, $y = 5.0$*, the matrix system for kriging is as follows, using the same semivariogram model (Table 6.2):

$$
\begin{vmatrix}
140.00 & 41.44 & 62.22 & 41.44 & 62.22 & 62.22 & 41.44 & 1 \\
41.44 & 140.00 & 62.22 & 17.78 & 10.68 & 10.68 & 17.78 & 1 \\
62.22 & 62.22 & 140.00 & 62.22 & 41.44 & 17.78 & 10.68 & 1 \\
41.44 & 17.78 & 62.22 & 140.00 & 62.22 & 10.68 & 0.58 & 1 \\
62.22 & 10.68 & 41.44 & 62.22 & 140.00 & 41.44 & 10.68 & 1 \\
62.22 & 10.68 & 17.78 & 10.68 & 41.44 & 140.00 & 62.22 & 1 \\
41.44 & 17.78 & 10.68 & 0.58 & 10.68 & 62.22 & 140.00 & 1 \\
1 & 1 & 1 & 1 & 1 & 1 & 1 & 0
\end{vmatrix}
\begin{vmatrix}
\lambda_1 \\ \lambda_2 \\ \lambda_3 \\ \lambda_4 \\ \lambda_5 \\ \lambda_6 \\ \lambda_7 \\ \mu
\end{vmatrix}
=
\begin{vmatrix}
140.00 \\ 41.44 \\ 62.22 \\ 41.44 \\ 62.22 \\ 62.22 \\ 41.44 \\ 1
\end{vmatrix}
$$

This system is solved for the kriging weights and Lagrangian multiplier to yield

$$
\begin{vmatrix}
\lambda_1 \\ \lambda_2 \\ \lambda_3 \\ \lambda_4 \\ \lambda_5 \\ \lambda_6 \\ \lambda_7 \\ \mu
\end{vmatrix}
=
\begin{vmatrix}
1 \\ 0 \\ 0 \\ 0 \\ 0 \\ 0 \\ 0 \\ 0
\end{vmatrix}
$$

Using these weights for estimation, of course, yields an estimate exactly equal to the data value at the sample location, which coincides with the estimation location. Moreover, because the Lagrangian multiplier is equal to zero, kriging variance is equal to zero (for exact interpolation).

6.8.2 A Note on Kriging Variance

As the previous calculation examples show, kriging variance is largely a function of sampling geometry in relationship to estimation location. The second term in the equation for kriging variance is the sum of products of kriging weights and point–sample covariances. The larger this sum is, the smaller is the kriging variance. The sum of kriging weights and point–sample covariances is proportional, obviously, to the magnitude of these covariances. As the exact interpolation example shows, the closer a data location is to the estimation location, the more similar at least one covariance value will be to the sill value of the semivariogram model. Therefore, the closer the sample locations are to the estimation location, the larger the magnitude of the covariance entries will be to the point-sample covariance matrix. In addition, the larger the sum of the products of these covariances with kriging weights, the smaller in magnitude will be kriging variance.

6.9 Sample Support

In the previous calculation, the estimate obtained from kriging is a point, or punctual, estimate. That is, the support of the estimate is punctual. This means that the estimate is valid only for the discrete point.

In some estimation applications, an estimate is desired that represents the average value of Z over a finite area. In mining engineering, for instance, kriging is used to estimate the average grade of ore over a volume of rock. In this case, the support of the estimate is referred to as an average, or block, support.

6.9.1 Block Kriging Versus Punctual Kriging

In block kriging the objective is to obtain an estimate that is an average over a finite area or that invokes a three-dimensional notion (although herein we describe a two-dimensional application). The kriging equations are modified as follows. The kriging system is the same as that for punctual kriging:

$$[\text{cov}_{ij}][\lambda\ \mu] = [\text{cov}_{0i}]$$

but, the right-hand vector contains point–block covariance entries, developed through integration as follows:

$$\text{cov}_{0i} = \frac{1}{\text{area}} \int_{x\min}^{x\max} \int_{y\min}^{y\max} \text{cov}(h_{xy,i})\, dy\, dx$$

This integral is written for a two-dimensional application, where area is the size of a region of size $(x\max - x\min)$ by $(y\max - y\min)$ (Figure 6.21), the center of this region is the point of estimation 0, and i represents one of the nearest-neighbor punctual samples to be used for estimation.

Most practitioners in geostatistics approximate the integral for point–block covariance using a procedure such as that described and implemented in David (1977, pp. 242–243):

$$\text{cov}_{0i} \approx \frac{1}{N^2} \sum_{k=1}^{N} \sum_{j=1}^{N} \text{cov}(h_{i,jk})$$

where N often is selected to be four. This approximation involves the following algorithm (Figure 6.21):

1. Determine a point within the block (area) surrounding the estimation location for each of the N^2 points i, j such that the coordinates of this point are

$$\text{xcoord} = \text{xmid} + ((i - ((N/2) + 1)) * \text{difx}/N) + \text{difx}/2N$$
$$\text{ycoord} = \text{ymid} + ((j - ((N/2) + 1)) * \text{dify}/N) + \text{dify}/2N$$

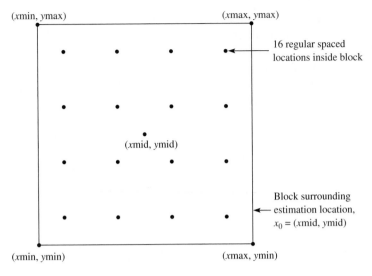

Figure 6.21 An example block with dimensions as shown, in this case divided into 16 discrete points for use with the standard procedure for block kriging.

where

$$\text{xmid} = (\text{xmax} - \text{xmin})/2$$
$$\text{ymid} = (\text{ymax} - \text{ymin})/2$$
$$\text{difx} = \text{xmax} - \text{xmin}$$
$$\text{dify} = \text{ymax} - \text{ymin}$$

and xmin, xmax, ymin, and ymax define the size of the area surrounding the estimation location.

2. Compute the distance between location (x_i, y_i), that is, the ith closest neighbor, and the location (xcoord, ycoord).
3. Determine the covariance for the distance computed in step 2.
4. Repeat steps 1–3 for all N^2 locations within the block.
5. Sum the N^2 covariance values.
6. Divide the sum of step 5 by N^2 to yield cov_{0i}.

A more accurate approximation of the integral for point–block covariance is obtained using Gaussian quadrature (Carr and Palmer, 1993; numerical integration to obtain point–block covariance is discussed also in Journel and Huijbregts (1978, pp. 98–108) and Davis and David (1978)). Gaussian quadrature (Chapter 2) is used to approximate the definite integral to obtain cov_{0i} as follows:

$$\text{cov}_{0i} = \frac{1}{\text{area}} \sum_{k=1}^{N} \sum_{j=1}^{N} W_k W_j \, \text{cov}(h_{kj,i})$$

where N is the number of Gauss points to use for the approximation.

The appropriate number of Gauss points N to use for the approximation of the definite integral depends on the order of the semivariogram model used to derive spatial autocovariance cov. Two of the covariance functions used in this chapter are exponential functions. At least five Gauss points are needed to accurately approximate the definite integral of an exponential function (Chapter 2). Optimal accuracy involves using $N = 4$ Gauss points for any of the four covariance models described in this chapter (Carr and Palmer, 1993). Some small error is tolerated when using the exponential or Gaussian semivariogram models, whereas a good degree of accuracy is obtained for the spherical and linear models.

Gaussian quadrature is implemented for computing point–block covariance by the following algorithm, assuming $N = 4$ Gauss points:

1. Determine the spatial location of Gauss points within the block, if $N = 4$, four rows (y) of points, each having four columns (x), represent a grid of 16 total points:

$$x_1 = \text{xmid} - (\text{difx}/2) * 0.8611363116$$
$$x_2 = \text{xmid} - (\text{difx}/2) * 0.3399810436$$
$$x_3 = \text{xmid} + (\text{difx}/2) * 0.3399810436$$
$$x_4 = \text{xmid} + (\text{difx}/2) * 0.8611363116$$
$$y_1 = \text{ymid} - (\text{dify}/2) * 0.8611363116$$
$$y_2 = \text{ymid} - (\text{dify}/2) * 0.3399810436$$
$$y_3 = \text{ymid} + (\text{dify}/2) * 0.3399810436$$
$$y_4 = \text{ymid} + (\text{dify}/2) * 0.8611363116$$

where $\text{xmid} = (\text{xmax} - \text{xmin})/2$, $\text{ymid} = (\text{ymax} - \text{ymin})/2$ $\text{difx} = \text{xmax} - \text{xmin}$, and $\text{dify} = \text{ymax} - \text{ymin}$, and xmin, xmax, ymin, and ymax define the area surrounding the estimation location 0.

2. Determine the weights W to be used in the calculation:

$$W_1 = 0.3478548452$$
$$W_2 = 0.6521451548$$
$$W_3 = 0.6521451548$$
$$W_4 = 0.3478548452$$

3. Determine the distance between each of the i nearest-neighboring locations and each of the 16 Gauss points; i.e., calculate $h_{i,kj}$, $k = 1$ to 4, $j = 1$ to 4.

4. Determine the covariance as

$$\text{cov}_{0i} = \frac{1}{\text{area}} \sum_{k=1}^{4} \sum_{j=1}^{4} \left(W_k \frac{\text{difx}}{2} \right) \left(W_j \frac{\text{dify}}{2} \right) \text{cov}(h_{i,kj})$$

where $\text{area} = (\text{difx})(\text{dify})$. A review of Chapter 2 reveals why the weights are multiplied by difx/2 and dify/2.

DEMONSTRATION

An example block and one known sample location are shown in Figure 6.22. Suppose the semivariogram model for this spatial estimation problem is a spherical one, with no nugget value, a sill equal to 10.0, and a range equal to 10.0 (distance units, generic). The covariance between the block and the point is estimated using the standard block kriging approach (e.g., David, 1977) for $N = 4$, 100, and 200 points, in comparison to the use of Gaussian quadrature for $N = 4$ Gauss points; see Table 6.6. Notice that the standard approach converges to the Gaussian quadrature solution, but that 40,000 points are required for the standard method to approach the accuracy of Gaussian quadrature using 4×4, or 16 total points!

This experiment is repeated using exponential and linear models (Table 6.6). Notice that Gaussian quadrature is associated with a small error for the exponential model, but is more accurate than using a 4×4 standard approach. Gaussian quadrature is accurate for the linear model.

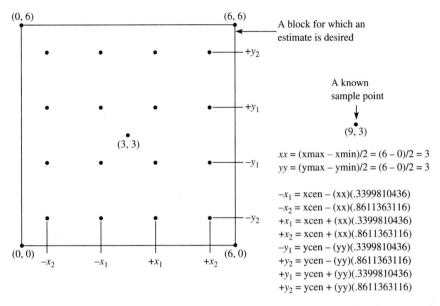

Figure 6.22 An example block showing discrete locations used for covariance calculations for the Gaussian quadrature approach to block kriging.

Table 6.6 Comparison of Gaussian Quadrature to the Standard Procedure for Block Kriging (see Figure 6.21)

Standard Approach	*4 × 4-Point Gaussian Quadrature*
For a spherical model: sill = 10, CO = 0, range = 10	
4 × 4, cov_{0i} = 2.097473	cov_{0i} = 2.099511
100 × 100 = 2.099537	
200 × 200 = 2.099535	
For an exponential model: sill = 10, CO = 0, range = 10/3	
4 × 4, cov_{0i} = 1.724465	cov_{0i} = 1.729156
100 × 100 = 1.729457	
200 × 200 = 1.729462	
For a linear model: sill = 10, CO = 0, range = 10	
4 × 4, cov_{0i} = 3.755674	cov_{0i} = 3.739456
100 × 100 = 3.739499	
200 × 200 = 3.739482	

6.10 JCBLOK: A Portable FORTRAN-77 Program for Punctual and Block Kriging

A portable FORTRAN-77 computer program, JCBLOK (with a user's guide), is presented for one- and two-dimensional punctual and block kriging (Figure 6.23, page 212). This program allows the use of up to five nested semivariogram models. Four semivariogram models are possible: spherical, exponential, Gaussian, and linear. Gaussian quadrature is used to approximate the point–sample covariance integral for block kriging. Lognormal and indicator transforms are performed by the program, if desired. And, the program can accommodate isotropic or anisotropic spatial structure.

In addition to the capabilities just mentioned, a general or quadrant search may be used for locating the N nearest-neighbor samples for estimation. A general search involves simply locating up to N closest data locations, regardless of where these locations are. A quadrant search on the other hand (Figure 6.24) locates N closest locations and assures that these locations occur in at least three of the four quadrants surrounding the estimation location. A quadrant search ensures that an estimate is reflective of spatial information surrounding the estimation location, whereas with a general search the N closest data locations may not surround the estimation location (they may be all on one side of the estimation location, for instance).

A quadrant search is more restrictive than a general search. Furthermore, when using kriging to obtain a grid of estimates, a quadrant search may not allow estimation to occur at the grid boundaries if all spatial samples $Z(x_i)$ are within the domain of the grid. Therefore, advantages and disadvantages are associated with either search strategy.

6.10.1 Using JCBLOK with the Lognormal and Indicator Transforms

JCBLOK will transform the spatial data Z to natural logarithms or indicator values. If a data transform is selected, the following estimates are obtained. In lognormal kriging, for instance, the data Z are transformed to their natural

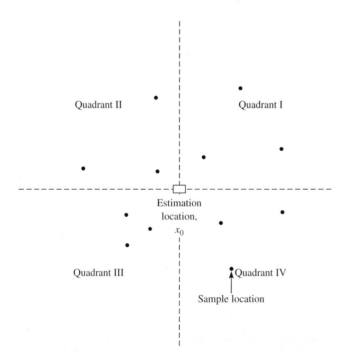

Figure 6.24   An example of a quadrant search showing the location of the three closest sample locations per quadrant for a total of NK = 12 closest data locations used for estimation.

logarithms, called Y herein. As kriging, punctual or block, proceeds, estimates are obtained of Y as follows (e.g., Journel and Huijbregts, 1978, p. 570):

$$Y^*(x_0) = \bar{Y} + \sum_{i=1}^{N} \lambda_i (Y(x_i) - \bar{Y})$$

where $\bar{Y}$ is the mean value of Y. Once this estimate is computed, an estimate of Z is obtained as (e.g., Journel and Huijbregts, 1978, p. 572; Journel, 1980)

$$Z^*(x_0) = XK_0 \exp\left(Y^*(x_0) + \frac{\text{krig. var.}}{2} \right)$$

where krig. var. is the kriging variance for the estimate of Y.

The parameter XK_0 is a correction factor. If not used, the estimate $Z^*(x_0)$ computed using $Y^*(x_0)$ may be biased; XK_0 corrects this bias.

To determine the value of XK_0 the following procedure is used:

1. Use the program JCBLOK in cross-validation mode. Cross-validation is the estimation of Z at a known data location, *without including that data location when finding the N nearest-neighbor locations for estimation.* Recall that kriging is an exact interpolator. If estimation is attempted at a known data location, the data value at that location is returned by kriging. In cross-validation, though, the objective is to see how closely kriging can estimate a known data value using surrounding neighbors for estimation. In cross-validation, an error is calculated at each data location: actual − estimated. Cross-validation is performed over all data locations to calibrate estimation accuracy.
2. In step 1, when prompted by JCBLOK to enter a value for XK_0, a value of 1.0 is used.
3. Once cross-validation is completed, summary statistics are printed. Two of these statistics are the average of the estimates and the average of the original sample data. XK_0 is recomputed to equal the ratio

$$XK_0 = \frac{\text{mean of original sample data}}{\text{mean of estimates}}$$

and step 1 is repeated to check the quality of XK_0.

If JCBLOK is used for the indicator transform, each estimate, punctual or block, yielded by the program is an estimate of the indicator function. This estimate is the probability that Z is less than or equal to the cutoff C at the estimation location.

In equation form, the estimate of the indicator function is obtained using a procedure analogous to that for estimating Z:

$$i^*(x_0) = \sum_{j=1}^{N} \lambda_j i(x_j)$$

except the subscript j is used to avoid confusion with the indicator function i.

6.11 Case Studies

6.11.1 An Application of JCBLOK to the NVSIM.DAT Data Set for Gridding

These data were shown previously to conform well to the normal distribution assumption implicit with kriging (Figures 6.7 and 6.8). The program JCBLOK

Table 6.7 Options for JCBLOK in
Application to NVSIM.DAT Data Set

IKRIG = 1; ITRAN = 0
IBEGC = 1; IENDC = 50
IBEGR = 1; IENDR = 50
SIZEX = SIZEY = 10.0
YMAX = 500.0
XMIN = 0.0
Semivariogram model parameters
  Model type: spherical
  Number of nested models: 1
  Nugget = 10.0
  Sill = 120.0
  Range = 100.0
  ANIS = 0.0; RATIO = 1.0
  ISERCH = 1; RADIUS = 100.0; NK = 10

is capable of producing estimates at intersections of a uniform rectangular or square grid. Many programs capable of generating contour maps require a regular grid of data values; hence, JCBLOK may be applied to a set of irregularly spaced spatial samples as a prelude to contour mapping.

Options for JCBLOK in application to the NVSIM.DAT data set for gridding are shown in Table 6.7. Semivariogram parameters for these data are interpreted from the omnidirectional semivariogram of Figure 6.14.

JCBLOK produces an output file containing results for estimation at each grid point. A contour map is developed from these estimates (Figure 6.25). A contour map may be a goal of geostatistical analysis. If so, this example serves as a demonstration of how to use JCBLOK for gridding as a prelude to contour mapping.

6.11.2 An Application of JCBLOK to the Block Estimation of Lognormal Data

A lognormal data set (NVSIMLOG.DAT, Appendix A) is used. The histogram of the untransformed data (Figure 6.26) reveals the lack of conformance of these data to a normal distribution. But the histogram of the log-transformed data appears to conform to a normal distribution (Figure 6.27). A probability plot of the log-transformed histogram results (Figure 6.28) indicates that a normal distribution assumption is appropriate for the log-transformed data.

Omnidirectional semivariograms for these data with no data transformation (Figure 6.29) and with lognormal transform (Figure 6.30) indicate that the semivariogram is more spherical in shape after transformation. This qualitatively demonstrates the necessity to consider data distribution characteristics in a geostatistical analysis.

Block kriging is attempted for these data. Options, including the size of the blocks, for the program JCBLOK are shown in Table 6.8. A cross-validation experiment is attempted first to evaluate the correction factor XK_0. Once this is finished, a gridding example is presented.

For the first application of JCBLOK to these data, the correction factor for

CONTOUR MAP FOR NVSIM.DAT SPATIAL DATA SET

Figure 6.25   Contour map for the NVSIM.DAT data set. This topographic map was drawn using the computer program TOPO, a *Golden Graphics* product (see reference list). A BASIC program must be run prior to using TOPO to render the output from JCBLOK compatible with this computer contouring algorithm. A complete listing of this BASIC program is given in Carr (1990). *Golden Graphics* is a product of Golden Software, Inc., P.O. Box 281, Golden, CO 80402 (USA).

lognormal kriging, XK_0, is set equal to 1.0. The summary statistics from cross-validation using this correction factor (Table 6.9) show that the mean of the estimates is approximately equal to that of the original data. This is what should happen for data that conform closely, or exactly, to the lognormal distribution. A larger difference in the two mean values occurs for data that are approximately, but not exactly, lognormally distributed.

Of course, the correction factor XK_0 is used to correct for any bias. On the basis of the summary statistics (Table 6.9), XK_0 is chosen to equal the ratio of the original data mean to that of the estimates, such that XK_0 is equal to 22.22454/22.61378, or 0.98. Using this correction factor yields the cross-validation summary statistics (Table 6.10) showing that the mean of the estimates is almost identical to the original data mean, a desired result for unbiased

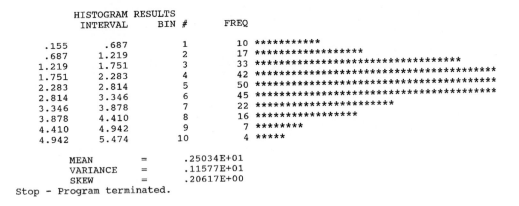

```
        HISTOGRAM RESULTS
           INTERVAL       BIN #        FREQ

    .000      23.833        1          181 ****************************************
  23.833      47.667        2           39 **********
  47.667      71.500        3           13 ****
  71.500      95.334        4            7 **
  95.334     119.167        5            2 *
 119.167     143.000        6            2 *
 143.000     166.834        7            0
 166.834     190.667        8            2 *
 190.667     214.501        9            0
 214.501     238.334       10            2 *

          MEAN      =      .22216E+02
          VARIANCE  =      .10097E+04
          SKEW      =      .12077E+06
Stop - Program terminated.
```

0;37;44m C:\TEXT=

Figure 6.26   Histogram for the NVSIMLOG.DAT data set.

```
        HISTOGRAM RESULTS
           INTERVAL       BIN #        FREQ

    .155       .687         1           10 **********
    .687      1.219         2           17 *****************
  1.219      1.751         3           33 *********************************
  1.751      2.283         4           42 ******************************************
  2.283      2.814         5           50 **************************************************
  2.814      3.346         6           45 *********************************************
  3.346      3.878         7           22 **********************
  3.878      4.410         8           16 ****************
  4.410      4.942         9            7 ********
  4.942      5.474        10            4 *****

          MEAN      =      .25034E+01
          VARIANCE  =      .11577E+01
          SKEW      =      .20617E+00
Stop - Program terminated.
```

0;37;44m C:\TEXT>

Figure 6.27   Histogram for the NVSIMLOG.DAT data set after log transformation.

Table 6.8   Options for JCBLOK in Application to the NVSIMLOG.DAT Data Set for Lognormal Block Kriging

---

IKRIG = 2; ITRAN = 1
IBEGC = IENDC = IBEGR = IENDR = 1 (for cross-validation to determine XK_0)
For gridding
   IBEGC = 1; IENDC = 50
   IBEGR = 1; IENDR = 50
   YMAX = 500.0; XMIN = 0.0
   SIZEX = SIZEY = 10.0
Lognormal semivariogram model parameters
   Number of nested models = 1
   Model type = spherical
   Nugget = 0.10
   Sill = 1.20
   Range = 80.0
   ANIS = 0.0; RATIO = 1.0
   ISERCH = 1; RADIUS = 80.0; NK = 10

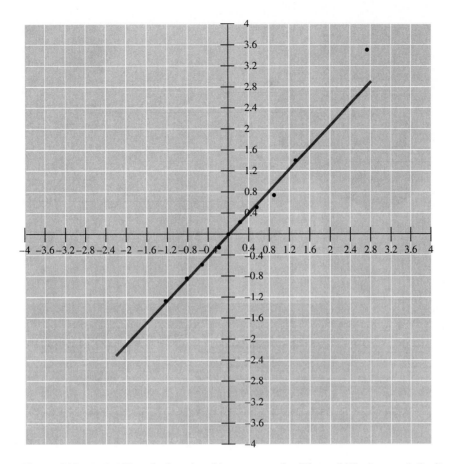

Figure 6.28 Probability plot based on histogram results (Figure 6.27). A normal distribution model represents the log-transformed data well.

estimation. Moreover, the average kriging error is closer to zero than for the previous result (Table 6.9).

Histograms show the distribution of the estimates after lognormal kriging (Figure 6.31) and after kriging without a data transform (Figure 6.32). When comparing these histograms to Figure 6.26, the histogram after lognormal kriging is more similar to that of the raw data than is the histogram after

Table 6.9 Summary Statistics from
JCBLOK, Cross-Validation Mode,
NVSIMLOG.DAT Data Set,
$XK_0 = 1.0$

1. Average kriging error	=	−0.389
2. Average squared error	=	762.048
3. Variance of error	=	764.981
4. Average kriging var.	=	0.603
5. Average estimate	=	22.614
6. Original data mean	=	22.225

UNTRANSFORMED DATA, OMNIDIRECTIONAL SEMI-VARIOGRAM CALCULATION

INTERVAL OF h		PAIRS	GAMMA-h	AVG. DIST. h
10.0-	20.0	82	.24553840E+03	.12121590E+02
20.0-	30.0	184	.78538960E+03	.23341430E+02
30.0-	40.0	208	.84244390E+03	.32823200E+02
40.0-	50.0	252	.74936960E+03	.42559940E+02
50.0-	60.0	407	.81825730E+03	.52760010E+02
60.0-	70.0	364	.82195840E+03	.63328210E+02
70.0-	80.0	486	.83187400E+03	.73501500E+02
80.0-	90.0	535	.74864260E+03	.83746350E+02
90.0-	100.0	508	.63115870E+03	.93758190E+02
100.0-	110.0	647	.75290530E+03	.10375110E+03
110.0-	120.0	576	.11570700E+04	.11330880E+03
120.0-	130.0	718	.10972500E+04	.12312160E+03
130.0-	140.0	803	.82515180E+03	.13388690E+03
140.0-	150.0	700	.99531560E+03	.14405360E+03
150.0-	160.0	724	.88980470E+03	.15380140E+03
160.0-	170.0	710	.10447260E+04	.16344830E+03
170.0-	180.0	865	.10250000E+04	.17364130E+03
180.0-	190.0	887	.10064000E+04	.18404320E+03
190.0-	200.0	815	.88950090E+03	.19404140E+03

```
                    SEMI-VARIOGRAM PLOT
      .13884840E+04  +
      .13422010E+04  +
      .12959190E+04  +
      .12496360E+04  +
      .12033530E+04  +
      .11570700E+04  +
      .11107870E+04  +                                    *
      .10645050E+04  +                                      *
      .10182220E+04  +                                              *  *
      .97193900E+03  +                                  *              *
      .92565620E+03  +
  G   .87937340E+03  +                                       *          *
  A   .83309060E+03  +          *
  M   .78680770E+03  +              *    *  *               *
  M   .74052500E+03  +      *    *              *     *
  A   .69424210E+03  +
      .64795930E+03  +
  h   .60167650E+03  +                         *
      .55539370E+03  +
      .50911090E+03  +
      .46282810E+03  +
      .41654530E+03  +
      .37026250E+03  +
      .32397960E+03  +
      .27769680E+03  +
      .23141400E+03  +   *
      .18513120E+03  +
      .13884840E+03  +
      .92565620E+02  +
      .46282810E+02  +
                      +----+----+----+----+----+----+----+----+----+----+
                      0    2    4    6    8   10   12   14   16   18   20
                              h * (           10.0 * 19)/20
                      * = > 30 PAIRS              o = < 30 PAIRS
Stop - Program terminated.
```

Figure 6.29 Omnidirectional semivariogram for the NVSIMLOG.DAT data set with no data transformation.

kriging without a data transform. Through the application of geostatistics, the distribution of estimates should be similar to that for the original data; therefore, the lognormal kriging model is more accurate for these data.

Now that the correction factor XK_0 has been determined for these data, JCBLOK is again applied to the NVSIMLOG.DAT data set to effect gridding.

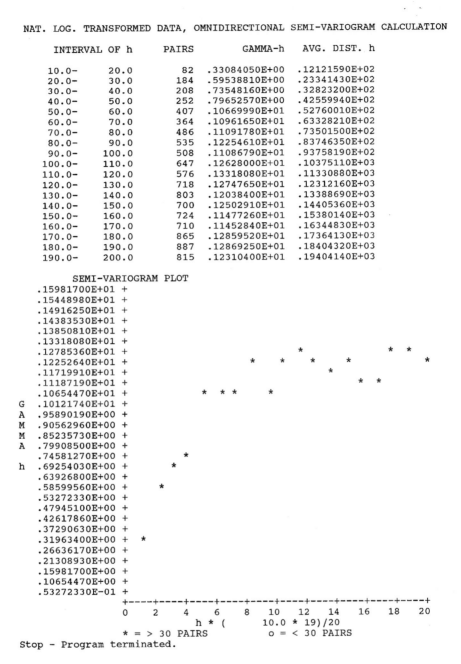

NAT. LOG. TRANSFORMED DATA, OMNIDIRECTIONAL SEMI-VARIOGRAM CALCULATION

INTERVAL OF h		PAIRS	GAMMA-h	AVG. DIST. h
10.0-	20.0	82	.33084050E+00	.12121590E+02
20.0-	30.0	184	.59538810E+00	.23341430E+02
30.0-	40.0	208	.73548160E+00	.32823200E+02
40.0-	50.0	252	.79652570E+00	.42559940E+02
50.0-	60.0	407	.10669990E+01	.52760010E+02
60.0-	70.0	364	.10961650E+01	.63328210E+02
70.0-	80.0	486	.11091780E+01	.73501500E+02
80.0-	90.0	535	.12254610E+01	.83746350E+02
90.0-	100.0	508	.11086790E+01	.93758190E+02
100.0-	110.0	647	.12628000E+01	.10375110E+03
110.0-	120.0	576	.13318080E+01	.11330880E+03
120.0-	130.0	718	.12747650E+01	.12312160E+03
130.0-	140.0	803	.12038400E+01	.13388690E+03
140.0-	150.0	700	.12502910E+01	.14405360E+03
150.0-	160.0	724	.11477260E+01	.15380140E+03
160.0-	170.0	710	.11452840E+01	.16344830E+03
170.0-	180.0	865	.12859520E+01	.17364130E+03
180.0-	190.0	887	.12869250E+01	.18404320E+03
190.0-	200.0	815	.12310400E+01	.19404140E+03

Figure 6.30 Omnidirectional semivariogram for the NVSIMLOG.DAT data set after log transformation of the data.

A contour map of the gridded results from lognormal kriging (Figure 6.33) is compared to one using kriging without the lognormal transformation (Figure 6.34). These two maps are different. More information than what is given in this chapter is required to determine which map is "better." Based on the fact that lognormal kriging yields estimates having a distribution more similar to the original data in comparison to kriging without the data transform, the map

Table 6.10 Summary Statistics
from JCBLOK, Cross-Validation
Mode, NVSIMLOG.DAT Data
Set, $XK_0 = 0.98$

1. Average kriging error	=	0.063
2. Average squared error	=	758.552
3. Variance of error	=	761.619
4. Average kriging var.	=	0.603
5. Average estimate	=	22.162
6. Original data mean	=	22.225

```
        HISTOGRAM RESULTS
          INTERVAL      BIN #        FREQ

  2.502    15.262        1           114 ****************************************
 15.262    28.022        2            74 *********************************
 28.022    40.781        3            28 *************
 40.781    53.541        4            11 *****
 53.541    66.301        5             8 ****
 66.301    79.061        6             5 ***
 79.061    91.821        7             5 ***
 91.821   104.580        8             0
104.580   117.340        9             2 *
117.340   130.100       10             1 *

        MEAN       =     .22614E+02
        VARIANCE   =     .42081E+03
        SKEW       =     .19486E+05
```

Figure 6.31 Histogram of estimation results from JCBLOK after lognormal kriging.

```
        HISTOGRAM RESULTS
          INTERVAL      BIN #        FREQ

  1.776    12.208        1            84 *****************************************
 12.208    22.641        2            85 *****************************************
 22.641    33.073        3            33 ********************
 33.073    43.506        4            22 *************
 43.506    53.938        5             7 *****
 53.938    64.370        6             6 ****
 64.370    74.803        7             5 ***
 74.803    85.235        8             1 *
 85.235    95.668        9             4 ***
 95.668   106.100       10             1 *

        MEAN       =     .21483E+02
        VARIANCE   =     .32077E+03
        SKEW       =     .11523E+05
```

Figure 6.32 Histogram of estimation results from JCBLOK after kriging with no data transformation.

Figure 6.33 Contour map of NVSIMLOG.DAT data developed using lognormal kriging. This map was drawn using the program TOPO, a *Golden Graphics* product (see reference list, and the caption to Figure 6.25).

based on lognormal kriging is expected to be better. Let this example simply serve to demonstrate how different maps can be obtained, depending on the assumptions (e.g., data transform) made in applying kriging for estimation.

6.11.3 Estimation of Indicator Functions Using JCBLOK

Suppose a nonparametric transform is chosen for the NVSIMLOG.DAT data set, one whose untransformed data values do not conform to a normal distribution. An experiment is conducted to apply an indicator transform to these data. The experiment proceeds by choosing an indicator transform such that the cutoff is equal to the median data value of 17.0.

In this case, gridding is attempted. Options for use with JCBLOK in application to the NVSIMLOG.DAT data set, for gridding and indicator transformation, are shown in Table 6.11. Semivariogram model parameters are interpreted from the semivariogram of the indicator transform (Figure 6.35). When prompted by the program JCBLOK the value 17.0 is entered for the cutoff to effect indicator kriging.

Figure 6.34 Contour map of NVSIMLOG.DAT data set developed using kriging without a data transformation. This map was drawn using the computer algorithm TOPO, a *Golden Graphics* product (see reference list, and captions to Figures 6.25 and 6.33).

Table 6.11 Options for JCBLOK Application to the NVSIMLOG.DAT Data Set for Indicator Block Kriging

---

IKRIG $= 2$, ITRAN $= 2$
IBEGC $= 1$; IENDC $= 50$
IBEGR $= 1$; IENDR $= 50$
SIZEX $=$ SIZEY $= 10.0$
YMAX $= 500.0$; XMIN $= 0.0$
Indicator semivariogram model parameters for $C = 17.0$
   Number of nested models $= 1$
   Model type $=$ spherical
   Nugget $= 0.125$
   Sill $= 0.230$
   Range $= 100.0$
   ANIS $= 0.0$; RATIO $= 1.0$
   ISEARCH $= 1$; RADIUS $= 100.0$; NK $= 10$

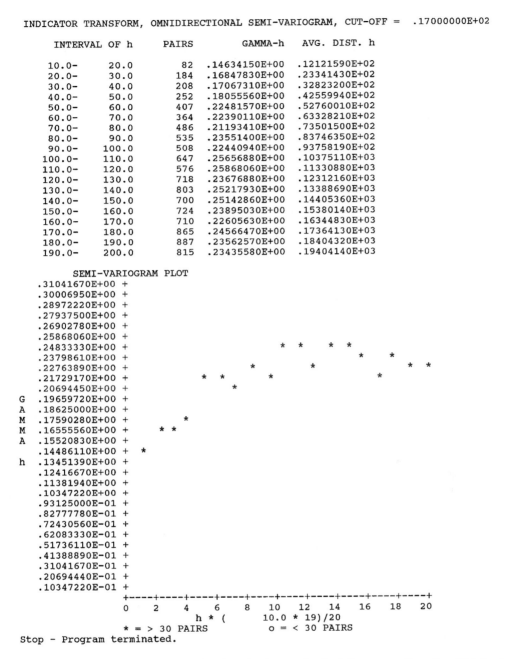

Figure 6.35 Omnidirectional semivariogram for the indicator transform $C = 17.0$ for the NVSIMLOG.DAT data set.

In this instance, a contour map is produced of the gridding results showing the spatial distribution of the indicator function (Figure 6.36). The contour lines show the probability of being less than or equal to the cutoff value. The larger the probability value is, the more likely is the possibility that Z is less than the cutoff value of 17.0.

Figure 6.36 Contour map of the indicator transform $C = 17.0$ for the NVSIMLOG.DAT data set. This map was drawn using the computer algorithm TOPO, a *Golden Graphics* product (see reference list, and captions to Figures 6.25 and 6.33).

Comparing this map to the one developed using the lognormal transform (Figure 6.33) reveals that contour lines for the indicator transform are similar to those from lognormal kriging. Small values of the indicator transform actually indicate a large probability of exceeding 17.0. Regions associated with small indicator values (Figure 6.36) match those regions associated with larger data values (Figure 6.33). This example serves as a demonstration for the use and interpretation of the indicator transform in kriging.

Chapter Summary

- Geostatistics encompasses generic methods for spatial analysis and interpolation.
- The semivariogram describes the spatial correlation inherent to a regionalized variable Z.

- Kriging is a least-squares, linear, unbiased estimator which uses information captured in the semivariogram for spatial interpolation.
- In developing a geostatistical model for a regionalized variable Z, the distribution for Z must be inspected prior to the application of semivariogram and/or kriging analysis. The reason for this is twofold: (1) to identify the presence, if any, of unusual (extreme) data values (outliers); and (2) to determine if a data transform is necessary prior to geostatistical analysis.

Exercises

(An asterisk indicates an advanced or graduate student problem)

1. Look up the word range in the dictionary. Speculate about the reasons why this word was chosen to represent that value of lag distance for which the semivariogram attains the sill.

2. At the range, the semivariogram is equal in value to the attribute called the sill. For lags larger than the range, the semivariogram remains constant, equal to the sill value. What does this behavior of the semivariogram imply? How does this answer relate to Problem 1?

3. Analyze the data sets in Appendix A, SANFER.DAT, AREA1.DAT, and AREA2.DAT.
   (a) Do a histogram analysis using the program HISTO (Chapter 4).
   (b) Do a semivariogram analysis using FGAM for (i) 4-directional calculations and (ii) omnidirectional calculations. Be sure to select an appropriate data transform if warranted.
   (c) Perform kriging using JCBLOK for cross-validation and gridding, using an appropriate data transform, if warranted.
   (d) Keep a detailed set of notes for steps (a) through (c) documenting (i) problems encountered; (ii) experiments to vary class size; (iii) experiments, if any, with indicator transforms; (iv) problems encountered with semivariogram analysis or kriging; and so on.

4. Compute by hand the kriging matrix required to yield an estimate at location $x = 1$, $y = 1$ (Figure 6.4), using the $N = 6$ closest data locations. Solve this matrix system using the program GAUSS (Chapter 2). Use the solutions for the kriging weights and Lagrangian multiplier to compute an estimate and kriging variance for this location. Verify your hand calculation using JCBLOK.

5. Redo Problem 4 using the $N = 3$ closest data locations. What is inferred about the influence of N on the calculation?

*6. Revise the block kriging procedure based on Gaussian quadrature to use 5-point approximation and repeat the experiment (Table 6.6) for the exponential model. Is an improvement in accuracy obtained using the Gaussian quadrature approach?

7. Obtain a set of spatial data and perform a geostatistical analysis, as in Problem 3. What did you learn?

8. Verify the results in Table 6.1 by hand and by using FGAM.

*9. Show that the expression for kriging variance given in Section 6.8 is equivalent to the error variance equation q by assuming that the sill of the semivariogram is equivalent to the variance of Z.

References and Suggested Readings

ARMSTRONG, M., ed. 1989. *Geostatistics, Parts 1 and 2.* Dordrecht, Netherlands: Kluwer Academic.

CARR, J.R. 1987. A comparison of FORTRAN, Pascal and C for variogram computation on a microcomputer. *Computers and Geosciences* **13**(6): 645–654.

CARR, J.R. 1990. UVKRIG: a FORTRAN-77 program for universal kriging. *Computers and Geosciences* **16**(2): 211–236.

CARR, J.R. 1992. A treatise on the use of geostatistics for the characterization of nonrenewable resources. *Nonrenewable Resources* **1**(1): 61–73.

CARR, J.R., and R.E. BAILEY. 1986. An indicator kriging model for the investigation of seismic hazard. *Mathematical Geology* **18**(4): 409–428.

CARR, J.R., and C.E. GLASS. 1985. Treatment of earthquake ground motion using regionalized variables. *Journal of the International Association for Mathematical Geology* **17**(3): 221–241.

CARR, J.R., and J.A. PALMER. 1993. Revisiting the accurate calculation of block-sample covariances using Gauss quadrature. *Mathematical Geology* **25**(5): 507–524.

CARR, J.R., R.E. BAILEY, and E.D. DENG. 1985. Use of indicator variograms for an enhanced spatial analysis. *Journal of the International Association for Mathematical Geology* **17**(8): 797–812.

CLARK, I. 1979. *Practical Geostatistics.* London: Applied Science.

DAVID, M. 1977. *Geostatistical Ore Reserve Estimation.* Amsterdam: Elsevier.

DAVIS, M., and M. DAVID. 1978. The numerical calculation of integrals of an isotropic variogram function on hypercubes. *Journal of the International Association for Mathematical Geology* **10**: 311–314.

DELHOMME, J.P. 1979. Spatial variability and uncertainty in groundwater flow parameters: a geostatistical approach. *Water Resources Research* **15**(2): 269–280.

DEVORE, J.L. 1991. *Probability and Statistics for Engineering and the Sciences,* 3d ed. Monterey, CA: Brooks/Cole.

Golden Graphics System, Golden Software, P.O. Box 281, Golden, CO 80402, USA.

GUARASCIO, M., M. DAVID, and C. HUIJBREGTS, eds. 1976. *Advanced Geostatistics for the Mining Industry.* Dordrecht, Netherlands: Reidel.

JOURNEL, A.G. 1980. The lognormal approach to predicting local distributions of selective mining unit grades. *Journal of Mathematical Geology* **12**(4): 285–303.

JOURNEL, A.G. 1983. Nonparametric estimation of spatial distributions. *Journal of the International Association for Mathematical Geology* **15**(3): 445–468.

JOURNEL, A.G., and C.J. HUIJBREGTS. 1978. *Mining Geostatistics.* London: Academic.

KNUDSEN, H.P., and Y.C. KIM. 1978. *A Short Course of Geostatistical Ore Reserve Estimation.* Tucson: University of Arizona, Department of Mining and Geological Engineering.

KRIGE, D.G. 1951. A statistical approach to some basic mine valuation problems on the Witwatersrand. *Journal of the Chemical and Metallurgical Mining Society of South Africa* **52**: 119–139.

MATHERON, G. 1963. Principles of geostatistics. *Economic Geology* **58**: 1246–1266.

OLEA, R.A., ed. 1991. *Geostatistical Glossary and Multilingual Dictionary.* New York: Oxford University Press.

SOARES, A., ed. 1993. *Proceedings of the 4th International Conference on Geostatistics.* Dordrecht, Netherlands: Kluwer Academic.

VERLY, G., M. DAVID, A. JOURNEL, and A. MARECHAL, eds. 1984. *Geostatistics for Natural Resources Characterization,* Parts 1 and 2. Dordrecht, Netherlands: Reidel.

Figure 6.6 A listing for the program FGAM.

```fortran
      PROGRAM FGAM
C
C     A PORTABLE FORTRAN-77 PROGRAM FOR SEMI-VARIOGRAM
C     CALCULATION
C
C     PROGRAM CAPABILITIES
C
C     1.  DIRECTIONAL VARIOGRAM CALCULATION FOR 4 SPATIAL
C         DIRECTIONS:
C         0.0   = EAST-WEST
C         45.0  = NE - SW
C         90.0  = NORTH-SOUTH
C         135.0 = NW - SE
C     2.  DIRECTIONAL VARIOGRAM CALCULATION FOR 8 SPATIAL
C         DIRECTIONS:
C         0.0   = EAST - WEST
C         22.5  = NE  - SW
C         45.0  = NE  - SW
C         67.5  = NE  - SW
C         90.0  = NORTH - SOUTH
C         112.5 = NW  - SE
C         135.0 = NW  - SE
C         157.5 = NW  - SE
C     3.  OMNIDIRECTIONAL (AVERAGE) CALCULATION
C     4.  LOG TRANSFORMATION
C     5.  INDICATOR TRANSFORMATION
C
C**********************************************************
C                 USER'S GUIDE, FGAM
C**********************************************************
C
C     LINE 1:  ENTER:  IDIR, ITRANS
C        WHERE:  IDIR = 1, OMNIDIRECTIONAL CALCULATION
C                     = 4, 4 SPATIAL DIRECTIONAL CALC.
C                     = 8, 8 SPATIAL DIRECTIONAL CALC.
C
C                ITRANS = 0, NO DATA TRANSFORMATION
C                       = 1, NATURAL LOG. TRANSFORMATION
C                       = 2, INDICATOR TRANSFORMATION
C                          (NOTE:  IF THIS OPTION IS
C                           CHOSEN, AN INTERACTIVE PROMPT
C                           ASKS FOR INPUT TO DEFINE CUT-OFF)
C
C     LINE 2:  ENTER:  XCLASS
C        WHERE:  XCLASS IS THE CLASS SIZE FOR h
C
C     LINES 3+:  ENTER SPATIAL DATA:
C
C        WHERE EACH LINE IS:
C
C        DATA VALUE, Y-COORD. (NORTHING), X-COORD. (EASTING)
C
C**********************************************************
C                 END OF USER'S GUIDE FOR FGAM
C**********************************************************
C
      PARAMETER (NN=4000, NDIR=8, INC=20)
      DIMENSION ZDAT(NN), X(NN), Y(NN)
      COMMON /SEMIV/ XGAM(INC,NDIR), XANGL(NDIR,2),
     2               NPAIR(INC,NDIR), GDIST(INC,NDIR)
C
C     OPEN FILES FOR INPUT (UNIT=5) AND OUTPUT (UNIT=1)
C
      OPEN(5, FILE = ' ')
      OPEN(1, FILE = ' ', STATUS = 'NEW')
C
C     READ OPTIONS, UNIT 5
C
      READ(5,*) IDIR, ITRANS
      READ(5,*) XCLASS
      CUT = 0.0
C
```

207

Figure 6.6 continued

```fortran
C
C      INITIALIZE DIRECTION ANGLES (IN RADIANS)
C
       XRAD = 0.01745329252
       IF (IDIR .EQ. 1) THEN
          XANGL(1,1) = 0.0
          XANGL(1,2) = 90.0 * XRAD
       ELSEIF (IDIR .EQ. 4) THEN
          DO I = 1,IDIR
             XANGL(I,1) = (0.0 + (I-1) * 45.0) * XRAD
             XANGL(I,2) = 22.5 * XRAD
          END DO
       ELSEIF (IDIR .EQ. 8) THEN
          DO I = 1,IDIR
             XANGL(I,1) = (0.0 + (I-1) * 22.5) * XRAD
             XANGL(I,2) = 11.25 * XRAD
          END DO
       END IF
C
C      READ SPATIAL DATA FROM UNIT 5
C
       NDAT = 0
10     CONTINUE
       READ(5,*,END=20) AZ, AY, AX
       NDAT = NDAT + 1
       ZDAT(NDAT) = AZ
       Y   (NDAT) = AY
       X   (NDAT) = AX
       GO TO 10
20     CONTINUE
C
C      TRANSFORM THE DATA IF DESIRED
C
       IF (ITRANS .NE. 0) THEN
       IF (ITRANS .EQ. 1) THEN
          DO I = 1, NDAT
             IF (ZDAT(I) .GT. 0.0) THEN
                ZDAT(I) = ALOG(ZDAT(I))
             END IF
          END DO

       ELSEIF (ITRANS .EQ. 2) THEN
          WRITE(*,*) 'ENTER THE VALUE OF THE CUT-OFF TO BE USED'
          WRITE(*,*) 'FOR THE INDICATOR TRANSFORMATION'
          READ(*,*) CUT
          DO I = 1,NDAT
             IF (ZDAT(I) .LE. CUT) THEN
                ZDAT(I) = 1.0
             ELSE
                ZDAT(I) = 0.0
             END IF
          END DO
       END IF
       END IF
C
C      INTIALIZE THE VARIOGRAM CALCULATION BY ZEROING ARRAYS
C
       DO I = 1,INC
       DO J = 1,NDIR
          XGAM(I,J)  = 0.0
          GDIST(I,J) = 0.0
          NPAIR(I,J) = 0
       END DO
       END DO
C
C      COMPLETE THE SEMI-VARIOGRAM CALCULATION
C
       DO I = 1,NDAT
       DO J = I+1, NDAT
C
C      CALCULATE THE DISTANCE BETWEEN LOCATIONS I AND J
C
       DIFX = X(I) - X(J)
       DIFY = Y(I) - Y(J)
       DIST = SQRT(DIFX * DIFX + DIFY * DIFY)
C
C      IF DIST > 20 * XCLASS, SKIP THIS PAIR
C
       IF (DIST .GT. 20.0 * XCLASS) THEN
          GO TO 100
       END IF
C
```

Figure 6.6 continued

```fortran
C
C     DETERMINE TO WHICH INCREMENT OF h THIS PAIR BELONGS
C
      JINC = INT(DIST / XCLASS) + 1
      IF (JINC .LT. 1) JINC = 1
      IF (JINC .GT. 20) GO TO 100
C
C     CALCULATE THE SQUARED DIFFERENCE IN DATA VALUE FOR PAIR
C
      DIFZSQ = (ZDAT(I) - ZDAT(J)) ** 2
C
C
C     IF IDIR .EQ. 4 .OR. 8, EVALUATE THE ORIENTATION OF THE PAIR
C
      IF (IDIR .EQ. 4 .OR. IDIR .EQ. 8) THEN
      IF (DIFX .NE. 0.0) THEN
      RATIO = DIFY / DIFX
      YANGL = ATAN(RATIO)
      IF (DIFX .LT. 0.0 .AND. DIFY .GT. 0.0) YANGL = YANGL +
   2                    3.141592654
      ELSE
      YANGL = 90.0 * 0.01745329252
      END IF
      DO IL = 1, IDIR
      ALESS = XANGL(IL,1) - XANGL(IL,2)
      AMORE = XANGL(IL,1) + XANGL(IL,2)
      IF (YANGL .GE. ALESS .AND. YANGL .LT. AMORE) THEN
      KI = IL
      ELSEIF(YANGL .GE. (ALESS+3.1416) .AND. YANGL .LT.
   3        (AMORE+3.1416)) THEN
      KI = IL
      END IF
      END DO
      END IF
      IF (IDIR .EQ. 1) KI = 1
C
C
C     PLACE SQUARE DIFFERENCE INTO PROPER ARRAY LOCATION
C
      XGAM(JINC,KI) = XGAM(JINC,KI) + DIFZSQ
      NPAIR(JINC,KI) = NPAIR(JINC,KI) + 1
      GDIST(JINC,KI) = GDIST(JINC,KI) + DIST
  100 CONTINUE
      END DO
      END DO
C
C
C     COMPLETE THE SEMI-VARIOGRAM CALCULATION
C
C
      DO I = 1,INC
      DO J = 1, IDIR
      IF (NPAIR(I,J) .NE. 0) THEN
      XGAM(I,J) = XGAM(I,J) / (2.0 * NPAIR(I,J))
      GDIST(I,J) = GDIST(I,J)/(         NPAIR(I,J))
      END IF
      END DO
      END DO
C
C     PLOT THE SEMI-VARIOGRAM(S)
C
      CALL PLOT(IDIR, XCLASS, CUT, ITRANS)
C
C     WRITE RESULTS TO UNIT 1
C
      DO J = 1, IDIR
      WRITE(1,50) J
   50 FORMAT(10X,'SEMI-VARIOGRAM FOR DIRECTION',I5)
      DO I = 1, INC
      WRITE(1,*) NPAIR(I,J), XGAM(I,J), GDIST(I,J)
      END DO
      END DO
C
C     TERMINATE THE PROGRAM
C
      CLOSE(5)
      CLOSE(1)
      STOP
      END
C
```

Figure 6.6 continued

```fortran
C
      SUBROUTINE PLOT(IDIR, XCLASS, CUT, ITRANS)
      PARAMETER(NN = 4000, NDIR = 8, INC = 20)
      COMMON /SEMIV/ XGAM(INC,NDIR), XANGL(NDIR,2),
     2               NPAIR(INC,NDIR), GDIST(INC,NDIR)
      CHARACTER*1 A(30,50), G(30)
C
C     SUBROUTINE TO PRINT AND PLOT SEMI-VARIOGRAM CALCULATION
C     RESULTS
C
      DO J = 1, IDIR
C
C     FIRST, TABULATE THE CALCULATION RESULTS
C
      IF (ITRANS .EQ. 0) THEN
        IF (IDIR .EQ. 1) THEN
          WRITE(*,10)
        ELSEIF (IDIR .EQ. 4) THEN
          XK = 0.0 + (J-1) * 45.0
          WRITE(*,11) XK
        ELSEIF (IDIR .EQ. 8) THEN
          XK = 0.0 + (J-1) * 22.5
          WRITE(*,11) XK
        END IF
      ELSEIF (ITRANS .EQ. 1) THEN
        IF (IDIR .EQ. 1) THEN
          WRITE(*,12)
        ELSEIF (IDIR .EQ. 4) THEN
          XK = 0.0 + (J-1) * 45.0
          WRITE(*,13) XK
        ELSEIF (IDIR .EQ. 8) THEN
          XK = 0.0 + (J-1) * 22.5
          WRITE(*,13) XK
        END IF
      ELSEIF (ITRANS .EQ. 2) THEN
        IF (IDIR .EQ. 1) THEN
          WRITE(*,14) CUT
        ELSEIF (IDIR .EQ. 4) THEN
          XK = 0.0 + (J-1) * 45.0
          WRITE(*,15) CUT, XK
        ELSEIF (IDIR .EQ. 8) THEN
          XK = 0.0 + (J-1) * 22.5
          WRITE(*,15) CUT, XK
        END IF
      END IF
10    FORMAT(1H1, 1X, 'UNTRANSFORMED DATA, OMNIDIRECTIONAL'
     2       ,' SEMI-VARIOGRAM CALCULATION',/)
11    FORMAT(1H1, 1X, 'UNTRANSFORMED DATA, DIRECTIONAL SEMI-'
     2       'VARIOGRAM CALCULATION, DIRECTION =', F7.2,/)
12    FORMAT(1H1, 1X, 'NAT. LOG. TRANSFORMED DATA, OMNIDIRECTIONAL'
     2       ,' SEMI-VARIOGRAM CALCULATION',/)
13    FORMAT(1H1, 1X, 'NAT. LOG. TRANSFORMED DATA, DIRECTIONAL'
     2       ,' SEMI-VARIOGRAM, DIRECTION =', F7.2,/)
14    FORMAT(1H1, 1X, 'INDICATOR TRANSFORM, OMNIDIRECTIONAL'
     2       ,' SEMI-VARIOGRAM, CUT-OFF =', E15.8,/)
15    FORMAT(1H1, 1X, 'INDICATOR TRANSFORM, DIRECT. SEMI-'
     2       'VARIO., CUT-OFF=', E15.8,' DIR.=', F7.2,/)
C
C     CREATE THE TABLE
C
      WRITE(*,40)
40    FORMAT(7X,'INTERVAL OF h',1X,'      PAIRS',8X,'GAMMA-h',
     2       3X,'AVG. DIST. h',//)
      JUSE = 0
      DO I = 1, INC
        IF (XGAM(I,J) .NE. 0.0) THEN
          JUSE = JUSE + 1
          XI = I * XCLASS
          XL = (I-1) * XCLASS
          WRITE(*,50) XL, XI, NPAIR(I,J), XGAM(I,J), GDIST(I,J)
        END IF
      END DO
50    FORMAT(1X,F9.1,'-',F9.1,I10,2E15.8)
C
```

Figure 6.6 continued

```fortran
C     NOW, CREATE THE PLOT.  FIRST, CALCULATE THE MAXIMUM
C     SEMI-VARIOGRAM VALUE FOR ALL INCREMENTS
C
      GMAX = XGAM(1,J)
      DISMAX = GDIST(1,J)
      DO I = 2, INC
      IF (XGAM(I,J) .GT. GMAX) THEN
      GMAX = XGAM(I,J)
      END IF
      IF (GDIST(I,J) .GT. DISMAX) THEN
      DISMAX = GDIST(I,J)
      END IF
      END DO
      GMAX = 1.2 * GMAX
C
C     FILL THE ARRAY, A, WITH BLANKS TO INITIALIZE
C
      DO IA = 1,30
      DO JA = 1,50
      A(IA,JA) = ' '
      END DO
      END DO
C
C     NOW, FILL THE ARRAY A WITH SYMBOLS FOR SEMI-VARIOGRAM
C
      DO I = 1, INC
      IF (XGAM(I,J) .NE. 0.0) THEN
      B = XGAM(I,J)
      JY = 1 + (30 - INT(B * 30.0 / GMAX))
      IF (JY .GT. 30) JY = 30
      JX = INT (50.0 * GDIST(I,J) / DISMAX)
      IF (JX .GT. 50) JX = 50
      IF (NPAIR(I,J) .GT. 30) THEN
      A(JY,JX) = '*'
      ELSE
      A(JY,JX) = 'o'
      END IF
      END IF
      END DO
C


C     FILL THE VECTOR, G
C
      DO IG = 1, 30
      G(IG) = ' '
      END DO
      G(12) = 'G'
      G(13) = 'A'
      G(14) = 'M'
      G(15) = 'M'
      G(16) = 'A'
      G(18) = 'h'
C
C     PLOT THE SEMI-VARIOGRAM
C
      WRITE(*,60)
60    FORMAT(/,10X,'SEMI-VARIOGRAM PLOT')
      DO K = 1,30
      XK = GMAX - (K-1) * (GMAX / 30.0)
      WRITE(*,70) G(K), XK, (A(K,JK), JK = 1,50)
      END DO
70    FORMAT(1X,A1,E15.8,' ',50A1)
      WRITE(*,80)
      WRITE(*,90)
      WRITE(*,100) XCLASS, JUSE
      WRITE(*,110)
80    FORMAT(18X,'+',5('---+---+'))
90    FORMAT(18X,'0',4X,'2',4X,'4',4X,'6',4X,'8',3X,'10',3X,'12',
     2       3X,'14',3X,'16',3X,'18',3X,'20')
100   FORMAT(30X,'h * (',F10.1,'*',I3,')/20')
110   FORMAT(18X,'* = > 30 PAIRS',10X,'o = < 30 PAIRS')
      END DO
      RETURN
      END
```

211

Figure 6.23 A listing for the program JCBLOK.

```
C     PROGRAM JCBLOK
C
C     A PORTABLE, FORTRAN-77 PROGRAM FOR ORDINARY-PUNCTUAL,
C     OR ORDINARY-BLOCK KRIGING.
C
C     PROGRAM CAPABILITIES:
C
C     1.  PUNCTUAL KRIGING
C     2.  BLOCK KRIGING
C     3.  GENERAL OR QUADRANT SEARCH
C     4.  UP TO FIVE NESTED VARIOGRAM MODELS
C     5.  SPHERICAL, EXPONENTIAL, GAUSSIAN, AND LINEAR
C         VARIOGRAM MODELS
C     6.  USE OF GAUSS QUADRATURE FOR INTEGRATING
C         BLOCK-SAMPLE COVARIANCES FOR BLOCK KRIGING
C     7.  ANISOTROPIC SPATIAL STRUCTURE MODELING
C     8.  CROSS-VALIDATION OR GRIDDING
C     9.  LOG-NORMAL AND INDICATOR TRANSFORMS
C
C     PROGRAM LIMITATIONS:
C
C     1.  ONE OR TWO-DIMENSIONAL ESTIMATION ONLY
C     2.  2000 SPATIAL DATA LOCATIONS, MAXIMUM
C     3.  USE OF UP TO 50 (MAXIMUM) CLOSEST LOCATIONS FOR
C         KRIGING
C*******************************************************************
C                   USER'S GUIDE FOR JCBLOK
C*******************************************************************
C
C     FREE FORMAT DATA ENTRY, ALL LINES:
C
C     LINE 1:  ENTER:  IKRIG, ITRAN
C              WHERE:  IKRIG = 1, PUNCTUAL KRIGING
C                            = 2, BLOCK KRIGING
C                      ITRAN = 0, NO DATA TRANSFORMATION
C                            = 1, LOG-NORMAL TRANSFORM
C                            = 2, INDICATOR TRANSFORM
C
C              NOTE:   IF ITRAN=1, PROGRAM
C                      WILL INTERACTIVELY REQUEST
C                      INFORMATION FOR THE CORRECTION
C                      FACTOR, XKO.  TO DETERMINE
C                      THIS VALUE, SET ITRAN=1,
C                      SET PROGRAM UP FOR CROSS-VAL.
C                      USE XKO=1.0, THEN EXAMINE
C                      THE SUMMARY STATISTICS AND SET
C                      XKO = ORIGINAL MEAN/MEAN OF EST.
C
C              NOTE:   IF ITRAN=2, AN
C                      INTERACTIVE MESSAGE
C                      WILL PROMPT ENTRY FOR
C                      C, THE CUT-OFF.
C
C     LINE 2:  ENTER:  IBEGC, IENDC, IBEGR, IENDR
C              WHERE:  IBEGC = BEGINNING COLUMN NO. FOR GRID
C                      IENDC = ENDING COLUMN NO. FOR GRID
C                      IBEGR = BEGINNING ROW NO. FOR GRID
C                      IENDR = ENDING ROW NO. FOR GRID
C              NOTE:   IF IBEGC=IENDC=IBEGR=IENDR=1, THEN
C                      CROSS VALIDATION MODE IS SELECTED
C
C     LINE 3:  ENTER:  SIZEX, SIZEY, YMAX, XMIN
C              WHERE:  SIZEX = DISTANCE BETWEEN COLUMNS IN
C                      GRID, WHICH IS ALSO THE X-
C                      DIMENSION OF BLOCK FOR BLOCK
C                      KRIGING
C                      SIZEY = DISTANCE BETWEEN ROWS IN GRID,
C                      WHICH IS ALSO THE Y-DIMENSION OF
C                      BLOCK FOR BLOCK KRIGING
C                      YMAX,XMIN = COORDINATE OF UPPER-LEFT
C                      HAND CORNER OF REGION WITHIN
C                      WHICH ESTIMATION IS PERFORMED.
C              NOTE:   YMAX-XMIN=0.0 FOR CROSS-
C                      VALIDATION MODE.
```

Figure 6.23 continued

```
C    LINES 4+:   VARIOGRAM PARAMETERS:
C         LINE 4A:   ENTER:  INEST
C                    WHERE:  INEST = NUMBER OF NESTED MODELS,
C                                    A NUMBER BETWEEN 1 AND 5
C
C         LINE 4B TO (UP TO) 4F:
C         DO I = 1, INEST
C         ENTER:  MODEL(I),CO(I),SILL(I),RANGE(I),ANIS(I),
C                 RATIO(I)
C         WHERE:  MODEL(I) = 1, SPHERICAL
C                            2, EXPONENTIAL
C                            3, GAUSSIAN
C                            4, LINEAR
C                 CO(I)    = NUGGET VALUE
C                 SILL(I)  = SILL VALUE
C                 RANGE(I) = RANGE FOR MODEL
C                 ANIS(I)  = ANGLE ORIENTATION, IN DEGREES,
C                            FOR MAJOR AXIS OF SPATIAL
C                            STRUCTURE ELLIPSE; USUALLY
C                            EQUAL TO ANGLE OF VARIOGRAM
C                            HAVING LARGEST RANGE; SET
C                            ANIS(I) = 0.0 FOR SPATIAL
C                            ISOTROPY
C                 RATIO(I) = RATIO OF LENGTHS OF:
C                            MAJOR AXIS/MINOR AXIS OF
C                            SPATIAL STRUCTURE ELLIPSE;
C                            USU. EQUAL TO RATIO OF
C                            RANGES FOR DIRECTIONAL
C                            VARIOGRAMS FOR THE ANGLES:
C                            ANIS(I) AND ANIS(I)+90.
C                            RATIO(I) = 1.0 FOR ISOTROPY.
C
C                 NOTE:  FOR A LINEAR VARIO.
C                        MODEL, SET THE SILL EQUAL
C                        TO GAMMA(h) FOR THE
C                        LARGEST INCREMENT OF h;
C                        SET RANGE=THIS INCREMENT
C
C    LINE 5:    SEARCH PARAMETERS:
C         ENTER:  ISERCH, RADIUS, NK
C         WHERE:  ISERCH = 1, GENERAL SEARCH
C                        = 2, QUADRANT SEARCH
C                 RADIUS = RADIUS OF SEARCH WINDOW TO USE
C                          FOR FINDING CLOSEST NEIGHBORING
C                          LOCATIONS
C                 NK = NUMBER OF CLOSEST LOCATIONS WITHIN
C                      THE SEARCH WINDOW OF RADIUS, RADIUS,
C                      TO USE; 50 = MAXIMUM.
C                 NOTE:  IF ISERCH = 2, THEN SET
C                        NK = MULTIPLE OF 4 IN THE
C                        RANGE: 4 TO 48.
C
C    LINES 6+:  SPATIAL DATA:  ENTER:
C         DATA VALUE, Y-COORD (NORTHING), X-COORD (EASTING)
C         ONE LINE FOR EACH DATA LOCATION
C
C*****************************************************
C              END OF USER'S GUIDE FOR JCBLOK
C*****************************************************
C
      PARAMETER(N=2000, NEST=5, JMAX=50)
      COMMON /SPADAT/ XDAT(N), X(N), Y(N), NTOT
      COMMON /MAT  /  COV1J(JMAX+1,JMAX+1), WT(JMAX+1),
     2                COV0I(JMAX+1), IKRIG
      COMMON /VARIO / INEST, MODEL(NEST), CO(NEST), SILL(NEST),
     2                RANGE(NEST), ANIS(NEST), RATIO(NEST)
      COMMON /SEARCH/ IUSE(JMAX), DIST(JMAX), ISERCH,RADIUS,NK,JUSE
      COMMON /XGRID / IBEGC, IENDC, IBEGR, IENDR, SIZEX, SIZEY,
     2                YMAX, XMIN
C
C   INITIALIZE UNIT 5 FOR INPUT, UNIT 1 FOR OUTPUT
C
      WRITE(*,*) 'ENTER FILENAMES FOR DATA INPUT AND OPTION ECHO'
      OPEN(5, FILE = ' ')
      OPEN(1, FILE = ' ', STATUS = 'NEW')
C
```

213

Figure 6.23 continued

```fortran
C
C       READ KRIGING OPTIONS
C
        READ(5,*) IKRIG, IKRIG, ITRAN
        IF (ITRAN .EQ. 1) THEN
          WRITE(*,*) 'ENTER VALUE FOR XKO FOR LOG-NORMAL KRIGING'
          READ(*,*) XKO
        ELSE
          XKO = 0.0
        END IF
        IF (ITRAN .EQ. 2) THEN
          WRITE(*,*) 'PLEASE ENTER THE VALUE OF THE CUT OFF FOR'
          WRITE(*,*) 'USE IN FORMING THE INDICATOR TRANSFORM'
          READ(*,*) CUT
        END IF
        READ(5,*) IBEGC, IENDC, IBEGR, IENDR
        READ(5,*) SIZEX, SIZEY, YMAX, XMIN
        READ(5,*) INEST
        IF (INEST .EQ. 1) THEN
          READ(5,*) MODEL(1), CO(1), SILL(1), RANGE(1), ANIS(1),
     2      RATIO(1)
        ELSE
          DO I = 1, INEST
            READ(5,*) MODEL(I), CO(I), SILL(I), RANGE(I), ANIS(I),
     2        RATIO(I)
          END DO
        END IF
        READ(5,*) ISERCH, RADIUS, NK
C
C       ACCESS SPATIAL DATA
C
        NTOT = 0
10      CONTINUE
        READ(5,*, END = 20) AD, AY, AX
        NTOT = NTOT + 1
        IF (ITRAN .EQ. 0) THEN
          XDAT(NTOT) = AD
        ELSEIF (ITRAN .EQ. 1) THEN
          IF (AD .GT. 0.0) THEN
            XDAT(NTOT) = ALOG(AD)
          ELSE
            XDAT(NTOT) = 0.0
          END IF
        ELSEIF (ITRAN .EQ. 2) THEN
          IF (AD .LE. CUT) THEN
            XDAT(NTOT) = 1.0
          ELSE
            XDAT(NTOT) = 0.0
          END IF
        END IF
        Y (NTOT) = AY
        X (NTOT) = AX
        GO TO 10
20      CONTINUE
C
C       ECHO THE DATA INPUT FOR CHECK
C
        CALL ECHO(ITRAN,XKO)
C
C       PERFORM KRIGING, CROSS-VALIDATION OR GRIDDING
C
        IF (IBEGC .EQ. 1 .AND. IENDC .EQ. 1 .AND.
     2    IBEGR .EQ. 1 .AND. IENDR .EQ. 1) THEN
          CALL CROSS(ITRAN,XKO)
        ELSE
          CALL GRID(ITRAN,XKO)
        END IF
C
C       TERMINATE THE PROGRAM
C
        CLOSE(5)
        CLOSE(1)
        STOP
        END
C
```

Figure 6.23 continued

```fortran
C
      SUBROUTINE CROSS(ITRAN,XKO)
      PARAMETER (N=2000, NEST=5, JMAX=50)
      COMMON /SPADAT/ XDAT(N), X(N), Y(N), NTOT
      COMMON /MAT   / COVIJ(JMAX+1,JMAX+1), WT(JMAX+1),
     2                COV0I(JMAX+1), IKRIG
      COMMON /VARIO / INEST, MODEL(NEST), CO(NEST), SILL(NEST),
     2                RANGE(NEST), ANIS(NEST), RATIO(NEST)
      COMMON /SEARCH/ IUSE(JMAX), DIST(JMAX), ISERCH,RADIUS,NK,JUSE
      REAL KRIGV
C
C     THIS SUBROUTINE PERFORMS CROSS VALIDATION FOR EITHER
C     PUNCTUAL OR BLOCK KRIGING
C
      WRITE(*,*) 'ENTER FILE NAME FOR OUTPUT OF CROSS-VAL RESULTS'
      OPEN(2, FILE = ' ', STATUS = 'NEW')
      WRITE(*,10)
      WRITE(2,10)
10    FORMAT(1H1, 10X, 'CROSS VALIDATION RESULTS',/)
      WRITE(*,15)
      WRITE(2,15)
15    FORMAT(/,1X,'LOC.',3X,'Y-COORD',3X,'X-COORD',5X,'ACTUAL',
     2       3X,'ESTIMATE',6X,'ERROR',2X,'KRIG VAR.',//)
C
C     KRIG EACH SAMPLE LOCATION IF POSSIBLE
C
      KOUNT = 0
      SUMERR = 0.0
      SUMSQ  = 0.0
      SUMK   = 0.0
      SUMEST = 0.0
      SUMDAT = 0.0
      SUMD2  = 0.0
C
C     COMPUTE THE MEAN OF THE SPATIAL DATA, Z (EVEN IF TRANS.)
C
      DO I = 1,NTOT
        SUMDAT = SUMDAT + XDAT(I)
        IF (ITRAN .EQ. 1) SUMD2 = SUMD2 + EXP(XDAT(I))
      END DO
      SUMDAT = SUMDAT / NTOT
      IF (ITRAN .EQ. 1) SUMD2 = SUMD2 / NTOT
      DO I = 1, NTOT
C
C     FIND CLOSEST LOCATIONS TO ITH LOCATION
C
        XX = X(I)
        YY = Y(I)
        CALL FIND(XX, YY, I)
        IF (JUSE .LT. 2) GO TO 100
        KOUNT = KOUNT + 1
C
C     FORM THE COVARIANCE MATRICES
C
        CALL CHAWK(XX,YY)
C
C     SOLVE FOR THE KRIGING WEIGHTS
C
        CALL GAUSIT(JUSE+1)
C
C     COMPUTE THE KRIGING ESTIMATE
C
        EST = 0.0
        DO I2 = 1, JUSE
          K = IUSE(I2)
          IF (ITRAN .EQ. 1) THEN
            EST = EST + WT(I2) * (XDAT(K) - SUMDAT)
          ELSE
            EST = EST + WT(I2) * XDAT(K)
          END IF
        END DO
C
```

215

Figure 6.23 continued

```fortran
C
C       COMPUTE KRIGING VARIANCE
C
        IF (INEST .EQ. 1) THEN
          SILTOT = SILL(1)
        ELSE
          SILTOT = 0.0
          DO J = 1, INEST
            SILTOT = SILTOT + SILL(J)
          END DO
        END IF
        SUM = 0.0
        DO J = 1, JUSE
          SUM = SUM + WT(J) * COVOI(J)
        END DO
        KRIGV = SILTOT - SUM - WT(JUSE + 1)
        IF (ITRAN .EQ. 1) THEN
          EST = XKO * EXP(EST + SUMDAT + KRIGV/2.0)
        END IF
        IF (ITRAN .EQ. 1) THEN
          ESTERR = EXP(XDAT(I)) - EST
        ELSE
          ESTERR = XDAT(I) - EST
        END IF
C
C       PRINT RESULTS TO SCREEN AND TO UNIT 1
C
        IF (ITRAN .EQ. 1) THEN
          ADAT = EXP(XDAT(I))
        ELSE
          ADAT = XDAT(I)
        END IF
        WRITE(*,20) I, Y(I), X(I), ADAT, EST, ESTERR, KRIGV
        WRITE(2,20) I, Y(I), X(I), ADAT, EST, ESTERR, KRIGV
20      FORMAT(I5,2E10.3,4E11.4)
C
C       UPDATE ERROR SUM, SQUARED ERROR SUM, AND KRIG VAR SUM
C
        SUMERR = SUMERR + ESTERR
        SUMSQ  = SUMSQ  + ESTERR * ESTERR
        SUMK   = SUMK   + KRIGV
        SUMEST = SUMEST + EST
100     CONTINUE
        END DO
C
C       SUMMARIZE CROSS VALIDATION STATISTICS
C
        IF (KOUNT .NE. 0) THEN
          AVGERR = SUMERR / KOUNT
          AVERSQ = SUMSQ / KOUNT
          VARERR = (SUMSQ - (SUMERR*SUMERR)/KOUNT) / (KOUNT-1)
          AVKRGV = SUMK / KOUNT
          AVGEST = SUMEST / KOUNT
          IF (ITRAN .EQ. 1) SUMDAT = SUMD2
          WRITE(*,200) AVGERR, AVERSQ, VARERR, AVKRGV, AVGEST, SUMDAT
          WRITE(2,200) AVGERR, AVERSQ, VARERR, AVKRGV, AVGEST, SUMDAT
200       FORMAT(//,10X,'AVERAGE KRIGING ERROR =',1E15.8,//,
     2           10X,'AVERAGE SQUARED ERROR =',1E15.8,//,
     3           10X,'VARIANCE OF THE ERROR =',1E15.8,//,
     4           10X,'AVERAGE KRIGING VAR.  =',1E15.8,//,
     5           10X,'AVERAGE ESTIMATE      =',1E15.8,//,
     6           10X,'ORIGINAL DATA MEAN    =',1E15.8)
        ELSEIF (KOUNT .EQ. 0) THEN
          WRITE(*,300)
          WRITE(2,300)
300       FORMAT(10X,'NO ESTIMATES WERE MADE; THE MOST LIKELY',/,
     2           10X,'CAUSE FOR THIS IS TOO SMALL A SEARCH RADIUS.',
     3      /,10X,'TRY INCREASING THE SEARCH RADIUS AND RETRY.')
        END IF
C
C       RETURN TO CALLING PROGRAM
C
        RETURN
        END
C
```

Figure 6.23 continued

```fortran
c
      SUBROUTINE GRID(ITRAN,XKO)
      PARAMETER (N=2000, NEST=5, JMAX=50)
      COMMON /SPADAT/ XDAT(N), X(N), Y(N), NTOT
      COMMON /MAT   / COVIJ(JMAX+1,JMAX+1), WT(JMAX + 1),
     2                COVOI(JMAX+1), IKRIG
      COMMON /VARIO / INEST, MODEL(NEST), CO(NEST), SILL(NEST),
     2                RANGE(NEST), ANIS(NEST), RATIO(NEST)
      COMMON /SEARCH/ IUSE(JMAX), DIST(JMAX), ISERCH,RADIUS,NK,JUSE
      COMMON /XGRID / IBEGC, IENDC, IBEGR, IENDR, SIZEX, SIZEY,
     2                YMAX, XMIN
      REAL KRIGV
c
c     THIS SUBROUTINE CREATES A GRID AND RETURNS ESTIMATES AT
c     GRID CELL CENTERS
c
      WRITE(*,*) 'ENTER FILE NAME FOR OUTPUT OF GRID RESULTS'
      OPEN(2, FILE = '   ', STATUS = 'NEW')
c
c     FORM GRID AND BEGIN ESTIMATION
c
c
c     FIRST, IF ITRAN =1, COMPUTE DATA MEAN
c
      IF (ITRAN .EQ. 1) THEN
      SUMDAT = 0.0
      DO I = 1,NTOT
      SUMDAT = SUMDAT + XDAT(I)
      END DO
      SUMDAT = SUMDAT / NTOT
      END IF
      YBEGIN = YMAX + SIZEY / 2.0
      XBEGIN = XMIN - SIZEX / 2.0
      DO I = IBEGR, IENDR
      YCOORD = YBEGIN - I * SIZEY
      DO J = IBEGC, IENDC
      XCOORD = XBEGIN + J * SIZEX
c
c     FIND CLOSEST LOCATIONS TO XCOORD,YCOORD
c

      CALL FIND(XCOORD,YCOORD,0)
      IF (JUSE .LT. 2) GO TO 100
c
c     FORM THE COVARIANCE MATRICES
c
      CALL CHAWK(XCOORD,YCOORD)
c
c     SOLVE FOR THE KRIGING WEIGHTS
c
      CALL GAUSIT(JUSE+1)
c
c     COMPUTE THE KRIGING ESTIMATE
c
      EST = 0.0
      DO IK = 1,JUSE
      KK = IUSE(IK)
      IF (ITRAN .EQ. 1) THEN
      EST = EST + WT(IK) * (XDAT(KK) - SUMDAT)
      ELSE
      EST = EST + WT(IK) * XDAT(KK)
      END IF
      END DO
c
c     COMPUTE THE KRIGING VARIANCE
c
      SUMK = 0.0
      DO IK = 1,JUSE
      SUMK = SUMK + WT(IK) * COVOI(IK)
      END DO
      IF (INEST .EQ. 1) THEN
      SILTOT = SILL(1)
      ELSE
      SILTOT = 0.0
      DO IK = 1,INEST
      SILTOT = SILTOT + SILL(IK)
      END DO
      END IF
      KRIGV = SILTOT - SUMK - WT(JUSE + 1)
      IF (ITRAN .EQ. 1) THEN
      EST = XKO * EXP(EST + SUMDAT + KRIGV/2.0)
      END IF
c
```

Figure 6.23 continued

```
C
C        PRINT ESTIMATION RESULTS
C
         WRITE(*,80) I, J, YCOORD, XCOORD, EST, KRIGV
         WRITE(2,80) I, J, YCOORD, XCOORD, EST, KRIGV
80       FORMAT(1X,I4,I4, 2E11.3,2E15.8)
100      CONTINUE
         END DO
         END DO
C
C        RETURN TO CALLING PROGRAM
C
         RETURN
         END
C
C
         SUBROUTINE FIND(XX,YY,J)
         PARAMETER (N=2000, NEST=5, JMAX=50)
         COMMON /SPADAT/ XDAT(N), X(N), Y(N), NTOT
         COMMON /SEARCH/ IUSE(JMAX), DIST(JMAX), ISERCH,RADIUS,NK,JUSE
         COMMON /VARIO / INEST, MODEL(NEST), CO(NEST), SILL(NEST),
        2                RANGE(NEST), ANIS(NEST), RATIO(NEST)
         INTEGER INIT(4)
         DIMENSION TMPDIS(4,JMAX), ITMUSE(4,JMAX), TEMPD(JMAX),
        2          ITEMP(JMAX)
C
C        A SUBROUTINE FOR LOCATING UP TO NK NEAREST NEIGHBORING
C        SAMPLE LOCATIONS FOR EACH ESTIMATION LOCATION
C
C


C
C        DETERMINE IF SPATIAL ANISOTROPY IS TO BE MODELED
C
         IF (INEST .EQ. 1) THEN
            IF (RATIO(1) .GT. 1.0) THEN
               KANIS = 1
            ELSE
               KANIS = 0
            END IF
         ELSE
            RMAX = RATIO(1)
            KANIS = 0
            DO J = 1, INEST
               IF (RATIO(J) .GT. 1.0) THEN
               IF (RATIO(J) .GE. RMAX) THEN
                  KANIS = J
               END IF
               END IF
            END DO
         END IF
         IF (KANIS .EQ. 0) THEN
            COSX = 1.0
            SINX = 0.0
            XMUL = 1.0
         ELSE
            XRAD = 0.017453292
            ARAD = ANIS(KANIS) * XRAD
            COSX = COS(ARAD)
            SINX = SIN(ARAD)
            XMUL = RATIO(KANIS)
         END IF
C
```

Figure 6.23 continued

```fortran
C       BEGIN SEARCH; COMPUTE DISTANCES BETWEEN XX,YY
C
      IF (ISERCH .EQ. 1) THEN
         JUSE = 0
         DO I = 1, NTOT
            IF (I .EQ. J) GO TO 100
            DX = XX - X(I)
            DY = YY - Y(I)
            XDIST = SQRT((DX*COSX + DY*SINX)**2 + XMUL*(
     2                    DY*COSX - DX*SINX)**2)
            IF (XDIST .GT. RADIUS) GO TO 100
            JUSE = JUSE + 1
            IF (JUSE .EQ. 1) THEN
               IUSE(JUSE) = I
               DIST(JUSE) = XDIST
            ELSEIF (JUSE .GT. 1 .AND. JUSE .LE. NK) THEN
               IUSE(JUSE) = I
               DIST(JUSE) = XDIST
               CALL HEAP(IUSE,DIST,JUSE)
            ELSEIF (JUSE .GT. NK) THEN
               JUSE = NK
               IF (XDIST .LT. DIST(NK)) THEN
                  IUSE(NK) = I
                  DIST(NK) = XDIST
                  CALL HEAP(IUSE,DIST,NK)
               END IF
            END IF
100         CONTINUE
         END DO

      ELSEIF (ISERCH .EQ. 2) THEN
         JUSE = 0
         IPRIME = NK / 4
         DO I = 1,4
            INIT(I) = 0
         END DO
         DO I = 1,NTOT
            IF (I .EQ. J) GO TO 200
            DX = XX - X(I)
            DY = YY - Y(I)
            XDIST = SQRT((DX*COSX + DY*SINX)**2 + XMUL*(
     2                    DY*COSX - DX*SINX)**2)
            IF (XDIST .GT. RADIUS) GO TO 200
C
C        DETERMINE TO WHICH SECTOR THIS DATA LOCATION BELONGS
C
C           XSECT = - DX
C           YSECT = - DY
C
C        CONVENTION:
C           XSECT AND YSECT POSITIVE, SECTOR = 1
C           XSECT NEGATIVE, YSECT POSITIVE, SECTOR = 2
C           XSECT NEGATIVE, YSECT NEGATIVE, SECTOR = 3
C           XSECT POSITIVE, YSECT NEGATIVE, SECTOR = 4
```

Figure 6.23 continued

```fortran
C
        IF (XSECT .GE. 0.0 .AND. YSECT .GE. 0.0) THEN
          KK = 1
        ELSEIF (XSECT .LE. 0.0 .AND. YSECT .GE. 0.0) THEN
          KK = 2
        ELSEIF (XSECT .LE. 0.0 .AND. YSECT .LE. 0.0) THEN
          KK = 3
        ELSEIF (XSECT .GE. 0.0 .AND. YSECT .LE. 0.0) THEN
          KK = 4
        END IF
        INIT(KK) = INIT(KK) + 1
        IF (INIT(KK) .LE. IPRIME) THEN
          TMPDIS(KK,INIT(KK)) = XDIST
          ITMUSE(KK,INIT(KK)) = I
          DO II = 1,INIT(KK)
            TEMPD(II) = TMPDIS(KK,II)
            ITEMP(II) = ITMUSE(KK,II)
          END DO
          JK = INIT(KK)
          IF (JK .GT. 1) THEN
            CALL HEAP(ITEMP,TEMPD,JK)
          END IF
          DO II = 1,INIT(KK)
            TMPDIS(KK,II) = TEMPD(II)
            ITMUSE(KK,II) = ITEMP(II)
          END DO
        ELSE
          INIT(KK) = IPRIME
          IF (XDIST .LT. TMPDIS(KK,IPRIME)) THEN
            TMPDIS(KK,IPRIME) = XDIST
            ITMUSE(KK,IPRIME) = I
            DO II = 1,IPRIME
              TEMPD(II) = TMPDIS(KK,II)
              ITEMP(II) = ITMUSE(KK,II)
            END DO
            CALL HEAP(ITEMP,TEMPD,IPRIME)
            DO II = 1,NK
              TMPDIS(KK,II) = TEMPD(II)
              ITMUSE(KK,II) = ITEMP(II)
            END DO
          END IF
        END IF
200     CONTINUE
        END IF
C
C       TAKE THE CLOSEST NK/4 LOCATIONS FROM EACH SECTOR
C
        IPRIME = NK / 4
C
C       RETURN IF A BOUNDARY CONDITION IS ENCOUNTERED
C
        IF (INIT(1) .EQ. 0 .AND. INIT(2) .EQ. 0) RETURN
        IF (INIT(2) .EQ. 0 .AND. INIT(3) .EQ. 0) RETURN
        IF (INIT(3) .EQ. 0 .AND. INIT(4) .EQ. 0) RETURN
        IF (INIT(1) .EQ. 0 .AND. INIT(4) .EQ. 0) RETURN
C
C       ELSE, DETERMINE HOW MANY (JUSE) CLOSEST POINTS WILL BE USED
C
        DO I = 1,4
          IF (INIT(I) .GE. IPRIME) THEN
            DO JI = 1, IPRIME
              KK = JUSE + JI
              IUSE(KK) = ITMUSE(I,JI)
              DIST(KK) = TMPDIS(I,JI)
            END DO
            JUSE = KK
          ELSEIF (INIT(I) .GT. 0 .AND. INIT(I) .LT. IPRIME) THEN
            DO JI = 1, INIT(I)
              KK = JUSE + JI
              IUSE(KK) = ITMUSE(I,JI)
              DIST(KK) = TMPDIS(I,JI)
            END DO
            JUSE = KK
          END IF
        END DO
C
C       RESORT FINAL ARRAY
C
        CALL HEAP(IUSE, DIST, JUSE)
C
C       RETURN TO CALLING PROGRAM
C
```

220

Figure 6.23 continued

```fortran
C
      RETURN
      END
C
C
      SUBROUTINE ECHO(ITRAN,XKO)
      PARAMETER (N=2000, NEST=5, JMAX=50)
      COMMON /SPADAT/ XDAT(N), X(N), Y(N), NTOT
      COMMON /MAT   / COVIJ(JMAX+1,JMAX+1), WT(JMAX+1),
     2                COVOI(JMAX+1), IKRIG
      COMMON /VARIO / INEST, MODEL(NEST), CO(NEST), SILL(NEST),
     2                RANGE(NEST), ANIS(NEST), RATIO(NEST)
      COMMON /SEARCH/ IUSE(JMAX), DIST(JMAX), ISERCH,RADIUS,NK,JUSE
      COMMON /XGRID / IBEGC, IENDC, IBEGR, IENDR, SIZEX, SIZEY,
     2                YMAX, XMIN
C
C     A SUBROUTINE TO PRINT INPUT DATA
C
C     ECHO OPTIONS FIRST
C
      IF (IKRIG .EQ. 1) THEN
         WRITE(*,10)
         WRITE(1,10)
      ELSEIF (IKRIG .EQ. 2) THEN
         WRITE(*,15)
         WRITE(1,15)
      END IF
10    FORMAT(1H1,   9X, 'PUNCTUAL KRIGING APPROACH',//)
15    FORMAT(1H1,   9X, 'BLOCK KRIGING APPROACH',//)
      IF (ITRAN .EQ. 0) THEN
         WRITE(*,16)
         WRITE(1,16)
      ELSEIF (ITRAN .EQ. 1) THEN
         WRITE(*,17) XKO
         WRITE(1,17) XKO
      ELSEIF (ITRAN .EQ. 2) THEN
         WRITE(*,18)
         WRITE(1,18)
      END IF
16    FORMAT(10X,'NO DATA TRANSFORMATION IN EFFECT',/)
17    FORMAT(10X,'LOG TRANSFORM IN EFFECT, CORRECTION =',E15.8,/)
18    FORMAT(10X,'INDICATOR TRANSFORM IN EFFECT',/)
      IF (IBEGC .EQ. 1 .AND. IENDC .EQ. 1 .AND. IBEGR .EQ. 1
     2             .AND. IENDR .EQ. 1) THEN
         WRITE(*,20)
         WRITE(1,20)
      ELSE
         WRITE(*,30) IBEGC, IENDC, IBEGR, IENDR, YMAX, XMIN
         WRITE(1,30) IBEGC, IENDC, IBEGR, IENDR, YMAX, XMIN
      END IF
20    FORMAT(//,10X,'CROSS VALIDATION RUN',/)
30    FORMAT(//,10X,'RESULTS FOR A GRID HAVING THE DIMENSIONS:',//,
     2      10X,'BEGINNING COLUMN NUMBER = ',I10,/,
     3      10X,'ENDING    COLUMN NUMBER = ',I10,/,
     4      10X,'BEGINNING ROW    NUMBER = ',I10,/,
     5      10X,'ENDING    ROW    NUMBER = ',I10,//,
     6      10X,'MAXIMUM Y-COORDINATE OF GRID = ',E15.8,//,
     7      10X,'MINIMUM X-COORDINATE OF GRID = ',E15.8,//)
      WRITE(*,40) SIZEX, SIZEY
      WRITE(1,40) SIZEX, SIZEY
40    FORMAT(10X,'SPACING IN GRID BETWEEN COLUMNS; ALSO, SIZE'
     2      ' OF BLOCK IN X: ',E10.3,/,
     3      10X,'SPACING IN GRID BETWEEN          ROWS; ALSO, SIZE'
     4      ' OF BLOCK IN Y: ',E10.3,/)
      WRITE(*,50) INEST
      WRITE(1,50) INEST
50    FORMAT(//,10X,'NUMBER OF NESTED VARIOGRAM MODELS = ',I3,/)
      WRITE(*,60)
      WRITE(1,60)
60    FORMAT(//,1X,'MODEL',4X,'NUGGET',6X,'SILL',5X,'RANGE',6X,
     2      'ANIS',5X,'RATIO',//)
      DO I = 1, INEST
         WRITE(*,70) MODEL(I),CO(I),SILL(I),RANGE(I),ANIS(I),RATIO(I)
         WRITE(1,70) MODEL(I),CO(I),SILL(I),RANGE(I),ANIS(I),RATIO(I)
      END DO
```

Figure 6.23 continued

```
70      FORMAT(1X,I5,5F10.3)
        IF (ISERCH .EQ. 1) THEN
            WRITE(*,80) RADIUS, NK
            WRITE(1,80) RADIUS, NK
        ELSEIF (ISERCH .EQ. 2) THEN
            WRITE(*,90) RADIUS, NK
            WRITE(1,90) RADIUS, NK
        END IF
80      FORMAT(/,10X,'A GENERAL SEARCH IS SELECTED WITH: ',/,
     2      10X,'SEARCH WINDOW RADIUS = ', F10.3,/,
     3      10X,'MAX. NO. OF NEAREST NEIGHBORS = ', I4,/)
90      FORMAT(/,10X,'A QUADRANT SEARCH IS SELECTED WITH: ',/,
     2      10X,'SEARCH WINDOW RADIUS = ', F10.3,/,
     3      10X,'MAX. NO. OF NEAREST NEIGHBORS = ', I4,/)
C
C       ECHO THE SPATIAL DATA NOW
C
        WRITE(*,100)
        WRITE(1,100)
100     FORMAT(1H1,10X,'SPATIAL DATA',//,
     2      10X,'DATA VALUE',13X,'Y-COORD',13X,'X-COORD',/)
        DO I = 1, NTOT
        IF (ITRAN .NE. 1) THEN
            WRITE(*,110) XDAT(I), Y(I), X(I)
            WRITE(1,110) XDAT(I), Y(I), X(I)
        ELSEIF (ITRAN .EQ. 1) THEN
            WRITE(*,110) EXP(XDAT(I)), Y(I), X(I)
            WRITE(1,110) EXP(XDAT(I)), Y(I), X(I)
        END IF
        END DO
110     FORMAT(3E20.7)
C
C       RETURN TO CALLING PROGRAM
C
        RETURN
        END

C
C
        SUBROUTINE HEAP(RB,RA,N)
        PARAMETER (JMAX = 50)
C
C       ADAPTED FROM PRESS, ET. AL., NUMERICAL RECIPES
C       PAGES 231 AND 232 (CALLED SUBROUTINE SORT THEREIN)
C
C       THIS SUBROUTINE SORTS AN ARRAY, RA, INTO
C       ASCENDING ORDER USING THE HEAPSORT ALGORITHM,
C       WHILE ALSO REARRANGING RB.
C
        DIMENSION RB(JMAX), RA(JMAX)
        INTEGER RB, RRB
        L = 0.5 * N + 1
        IR = N
10      CONTINUE
        IF (L .GT. 1) THEN
            L = L - 1
            RRA = RA(L)
            RRB = RB(L)
        ELSE
            RRA = RA(IR)
            RRB = RB(IR)
            RA(IR) = RA(1)
            RB(IR) = RB(1)
            IR = IR - 1
            IF (IR .EQ. 1) THEN
                RA(1) = RRA
                RB(1) = RRB
                RETURN
            END IF
        END IF
        I = L
        J = L + L
```

222

Figure 6.23 continued

```fortran
20    IF (J .LE. IR) THEN
        IF (J .LT. IR) THEN
          IF (RA(J) .LT. RA(J+1)) J = J + 1
        END IF
        IF (RRA .LT. RA(J)) THEN
          RA(I) = RA(J)
          RB(I) = RB(J)
          I = J
          J = J + J
        ELSE
          J = IR + 1
        END IF
        GO TO 20
      END IF
      RA(I) = RRA
      RB(I) = RRB
      GO TO 10
C
C     RETURN TO CALLING PROGRAM
C
      RETURN
      END
C
C
      SUBROUTINE CHAWK(X0,Y0)
      PARAMETER (N=2000, NEST=5, JMAX=50)
      COMMON /SPADAT/ XDAT(N), X(N), Y(N), NTOT
      COMMON /MAT  / COVIJ(JMAX+1,JMAX+1), WT(JMAX+1),
     2               COVOI(JMAX+1), IKRIG
      COMMON /VARIO / INEST, MODEL(NEST), CO(NEST), SILL(NEST),
     2               RANGE(NEST), ANIS(NEST), RATIO(NEST)
      COMMON /SEARCH/ IUSE(JMAX), DIST(JMAX), ISERCH,RADIUS,NK,JUSE
      COMMON /XGRID / IBEGC, IENDC, IBEGR, IENDR, SIZEX, SIZEY,
     2               YMAX, XMIN
      DIMENSION XLOC(4), YLOC(4), XWT(4), COVTMP(16)

C
C     A SUBROUTINE TO CONSTRUCT THE INTERSAMPLE COVARIANCE
C     MATRIX, COVIJ, AND THE POINT-SAMPLE COVARIANCE VECTOR, COVOI
C
C     FIRST, REGARDLESS OF THE VALUE OF IKRIG, FORM THE MATRIX,
C     COVIJ, BECAUSE ITS FORMATION DOES NOT REQUIRE INTEGRATION,
C     EVEN FOR BLOCK KRIGING.
C
      DO I = 1, JUSE
        IF (INEST .EQ. 1) THEN
          COVIJ(I,I) = SILL(1)
        ELSE
          SILTOT = 0.0
          DO J = 1,INEST
            SILTOT = SILTOT + SILL(J)
          END DO
          COVIJ(I,I) = SILTOT
        END IF
        COVIJ(I,JUSE+1) = 1.0
        COVIJ(JUSE+1,I) = 1.0
        K = IUSE(I)
        DO J = I+1, JUSE
          L = IUSE(J)
          DX = X(K) - X(L)
          DY = Y(K) - Y(L)
          B = SQRT(DX*DX + DY*DY)
          COVIJ(I,J) = XGAM(B)
          COVIJ(J,I) = COVIJ(I,J)
        END DO
      END DO
      COVIJ(JUSE+1,JUSE+1) = 0.0
C
C     FORM THE POINT-SAMPLE COVARIANCE VECTOR, COVOI.
C     USE GAUSS QUADRATURE INTEGRATION FOR BLOCK KRIGING OPTION
```

Figure 6.23 continued

```fortran
c

      IF (IKRIG .EQ. 1) THEN
      DO I = 1, JUSE
        B = DIST(I)
        COVOI(I) = XGAM(B)
      END DO
      COVOI(JUSE+1) = 1.0
      ELSEIF (IKRIG .EQ. 2) THEN
      X2 = SIZEX / 2.0
      Y2 = SIZEY / 2.0
      XLOC(1) = X0 - X2 * 0.8611363116
      XLOC(4) = X0 + X2 * 0.8611363116
      YLOC(4) = Y0 + Y2 * 0.8611363116
      YLOC(1) = Y0 - Y2 * 0.8611363116
      XLOC(2) = X0 - Y2 * 0.3399810436
      YLOC(3) = Y0 + Y2 * 0.3399810436
      XLOC(3) = X0 + X2 * 0.3399810436
      XLOC(2) = X0 - X2 * 0.3399810436
      XWT(1) =            0.3478548452
      XWT(4) =            0.3478548452
      XWT(2) =            0.6521451548
      XWT(3) =            0.6521451548
      DO K = 1, JUSE
        LM = 0
        XN = X(IUSE(K))
        YN = Y(IUSE(K))
        DO L = 1,4
          DO M = 1,4
            LM = LM + 1
            XX = XLOC(L)
            YY = YLOC(M)
            XYDIST = SQRT((XN-XX)**2 + (YN-YY)**2)
            COVTMP(LM) = XGAM(XYDIST)*XWT(L)*X2*XWT(M)*Y2
          END DO
        SUM = 0.0
        DO LI = 1,16
          SUM = SUM + COVTMP(LI)
        END DO
        COVOI(K) = SUM / (SIZEX * SIZEY)
      END DO
      COVOI(JUSE+1) = 1.0
      END IF
c
c     RETURN TO CALLING PROGRAM
c
      RETURN
      END
c
c
      FUNCTION XGAM(D)
      PARAMETER (NEST=5)
      COMMON /VARIO / INEST, MODEL(NEST), CO(NEST), SILL(NEST),
     2                RANGE(NEST), ANIS(NEST), RATIO(NEST)
c
c     FUNCTION TO EVALUATE SEMI-VARIOGRAM MODEL AND CONVERT
c     THE VALUE FROM THE MODEL TO A COVARIANCE
```

Figure 6.23 continued

```fortran
C
      TMP = 0.0
      SUM = 0.0
      DO I = 1, INEST
      SUM = SUM + SILL(I)
      IF (MODEL(I) .EQ. 1) THEN
      IF (D .LT. RANGE(I)) THEN
      C = SILL(I) - CO(I)
      TMP = TMP + CO(I) + C*((1.5*D/RANGE(I)) - (0.5*D**3/
     RANGE(I)**3))
    2 IF (D .EQ. 0.0) TMP = 0.0
      ELSE
      TMP = TMP + SILL(I)
      END IF
      ELSEIF (MODEL(I) .EQ. 2) THEN
      C = SILL(I) - CO(I)
      R = RANGE(I) / 3.0
      TMP = TMP + CO(I) + C*(1.0 - EXP(-D/R))
      IF (D .EQ. 0.0) TMP = 0.0
      ELSEIF (MODEL(I) .EQ. 3) THEN
      C = SILL(I) - CO(I)
      R = RANGE(I) / SQRT(3.0)
      TMP = TMP + CO(I) + C*(1.0 - EXP(-D**2/R**2))
      IF (D .EQ. 0.0) TMP = 0.0
      ELSEIF (MODEL(I) .EQ. 4) THEN
      C = SILL(I) - CO(I)
      SLOPE = C/RANGE(I)
      TMP = TMP + CO(I) + SLOPE * D
      IF (D .EQ. 0.0) TMP = 0.0
      END IF
      END DO
      XGAM = SUM - TMP
C
C     RETURN TO SUBROUTINE CHAWK
C
      RETURN
      END
C
```

```fortran
      SUBROUTINE GAUSIT(N)
C
C     A STANDARD FORTRAN-77 PROGRAM FOR GAUSS ELIMINATION
C     USING A MAXIMUM PIVOT ELEMENT SEARCH FOR EACH
C     UNKNOWN
C
C     THIS VERSION OF GAUSS ELIMINATION IS USED WITH
C     ITERATIVE IMPROVEMENT TO YIELD CALCULATIONS OF
C     SUPERIOR ACCURACY
      PARAMETER (JMAX = 50)
      COMMON /MAT  / COVIJ(JMAX+1,JMAX+1), WT(JMAX+1),
     2               COVOI(JMAX+1), IKRIG
      DIMENSION A(JMAX+1,JMAX+2), TEMP(JMAX+1,JMAX+2)
C
C     PLACE THE MATRIX, COVIJ, INTO THE MATRIX, A; ALSO, PLACE
C     THE VECTOR, COVOI, INTO THE LAST COLUMN OF A
C
      NITT = 3
      DO I = 1,N
      DO J = 1,N
      A(I,J) = COVIJ(I,J)
      END DO
      END DO
      DO I = 1,N
      A(I, N+1) = COVOI(I)
      END DO
C
C     INITIALIZE MATRIX SIZES
C
      M = N + 1
      L = N - 1
C
C     PRESERVE THE MATRIX, A, BY PLACING IT IN THE MATRIX, TEMP
C
      DO I = 1,N
      DO J = 1,M
      TEMP(I,J) = A(I,J)
      END DO
      END DO
```

Figure 6.23 continued

```fortran
C
C      CALL THE SUBROUTINE SOLVE FOR A SOLUTION
C
       CALL SOLVE (TEMP, WT, M, L)
C
C      USE ITERATIVE IMPROVEMENT FOR GREATER ACCURACY
C
       DO I = 1, NITT
          CALL GPROVE(A, TEMP, WT, M, L)
       END DO
C
C      RETURN TO CALLING PROGRAM
C
       RETURN
       END
C
C
       SUBROUTINE SOLVE(A, X, M, L)
C
C      SUBROUTINE TO PERFORM GAUSS ELIMINATION
C
       PARAMETER (JMAX = 50)
       DIMENSION A(JMAX+1,JMAX+2), X(JMAX+1)
C
C      BEGIN THE FORWARD REDUCTION (ELIMINATION)
C
       N = L + 1
       DO K = 1,L
       JJ = K
          XLARG = ABS(A(K,K))
          K1 = K + 1
C
C      SEARCH FOR LARGEST PIVOT ELEMENT
C
          DO I = K1, N
             CA = ABS(A(I,K))
             DIFF = XLARG - CA
             IF (DIFF .LT. 0) THEN
                XLARG = CA
                JJ = I
             END IF
          END DO
          IDIFF = JJ - K
C
C      SWAP ROWS, IF NECESSARY, DEPENDING ON THE
C      POSITION OF THE MAXIMUM PIVOT ELEMENT.
C
          IF (IDIFF .NE. 0) THEN
             DO J = K,M
                TEMP = A(JJ,J)
                A(JJ,J) = A(K,J)
                A(K,J) = TEMP
             END DO
          END IF
C
C      THIS IS THE MAJOR ASPECT OF GAUSS ELIMINATION:
C      DIVIDE REMAINING ELEMENTS OF A (AND B WHICH
C      IS IN THE LAST COLUMN OF A) BY THE PIVOT
C      ELEMENT, A(K,K)
C
          DO I = K1, N
             QUOT = A(I,K) / A(K,K)
             DO J = K1, M
                A(I,J) = A(I,J) - QUOT * A(K,J)
             END DO
          END DO
          DO I = K1, N
             A(I,K) = 0.0
          END DO
       END DO
C
C      SOLVE, NOW, FOR LAST UNKNOWN
C
       X(N) = A(N,M) / A(N,N)
```

Figure 6.23 continued

```fortran
C
C     USING THIS SOLUTION, BACK SUBSTITUTE TO FIND
C     THE REMAINING UNKNOWNS

      DO IN = 1,L
        SUM = 0.0
        I = N - IN
        I1 = I + 1
        DO J = I1, N
          SUM = SUM + A(I,J) * X(J)
        END DO
        X(I) = (A(I,M) - SUM) / A(I,I)
      END DO

C
C     RETURN TO MAIN CALLING PROGRAM

      RETURN
      END

C
C
      SUBROUTINE GPROVE (A, TEMP, X, M, L)

C
C     PROGRAM TO PROVIDE BETTER ACCURACY IN GAUSS
C     ELIMINATION THROUGH ITERATIVE IMPROVEMENT

      PARAMETER (JMAX = 50)
      DIMENSION A(JMAX+1,JMAX+2), TEMP(JMAX+1,JMAX+2), X(JMAX+1)
      DIMENSION R(JMAX+1), RR(JMAX+1)

C
C     STEP 1:  MULTIPLY THE SOLUTION, X, BY ORIGINAL
C              MATRIX, A

      N = M - 1
      DO I = 1,N
        R(I) = 0.0
        DO J = 1,N
          R(I) = R(I) + A(I,J) * X(J)
        END DO
      END DO

C
C     STEP 2:  SUBTRACT THE VECTOR, B, FROM THE VECTOR, R.
C

      DO I = 1,N
        R(I) = R(I) - A(I,M)
      END DO

C
C     STEP 3:  PLACE THE MATRIX, A, INTO THE MATRIX, TEMP
C

      DO I = 1,N
        DO J = 1, N
          TEMP(I,J) = A(I,J)
        END DO
      END DO

C
C     STEP 3A:  PLACE THE VECTOR, R, INTO THE N+1 COLUMN
C               OF THE MATRIX, TEMP.

      DO I = 1,N
        TEMP(I,M) = R(I)
      END DO

C
C     STEP 4:  SOLVE FOR THE ERROR, E (PLACED IN THE VECTOR,
C              RR)

      CALL SOLVE(TEMP, RR, M, L)

C
C     STEP 5 (FINAL STEP):  UPDATE THE SOLUTION, X.

      DO I = 1,N
        X(I) = X(I) - RR(I)
      END DO

C
C     RETURN TO CALLING PROGRAM

      RETURN
      END
```

7

Computer Graphics: Visualizing Quantitative Information

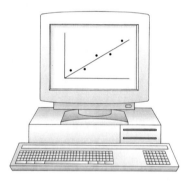

In his book, *The Visual Display of Quantitative Information,* Edward Tuft presents a wide variety of figures, graphs, and charts to document the creation of visual images from numbers. Some of these images are historic and some are recent. They clearly demonstrate the improved ability that graphics presentations offer for data interpretation.

The phrase *a picture is worth a thousand words* is often taken for granted because it is cliché. Much of our learning, though, comes through sensory perception. At first, a child responds mostly to touch and sound. As sight improves, the child soon develops the ability to recognize patterns; imagery then stimulates development. With sight alone, the brain recognizes, evaluates, and analyzes images that can be described only tediously with words and numbers.

This chapter focuses specifically on the conversion of numbers to pictures. The objective is to demonstrate improved ability to interpret data by converting them to images. This chapter supplements many of the others in this text. For example, computer graphics are used to revisit portions of Chapters 5 and 6. Computer graphics will also be used in Chapter 8 on finite element analysis, in Chapter 10 on image processing, and in Chapter 11 on fractals.

We first discuss what is meant by computer graphics in the context of this book. Our definition for computer graphics is patterned after one presented in Foley and Van Dam (1982, p. 3): Computer graphics is the creation of pictures, via computers, from quantitative information. However, this definition excludes hardware differences among computers. Not all computers are capable of creating high-resolution, publication-quality graphical images. But many microcomputers are capable of displaying graphical images of reasonable quality, which is sufficient for the purposes of this text.

7.1 Computer Graphics and Hardware

Good quality computer-generated graphics are possible with existing microcomputer technology. This technology is expected to improve substantially in resolution over the next ten years due to improved microcomputer hardware technology, so what is covered in this chapter will only become more exciting.

Many of the applications within this chapter are presented in a manner that to a large extent is hardware independent; that is, with the exception of specifying where on the screen to draw a figure, the graphical construction of the figure using lines is independent of the hardware. Therefore, many of the graphical applications presented herein can be adapted easily to more advanced hardware. Herein, graphics applications are designed to take advantage of the VGA graphics environment available with many MS-DOS-compatible microcomputers.

7.1.1 VGA Computer Graphics Environment

VGA graphics technology allows a palette of 16 colors to be used for high pixel resolution—640 horizontal × 480 vertical—or a palette of 256 colors to be used for medium pixel resolution—320 horizontal × 200 vertical. A palette is a collection of colors; the word is chosen to represent the same notion as that of an artist's palette of paints. A pixel, also known as a pel (picture element), is one element of a digital raster. It is the smallest dot size on a computer monitor screen. A raster is a rectangular matrix of numbers; that is, a raster is a digital image.

Software programs may be used to design palettes. The use of the plural is significant and is commensurate with the fact that, whereas only 16 different colors may be displayed at one time in high-resolution mode using VGA hardware, a large number of different 16 color combinations (palettes) may be designed for display. This notion is important for our discussion of image processing, which is presented later in this text. Experiments to design palettes are possible using the program PALDSGN (Figure 7.1, page 264).

With the VGA hardware, one 8-bit byte is used to represent each of the primary colors: red, green, or blue. Of the 8 bits used for each primary color, the two highest-order bits are zero, leaving the 6 lower order bits to represent the intensity of each color.

An 8-bit byte might look like this:

00101101

where each bit is a number, 0 or 1. The entire collection of 8 bits is a byte, or computer word. The byte represents a number, which is explicitly determined using the following summation:

$$number = \sum_{i=1}^{8} b_i 2^{(i-1)}$$

where b_i is a bit, 0 or 1; the subscript i increments from right to left for the byte. In the case of the byte shown above, 00101101, the number is

$$1(2^0) + 0(2^1) + 1(2^2) + 1(2^3) + 0(2^4) + 1(2^5) + 0(2^6) + 0(2^7) = 45$$

For this byte, b_1 is equal to 1 and b_8 is equal to 0. Because the VGA graphics environment uses only the 6 lowest-order bits to represent intensity, there are 2^6, or 64, possible intensity values for each primary color.

The program PALDSGN mixes the primary colors, depending on the intensity specified for each, into a wide range of colors using the function RGB (Microsoft FORTRAN-77 Version 5.1 Compiler Documentation, Advanced Topics, p. 300). This function mixes colors as described by the color cube (Figure 7.2). Each primary color is associated with an intensity that ranges between 0 and 63, for a total of 64 intensity levels.

This color cube describes additive color combinations. For instance, red + green gives yellow; red + blue gives magenta; blue + green gives cyan; and so on. Additive color combinations are reproduced by the cathode ray tube (CRT) of a color television or a color computer monitor. The concept of additive color contrasts that of subtractive color. With subtractive color, combinations are obtained, for instance, by mixing different colored paints. For example, the mixing of equal parts of blue and yellow paints yields a green paint. Subtractive color is mentioned only to avoid confusion when attempting to understand the color combinations described by the color cube (Figure 7.2).

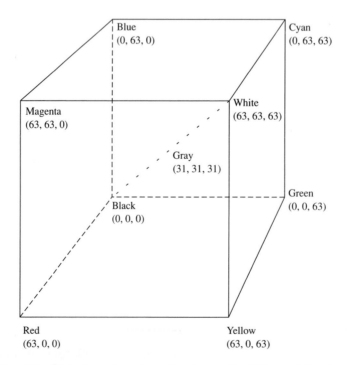

Figure 7.2   The color cube, a figure developed using Foley and Van Dam (1982, p. 611).

Suppose the program PALDSGN is used to create an additive color palette having cyan (at full intensity) as one of its 16 colors. Notice (Figure 7.2) that cyan is obtained by mixing equal intensities of blue and green primary colors. To obtain cyan at full intensity, PALDSGN is used with the following intensities for red, blue, and green: intensity of red = 0; intensity of blue = 63; intensity of green = 63.

Design of two complete palettes further demonstrates the use of the color cube. In Chapter 10, a palette is used consisting of 16 shades of gray, ranging from black to white. Equal intensities of all three primary colors yields varying shades of gray. The color black is obtained if the intensity values for red, blue, and green are zero; bright white is obtained if the intensity values for red, blue, and green are 63. Table 7.1 shows the design of a palette whose 16 colors are varying shades of gray, ranging from black for color 0 to bright white for color 15.

A second palette (Table 7.2) designed for use later in this chapter has cold colors (shades of violet and blue) associated with smaller color numbers (0 to 7), and warm colors (shades of yellow and red) associated with larger color numbers (8 to 15). This particular palette is used in Section 7.6 to demonstrate the pseudo-contour mapping of kriging results wherein cold colors represent smaller data values and warm colors represent larger data values.

Exercises at the end of this chapter challenge you to create your own palette of colors using the program PALDSGN. On the basis of these exercises, you will discover the tremendous creative power you have for designing colors using VGA graphics hardware.

Table 7.1 Intensity Values for Designing a 16-Color Palette of Varying Intensities of the Color Gray

Intensities			
Red	Blue	Green	Color
0	0	0	Black
4	4	4	Gray (4)
8	8	8	Gray (8)
12	12	12	Gray (12)
16	16	16	Gray (16)
20	20	20	Gray (20)
24	24	24	Gray (24)
28	28	28	Gray (28)
32	32	32	Gray (32)
36	36	36	Gray (36)
40	40	40	Gray (40)
44	44	44	Gray (44)
48	48	48	Gray (48)
52	52	52	Gray (52)
56	56	56	Gray (56)
60	60	60	White

Note. The numbers in parentheses following the color gray indicate the intensity for gray.

Table 7.2   Design of a Palette,
COLORDEN.PAL, for Color Density Slicing

	Intensity		
Red	Blue	Green	Color
0	0	0	Black
20	37	0	Dark violet
31	48	0	Violet
42	59	0	Bright violet
0	30	0	Dark blue
0	45	0	Blue
0	60	0	Bright blue
0	0	30	Dark green
0	0	45	Green
0	0	60	Bright green
30	0	30	Dark yellow
45	0	45	Yellow
60	0	60	Bright yellow
30	0	0	Dark red
45	0	0	Red
60	0	0	Bright red

Aside from the ability to use color to generate graphical images, the more pixels allowed by a graphics adaptor for the representation of images, and hence the smaller each pixel is, the better is the resolution of the resultant graphical images. Hence, higher-resolution graphics can be obtained using VGA hardware and a 16-color palette, but the range of colors that may be used for drawing the graphics is sacrificed relative to lower-resolution graphics for which a 256-color palette is available. As technology evolves, the ability to produce high-quality graphics using a larger palette of colors will improve. For example, the Super VGA graphics environment allows the use of a 256-color palette for the 640 $\times$ 480-pixel resolution mode.

7.2 Viewing and Windowing

A hypothetical situation of a graph drawn on a computer monitor is shown in Figure 7.3. As stated previously, the computer monitor screen is a large field of pixels. In a high-resolution VGA graphics mode, the size of this field is 640 (horizontal) by 480 (vertical) pixels, which represents 307,200 total pixels.

This pixel domain is what is available for drawing graphical images (plotters and printers also may be used for drawing graphics to yield what is known as a hard copy, or paper copy, of the graphical image displayed on a monitor screen). Viewing a graphical image on the monitor screen requires software that can draw the image by setting pixels equal to one of the colors available in the palette, where the color differs from what is known as the background color (often black).

For example, suppose a drawing is generated based on some data, such as the drawing shown on the monitor screen in Figure 7.3. This drawing is one

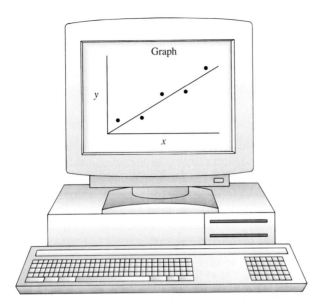

Figure 7.3 A hypothetical two-dimensional graph shown on a computer monitor.

of a two-dimensional x-y plot of bivariate data. A hypothetical regression line is drawn with the data to show the best-fit line for these data. Such data are usually in units, often real numbers, other than pixel coordinates required to plot the data on the screen. How is this problem of the usual lack of data coordinate compatibility with pixel coordinates reconciled?

One approach is to painstakingly compute pixel coordinates by hand based on the data values. For instance, suppose the point (4.5, 7.9) is to be drawn on the screen using the high-resolution, 640×480-pixel VGA graphics mode. If the maximum value xmax and minimum value xmin of the x coordinates for the data set to which the point (4.5, 7.9) belongs are known, and likewise the maximum and minimum y coordinates, ymax and ymin, are known, then, assuming that the full monitor screen is used for plotting, the pixel coordinates used for plotting the point on the screen are

$$\text{xpixel} = 640((4.5 - \text{xmin})/(\text{xmax} - \text{xmin}))$$
$$\text{ypixel} = 480(1 - ((7.9 - \text{ymin})/(\text{ymax} - \text{ymin})))$$

Notice that in the calculation of the y-pixel coordinate, the fact that ypixel = 0 is at the top of the monitor screen and ypixel = 480 is at the bottom of the monitor screen must be accounted for, whereas in plotting x-y data, $y = 0$ is often at the bottom of the plot.

This task of converting data coordinates to pixel coordinates is an arduous one. A more convenient and consequently more elegant method involves using software routines to automatically convert data coordinates to pixel (screen) coordinates. This is easily accomplished using commands for windowing and viewing.

A windowing function establishes or defines within the computer algorithm the range of actual coordinate values associated with the data to be plotted. For instance, a windowing command may have the following syntax:

$$\text{WINDOW(xmin, ymin, xmax, ymax)}$$

where xmin, ymin, xmax, and ymax define the ranges of the actual x and y coordinates.

A viewing function is used in conjunction with the windowing function to map the data coordinate range defined by the windowing function to pixel (screen) coordinates. For example, a viewing function may have the following syntax:

VIEW(xpixmin, ypixmin, xpixmax, ypixmax)

where xpixmin and xpixmax are, respectively, the minimum and maximum x-direction pixel numbers for the plotting domain; ypixmin and ypixmax are, respectively, the minimum and maximum y-direction pixel numbers for the plotting domain. Once the viewing command is used, graphical images can be drawn only inside the pixel region specified by this command. For instance, although the high-resolution VGA graphics mode offers a total 640 × 480-pixel field, if the viewing command used is

VIEW(1, 1, 300, 200)

then graphical images can be drawn only in the upper left 300 × 200-pixel region of the total 640 × 480-pixel field.

Once the viewing function is activated within a computer algorithm, thereafter all calls to graphics functions are made using actual data coordinates (as defined by the windowing function). The viewing function automatically converts these coordinates to pixel coordinates, within the pixel domain specified in the call to the viewing function, for use with the graphics functions. This is demonstrated later in this chapter. A useful review of windowing and viewing functions offered by several languages is shown in Table 7.3.

At first glance, the power represented by the windowing and viewing combination of functions is not necessarily obvious. Yet these are two of the three most important functions available for computer graphics. The third function is one that draws lines and is discussed in the next section. Once a windowing command is activated in a computer algorithm, only the viewing command needs to be changed to change the scale of the graphical image.

Table 7.3   Example Window and View Commands Used with Various Computer Languages/Compilers

	Commands	
Language	Windowing	Viewing
BASIC[a]	WINDOW	VIEW
FORTRAN-77[b]	SETWINDOW	CALL SETVIEWPORT
C[c]	(custom algorithms; see footnote c)	
Pascal[d]	WINDOW	SETVIEWPORT

[a] Such as is allowed with the IBM BASICA interpreter.

[b] These are functions built into the Microsoft FORTRAN Version 5.0 and 5.1 compilers; these functions are C language routines which are linked automatically by the compiler to FORTRAN programs.

[c] For example, see Hunt (1989).

[d] Functions used, for example, with the Borland Turbo Pascal Version 5.0 compiler.

For example, suppose the range of data coordinates is as defined by the following windowing command:

$$\text{WINDOW}(-1, -1, 10, 10)$$

and is used with the viewing command

$$\text{VIEW}(1, 1, 639, 479)$$

which is commensurate with the high-resolution VGA graphics mode and makes use of the entire screen. In using this combination of functions, the real coordinate $(-1, -1)$ is automatically associated with the pixel $(1, 479)$ at the lower left corner of the screen, which implicitly assumes that the viewing function automatically accounts for the fact that real y coordinates increase from the bottom of the screen (plot) to the top. The real coordinate $(10, 10)$ is associated with the upper right portion of the screen at pixel coordinate $(639, 1)$. Moreover, the real coordinate $(10, -1)$ is associated with the lower right corner of the screen at pixel coordinate $(639, 479)$. These are but a few examples to show how the windowing and viewing functions are used to complement one another.

Suppose a graphical image having real coordinates as defined by the windowing function just shown is to be plotted in the upper left quadrant of the screen so that other images may be drawn on the screen, but in different positions. In this case, the windowing command does not change because this command simply defines the range of data coordinates. Instead, only the viewing command is changed:

$$\text{VIEW}(1, 1, 319, 239)$$

and the graphical image is redrawn. The real coordinate $(-1, -1)$ is automatically associated with the pixel $(1, 239)$, the real coordinate $(10, 10)$ is associated with the pixel $(319, 1)$, and so on. Notice that in addition to changing the position on the screen where a plot is made, this change to the viewing command also reduces the size of the plotting region, hence reduces the size of the plot. This demonstrates how the scale of a plot can be changed simply by changing the viewing command.

By this example, the true power of the windowing/viewing combination is realized for (1) automatic coordinate conversion from physical to screen coordinates, and (2) convenient rescaling of graphical images. This is accomplished simply by using the viewing function to change the size of the pixel domain used for plotting. In this manner, a graph can easily be enlarged or reduced without the need to change the windowing function.

7.3 Vector Generation

As mentioned earlier, the three most important graphics functions are those for windowing, viewing, and vector generation. A vector generation function is one that draws a line, in a specified color, between two points on the monitor screen. Such a line can be of any length within the domain of the viewing screen. Therefore, any shape, picture, font (lettering style), and so on, may be generated on the screen using vector generation.

A vector generation function becomes a point, or dot, function if a line is drawn on the screen between a pixel and itself. Therefore, many software languages capable of generating graphics offer a point-setting function, which enables a single pixel to be set to a particular color and which is simpler to use for such a purpose than the vector generation function. However, given that the point-setting function is simply one form of vector generation, the statement made earlier still holds: Any shape may be drawn on the screen using a vector generation function.

Example 7.1   Drawing a Box on the Screen

This example is used to demonstrate windowing, viewing, and vector generation. The box shown in Figure 7.4 is to be drawn on the screen three times, each time using a different primary color, red, green, or blue. The viewing function is to be used such that the three boxes are displayed simultaneously, side by side, on the screen.

A computer algorithm called BOXDRAW (Figure 7.5, page 266) is designed for this purpose. A black-and-white print of the resultant output is shown in Figure 7.6. In developing this algorithm, notice that the following functions are performed, as documented by comment statements within the program listing:

1. A windowing function is used to define the actual coordinate range for the box, e.g., window(-1.0, -1.0, 4.0, 4.0). The physical coordinate range is chosen to be larger than what is shown in Figure 7.4 to ensure the box fits entirely within the viewing domain as specified using the viewing function.
2. A viewing command is used to define this pixel domain for each of the three drawings, e.g.,

for drawing 1: view(10, 10, 190, 190)
for drawing 2: view(210, 10, 390, 190)
for drawing 3: view(410, 10, 590, 190)

In other words, the first box is drawn within the pixel region: xpixmin $= 10$; xpixmax $= 190$; ypixmin $= 10$; ypixmax $= 190$; in fact, for all three boxes,

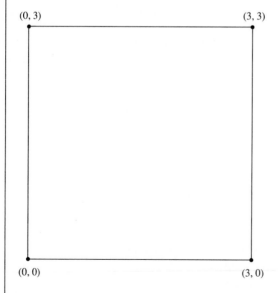

Figure 7.4   An example box; each corner is defined by coordinates in a two-dimensional system.

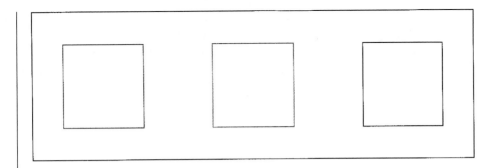

PUSH THE ENTER KEY TO FINISH PROGRAM

Figure 7.6 A black and white print of the display produced using the program BOXDRAW.

ypixmin and ypixmax are the same. For the second box, the xpixel region is changed: xpixmin = 210; xpixmax = 390. And, for the third box, the xpixel range is changed: xpixmin = 410; xpixmax = 590. This demonstrates how to change the position of a drawing on the screen without changing the windowing function (therefore avoiding the need for recomputing coordinates of the corners of the box for each drawing).

3. A primary color, red, green, or blue, is designed using the default palette of colors as listed at the beginning of the program listing: color 4 = red; color 2 = green; color 1 = blue.

4. The box is drawn by starting at one of the corners and proceeding to the other three corners, finally returning to the starting corner.

   This simple exercise shows how to use a viewing function to change the position on the screen for drawing. A windowing function is used only once for this purpose. A viewing function is used to change the position on the screen at which each box is drawn. This program also demonstrates the assignment of color using one of 16 colors to draw each box. Vector generation is also demonstrated for drawing the sides of the box. The principles of graphics drawing applied in this program lead to a sophisticated two-dimensional graphics program which is introduced and demonstrated in the next section. ∎

7.3.1 Computer-Aided Drawing of Two-Dimensional Graphs

In the previous example, a demonstration is presented showing the use of a vector generator for drawing a shape. On the basis of this demonstration, a computer program is designed for drawing two-dimensional graphs. It will also have the capability of drawing graphs for linear least-squares regression, multivariate data analysis, and geostatistics, as well as graphs showing the results of Fourier analysis, image processing, and fractals. This program, once modified, will be useful for displaying results developed using finite element analysis.

Figure 2.3 is revisited to observe the essential components of a two-dimensional graph. The following capabilities must be designed into the program: viewing and windowing for data to pixel coordinate transformation; vector generation to draw the axes, put tick marks on the axes, and draw the best fit line, $y = mx + b$; and point generation in conjunction with vector generation to plot the points x_i, y_i. In addition to these capabilities, text generation to label the graph and the axes is desirable.

This overview of the design of the algorithm is implemented in a computer program as follows:

1. The mode for data entry into the program is determined. Six options are allowed: (1) a general data file of x, y values; (2) interactive input through the keyboard; (3) the display of cross-validation results from the program JCBLOK (Chapter 6); (4) the display of correspondence analysis results from Chapter 5; (5) the display of semivariogram results from the program FGAM (Chapter 6); or (6) the display of a one-dimensional time series or function (for which the x dimension is not explicitly present in the data file).
2. The data are accessed for plotting.
3. Interactive keyboard entry is used to determine whether the x, y data are to be converted to logarithms or displayed as is (this is useful for displaying the types of analyses discussed in Chapter 11 on fractals); whether the y values and/or x values are to be rescaled (to obtain square graphs, the y-axis range is equated to the x-axis range); and the ranges of the x and y axes (minimum and maximum values) to be used to plot the data, which are useful for displaying the entire data set or a subset thereof.
4. Interactive query is used to define the option for fitting a function to the x, y data. If a function is to be plotted with the data, then it is defined as follows:

$$y = a_0 + a_1 x + a_2 x^2 + a_3 x^3 + a_4 x^4 + a_5 \exp(a_6 x^{a_7})$$

If a semivariogram is plotted, and a spherical model is fit to it, the sill and range of this model also must be specified. In using the function shown above for semivariogram model fitting, the following is used to assign the coefficients a:

(a) for a spherical model:
   $a_0 =$ nugget value
   $a_1 = 1.5(\text{sill} - \text{nugget})/\text{range}$
   $a_2 = 0.0$
   $a_3 = -0.5(\text{sill} - \text{nugget})^3/\text{range}^3$
   Coefficients a_4 through a_7 are all set to zero.

(b) for an exponential model:
   $a_0 =$ nugget value
   $a_1 = a_2 = a_3 = a_4 = 0$
   $a_5 = -(\text{sill} - \text{nugget})$
   $a_6 = -1/\text{range}$ (remember, in this case, the range is $1/3$ of the practical range)
   $a_7 = 1.0$

(c) for a Gaussian model:
a_0 = nugget value
$a_1 = a_2 = a_3 = a_4 = 0$
$a_5 = -(\text{sill} - \text{nugget})$
$a_6 = -1/\text{range}^2$ (remember, in this case, the range is the practical range divided by the square root of 3)
$a_7 = 2.0$
(d) for a linear model:
a_0 = nugget value
a_1 = slope
a_2 through a_7 are equal to zero
5. Interactive query is used to determine if the data points x, y are to be connected by line segments, or simply plotted as points.
6. A title for the graph is specified up to 50 characters in length.
7. The graphical image is drawn.

The computer algorithm based on these seven steps is called GRAPH2D and is described in the next section.

7.3.2 GRAPH2D: An Interactive Program for Plotting Two-Dimensional Graphical Images

A computer algorithm called GRAPH2D (Figure 7.7, page 269) has been designed based on these seven steps. A detailed user's guide is provided at the beginning of this program. Several examples demonstrate how this program is used.

Example 7.2 Application of GRAPH2D to Reproduce Figure 2.3

A simple two-dimensional graph showing a linear regression for a small, bivariate data set was shown in Figure 2.3. The program GRAPH2D may be used to draw this figure. Option 2 for keyboard data entry is chosen. For a small data set, such as this one, this option is perhaps the simplest one to use. The resultant plot is shown in Figure 7.8. The option for plotting a function is chosen to fit a line to these data using the following coefficients: $a_0 = 3.66$; $a_1 = -1.27$; a_2 through a_7 are set equal to zero. All interactive options used in GRAPH2D to obtain this plot are documented in the caption to Figure 7.8. ∎

Example 7.3 Using GRAPH2D to Display the Results from Correspondence Analysis

GRAPH2D has been designed to allow an option for reading information from the output file created by CORSPOND and to display this information graphically. The result is a plot that is, perhaps, more aesthetically appealing than the line printer graphs produced by CORSPOND.

In this example, the correspondence analysis results for the September 1986 Stillwater data set given in Chapter 5 are plotted using the first two significant factors. CORSPOND (Chapter 5) is first applied to these data to create an output file. Once this is accomplished, the fourth data entry option is selected in GRAPH2D to access this

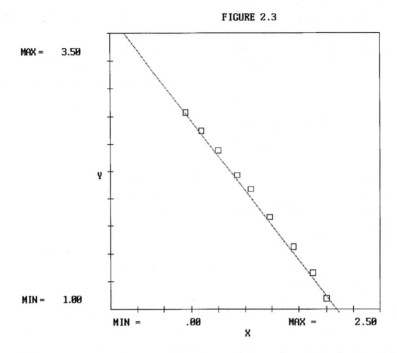

FIGURE 2.3

Figure 7.8   A reproduction of Figure 2.3 using GRAPH2D. Options used with GRAPH2D to produce this figure are keyboard data entry, no log transformation (these data come from Example 2.5), no data rescaling, minimum x value is 0.0, maximum x value is 2.5, minimum y value is 0.0, maximum y value is 3.5, a function is plotted with $a_0 = 3.66$, $a_1 = -1.27$, and a_2 through a_7 are all 0.0, points only are plotted, and the title is as shown.

file. The resultant graph (Figure 7.9) is comparable to that shown in Figure 5.19. All options used with GRAPH2D to develop this plot are listed in the caption to Figure 7.9. *Note:* The overprint summary from CORSPOND must be used when studying the graph of Figure 7.9 to understand which symbols overprint others.   ∎

Example 7.4   Using GRAPH2D to Display Semivariograms

Another option designed into GRAPH2D is the ability to read the output file storing semivariogram results created by the program FGAM (Chapter 6). *Note*: If the directional semivariogram option is used with FGAM, each directional semivariogram is stored in a separate file; these must be plotted one at a time, as is demonstrated momentarily.

Examples are presented to display both omnidirectional and directional semivariograms. An omnidirectional semivariogram is shown in Figure 7.10, computed using the NVSIM.DAT (Chapter 6; Appendix A) data set. Notice that a spherical semivariogram model is displayed with these data; in fact, the model parameters are those listed in Table 6.7 (see Chapter 6). In using GRAPH2D to fit this particular model to the calculated semivariogram values, the following coefficients are used: $a_0 = 10.0$; $a_1 = 1.65$; $a_2 = 0.0$; $a_3 = -5.5 \times 10^{-5}$; coefficients a_4 through a_7 are zero. A plot (Figure 7.10) is useful for visually assessing the quality of the semivariogram model fit

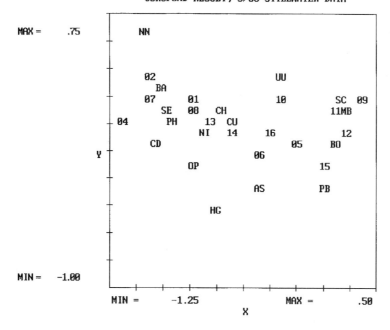

Figure 7.9 GRAPH2D plot of correspondence analysis results for the September 1986 Stillwater data (Chapter 5). Factor 1 (horizontal axis) is plotted versus factor 2 (vertical axis). The following options are used with GRAPH2D to obtain this plot: correspondence analysis data entry option is chosen; 3 factors were calculated by CORSPOND (Chapter 5), factors 1 and 2 are chosen for plotting; no log transformation and no data rescaling are used; minimum $x = -1.25$, maximum $x = 0.5$, minimum $y = -1.0$, and maximum $y = 0.75$; no function is plotted and the title is as shown.

to the calculated values of the semivariogram. In this case (Figure 7.10), the model fits the data quite well.

Directional semivariograms are useful for assessing whether the spatial correlation for a particular regionalized phenomenon is isotropic or anisotropic. The LONGB.DAT earthquake data set (Appendix A) is used for this example. Four directional semivariograms are calculated using FGAM and displayed in Figure 7.11 using GRAPH2D.

Spatial anisotropy is manifested in directional semivariograms as different ranges for orthogonal spatial directions. In the example (Figure 7.11), a weak spatial anisotropy is detected; the north–south direction does seem to show a longer range in comparison to the east–west direction. Because the data are only weakly anisotropic, an isotropic model will adequately represent the data and is more easily implemented in kriging.

Example 7.5 Using GRAPH2D to Display the Results from Cross-Validation Using the Program JCBLOK

One option available when using the program JCBLOK (Chapter 6) is cross-validation. In cross-validation, estimates are obtained at spatial locations associated with

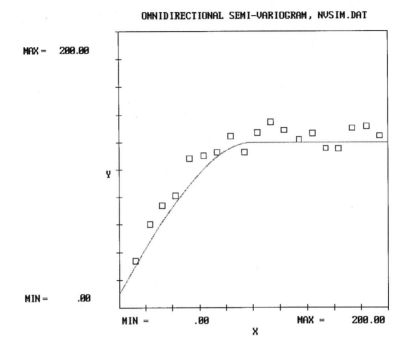

Figure 7.10 GRAPH2D plot of an omnidirectional semivariogram for the NVSIM.DAT (Appendix A) data set. The following options are used with GRAPH2D to obtain this plot: semivariogram data entry option is chosen; no log transformation or data rescaling is used; minimum $x = 0.0$, maximum $x = 200.0$, minimum $y = 0.0$, and maximum $y = 200.0$; a function is plotted with these data such that $a_0 = 10.0$, $a_1 = 1.65$, $a_2 = 0.0$, $a_3 = -5.5E-05$, and a_4 through a_7 are equal to 0.0; points only are plotted and the title is as shown. Lag distance (x axis) is plotted versus semivariance (y axis).

measured values. Of course, the measured value at a spatial location is not used in computing the estimated value at that location. Therefore, one objective of cross-validation is to assess the quality of estimation for a particular spatial data set using a chosen semivariogram model.

A typical visual way in which the results of cross-validation are displayed is through the use of a scatter diagram. Such a display is nothing more than a two-dimensional graph for which the x axis is associated with actual data values and the y axis is associated with estimated values. If all estimates are perfect, such a graph is a straight line having a slope of 1.0 and an intercept of 0. In practice, estimation is rarely perfect. Therefore, a certain amount of scatter about this 1:1 line will often occur.

GRAPH2D has an option for creating a scatter diagram. For example, the quality of estimates is assessed using the NVSIM.DAT data (Appendix A). For these data, the following semivariogram model is used: model type = spherical; nugget = 10.0; sill = 120.0; range = 100 units; anis = 0.0; ratio = 0.0; maximum number of nearest-neighbor values used for estimation = 10; search window radius = 100. The program JCBLOK is used to create an output file that stores the results for the cross-validation.

The resultant plot developed using GRAPH2D and the cross-validation display option is shown in Figure 7.12. A function is plotted for these data and is the line having

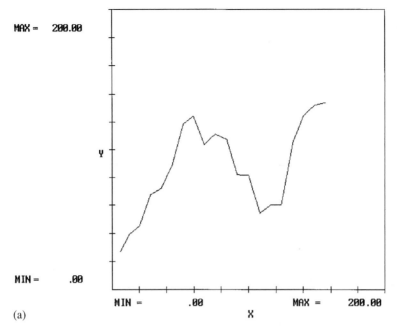

(a)

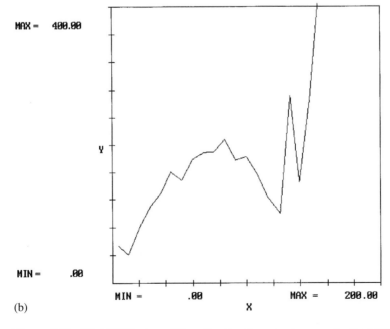

(b)

Figure 7.11 GRAPH2D plots of four directional semivariograms for the LONGB.DAT
(Appendix A) data set. Options used with GRAPH2D to obtain these plots can be inferred
from the plots, except in the case of this data set, the *y* values for all four plots were
rescaled by a factor of 50.0 (rescaling is an option available in GRAPH2D). For each plot,
lag distance (*x* axis) is plotted versus semivariance (*y* axis). Directions are (a) east–west,
(b) northeast–southwest.

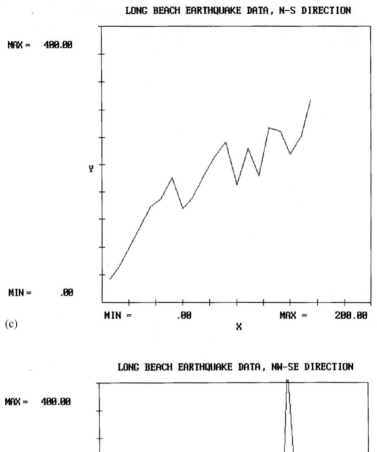

(c)

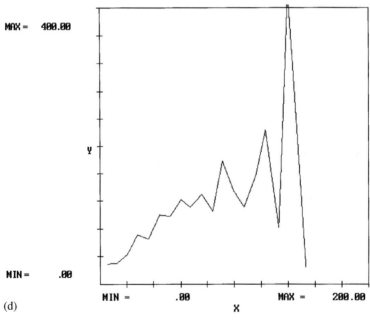

(d)

Figure 7.11 continued   Directions are (c) north–south, and (d) northwest–southeast.

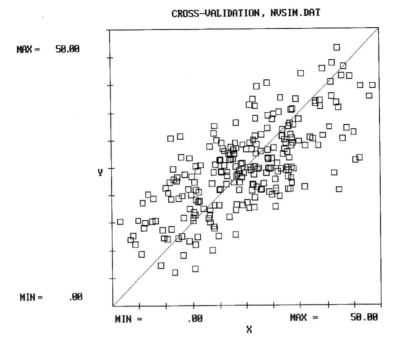

Figure 7.12   GRAPH2D display of cross-validation results for the NVSIM.DAT (Appendix A) data set. Actual data values (x axis) are plotted versus estimated data values (y axis). The following options are used with GRAPH2D to obtain this plot: cross-validation data entry; no log-transform or data rescaling is used; minimum $x = 0.0$, maximum $x = 50.0$, minimum $y = 0.0$, and maximum $y = 50.0$; a function is plotted having the coefficients $a_0 = 0.0$, $a_1 = 1.0$, and a_2 through $a_7 = 0.0$; points only are plotted and the title is as shown.

a $1:1$ slope, the basis for assessing the quality of estimation. This line is plotted by selecting the following coefficients: $a_0 = 0.0$; $a_1 = 1.0$; and coefficients a_2 through a_7 are zero.

This plot provides a qualitative assessment of estimation quality. Parameters of the semivariogram model may be varied to monitor their impact on estimation quality vis-à-vis the scatter diagram, as is demonstrated in Figure 7.13. A preferred model is one yielding the least amount of scatter with respect to the $1:1$ line. The scatter diagram also provides way to compare different types of interpolation models. Again, the one yielding the least amount of scatter about the $1:1$ line may be preferred for the data set.   ■

7.4 Two-Dimensional Coordinate Rotations

With some two-dimensional graphics applications, a coordinate rotation, or transformation, is warranted. This transformation is a simple counterclockwise rotation of the x and y axes about the origin.

Effecting such a transformation is accomplished by multiplying a rotation matrix $[R]$ by each coordinate vector (x, y) (the word vector in this case, represents the matrix notion). For two-dimensional rotations, this matrix is a

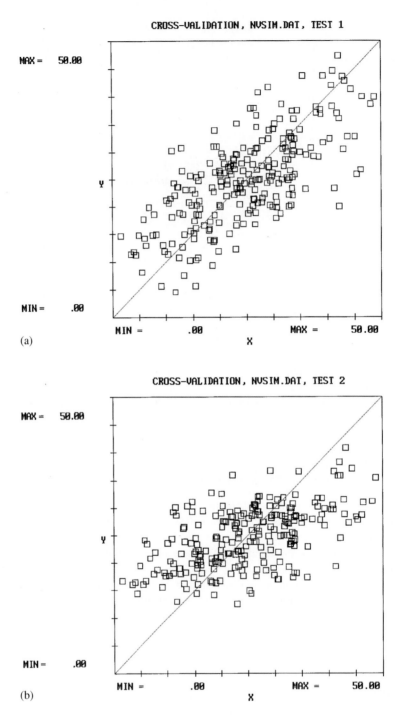

Figure 7.13 Two cross-validation experiments to show the influence the semivariogram model has on data estimation. (a) GRAPH2D is used to display the cross validation for the NVSIM.DAT data set using a nugget equal to 0.0, a sill of 120.0, and a range of 100. (b) GRAPH2D is used to display the cross validation for the NVSIM.DAT data set using a nugget of 120.0, a sill of 120.0, and a range of 100.0; this yields a gross smoothing of the data. The scatter in each of these two results is more severe than what is shown in Figure 7.12. In each of these plots, actual data values (*x* axis) are plotted versus estimated values (*y* axis).

function of the single rotation angle θ about the origin:

$$[R] = \begin{vmatrix} \cos\theta & \sin\theta \\ -\sin\theta & \cos\theta \end{vmatrix}$$

Coordinate transformation is accomplished using the following matrix multiplication:

$$\{x'\ y'\} = \{x\ y\}[R]$$

where $\{x'\ y'\}$ is the vector containing the transformed coordinates. A graphical plot showing the results from rotation is developed using the transformed vectors $\{x'\ y'\}$ rather than the original untransformed vectors $\{x\ y\}$.

7.5 Three-Dimensional Graphical Images

Of the many graphical displays possible on a computer monitor screen, those representing three-dimensional images are perhaps the most exciting. Coincidentally, these images are an enigma: How is a three-dimensional image plotted on a two-dimensional screen? But, this enigma is no less than that faced by an artist who wishes to represent a three-dimensional image on a two-dimensional canvas. Thus, we shall mimic the artist when developing computer algorithms for the display of three-dimensional images on a two-dimensional screen.

7.5.1 The Forgetting Transform

A three-dimensional object is plotted on a two-dimensional screen using a projection. For example, a line drawn in three-dimensional space is simply projected onto a two-dimensional surface. Physically, such a projection is accomplished easily by placing the line, for example, represented by a rod, between a sheet of paper and a light bulb, as shown in Figure 7.14. In terms of computer-generated graphics, though, how is such a projection accomplished?

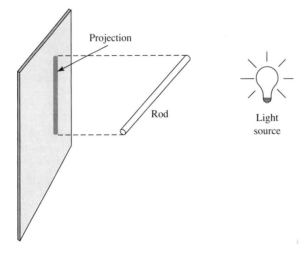

Figure 7.14   A drawing showing the projection of a rod onto a two-dimensional surface using a light source. What is shown is the rod's shadow, which is a way of showing how an object oriented in three-dimensional space will appear if depicted on a two-dimensional surface.

This is an important and provocative question. Projecting three-dimensional objects onto a two-dimensional screen is accomplished simply by forgetting one of the coordinate directions when plotting points or drawing lines. This is the definition of the forgetting transform. For example, with respect to the cube in Figure 7.15, by forgetting the x-coordinate direction, a two-dimensional projection of the cube onto the y-z plane is obtained; by forgetting the y-coordinate direction, a projection onto the x-z plane is obtained; and by forgetting the z-coordinate direction, a projection onto the x-y plane is obtained. Applications of the forgetting transform are demonstrated after the next two sections.

7.5.2 Three-Dimensional Coordinate Rotations

Two-dimensional rotation was introduced in Section 7.4. Now coordinate rotation is discussed relative to three coordinate axes. Whereas two-dimensional coordinate rotation involves only one angle of revolution, three-dimensional coordinate rotation involves up to three angles of revolution, one for each of the three coordinate directions. For example, a three-dimensional object may simply spin about the z axis; or it may spin about the z axis while tipping about the x axis; or it may spin about the z axis, tip about the x axis, and twirl about the y axis. Any rotation or combination of rotations leads to a different position of the three-dimensional object in space.

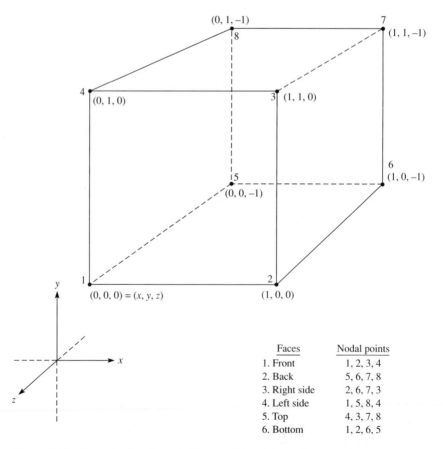

Faces	Nodal points
1. Front	1, 2, 3, 4
2. Back	5, 6, 7, 8
3. Right side	2, 6, 7, 3
4. Left side	1, 5, 8, 4
5. Top	4, 3, 7, 8
6. Bottom	1, 2, 6, 5

Figure 7.15   An example cube; coordinates of the eight vertices are shown.

Because each point associated with a three-dimensional object is defined by a 3×1 coordinate vector $\{x\ y\ z\}$, a rotation matrix $[R]$, itself a 3×3 matrix, is required to perform the following coordinate transformation:

$$\{x'\ y'\ z'\} = \{x\ y\ z\}[R]$$

The 3×3 matrix $[R]$ can be developed by considering three two-dimensional rotations. For example, rotation about the x axis is accomplished using $[R_x]$ as

$$[R_x] = \begin{vmatrix} 1 & 0 & 0 \\ 0 & \cos\beta & \sin\beta \\ 0 & -\sin\beta & \cos\beta \end{vmatrix}$$

where β is the rotation about the x axis. If a rotation about the y axis is considered, the rotation matrix $[R_y]$ is

$$[R_y] = \begin{vmatrix} \cos\alpha & 0 & -\sin\alpha \\ 0 & 1 & 0 \\ \sin\alpha & 0 & \cos\alpha \end{vmatrix}$$

where α is the rotation about the y axis. And a rotation about the z axis is accomplished using $[R_z]$ such that

$$[R_z] = \begin{vmatrix} \cos\theta & \sin\theta & 0 \\ -\sin\theta & \cos\theta & 0 \\ 0 & 0 & 1 \end{vmatrix}$$

where θ is the rotation about the z axis.

Of course, a rotation may be considered relative to only one axis, or any combination of the three axes. For each axis associated with a rotation, the following multiplication is used:

$$\{x'\ y'\ z'\} = \{x\ y\ z\}[R]$$

For instance, to consider a rotation about all three axes, the following multiplications must be performed:

$$\{x'\ y'\ z'\} = \{x\ y\ z\}[R_x]$$
$$\{x''\ y''\ z''\} = \{x'\ y'\ z'\}[R_y]$$
$$\{x'''\ y'''\ z'''\} = \{x''\ y''\ z''\}[R_z]$$

Rather than performing three matrix multiplications, the three matrices $[R_x]$, $[R_y]$, and $[R_z]$ may be combined by realizing from the previous three multiplications that

$$\{x'''\ y'''\ z'''\} = \{x\ y\ z\}[R_x][R_y][R_z]$$

Therefore, the combination $[R_x][R_y][R_z]$ is equal to $[R_{xyz}]$:

$$[R_{xyz}] = \begin{vmatrix} \cos\alpha\cos\theta & \cos\alpha\sin\theta & -\sin\alpha \\ (\sin\alpha\sin\beta\cos\theta - \cos\beta\sin\theta) & (\cos\beta\cos\theta + \sin\beta\sin\alpha\sin\theta) & \sin\beta\cos\alpha \\ (\sin\theta\sin\beta + \cos\beta\sin\alpha\cos\theta) & (\cos\beta\sin\alpha\sin\theta - \sin\beta\cos\theta) & \cos\beta\cos\alpha \end{vmatrix}$$

In this discussion, all rotations are counterclockwise about a coordinate axis for positive-valued angles. If, however, the order of multiplication is $[R]\{x\ y\ z\}$, rotations are clockwise for positive-valued angles.

Example 7.6   Rotation of a Cube

Rotate the cube in Figure 7.15 about the z axis using an angle of 30°. Eight matrix multiplications must be performed to complete this problem as follows:

$$\{x'\ y'\ z'\} = \{x\ y\ z\}[R_{xyz}]$$

But for this problem $[R_{xyz}]$ simplifies to

$$[R_{xyz}] = \begin{vmatrix} \cos 30 & \sin 30 & 0 \\ -\sin 30 & \cos 30 & 0 \\ 0 & 0 & 1 \end{vmatrix}$$

The transformed coordinates therefore are

$$\{x_1'\ y_1'\ z_1'\} = \{0\ 0\ 0\}[R_{xyz}] = \{0\ 0\ 0\}$$
$$\{x_2'\ y_2'\ z_2'\} = \{1\ 0\ 0\}[R_{xyz}] = \{\cos 30, \sin 30, 0\}$$
$$\{x_3'\ y_3'\ z_3'\} = \{1\ 0\ 1\}[R_{xyz}] = \{\cos 30, \sin 30, 1\}$$
$$\{x_4'\ y_4'\ z_4'\} = \{0\ 0\ 1\}[R_{xyz}] = \{0, 0, 1\}$$
$$\{x_5'\ y_5'\ z_5'\} = \{0\ -1\ 0\}[R_{xyz}] = \{\sin 30, -\cos 30, 0\}$$
$$\{x_6'\ y_6'\ z_6'\} = \{1\ -1\ 0\}[R_{xyz}] = \{\cos 30 + \sin 30, \sin 30 - \cos 30, 0\}$$
$$\{x_7'\ y_7'\ z_7'\} = \{1\ -1\ 1\}[R_{xyz}] = \{\cos 30 + \sin 30, \sin 30 - \cos 30, 1\}$$
$$\{x_8'\ y_8'\ z_8'\} = \{0\ -1\ 1\}[R_{xyz}] = \{\sin 30, -\cos 30, 1\}$$

7.5.3 Vanishing Points

In representing the true perspective of a three-dimensional object on a two-dimensional medium, a vanishing point must be considered. Notice in Figure 7.16 that the size of objects decreases as the vanishing point is approached. Of course, careful study of a photograph, which is a two-dimensional image of a three-dimensional realm, reveals that objects in the photograph do decrease in size toward a vanishing point. The instructor of an art class will emphasize the necessity of considering a vanishing point in order for drawings to appear realistic once completed.

In the application of the forgetting transform herein, the vanishing point is considered to be at infinity. Parallel lines remain parallel in all perspectives using such a vanishing point. This is done for simplicity; otherwise, subsequent to rotation, a second operation on the transformed coordinates would be required to achieve the effect of the object decreasing in size to the vanishing point. Such an operation is described in Foley and Van Dam (1982, p. 290). For the purpose of this chapter, a vanishing point at infinity is adequate.

7.5.4 Plotting Three-Dimensional Grids: Hidden Line Removal and Shading Algorithms

A common application in the earth sciences for three-dimensional computer graphics is the display of three-dimensional grids. Given topography as a common example, grids are displayed consisting of grid points defined by x and y coordinates and an elevation (the third dimension). These grids are often displayed with some sort of spin and tip; the tip gives an effect as if the viewer is above the grid and peering down.

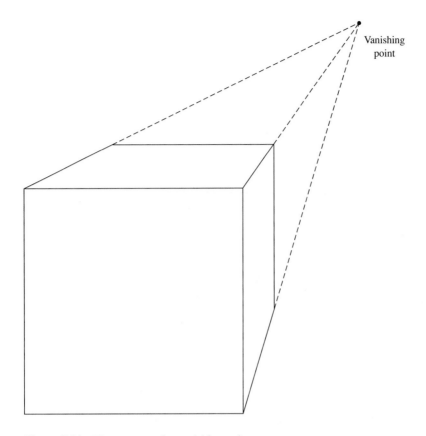

Figure 7.16   The concept of a vanishing point.

Two methods for displaying three-dimensional grids are discussed. First, the grid may be displayed by drawing the outlines of polygons; each polygon is defined by four grid points (Figure 7.17, a display of actual topography). Second, a shading algorithm may be used to display the polygons as solids; the intensity of the color of each polygon depends on the orientation of the polygon relative to some light source (Figure 7.18, a shaded relief drawing of the topography of Figure 7.17). A shading algorithm may be used with a low sun angle (light source), hence a display of this type is called shaded relief.

Figures 7.17 and 7.18 are displayed using a computer graphics algorithm that is capable of (1) three-dimensional rotations; (2) implementing a forgetting transform to forget z (the coordinate direction perpendicular to the screen), hence plotting only the xy plane; (3) hidden line removal; and (4) shading. This computer algorithm is called GRID3D (Figure 7.19, page 276).

Hidden line removal is used to obtain a display (Figure 7.17). Some of the polygons comprising the grid cannot be displayed because they face away from the viewer, as is the case with actual topography (all sides of a mountain cannot be seen from a given perspective, for instance). Moreover, some polygons are closer to the viewer than others and, whereas they all might face the viewer, the closer polygons may partially or completely obscure those that are farther away. Because of these problems associated with displaying three-dimensional

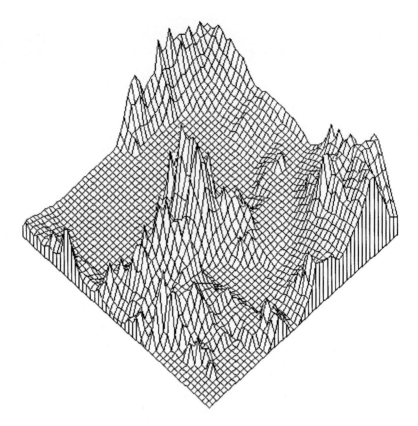

Figure 7.17   A display of actual topography (CALIENTE.DAT) using the GRID3D
program. A tip angle of −70 degrees and a spin angle of 225 degrees is used to obtain this
plot.

grids, a hidden line algorithm must be used to remove parts of those polygons
that are partially or totally hidden by other polygons. This algorithm should
also identify those polygons that face away from a viewer to avoid drawing
them at all. By using an algorithm to remove hidden lines, the drawing looks
like that of a solid and is cleaner in appearance because fewer lines are drawn.

Several algorithms are described in the literature for hidden line (hidden
surface) removal. One such algorithm is described in Foley and Van Dam
(1982). Considerable detail on hidden line removal is given in Ammeraal
(1986, pp. 88–130; a program is given on pp. 119–129). Ammeraal (1988, pp.
237–249) improves the 1986 algorithm by allowing for curved (polygonal)
surfaces. The Foley and Van Dam (1982) and Ammeraal (1986, 1988) algorithms are especially well suited for computer-aided-design (CAD) drawings
and are capable of drawing complex objects.

A simpler hidden line algorithm is found in Angell (1985, pp. 189–206).
The word simpler implies an ease of programming and implementation on
microcomputer systems. Furthermore, this algorithm works extremely well for
displaying three-dimensional grids. The algorithm is as follows:

1. Apply any desirable spin (rotation) and tip to the three-dimensional grid.
2. Draw the grid, polygon by polygon, from the part of the grid that is farthest

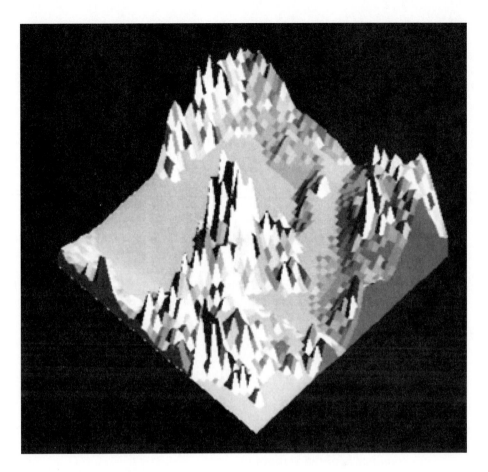

Figure 7.18   A shaded relief display of the grid shown in Figure 7.17. The shaded relief is also produced using GRID3D and, in this case, a low sun angle defined as dix $= -100.0$, diy $= -0.1$, diz $= 50.0$.

from the viewer to the part that is closest to the viewer. Fill each polygon with the background color as it is being drawn. In this manner, a closer polygon overwrites one more distant for a correct hidden line removal.

This is the complete and simple algorithm. Because this algorithm is the one used in GRID3D and because the algorithms of Foley and Van Dam (1982) and Ammeraal (1986, 1988) are much more complex and exceed the scope of this text, only the Angell (1985) algorithm is described in detail. That this algorithm does yield correct hidden line removal can be inferred (Figure 7.17).

A shading algorithm is also given by Angell (1985, pp. 189–206). In fact, as implemented by Angell, the shading algorithm can be used simultaneously with the hidden line removal algorithm. Instead of filling each polygon with the background color (as is done in the algorithm described above), each polygon is filled with a color whose intensity depends on the orientation of the polygon relative to some light source. The direction of the light source is defined by a vector oriented in three-dimensional space. This vector is assumed to pass through the origin ($x = 0$, $y = 0$, $z = 0$). Therefore, the source of the illumination is defined by three values: dix, diy, and diz; these three values define a point in three-dimensional space. The light source vector passes through this point and the origin.

For example, a very low sun angle (the elevation of the sun is defined by the y-coordinate direction diy) can be designed as follows: dix $= -100.0$; diy $= -0.1$; diz $= 50.0$. This is the sun angle, for instance, that is used to obtain the previous display (Figure 7.18). Notice that the value diy is small relative to dix and diz. Furthermore, varying dix and diz causes the position of the sun to move; for example, setting diz $= 0$ and dix > 0 puts the sun to the west; setting diz $= 0$ and dix < 0 puts the sun to the east; setting dix $= 0$ and diz > 0 puts the sun to the south; and setting dix $= 0$ and diz < 0 puts the sun to the north; the values used above, dix $= -100$ and diz $= 50$, put the sun to the southeast.

The Angell (1985) shading algorithm computes the vector normal to each polygon. This normal (perpendicular, orthogonal) vector is computed as follows. First, the normal vector passes through the origin and a point defined by the coordinates Nx, Ny, and Nz. Second, assuming that the polygon is defined by four vertices, numbered 1 through 4 in a counterclockwise fashion,

$$Nx = dz_2 dy_1 - dz_1 dy_2$$
$$Ny = dz_1 dx_2 - dz_2 dx_1$$
$$Nz = dx_1 dy_2 - dx_2 dy_1$$

where

$$dx_1 = x_2 - x_1, \quad dy_1 = y_2 - y_1, \quad dz_1 = z_2 - z_1$$
$$dx_2 = x_3 - x_2, \quad dy_2 = y_3 - y_2, \quad dz_2 = z_3 - z_2$$

for which the subscripts describe one of the vertices, 1, 2, or 3, and x, y, z are coordinates of the vertices.

If this normal vector points away from the viewer, the polygon is not drawn; otherwise, the cosine between this normal vector and the sun angle is computed. If, for example, the polygon is orthogonal to the sun direction, its normal vector is parallel to the sun angle and the cosine is 1.0; such a polygon should be brightly illuminated. On the other hand, if a polygon is parallel to the sun angle, its normal is orthogonal to the sun angle and the cosine is zero; such a polygon should not be illuminated. Therefore, the intensity of color used for shading each polygon is the maximum allowable intensity multiplied by the cosine between the sun angle and the polygon normal. This is the shading algorithm used in GRID3D.

Options used with GRID3D to obtain both Figures 7.17 and 7.18 are described in the captions for these figures. Another example is presented to show how GRID3D may be used to display the results from the kriging algorithm JCBLOK (Chapter 6). In this application, only polygon outlines will be displayed; a result using shading will not be shown (although the program GRID3D always displays both the polygon outline and shaded relief images). In Chapter 6, an application to the data set NVSIM.DAT was made. A contour map was displayed (Figure 6.26) which showed the application of kriging to these data. This kriged result may also be displayed as a three-dimensional grid (Figure 7.20). Options used to obtain this display are described in the caption for this figure. Perhaps larger and smaller data regions are more easily inferred when the kriged result is displayed as a three-dimensional grid rather than as a contour map.

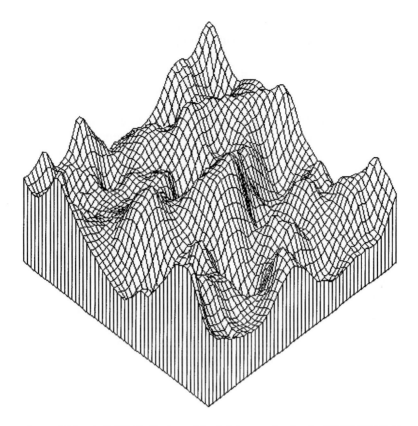

Figure 7.20   A GRID3D display of the kriging results for the NVSIM.DAT (Appendix A) data set; this kriging result is that produced by JCBLOK in Chapter 6; a tip angle of -60 and a spin angle of 135 are used to obtain this display. Notice, in contrast to the contour maps shown in Chapter 6, how readily apparent larger and smaller data values are in this GRID3D plot.

7.6 Mapping Using Color Density Slicing

Another way in which kriged results may be displayed is by using color density slicing. This term means nothing more than assigning to each estimated value in the grid an integer value that represents one of the colors in a palette. Then the grid is displayed as a digital image using these colors. The result of this display is a pseudo-contour map based on the use of colors, rather than contour lines.

To create a pseudo-contour map using color density slicing, the following algorithm is used:

1. Assuming that the high-resolution VGA graphics mode is used, design a 16-color palette using the program PALDSGN. For example, the palette for Table 7.2 is designed specifically for use in color density slicing for the display of kriging results. As an alternative, the default palette (listed at the beginning of the BOXDRAW program) may be used.
2. Determine the minimum and maximum values in the grid of data to be contoured; call these datamin and datamax.

3. Assign to each grid intersection an integer between 0 and 15, representing one of the colors in the palette using the following procedure:
(a) calling the grid intersection i, j, then
(b) $color_{i,j} = 15((\text{data value}_{i,j} - \text{datamin})/(\text{datamax} - \text{datamin}))$
(c) if $color_{i,j} < 0$, then set $color_{i,j} = 0$
(d) if $color_{i,j} > 15$, then set $color_{i,j} = 15$
4. Generate the display by assigning a color to each pixel coinciding with a grid intersection based on the integer color value. A computer algorithm (Figure 7.21, page 286) based on this procedure is called COLORDEN. An application of this computer algorithm is shown in the next section.

7.6.1 Color Density Mapping Using the Program COLORDEN

An application of the program COLORDEN to the display of gridding results from the program JCBLOK (Chapter 6) is made with respect to the NVSIM.DAT data set. Results from COLORDEN are shown in Figures 7.22 through 7.24. This program first displays the pseudo-contour map using the 16 colors of the palette. A legend is shown to the left of the map which associates each color with a numerical range (see Figure 7.22). Once this map is finished (and a hard copy is made if desired), the enter key is pushed to go on to the next contouring step.

The program is started by typing the command COLORDEN. The program first requests the name of the file, created by JCBLOK, that stores the gridding results. Then, the program asks if the default 16-color palette will be used for display or if a custom palette designed using PALDSGN will be used. If a

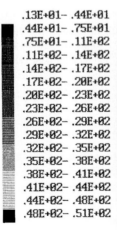
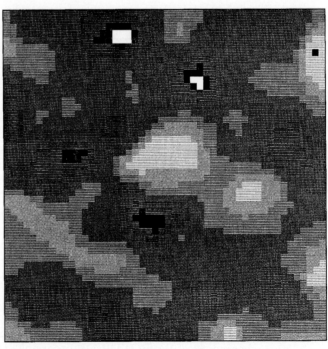

.13E+01– .44E+01
.44E+01– .75E+01
.75E+01– .11E+02
.11E+02– .14E+02
.14E+02– .17E+02
.17E+02– .20E+02
.20E+02– .23E+02
.23E+02– .26E+02
.26E+02– .29E+02
.29E+02– .32E+02
.32E+02– .35E+02
.35E+02– .38E+02
.38E+02– .41E+02
.41E+02– .44E+02
.44E+02– .48E+02
.48E+02– .51E+02

Figure 7.22   A COLORDEN display of the NVSIM.DAT kriging results; the palette used is that listed in Table 7.2.

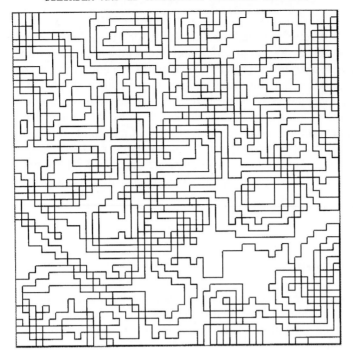

Figure 7.23 A COLORDEN contour map of the NVSIM.DAT kriging results produced using 16 contouring intervals.

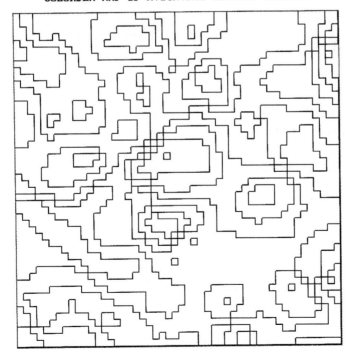

Figure 7.24 A COLORDEN contour map of the NVSIM.DAT kriging results produced using 8 contouring intervals.

PALDSGN palette is used, the file name for this palette must be specified when requested by the program. Subsequent to the specification of the palette, COLORDEN asks the user to specify the number of contouring intervals desired: 4, 8, 12, or 16. After the initial color density map is displayed (e.g., Figure 7.22) and the enter key is pushed, the program creates a contour map by outlining the boundaries between the colors in the first map. If fewer than 16 contour intervals are requested, the program first replots the color map using a fewer number of colors; then the contour map is created. The ultimate result of the contour map is shown in Figures 7.23 and 7.24. These maps are not as elegant as those shown in Chapter 6, but they do offer one approach to contour mapping. The two figures contrast the use of 16 and 8, respectively, contouring intervals.

7.7 Contour Mapping

The foregoing discussion presents the notion of pseudo-contour mapping using color density slicing. This section presents an algorithm for drawing actual contours. This algorithm is implemented in a computer program for contour map drawing.

7.7.1 Algorithm

A detailed treatise on contouring geologic surfaces is given by Jones et al. (1986). Useful algorithms are given therein for contour line drawing. Two basic steps are followed by most contouring algorithms (Jones et al., 1986, p. 58). For a regular grid of data (such as what is produced by the computer algorithm JCBLOK, Chapter 6), (1) find where a given contour line passes between two grid points, and (2) find where the line goes through a grid square and draw it.

Interestingly, a useful diagram of such an algorithm is given in Robinson and Coruh (1988). This algorithm is shown diagrammatically in Figure 7.25. Grid cells are emphasized, along with diagonals across them (as shown). The algorithm is as follows:

1. Determine a contour interval. Each interval is equal to the minimum gridded value plus the gridded data range (maximum minus minimum) divided by the number of desired contour intervals:

$$\text{min grid value} + (\text{grid range})/\text{NTOUR}$$

2. Search through all grid cells for intersections of the contour interval determined in step 1. Examples of grid cells are shown in Figure 7.25. Each grid cell is defined by two grid rows, I and $I + 1$, and two grid columns, $J - 1$ and J. First, determine if a contour line passes through the top of the grid cell. If so, use linear interpolation to determine the exact intersection (Figure 7.25). If the line does enter the top of the cell, determine if it crosses the diagonal. If so, use linear interpolation to determine the exact intersection (Figure 7.25). If the line crosses the diagonal, determine where it exits the cell. If the line does not cross the diagonal, determine where it exits on the right side of the cell. If the line does not intersect the top of the cell, check for an entry on the left side of the cell. If entry is found here, see if the

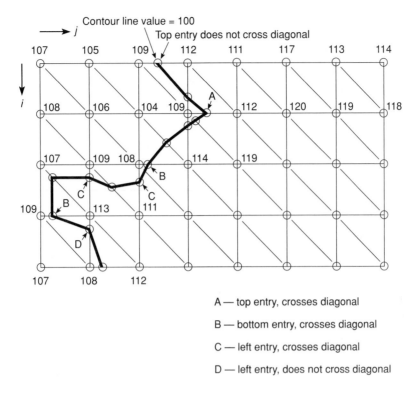

A — top entry, crosses diagonal

B — bottom entry, crosses diagonal

C — left entry, crosses diagonal

D — left entry, does not cross diagonal

Figure 7.25 An algorithm for drawing contour lines based on using grid cells and diagonals across them, as shown. The algorithm proceeds by determining first if and then where a contour line intersects a grid cell and passes through it.

line crosses the diagonal. If the line does not enter on the left side, see if it enters the cell from the bottom. If no entries are found, proceed to the next cell (and so on). This procedure is repeated for each and every grid cell for each and every contour interval.

3. Draw the contour lines as the algorithm proceeds.

7.7.2 CTOUR: A Computer Algorithm for Contour Mapping

A computer program, CTOUR is presented in Figure 7.26 (page 289) for contour line drawing (mapping). A user's guide is presented at the beginning of the listing. This computer program implements the algorithm described in the previous section and shown in Figure 7.25.

Up to 10 contouring steps are allowed by the program. Furthermore, the program is flexible and can read a variety of files, including those created by JCBLOK (Chapter 6), as well as files compatible with the program GRID3D. Once CTOUR requests the name of the input data file, it requests the number of values appearing on each line in the input file. For JCBLOK-created files, this number is 6; for a general input file to GRID3D, such as the CALI-ENTE.DAT file, this number is 3. Once this value is entered, the program requests information specifying which of these values is the one to be contoured. If the kriging estimates from JCBLOK are to be contoured, the value 5 is entered to show that the 5th value of each line in the data value is the one to be contoured.

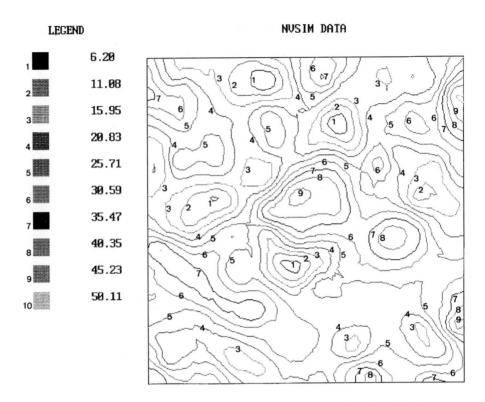

LEGEND   **NVSIM DATA**

1	6.20
2	11.08
3	15.95
4	20.83
5	25.71
6	30.59
7	35.47
8	40.35
9	45.23
10	50.11

Figure 7.27   A contour map of the NVSIM.DAT data set drawn using CTOUR. The program JCBLOK (Chapter 6) was used first to obtain a regular 50 × 50 grid. Six values appear on each line of the output file from JCBLOK; the fifth value on each line (the estimate from kriging) was chosen for contouring. Because this is a black and white print of an original color display, numbers have been added post program execution to render the map more interpretable.

Once the interactive data entry session is over, CTOUR draws the contour map, using different colors for each contour interval. A legend is drawn to show the data value associated with each contour color.

Examples are shown for the NVSIM.DAT data set (Figure 7.27) and the CALIENTE.DAT data set (Figure 7.28). These are black and white displays of the original color screen drawings, hence interpreting these contour maps is difficult. However, numbers have been artificially added (post program execution) to make the contour intervals more understandable. Comparing the contour map for the NVSIM.DAT data set (Figure 7.27) with the map shown in Chapter 6 (Figure 6.25) reveals the accuracy of the CTOUR computer algorithm.

Chapter Summary

- Converting numbers to pictures dramatically improves the ability to interpret quantitative information.
- Computer graphics are generated by turning on screen pixels to various colors.

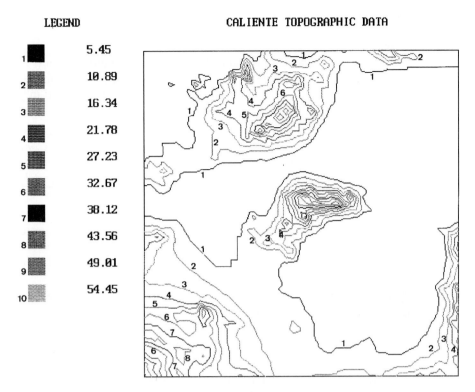

LEGEND

CALIENTE TOPOGRAPHIC DATA

1	5.45
2	10.89
3	16.34
4	21.78
5	27.23
6	32.67
7	38.12
8	43.56
9	49.01
10	54.45

Figure 7.28   A contour map of the CALIENTE.DAT data set. These data are already in grid format, hence no prior processing (such as using JCBLOK) is required for this data file. These data represent a 50 × 50 grid of topography. Three values appear on each line of the file; of these, the third value on each line was chosen for contouring. Numbers have been added to enhance interpretation because this figure is a black and white print of an original color display.

- The three most important computer graphics functions are windowing, viewing, and vector generation (one form of which is point (pixel) setting).
- GRAPH2D is an interactive, two-dimensional graph drawing program which supplements many of the chapters composing this text.
- The plotting of three-dimensional graphics images involves forgetting (discarding) one of the spatial directions and using the remaining two for plotting.
- The display of three-dimensional grids is a common use of computer graphics in the earth sciences. Hidden lines may be removed by drawing the grid from the point farthest from the viewer to the point closest to the viewer. A shading algorithm creates shaded relief images of three-dimensional grids.
- Color density mapping is a simple alternative to the plotting of contour maps.
- Outlining the color boundaries in a color density map yields a simple contour map.
- Algorithms for drawing actual contour maps are based on (1) finding where contour lines cross grid cells, and (2) following the lines through the cells.
- A picture is worth a thousand words.

Exercises

(An asterisk indicates an advanced or graduate student problem)

1. Design four different sets of 16-color palettes using the program PALDSGN. Use the color cube shown in Figure 7.2 to aid in your design of different colors.

2. Locate one or more two-dimensional graphs by searching through recent issues of the following journals: *Mathematical Geology, Computers and Geosciences, Nonrenewable Resources, International Journal of Mining and Geological Engineering, Bulletin of the Association of Engineering Geologists,* and *Water Resources Research.* Attempt to reproduce the graph(s) using the program GRAPH2D.

3. Use the program GRAPH2D to plot the results of the correspondence analysis of the June and September 1986, Stillwater data sets (Tables 5.1 and 5.2). Create the following plots for both data sets: factors 1 and 2, 1 and 3, and 2 and 3.

4. Suppose you wish to obtain more detail in your correspondence analysis plots. For instance, suppose you wish to focus only on the lower right quadrant of Figure 5.19 (see Chapter 5). How would you do this using the program GRAPH2D? Please experiment with your ideas regarding a solution to this problem.

5. Run the program JCBLOK (Chapter 6) to obtain a cross-validation of the NVSIMLOG.DAT data set using lognormal kriging. Plot the results of the cross-validation using the program GRAPH2D.

6. Run the program JCBLOK (Chapter 6) to grid the NVSIMLOG.DAT data set. Use the same grid parameters as were used for these data in Chapter 6. Use the program COLORDEN to display this grid, then contour the data using 8 contour levels. Compare this contour map to that presented in Chapter 6. Identify similarities and differences.

7. Display the grid results developed in Problem 6 using GRID3D. Experiment with several spin and tip angles to achieve a display that appears to be useful for interpretation.

8. Experiment with GRID3D for displaying the CALIENTE.DAT data set using several different tip and spin angles. Experiment with several different sun angles to see how they affect the shading algorithm.

9. Compute the integer value (number) represented by each of the following 8-bit bytes: (a) 00111111, (b) 00000001, and (c) 00101011.

10. Apply the program JCBLOK (Chapter 6) to obtain regular grids of size 50 × 50 for the following data sets found in Appendix A: (a) LONGB.-DAT, (b) SANFER.DAT, and (c) AREA1.DAT. Use the program CTOUR to display contour maps of the resultant grids. Use the program COLORDEN to display these grids. Use the COLORDEN displays to evaluate the accuracy of the contour maps derived using CTOUR.

*11. Modify CTOUR to change the contouring algorithm slightly such that the diagonals across grid cells are from the top right to the bottom left of each cell (note how this differs from Figure 7.25). Does this modification change resultant contour maps?

References and Suggested Readings

AMMERAAL, L. 1986. *Programming Principles in Computer Graphics.* Chichester, UK: Wiley.

AMMERAAL, L. 1988. *Interactive 3D Computer Graphics.* Chichester, UK: Wiley.

ANGELL, I.O. 1985. *Advanced Graphics with the IBM Personal Computer.* London: Macmillan.

FOLEY, J.D., and A. VAN DAM. 1982. *Fundamentals of Interactive Computer Graphics.* Reading, MA: Addison-Wesley.

HUNT, W.J. 1989. *The C Toolbox,* 2d ed. Reading, MA: Addison-Wesley.

JONES, T.A., D.E. HAMILTON, and C.R. JOHNSON. 1986. *Contouring Geologic Surfaces With the Computer.* New York: Van Nostrand Reinhold.

Microsoft Corporation. 1991. *Microsoft FORTRAN Version 5.1: Advanced Topics,* Redmond, WA: Microsoft Corporation.

ROBINSON, E.S. and C. CORUH. 1988. *Basic Exploration Geophysics.* New York: Wiley.

TUFT, E.R. 1983. *The Visual Display of Quantitative Information.* Chesire, CT: Graphics Press.

Figure 7.1 A listing of the program PALDSGN.

```fortran
C     PALDSGN
C     A PROGRAM TO EXPERIMENT WITH PALETTE DESIGN
C     FOR HIGH RESOLUTION, VGA GRAPHICS MODE
C     I.E., VRES16COLOR MODE
C     *************************************************
C                 USER'S GUIDE FOR PALDSGN
C     *************************************************
C     YOU WILL BE ASKED A SERIES OF INTERACTIVE QUESTIONS TO
C     SPECIFY THE INTENSITIES FOR EACH PRIMARY COLOR, RED,
C     BLUE, AND GREEN, FOR USE IN DESIGNING A COLOR.
C
C     FOR EXAMPLE, IF YOU DESIRE THE COLOR, BRIGHT RED, THEN
C     SET RED = 63, BLUE = 0, GREEN = 0.
C
C     USE THE COLOR CUBE TO DESIGN ANY COLOR.
C
C     THE PROGRAM ASKS YOU TO DESIGN 16 COLORS TO CREATE A
C     PALETTE. YOU WILL THEN BE GIVEN THE OPTION TO SAVE THE
C     PALETTE FOR USE IN OTHER PROGRAMS IN THIS TEXT.
C     *************************************************
C              END OF USER'S GUIDE AND DISCUSSION
C     *************************************************
C
      INCLUDE 'FGRAPH.FI'
      INCLUDE 'FGRAPH.FD'
      INTEGER*2   DUMMY2,  DUMMY
      INTEGER*4   DUMMY4, IBLUE, IRED, IGREEN
      INTEGER*4   RGB, TMP, IPAL(16,3), TMPMAX
      CHARACTER*2 BCOLOR(16)
      RECORD /VIDEOCONFIG/ VC
      RECORD /RCCOORD/ CURPOS
      DATA BCOLOR /' 1',' 2',' 3',' 4',' 5',' 6',' 7',' 8',' 9',
     2             '10','11','12','13','14','15','16' /
C
      TMPMAX = 0

C     SELECT THE HIGH RESOLUTION, 16 COLOR PALETTE VGA MODE
C
      DUMMY2 = SETVIDEOMODE($VRES16COLOR)
C
C     GET CONFIGURATION VARIABLES FOR CURRENT MODE
C
      CALL GETVIDEOCONFIG (VC)
C
C     BEGIN PALETTE DESIGN
C
      DO I = 1, 16
C
C     CLEAR THE SCREEN FOR THIS COLOR DESIGN
C
      CALL CLEARSCREEN ($GCLEARSCREEN)
C
C     SEEK INPUT TO DEFINE INTENSITIES FOR PRIMARY COLORS
C
20    CONTINUE
      WRITE(*,*) 'COLOR BEING DESIGNED = ', I
      WRITE(*,*)
      WRITE(*,*) 'ENTER A VALUE FOR RED:    0 TO 63'
      READ(*,*)  IRED
      WRITE(*,*) 'ENTER A VALUE FOR BLUE:   0 TO 63'
      READ(*,*)  IBLUE
      WRITE(*,*) 'ENTER A VALUE FOR GREEN:  0 TO 63'
      READ(*,*)  IGREEN
C
C     USE RGB FUNCTION TO DESIGN COLOR USING COLOR CUBE
C
      TMP = RGB(IRED, IGREEN, IBLUE)
      IF (TMP .GT. TMPMAX) THEN
         TMPMAX = TMP
         J      = I
      END IF
C
```

Figure 7.1 continued

```fortran
C     PUT COLOR INTO PALETTE
C
      DUMMY4 = REMAPPALETTE(I-1,TMP)
C
C     DRAW A COLORED SQUARE OF THE CHOSEN COLOR
C
      DUMMY2 = SETCOLOR(I-1)
      DUMMY2 = RECTANGLE ($GFILLINTERIOR, 500, 100, 600, 200)
C
C     DESIGN ANOTHER COLOR OR CONTINUE TO NEXT COLOR
C
      WRITE(*,*) 'ARE YOU HAPPY WITH THIS COLOR?'
      WRITE(*,*) 'ENTER 1 FOR YES; 0 TO CHANGE THIS COLOR'
      READ(*,*) IQUIT
      IF (IQUIT .EQ. 0) GO TO 20
C
C     PLACE THIS COLOR INTO IPAL ARRAY, THEN DESIGN NEXT COLOR
C
      IPAL(I,1) = IRED
      IPAL(I,2) = IGREEN
      IPAL(I,3) = IBLUE
      END DO
C
C     DISPLAY THE FINAL PALETTE
C
      CALL CLEARSCREEN ($GCLEARSCREEN)
      CALL SETTEXTPOSITION(1,1,CURPOS)
      DUMMY = SETTEXTCOLOR(J)
      CALL OUTTEXT('FINAL PALETTE')
      CALL SETTEXTPOSITION(1,68,CURPOS)
      CALL OUTTEXT('COLOR')
      DO I = 1, 16
      DUMMY2 = SETCOLOR(I-1)
      DUMMY2 = RECTANGLE ($GFILLINTERIOR, 501, (16*(I)+1), 516,
    2                                        (16*(I+1)))
      CALL SETTEXTPOSITION(I+1,70,CURPOS)
      CALL OUTTEXT(BCOLOR(I))
      END DO
C
C     SAVE THIS PALETTE TO AN OUTPUT FILE IF DESIRED
C
      CALL SETTEXTPOSITION(22,1,CURPOS)
      WRITE(*,*) 'DO YOU WISH TO SAVE THIS PALETTE TO A FILE?'
      WRITE(*,*) 'ENTER 1 FOR YES; 0 FOR NO'
      READ(*,*) ISAVE
      IF (ISAVE .EQ. 1) THEN
      OPEN (1, FILE = ' ', STATUS = 'NEW')
      DO I = 1, 16
      WRITE(1,*) (IPAL(I,JK), JK = 1,3)
      END DO
      END IF
C
C     TERMINATE THE PROGRAM
C
      DUMMY2 = SETVIDEOMODE($DEFAULTMODE)
      STOP
      END
C
C     RGB FUNCTION
C
      INTEGER*4 FUNCTION RGB(R, G, B)
      INTEGER*4 R, G, B
      RGB = ISHL(ISHL(B,8) .OR. G,8) .OR. R
      RETURN
      END
```

Figure 7.5 A listing of the program BOXDRAW.

```fortran
C     PROGRAM BOXDRAW
C
C     A PROGRAM TO DEMONSTRATE:
C        1.  WINDOWING
C        2.  VIEWING
C        3.  VECTOR DRAWING
C        4.  CHANGE OF VIEW DOMAIN
C
C     ************************************************************
C                         USER'S GUIDE
C     ************************************************************
C
C     ALL THAT IS NECESSARY TO RUN THIS PROGRAM IS TO TYPE
C     THE COMMAND:  BOXDRAW.
C
C        E.G.:  A>BOXDRAW
C
C     NO DATA INPUT IS REQUIRED.
C
C     THE DEFAULT PALETTE OF COLORS, AS SHOWN BELOW,
C     IS USED FOR THIS GRAPHICS DRAWING PROGRAM:
C
C     DEFAULT PALETTE:
C
C        COLOR NUMBER          COLOR
C             0                BLACK
C             1                BLUE
C             2                GREEN
C             3                CYAN
C             4                RED
C             5                MAGENTA
C             6                BROWN
C             7                WHITE
C             8                DARK GREY
C             9                LIGHT BLUE
C            10                LIGHT GREEN
C            11                LIGHT CYAN
C            12                LIGHT RED
C            13                LIGHT MAGENTA
C            14                YELLOW
C            15                BRIGHT WHITE
C     ************************************************************
C               END OF USER'S GUIDE AND DISCUSSION
C     ************************************************************
C
      INCLUDE 'FGRAPH.FI'
      INCLUDE 'FGRAPH.FD'
      INTEGER*2  DUMMY2
      DOUBLE PRECISION XX(4), YY(4)
      RECORD /WXYCOORD/ XY
      RECORD /VIDEOCONFIG/ VC
      RECORD /RCCOORD/ CURPOS
C
C     1.  SELECT THE HIGH RESOLUTION, VGA GRAPHICS MODE
C
      DUMMY2 = SETVIDEOMODE ($VRES16COLOR)
C
C     2.  GET CONFIGURATION VARIABLES FOR CURRENT MODE
C
      CALL GETVIDEOCONFIG (VC)
C
C     3.  DEFINE THE WINDOW REGION; SET THE .TRUE./.FALSE.
C         LOGICAL SWITCH TO .TRUE. TO SHOW THAT THE PHYSICAL
C         Y-COORDINATES OF THE BOX INCREASE FROM BOTTOM TO TOP;
C         MOREOVER, DEFINE THE WINDOW DOMAIN TO BE SLIGHTLY
C         LARGER THAN THE ACTUAL BOX SIZE.
C
      DUMMY2 = SETWINDOW (.TRUE., -1.0, -1.0, 4.0, 4.0)
C
C     4.  CLEAR THE SCREEN IN PREPARATION FOR DRAWING
C
      CALL CLEARSCREEN ($GCLEARSCREEN)
C
```

```
C     5A.   PREPARE FOR DRAWING BOX 1.   SET VIEWING DOMAIN
C
      K = 1
      CALL SETVIEWPORT( 10, 10, 190, 190)
C     5B.   SET COLOR FOR BOX TO RED
C
      DUMMY2 = SETCOLOR(4)
C     5C.   DRAW BOX 1
C
      GO TO 1000
C     6A.   PREPARE FOR DRAWING BOX 2.   SET VIEWING DOMAIN
C
10    CONTINUE
      K = 2
      CALL SETVIEWPORT( 210, 10, 390, 190)
C     6B.   SET COLOR FOR BOX TO GREEN
C
      DUMMY2 = SETCOLOR(2)
C     6C.   DRAW BOX 2
C
      GO TO 1000
C     7A.   PREPARE FOR DRAWING BOX 3.   SET VIEWING DOMAIN
C
20    CONTINUE
      K = 3
      CALL SETVIEWPORT( 410, 10, 590, 190)
C

C     7B.   SET COLOR FOR BOX TO BLUE
C
      DUMMY2 = SETCOLOR(1)
C     7C.   DRAW BOX 3
C
      GO TO 1000
C     FINISH THE DRAWING BY DRAWING A WHITE BORDER AROUND
C     THE THREE BOXES
C
30    CONTINUE
      DUMMY2 = SETWINDOW(.TRUE., 0.0, 0.0, 3.0, 1.5)
      CALL SETVIEWPORT (4, 4, 596, 196)
      DUMMY2 = SETCOLOR(15)
      YY(3) = 1.5D0
      YY(4) = 1.5D0
      CALL MOVETO_W( XX(1), YY(1), XY)
      DUMMY2 = LINETO_W(XX(2), YY(2))
      DUMMY2 = LINETO_W(XX(3), YY(3))
      DUMMY2 = LINETO_W(XX(4), YY(4))
      DUMMY2 = LINETO_W( XX(1), YY(1))
C     PAUSE FOR PRINT SCREEN OPTION
C
      CALL SETTEXTPOSITION(22,3,CURPOS)
      CALL OUTTEXT(' PUSH THE ENTER KEY TO FINISH PROGRAM')
C     RETURN SCREEN TO DEFAULT MODE
C
      READ(*,*)
      DUMMY2 = SETVIDEOMODE ($DEFAULTMODE)
      GO TO 2000
1000  CONTINUE
C
```

Figure 7.5 continued

```
C     BOX DRAWING ROUTINE
C

      XX(1) = 0.D0
      YY(1) = 0.D0
      XX(2) = 3.D0
      YY(2) = 0.D0
      XX(3) = 3.D0
      YY(3) = 3.D0
      YY(4) = 3.D0
      XX(4) = 0.D0

C
C     MOVE TO STARTING CORNER (LOWER, LEFT) OF BOX
C
      CALL MOVETO_W( XX(1), YY(1), XY)
C
C     DRAW THE FOUR SIDES OF THE BOX
C
      DO I = 2, 4
        DUMMY2 = LINETO_W( XX(I), YY(I))
      END DO
      DUMMY2 = LINETO_W( XX(1), YY(1))
      IF (K .EQ. 1) THEN
        GO TO 10
      ELSEIF (K .EQ. 2) THEN
        GO TO 20
      ELSEIF (K .EQ. 3) THEN
        GO TO 30
      END IF
2000  STOP
      END
```

Figure 7.7 A listing of the program GRAPH2D.

```
C    PROGRAM GRAPH2D
C
C    AN INTERACTIVE PROGRAM FOR THE DESIGN OF 2D GRAPHS
C
C    *******************************************************
C              USER'S GUIDE:  GRAPH2D
C    *******************************************************
C
C    THIS PROGRAM IS AN INTERACTIVE ONE; QUESTIONS APPEAR ON
C    THE SCREEN AS FOLLOWS:
C
C    QUESTION 1:  HOW ARE DATA TO BE ENTERED INTO PROGRAM?
C              ENTER 1 FOR DATA FILE;
C              ENTER 2 FOR KEYBOARD;
C              ENTER 3 IF DATA ARE RESULTS FOR CROSS-VALIDATION
C                   FROM THE PROGRAM, JCBLOK (CHAPTER 6)
C              ENTER 4 IF DATA ARE OUTPUT FROM
C                   CORSPOND (CHAPTER 5).
C              ENTER 5 IF DATA ARE OUTPUT FROM FGAM (CHAP 6)
C              ENTER 6 IF A ONE-DIMENSIONAL PROFILE IS TO
C                   BE GRAPHED
C
C    IF OPTION SPECIFIED IS 1, THEN:
C
C    DATA FILE ENTRY:   THE FILE SHOULD LOOK LIKE THIS:
C
C              X1, Y1
C              X2, Y2
C                .  .
C              XN, YN;    WHERE N CAN BE UP TO 1000
C
C    IF OPTION SPECIFIED IS 2, THEN:

C    KEYBOARD ENTRY:  THE PROGRAM PROMPTS THE QUESTION:
C              HOW MANY DATA PAIRS DO YOU HAVE?
C              RESPOND WITH AN INTEGER FROM 1 TO 1000
C
C              THEN, THE  PROGRAM RESPONDS WITH N QUESTIONS:
C              ENTER X,Y FOR PAIR I:
C
C              ENTER THE DATA AS FOLLOWS:
C                   X1, Y1 <ENTER>
C                   X2, Y2 <ENTER>
C                   UP TO XN, YN <ENTER>
C
C    IF OPTION SPECIFIED IS 3, THEN THE FILE NAME STORING
C    THE RESULTS FROM CROSS VALIDATION (JCBLOK) IS REQUESTED
C
C    IF OPTION SPECIFIED IS 4, THEN THE FILE NAME STORING
C    THE CORRESPONDENCE ANALYSIS FACTORS IS REQUESTED;
C    THEN, AN INTERACTIVE QUESTION MUST BE ANSWERED STATING
C    HOW MANY FACTORS APPEAR IN THE FILE; ANOTHER QUESTION
C    THEN ASKS WHICH TWO FACTORS ARE TO BE USED FOR PLOTTING.
C
C    IF OPTION SPECIFIED IS 5, THEN THE FILE NAME STORING
C    THE SEMI-VARIOGRAM RESULTS FROM FGAM (CHAP 6) IS REQUESTED
C
C    IF OPTION 6 IS SPECIFIED, THEN THE NAME OF THE FILE
C    STORING THE ONE DIMENSIONAL PROFILE IS REQUESTED;
C    THE ONE DIMENSIONAL PROFILE IS CONSIDERED TO REPRESENT
C    THE "Y" DIRECTION; A VALUE FOR THE X-DIRECTION
C    INCREMENT IS REQUESTED ENABLING THE PROGRAM TO
C    AUTOMATICALLY GENERATE X-COORDINATES FOR PLOTTING.
```

269

Figure 7.7 continued

```fortran
C
C      QUESTION 2.   RESPOND TO QUESTIONS TO DEFINE GRAPH PARAMETERS
C
C         2A.   DO YOU WISH TO CONVERT X OR Y TO LOGARITHMS?
C               (THIS IS USEFUL FOR ANALYSES DESCRIBED IN
C               CHAPTER 11).  ENTER 1 FOR YES, 0 FOR NO.
C         2B.   DO YOU WISH TO RESCALE THE X-DIRECTION VALUES
C               AND/OR THE Y-DIRECTION VALUES?  IF SO:
C               IF X IS TO BE RESCALED, ENTER THE RESCALING
C                  VALUE, XMAG;
C               IF Y IS TO BE RESCALED, ENTER THE RESCALING
C                  VALUE, YMAG;
C         2C.   WHAT ARE THE MINIMUM AND MAXIMUM VALUES FOR THE
C               X-AXIS?
C               ENTER LIKE THIS:  XMIN, XMAX <ENTER>
C               ATTENTION:  IF THE X-AXIS IS RESCALED IN 2A, BE
C                           SURE TO ACCOUNT FOR THIS WHEN
C                           ENTERING XMIN, XMAX
C         2D.   WHAT ARE THE MINIMUM AND MAXIMUM VALUES FOR THE
C               Y-AXIS?
C               ENTER LIKE THIS:  YMIN, YMAX <ENTER>
C               ATTENTION:  IF THE Y-AXIS IS RESCALED IN 2A, BE
C                           SURE TO ACCOUNT FOR THIS WHEN
C                           ENTERING YMIN, YMAX
C      QUESTION 3.   WILL A FUNCTION BE FIT TO YOUR DATA OF THE FORM:
C
C           Y = A(0) + A(1)*X + A(2) *X*X + A(3) *X*X*X +
C               A(4) *X*X*X*X + A(5) *EXP(A(6)*X**(A(7)))
C
C               RESPOND:  1 = YES; 0 = NO
C
C      QUESTION 4.   IF ANSWER TO QUESTION 3 IS YES, THEN
C
C               ENTER THE COEFFICIENTS, A, LIKE THIS:
C
C               A0, A1, A2, A3, A4, A5, A6, A7 <ENTER>
C
C               EXAMPLE, IF Y = 3X + 4, THEN ENTER:
C
C               4.0, 3.0, 0.0, 0.0, 0.0, 0.0, 0.0
C
C      QUESTION 5.   ARE THE POINTS, (X,Y) TO BE CONNECTED BY
C                    LINES OR SIMPLY PLOTED AS POINTS?
C
C                    ENTER 1, FOR LINES; 0 FOR POINTS
C
C      QUESTION 6.   WHAT IS THE TITLE FOR THE GRAPH?
C                    ENTER THE TITLE
C
C      ********************************************************
C                    END OF USER'S GUIDE AND DISCUSSION
C      ********************************************************
C
       INCLUDE 'FGRAPH.FI'
       INCLUDE 'FGRAPH.FD'
       INTEGER*2 DUMMY2
       INTEGER KTEMP(400,2)
       REAL*8 X(1000), Y(1000), XMOD, YMOD, XINC, YINC
       REAL*8 XRANGE, YRANGE, A(8), FACT(10)
       CHARACTER*80 HEADER
       CHARACTER*50 TITLE
       CHARACTER*1 XLABEL, YLABEL
       CHARACTER*2 CORLAB(400)
       RECORD /WXYCOORD/ XY
       RECORD /RCCOORD / CURPOS
       XLABEL = 'X'
       YLABEL = 'Y'
       DUMMY2 = SETVIDEOMODE($VRES16COLOR)
C
```

Figure 7.7 continued

```fortran
C      ACCESS ALL DATA FOR THE GRAPH THROUGH INTERACTIVE QUERY
C
       CALL CLEARSCREEN($GCLEARSCREEN)
       CALL SETTEXTPOSITION(1,10,CURPOS)
       WRITE(*,*) 'WELCOME TO THE GRAPH2D PROGRAM'
       CALL SETTEXTPOSITION(3,5,CURPOS)
       WRITE(*,*) 'ARE DATA TO BE READ FROM AN INPUT FILE, OR'
       CALL SETTEXTPOSITION(4,5,CURPOS)
       WRITE(*,*) 'ENTERED THROUGH THE KEYBOARD?  ENTER AS FOLLOWS:'
       CALL SETTEXTPOSITION(8,10,CURPOS)
       WRITE(*,*) 'ENTER 1 FOR GENERAL DATA FILE'
       CALL SETTEXTPOSITION(9,10,CURPOS)
       WRITE(*,*) 'ENTER 2 FOR KEYBOARD ENTRY'
       CALL SETTEXTPOSITION(10,10,CURPOS)
       WRITE(*,*) 'ENTER 3 FOR CROSS-VALIDATION DISPLAY'
       CALL SETTEXTPOSITION(11,10,CURPOS)
       WRITE(*,*) 'ENTER 4 FOR CORRESPONDENCE ANALYSIS DISPLAY'
       CALL SETTEXTPOSITION(12,10,CURPOS)
       WRITE(*,*) 'ENTER 5 FOR SEMI-VARIOGRAM DISPLAY'
       CALL SETTEXTPOSITION(13,10,CURPOS)
       WRITE(*,*) 'ENTER 6 FOR A ONE DIMENSIONAL PROFILE'
       CALL SETTEXTPOSITION(15,10,CURPOS)
       WRITE(*,*) 'ENTER YOUR RESPONSE NOW'
       READ(*,*) IDAT
       IF (IDAT .EQ. 1) THEN
       OPEN(5, FILE = ' ')
       N = 1
10     CONTINUE
       READ(5,*,END=20) X(N), Y(N)
       N = N + 1
       GO TO 10
20     N = N - 1
       ELSEIF (IDAT .EQ. 2) THEN
       WRITE(*,*) 'HOW MANY DATA PAIRS DO YOU HAVE?'
       READ(*,*) N
       DO I = 1,N
       WRITE(*,*) 'ENTER X,Y VALUES FOR PAIR ', I
       READ(*,*) X(I), Y(I)
       END DO
       ELSEIF (IDAT .EQ. 3) THEN

       WRITE(*,*) 'ENTER FILENAME FOR CROSS-VAL RESULTS'
       OPEN(5, FILE = ' ')
       N = 1
C
C            READ INITIAL HEADER LINES
C
       DO I = 1,5
       READ(5,24) HEADER
       END DO
24     FORMAT(1A80)
25     CONTINUE
       READ(5,*,ERR=26) LL,YYY,XXX,X(N),Y(N),ERR,ZZZ
       N = N + 1
       GO TO 25
26     N = N - 1
       ELSEIF (IDAT .EQ. 4) THEN
       WRITE(*,*) 'ENTER FILENAME FOR CORSPOND RESULTS'
       OPEN(5, FILE = ' ')
       WRITE(*,*) 'HOW MANY FACTORS ARE THERE IN YOUR FILE?'
       READ(*,*) IFACT
       WRITE(*,*) 'WHICH TWO FACTORS DO YOU WISH TO USE FOR'
       WRITE(*,*) 'PLOTTING?  ENTER LIKE THIS:  1,2; OR  3,4'
       READ(*,*) ILOC, JLOC
       N = 1
27     CONTINUE
       READ(5,29,END=28) CORLAB(N), (FACT(JK),  JK = 1,  IFACT)
       X(N) = FACT(ILOC)
       Y(N) = FACT(JLOC)
       N = N + 1
       GO TO 27
28     N = N - 1
29     FORMAT(1X,1A2, 1E15.3, 9 (1X,1E15.3))
       ELSEIF (IDAT .EQ. 5) THEN
C
C      SPECIFY INPUT FILE
C
       WRITE(*,*) 'ENTER THE FILE NAME STORING THE SEMI-VARIOGRAM'
       WRITE(*,*) 'RESULTS'
       OPEN(5, FILE = ' ')
C
```

Figure 7.7 continued

```
C     READ THE INITIAL HEADER FROM THE FILE
C
      READ(5,24) HEADER
C
C     READ THE DATA
C
      N = 1
31    CONTINUE
      READ(5,*,END=32) NN, Y(N), X(N)
      IF (NN .EQ. 0) GO TO 31
      N = N + 1
      GO TO 31
32    CONTINUE
      N = N - 1
      ELSEIF (IDAT .EQ. 6) THEN
C
C     SPECIFY FILE NAME STORING ONE-DIMENSIONAL PROFILE
C
      WRITE(*,*) 'ENTER THE NAME OF THE FILE STORING THE 1D PROFILE'
      OPEN(5, FILE = ' ')
      WRITE(*,*) 'ENTER THE SAMPLING INTERVAL FOR X-DIRECTION'
      READ(*,*) XINC
      N = 1
33    CONTINUE
      READ(5,*,END=34) Y(N)
      X(N) = XINC * FLOAT(N)
      N = N + 1
      GO TO 33
34    CONTINUE
      N = N - 1
      END IF
      WRITE(*,*) 'DO YOU WISH TO CONVERT X OR Y TO LOGARITHMS?'
      WRITE(*,*) 'ENTER 1 FOR YES, 0 FOR NO'
      READ(*,*) ILOG
      IF (ILOG .EQ. 1) THEN
      WRITE(*,*) 'CONVERT X VALUES?    1 = YES,  0 = NO'
      READ(*,*) IXLOG
      WRITE(*,*) 'CONVERT Y VALUES?    1 = YES,  0 = NO'
      READ(*,*) IYLOG
      DO I = 1,N
      IF (IXLOG .EQ. 1 .AND. X(I) .GT. 0.) X(I) = ALOG10(X(I))
      IF (IYLOG .EQ. 1 .AND. Y(I) .GT. 0.) Y(I) = ALOG10(Y(I))
      END DO
      END IF
      WRITE(*,*) 'DO YOU WISH TO RESCALE X AND/OR Y VALUES?'
      WRITE(*,*) 'ENTER 1 FOR YES, 0 FOR NO'
      READ(*,*) IRESCL
      IF (IRESCL .EQ. 1) THEN
      WRITE(*,*) 'IS THE X-AXIS TO BE RESCALED?   1 = YES;  0=NO'
      READ(*,*) IXSCL
      IF (IXSCL .EQ. 1) THEN
      WRITE(*,*) 'ENTER RESCALING PARAMETER FOR X-DIRECTION'
      READ(*,*) XMAG
      END IF
      WRITE(*,*) 'IS THE Y-AXIS TO BE RESCALED? 1=YES;  0=NO'
      READ(*,*) IYSCL
      IF (IYSCL .EQ. 1) THEN
      WRITE(*,*) 'ENTER RESCALING PARAMETER FOR Y-DIRECTION'
      READ(*,*) YMAG
      END IF
      DO I = 1,N
      IF (IXSCL .EQ. 1) THEN
      X(I) = X(I) * XMAG
      END IF
      IF (IYSCL .EQ. 1) THEN
      Y(I) = Y(I) * YMAG
      END IF
      END DO
      END IF
      WRITE(*,*) 'WHAT ARE THE MINIMUM AND MAXIMUM X-AXIS VALUES'
      WRITE(*,*) 'YOU WISH TO USE FOR THE GRAPH?'
      READ(*,*) XMIN, XMAX
      WRITE(*,*) 'WHAT ARE THE MINIMUM AND MAXIMUM Y-AXIS VALUES'
      WRITE(*,*) 'YOU WISH TO USE FOR THE GRAPH?'
      READ(*,*) YMIN, YMAX
      WRITE(*,*) 'DO YOU WISH TO PLOT A FUNCTION WITH THESE DATA?'
      WRITE(*,*) 'ENTER 1 FOR YES, 0 FOR NO'
```

Figure 7.7 continued

```fortran
      READ(*,*) IFUN
      IF (IFUN .EQ. 1) THEN
      WRITE(*,*) 'PLEASE ENTER VALUES FOR THE COEFFICIENTS, A'
      READ(*,*) (A(JK), JK = 1, 8)
      IF (IDAT .EQ. 5) THEN
      WRITE(*,*) 'IF YOUR SEMI-VARIOGRAM MODEL IS SPHERICAL'
      WRITE(*,*) 'THEN ENTER THE SILL AND THE RANGE NOW'
      WRITE(*,*) 'ENTER 1 FOR SPHERICAL MODEL; 0 FOR OTHER'
      READ(*,*) ISPHER
      IF (ISPHER .EQ. 1) THEN
      WRITE(*,*) 'ENTER SILL AND RANGE NOW:   SILL,RANGE'
      READ(*,*) SILL, RANGE
      ELSE
         SILL = 0.0
         RANGE = 1.0E+23
      END IF
      END IF
      WRITE(*,*) 'DO YOU WISH TO CONNECT THE POINTS, X, Y, WITH'
      WRITE(*,*) 'LINES, OR JUST PLOT THEM AS POINTS?'
      WRITE(*,*) 'ENTER 1 FOR LINES, 0 FOR POINTS'
      READ(*,*) ILINE
      WRITE(*,*) 'WHAT IS THE TITLE FOR THE GRAPH?'
      READ(*,30) TITLE
30    FORMAT(1A50)
C
C     DRAW THE GRAPH
C
C     1.  SET THE VIDEOMODE TO HIGH RESOLUTION, VGA MODE
C
      DUMMY2 = SETVIDEOMODE($VRES16COLOR)
C
C     2.  CLEAR THE SCREEN
C
      CALL CLEARSCREEN ($GCLEARSCREEN)
C
C     3.  SET THE COLOR FOR DRAWING THE GRAPH TO BRIGHT WHITE
C
      DUMMY2 = SETCOLOR(15)
C

C     4.  DEFINE THE COORDINATE WINDOW
C         MAKE THE WINDOW SLIGHTLY LARGER THAN THE X-AXIS AND
C         Y-AXIS RANGES
C
      XRANGE = XMAX - XMIN
      YRANGE = YMAX - YMIN
      DUMMY2 = SETWINDOW(.TRUE., (XMIN-XRANGE/100.D0), (YMIN-YRANGE/
     2 100.D0) , (XMAX+XRANGE/100.D0), (YMAX+YRANGE/100.D0))
C
C     5.  DEFINE THE VIEWPORT FOR THE GRAPH
C
      IX = 400
      IY = DINT(IX * YRANGE/XRANGE)
      IF (IY .GT. IX) THEN
         IY = 400
         IX = 400
      END IF
      CALL SETVIEWPORT(200, 25, IX+200, IY+25)
      KROW = (INT((IY+25)/16))/2 + 2
      KCOL = 23
      CALL SETTEXTPOSITION(KROW,KCOL,CURPOS)
      CALL OUTTEXT(YLABEL)
C
C     6.  DRAW THE BORDER FOR THE GRAPH
C
      CALL MOVETO_W(XMIN, YMIN, XY)
      DUMMY2 = LINETO_W(XMAX, YMIN)
      DUMMY2 = LINETO_W(XMAX, YMAX)
      DUMMY2 = LINETO_W(XMIN, YMAX)
      DUMMY2 = LINETO_W(XMIN, YMIN)
C
C     7.  DRAW THE TIC MARKS ON THE X AND Y AXES
C
      XINC = XRANGE / 10.D0
      YINC = YRANGE / 10.D0
      DO I = 1,10
      CALL MOVETO_W(XMIN+I*XINC, YMIN-YINC/10.D0, XY)
      DUMMY2 = LINETO_W(XMIN+I*XINC, YMIN+YINC/10.D0)
      CALL MOVETO_W(XMIN-XINC/10.D0, YMIN+I*YINC, XY)
      DUMMY2 = LINETO_W(XMIN+XINC/10.D0, YMIN+I*YINC)
      END DO
C
```

Figure 7.7 continued

```fortran
C     8.  PLACE LABELS ON GRAPH AND AXES
C
      KROW = 1
      KCOL = 27 + (50 - LEN_TRIM(TITLE))/2
      CALL SETTEXTPOSITION(KROW, KCOL, CURPOS)
      CALL OUTTEXT(TITLE)
      KROW = INT ((IY+25)/16) + 2
      IF (KROW .GT. 30) KROW = 30
      CALL SETTEXTPOSITION(KROW,27,CURPOS)
      WRITE(*,100) XMIN, XMAX
100   FORMAT(' MIN = ',F10.2,16X,'MAX = ',1F10.2)
      KCOL = 27 + (50 - LEN_TRIM(XLABEL))/2
      CALL SETTEXTPOSITION(KROW+1,KCOL,CURPOS)
      CALL OUTTEXT(XLABEL)
      CALL SETTEXTPOSITION(3,7,CURPOS)
      WRITE(*,101) YMAX
101   FORMAT(' MAX = ',1F10.2)
      KROW = KROW - 3
      CALL SETTEXTPOSITION(KROW, 7, CURPOS)
      WRITE(*,102) YMIN
102   FORMAT(' MIN = ',1F10.2)
C
C     9.  PLOT THE POINTS
C
      IF (IDAT .NE. 4) THEN
         IF (ILINE .EQ. 1) THEN
            CALL MOVETO_W(X(1), Y(1), XY)
            DO I = 2,N
               DUMMY2 = LINETO_W(X(I), Y(I))
            END DO
         ELSEIF (ILINE .EQ. 0) THEN
C
C     PLOT THE POINTS AS SMALL RECTANGLES
C
            DO I = 1,N
               DUMMY2 = RECTANGLE_W($GBORDER,  (X(I)-XINC/10.D0),
2                                   (Y(I)-YINC/10.D0),(X(I)+XINC/10.D0),
3                                   (Y(I)+YINC/10.D0))
            END DO
      ELSEIF (IDAT .EQ. 4) THEN
         KMAX = INT((IY+25)/16)  - 4
         DO I = 1, N
            KROW = 3 + INT(KMAX * (YMAX-Y(I))/YRANGE)
            IF (KROW .GT. KMAX+3) THEN
               KROW = KMAX + 3
            END IF
            JMAX = 47
            KCOL = 28 + INT(JMAX*(X(I) -  XMIN)/XRANGE)
            IF (KCOL .GT. 75) THEN
               KCOL = 75
            END IF
            IF (I .GT. 1) THEN
               IT = 0
               DO J = 1, I-1
                  IF (KROW .EQ. KTEMP(J,1)) THEN
                     IF (KCOL .EQ. KTEMP(J,2)-1.OR.KCOL.EQ.KTEMP(J,2)+1) THEN
                        IT = 1
                     END IF
                  END IF
               END DO
            ELSE
               IT = 0
            END IF
            IF (IT .EQ. 0) THEN
               KTEMP(I,1) = KROW
               KTEMP(I,2) = KCOL
            END IF
            IF (IT .EQ. 1) GO TO 200
            CALL SETTEXTPOSITION(KROW, KCOL,  CURPOS)
            CALL OUTTEXT(CORLAB(I))
200         CONTINUE
         END DO
      END IF
C
```

Figure 7.7 continued

```fortran
C       10.  PLOT THE FUNCTION IF DESIRED
C
        IF (IFUN .EQ. 1) THEN
          DO I = 1, 1000
            XMOD = XMIN + ((I-1)*XRANGE/1000.D0)
            YMOD = A(1) + A(2)*XMOD + A(3)*XMOD**2 + A(4)*XMOD**3 +
     2             A(5)*XMOD**4 + A(6)*EXP(A(7)*XMOD**(A(8)))
            IF (IDAT .EQ. 5 .AND. XMOD .GE. RANGE) YMOD = SILL
            IF (XMOD .LE. XMAX .AND. YMOD .LE. YMAX) THEN
              DUMMY2 = SETCOLOR(2)
              DUMMY2 = SETPIXEL_W(XMOD,YMOD)
            END IF
          END DO
        END IF
C
C       PAUSE TO PRINT THE SCREEN IF DESIRED
C
        READ(*,*)
        DUMMY2 = SETVIDEOMODE($DEFAULTMODE)
        STOP
        END
```

275

Figure 7.19 A listing of the program GRID3D.

```
C    PROGRAM GRID3D
C
C    GRID3D IS A GENERAL GRAPHICS PROGRAM FOR DISPLAYING THREE
C    DIMENSIONAL GRIDS USING HIDDEN LINE AND SHADING ALGORITHMS.
C    THE RESULT IS THE DISPLAY OF SOLID, SHADED FACETS.   THE GRID
C    IS CONSIDERED TO OCCUPY THE X-Z PLANE; ELEVATIONS, OR VALUES
C    IN GENERAL AT THE GRID INTERSECTIONS, ARE CONSIDERED TO BE
C    Y COORDINATES; AFTER ROTATING THE GRID WITH A TIP AND SPIN,
C    A FORGETTING TRANSFORM IS USED TO PLOT THE XY PLANE BY
C    FORGETTING THE Z-COORDINATE DIRECTION.
C
C    THE HIDDEN LINE ALGORITHM IS A VERY SIMPLE ONE AS GIVEN
C    IN ANGELL (1985), "ADVANCED GRAPHICS WITH THE IBM PERSONAL
C    COMPUTER."  THIS ALGORITHM DRAWS THE GRID POLYGONS FROM
C    BACK (FARTHEST) TO FRONT (NEAREST), FILLING IN EACH
C    POLYGON WITH FOREGROUND SHADE; IN THIS FASHION, A CLOSER
C    POLYGON MASKS A MORE DISTANT ONE FOR A CORRECT HIDDEN LINE
C    REMOVAL.
C
C    THIS PROGRAM DISPLAYS GRIDS CREATED BY JCBLOK (CHAPTER 6) AND
C    FORINV2 (CHAPTER 11); OR, WILL DISPLAY ANY GENERAL GRID
C    WHERE, FOR AN N-ROW BY M-COLUMN GRID, NM TOTAL GRID POINTS
C    ARE DEFINED, GRID POINT BY GRID POINT, AS FOLLOWS:
C
C           LINE 1:  X1, Z1, Y1  [WHERE Y = ELEV. OR VALUE]
C                               TO
C           LINE NM:  XNM, ZNM, YNM [FOR THE NM-TH GRID POINT]
C
C    WHERE LINE 1 IS FOR THE UPPER, LEFT CORNER OF THE GRID;
C    THEN, LINES 2-NM ARE WRITTEN TO DEFINE THE GRID, ROW
C    BY ROW; I.E., LINE 2 IS FOR THE SECOND GRID POINT ON
C    ROW 1; LINE M+1 IS FOR THE 1ST GRID POINT ON THE SECOND
C    LINE; AND SO ON.
C
C    ***********************************************************
C    ***********************************************************
C                 USER'S GUIDE:  GRID3D
C    ***********************************************************
C    ***********************************************************
C
C    A FEW PARAMETERS ARE ENTERED INTO THIS PROGRAM
C    INTERACTIVELY AS FOLLOWS:
C
C    1.   'ENTER THE NUMBER OF ROWS COMPRISING THE GRID';
C         THIS IS THE VALUE, N, DISCUSSED ABOVE
C    2.   'ENTER THE NUMBER OF COLUMNS COMPRISING THE GRID';
C         THIS IS THE VALUE, M, DISCUSSED ABOVE
C    3.   'ENTER THE OPTION FOR DATA FILE INPUT:'
C         'ENTER:   1 = FILE FROM JCBLOK (CHAP 6);
C         'ENTER:   2 = FILE FROM FORINV2 (CHAP 11);
C         'ENTER:   3 = GENERAL FILE AS DESCRIBED ABOVE
C    4.   'ENTER THE NAME OF THE DATA FILE STORING THE GRID';
C         THIS FILE IS CREATED BY JCBLOK, FORINV2, OR IS
C         A GENERAL FILE AS DESCRIBED ABOVE.
C    5.   'ENTER THE TIP ANGLE'; SUGGESTION:  ENTER A
C         NEGATIVE NUMBER; TRY -2 AND -30 TO SEE THE
C         DIFFERENCE, THEN CHOOSE A VALUE MOST APPROPRIATE
C         FOR YOUR DISPLAY
C    6.   'ENTER THE SPIN ANGLE'; ENTER ANY ANGLE, 1-360; OR,
C         -1 TO -360; EXPERIMENT WITH THIS TO DETERMINE
C         A PREFERENCE
C    7.   'ENTER THE SUN DIRECTION COORDINATES'
C         DIX, DIY, AND DIZ; THESE COORDINATES FORM
C         A VECTOR PASSING THROUGH THE ORIGIN,
C         0,0,0.   TRY DIX = -100.0, DIY = -0.1, DIZ = 50.0
C
C    *** IMPORTANT NOTE ***
C    AFTER THE SPIN ANGLE IS ENTERED, GRID3D FIRST
C    DRAWS ONLY THE OUTLINE OF POLYGONS USING A
C    HIDDEN LINE ALGORITHM.  THEN, THE PROGRAM
C    PAUSES, ***WAITING FOR THE ENTER KEY TO BE
C    PUSHED***.   ONCE THE ENTER KEY IS PUSHED, GRID3D
C    REDRAWS THE GRID USING A SHADING ALGORITHM.
C    THE PAUSE IS USED SHOULD THE PROGRAM USER DESIRE
C    A HARD COPY OF THE POLYGON OUTLINE GRID.
C
C    ***********************************************************
C         END OF USER'S GUIDE FOR GRID3D
C    ***********************************************************
C    ***********************************************************
C
```

276

Figure 7.19 continued

```fortran
      INCLUDE 'FGRAPH.FI'
      INCLUDE 'FGRAPH.FD'
      INTEGER*2 DUMMY2
      INTEGER*4 KROW(2500), JCOL(2500), KOORD(4,2), KSKRT(2500,3)
      INTEGER*4 KOLOR, DUMMY4, IB, IG, IR, RGB, KEDGE1, KEDGE2
      INTEGER*4 XMIN, YMIN, XMAX, YMAX
      REAL*4 TIP, SPIN, MX, KX
      REAL*4 DAT(2500,4), R(3,3), Z(2500)
      RECORD /XYCOORD/ XY
C
C     ACCESS THE GRID INFORMATION
C
      WRITE(*,*) 'ENTER NUMBER OF ROWS COMPRISING THE GRID'
      READ (*,*) NROW
      WRITE(*,*) 'ENTER NUMBER OF COLUMNS COMPRISING THE GRID'
      READ (*,*) NCOL
      WRITE(*,*) 'ENTER THE OPTION SPECIFYING THE TYPE OF FILE:'
      WRITE(*,*) 'ENTER:  1 = FILE FROM JCBLOK (CHAP 6)'
      WRITE(*,*) 'ENTER:  2 = FILE FROM FORINV2 (CHAP 11)'
      WRITE(*,*) 'ENTER:  3 = GENERAL FILE, Z,X,Y FORMAT'
      READ(*,*) IFILE
      WRITE(*,*) 'ENTER THE NAME OF THE DATA FILE STORING THE GRID'
      OPEN(5, FILE = ' ')
      DATMIN = 0.0
      DATMAX = 0.0
      DO I = 1,NROW
      DO J = 1, NCOL
      K = (I-1) * NCOL + J
      IF (IFILE .EQ. 1) THEN
      READ(5,*) DAT(K,3), DAT(K,1), XDUM1, ZDUM1, DAT(K,2), ADUM
      ELSEIF (IFILE .NE. 1) THEN
      READ(5,*) DAT(K,3), DAT(K,1), DAT(K,2)
      END IF
      IF (DAT(K,2) .LT. DATMIN) DATMIN = DAT(K,2)
      IF (DAT(K,2) .GT. DATMAX) DATMAX = DAT(K,2)
      END DO
      END DO
      MM = NROW * NCOL
      IF (DATMIN .LT. 0.) THEN
      DO I = 1, MM
      DAT(I,2) = DAT(I,2) - DATMIN
      END DO
      END IF
C
C     DETERMINE COORDINATES OF SKIRT (BOTTOM BORDER OF GRID)
C
      K = 0
      DO I = 1,NROW
      DO J = 1,NCOL
      IF ((I.EQ.1.OR.I.EQ.NROW) .OR. (J.EQ.1.OR.J.EQ.NCOL)) THEN
      K = K + 1
      L = (I-1)*NCOL + J
      KSKRT(K,1) = NINT(DAT(L,1))
      KSKRT(K,2) = 0
      KSKRT(K,3) = NINT(DAT(L,3))
      END IF
      END DO
      END DO
C
C     ROTATE AND TIP THE GRID AND SKIRT
C
      WRITE(*,*) 'ENTER TIP ANGLE'
      READ (*,*) TIP
      WRITE(*,*) 'ENTER SPIN ANGLE'
      READ (*,*) SPIN
      WRITE(*,*) 'ENTER SUN ANGLE COORDINATES, DIX,DIY,DIZ'
      READ (*,*) DIX, DIY, DIZ
C
C     CHANGE ROWS AND COLUMNS OF DATA DEPENDING ON SPIN TO
C     FACILITATE THE CORRECTNESS OF THE HIDDEN LINE ALGORITHM
C
      IF ((SPIN.GT.90.AND.SPIN.LE.180).OR.
   2  (SPIN.LT.-180.AND.SPIN.GE.-270)) THEN
      DO J = 1, NCOL
      DO I = NROW, 1, -1
      KK = (I-1)*NCOL + J
      LL = (J-1)*NROW + (NROW-I) + 1
      DAT(LL,4) = DAT(KK,2)
      DAT(LL,5) = DAT(KK,2)
      DAT(LL,6) = DAT(KK,3)
      END DO
      END DO
      DO I = 1, MM
      DAT(I,2) = DAT(I,4)
      DAT(I,2) = DAT(I,5)
      DAT(I,3) = DAT(I,6)
      END DO
C
C
```

Figure 7.19 continued

```fortran
      ELSEIF ((SPIN .GT. 180 .AND. SPIN .LE. 270) .OR.
     2        (SPIN .LT. - 90 .AND. SPIN .GE. -180)) THEN
        DO I = NROW,1,-1
          DO J = NCOL,1,-1
            KK = (I-1)*NCOL + J
            LL = (NROW-I)*NCOL + (NCOL-J) + 1
            DAT(LL,4) = DAT(KK,2)
          END DO
        DO I = 1,MM
        DAT(I,2) = DAT(I,4)
      END DO
      ELSEIF ((SPIN .GT. 270 .AND. SPIN .LE. 360.) .OR.
     2        (SPIN .LT. 0. .AND. SPIN .GE. -90.)) THEN
        DO J = NCOL,1,-1
          DO I = 1,NROW
            KK = (I-1)*NCOL + J
            LL = (I-1)*NCOL + (NCOL-J) + 1
            DAT(LL,4) = DAT(KK,2)
          END DO
        DO I = 1,MM
        DAT(I,2) = DAT(I,4)
      END DO
      END IF
      IF (SPIN .GT. 90 .AND. SPIN .LE. 180) THEN
        SPIN = SPIN - 90.0
      ELSEIF (SPIN .GT. 180 .AND. SPIN .LE. 270) THEN
        SPIN = SPIN - 180.0
      ELSEIF (SPIN .LT. -90 .AND. SPIN .GE. -180) THEN
        SPIN = SPIN + 180.0
      ELSEIF (SPIN .LT. -180 .AND. SPIN .GE. -270) THEN
        SPIN = SPIN + 270.0
      ELSEIF (SPIN .LT. 0. .AND. SPIN .GE. -90.) THEN
        SPIN = SPIN + 90.
      ELSEIF (SPIN .GT. 270 .AND. SPIN .LE. 360.) THEN
        SPIN = SPIN - 270.0
      END IF

C     DEFINE THE ROTATION MATRIX, R
      TIP = TIP / 57.29577951
      SPIN=SPIN / 57.29577951
      R(1,1) = COS(SPIN)
      R(1,2) = 0.
      R(1,3) = - SIN(SPIN)
      R(2,1) = SIN(SPIN)*SIN(TIP)
      R(2,2) = COS(TIP)
      R(2,3) = SIN(TIP) * COS(SPIN)
      R(3,1) = COS(TIP)*SIN(SPIN)
      R(3,2) = - SIN(TIP)
      R(3,3) = COS(TIP)*COS(SPIN)
C     TRANSFORM GRID VALUES
      DO I = 1,MM
        X1 = R(1,1)*DAT(I,1) + R(1,2)*DAT(I,2) + R(1,3)*DAT(I,3)
        X1 = X1 * 100.0
        Y1 = R(2,1)*DAT(I,1) + R(2,2)*DAT(I,2) + R(2,3)*DAT(I,3)
        Y1 = Y1 * 100.0
        Z(I) = R(3,1)*DAT(I,1) + R(3,2)*DAT(I,2) + R(3,3)*DAT(I,3)
        KROW(I) = NINT(Y1)
        JCOL(I) = NINT(X1)
      END DO
C     TRANSFORM SKIRT VALUES
      K = 0
      DO I = 1,NROW
        DO J = 1, NCOL
          IF ((I.EQ.1.OR.I.EQ.NROW) .OR. (J.EQ.1.OR.J.EQ.NCOL)) THEN
            K = K + 1
            XDUM = FLOAT(KSKRT(K,1))*R(1,1)+FLOAT(KSKRT(K,2))*R(1,2) +
     2             FLOAT(KSKRT(K,3))*R(1,3)
            YDUM = FLOAT(KSKRT(K,1))*R(2,1)+FLOAT(KSKRT(K,2))*R(2,2) +
     2             FLOAT(KSKRT(K,3))*R(2,3)
            XDUM = XDUM * 100.
            YDUM = YDUM * 100.
            KSKRT(K,1) = NINT(XDUM)
            KSKRT(K,2) = NINT(YDUM)
          END IF
        END DO
      END DO
C
```

Figure 7.19 continued

```fortran
C     PLOT THE GRID
C
      DUMMY2 = SETVIDEOMODE($VRES16COLOR)
      CALL CLEARSCREEN($GCLEARSCREEN)
C
C     ESTABLISH THE VIEWING FIELD TO BE EQUAL TO THE ENTIRE SCREEN
C
      CALL SETVIEWPORT(1,1,640,480)
C
C     TRANSFORM COORDINATES TO INTEGERS
C
      XMIN = JCOL(1)
      YMIN = KROW(1)
      XMAX = JCOL(1)
      YMAX = KROW(1)
      DO KI = 2,MM
      IF (JCOL(KI) .LT. XMIN)  XMIN = JCOL(KI)
      IF (KROW(KI) .LT. YMIN)  YMIN = KROW(KI)
      IF (JCOL(KI) .GT. XMAX)  XMAX = JCOL(KI)
      IF (KROW(KI) .GT. YMAX)  YMAX = KROW(KI)
      END DO
C     CHECK SKIRT TRANSFORMED COORDINATES
      K = 0
      DO I = 1,NROW
      DO J = 1,NCOL
      IF((I.EQ.1.OR.I.EQ.NROW).OR.(J.EQ.1.OR.J.EQ.NCOL)) THEN
      K = K + 1
      IF (KSKRT(K,1) .LT. XMIN) XMIN = KSKRT(K,1)
      IF (KSKRT(K,1) .GT. XMAX) XMAX = KSKRT(K,1)
      IF (KSKRT(K,2) .GT. YMAX) YMAX = KSKRT(K,2)
      IF (KSKRT(K,2) .LT. YMIN) YMIN = KSKRT(K,2)
      END IF
      END DO
      END DO
C

C     TRANSFORM COORDINATES TO PIXEL COORDINATES
C
      K = 0
      DO I = 1,NROW
      DO J = 1,NCOL
      IF ((I.EQ.1.OR.I.EQ.NROW).OR.(J.EQ.1.OR.J.EQ.NCOL)) THEN
      K = K + 1
      KDUM1 = 100+ NINT(400.*(FLOAT(KSKRT(K,1))-FLOAT(XMIN))/
     2          (FLOAT(XMAX)-FLOAT(XMIN)))
      KDUM2   = 10 + NINT(400.*(FLOAT(KSKRT(K,2))-FLOAT(YMIN))
     2          )/(FLOAT(YMAX)-FLOAT(YMIN)))
      KSKRT(K,2) = 410 - KDUM2
      KSKRT(K,1) = KDUM1
      END IF
      END DO
      DO I = 1,MM
      KDUM1= 10 + NINT(400.*(FLOAT(KROW(I))-FLOAT(YMIN))/
     2          (FLOAT(YMAX)-FLOAT(YMIN)))
      KROW(I) = 410 - KDUM1
      JDUM1= 100 + NINT(400.*(FLOAT(JCOL(I))-FLOAT(XMIN))/
     2          (FLOAT(XMAX)-FLOAT(XMIN)))
      JCOL(I) = JDUM1
      END DO
C
C     DRAW THE GRID, POLYGON BY POLYGON, FROM THE REAR FORWARD
C
      KOUNT = 1
      DXYZ = SQRT(DIX*DIX+DIY*DIY+DIZ*DIZ)
C
C     REMAP PALETTE
```

Figure 7.19 continued

```
C
15        CONTINUE
          IF (KOUNT .EQ. 1) THEN
             KEDGE1 = 0
             KEDGE2 = 0
             KOLOR  = 0
          ELSE
             KEDGE1 = 6
             KEDGE2 = 8
          DO I = 1,16
             IR = (I-1) * 4
             IG = (I-1) * 4
             IB = (I-1) * 4
             KOLOR = RGB(IR, IG, IB)
             DUMMY4 = REMAPPALETTE(I-1,KOLOR)
          END DO
          END IF
          DO I = 1,NROW-1
          DO J = 1,NCOL-1
             JJ = I * NCOL + J
             KK = (I-1) * NCOL + J
             LSWIT = 0
             KPROD1 = 0
             KPROD2 = 0
             IF (I.EQ. 1 .AND. J.EQ.1) THEN
                KOORD(1,1) = KSKRT(1,1)
                KOORD(1,2) = KSKRT(1,2)
                KOORD(2,1) = KSKRT(NCOL+1,1)
                KOORD(2,2) = KSKRT(NCOL+1,2)
                KOORD(3,1) = JCOL(NCOL+1)
                KOORD(3,2) = KROW(NCOL+1)
                KOORD(4,1) = JCOL(1)
                KOORD(4,2) = KROW(1)
                LSWIT = 1
                DUMMY2 = SETCOLOR(KEDGE2)
                GO TO 10
             END IF
             IF (I. EQ. 1 .AND. J.EQ.NCOL-1) THEN
                KOORD(1,1) = KSKRT(NCOL,1)
                KOORD(1,2) = KSKRT(NCOL,2)
                KOORD(2,1) = KSKRT(NCOL+2,1)
                KOORD(2,2) = KSKRT(NCOL+2,2)
                KOORD(3,1) = JCOL(2*NCOL)
                KOORD(3,2) = KROW(2*NCOL)
                KOORD(4,1) = JCOL(NCOL)
                KOORD(4,2) = KROW(NCOL)
                LSWIT = 1
                DUMMY2 = SETCOLOR(KEDGE2)
                GO TO 10
             END IF
             IF (J.EQ.1.AND.I.NE.1) THEN
                KB = NCOL + (I-2)*2 + 1
                KOORD(1,1) = KSKRT(KB+2,1)
                KOORD(1,2) = KSKRT(KB+2,2)
                KOORD(2,1) = KSKRT(KB,1)
                KOORD(2,2) = KSKRT(KB,2)
                KOORD(3,1) = JCOL(KK)
                KOORD(3,2) = KROW(KK)
                KOORD(4,1) = JCOL(JJ)
                KOORD(4,2) = KROW(JJ)
                LSWIT = 3
                DUMMY2 = SETCOLOR(KEDGE2)
                GO TO 10
             END IF
             IF (J.EQ.NCOL-1 .AND. I.NE.1) THEN
                LSWIT = 7
                GO TO 5
7            CONTINUE
                KPROD1 = 0
                KPROD2 = 0
                KB = NCOL+(I-1)*2
                IF (I. NE. NROW-1) THEN
                   KOORD(1,1) = JCOL(KK+1)
                   KOORD(1,2) = KROW(KK+1)
                   KOORD(2,1) = KSKRT(KB,1)
                   KOORD(2,2) = KSKRT(KB,2)
                   KOORD(3,1) = KSKRT(KB+2,1)
                   KOORD(3,2) = KSKRT(KB+2,2)
                   KOORD(4,1) = JCOL(JJ+1)
                   KOORD(4,2) = KROW(JJ+1)
                   DUMMY2 = SETCOLOR(KEDGE2)
                   LSWIT = 6
```

Figure 7.19 continued

```fortran
      ELSEIF ( I.EQ. NROW-1) THEN
        KOORD(1,1) = JCOL(KK+1)
        KOORD(1,2) = KROW(KK+1)
        KOORD(2,1) = KSKRT(KB,1)
        KOORD(2,2) = KSKRT(KB,2)
        KOORD(3,1) = KSKRT(KB+NCOL,1)
        KOORD(3,2) = KSKRT(KB+NCOL,2)
        KOORD(4,1) = JCOL(JJ+1)
        KOORD(4,2) = KROW(JJ+1)
        LSWIT = 3
      END IF
      DUMMY2 = SETCOLOR(KEDGE2)
      GO TO 10
      END IF
3     CONTINUE
      KPROD1 = 0
      KPROD2 = 0
      IF (I .EQ. NROW-1) THEN
        LSWIT = 2
        GO TO 5
      END IF
4     CONTINUE
      KPROD1 = 0
      KPROD2 = 0
      KB = NCOL + (NROW-2)*2 + J
        KOORD(1,1) = JCOL(JJ)
        KOORD(1,2) = KROW(JJ)
        KOORD(2,1) = JCOL(JJ+1)
        KOORD(2,2) = KROW(JJ+1)
        KOORD(3,1) = KSKRT(KB+1,1)
        KOORD(3,2) = KSKRT(KB+1,2)
        KOORD(4,1) = KSKRT(KB,1)
        KOORD(4,2) = KSKRT(KB,2)
        LSWIT = 6
      DUMMY2 = SETCOLOR(KEDGE1)
      GO TO 10
      END IF
5     CONTINUE
        KOORD(1,1) = JCOL(KK)
        KOORD(1,2) = KROW(KK)
        KOORD(2,1) = JCOL(KK+1)
        KOORD(2,2) = KROW(KK+1)
        KOORD(3,1) = JCOL(JJ+1)
        KOORD(3,2) = KROW(JJ+1)
        KOORD(4,1) = JCOL(JJ)
        KOORD(4,2) = KROW(JJ)
        KDX1 = JCOL(KK+1) - JCOL(KK)
        KDY1 = KROW(KK+1) - KROW(KK)
        KDX2 = JCOL(JJ+1) - JCOL(KK+1)
        KDY2 = KROW(JJ+1) - KROW(KK+1)
        KPROD1 = KDX1 * KDY2 - KDX2 * KDY1
        KDX1 = JCOL(JJ) - JCOL(JJ+1)
        KDY1 = KROW(JJ) - KROW(JJ+1)
        KDX2 = JCOL(KK) - JCOL(JJ)
        KDY2 = KROW(KK) - KROW(JJ)
        KPROD2 = KDX1*KDY2 - KDX2*KDY1
      IF (KOUNT .GT. 1) THEN
        NZ = -KPROD1
        DZ1 = -(Z(KK+1) - Z(KK))
        DZ2 = -(Z(JJ+1) - Z(KK+1))
        NY = NINT(DZ1)*KDX2 - NINT(DZ2)*KDX1
        NX = NINT(DZ2)*KDY1 - NINT(DZ1)*KDY2
        NXYZ = NINT(SQRT(FLOAT(NX*NX)+FLOAT(NY*NY)+FLOAT(NZ*NZ)))
        IF (NXYZ .GT. 0) THEN
          KSHAD = 50-NINT(50.*(DIX*FLOAT(NX)+DIY*FLOAT(NY)+DIZ*
     2            FLOAT(NZ))/(DXYZ*FLOAT(NXYZ)))
        ELSE
          KSHAD = 0
        END IF
        IF (KSHAD .GT. 100) KSHAD = 100
        IF (KSHAD .LT.  0) KSHAD = 0
        KOLOR = NINT(16.*FLOAT(KSHAD)/100.)
        KOLOR = KOLOR - 1
        IF (KOLOR .LT. 1) KOLOR = 1
        IF (KOLOR .GT.15) KOLOR = 15
      END IF
      DUMMY2 = SETCOLOR(KOLOR)
10    CONTINUE
      IF ((KPROD1.LT.0.AND.KPROD2.LT.0) .OR.
     2    (KPROD1.GE.0.AND.KPROD2.GE.0)) THEN
C
```

Figure 7.19 continued

```
C      FIND MIN AND MAX Y VALUES FOR POLYGON
C
       KMIN = 500
       KMAX = -10
       DO KI = 1,4
         IF (KOORD(KI,2) .LT. KMIN) KMIN = KOORD(KI,2)
         IF (KOORD(KI,2) .GT. KMAX) KMAX = KOORD(KI,2)
       END DO
       DO IROW = KMIN, KMAX
C      FIND X INTERCEPTS FOR THE IROW PIXEL LINE
         JMIN = 1000
         JMAX = -10
         DO KI = 1,4
           KJ = MOD(KI,4) + 1
           KDIFF = KOORD(KI,2) - KOORD(KJ,2)
           IF (KDIFF .NE. 0) THEN
             JDIFF = IROW - KOORD(KI,2)
             MDIFF = KOORD(KJ,2) - KOORD(KI,2)
             MX = FLOAT(JDIFF) / FLOAT(MDIFF)
             IF (MX .GE. 0..AND. MX .LE. 1.) THEN
               KX = KOORD(KI,1) + MX*(KOORD(KJ,1)-KOORD(KI,1))
               JX = NINT(KX)
               IF (JX .LT. JMIN) JMIN = JX
               IF (JX .GT. JMAX) JMAX = JX
             END IF
           END IF
         END DO
         IF (JMIN .LT. JMAX) THEN
           CALL MOVETO(JMIN,IROW,XY)
           DUMMY2 = LINETO(JMAX,IROW)
         END IF
       END DO
       IF (KOUNT .EQ. 1) THEN
         DUMMY2 = SETCOLOR(15)
         CALL MOVETO(KOORD(1,1),KOORD(1,2),XY)
         DUMMY2 = LINETO(KOORD(2,1), KOORD(2,2))
         DUMMY2 = LINETO(KOORD(3,1), KOORD(3,2))
         DUMMY2 = LINETO(KOORD(4,1), KOORD(4,2))
         DUMMY2 = LINETO(KOORD(1,1), KOORD(1,2))
       END IF

       IF (LSWIT .EQ. 1) THEN
         LSWIT = 0
         GO TO 5
       ELSEIF (LSWIT .EQ. 2) THEN
         LSWIT = 0
         GO TO 4
       ELSEIF (LSWIT .EQ. 3) THEN
         LSWIT = 0
         GO TO 3
       ELSEIF (LSWIT .EQ. 6) THEN
         LSWIT = 0
         GO TO 20
       ELSEIF (LSWIT .EQ. 7) THEN
         LSWIT = 0
         GO TO 7
       END IF
       ELSE
         KA1 = JCOL(KK)
         KB1 = KROW(KK)
         KA2 = JCOL(KK+1)
         KB2 = KROW(KK+1)
         KA3 = JCOL(JJ+1)
         KB3 = KROW(JJ+1)
         KA4 = JCOL(JJ)
         KB4 = KROW(JJ)
         DO L = 1,2
           KC11 = KA2 - KA1
           KC12 = KA3 - KA4
           KC21 = KB2 - KB1
           KC22 = KB3 - KB4
           KD1 = KA3 - KA1
           KD2 = KB3 - KB1
           KDET = KC11*KC22 - KC21*KC12
           IF (KDET .NE. 0) THEN
             MX = FLOAT((KD1*KC22-KD2*KC12))/FLOAT(KDET)
             IF (MX .GE. 0..AND. MX .LE. 1.) THEN
               KA5 = KA1 + MX*KC11
               KB5 = KB1 + MX*KC21
C
```

Figure 7.19 continued

```fortran
C
C          FILL IN THE TRIANGLE WITH FOREGROUND SHADE

           KOORD(1,1) = KA1
           KOORD(1,2) = KB1
           KOORD(2,1) = KA4
           KOORD(2,2) = KB4
           KOORD(3,1) = KA5
           KOORD(3,2) = KB5
           IF (KOUNT .GT. 1) THEN
           KDX1 = KA4 - KA1
           KDY1 = KB4 - KB1
           KDZ1 = -NINT(Z(JJ)-Z(KK))
           KDX2 = KA5 - KA4
           KDY2 = KB5 - KB4
           KDZ2 = NINT(-Z(KK) - MX*(Z(JJ)-Z(KK)))
           NZ = -(KDX1*KDY2-KDX2*KDY1)
           NY = KDZ1*KDX2-KDZ2*KDX1
           NX = KDY1*KDZ2-KDY2*KDZ1
           NXYZ = NINT(SQRT(FLOAT(NX*NX)+FLOAT(NY*NY)+FLOAT(NZ*NZ
      2          )))
           IF (NXYZ .GT. 0) THEN
           KSHAD = 50 - NINT(50.*(DIX*FLOAT(NX)+DIY*FLOAT(NY)+
      2          DIZ * FLOAT(NZ))/(DXYZ*FLOAT(NXYZ)))
           KSHAD = IABS(KSHAD)
           ELSE
           KSHAD = 0
           END IF
           IF (KSHAD .GT. 100) KSHAD = 100
           IF (KSHAD .LT.   0) KSHAD = 0
           KOLOR = NINT(16. * FLOAT(KSHAD)/100.)
           KOLOR = KOLOR - 1
           IF (KOLOR .GT. 15) KOLOR = 15
           IF (KOLOR .LT. 1) KOLOR = 1
           END IF
           DUMMY2 = SETCOLOR(KOLOR)
           KMIN = 500
           KMAX = -10

           DO KI = 1,3
           IF (KOORD(KI,2) .LT. KMIN) KMIN = KOORD(KI,2)
           IF (KOORD(KI,2) .GT. KMAX) KMAX = KOORD(KI,2)
           END DO
           DO IROW = KMIN, KMAX
           JMIN = 1000
           JMAX = -10
           DO KI = 1,3
           KJ = MOD(KI,3) + 1
           KDIFF = KOORD(KI,2) - KOORD(KJ,2)
           IF (KDIFF .NE. 0) THEN
           JDIFF = IROW - KOORD(KI,2)
           MDIFF = - KDIFF
           MX = FLOAT(JDIFF) / FLOAT(MDIFF)
           IF (MX .GE. 0..AND. MX .LE. 1.) THEN
           KX = KOORD(KI,1) + MX*(KOORD(KJ,1) - KOORD(KI,1))
           JX = NINT(KX)
           IF (JX .LT. JMIN) JMIN = JX
           IF (JX .GT. JMAX) JMAX = JX
           END IF
           END IF
           END DO
           IF (JMIN .LT. JMAX) THEN
           CALL MOVETO(JMIN,IROW,XY)
           DUMMY2 = LINETO(JMAX,IROW)
           END IF
           END DO
           IF (KOUNT .EQ. 1) THEN
           DUMMY2 = SETCOLOR(15)
           CALL MOVETO(KOORD(1,1), KOORD(1,2), XY)
           DUMMY2 = LINETO(KOORD(2,1), KOORD(2,2))
           DUMMY2 = LINETO(KOORD(3,1), KOORD(3,2))
           DUMMY2 = LINETO(KOORD(1,1), KOORD(1,2))
           END IF
C
```

283

Figure 7.19 continued

```
C
C       LIKEWISE FOR THE SECOND TRIANGLE, FOREGROUND SHADING

        KOORD(1,1) = KA2
        KOORD(1,2) = KB2
        KOORD(2,1) = KA3
        KOORD(2,2) = KB3
        KOORD(3,1) = KA5
        KOORD(3,2) = KB5
        IF (KOUNT .GT. 1) THEN
        KDX1 = KA3 - KA2
        KDY1 = KB3 - KB2
        KDZ1 = -NINT(Z(JJ+1)-Z(KK+1))
        KDX2 = KA5 - KA3
        KDY2 = KB5 - KB3
        KDZ2 = NINT(-Z(KK+1)-MX*(Z(JJ+1)-Z(KK+1)))
        NZ = -(KDX1*KDY2-KDX2*KDY1)
        NY = KDZ1*KDX2-KDZ2*KDX1
        NX = KDY1*KDZ2-KDY2*KDZ1
        NXYZ = NINT(SQRT(FLOAT(NX*NX)+FLOAT(NY*NY)+FLOAT(NZ*NZ
     )))
        IF (NXYZ .GT. 0) THEN
     2  KSHAD = 50-NINT(50.*(DIX*FLOAT(NX)+DIY*FLOAT(NY)+
     2         DIZ*FLOAT(NZ))/(DXYZ*FLOAT(NXYZ)))
        KSHAD = IABS(KSHAD)
        ELSE
        KSHAD = 0
        END IF
        IF (KSHAD .GT. 100) KSHAD = 100
        IF (KSHAD .LT.   0) KSHAD =   0
        KOLOR = NINT(16.*FLOAT(KSHAD)/100.)
        KOLOR = KOLOR - 1
        IF (KOLOR .GT. 15) KOLOR = 15
        IF (KOLOR .LT.  1) KOLOR =  1
        END IF
        DUMMY2 = SETCOLOR(KOLOR)
        KMIN = 500
        KMAX = -10
        DO KI = 1,3
        IF (KOORD(KI,2) .LT. KMIN) KMIN = KOORD(KI,2)
        IF (KOORD(KI,2) .GT. KMAX) KMAX = KOORD(KI,2)
        END DO

        DO IROW = KMIN, KMAX
        JMIN = 1000
        JMAX =  -10
        DO KI = 1,3
        KJ = MOD(KI,3) + 1
        KDIFF = KOORD(KI,2) - KOORD(KJ,2)
        IF (KDIFF .NE. 0) THEN
        JDIFF = IROW - KOORD(KI,2)
        MDIFF = - KDIFF
        MX = FLOAT(JDIFF) / FLOAT(MDIFF)
        IF (MX .GE. 0..AND. MX .LE. 1.) THEN
        KX = KOORD(KI,1) + MX*(KOORD(KJ,1) - KOORD(KI,1))
        JX = NINT(KX)
        IF (JX .LT. JMIN) JMIN = JX
        IF (JX .GT. JMAX) JMAX = JX
        END IF
        END IF
        END DO
        IF (JMIN .LT. JMAX) THEN
        CALL MOVETO(JMIN,IROW,XY)
        DUMMY2 = LINETO(JMAX,IROW)
        END IF
        END DO
        IF (KOUNT .EQ. 1) THEN
        DUMMY2 = SETCOLOR(15)
        CALL MOVETO(KOORD(1,1), KOORD(1,2), XY)
        DUMMY2 = LINETO(KOORD(2,1), KOORD(2,2))
        DUMMY2 = LINETO(KOORD(3,1), KOORD(3,2))
        DUMMY2 = LINETO(KOORD(1,1), KOORD(1,2))
        END IF
        ELSE
        KAA = KA2
        KA2 = KA4
        KA4 = KAA
        KBB = KB2
        KB2 = KB4
        KB4 = KBB
        END IF
```

284

Figure 7.19 continued

```
              ELSE
                 KAA = KA2
                 KA2 = KA4
                 KA4 = KAA
                 KBB = KB2
                 KB2 = KB4
                 KB4 = KBB
              END IF
           END DO
        END IF
        IF (LSWIT .EQ. 2) GO TO 4
        IF (LSWIT .EQ. 7) GO TO 7
20      CONTINUE
        END DO
     END DO
     IF (KOUNT .EQ. 1) THEN
        KOUNT = 2
        READ(*,*)
        GO TO 15
     END IF

C
C    PAUSE TO ALLOW SCREEN TO BE PRINTED
C
     READ(*,*)
     DUMMY2 = SETVIDEOMODE($DEFAULTMODE)
     STOP
     END

C
C
     INTEGER*4 FUNCTION RGB(R,G,B)
     INTEGER*4 R, G, B
     RGB = ISHL(ISHL(B,8) .OR. G,8) .OR. R
     RETURN
     END
```

285

Figure 7.21 A listing of the program COLORDEN.

```fortran
C     PROGRAM COLORDENSITY MAPPING
C
C     AN INTERACTIVE GRAPHICS PROGRAM FOR THE PREPARATION
C     OF PSEUDO-CONTOUR MAPS USING COLOR DENSITY SLICING
C     OF KRIGING RESULTS.
C
C***********************************************
C              USER'S GUIDE:  COLORDEN
C***********************************************
C
C     THE PROGRAM REQUESTS INFORMATION DEFINING THE INPUT DATA
C     FILE CREATED BY THE PROGRAM, JCBLOK (CHAPTER 6), WHICH
C     CONTAINS THE RESULTS OF THE GRID ESTIMATION
C
C     THIS PROGRAM CAN ACCEPT UP TO 100 X 100 GRIDS
C
C     AFTER THE FILE NAME IS SPECIFIED, SEVERAL QUESTIONS
C     ARE ASKED:
C
C     QUESTION 1:  DO YOU WISH TO USE THE DEFAULT COLOR PALETTE
C                  OR ONE CREATED BY THE PROGRAM, PALDSGN
C
C                  ENTER 1, FOR ONE CREATED BY PALDSGN; ENTER
C                        0, FOR THE DEFAULT PALETTE
C
C                  IF A PALDSGN PALETTE IS CHOSEN, THEN THE
C                  FOLLOWING QUESTION IS ASKED:
C                  SPECIFY THE INPUT FILE NAME FOR THE PALETTE
C                  (UNIT 1)
C
C     QUESTION 2:  ENTER THE TITLE FOR THE PSEUDO CONTOUR MAP
C                  UP TO 50 CHARACTERS IN LENGTH
C
C     QUESTION 3:  ENTER THE NUMBER OF CONTOUR INTERVALS
C                  DESIRED; NOTE:  AFTER THE PSEUDO-CONTOUR
C                  MAP IS CREATED USING THE PALETTE CHOSEN
C                  IN QUESTION 1, A SECOND MAP SHOWING
C                  CONTOUR LINES IS CREATED; IT IS THIS SECOND
C                  MAP TO WHICH QUESTION 3 PERTAINS.
C
C     ATTENTION:   AFTER THE FIRST MAP IS DRAWN, THE ENTER
C                  KEY MUST BE PUSHED BEFORE THE PROGRAM WILL
C                  BEGIN TO DRAW THE CONTOUR MAP
C
C***********************************************
C           END OF USER'S GUIDE AND DISCUSSION
C***********************************************
C
      INCLUDE 'FGRAPH.FI'
      INCLUDE 'FGRAPH.FD'
      INTEGER*2 DUMMY2, DUMC, DUMR, DUMM
      INTEGER*4 IR, IB, IG, DUMMY4, IBUF(400), RGB, TMP, TMPMAX
      CHARACTER*50 TITLE
      REAL*4 XBUF(100,100)
      RECORD /RCCORD/ CURPOS
      RECORD /XYCOORD/ XY
C
C     ENTER INPUT FILE NAME, UNIT 5, CONTAIN GRID RESULTS
C
      OPEN (5, FILE = ' ')
C
C     SET VIDEO MODE
C
      DUMMY2 = SETVIDEOMODE($VRES16COLOR)
C
C     DEFINE THE PALETTE
C
      WRITE(*,*) 'DO YOU WISH TO USE A PALETTE CREATED BY PALDSGN?'
      WRITE(*,*) 'ENTER 1 FOR YES, 0 FOR NO'
      READ(*,*) IPAL
      IF (IPAL .EQ. 1) THEN
        OPEN (1, FILE = ' ')
        TMPMAX = 0
        DO I = 1,16
          READ(1,*) IR, IG, IB
          TMP = RGB(IR, IG, IB)
          IF (TMP .GT. TMPMAX) THEN
            TMPMAX = TMP
            JK = I
          END IF
          DUMMY4 = REMAPPALETTE(I-1, TMP)
        END DO
      END IF
C
```

286

Figure 7.21 continued

```fortran
C     FILL XBUF ARRAY WITH GRID
C
      DATMAX = -10000000000.
      DATMIN = 10000000000.
      K = 0
      L = 0
10    CONTINUE
      READ(5,*,END=20) I, J, YYY, XXX, XBUF(I,J), ZZZ
      IF (I .GT. K) THEN
        K = I
      END IF
      IF (J .GT. L) THEN
        L = J
      END IF
      IF (XBUF(I,J) .GT. DATMAX) THEN
        DATMAX = XBUF(I,J)
      END IF
      IF (XBUF(I,J) .LT. DATMIN) THEN
        DATMIN = XBUF(I,J)
      END IF
      GO TO 10
20    CONTINUE
C
C     DETERMINE THE DATA RANGE
C
      DATRNG = DATMAX - DATMIN
C
C     ENTER THE TITLE FOR THE CONTOUR MAP
C
      WRITE(*,*) 'ENTER THE TITLE FOR THE CONTOUR MAP, UP TO'
      WRITE(*,*) '50 CHARACTERS IN LENGTH.'
      READ(*,25) TITLE
25    FORMAT(1A50)
C
C     DEFINE THE NUMBER OF CONTOUR LINES DESIRED
C
      WRITE(*,*) 'HOW MANY CONTOUR INTERVALS DO YOU DESIRE'
      WRITE(*,*) 'FOR THE SECOND CONTOUR MAP?'
      WRITE(*,*) 'ENTER 4, 8, 12, OR 16'
      READ(*,*) ICONT
C

C     CLEAR THE SCREEN
C
      CALL CLEARSCREEN ($GCLEARSCREEN)
C
C     DEFINE THE VIEWING REGION
C
      CALL SETVIEWPORT(1,1,640,480)
C
C     DRAW A BORDER AROUND THE MAP IN THE BRIGHTEST PALETTE COLOR
C
      IF (IPAL .EQ. 0) JK = 15
      DUMMY2 = SETCOLOR(JK)
      CALL MOVETO(200, 25, XY)
      DUMMY2 = LINETO(601,25)
      DUMMY2 = LINETO(601,426)
      DUMMY2 = LINETO(200,426)
      DUMMY2 = LINETO(200,25)
C
C     CREATE THE PSEUDO CONTOUR MAP
C
      IROW = INT(400 / K)
      ICOL = INT(400 / L)
      DO ITOT = 1, K
        DO JTOT = 1, L
          ICOLOR = INT(15 * (XBUF(ITOT,JTOT)-DATMIN)/DATRNG)
          IF (ICOLOR .LT. 0) ICOLOR = 0
          JPOS = (JTOT-1) * ICOL
          DO KTOT = 1, ICOL
            IBUF(JPOS+KTOT) = ICOLOR
          END DO
        END DO
        IPOS = (ITOT - 1) * IROW
        DO IRIT = 1, IROW
          III = IPOS + IRIT
          DO JRIT = 1, ICOL * L
            DUMMY2 = SETCOLOR(IBUF(JRIT))
            DUMMY2 = SETPIXEL(JRIT+200, III+25)
          END DO
        END DO
      END DO
C
```

287

Figure 7.21 continued

```fortran
C
C      DRAW LEGEND FOR MAP
C
       DO I = 1, 16
          DUMMY2 = SETCOLOR(I-1)
          DUMMY2 = RECTANGLE($GFILLINTERIOR, 11, (16*I+1), 26, (16*(I+1)))
          DATINC = DATRNG/16.
          DATNOW = DATMIN + (I-1) * DATINC
          DAT2 = DATMIN + I * DATINC
          IF (I.EQ.1) CALL SETTEXTPOSITION(I+1, 7, CURPOS)
          IF (I .GT. 1) CALL SETTEXTPOSITION(I+1, 6, CURPOS)
          WRITE(*,100) DATNOW, DAT2
100       FORMAT(E8.2,'-',E8.2, )
       END DO
C
C      PRINT THE TITLE
C
       KCOL = 27 + (50 - LEN TRIM(TITLE))/2
       CALL SETTEXTPOSITION(1, KCOL, CURPOS)
       CALL OUTTEXT(TITLE)
C
C      PAUSE; HIT ENTER TO DRAW A CONTOUR MAP
C
       READ(*,*)
C
C      REMAP COLOR 15 TO BRIGHT WHITE
C
       TMP = RGB(63, 63, 63)
       DUMMY4 = REMAPPALETTE(15,TMP)
C
C      CHANGE THE RANGE OF COLORS IN MAP IF FEWER THAN 16
C      CONTOUR INTERVALS ARE DESIRED FOR SECOND MAP
C
       IF (ICONT .NE. 16) THEN
          DO ITOT = 1,K
             IPOS = (ITOT - 1) * IROW
             DO IRIT = 1, IROW
                III = IPOS + IRIT
                DO JRIT = 1, ICOL*L
                   DUMM = ICONT * GETPIXEL(JRIT+200,III+25) / 15
                   DUMMY2 = SETCOLOR(DUMM)
                   DUMMY2 = SETPIXEL(JRIT+200, III+25)
                END DO
             END DO
          END DO
       END IF
C
C      DRAW THE CONTOUR MAP
C
       DUMMY2 = SETCOLOR(0)
       DUMMY2 = RECTANGLE($GFILLINTERIOR,  1,1,199,426)
       DO ITOT = 1, K
          IPOS = (ITOT - 1) * IROW
          DO IRIT = 1, IROW
             III = IPOS + IRIT
             DO JRIT = 1, ICOL*L
                DUMC = GETPIXEL(JRIT+201,III+25)
                DUMR = GETPIXEL(JRIT+200,III+26)
                DUMM = GETPIXEL(JRIT+200,III+25)
                IF (DUMM .NE. DUMC .OR. DUMM .NE. DUMR) THEN
                   DUMMY2 = SETCOLOR(15)
                   DUMMY2 = SETPIXEL(JRIT+200,III+25)
                ELSE
                   DUMMY2 = SETCOLOR(0)
                   DUMMY2 = SETPIXEL(JRIT+200,III+25)
                END IF
             END DO
          END DO
       END DO
C
C      PRINT THE TITLE FOR THE MAP
C
       KCOL = 27 + (50 - LEN TRIM(TITLE))/2
       CALL SETTEXTPOSITION(1, KCOL, CURPOS)
       CALL OUTTEXT(TITLE)
C
C      PAUSE; HIT ENTER TO EXIT PROGRAM
C
       READ(*,*)
       DUMMY2 = SETVIDEOMODE($DEFAULTMODE)
       STOP
       END
C
C      RGB FUNCTION
C
       INTEGER*4 FUNCTION RGB(R, G, B)
       INTEGER*4 R, G, B
       RGB = ISHL(ISHL(B,8) .OR.  G, 8) .OR. R
       RETURN
       END
```

288

Figure 7.26 A listing of the computer algorithm CTOUR for contour line drawing. A complete user's guide is given.

```fortran
C     PROGRAM CTOUR
C
C     A PROGRAM FOR DRAWING CONTOURS USING A REGULAR GRID OF DATA VALUES;
C     UP TO TEN (10) CONTOUR INTERVALS ARE ALLOWED; THE MINIMUM NUMBER IS
C     TWO (2).   CONTOURS ARE DRAWN IN DIFFERENT COLORS USING THE DEFAULT
C     PALETTES; A LEGEND ASSOCIATES CONTOUR COLORS WITH DATA VALUES.
C
C     ****************************************************
C                    USER'S GUIDE:   CTOUR
C     ****************************************************
C
C     A FEW INTERACTIVE QUESTIONS COMPOSE THE TOTAL USER SPECIFIED
C     INPUT TO THIS PROGRAM:
C
C     1.   ENTER NUMBER OF ROWS FOR THE GRID TO BE CONTOURED;
C     2.   ENTER NUMBER OF COLUMNS FOR THIS GRID;
C     3.   ENTER THE NAME OF THE INPUT FILE STORING THE GRID
C     4.   ENTER NUMBER OF VALUES ON EACH LINE OF THIS GRID FILE;
C          FOR INSTANCE, IF EACH LINE OF THE FILE STORING THE GRID
C          LOOKS LIKE THIS:
C
C               1    2   25.9    36.4    92.9
C
C          ENTER THE NUMBER, 5, FOR THIS FOURTH QUESTION.
C     5.   OF THE NUMBER ENTERED FOR QUESTION 4, SPECIFY WHICH OF THE
C          VALUES IS TO BE CONTOURED; FOR EXAMPLE, IF THE THIRD NUMBER
C          ON EACH LINE OF THE GRID FILE IS THE ONE TO BE CONTOURED,
C          ENTER THE NUMBER, 3.
C     6.   SPECIFY THE NUMBER OF CONTOUR INTERVALS; THE MAXIMUM NUMBER
C          IS TEN (10); THE MINIMUM NUMBER IS TWO (2).
C     7.   ENTER A TITLE FOR THE CONTOUR MAP, UP TO FIFTY SPACES IN
C          LENGTH
C
C     ****************************************************
C                  END OF USER'S GUIDE:  CTOUR
C     ****************************************************
C
      INCLUDE 'FGRAPH.FI'
      INCLUDE 'FGRAPH.FD'


      INTEGER*2 DUMMY2
      DIMENSION GRID(100,100), XLINE(100)
      CHARACTER*50 TITLE
      RECORD /WXYCOORD/ XY
      RECORD /RCCOORD / CURPOS
C
C     ACCESS THE GRID INFORMATION
C
      WRITE(*,*)' ENTER NUMBER OF ROWS FOR GRID TO BE CONTOURED'
      READ (*,*)   NROW
      WRITE(*,*)' ENTER NUMBER OF COLUMNS FOR THIS GRID'
      READ (*,*)   NCOL
      WRITE(*,*)' ENTER THE FILE NAME STORING THE GRID'
      OPEN(5, FILE = ' ')
      WRITE(*,*)' HOW MANY VALUES ARE THERE, EACH LINE OF GRID FILE?'
      READ (*,*) NLINE
      WRITE(*,*)' OF THESE VALUES, WHICH ONE IS TO BE CONTOURED?'
      READ (*,*) NVAL
      DO I = 1, NROW
         READ(5,*) (XLINE(JK), JK = 1, NLINE)
         GRID(I,J) = XLINE(NVAL)
      END DO
C
C     DETERMINE DATA RANGE
C
      GMIN = GRID(1,1)
      GMAX = GRID(1,1)
      DO I = 1, NROW
      DO J = 1, NCOL
         IF (GRID(I,J) .GT. GMAX) THEN
            GMAX = GRID(I,J)
         END IF
         IF (GRID(I,J) .LT. GMIN) THEN
            GMIN = GRID(I,J)
         END IF
      END DO
      END DO
C
```

289

Figure 7.26 continued

```fortran
C
C     DETERMINE CONTOUR INTERVALS
C
      WRITE(*,*) 'ENTER NUMBER OF CONTOUR INTERVALS DESIRED; 10 MAX'
      READ (*,*) NTOUR
      WRITE(*,*) 'ENTER TITLE FOR MAP, UP TO FIFTY SPACES IN LENGTH'
      READ(*,10) TITLE
10    FORMAT(1A50)
      IF (NTOUR .GT. 10) NTOUR = 10
      IF (NTOUR .LT. 2 ) NTOUR = 2
      RANGE = GMAX - GMIN
      STEP = RANGE / FLOAT(NTOUR)
      GRDMIN = GMIN
C
C     CONTOUR ALGORITM AND DRAWING
C
C
      DUMMY2 = SETVIDEOMODE($VRES16COLOR)
      CALL CLEARSCREEN($CLEARSCREEN)
      DUMMY2 = SETCOLOR(15)
C     WINDOWING AND VIEWING:
      DUMMY2 = SETWINDOW(.FALSE., 1, 1,NCOL+1,NROW+1)
      CALL SETVIEWPORT(201,41,600,440)
C     BORDER DRAWING
      CALL MOVETO W(1,1,XY)
      DUMMY2 = LINETO_W(NCOL,1)
      DUMMY2 = LINETO_W(NCOL,NROW)
      DUMMY2 = LINETO_W(1,NROW)
      DUMMY2 = LINETO_W(1,1)
C     BEGIN CONTOURING ALGORITHM, ONE CONTOUR STEP AT A TIME
      DO K = 1, NTOUR
      VALUE = STEP * FLOAT(K) * 0.99 + GRDMIN
C     SET DRAWING COLOR TO COLOR NUMBER, K (SEE LISTING OF PROGRAM,
C     BOXDRAW, FOR LIST OF DEFAULT COLORS)
      DUMMY2 = SETCOLOR(K)
      DO I = 1, NROW - 1
        DO J = 2, NCOL
C       CHECK WHETHER CONTOUR LINE APPEARS IN CELL, I,J:
          KTOP = 0

          GMAX = GRID(I,J)
          GMIN = GRID(I,J-1)
          KLAATU = 1
          IF (GMAX .LT. GMIN) THEN
            KLAATU = -1
            DUM = GMAX
            GMAX = GMIN
            GMIN = DUM
          END IF
          IF (GMIN .LT. VALUE .AND. GMAX .GT. VALUE) THEN
            KTOP = 1
            IF (GMAX .NE. GMIN) THEN
            IF (KLAATU .LT. 0) THEN
              POS = FLOAT(J) - (VALUE-GMIN)/(GMAX-GMIN)
            ELSE
              POS = FLOAT(J-1) + (VALUE-GMIN)/(GMAX-GMIN)
            END IF
            END IF
            CALL MOVETO W(POS,I,XY)
C           SEE WHERE THE LINE GOES
            GMAX = GRID(I+1,J)
            GMIN = GRID(I,J-1)
            KLAATU = 1
            IF (GMAX .LT. GMIN) THEN
              KLAATU = -1
              DUM = GMAX
              GMAX = GMIN
              GMIN = DUM
            END IF
            IF (GMIN .LT. VALUE .AND. GMAX .GT. VALUE) THEN
            IF (GMAX .NE. GMIN) THEN
            IF (KLAATU .LT. 0) THEN
              XPOS = J - (VALUE-GMIN)/(GMAX-GMIN)
              YPOS = (I+1) - (VALUE-GMIN)/(GMAX-GMIN)
            ELSE
              XPOS = (J-1) + (VALUE-GMIN)/(GMAX-GMIN)
              YPOS = I + (VALUE-GMIN)/(GMAX-GMIN)
            END IF
            END IF
            DUMMY2 = LINETO_W(XPOS,YPOS)
```

Figure 7.26 continued

```fortran
c     FIND WHERE THE LINE FINISHES, THIS CELL
      GMAX = GRID(I+1,J-1)
      GMIN = GRID(I,J-1)
      KLAATU = 1
      IF (GMAX .LT. GMIN) THEN
        KLAATU = -1
        DUM = GMAX
        GMAX = GMIN
        GMIN = DUM
      END IF
      IF (GMIN .LT. VALUE .AND. GMAX .GT. VALUE) THEN
        IF (GMAX .NE. GMIN) THEN
          IF (KLAATU .LT. 0) THEN
            POS = (I+1) - (VALUE-GMIN)/(GMAX-GMIN)
          ELSE
            POS = I + (VALUE-GMIN)/(GMAX-GMIN)
          END IF
        END IF
        DUMMY2 = LINETO_W(J-1,POS)
      ELSE
        IF (GRID(I+1,J) .GT. GRID(I+1,J-1)) THEN
          POS=(J-1)+(VALUE-GRID(I+1,J-1))/
     2        (GRID(I+1,J)-GRID(I+1,J-1))
        ELSEIF (GRID(I+1,J) .LT. GRID(I+1,J-1)) THEN
          POS = J - (VALUE-GRID(I+1,J))/(GRID(I+1,J-1)-GRID(I+1,J))
        END IF
        DUMMY2 = LINETO_W(POS,I+1)
      END IF
      ELSE
        KTOP = 0
        IF(GRID(I+1,J) .GT. GRID(I,J)) THEN
          POS = I + (VALUE-GRID(I,J))/(GRID(I+1,J)-GRID(I,J))
        ELSEIF (GRID(I+1,J) .LT. GRID(I,J)) THEN
          POS = (I+1)-(VALUE-GRID(I+1,J))/(GRID(I,J)-GRID(I+1,J))
        END IF
        DUMMY2 = LINETO_W(J,POS)
      END IF
      END IF
c     TOP DOWN SEARCH COMPLETED; IF NO LINE CROSSING FOUND, START
c     SEARCH FROM LEFT SIDE

      KSIDE = 0
      IF (KTOP .EQ. 0) THEN
      GMAX = GRID(I+1,J-1)
      GMIN = GRID(I,J-1)
      KLAATU = 1
      IF (GMAX .LT. GMIN) THEN
        KLAATU = -1
        DUM = GMAX
        GMAX = GMIN
        GMIN = DUM
      END IF
      IF (GMIN .LT. VALUE .AND. GMAX .GT. VALUE) THEN
        KSIDE = 1
        IF (GMAX .NE. GMIN) THEN
          IF (KLAATU .LT. 0) THEN
            POS = (I+1) - (VALUE-GMIN)/(GMAX-GMIN)
          ELSE
            POS = I + (VALUE-GMIN)/(GMAX-GMIN)
          END IF
        END IF
      END IF
c     CALL MOVETO_W(J-1,POS,XY)
c     FIND WHERE THIS LINE GOES
      GMAX = GRID(I+1,J)
      GMIN = GRID(I,J-1)
      KLAATU = 1
      IF (GMAX .LT. GMIN) THEN
        KLAATU = -1
        DUM = GMAX
        GMAX = GMIN
        GMIN = DUM
      END IF
      IF (GMIN .LT. VALUE .AND. GMAX .GT. VALUE) THEN
        IF (GMAX .NE. GMIN) THEN
          IF (KLAATU .LT. 0) THEN
            XPOS = J - (VALUE-GMIN)/(GMAX-GMIN)
            YPOS = (I+1) - (VALUE-GMIN)/(GMAX-GMIN)
          ELSE
            XPOS = (J-1)+(VALUE-GMIN)/(GMAX-GMIN)
            YPOS = I + (VALUE-GMIN)/(GMAX-GMIN)
          END IF
        END IF
        DUMMY2 = LINETO_W(XPOS,YPOS)
```

291

Figure 7.26 continued

```fortran
C     CONTINUE LINE ACROSS.
      IF (GRID(I+1,J) .GT. GRID(I,J)) THEN
        POS = I + (VALUE-GRID(I,J))/(GRID(I+1,J)-GRID(I,J))
      ELSEIF (GRID(I+1,J) .LT. GRID(I,J)) THEN
        POS = (I+1)-(VALUE-GRID(I+1,J))/(GRID(I,J)-GRID(I+1,J))
      END IF
      DUMMY2 = LINETO_W(J,POS)
      ELSE
      IF (GRID(I+1,J) .GT. GRID(I+1,J-1)) THEN
        POS=(J-1)+(VALUE-GRID(I+1,J-1))/
2              (GRID(I+1,J)-GRID(I+1,J-1))
      ELSEIF (GRID(I+1,J) .LT. GRID(I+1,J-1)) THEN
        POS = J-(VALUE-GRID(I+1,J))/(GRID(I+1,J-1)-GRID(I+1,J))
      END IF
      DUMMY2 = LINETO_W(POS,I+1)
      END IF
      END IF
C     SEE IF CONTOUR LINE ENTERS FROM BOTTOM TO RIGHT SIDE IF
A
C     LEFT SIDE ENTRY WAS NOT FOUND
      IF (KSIDE .EQ. 0) THEN
      GMAX = GRID(I+1,J)
      GMIN = GRID(I+1,J-1)
      KLAATU = 1
      IF (GMAX .LT. GMIN) THEN
        KLAATU = -1
        DUM = GMAX
        GMAX = GMIN
        GMIN = DUM
      END IF
      IF (GMIN .LT. VALUE .AND. GMAX .GT. VALUE) THEN
      IF (GMIN .NE. GMAX) THEN
      IF (KLAATU .LT. 0) THEN
        POS = J - (VALUE - GMIN)/(GMAX-GMIN)
      ELSE
        POS = (J-1) + (VALUE-GMIN)/(GMAX-GMIN)
      END IF
      END IF
      CALL MOVETO_W(POS,I+1,XY)
      GMIN = GRID(I,J-1)
      GMAX = GRID(I+1,J)
      KLAATU = 1

      IF (GMAX .LT. GMIN) THEN
        KLAATU = -1
        DUM = GMAX
        GMAX = GMIN
        GMIN = DUM
      END IF
      IF (GMIN .LT. VALUE .AND. GMAX .GT. VALUE) THEN
      IF (GMAX .NE. GMIN) THEN
      IF (KLAATU .LT. 0) THEN
        XPOS = J - (VALUE-GMIN)/(GMAX-GMIN)
        YPOS = (I+1)-(VALUE-GMIN)/(GMAX-GMIN)
      ELSE
        XPOS = (J-1)+(VALUE-GMIN)/(GMAX-GMIN)
        YPOS = I + (VALUE-GMIN)/(GMAX-GMIN)
      END IF
      END IF
      DUMMY2 = LINETO_W(XPOS,YPOS)
      GMAX = GRID(I+1,J)
      GMIN = GRID(I,J)
      KLAATU = 1
      IF (GMAX .LT. GMIN) THEN
        KLAATU = -1
        DUM = GMAX
        GMAX = GMIN
        GMIN = DUM
      END IF
      IF (GMIN .LT. VALUE .AND. GMAX .GT. VALUE) THEN
      IF (GMAX .NE. GMIN) THEN
      IF (KLAATU .LT. 0) THEN
        POS = (I+1) - (VALUE-GMIN)/(GMAX-GMIN)
      ELSE
        POS = I + (VALUE-GMIN)/(GMAX-GMIN)
      END IF
      END IF
      DUMMY2 = LINETO_W(J,POS)
      END IF
      END IF
      END IF
      END IF
      END DO
      END DO
C
```

292

Figure 7.26 continued

```
C
C        PRINT TITLE ON MAP
C
         KROW = 1
         KCOL = 27 + (50 - LEN_TRIM(TITLE))/2
         CALL SETTEXTPOSITION(KROW,KCOL,CURPOS)
         CALL OUTTEXT(TITLE)
C
C        PRINT THE LEGEND
C
         KROW = 1
         KCOL = 5
         CALL SETTEXTPOSITION(KROW,KCOL,CURPOS)
         CALL OUTTEXT('LEGEND')
         CALL SETVIEWPORT(1,1,640,480)
         DO I = 1, NTOUR
            VALUE = STEP * I * .99 + GRDMIN
            DUMMY2 = SETCOLOR(I)
            DUMMY2 = RECTANGLE($GFILLINTERIOR,10,I*32,30,I*32+20)
            CALL SETTEXTPOSITION(I*2+1,8, CURPOS)
            WRITE(*,100) VALUE
100         FORMAT(F10.2, )
         END DO
C
C        PAUSE FOR PRINTING, IF DESIRED
C
         READ(*,*)
         DUMMY2 = SETVIDEOMODE($DEFAULTMODE)
         STOP
         END
```

293

8

Finite Element Analysis

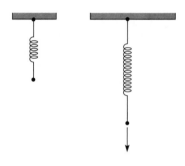

How does a system respond to a disturbance? Of course, the answer to this question depends on what the system is. If the system is an economic one, a disturbance may have a complicated, perhaps unpredictable effect. One example of such a disturbance might be the announced merger between two major and formerly rival corporations. On the other hand, if the system is a collection of springs, a disturbance has a measurable and predictable effect. In this case, a disturbance might be a force or forces applied to one or several of the springs that compose the system. The same force or forces applied in the same way to the system will in theory, and usually in practice, produce the same effect.

Often systems are complex. An economic system is complex due to the number of entities that comprise the system and because humans control the entities. A system of springs, on the other hand, is a physical, inorganic system of finite pieces and is not as complex as an economic system. Therefore, if the system under study is discretized easily as a collection of interconnected pieces of finite size, finite element analysis may be used to assess the effect that a disturbance has on such a system.

Each piece composing the system is called an element, and each element is of finite size; hence the term finite element analysis. As applied in this chapter for the analysis of stress and strain, finite element analysis is based on Hooke's law. This law explains, for example, the effect a force has on a spring, also the relationship between stress and strain.

8.1 Hooke's Law

A spring (Figure 8.1) is a device that displaces, or deflects, a certain amount if a force, either compressive or tensile, is applied to it. For instance, if a tensile force F is applied to the spring, the spring will elongate an amount d (Figure 8.1).

A spring has a finite stiffness depending on a number of factors: the material comprising the spring, its mass, and its length. If this stiffness is represented by K, the amount the spring displaces by d, and the applied force by F, then

$$F = Kd$$

from which the displacement d may be determined as F/K.

This relationship is known as Hooke's law, named for Robert Hooke, an English scientist who proposed the anagram ceiiinosssttuv, which in Latin is *Ut Tensio sic Vis* and means "the force varies as the stretch" (Popov, 1976, p. 35). Hooke's law is a simple one and shows that the amount of displacement the spring sustains is directly proportional to the amount of force applied to the spring; moreover, the displacement is inversely proportional to the stiffness of the spring.

8.2 Strain and Stress

How is the strength of a spring determined relative to the amount of displacement it can sustain before breaking? This question is answered by exploring the relationship between strain and stress. Strain is the change in length (for a given dimension) in relationship to the original length. For example, given the spring in Figure 8.1 which lengthens by an amount d,

$$\text{strain } \varepsilon = \frac{d}{L}$$

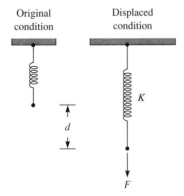

Figure 8.1   A simple spring showing the effect d of a tensile force F.

Figure 8.2   A bar element, which can be thought of as a spring.

where L is the original length of the spring, the length prior to the application of the force F.

Stress is the force acting on a material over a given area. Mathematically, this is written as

$$\text{stress } \sigma = \frac{\text{force}}{\text{area}}$$

For example, the bar in Figure 8.2 can be thought of as a spring having a cross-sectional area A and subjected to a tensile force F. The (tensile) stress in the bar, due to F, is F/A.

Strain and stress are related, as can be seen in Figure 8.3. This plot is known as a stress–strain diagram. Each material has a unique stress–strain diagram. For instance, a ductile material has a diagram (Figure 8.3a) showing, beyond a certain limit of stress, that small additional increments of stress yield large strains. In contrast, brittle materials break at a certain limit of stress (Figure 8.3b).

For either stress–strain diagram and for most materials, the slope at the origin is a constant for a material and is known as Young's modulus (Thomas

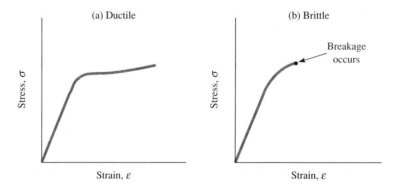

Figure 8.3   Strain–stress diagrams for (a) a ductile material showing severe distortions (strains) beyond a particular limiting stress, and (b) a brittle material, which breaks at a particular limiting stress.

Young, *Lectures on Natural Philosophy,* 1807). Young's modulus is referred to by the letter E. By examining Figure 8.3, the following relationship is realized at and near the origin of the stress–strain diagram:

$$\sigma = E\varepsilon$$

This relationship between stress and strain is also known as Hooke's law. In summary, two forms of Hooke's law have been introduced:

$$F = Kd$$

and

$$\sigma = E\varepsilon$$

Both forms, if used together, lead to a solution for K, the axial stiffness of a bar. First, recall that stress is equal to F/A (force per area). If F is an axial force, axial stress is determined by dividing this force by the cross-sectional area of the spring or bar. Recalling that $F = Kd$, then

$$\sigma A = Kd$$

Further recalling that the displacement d is equal to the strain multiplied by the length of the spring or bar, we have

$$\sigma A = K\varepsilon L \qquad \text{or} \qquad K = \frac{\sigma A}{\varepsilon L}$$

But stress divided by strain is equal to Young's modulus E. Therefore, K is equal to AE/L. In other words, the stiffness (in this case axial stiffness) of the bar or spring is directly proportional to the cross-sectional area (mass) and Young's modulus, but is inversely proportional to the length of the bar or spring.

Returning to the question posed at the beginning of this section regarding the capacity of a spring to sustain displacement, how much displacement a spring can sustain without breaking is described by the stress–strain curve. If the spring is composed of a ductile material, it will not break immediately. But beyond a particular stress, it will behave plastically, sustaining considerable displacement for small increases in stress, and eventually the ductile spring will break. If the spring is composed of a brittle material, it will not behave plastically before breaking at a particular limiting stress. Whether a material is brittle or ductile and precise knowledge of the critical stress causing the ductile behavior or breakage can be determined from laboratory experiments designed to apply variable stress and record resultant strains to develop a stress–strain diagram for the material.

8.3 Poisson's Ratio

An additional property of solid materials is Poisson's ratio, named for S.D. Poisson, a French scientist, who first recognized and reported this concept. This ratio describes the absolute value of the magnitude of lateral strain relative to axial strain:

$$\text{Poisson's ratio } \nu = \left| \frac{\text{lateral strain}}{\text{axial strain}} \right|$$

If a solid material is subjected to axial tension (Figure 8.4), it will elongate in its axial direction and decrease in width perpendicular to the axis, that is,

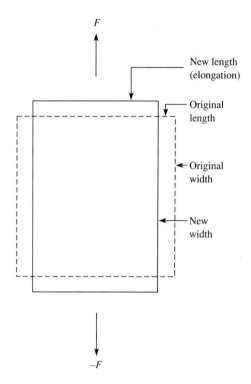

Figure 8.4   The concept of Poisson's ratio.

laterally. Conversely, if compressed axially, the material will tend to shorten axially and expand laterally.

Poisson's ratio ranges between 0 and 0.5. Cork has a Poisson's ratio of approximately zero; rubber is a material for which Poisson's ratio approaches 0.5. For many materials, including many types of rock, Poisson's ratio is in the range 0.25 to 0.35 (Popov, 1976, p. 42). Poisson's ratio will be used later in this chapter to derive the concept of stiffness for two-dimensional solid finite elements.

8.4 Truss Systems: Finite Element Analysis in Its Most Simple Form

A finite element (cf. Cook, 1974, p. 2) is a single, unique constituent of a finite element mesh. A mesh is the system of all finite elements used to analyze stress and strain for a particular physical system. For example, a system of bar elements, also called a truss system, is shown in Figure 8.5. Each bar is a finite element, and with respect to truss systems these elements are called bar elements.

A given finite element in a mesh is defined by its nodal points (indicated by numbers in Figure 8.5). Each nodal point is part of the finite element mesh and is associated with unique coordinates, either in two- or three-dimensional space. The finite elements are also shown in this figure by a number enclosed in a circle.

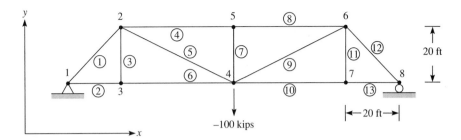

Figure 8.5 A simple truss system consisting of 8 nodal points and 13 elements. A kip is equal to 1000 pounds (454 kilograms). Numbers inside circles represent elements; numbers alone represent nodal points.

In the case of the truss system, the objective of the finite element analysis of this system is to assess the affect that the force F has on the entire system. The affect is assessed at the nodal points and measured as displacement. Suppose, for instance, that a two-dimensional finite element analysis is to be performed. In this case, each nodal point is free to move in two directions, x and y, unless physically restricted from doing so. Each nodal point, therefore, has two degrees of freedom in which to move (displace) for a two-dimensional finite element analysis.

Therefore, given that force F is acting on the system at nodal point 4 (Figure 8.5), every nodal point displaces in the x and/or y directions in response to this force, with the exception of nodal point 1, which is physically restricted from moving in either the x or y directions. (Such a restriction is known as a boundary condition; boundary conditions are discussed later in this chapter.) The objective of finite element analysis is to assess the magnitudes of these displacements.

Recall from Hooke's law that $F = Kd$, where d is a displacement. This law in scalar form pertains to a single spring. However, the truss system represents a system of springs. Hooke's law, though, pertains to each element of this system and is written for this system using matrix algebra:

$$\{F\} = [K]\{d\}$$

where $\{F\}$ is the collection of all forces acting over all degrees of freedom at the nodal points, and $\{d\}$ is the vector of displacements for all degrees of freedom at all nodal points. The matrix $[K]$ is the system stiffness matrix and describes the stiffness for the entire system. This is a square, symmetrical, and diagonalized matrix having a dimension equal to the total number of nodal points times the number of degrees of freedom per nodal point.

For a two-dimensional truss system associated with eight nodal points (Figure 8.5), the size of $[K]$ is 16×16, because 16 is the product of 8 (nodal points) and 2 (degrees of freedom per nodal point). Moreover, the size of $\{F\}$ and of $\{d\}$ is 16×1. True, nodal point 1 is restricted from movement, but for the sake of programming simplicity, its degrees of freedom, although restricted, are included in the size of the matrix system. Large values for stiffness are associated with these freedoms in the matrix $[K]$ to result in restriction (that is, displacement d is infinitely small if stiffness K is infinitely large).

The vector $\{d\}$ is that which is solved. The vector $\{F\}$ contains known quantities, the active forces for which displacements result. What is needed to allow us to solve for the vector $\{d\}$ is the system stiffness matrix $[K]$. This matrix is developed as the sum of all element stiffness matrices. Therefore, what is needed first is a formula for the element stiffness matrix, in this case for bar elements.

8.4.1 Element Stiffness Matrix: Bar Elements

In this section, the simplest of the finite elements, the bar element, is used to model truss systems. To develop the system stiffness matrix $[K]$, information is needed regarding the element stiffness matrices, as was discussed previously.

To derive the element stiffness matrix for a bar element (Figure 8.6) an arbitrary x-direction displacement u_A is shown applied to nodal point A whereas nodal point B is fixed (letter designation of nodal points is used to emphasize the generic nature of this derivation). The consequence of this arbitrary displacement is an axial shortening, and the amount of this shortening is

$$\Delta \text{ axial length} = \Delta L = u_A \cos \alpha$$

In terms of Hooke's law, $F = Kd$, the amount of axial shortening, which is a consequence of u_A, is equal to d. This displacement must be the result of a force, in this case called F_{ax} (axial force), that is equal to

$$Ku_A \cos \alpha$$

which is, of course, Kd.

This force is resolved into its x and y direction components as follows:

$$F_{ax,x} = F_{ax} \cos \alpha, \qquad F_{ax,y} = F_{ax} \sin \alpha$$

Note: The finite element procedure provides an approximate solution for displacement. Why this is so is evident in Figure 8.6. Assume that u_A is so small that the change in the orientation angle α of the bar element is negligible. In reality, this change in orientation of an element does occur and the magnitude of this change is dependent on the magnitude of the force acting on the element. But this change in orientation can only be known once displacements are calculated and these, of course, are to be solved by the finite element procedure. This is the dilemma. What is known prior to the application of finite element

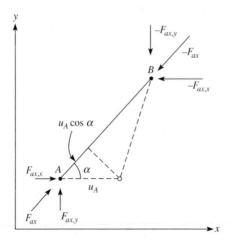

Figure 8.6 A single oriented bar element (solid line) showing the consequences (dashed line) of an arbitrary x-direction displacement u_A.

analysis is the predeformation orientation angle of each element, and this is what must be used.

A static condition is assumed for all applications of finite element analysis presented in this text. This assumes that all forces acting on systems are balanced (that is, are in equilibrium). For every action (force), an equal but opposite reaction (force) cancels the action (force) for no net force. This means that the following summations hold for subsequent finite element analyses:

$$\Sigma F_x = 0, \qquad \Sigma F_y = 0$$

Because a static condition is assumed for the bar element (Figure 8.6), the following condition holds:

$$\Sigma F_x = F_{ax,x} - F_{ax,x} = 0$$

where $-F_{ax,x}$ is the reaction (force) at nodal point B to the force $F_{ax,x}$ at nodal point A. Likewise, for the y direction,

$$\Sigma F_y = F_{ax,y} - F_{ax,y} = 0$$

Combining the x- and y-direction equations into a single equation leads to

$$\Sigma F_x + \Sigma F_y = F_{ax,x} + F_{ax,y} - F_{ax,x} - F_{ax,y} = 0$$

Substituting for $F_{ax,x}$ and $F_{ax,y}$ gives

$$F_{ax} \cos \alpha + F_{ax} \sin \alpha - F_{ax} \cos \alpha - F_{ax} \sin \alpha = 0$$

Substituting the solution given above for F_{ax} yields

$$Ku_A(\cos^2\alpha + \cos \alpha \sin \alpha - \cos^2\alpha - \cos \alpha \sin \alpha) = 0$$

Finally, recalling that K is equal to AE/L, gives

$$\frac{AEu_A}{L}(\cos^2\alpha + \cos \alpha \sin \alpha - \cos^2\alpha - \cos \alpha \sin \alpha) = 0$$

This completes the analysis of an arbitrary displacement applied to the x-direction degree of freedom at nodal point A (Figure 8.6). Actually, this is also the solution for the x-direction degree of freedom at nodal point B, except the signs are reversed to acknowledge the equal but opposite reaction at this nodal point (in this case, recall the situation in Figure 8.6, but assume that an arbitrary x displacement exists at nodal point B, whereas nodal point A is fixed). Then, at nodal point B, the following holds for the x-direction degree of freedom:

$$\frac{AEu_B}{L}(-\cos^2\alpha - \cos \alpha \sin \alpha + \cos^2\alpha + \cos \alpha \sin \alpha) = 0$$

Now y-direction arbitrary displacements are analyzed. In doing so, the situation in Figure 8.7 is used, which is almost identical to that for x-direction arbitrary displacements (Figure 8.6), except an arbitrary y-direction displacement, v_A, is associated with nodal point A (whereas nodal point B is fixed). This arbitrary displacement results in an axial shortening by an amount

$$\text{axial shortening } \Delta L = v_A \sin \alpha$$

This shortening also implies an axial force:

$$F_{ax} = Kv_A \sin \alpha = \frac{AE}{L} v_A \sin \alpha$$

Resolving this axial force into its x and y components, $F_{ax,x}$ and $F_{ax,y}$, and realizing that equal but opposite reaction forces act at nodal point B yields on

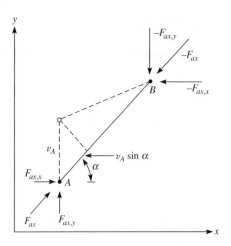

Figure 8.7   A single oriented bar element showing the consequences of an arbitrary y-direction displacement v_A.

the assumption of static conditions

$$\frac{AE}{L} v_A(\cos \alpha \sin \alpha + \sin^2\alpha - \cos \alpha \sin \alpha - \sin^2\alpha) = 0$$

Because of symmetry, the following is written for nodal point B (that is, assuming a situation similar to Figure 8.7, but where the arbitrary displacement occurs at nodal point B, while nodal point A is fixed):

$$\frac{AE}{L} v_B(-\cos \alpha \sin \alpha - \sin^2\alpha + \cos \alpha \sin \alpha + \sin^2\alpha) = 0$$

Four equations have now been derived to describe the four total degrees of freedom for the bar element: The element is defined by two nodal points, each of which has two degrees of freedom for a two-dimensional analysis; therefore, four total degrees of freedom exist for the element. The four equations are assembled as the columns of a matrix, resulting in the following matrix system:

$$\frac{AE}{L}
\begin{vmatrix}
\cos^2\alpha & \cos \alpha \sin \alpha & -\cos^2\alpha & -\cos \alpha \sin \alpha \\
\cos \alpha \sin \alpha & \sin^2\alpha & -\cos \alpha \sin \alpha & -\sin^2\alpha \\
-\cos^2\alpha & -\cos \alpha \sin \alpha & \cos^2\alpha & \cos \alpha \sin \alpha \\
-\cos \alpha \sin \alpha & -\sin^2\alpha & \cos \alpha \sin \alpha & \sin^2\alpha
\end{vmatrix}
\begin{vmatrix}
u_A \\ v_A \\ u_B \\ v_B
\end{vmatrix}
=
\begin{vmatrix}
F_{A,x} \\ F_{A,y} \\ F_{B,x} \\ F_{B,y}
\end{vmatrix}$$

This system has the form $[k]\{d\} = \{f\}$, which is, of course, Hooke's law. Moreover, the element stiffness matrix $[k]$ is

$$\underset{4 \times 4}{[k]} = \frac{AE}{L}
\begin{vmatrix}
C2 & CS & -C2 & -CS \\
CS & S2 & -CS & -S2 \\
-C2 & -CS & C2 & CS \\
-CS & -S2 & CS & S2
\end{vmatrix}$$

where $C2 = \cos^2\alpha$; $CS = \cos \alpha \sin \alpha$; and $S2 = \sin^2\alpha$.

This matrix is the generic formula for element stiffness matrix $[k]$ for any bar element oriented in two-dimensional space. A bar element is defined by two nodal points. In a two-dimensional space, each nodal point is defined by an x and a y coordinate. For instance, the two nodal points in Figures 8.6 and 8.7 are defined by coordinates: (x_A, y_A) and (x_B, y_B).

The length of any bar element is computed from nodal coordinates:

$$L = \sqrt{(x_A - x_B)^2 + (y_A - y_B)^2}$$

Moreover,

$$\cos \alpha = \frac{x_B - x_A}{L}, \qquad \sin \alpha = \frac{y_B - y_A}{L}$$

Therefore, the total information that is required to compute $[k]$ for a bar element consists of (1) coordinates of two nodal points, (2) Young's modulus for the material composing the bar element, and (3) the cross-sectional area for the bar element. On the basis of this list, the stiffness of a bar element oriented in two-dimensional space depends on this orientation (the orientation angle) and on the axial stiffness of the element.

8.4.2 Computing the System Stiffness Matrix

For the finite element system of bar elements in Figure 8.8, a demonstration is presented for assembling the system stiffness matrix $[K]$ for this mesh. Three nodal points define this mesh and each nodal point is associated with two degrees of freedom. In this case, the system stiffness matrix $[K]$ is a 6×6 matrix.

The skeletal form of this matrix (Figure 8.9) has a row and a column associated with one of the degrees of freedom in the system: $1x$, $1y$, $2x$, $2y$, $3x$, and $3y$. All entries in the matrix $[K]$ are initialized to zero to begin this process.

The system stiffness matrix $[K]$ is assembled by disassembling the element stiffness matrices and placing their entries into the matrix $[K]$ in a summation process. In the case wherein two elements compose the mesh (Figure 8.8), one of the elements is defined by nodal points 1 and 2. These nodal points have, respectively, the coordinates $(0, 0)$ and $(1, 0)$. The length of this element is therefore equal to 1.0. If, for simplicity, Young's modulus E and cross-sectional area A are assumed to both equal to 1.0, the element stiffness matrix for this element is

$$[k] = \frac{1}{1} \begin{vmatrix} 1.0 & 0.0 & -1.0 & 0.0 \\ 0.0 & 0.0 & 0.0 & 0.0 \\ -1.0 & 0.0 & 1.0 & 0.0 \\ 0.0 & 0.0 & 0.0 & 0.0 \end{vmatrix}$$

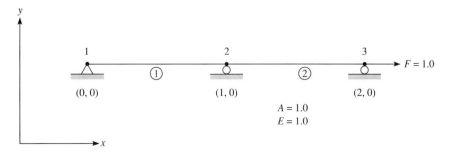

Figure 8.8 A simple truss system consisting of 3 nodal points and 2 elements.

	1x	1y	2x	2y	3x	3y
1x	0	0	0	0	0	0
1y	0	0	0	0	0	0
2x	0	0	0	0	0	0
2y	0	0	0	0	0	0
3x	0	0	0	0	0	0
3y	0	0	0	0	0	0

Figure 8.9 The skeletal form of the system stiffness matrix for the mesh shown in Figure 8.8.

because $\cos \alpha = \sin \alpha = 1$ and $\cos^2\alpha = \sin^2\alpha = 1$. This matrix is disassembled and its entries placed into the system stiffness matrix $[K]$ using the formula

$$K_{ix,jy} = K_{ix,jy} + k_{ix,jy}$$

This relationship is demonstrated as follows. First, let the element stiffness matrix for element 1 be written as

$$[k] = \begin{array}{c} \\ 1x \\ 1y \\ 2x \\ 2y \end{array} \begin{array}{|cccc|} 1x & 1y & 2x & 2y \\ 1.0 & 0.0 & -1.0 & 0.0 \\ 0.0 & 0.0 & 0.0 & 0.0 \\ -1.0 & 0.0 & 1.0 & 0.0 \\ 0.0 & 0.0 & 0.0 & 0.0 \end{array}$$

which associates each line and each column with a nodal point and one of its freedoms, x or y. For example, to disassemble this matrix, the entry for row $1x$, column $1x$ is placed into row $1x$, column $1x$ in the system stiffness matrix $[K]$ (Figure 8.10). Likewise, the entry for row $1x$, column $1y$ is placed into row $1x$, column $1y$ of the matrix $[K]$, and so on.

All that is required to complete the matrix $[K]$ is to compute the element stiffness matrix for element 2 and place it into the matrix $[K]$. The element stiffness matrix for element 2 is identical to that for element 1 by virtue of identical geometry. But the rows and columns for this matrix are now associated with nodal points 2 and 3. Therefore, the entry in row 1, column 1 of the matrix $[k]$ is placed into row $2x$, column $2x$ of the matrix $[K]$ (Figure 8.11). Row $2x$, column $2x$ of matrix $[K]$ is already equal to 1.0 because of the disassembly of the element stiffness matrix for element 1. To this value is

	1x	1y	2x	2y	3x	3y
1x	1	0	−1	0	0	0
1y	0	0	0	0	0	0
2x	−1	0	1	0	0	0
2y	0	0	0	0	0	0
3x	0	0	0	0	0	0
3y	0	0	0	0	0	0

Figure 8.10   The system stiffness matrix after the addition of the element stiffness matrix for element number 1.

	1x	1y	2x	2y	3x	3y
1x	∞*	0	−1	0	0	0
1y	0	∞*	0	0	0	0
2x	−1	0	2	0	−1	0
2y	0	0	0	∞*	0	0
3x	0	0	−1	0	1	0
3y	0	0	0	0	0	∞*

∞* — boundary condition

Figure 8.11   The final, actual system stiffness matrix for the truss shown in Figure 8.8.

added another value of 1.0 from the disassembly of the element stiffness matrix for element 2. This brings the total value in row $2x$, column $2x$ of matrix $[K]$ to 2.0.

Nodal points common to more than one element derive additive stiffness from all of these elements. The result of disassembling the element stiffness matrix for element 2 and placing it into the matrix $[K]$ (Figure 8.11) shows the final, complete system stiffness matrix for the two-element mesh (Figure 8.8).

8.5   A Rayleigh–Ritz Derivation of the Element Stiffness Matrix for Bar Elements

A more complex approach to deriving the element stiffness matrix for bar elements than that just presented is given here. One objective of this section is to show that this complex method yields the same stiffness matrix as that yielded by the approach just presented. The major objective of this section is to establish the procedure for determining the element stiffness matrix for the solid two-dimensional element applied for geological analysis later in this chapter.

8.5.1   An Overview

A finite element approach to the analysis of assumed displacement fields is used herein to derive stiffness matrices for elements. A Rayleigh–Ritz approach to this derivation is taken (Cook, 1974). In the Rayleigh–Ritz implementation used in this chapter, a function is used to describe the potential energy for single finite elements and consequently systems of finite elements.

Potential energy is defined for a single spring (e.g., Figure 8.1) as (Cook, 1974, p. 54)

$$\Pi_P = \frac{kd^2}{2} - Fd = \text{strain energy} - \text{work}$$

which is equal to zero at $d = 0$ (no displacement). In other words, potential energy of the spring, once displaced an amount d, is equal to one-half the product of stiffness and square of the displacement less the amount of work done against the force. This work is equal to the product of the force and displacement: Fd.

For a system of M finite elements, the potential energy of the system is written as (Cook, 1974, p. 77)

$$\Pi_P = \left(\sum_{i=1}^{M} \Pi_{PE,i} \right) - [D]^T [F]$$

for which

$$\Pi_{PE,i}$$

is the potential energy for one of the M elements, $[D]$ is the vector of displacements for all nodal point degrees of freedom for the system, and $[F]$ is the vector of forces acting on all nodal point degrees of freedom comprising the system.

For a single element (Cook, 1974, p. 76),

$$\Pi_{\text{PE}} = \frac{1}{2}[d]^{\text{T}} \left(\int_{\text{VOL}} [B]^{\text{T}}[E][B] \, dV \right) [d] + [d]^{\text{T}} \int_{\text{VOL}} [B]^{\text{T}}[\sigma_0] \, dV$$

$$- [d]^{\text{T}} \int_{\text{VOL}} [N]^{\text{T}}[\text{BF}] \, dV - [d]^{\text{T}} \int_{\text{SURFACE}} [N]^{\text{T}}[\Phi] \, dS$$

for which the matrix $[B]$ is a strain–displacement matrix for the element, which relates strains in the element to nodal point displacements as

$$[\varepsilon] = [B][d]$$

The matrix $[N]$ describes the assumed displacement field for the element, such that displacements anywhere within the element may be determined from the displacements at the nodal points $[d]$ as $[N][d]$. The matrices $[B]$ and $[N]$ are related as follows:

$$[B] = \begin{vmatrix} \dfrac{\partial}{\partial x} & 0 \\ 0 & \dfrac{\partial}{\partial y} \\ \dfrac{\partial}{\partial y} & \dfrac{\partial}{\partial x} \end{vmatrix} [N]$$

In other words, the strain–displacement matrix $[B]$ is developed by differentiating $[N]$.

In the equation for the potential energy of a single element, the second term is a function of the initial stress in the element, the third term is a function of initial body forces $[\text{BF}]$ in the element, and the fourth term is a function of initial surface tractions present on the element.

Substituting the formula for the potential energy of an element into that for the system yields (from Cook, 1974, p. 77)

$$\Pi_{\text{P}} = \frac{1}{2}[D]^{\text{T}} \left(\sum_{i=1}^{M} \int_{\text{VOL}} [B]^{\text{T}}[E][B] \, dV \right) [D] + [D]^{\text{T}} \sum_{i=1}^{M} \left(\int_{\text{VOL}} [B]^{\text{T}}[\sigma_0] \, dV \right.$$

$$\left. - \int_{\text{VOL}} [N]^{\text{T}}[\text{BF}] \, dV - \int_{\text{SURFACE}} [N]^{\text{T}}[\Phi] \, dS \right) - [D]^{\text{T}}[F]$$

In this expression, the component

$$\int_{\text{VOL}} [B]^{\text{T}}[E][B] \, dV$$

is the element stiffness matrix, considering that the first term of the potential energy equation

$$\frac{1}{2}[D]^{\text{T}} \left(\sum_{i=1}^{M} \int_{\text{VOL}} [B]^{\text{T}}[E][B] \, dV \right) [D] = \frac{1}{2}[D]^{\text{T}}[D] \left(\sum_{i=1}^{M} \int_{\text{VOL}} [B]^{\text{T}}[E][B] \right)$$

is analogous to $kd^2/2$. How the strain–displacement matrix $[B]$ is determined is shown next for a bar element.

8.5.2 Use of Assumed Displacement Fields in the Rayleigh–Ritz Method

The finite element method may be viewed as a form of the Rayleigh–Ritz potential energy method (cf. Cook, 1974). This method is a technique for approximating the equilibrium configuration of a structure under study. Usually, this is done using a polynomial to describe the approximate displacement field at every point within an element.

For example, derivation of the element stiffness matrix for a bar element is revisited. Such an element (Figure 8.12) is drawn parallel to the x axis (i.e., with an orientation angle equal to zero degrees). If x-direction displacement u is considered to vary linearly within the element, the following polynomial is proposed:

$$u = a_1 + a_2 x = [1 \quad x] \begin{vmatrix} a_1 \\ a_2 \end{vmatrix}$$

This is a general, linear polynomial describing the displacement u in the x direction at any point x within the element. Because this polynomial is representative of the x-direction displacement field for any position x, the equation may be written specifically at each nodal point A and B:

$$u_A = a_1 + a_2 x_A = [1 \quad x_A] \begin{vmatrix} a_1 \\ a_2 \end{vmatrix}$$

and

$$u_B = a_1 + a_2 x_B = [1 \quad x_B] \begin{vmatrix} a_1 \\ a_2 \end{vmatrix}$$

The coefficients a_1 and a_2 are common to these two equations because the polynomial describing u represents the x-direction displacement everywhere within the element. Hence the two equations can be combined such that (using notation consistent with Cook (1974))

$$\begin{vmatrix} u_A \\ u_B \end{vmatrix} = \begin{vmatrix} 1 & x_A \\ 1 & x_B \end{vmatrix} \begin{vmatrix} a_1 \\ a_2 \end{vmatrix} = [DA][a]$$

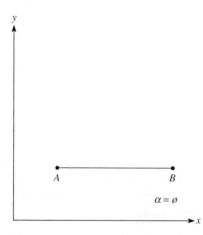

Figure 8.12   A single bar element oriented parallel to the x axis.

Writing this as a solution for $[a]$ gives

$$[a] = [DA]^{-1}[u]$$

Because $[a] = [a_1 \quad a_2]$, the general polynomial for u may be written as

$$u = [1 \quad x][DA]^{-1}[u]$$

Using the procedure (Chapter 3) for inverting a 2×2 matrix, the inverse of the matrix $[DA]$ is computed:

$$[DA]^{-1} = \frac{1}{\det[DA]} \begin{vmatrix} x_B & -x_A \\ -1 & 1 \end{vmatrix}$$

where $\det[DA] = x_B - x_A =$ length of the bar element L (for the unique orientation shown in Figure 8.12).

For simplicity, let x_A be at $x = 0$. Because $x_B - x_A$ equals the length L, then $x_B = L$. Hence,

$$[DA]^{-1} = \frac{1}{L} \begin{vmatrix} L & 0 \\ -1 & 1 \end{vmatrix} = \begin{vmatrix} 1 & 0 \\ -\dfrac{1}{L} & \dfrac{1}{L} \end{vmatrix}$$

Substituting this result into the solution for u gives

$$u = [1 \quad x] \begin{vmatrix} 1 & 0 \\ -\dfrac{1}{L} & \dfrac{1}{L} \end{vmatrix} [u] = \begin{vmatrix} \left(1 - \dfrac{x}{L}\right) & \dfrac{x}{L} \end{vmatrix} [u]$$

Knowing that $[u] = [u_A \quad u_B] = \{d\}$, using notation from Cook (1974), the following may be written:

$$u = \begin{vmatrix} \left(1 - \dfrac{x}{L}\right) & \dfrac{x}{L} \end{vmatrix} [u] = [N][d]$$

where the matrix $[N]$ describes the nature of the assumed displacement field.

Because strain is the change in length (size of a dimension) as assessed against the original dimension (Section 8.2), differential calculus is used to write the following:

$$\varepsilon_x = \frac{\partial u}{\partial x}$$

for x-directional strain. For a two-dimensional analysis of strain, two additional strains, strain in the y direction and a shear strain component, are written as

$$\varepsilon_y = \frac{\partial v}{\partial y}, \qquad \gamma_{xy} = \frac{\partial u}{\partial y} + \frac{\partial v}{\partial x}$$

These three components of strain may be computed from displacements $\{d\}$ using a strain–displacement matrix $[B]$, such that

$$[\varepsilon] = [B][d], \qquad [\varepsilon] = [\varepsilon_x \quad \varepsilon_y \quad \gamma_{xy}]$$

The strain–displacement matrix $[B]$ is found by taking partial derivatives of the matrix $[N]$ with respect to x and y. But, when only x-direction displacement is considered (Figure 8.12), the strain–displacement matrix $[B]$ is determined uniquely for x such that

$$[B_x] = \frac{\partial}{\partial x}[N] = \begin{vmatrix} \dfrac{-1}{L} & \dfrac{1}{L} \end{vmatrix}$$

Now the element stiffness matrix $[k]$ is solved using integral calculus such that

$$[k] = EA \int_0^L [B_x]^T [B_x] \, dx$$

where E is Young's modulus of the bar element, A is its cross section, and $[B_x]$ is its x-direction strain–displacement matrix. This is identical to the formula for $[k]$ developed in Section 8.5.1 provided that Young's modulus E is acknowledged to be a scalar for a bar element rather than a matrix. Further, the product AL is the volume of the bar element. Because the cross-sectional area A is assumed to be constant for the bar element, the integration is performed over the length L of the bar element, which in this case is tantamount to integrating over the volume of the element.

Solving this integral gives

$$[k] = AE \int_0^L [B_x]^T [B_x] \, dx = AE \int_0^L \begin{vmatrix} \dfrac{-1}{L} \\[2mm] \dfrac{1}{L} \end{vmatrix} \begin{vmatrix} \dfrac{-1}{L} & \dfrac{1}{L} \end{vmatrix} dx$$

$$= AE \int_0^L \begin{vmatrix} \dfrac{1}{L^2} & \dfrac{-1}{L^2} \\[2mm] \dfrac{-1}{L^2} & \dfrac{1}{L^2} \end{vmatrix} dx = \frac{AE}{L^2} \int_0^L \begin{vmatrix} 1 & -1 \\ -1 & 1 \end{vmatrix} dx$$

$$= \frac{AE}{L^2} \begin{vmatrix} x & -x \\ -x & x \end{vmatrix}_0^L = \frac{AE}{L} \begin{vmatrix} 1 & -1 \\ -1 & 1 \end{vmatrix}$$

In part, this is what is determined in Section 8.4; that is, the matrix is multiplied by axial stiffness: AE/L. However, the matrix differs from what is determined in Section 8.4, because the one from the Rayleigh–Ritz derivation is only a 2×2 matrix, whereas the one in Section 8.4 is a 4×4 matrix. In fact, element stiffness matrix $[k]$ in Section 8.4 must be a 4×4 matrix to account for the four total degrees of freedom each bar element has.

To resolve this discrepancy, the concept of coordinate transformation is introduced vis-à-vis the finite element formulation. To orient a one-dimensional object in a two-dimensional field, the following expression is used (following the logic presented in Chapter 7, this time assuming that the one-dimensional bar element pivots about nodal point A; cf. Figure 8.12):

$$[x' \quad y'] = x_B[\cos \alpha \quad \sin \alpha] = x_B[C \quad S]$$

which assumes, of course, $x_A = 0$. Using this formula, a transformation matrix $[T]$ is defined such that

$$[T] = \begin{vmatrix} C & S & 0 & \\ 0 & C & S \end{vmatrix} = \begin{vmatrix} \cos \alpha & \sin \alpha & 0 & 0 \\ 0 & 0 & \cos \alpha & \sin \alpha \end{vmatrix}$$

Performing the following multiplication

$$[k] \quad = \quad [T]^T \quad [k] \quad [T]$$
$$4 \times 4 \quad 4 \times 2 \ 2 \times 2 \ 2 \times 4$$

leads to

$$[k] = [T]^\mathrm{T} \left(\frac{AE}{L} \begin{vmatrix} 1 & -1 \\ -1 & 1 \end{vmatrix} \right) [T] = \frac{AE}{L} \begin{vmatrix} \text{C2} & \text{CS} & -\text{C2} & -\text{CS} \\ \text{CS} & \text{S2} & -\text{CS} & -\text{S2} \\ -\text{C2} & -\text{CS} & \text{C2} & \text{CS} \\ -\text{CS} & -\text{S2} & \text{CS} & \text{S2} \end{vmatrix}$$

for which C2 = $\cos^2\alpha$, CS = $\cos \alpha \sin \alpha$, and S2 = $\sin^2\alpha$, which is what is derived in Section 8.4. In summary, the Rayleigh–Ritz method, upon coordinate transformation and using the assumed displacement field as defined earlier in this section, is identical to the procedure presented in Section 8.4.

8.6 TRUSS: A FORTRAN-77 Program for Two-Dimensional Finite Element Analysis of Truss Systems

The theory of finite element analysis is one thing; the practice of finite element analysis is another. In fact, depending on the complexity of the finite element mesh, the practice of finite element analysis can be one of the most frustrating computer applications a person will ever encounter. An important aspect of finite element analysis education, therefore, is the exposure to implementing finite element analysis on a computer.

A successful finite element analysis is a rewarding product, and the foregoing paragraph is not meant to discourage its use. On the contrary, the paragraph is meant only to acknowledge what students will discover on their own. In this section, experience is gained, in small measure, regarding the frustration of entering data into a finite element analysis computer program. But, in utilizing computer graphics, the joy of a successful analysis is also experienced.

The truss system in Figure 8.5 is analyzed as a demonstration. Before a computer program is presented for this analysis, the information required for a complete finite element analysis of a truss system is reviewed. This information consists of the following:

1. The cross-sectional area A, Young's modulus E, and $x - y$ coordinates for each of the two nodal points of each element in the truss
2. Element connectivity, which is simply defining each element by its nodal point numbers (as demonstrated below)
3. Active forces acting at nodal point degrees of freedom. Only active forces, not reaction forces, are required. The reactions are implicitly accounted for in the formulation of the system stiffness matrix, provided boundary conditions, as discussed subsequently, are specified. For example, in the case of Figure 8.5, only one active force is applied to the y-direction degree of freedom at nodal point 4. This force is equal to -100 kips $= -445,000$ N[ewtons]; the negative sign indicates that the force acts downward. (*Note:* Designating this force as a negative value is at the discretion of the program user; this value may be entered as a positive quantity if desired. If the value is entered as a negative value, then negative y displacements represent downward movement and positive values represent upward movement. If the force is entered as a positive number, then positive y displacements

represent downward movement and negative values represent upward movement. This discussion follows from the fact that the force applied at nodal point 4 acts downward.) Because in Figure 8.5 only one degree of freedom is associated with an active force, the force vector $\{F\}$, in this case a 16×1 vector, contains all zero entries except for the eighth position in the vector, the position associated with the y degree of freedom at nodal point 4. This entry is equal to -100 kips or, if the metric system is used, $-445,000$ N.
4. Data entry is required to specify boundary conditions. Such conditions describe restraints placed on specific nodal points, preventing them from displacing in one or both spatial directions. For instance, nodal point 1 in Figure 8.5 is associated with a physical constraint preventing its movement in both spatial directions. Nodal point 8 is associated with a roller, allowing it to move in the horizontal (x) direction, but preventing it from moving in the y direction (even in the upward direction). The remaining six nodal points are unrestricted.

Two approaches to accounting for boundary conditions are possible. One is simple in concept and simple to implement in a computer program; the other is simple in concept, but difficult to implement in a computer program. This second method involves removing from the system stiffness matrix $[K]$ the rows and columns associated with restricted degrees of freedom at specified nodal points.

This second method entails the removal of the first and second rows and the first and second columns of the matrix $[K]$, because these are associated with the x and y degrees of freedom at nodal point 1. The 16th row and column must also be removed, because these are associated with the y degree of freedom at nodal point 8. The result of this modification is a revised matrix $[K]$, a 13×13 matrix. Likewise, the force vector $\{F\}$ is modified to remove its first, second, and 16th entries, because forces acting at restricted freedoms are not necessary once the matrix $[K]$ is modified. Once the force vector is modified, Gaussian elimination (or a related equation solution algorithm) is used to solve for the vector $\{d\}$ in the system $\{F\} = [K]\{d\}$. Because $[K]$ and $\{F\}$ are modified, the vector $\{d\}$ contains displacements only for unrestricted freedoms; restricted freedoms are automatically associated with displacements equal to zero.

This concept is fairly simple to describe verbally, but involves careful planning when developing a computer algorithm to perform the appropriate modifications for the requisite boundary conditions. An alternative method for use in accounting for boundary conditions that is easy to implement in a computer algorithm is based on the notion that in Hooke's law relative to a single spring, if the stiffness K is infinitely large, the resultant displacement d is infinitely small for any force F. In implementing this method for accounting for boundary conditions, the matrix $[K]$ and the vector $\{F\}$ are not modified by removing rows and columns; in fact, the vector $\{F\}$ is not change at all. Instead, entries on the diagonal of the system stiffness matrix $[K]$ are changed to equal large numbers for rows associated with nodal degrees of freedom to be restricted. Three diagonal entries in $[K]$ in Figure 8.11, associated with the first, second, and 16th rows, are set equal to a large number, such as $1.0E + 15$ or some other large number (up to the limit of the computer on which the analysis is to be performed). Once these large numbers are entered at appropriate diagonal positions within the matrix $[K]$, the system $\{F\} = [K]\{d\}$ is solved for $\{d\}$ using Gaussian elimination. Displacements associated with restricted freedoms are found to be extremely small numbers in this procedure because of large numbers entered into the matrix $[K]$.

One additional property of the system stiffness matrix $[K]$ is discussed. If nodal points comprising a mesh are numbered efficiently, the matrix $[K]$ is banded (Figure 8.13). Efficient nodal point numbering is used (Figure 8.5), numbering systematically from left to right, up and down; numbering is not random or haphazard. In this manner, the maximum difference in nodal point numbers for any one element is kept to a minimum (this maximum difference is 3 for Figure 8.5). In this case, only a symmetrical region, or band, around the diagonal of the matrix $[K]$ contains nonzero entries; outside the band, all entries in the matrix $[K]$ are zero. The size of this band on either side of the diagonal is called the bandwidth. The Gaussian elimination equation solution technique is modified for such a matrix to use only that portion of $[K]$ that includes the diagonal and the bandwidth region above the diagonal (see Figure 8.13), which greatly enhances the efficiency of Gaussian elimination for finite element solutions.

This is an overview of the information required to perform finite element analysis on a system of bar elements. A computer algorithm is introduced for finite element analysis of truss systems. This program is called TRUSS (Figure 8.14, page 364). A detailed user's guide is presented at the beginning of this listing. A brief outline of this program is as follows:

1. Access all data for the mesh and for active forces.
2. Form the system stiffness matrix $[K]$:
   (a) Form the element stiffness matrices $[k]$.
   (b) Put element stiffness matrices into the system stiffness matrix $[K]$.
3. Compute displacements $\{d\}$ using Gaussian elimination.
4. Compute strains in all elements using the displacements computed for each element.
5. Compute stresses from strains for all elements.
6. Print the results.

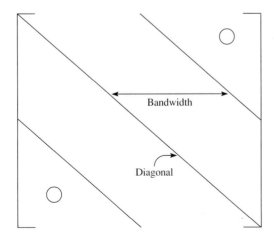

Figure 8.13 The concept of a banded matrix.

Note: When entering data into the program TRUSS, units for Young's modulus E must be consistent with the units for both active forces and nodal coordinates. For example, if forces are given in pounds and nodal coordinates are given in feet, then E must be expressed in units of pounds per square foot (lb/ft^2). If forces are expressed in pounds and nodal coordinates are in inches, then E must be expressed as pounds per square inch ($lb/in.^2$). If forces are expressed in newtons (N) or meganewtons (MN), and nodal coordinates are in meters, then E must be expressed as megapascals (MPa).

8.6.1 Error Trapping

As noted, finite element analysis can be frustrating. The source of this frustration lies in preparation of input data to a finite element analysis computer program. Mistakes are made easily when entering data for nodal coordinates and element connectivity. This is a common reason why incorrect finite element analyses are obtained.

A visually stimulating way to trap (find) errors is to use computer graphics to draw a mesh based on the nodal point coordinates and element connectivity information. If errors exist in this data base, they are spotted easily when drawing the mesh on the screen because the picture of the mesh will not be correct.

Herein is presented a computer graphics program for drawing and checking finite element meshes. This program is called MESHDRAW (Figure 8.15, page 371); a detailed user's guide is presented at the beginning of this listing. This program is able to draw meshes consisting of bar elements and can draw meshes consisting of two-dimensional solid elements, which are used later in this chapter.

Use of this program is straightforward. Once the command MESHDRAW is typed, the program seeks the following information interactively:

1. Whether (a) a truss system of bar elements or (b) a system of 2D solid elements is to be plotted.
2. The name of the file that contains the nodal point coordinate and element connectivity data base. This file is that used for TRUSS or QUAD (presented later in this chapter), depending on option.
3. Whether to plot the original mesh or the deformed mesh subsequent to finite element analysis. If the deformed mesh is to be plotted, then we must know (a) what magnification factor will be used to show the deformations and (b) the name of the output file storing results from the finite element analysis.

If the original, undeformed mesh is plotted, then MESHDRAW draws the mesh using a green color. If the deformed mesh is plotted, MESHDRAW draws each element in a color that represents the amount of stress in the element: Red shows the largest compressive stress, yellow indicates intermediate compressive stress, white indicates the smallest compressive stress, green indicates the largest tensile stress, magenta indicates intermediate tensile stress, and blue indicates the smallest tensile stress. Using these colors for representing stress in elements reveals not only the deformed shape, but also where in the mesh the greatest stress in elements occurs and whether the stress is tensile or compressive.

A few examples are presented to demonstrate the use of TRUSS and MESH-DRAW.

8.6.2 Finite Element Analysis of a Truss System

A small, simple truss system consisting of 8 nodal points and 13 elements (Figure 8.5) is analyzed. The affect of one active force acting in the y direction at nodal point 4 is assessed for this system.

Data are entered for the truss system into a data file called FIG85.DAT (Figure 8.16). Compare this data file to the user's guide for TRUSS (Figure 8.14). Material properties are for steel (Table 8.1). A drawing produced using the program MESHDRAW (Figure 8.17) verifies that the data, at least for nodal point coordinates and element connectivity, are error free.

Because no errors are found from the drawing, finite element analysis is completed by running the program TRUSS. When using this program, an interactive statement appears on the screen seeking information for the input data file name (unit 5). In this example, the name FIG85.DAT is entered. Next, an interactive statement appears asking that the file name for unit 1 be specified. This is an output file to contain results of the finite element analysis. It is a new file to be created by the program, hence it must not already exist. In this example, the name FIG85.OUT is used. Results from the TRUSS analysis are listed in Figure 8.18 for the truss in Figure 8.5.

The drawing in Figure 8.19 is created by MESHDRAW to show the deformed shape of the mesh; a magnification factor of 400 is used. The mesh has displaced downward, in the direction the one active force is acting; this deformed shape is expected. Deformations are symmetrical with respect to the center of the

```
8,13,1,1,3,2
1,0.0,0.0
2,20.0,20.0
3,20.0,0.0
4,50.0,0.0
5,50.0,20.0
6,80.0,20.0
7,80.0,0.0
8,100.0,0.0
1,1,2,1
2,1,3,1
3,2,3,1
4,2,5,1
5,2,4,1
6,3,4,1
7,4,5,1
8,5,6,1
9,4,6,1
10,4,7,1
11,6,7,1
12,6,8,1
13,7,8,1
0.333,4320000000.0
4
0.0,-100000.0
1,1,1
8,0,1
```

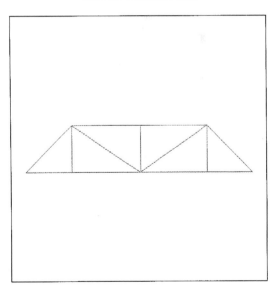

PLOT OF ORIGINAL MESH

Figure 8.16 A listing of the data file, FIG85.DAT (which is used to analyze the situation shown in Figure 8.5).

Figure 8.17 The original, predeformed mesh for Figure 8.5, drawn by the program MESHDRAW.

Table 8.1 Material Properties for the Mesh of Figure 8.5

Young's modulus E:	4,320,000,000 lb/ft^2
Cross-sectional area A:	0.333 ft^2

```
FINITE ELEMENT ANALYSIS:  TRUSS SYSTEMS

PARAMETERS:

NUMBER OF NODES                =     8
NUMBER OF ELEMENTS             =    13
NUMBER OF MATERIALS            =     1
NUMBER OF NODE POINTS W. FORCES =    1


NODAL POINT COORDINATES
     NODE                 X                      Y

       1         .00000000E+00          .00000000E+00
       2         .20000000E+02          .20000000E+02
       3         .20000000E+02          .00000000E+00
       4         .50000000E+02          .00000000E+00
       5         .50000000E+02          .20000000E+02
       6         .80000000E+02          .20000000E+02
       7         .80000000E+02          .00000000E+00
       8         .10000000E+03          .00000000E+00


ELEMENT CONNECTIVITY
   ELEMENT          NODE-A          NODE-B          MATERIAL

       1               1               2               1
       2               1               3               1
       3               2               3               1
       4               2               5               1
       5               2               4               1
       6               3               4               1
       7               4               5               1
       8               5               6               1
       9               4               6               1
      10               4               7               1
      11               6               7               1
      12               6               8               1
      13               7               8               1


ACTIVE FORCE INFORMATION:
     NODE          FORCE-X              FORCE-Y

       1        .00000000E+00         .00000000E+00
       2        .00000000E+00         .00000000E+00
       3        .00000000E+00         .00000000E+00
       4        .00000000E+00        -.10000000E+06
       5        .00000000E+00         .00000000E+00
       6        .00000000E+00         .00000000E+00
       7        .00000000E+00         .00000000E+00
       8        .00000000E+00         .00000000E+00
```

Figure 8.18 A listing of the output from the program TRUSS for the mesh of Figure 8.5, as described by the data file listed in Figure 8.16.

```
BOUNDARY CONDITIONS:
            NODE              X-BOUND              Y-BOUND

             1                   1                    1
             8                   0                    1

                 NODAL DISPLACEMENTS

NODAL POINT          X-DISP              Y-DISP

     1            -.484E-16           -.500E-10
     2             .434E-02           -.631E-02
     3             .695E-03           -.631E-02
     4             .174E-02           -.143E-01
     5             .174E-02           -.143E-01
     6            -.869E-03           -.631E-02
     7             .278E-02           -.631E-02
     8             .348E-02           -.500E-10
                 STRESS IN ELEMENTS

        ELEMENT               STRESS

           1                 .212E+06
           2                -.150E+06
           3                 .000E+00
           4                 .375E+06
           5                -.271E+06
           6                -.151E+06
           7                 .000E+00
           8                 .375E+06
           9                -.271E+06
          10                -.150E+06
          11                 .000E+00
          12                 .212E+06
          13                -.152E+06
Stop - Program terminated.


0;37;44m C:\TEXT\CHAPTER8>
```

Figure 8.18 continued

mesh, and this, too, is expected because the one active force acts at the center of the mesh; all nodal points have moved closer to the one active force, as expected. Results from TRUSS seem to verify intuition and a correct analysis seems to have been obtained.

A black and white reproduction of the color graph created by the program MESHDRAW is shown in Figure 8.19. If a deformed mesh is plotted using MESHDRAW, as in this figure, elements are drawn in color based on the level of stress calculated for each element by the program TRUSS, according to the color scheme is described earlier. Up to now, no mention has been made of how the program TRUSS computes stress and strain.

Stress is obtained from strain through a multiplication by Young's modulus E: stress = E(strain). Therefore, to compute the axial stress in a bar element, the axial strain is first required. Recall that axial strain is the ratio of the change in axial length to the original length. Therefore, the program TRUSS makes the following calculations:

1. Original nodal point coordinates are updated to include the computed displacements dx and dy for the nodal point, such that

$$X' = X\text{(original)} + dx, \qquad Y' = Y\text{(original)} + dy$$

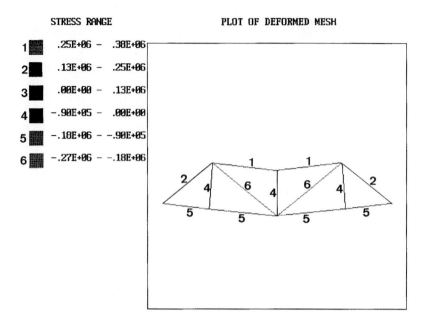

Figure 8.19   The deformed mesh for the situation shown in Figure 8.5, drawn by the program MESHDRAW.

2. The deformed length of the element is calculated as follows (assuming two nodal points of the element are called A and B):

$$L' = \sqrt{(X_B' - X_A')^2 + (Y_B' - Y_A')^2}$$

3. The change in length is calculated as $L - L'$; therefore, strain is computed as

$$\varepsilon_{\text{axial}} = \frac{L - L'}{L}$$

Relative to the numerator of this expression, if L' is greater than L, then an elongation has occurred, indicating the presence of a tensile strain in the element. In this case, the strain has a negative value. If L' is less than L, compression is indicated and strain is positive. Consequently, stress, which is computed from strain, is negative if tensile and positive if compressive. When using the program MESHDRAW to draw the deformed mesh, the following convention is used for displaying stress: Negative values indicate tension, and positive values indicate compression.

8.6.3   A Simple Test Problem

Suppose the mesh in Figure 8.8 is analyzed to determine the influence of an axial, tensile force. If the magnitude of this force is equal to 1.0 and if cross-sectional area A and Young's modulus E are both equal to 1.0, then nodal point 3 is expected to move in the axial direction by an amount equal to $F/K = F/AE/L = FL/AE = L$. The total length of this mesh, from nodal point 1 to nodal point 3, is 2.0, so this is the total axial displacement to be expected. Consequently, the x-direction displacement at nodal point 3 is expected to be

```
3,2,1,1,1,3
1,0.0,0.0
2,1.0,0.0
3,2.0,0.0
1,1,2,1
2,2,3,1
1.0,1.0
3
1.0,0.0
1,1,1
2,0,1
3,0,1

0;37;44m C:\TEXT=
```

Figure 8.20   A listing of the data file used to analyze the situation shown in Figure 8.8.

2.0. Likewise, nodal point 2 is expected to displace by an amount equal to its distance from the origin, 1.0.

A listing of the TRUSS data file for this analysis is shown in Figure 8.20. For such a simple mesh, MESHDRAW is not really needed to verify the data base; this is easy to do by inspection. Results from TRUSS (Figure 8.21) indicate x-direction displacements for nodal points 2 and 3 are, respectively, 1.0 and 2.0, as expected. Although this is a simple analysis, an excellent test problem is obtained. The mesh is small and manageable. Displacements are computed easily by hand using Hooke's law. Moreover, if the program computes the same displacements as computed by hand, correct program performance is verified. The program TRUSS notwithstanding, this simple test problem is a standard one for any finite element computer program capable of analyzing truss systems.

8.7 Two-Dimensional, Solid, Isoparametric, Four-Node Elements: An Advanced Application of Gaussian Quadrature for Deriving the Element Stiffness Matrix

In previous sections, our discussion focused on the one-dimensional bar element. A formula was derived for the element stiffness matrix $[k]$. Information from all element stiffness matrices comprises the system stiffness matrix $[K]$ for a given mesh. Once this matrix is formed and the force vector $\{F\}$ is defined, the system $\{F\} = [K]\{d\}$ is solved for $\{d\}$ using Gaussian elimination. Recall that the fundamental equation of finite element analysis is $\{F\} = [K]\{d\}$. In this section, a four-node solid element is introduced. The only difference between this section and the previous ones is found in the formula for the element stiffness matrix for the four-node solid element. The manner in which the element stiffness is placed into the system stiffness matrix, boundary conditions are specified, and active forces are specified is identical to what is used for the bar element. Once the system stiffness matrix $[K]$ is determined for a mesh of solid four-node elements, the equation system $\{F\} = [K]\{d\}$ is solved using Gaussian elimination to determine displacements at nodal point degrees of freedom, just as was done in earlier sections.

```
FINITE ELEMENT ANALYSIS:   TRUSS SYSTEMS

PARAMETERS:

NUMBER OF NODES                =    3
NUMBER OF ELEMENTS             =    2
NUMBER OF MATERIALS            =    1
NUMBER OF NODE POINTS W. FORCES =   1



NODAL POINT COORDINATES
      NODE                 X                        Y

        1        .00000000E+00           .00000000E+00
        2        .10000000E+01           .00000000E+00
        3        .20000000E+01           .00000000E+00



ELEMENT CONNECTIVITY
   ELEMENT            NODE-A              NODE-B            MATERIAL

        1               1                   2                 1
        2               2                   3                 1



ACTIVE FORCE INFORMATION:
      NODE          FORCE-X              FORCE-Y

        1        .00000000E+00           .00000000E+00
        2        .00000000E+00           .00000000E+00
        3        .10000000E+01           .00000000E+00



BOUNDARY CONDITIONS:
      NODE           X-BOUND              Y-BOUND

        1               1                    1
        2               0                    1
        3               0                    1


               NODAL DISPLACEMENTS

NODAL POINT        X-DISP           Y-DISP

        1        .100E-14         .000E+00
        2        .100E+01         .000E+00
        3        .200E+01         .000E+00
               STRESS IN ELEMENTS

      ELEMENT              STRESS

        1                 -.100E+01
        2                 -.100E+01
Stop - Program terminated.


0;37;44m C:\TEXT=
```

Figure 8.21 A listing of the output from the program TRUSS for the situation shown in Figure 8.8.

8.7.1 Introduction to Isoparametric, Solid, Four-Node Elements

Isoparametric means that the same interpolation functions used to define the shape of an element are used also to define the assumed displacement field within the element. A Rayleigh–Ritz approach was used in Section 8.5 to derive the stiffness matrix for a bar element. In that derivation, the assumed displacement field $u = a_1 + a_2x$ was used, but the shape of the bar element was not considered in this derivation. Now, shape is considered when deriving the element stiffness matrix for the four-node, solid, isoparametric element.

Derivation of this element stiffness matrix begins by studying a two-coordinate system (Figure 8.22): a global x-y coordinate system, and a local ξ-η coordinate system. In the discussion to follow, the local coordinate system is the focus for deriving the element stiffness matrix. A Jacobian magnification matrix is used to transform the element stiffness matrix from the local to the global coordinate system.

8.7.2 Rayleigh–Ritz Derivation of Element Stiffness

Before a rigorous derivation of this element stiffness matrix is presented, some qualitative aspects of the four-node, solid, two-dimensional element are discussed. Because this element is two dimensional, each of its four nodal points is associated with two degrees of freedom, giving the element eight total degrees of freedom. The element stiffness matrix for such an element must be 8×8 in size.

A local coordinate system (Figure 8.22) having an origin at the center of the element is used to derive element stiffness. For the four nodal points 1 through 4, the local coordinates are node 1, $(-1, -1)$; node 2, $(1, -1)$; node 3, $(1, 1)$; and node 4, $(-1, 1)$.

Derivation of element stiffness for the four-node element follows in form and substance that which is presented for the bar element in Section 8.5.

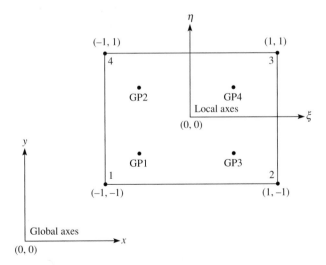

Figure 8.22 A four-node, quadrilateral, isoparametric, variable strain element. Global and local coordinate systems are shown. The locations of Gaussian points are also shown. Coordinates shown for the four nodal points are relative to the local coordinate system.

Expressing the relationship for the assumed displacement field for the four-node element for both horizontal displacement u and vertical displacement v gives

$$u = a_1 + a_2\xi + a_3\eta + a_4\xi\eta = [1\ \xi\ \eta\ \xi\eta][a_{1-4}]$$
$$v = a_5 + a_6\xi + a_7\eta + a_8\xi\eta = [1\ \xi\ \eta\ \xi\eta][a_{5-8}]$$

These relationships are written as functions of the local coordinates. Each of these two polynomials has four terms, one for each horizontal (u) and vertical (v) degree of freedom at each nodal point. For u and v combined, eight total terms (one for each of the eight total degrees of freedom) are present.

Because the four-node element is isoparametric, functions for shape and displacement are coincidental. Hence, for shape,

$$x = a_1 + a_2\xi + a_3\eta + a_4\xi\eta = [1\ \xi\ \eta\ \xi\eta][a_{1-4}]$$

$$y = a_5 + a_6\xi + a_7\eta + a_8\xi\eta = [1\ \xi\ \eta\ \xi\eta][a_{5-8}]$$

for which the coefficients a are the same for both $\{u\ v\}$ and $\{x\ y\}$. Again, the term isoparametric refers to the characterization of element shape and displacement field by the same polynomial(s).

Polynomials for displacement/shape may be written at each of the four nodal points, such that

$$
\begin{vmatrix} u_1 \\ u_2 \\ u_3 \\ u_4 \end{vmatrix} =
\begin{vmatrix} 1 & \xi_1 & \eta_1 & \xi_1\eta_1 \\ 1 & \xi_2 & \eta_2 & \xi_2\eta_2 \\ 1 & \xi_3 & \eta_3 & \xi_3\eta_3 \\ 1 & \xi_4 & \eta_4 & \xi_4\eta_4 \end{vmatrix}
\begin{vmatrix} a_1 \\ a_2 \\ a_3 \\ a_4 \end{vmatrix}
$$

$$
\begin{vmatrix} v_1 \\ v_2 \\ v_3 \\ v_4 \end{vmatrix} =
\begin{vmatrix} 1 & \xi_1 & \eta_1 & \xi_1\eta_1 \\ 1 & \xi_2 & \eta_2 & \xi_2\eta_2 \\ 1 & \xi_3 & \eta_3 & \xi_3\eta_3 \\ 1 & \xi_4 & \eta_4 & \xi_4\eta_4 \end{vmatrix}
\begin{vmatrix} a_5 \\ a_6 \\ a_7 \\ a_8 \end{vmatrix}
$$

which are simply written as $\{u\} = [DA]\{a_{1-4}\}$ and $\{v\} = [DA]\{a_{5-8}\}$.

Because $\{a_{1-4}\} = [DA]^{-1}\{u\}$ and $\{a_{5-8}\} = [DA]^{-1}\{v\}$,

$$u = [1\ \xi\ \eta\ \xi\eta][DA]^{-1}[u]$$
$$v = [1\ \xi\ \eta\ \xi\eta][DA]^{-1}[v]$$

and

$$x = [1\ \xi\ \eta\ \xi\eta][DA]^{-1}[x]$$
$$y = [1\ \xi\ \eta\ \xi\eta][DA]^{-1}[y]$$

where $[x]$ is the vector of global x coordinates for the four nodal points of the element $[x] = [x_1\ x_2\ x_3\ x_4]$, and $[y]$ is the vector of global y coordinates for the four nodal points of the element $[y] = [y_1\ y_2\ y_3\ y_4]$.

In these expressions, the product

$$[1\ \xi\ \eta\ \xi\eta][DA]^{-1}$$

composes the matrix $[N]$, a 1×4 matrix. To derive this matrix, information is required for $[DA]^{-1}$. Using the *local* coordinates for the four nodal points, we find

$$[DA] = \begin{vmatrix} 1 & \xi_1 & \eta_1 & \xi_1\eta_1 \\ 1 & \xi_2 & \eta_2 & \xi_2\eta_2 \\ 1 & \xi_3 & \eta_3 & \xi_3\eta_3 \\ 1 & \xi_4 & \eta_4 & \xi_4\eta_4 \end{vmatrix} = \begin{vmatrix} 1 & -1 & -1 & 1 \\ 1 & 1 & -1 & -1 \\ 1 & 1 & 1 & 1 \\ 1 & -1 & 1 & -1 \end{vmatrix}$$

From this knowledge,

$$[DA]^{-1} = \frac{1}{4}\begin{vmatrix} 1 & 1 & 1 & 1 \\ -1 & 1 & 1 & -1 \\ -1 & -1 & 1 & 1 \\ 1 & -1 & 1 & -1 \end{vmatrix}$$

Acknowledging that $[N] = [N_1\, N_2\, N_3\, N_4]$, knowledge of $[DA]^{-1}$ leads to

$$N_1 = \frac{1 - \xi - \eta + \xi\eta}{4}, \qquad N_2 = \frac{1 + \xi - \eta - \xi\eta}{4},$$
$$N_3 = \frac{1 + \xi + \eta + \xi\eta}{4}, \qquad N_4 = \frac{1 - \xi + \eta - \xi\eta}{4}$$

Combining the horizontal and vertical displacements into a single vector and using the entries for the matrix $[N]$ gives

$$\begin{vmatrix} u \\ v \end{vmatrix} = \begin{vmatrix} N_1 & 0 & N_2 & 0 & N_3 & 0 & N_4 & 0 \\ 0 & N_1 & 0 & N_2 & 0 & N_3 & 0 & N_4 \end{vmatrix} \begin{vmatrix} u_1 \\ v_1 \\ u_2 \\ v_2 \\ u_3 \\ v_3 \\ u_4 \\ v_4 \end{vmatrix} = [N'][d]$$

Unlike the Rayleigh–Ritz derivation of the element stiffness matrix for the bar element, which did not make use of a local coordinate system, a relationship between the local and global coordinate systems must now be introduced. The notion of the Jacobian matrix $[J]$ is used to relate derivatives in the local domain to those in the global domain, such that

$$\begin{vmatrix} \dfrac{\partial}{\partial\xi} \\ \dfrac{\partial}{\partial\eta} \end{vmatrix} = \begin{vmatrix} \dfrac{\partial x}{\partial\xi} & \dfrac{\partial y}{\partial\xi} \\ \dfrac{\partial x}{\partial\eta} & \dfrac{\partial y}{\partial\eta} \end{vmatrix} \begin{vmatrix} \dfrac{\partial}{\partial x} \\ \dfrac{\partial}{\partial y} \end{vmatrix} = [J]\begin{vmatrix} \dfrac{\partial}{\partial x} \\ \dfrac{\partial}{\partial y} \end{vmatrix}$$

for which

$$[J] = \begin{vmatrix} N_{1,\xi} & N_{2,\xi} & N_{3,\xi} & N_{4,\xi} \\ N_{1,\eta} & N_{2,\eta} & N_{3,\eta} & N_{4,\eta} \end{vmatrix} \begin{vmatrix} x_1 & y_1 \\ x_2 & y_2 \\ x_3 & y_3 \\ x_4 & y_4 \end{vmatrix}$$

which borrows notation from Cook (1974, p. 102) such that

$$N_{i,\xi} = \frac{\partial N_i}{\partial \xi}, \qquad N_{i,\eta} = \frac{\partial N_i}{\partial \eta}$$

Before proceeding further, a brief respite allows reflection on the objective of this section: the derivation of the element stiffness matrix for the four-node, solid element. The following integral was introduced (Section 8.5):

$$[k] = \int_{VOL} [B]^T [E][B] \, dV$$

In the present case, matrix $[k]$ for the four-node, solid element is an 8×8 matrix. Matrix $[E]$ for the four-node, solid element is a 3×3 matrix and is a function of Young's modulus and Poisson's ratio. This matrix is discussed in detail momentarily. Because the matrix $[E]$ has a dimension of 3×3, the matrix $[B]$, which is the strain–displacement matrix, must be a 3×8 matrix to yield a matrix $[k]$ having a size 8×8.

Because matrix $[B]$ relates strain in the four-node element to displacements at nodal points, derivation of this matrix begins with a discussion of strain. Three components of strain are associated with a solid, two-dimensional element: strain in the x direction, strain in the y direction, and shear strain. This is written as

$$[\varepsilon] = \begin{vmatrix} \varepsilon_x \\ \varepsilon_y \\ \gamma_{xy} \end{vmatrix} = \begin{vmatrix} \dfrac{\partial u}{\partial x} \\[2mm] \dfrac{\partial v}{\partial y} \\[2mm] \dfrac{\partial u}{\partial y} + \dfrac{\partial v}{\partial x} \end{vmatrix}$$

$$= \begin{vmatrix} 1 & 0 & 0 & 0 \\ 0 & 0 & 0 & 1 \\ 0 & 1 & 1 & 0 \end{vmatrix} \begin{vmatrix} \dfrac{\partial u}{\partial x} \\[2mm] \dfrac{\partial u}{\partial y} \\[2mm] \dfrac{\partial v}{\partial x} \\[2mm] \dfrac{\partial v}{\partial y} \end{vmatrix}$$

Recalling the definition of the Jacobian matrix $[J]$ presented earlier, global strain is related to local strain as

$$[\partial \text{ local}] = [J][\partial \text{ global}], \therefore [\partial \text{ global}] = [J]^{-1}[\partial \text{ local}]$$

Therefore,

$$
\begin{vmatrix} \dfrac{\partial u}{\partial x} \\[6pt] \dfrac{\partial u}{\partial y} \\[6pt] \dfrac{\partial v}{\partial x} \\[6pt] \dfrac{\partial v}{\partial y} \end{vmatrix} = \begin{vmatrix} [J]^{-1} & 0 \\ 0 & [J]^{-1} \end{vmatrix} \begin{vmatrix} \dfrac{\partial u}{\partial \xi} \\[6pt] \dfrac{\partial u}{\partial \eta} \\[6pt] \dfrac{\partial v}{\partial \xi} \\[6pt] \dfrac{\partial v}{\partial \eta} \end{vmatrix}
$$

Global strain may also be related to local displacements, such that

$$
\begin{vmatrix} \varepsilon_x \\[6pt] \varepsilon y \\[6pt] \gamma_{xy} \end{vmatrix} = \begin{vmatrix} 1 & 0 & 0 & 0 \\ 0 & 0 & 0 & 1 \\ 0 & 1 & 1 & 0 \end{vmatrix} \begin{vmatrix} [J]^{-1} & 0 \\ 0 & [J]^{-1} \end{vmatrix} \begin{vmatrix} \dfrac{\partial u}{\partial \xi} \\[6pt] \dfrac{\partial u}{\partial \eta} \\[6pt] \dfrac{\partial v}{\partial \xi} \\[6pt] \dfrac{\partial v}{\partial \eta} \end{vmatrix}
$$

All that is needed now is to describe the relationship between the local and the global displacements. Recall that

$$
\begin{vmatrix} u \\ v \end{vmatrix} = \begin{vmatrix} N_1 & 0 & N_2 & 0 & N_3 & 0 & N_4 & 0 \\ 0 & N_1 & 0 & N_2 & 0 & N_3 & 0 & N_4 \end{vmatrix} \begin{vmatrix} u_1 \\ v_1 \\ u_2 \\ v_2 \\ u_3 \\ v_3 \\ u_4 \\ v_4 \end{vmatrix}
$$

It follows that

$$
\begin{vmatrix} \dfrac{\partial u}{\partial \xi} \\[6pt] \dfrac{\partial u}{\partial \eta} \\[6pt] \dfrac{\partial v}{\partial \xi} \\[6pt] \dfrac{\partial v}{\partial \eta} \end{vmatrix} = \begin{vmatrix} N_{1,\xi} & 0 & N_{2,\xi} & 0 & N_{3,\xi} & 0 & N_{4,\xi} & 0 \\ N_{1,\eta} & 0 & N_{2,\eta} & 0 & N_{3,\eta} & 0 & N_{4,\eta} & 0 \\ 0 & N_{1,\xi} & 0 & N_{2,\xi} & 0 & N_{3,\xi} & 0 & N_{4,\xi} \\ 0 & N_{1,\eta} & 0 & N_{2,\eta} & 0 & N_{3,\eta} & 0 & N_{4,\eta} \end{vmatrix} \begin{vmatrix} u_1 \\ v_1 \\ u_2 \\ v_2 \\ u_3 \\ v_3 \\ u_4 \\ v_4 \end{vmatrix}
$$

in which the notation for partial derivatives (Cook, 1974, p. 102) is used. This system is written in shorthand to facilitate subsequent discussion, such that

$$
\begin{vmatrix} \dfrac{\partial u}{\partial \xi} \\[2mm] \dfrac{\partial u}{\partial \eta} \\[2mm] \dfrac{\partial v}{\partial \xi} \\[2mm] \dfrac{\partial v}{\partial \eta} \end{vmatrix} = \begin{vmatrix} N_{i,\xi} & N_{i,\eta} \end{vmatrix} \begin{vmatrix} u_1 \\ v_1 \\ u_2 \\ v_2 \\ u_3 \\ v_3 \\ u_4 \\ v_4 \end{vmatrix}
$$

Substituting this solution into that for global strain yields

$$
\begin{vmatrix} \varepsilon_x \\ \varepsilon y \\ \gamma_{xy} \end{vmatrix} = \begin{vmatrix} 1 & 0 & 0 & 0 \\ 0 & 0 & 0 & 1 \\ 0 & 1 & 1 & 0 \end{vmatrix} \begin{vmatrix} [J]^{-1} & 0 \\ 0 & [J]^{-1} \end{vmatrix} \begin{vmatrix} N_{i,\xi} & N_{i,\eta} \end{vmatrix} \begin{vmatrix} u_1 \\ v_1 \\ u_2 \\ v_2 \\ u_3 \\ v_3 \\ u_4 \\ v_4 \end{vmatrix}
$$

Further,

$$
[\varepsilon] = [B][d] = [B] \begin{vmatrix} u_1 \\ v_1 \\ u_2 \\ v_2 \\ u_3 \\ v_3 \\ u_4 \\ v_4 \end{vmatrix}
$$

wherein

$$
[B] = \begin{vmatrix} 1 & 0 & 0 & 0 \\ 0 & 0 & 0 & 1 \\ 0 & 1 & 1 & 0 \end{vmatrix} \begin{vmatrix} [J]^{-1} & 0 \\ 0 & [J]^{-1} \end{vmatrix} \begin{vmatrix} N_{i,\xi} & N_{i,\eta} \end{vmatrix}
$$

which is the strain–displacement matrix.

If partial derivatives are taken of N_i with respect to the two local axes, eight equations are obtained:

$$N_{1,\xi} = \frac{\eta - 1}{4}, \qquad N_{1,\eta} = \frac{\xi - 1}{4}$$

$$N_{2,\xi} = \frac{1 - \eta}{4}, \qquad N_{2,\eta} = \frac{-1 - \xi}{4}$$

$$N_{3,\xi} = \frac{1 + \eta}{4}, \qquad N_{3,\eta} = \frac{1 + \xi}{4}$$

$$N_{4,\xi} = \frac{-1 - \eta}{4}, \qquad N_{4,\eta} = \frac{1 - \xi}{4}$$

With the strain–displacement matrix $[B]$ defined, the integral for the element stiffness matrix is recalled:

$$[k] = \int_{\text{VOL}} [B]^{\text{T}}[E][B] \, dV$$

which may be written as a double integral:

$$[k] = \int_X \int_Y [B]^{\text{T}}[E][B] \, t \, dY \, dX$$

where t is the thickness of the two-dimensional, solid, four-node element. If t is assumed equal to 1, then its presence in the calculation can be implicit rather than explicit. This is done to simplify the remainder of this discussion.

Expressing the integral for the element stiffness matrix in terms of the local coordinate system yields

$$[k] = \int_{-1}^{1} \int_{-1}^{1} [B]^{\text{T}}[E][B] \, (\det[J]) \, d\eta \, d\xi$$

where $\det[J]$ is the determinant of the Jacobian matrix. This determinant is a scaling factor yielding the global area of the four-node element from its local area.

Again, a situation is encountered wherein an integral equation must be solved, yet computer languages often cannot be used for evaluating the integral. Instead, the integral must be approximated. In the case of the stiffness formula for the four-node element, Gaussian quadrature is used.

If N is the number of Gauss points used for each coordinate direction, the 8×8 element stiffness matrix $[k]$ for the four-node quadrilateral element is found as a double summation:

$$[k] = \sum_{i=1}^{N} \sum_{j=1}^{N} w_i w_j [B_{ij}]^{\text{T}}[E][B_{ij}] \det[J_{ij}]$$

The limits of definite integration for both coordinate directions are $[-1, 1]$, and if N is equal to 2, which is a sufficient condition given that the polynomials used to describe element shape and displacement field are of order less than 3, then the Gaussian quadrature weights w_i and w_j are equal to 1.0. Moreover,

$$\xi = \pm\frac{1}{\sqrt{3}}, \qquad \eta = \pm\frac{1}{\sqrt{3}}$$

These locations are called Gauss points (Figure 8.22). The equation for element stiffness simplifies to

$$[k] = \sum_{i=1}^{2} \sum_{j=1}^{2} [B_{ij}]^{T}[E][B_{ij}](\det[J_{ij}])$$

Before proceeding to an example showing the explicit calculation of the matrices $[B]$ and $[J]$, a detailed discussion of the material constitutive matrix $[E]$ is presented.

8.7.3 The Material Constitutive Matrix $[E]$

We have been using matrix $[E]$ in equations for deriving the element stiffness matrix $[k]$ for a solid, two-dimensional element. Its definition has been deferred until now.

Matrix $[E]$ is known as the material constitutive matrix. This matrix describes the relationship between strain and stress for a given material. Several different formulas are available for this matrix, depending on stress assumptions. If we wish to consider plane stress conditions for a solid, two-dimensional element, strain and stress are analyzed within the solid element by assuming that stress normal (orthogonal) to the element is equal to zero.

Note: The fact that stress normal to the element is assumed to be zero does not mean that strain normal to the element is zero! In fact, strain normal to the element definitely is not zero and the thickness t of the element changes. This appears to contradict the information given in Section 8.2. If, for plane stress assumptions, the strain normal to an element is not equal to zero, how can stress normal to the element be considered to be zero? This is explained easily by realizing that Section 8.2 pertains to one-dimensional strain–stress relationships. Herein, discussion concerns two-dimensional strain–stress relationships. Moreover, recall the notion of Poisson's ratio: For a particular axial stress, strain normal to the axis will be nonzero, even though stress normal to the axis is zero. In the case of the solid two-dimensional element, stress within the plane of the element will cause a change in element thickness because of the Poisson relationship for the material comprising the element.

If plane-stress assumptions are used for a finite element analysis using solid, two-dimensional elements, the material constitutive matrix $[E]$ is

$$[E] = \frac{E}{1 - \nu^2}\begin{vmatrix} 1 & \nu & 0 \\ \nu & 1 & 0 \\ 0 & 0 & \dfrac{1-\nu}{2} \end{vmatrix}$$

for which ν is Poisson's ratio and E is Young's modulus. For a two-dimensional strain–stress analysis,

$$[\sigma] = [E][\varepsilon]$$

for which

$$[\sigma] = \begin{vmatrix} \sigma_x \\ \sigma_y \\ \tau_{xy} \end{vmatrix}, \qquad [\varepsilon] = \begin{vmatrix} \varepsilon_x \\ \varepsilon_y \\ \gamma_{xy} \end{vmatrix}$$

where the three components of stress are stress in the x direction, stress in the y direction, and shear stress. For a two-dimensional plane-stress analysis,

$$\sigma_x = \frac{E(\varepsilon_x + \nu\varepsilon_y)}{1 - \nu^2}, \qquad \sigma_y = \frac{E(\nu\varepsilon_x + \varepsilon_y)}{1 - \nu^2}, \qquad \tau_{xy} = \frac{\gamma_{xy}E(1 - \nu)}{2(1 - \nu^2)} = \frac{\gamma_{xy}E}{2(1 + \nu)}$$

These three relationships are known as Hooke's law for two dimensions (and plane-stress assumptions).

In this text, earth sciences applications of finite element analysis are emphasized. For these analyses, meshes are designed to model discrete slices in the earth's crust. The assumption in these analyses is that the earth's crust is infinitely the same as the slice in the direction orthogonal to the slice, where the word infinitely is used subjectively.

The assumption of plane stress is not appropriate for such earth sciences analyses. Elements cannot be allowed to change in thickness for such a discrete slice of earth. This slice is modeled in a finite element analysis as if earth is not disturbed; hence thickness of the slice must not change. This suggests another type of assumption for the material constitutive matrix: plane strain.

In this assumption, strain normal to the element is assumed to be zero. This means that stress normal to the element is likely to be nonzero. For earth sciences analyses, this makes sense because stress in earth is a three-dimensional phenomenon. To effect a plane-strain finite element analysis, the following material constitutive matrix is used to calculate the element stiffness matrix:

$$[E] = \frac{E}{(1 + \nu)(1 - 2\nu)} \begin{vmatrix} 1 - \nu & \nu & 0 \\ \nu & 1 - \nu & 0 \\ 0 & 0 & \dfrac{1 - 2\nu}{2} \end{vmatrix}$$

This plane-strain form of the material constitutive matrix is used for two-dimensional finite element analyses in this chapter.

For a plane-strain analysis, the following relationships exist:

$$\sigma_x = \frac{E(\varepsilon_x(1 - \nu) + \nu\varepsilon_y)}{(1 + \nu)(1 - 2\nu)}, \qquad \sigma_y = \frac{E(\nu\varepsilon_x + \varepsilon_y(1 - \nu))}{(1 + \nu)(1 - 2\nu)}, \qquad \tau_{xy} = \frac{\gamma_{xy}E}{2(1 + \nu)}$$

which collectively comprise Hooke's law for two dimensions.

8.7.4 An Example of Element Stiffness Matrix Calculation

An example calculation is presented to clarify the foregoing discussion. The element stiffness matrix $[k]$ is to be calculated for an element (Figure 8.23), assuming that thickness t of this element is equal to 1 (unit). Use two Gauss points per coordinate direction for the calculations and assume a plane-strain analysis. Assume also that Young's modulus E is equal to 10.0 (units) and Poisson's ratio is equal to 0.3 (unitless).

Because two Gauss points are used per coordinate direction, four total Gauss points (Figure 8.23), defined using their local coordinates, are used to derive $[k]$. At each of these four Gauss points, the equation

$$[k_{ij}] = [B_{ij}]^{T}[E][B_{ij}] \det[J_{ij}]$$

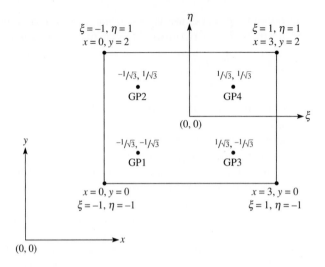

Figure 8.23   A four-node quadrilateral element used for the calculations presented in Section 8.7.3.

must be solved. Once all four of these equations are solved, the element stiffness matrix is obtained as

$$[k] = \sum_{i=1}^{2} \sum_{j=1}^{2} [k_{ij}]$$

In proceeding through this solution, a list is first compiled of the required steps leading to the complete calculation of the element stiffness matrix:

1. Calculate the matrix $[E]$. This can be done before or after the next two steps.
2. At each of the four Gauss points, compute the Jacobian matrix $[J_{ij}]$ and the inverse of this matrix $[J_{ij}]^{-1}$, which entails the calculation of $\det[J_{ij}]$.
3. Calculate the strain–displacement matrix $[B_{ij}]$.

Calculation now proceeds through the Gauss points in this order: GP1, GP2, GP3, and GP4. Beginning at Gauss point 1 (GP1), the matrix $[J_{11}]$ and its inverse are computed. This first Gauss point is defined by the local coordinates:

$$\xi = -\frac{1}{\sqrt{3}}, \qquad \eta = -\frac{1}{\sqrt{3}}, \qquad -\frac{1}{\sqrt{3}} = -0.577$$

Using these local coordinates and the global coordinates of the four nodal points (Figure 8.23), matrix $[J_{11}]$ is computed as

$$[J_{11}] = \begin{vmatrix} N_{1,\xi} & N_{2,\xi} & N_{3,\xi} & N_{4,\xi} \\ N_{1,\eta} & N_{2,\eta} & N_{3,\eta} & N_{4,\eta} \end{vmatrix} \begin{vmatrix} x_1 & y_1 \\ x_2 & y_2 \\ x_3 & y_3 \\ x_4 & y_4 \end{vmatrix}$$

In the case of this example and this Gauss point, the following derivatives are explicitly calculated:

$$N_{1,\xi} = \frac{\eta - 1}{4} = \frac{-0.577 - 1}{4} = -0.394, \qquad N_{1,\eta} = \frac{-1.577}{4} = -0.394$$

$$N_{2,\xi} = \frac{1 - \eta}{4} = 0.394, \qquad N_{2,\eta} = \frac{-1 - \xi}{4} = -0.106$$

$$N_{3,\xi} = \frac{1 + \eta}{4} = 0.106, \qquad N_{3,\eta} = \frac{1 + \xi}{4} = 0.106$$

$$N_{4,\xi} = \frac{-1 - \eta}{4} = -0.106, \qquad N_{4,\eta} = \frac{1 - \xi}{4} = 0.394$$

The Jacobian matrix $[J_{11}]$ is calculated using these derivatives, such that

$$[J_{11}] = \begin{vmatrix} -0.394 & 0.394 & 0.106 & -0.106 \\ -0.394 & -0.106 & 0.106 & 0.394 \end{vmatrix} \begin{vmatrix} 0 & 0 \\ 3 & 0 \\ 3 & 2 \\ 0 & 2 \end{vmatrix} = \begin{vmatrix} 1.5 & 0.0 \\ 0.0 & 1.0 \end{vmatrix}$$

The determinant of this matrix, $\det[J]$, is equal to $1.5(1.0) = 1.5$. The global area of the lower left rectangular quadrant of the element (Figure 8.23) is 1.5; the local area of this quadrant is equal to 1.0. This underscores the fact that the number $\det[J]$ yields the global area from the local area.

The inverse of this Jacobian matrix is found to be

$$[J_{11}]^{-1} = \frac{1}{1.5} \begin{vmatrix} 1.0 & 0.0 \\ 0.0 & 1.5 \end{vmatrix} = \begin{vmatrix} 0.67 & 0.00 \\ 0.00 & 1.00 \end{vmatrix}$$

The strain–displacement matrix for the region of the element surrounding this first Gauss point is calculated in two parts. First, the product

$$\begin{vmatrix} [J_{11}]^{-1} & 0 \\ 0 & [J_{11}]^{-1} \end{vmatrix} \begin{vmatrix} N_{i,\xi} & N_{i,\eta} \end{vmatrix}$$

is formed, which for this example and the first Gauss point is expanded as

$$\begin{vmatrix} 0.67 & 0 & 0 & 0 \\ 0 & 1.00 & 0 & 0 \\ 0 & 0 & 0.67 & 0 \\ 0 & 0 & 0 & 1.00 \end{vmatrix}$$

$$\times \begin{vmatrix} -0.394 & 0 & 0.394 & 0 & 0.106 & 0 & -0.106 & 0 \\ -0.394 & 0 & -0.106 & 0 & 0.106 & 0 & 0.394 & 0 \\ 0 & -0.394 & 0 & 0.394 & 0 & 0.106 & 0 & -0.106 \\ 0 & -0.394 & 0 & -0.106 & 0 & 0.106 & 0 & 0.394 \end{vmatrix}$$

the product of which is

$$
\begin{vmatrix}
-0.262 & 0 & 0.262 & 0 & 0.070 & 0 & -0.070 & 0 \\
-0.394 & 0 & -0.106 & 0 & 0.106 & 0 & 0.394 & 0 \\
0 & -0.262 & 0 & 0.262 & 0 & 0.070 & 0 & -0.070 \\
0 & -0.394 & 0 & -0.106 & 0 & 0.106 & 0 & 0.394
\end{vmatrix}
$$

The calculation of matrix $[B_{11}]$ is completed by multiplying this matrix by

$$
\begin{vmatrix}
1 & 0 & 0 & 0 \\
0 & 0 & 0 & 1 \\
0 & 1 & 1 & 0
\end{vmatrix}
$$

to yield

$$
[B_{11}] =
\begin{vmatrix}
-0.262 & 0 & 0.262 & 0 & 0.070 & 0 & -0.070 & 0 \\
0 & -0.394 & 0 & -0.106 & 0 & 0.106 & 0 & 0.394 \\
-0.394 & -0.262 & -0.106 & 0.262 & 0.106 & 0.070 & 0.394 & -0.070
\end{vmatrix}
$$

With the strain–displacement matrix computed, solution proceeds to determining the material constitutive matrix $[E]$. For plane-strain assumptions, and for the material properties as specified for this problem, this matrix is

$$
[E] = \frac{10}{0.52}
\begin{vmatrix}
0.7 & 0.3 & 0.0 \\
0.3 & 0.7 & 0.0 \\
0.0 & 0.0 & 0.2
\end{vmatrix}
=
\begin{vmatrix}
13.46 & 5.77 & 0.00 \\
5.77 & 13.46 & 0.00 \\
0.00 & 0.00 & 3.85
\end{vmatrix}
$$

Finally, the matrix $[k_{11}]$ is calculated as

$$
[B_{11}]^{\mathrm{T}}[E][B_{11}] \det[J_{11}]
$$

which is an 8×8 matrix equal to

$$
1.5
\begin{vmatrix}
1.528 & 0.997 & -0.769 & -0.239 & -0.410 & -0.267 & -0.349 & -0.491 \\
0.997 & 2.358 & -0.491 & 0.295 & -0.267 & -0.633 & -0.239 & -2.021 \\
-0.769 & -0.491 & 0.973 & -0.267 & 0.206 & 0.132 & -0.410 & 0.627 \\
-0.239 & 0.295 & -0.267 & 0.416 & 0.064 & -0.079 & 0.442 & -0.633 \\
-0.410 & -0.267 & 0.206 & 0.064 & 0.110 & 0.072 & 0.094 & 0.132 \\
-0.267 & -0.633 & 0.132 & -0.079 & 0.072 & 0.170 & 0.064 & 0.542 \\
-0.349 & -0.239 & -0.410 & 0.442 & 0.094 & 0.064 & 0.665 & -0.267 \\
-0.491 & -2.021 & 0.627 & -0.633 & 0.132 & 0.542 & -0.267 & 2.111
\end{vmatrix}
$$

Following the same steps used to compute $[k_{11}]$, the remaining three matrices, $[k_{12}]$, $[k_{21}]$, and $[k_{22}]$, are computed. The result for $[k_{12}]$ is

$$
1.5
\begin{vmatrix}
0.665 & 0.267 & 0.094 & -0.064 & -0.410 & -0.442 & -0.349 & 0.239 \\
0.267 & 2.111 & -0.132 & 0.542 & -0.627 & -0.633 & 0.491 & -2.021 \\
0.094 & -0.132 & 0.110 & -0.072 & 0.206 & -0.064 & -0.410 & 0.267 \\
-0.064 & 0.542 & -0.072 & 0.170 & -0.132 & -0.079 & 0.267 & -0.633 \\
-0.410 & -0.627 & 0.206 & -0.132 & 0.973 & 0.267 & -0.769 & 0.491 \\
-0.442 & -0.633 & -0.064 & -0.079 & 0.267 & 0.416 & 0.239 & 0.295 \\
-0.349 & 0.491 & -0.410 & 0.267 & -0.769 & 0.239 & 1.528 & -0.997 \\
0.239 & -2.021 & 0.267 & -0.633 & 0.491 & 0.295 & -0.997 & 2.358
\end{vmatrix}
$$

The result for $[k_{21}]$ is

$$
1.5
\begin{vmatrix}
0.973 & 0.267 & -0.769 & 0.491 & -0.410 & -0.627 & 0.206 & -0.132 \\
0.267 & 0.416 & 0.239 & 0.295 & -0.442 & -0.633 & -0.064 & -0.079 \\
-0.769 & 0.239 & 1.528 & -0.997 & -0.349 & 0.491 & -0.410 & 0.267 \\
0.491 & 0.295 & -0.997 & 2.358 & 0.239 & -2.021 & 0.267 & -0.633 \\
-0.410 & -0.442 & -0.349 & 0.239 & 0.665 & 0.267 & 0.094 & -0.064 \\
-0.627 & -0.633 & 0.491 & -2.021 & 0.267 & 2.111 & -0.132 & 0.542 \\
0.206 & -0.064 & -0.410 & 0.267 & 0.094 & -0.132 & 0.110 & -0.072 \\
-0.132 & -0.079 & 0.267 & -0.633 & -0.064 & 0.542 & -0.072 & 0.170
\end{vmatrix}
$$

and for $[k_{22}]$ is

$$
1.5
\begin{vmatrix}
0.110 & 0.072 & 0.094 & 0.132 & -0.410 & -0.267 & 0.206 & 0.064 \\
0.072 & 0.170 & 0.064 & 0.542 & -0.267 & -0.633 & 0.132 & -0.079 \\
0.094 & 0.064 & 0.665 & -0.267 & -0.349 & -0.239 & -0.410 & 0.442 \\
0.132 & 0.542 & -0.267 & 2.111 & -0.491 & -2.021 & 0.627 & -0.633 \\
-0.410 & -0.267 & -0.349 & -0.491 & 1.528 & 0.997 & -0.769 & -0.239 \\
-0.267 & -0.633 & -0.239 & -2.021 & 0.997 & 2.358 & -0.491 & 0.295 \\
0.206 & 0.132 & -0.410 & 0.627 & -0.769 & -0.491 & 0.973 & -0.267 \\
0.064 & -0.079 & 0.442 & -0.633 & -0.239 & 0.295 & -0.267 & 0.416
\end{vmatrix}
$$

Summing these four matrices yields the total element stiffness matrix:

$$
1.5
\begin{vmatrix}
3.276 & 1.603 & -1.350 & 0.320 & -1.640 & -1.603 & -0.286 & -0.320 \\
1.603 & 5.055 & -0.320 & 1.674 & -1.603 & -2.532 & 0.320 & -4.200 \\
-1.350 & -0.320 & 3.276 & -1.603 & -0.286 & 0.320 & -1.640 & 1.603 \\
0.320 & 1.674 & -1.603 & 5.055 & -0.320 & -4.200 & 1.603 & -2.532 \\
-1.640 & -1.603 & -0.286 & -0.320 & 3.276 & 1.603 & -1.350 & 0.320 \\
-1.603 & -2.532 & 0.320 & -4.200 & 1.603 & 5.055 & -0.320 & 1.674 \\
-0.286 & 0.320 & -1.640 & 1.603 & -1.350 & -0.320 & 3.276 & -1.603 \\
-0.320 & -4.200 & 1.603 & -2.532 & 0.320 & 1.674 & -1.603 & 5.055
\end{vmatrix}
$$

8.7.5 Computing Strain and Stress for Quadrilateral Elements

Once calculated, the element stiffness matrices are disassembled and placed into the system stiffness matrix in the same manner as was done for the bar elements. Nodal displacements $\{d\}$ are calculated once the system stiffness matrix is derived. Once the nodal point displacements $\{d\}$ are calculated, strains and stresses may be calculated at the Gauss points internal to each element. This is accomplished using the following steps:

1. Isolate the eight nodal point displacements for each quadrilateral element. Put these eight values into a vector $\{d'\}$.
2. For each Gauss point internal to the element, recompute the strain–displacement matrix $[B_{ij}]$.
3. Compute the vector of three strains at each Gauss point as

$$
\begin{vmatrix}
\varepsilon_x \\
\varepsilon_y \\
\gamma_{xy}
\end{vmatrix}
= [B_{ij}]
\begin{vmatrix}
d_{1,x} \\
d_{1,y} \\
d_{2,x} \\
d_{2,y} \\
d_{3,x} \\
d_{3,y} \\
d_{4,x} \\
d_{4,y}
\end{vmatrix}
= [B_{ij}][d']
$$

4. Using the strain vector, the three components of stress are calculated for the region surrounding each Gauss point as

$$
\begin{vmatrix}
\sigma_x \\
\sigma_y \\
\tau_{xy}
\end{vmatrix}
= [E]
\begin{vmatrix}
\varepsilon_x \\
\varepsilon_y \\
\gamma_{xy}
\end{vmatrix}
$$

The vectors for strain and stress are unique to each Gauss point. Therefore, the isoparametric, four-node element is a variable strain element. This property is attractive for earth sciences analyses in which strain and stress likely vary rapidly over space.

8.8 Nonlinear, Elastic Finite Element Analysis: An Application for the Analysis of Fractured Rock Systems

Up to now, discussions on both the bar element and quadrilateral element have assumed that a linear, elastic relationship (Figure 8.24) exists between strain and stress (likewise force and displacement). For some materials, a linear, elastic relationship exists between force and displacement up to a limiting value of force, beyond which the relationship is nonlinear (Figure 8.25).

Fractured rock is a material for which the relationship between force and displacement is similar (Figure 8.25). For this discussion, assume the unloading curve is identical to the loading curve in Figure 8.25; that is, the removal of load causes a decrease in displacement which follows exactly the curve produced as load is applied.

Several approaches exist for modeling the behavior of fractures using finite element analysis. A sophisticated isoparametric ''joint'' finite element was developed by Goodman and others (1968), which is described clearly and applied in Goodman (1976). This element is defined using four nodal points, hence its element stiffness matrix has dimensions 8×8 for a two-dimensional analysis. This element is a bit odd in that it has no area; it simply separates two isoparametric, quadrilateral, solid elements.

An alternative continuum method for assessing the nonlinear, elastic behavior of fractured materials is developed by Zienkiewicz and Pande (1977). This method is useful for indicating places in a finite element mesh where fractures may open or where the fractures may fail in shear based on the state of stress in the rock. This method is an excellent educational tool for use in understanding the behavior of jointed rock masses in response to stress.

The term continuum refers to the type of finite element mesh designed for this analysis. A continuum mesh is one composed of connected finite elements modeling an unbroken, or unfractured, medium. Obviously, this statement contradicts the objective of this section: the finite element analysis of jointed

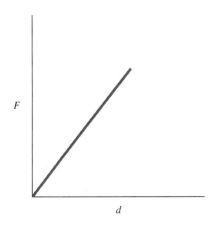

Figure 8.24 Linear elastic material behavior.

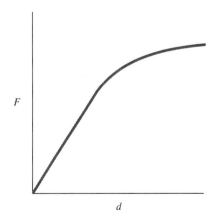

Figure 8.25 Nonlinear elastic material behavior.

(fractured, discontinuous) rock masses. This statement also reveals that the Zienkiewicz and Pande method is only an approximate one. Its primary utility is the indication in a continuum mesh of where rock fractures are likely to fail. Again, a more correct analysis of discontinuous rock masses is afforded using the Goodman joint element. The use of this element yields a mesh that is a discontinuum. The objective of this section is to develop a simple, educational tool for indicating where fractures are likely to fail. This explains why a continuum analysis is deemed acceptable for this chapter.

8.8.1 A Summary of the Continuum Analysis of Nonlinear, Elastic Behavior of Fractured Rock Masses

For this analysis, a mesh of four-node, isoparametric, quadrilateral elements is used (Figure 8.26). One hypothetical system of fractures is oriented in the horizontal direction, bisecting the rock mass that is modeled by the mesh. The disturbance affecting the rock mass is a rectangular tunnel (Figure 8.26). Finite element analysis is used to assess the affect that this disturbance has on the surrounding rock mass, including the system of fractures.

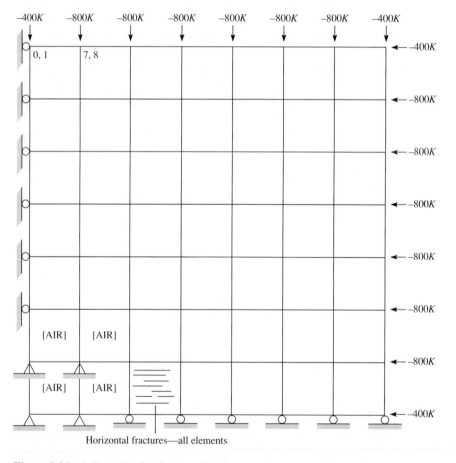

Figure 8.26   A 7-row by 7-column mesh of square elements used to model a square tunnel opening as shown (the air elements).

In using the continuum method for analyzing the affect that this disturbance has on the fracture system, the following algorithm is employed:

1. Assemble the system stiffness matrix for this mesh of four-node, quadrilateral elements.
2. Analyze displacements $\{d\}$ using the system $\{F\} = [K]\{d\}$, where the forces are the active forces (Figure 8.26).
3. Using the computed displacements, analyze strain and stress at each Gauss point of each element.
4. Assume that the fractures whose behavior in this stress field is to be assessed intersect each Gauss point.
5. Rotate the stress vector at a Gauss point to the orientation of each fracture to be analyzed. This rotation is as follows:

$$[\sigma'] = [\sigma][R_\sigma], \qquad [R_\sigma] = \begin{vmatrix} C2 & S2 & 2CS \\ S2 & C2 & -2CS \\ -CS & CS & (C2 - S2) \end{vmatrix}$$

   which uses a 3×3 stress rotation matrix, for which $C2 = \cos^2\alpha$, $S2 = \sin^2\alpha$, $CS = \cos\alpha \sin\alpha$, and $\alpha = $ fracture orientation.
6. Rotate the material constitutive matrix $[E]$ to the orientation of each fracture to be analyzed. This rotation is as follows:

$$[E'] = [R_\sigma][E][R_\sigma]^T$$

7. Analyze the behavior of each fracture:
   (a) If stress normal (perpendicular) to the fracture is tensile, then the fracture is deemed to be open. In this case, the material constitutive matrix $[E']$ is modified such that

$$E'(1, 1) = E'(1, 1) - E'(1, 2)E'(2, 1)/E'(2, 2)$$

   Then $E'(1, 2)$, $E'(2, 1)$, $E'(2, 2)$, and $E'(3, 3)$ are each mulitplied by 0.01 (i.e., set equal to 1% of the original values).
   (b) If stress normal to the fracture is compressive, a Mohr–Coulomb criterion is used to assess whether the fracture is in a shear failure mode. The shear strength of the fracture is a function of cohesion and friction such that

$$\tau_{max} = C + \sigma_N \tan(\phi)$$

   where C is cohesion, and ϕ is the friction angle, the tangent of which is multiplied by the normal stress. If the shear stress at the Gauss point (the third component of the stress vector) is greater than this Mohr–Coulomb criterion, that is, if

$$\tau_{xy} > \tau_{max}$$

   then the fracture is deemed to be in shear failure at the Gauss point. In this case, the material constitutive matrix $[E]$ is modified such that

$$E'(3, 3) = 0.01E'(3, 3)$$

8. Once the analysis is complete, rotate the material constitutive matrix $[E']$ back to its original orientation:

$$[E] = [R_\varepsilon]^T[E'][R_\varepsilon], \qquad [R_\varepsilon] = \begin{vmatrix} C2 & S2 & 2CS \\ S2 & C2 & -2CS \\ -2CS & 2CS & (C2 - S2) \end{vmatrix}$$

where $C2 = \cos^2\alpha$, $S2 = \sin^2\alpha$, $CS = \cos\alpha\sin\alpha$, and α = fracture orientation.

9. Recompute the element stiffness matrix $[k]$ for the element using the modified material constitutive matrices, once fracture behavior at all Gauss points internal to the element is analyzed.
10. Reassemble the system stiffness matrix $[K]$ and compute the displacements again using the original force vector $\{F\}$.
11. Repeat steps 1 through 10 until some convergence criterion is met (i.e., until no new fracture failures are found). This achieves the nonlinear analysis as depicted in Figure 8.25.

In proceeding through these 11 steps, a record is kept by the finite element computer program of which Gauss points are associated with fractures in failure. Once the program is finished, a computer graphics program (MESHDRAW) is used to display the mesh and show by color the locations of Gauss points associated with failed fractures. This achieves a most useful visual analysis of fracture response to a disturbance.

8.9 QUAD: A FORTRAN-77 Program for Linear, Elastic and Nonlinear, Elastic Finite Element Analysis Using Four-Node, Isoparametric Elements

A valuable suite of programs for finite element analysis is presented in Smith (1982). Many valuable main programs and supporting subroutines are presented in this reference for two- and three-dimensional analyses. A program called QUAD is developed based on Smith's program 5.0: plane-strain analysis of a linear, elastic solid using four-node, isoparametric elements. QUAD is a substantial modification of Smith's program to include an option for analyzing the nonlinear, elastic behavior of jointed rock masses.

Another attractive aspect of Smith's programs is automatic mesh generation. Finite element mesh generators are used for regular meshes. Such generators automatically assign nodal point coordinates and specify element connectivity, simply using some initial parameters: x and y dimensions of each element, and the total number of rows and columns of elements in the mesh. The program QUAD is designed to have two of Smith's generators: one to generate rectangular grids of elements, and the other to generate meshes to analyze circular or elliptical tunnel openings. Each of these generators is described in Section 8.10.

With automatic mesh generation, a lesser chance that error is present in the mesh is obtained. Therefore, error trapping is not as critical when using mesh generation as when entering nodal point coordinates and element connectivity by hand. Moreover, creating a data file for finite element analysis in which automatic mesh generation is used is a much simpler process than creating a file when mesh generation is not used. Therefore, the program QUAD, although using a more complicated four-node element, will perhaps be perceived as being easier to use than the program TRUSS (Section 8.6).

Although automatic mesh generation is used by QUAD, the graphics program MESHDRAW is useful for displaying the initial, undeformed mesh, and certainly for displaying the deformed mesh. Therefore, QUAD creates two output files: (1) The nodal point coordinates and element connectivity are saved in a file as they are generated by the mesh generator, and (2) displacements, strains, and stresses (and fracture failure information if a nonlinear, elastic analysis is performed) are saved in a separate file. These files are compatible with the program MESHDRAW, as will be demonstrated in the next section.

A complete listing of QUAD (Figure 8.27, page 378) includes a detailed user's guide. Data input is straightforward, as is demonstrated in several examples in Section 8.10.

8.10 Applications of QUAD for Finite Element Analysis Using the Two-Dimensional, Solid, Four-Node, Isoparametric Element

Several examples are presented to demonstrate (1) the use of the program QUAD for finite element analysis using a solid two-dimensional element; (2) the use of QUAD with automatic mesh generation; (3) the use of QUAD to perform a nonlinear, elastic analysis to assess the behavior of fractures in rock; and (4) the display of these results using the program MESHDRAW.

8.10.1 Analysis of Displacement Beneath a Spread Footing on Soil

An example is presented to show how to use the program QUAD (1) for automatic, rectangular mesh generation, and (2) to obtain all the input information required by this program. A situation (Figure 8.28) of a spread footing

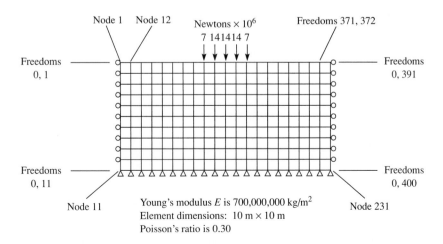

Figure 8.28   A mesh used to assess the effect of a spread footing bearing on soil.

Table 8.2 Properties for the Mesh Shown in Figure 8.28

NXE = 20	NYE = 10	N = 400
W = 23	NN = 231	RN = 41
NL = 5	AA = 10.0	BB = 10.0
NAIR = 0	NANGLE = 0	MESH = 1

Young's modulus = 144,000,000 lb/ft^2
Poisson's ratio = 0.30
Bearing pressure of footing (uniform): 288,000 lb/ft^2

bearing on soil was used in Carr (1992). A rectangular array of elements is used; 10 rows of elements are shown in this array, each containing 20 elements, for 200 total elements along with 231 total nodal points. Fortunately, automatic mesh generation is employed for this analysis, hence avoiding the tedium of entering nodal point coordinates and element connectivity information by hand. All properties of this mesh are listed in Table 8.2.

A complete data file (Figure 8.29) is presented for this analysis, prepared by following the user's guide shown at the beginning of the listing of the

```
20,10,400,23,231,41,5       11    132        223
700000000.0,0.30,0,0,1       1    1          1
10.0,10.0                    1    1          0
1                           22    143        224
1                            1    1          1
0                            1    1          0
2                           33    154        225
1                            1    1          1
0                            1    1          0
3                           44    165        226
1                            1    1          1
0                            1    1          0
4                           55    176        227
1                            1    1          1
0                            1    1          0
5                           66    187        228
1                            1    1          1
0                            1    1          0
6                           77    198        229
1                            1    1          1
0                            1    1          0
7                           88    209        230
1                            1    1          1
0                            1    1          0
8                           99    220        231
1                            1    1          1
0                            1    1          1
9                          110    221        152,-7000000.0
1                            1    1          172,-14000000.0
0                            1    0          192,-14000000.0
10                         121    222        212,-14000000.0
1                            1    1          232,-7000000.0
0                            1    0
```

Figure 8.29 A data file used with the program QUAD to analyze the situation shown in Figure 8.28.

program QUAD (see Figure 8.27). The first two lines of the data file define options for mesh generation and subsequent finite element analysis.

To prepare a data file for use with QUAD, the mesh, even though it will be generated by the program, must be drawn on paper (Figure 8.28). Such a drawing shows the total number of elements and nodal points. Once the drawing is complete, the nodal points must be numbered, starting at the upper left corner of the mesh and proceeding downward along the first column of nodal points, then up to the top of the next column, and so on.

Subsequent to nodal point numbering, all nodal freedoms must be numbered. This, too, begins at the upper left nodal point. A distinct freedom number must be assigned to each *x*- and *y*-direction freedom at each nodal point. If *x*- and/ or *y*-direction movements are restricted, no freedom number is assigned to the restricted freedom. Freedom numbers are shown in Figure 8.28 as a pair of numbers *x,y*. Notice how these are numbered. The largest freedom number in the mesh is equal to the total number of nodal points (231) multiplied by the total number of freedoms per node (2) minus the total number of re- stricted freedoms (62). Therefore, in this example the largest freedom number is 400.

Boundary conditions specified in the data file (Figure 8.29) follow the input guide shown at the beginning of the listing of QUAD. Each nodal point associ- ated with a boundary condition is specified, followed by two lines: one to show the *x*-direction boundary condition—0 if free to move, 1 if restricted—and the other to show the *y*-direction boundary condition—0 if free to move, 1 if restricted. Comparing the listing of the data file to the mesh (Figure 8.28) shows how the boundary conditions are specified.

Once the boundary conditions are specified, the active forces operating on the system are entered. Negative forces are compressive and positive forces are tensile. In this analysis, compressive forces are analyzed (Figure 8.28). When entering data describing forces, the freedom number associated with an active force is specified, along with the magnitude of the force for that freedom number. Freedom numbers and forces are shown in Figure 8.28.

Using the data file in Figure 8.29, the program QUAD yields the analysis shown in Figure 8.30. This figure is a drawing from the program MESHDRAW. When running the program QUAD for this analysis, the program requests information for the name of the input data file, unit 5 (that containing the file listed in Figure 8.29). The program then asks the user to specify the name of the output file, unit 1, storing nodal point coordinates and element connectivity (called FOOTINGC.OUT in this example). Finally, the program asks the user to specify the name of the output file, unit 2, storing the displacements, strains, and stresses for the analysis (called FOOTING.OUT for this example). Note that both output files are newly created by the program and must not exist prior to running the program.

Figure 8.30 was created using MESHDRAW as described by the interactive session shown in Figure 8.31. Note that the option for 2D solid elements is selected first; then, the name of the nodal point coordinates/mesh connectivity file is specified (FOOTINGC.OUT in this case). Following this, the option to display the deformed mesh is chosen. A magnification factor of 200.0 is chosen for displaying the displacements (displacements may be small, so a magnifica- tion factor is used to render them visible on the screen). After the entry of the magnification factor, the name of the output file that stores the displacements,

PLOT OF DEFORMED MESH

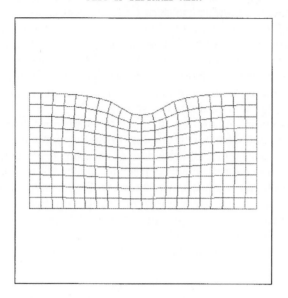

Figure 8.30   The deformed mesh for the situation shown in Figure 8.28, drawn by the program MESHDRAW.

```
IS THE MESH TO BE DRAWN COMPRISED OF BAR ELEMENTS
OR 2D SOLID ELEMENTS?
ENTER 1 FOR BAR ELEMENTS; 2 FOR 2D SOLID ELEMS
2


PLEASE ENTER THE NAME OF THE DATA FILE, UNIT 5,
WHICH STORES THE ORIGINAL NODAL COORDINATES
AND ELEMENT CONNECTIVITY
File name missing or blank - please enter file name
UNIT 5? FOOTINGC.OUT


WILL THE ORIGINAL OR THE DEFORMED MESH BE DRAWN?
ENTER 1, FOR THE ORIGINAL, OR 2 FOR THE DEFORMED
2


ENTER THE MAGNIFICATION FACTOR FOR THE PLOTTING
OF THE DEFORMED MESH:   .5, 1.0, 1000., ETC.
200.


ENTER THE NAME OF THE FILE, UNIT 1, WHICH
STORES THE DISPLACEMENTS
File name missing or blank - please enter file name
UNIT 1? FOOTING.OUT

WHICH STRESS MAP WOULD YOU LIKE?
ENTER:   1, FOR X-DIRECTION STRESS
         2, FOR Y-DIRECTION STRESS
         3, FOR SHEAR STRESS
         4, FOR FRACTURE FAILURE DIAGRAM
         5, FOR OUTLINE OF DEFORMED MESH ONLY
5
```

Figure 8.31   The interactive session for the program MESHDRAW, which results in the drawing shown in Figure 8.30.

strains, and stresses is specified (FOOTING.OUT in this case). Finally, the option (No. 5) is chosen to display the deformed mesh only.

8.10.2 Analysis of a Rectangular Opening

The rectangular opening in Figure 8.26 is analyzed. Mesh parameters and material parameters are listed in Table 8.3. Four air elements are used to represent the opening. The reason for using such elements is due to using automatic mesh generation. The rectangular mesh generator GEOMET used in the program QUAD yields a regular rectangular array of elements without allowing any gaps in the elements. Therefore, to design voids into the mesh, certain elements representing voids are designated as air elements, whose Young's moduli are set equal to zero.

A complete data set for this analysis is shown in Figure 8.32. The option NAIR is set equal to 4. The four air elements are designated as follows:

IXAIR	IYAIR
1	6
1	7
2	6
2	7

In other words, IXAIR is the column number and IYAIR is the row number on which the air elements occur.

Further note that NANGLE, the option specifying the number of fracture orientations to be analyzed, is set equal to 1. The fracture orientation angle is set equal to 0.0. No cohesion is assumed for the fractures whose friction angle is 30 degrees. The reader must determine the freedom numbers for the situation shown in Figure 8.26.

The program QUAD is used with this data set to analyze the behavior of the horizontal fractures around the square opening. As in the previous example, this program asks the user to specify the name of the input file (unit 5), the name of the output file storing nodal point coordinates and element connectivity

---

Table 8.3 Properties for the Mesh Shown in
  Figure 8.26

NXE = 7	NYE = 7	N = 108
W = 18	NN = 64	RN = 16
NL = 16	AA = 5.0	BB = 5.0
NAIR = 4	NANGLE = 1	MESH = 1

Air element positions		Fracture properties
IXAIR	IYAIR	Angle = 0.0 degrees
1	6	Cohesion = 0.0
1	7	Friction = 0.58
2	6	
2	7	

Young's modulus = 144,000,000 lb/ft^2
Poisson's ratio = 0.30
Uniform isotropic pressure = 160,000 lb/ft^2

7,7,108,18,64,16,16	4	16	64
144000000.0,0.3,4,1,1	1	1	0
5.0,5.0	0	1	1
1,6	5	24	1,-400000.0
1,7	1	0	8,-800000.0
2,6	0	1	20,-800000.0
2,7	6	32	35,-800000.0
0.0	1	0	50,-800000.0
0.0	0	1	65,-800000.0
0.58	7	40	80,-800000.0
1	1	0	94,-400000.0
1	1	1	95,-400000.0
0	8	48	96, -800000.0
2	1	0	98, -800000.0
1	1	1	100,-800000.0
0	15	56	102,-800000.0
3	1	0	104,-800000.0
1	1	1	106,-800000.0
0			108,-400000.0

Figure 8.32   The data set used with the program QUAD for the example in Section 8.10.2.

information (unit 1), and the name of three additional output files (units 2, 3, and 4) storing the results of the nonlinear, elastic analysis for three load iterations.

MESHDRAW is used to display the original mesh for error trapping. The interactive session for MESHDRAW for this purpose is shown in Figure 8.33, yielding the final plot (Figure 8.34).

Once error trapping is finished, and in this case no errors are found, results for the nonlinear, elastic analysis are computed and displayed (Figures 8.35 through 8.37). The interactive session for obtaining the plot of Figure 8.35 is shown in Figure 8.38; for the other two plots (Figures 8.36 and 8.37), the interactive sessions are the same (Figure 8.38), except the output files are LARGE4F2.OUT and LARGE4F3.OUT. QUAD models the propagation of failure of the horizontal fractures upward above the square opening through the three iterations. This is an excellent demonstration of how useful the continuum method is for indicating how fractures behave in response to a disturbance.

```
IS THE MESH TO BE DRAWN COMPRISED OF BAR ELEMENTS
OR 2D SOLID ELEMENTS?
ENTER 1 FOR BAR ELEMENTS; 2 FOR 2D SOLID ELEMS
2


PLEASE ENTER THE NAME OF THE DATA FILE, UNIT 5,
WHICH STORES THE ORIGINAL NODAL COORDINATES
AND ELEMENT CONNECTIVITY
File name missing or blank - please enter file name
UNIT 5? LARGE4FC.OUT


WILL THE ORIGINAL OR THE DEFORMED MESH BE DRAWN?
ENTER 1, FOR THE ORIGINAL, OR 2 FOR THE DEFORMED
1
```

Figure 8.33   The interactive session for the program MESHDRAW, which results in the drawing shown in Figure 8.34.

PLOT OF ORIGINAL MESH

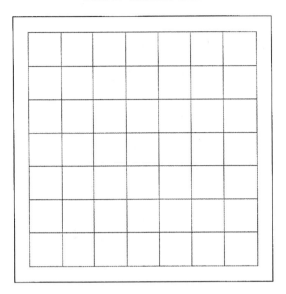

Figure 8.34   The original, undeformed mesh for the situation shown in Figure 8.26 (Section 8.10.2).

PLOT OF DEFORMED MESH

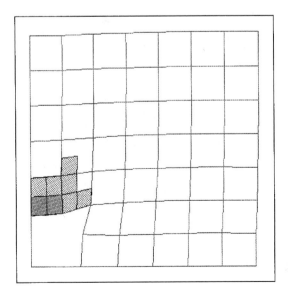

Figure 8.35   A nonlinear elastic analysis of the situation shown in Figure 8.26 after one load iteration. Each element is subdivided into its four (Gaussian point) quadrants. If a joint is deemed in failure at a Gaussian point, its quadrant is outlined. This explains why this figure appears to have more elements than Figure 8.26.

PLOT OF DEFORMED MESH

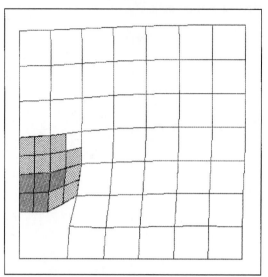

Figure 8.36   The same as Figure 8.35, but after two load iterations. Notice how a vertically propagating fracture failure is being modeled.

PLOT OF DEFORMED MESH

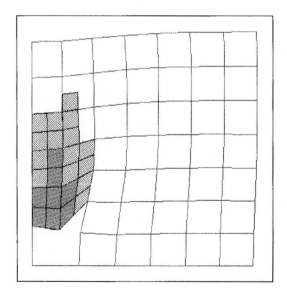

Figure 8.37   The same as Figure 8.35, but after three load iterations. Fracture failure has propagated upward considerably.

8.10.3 Analysis of Stress Around an Elliptical Tunnel

The primary purpose of this example is to demonstrate the use of the program QUAD for generating circular and elliptical meshes for the analysis of the behavior of circular and elliptical openings. In this case, the example from

```
IS THE MESH TO BE DRAWN COMPRISED OF BAR ELEMENTS
OR 2D SOLID ELEMENTS?
ENTER 1 FOR BAR ELEMENTS; 2 FOR 2D SOLID ELEMS
2


PLEASE ENTER THE NAME OF THE DATA FILE, UNIT 5,
WHICH STORES THE ORIGINAL NODAL COORDINATES
AND ELEMENT CONNECTIVITY
File name missing or blank - please enter file name
UNIT 5? LARGE4FC.OUT



WILL THE ORIGINAL OR THE DEFORMED MESH BE DRAWN?
ENTER 1, FOR THE ORIGINAL, OR 2 FOR THE DEFORMED
2


ENTER THE MAGNIFICATION FACTOR FOR THE PLOTTING
OF THE DEFORMED MESH:   .5, 1.0, 1000., ETC.
100.

ENTER THE NAME OF THE FILE, UNIT 1, WHICH
STORES THE DISPLACEMENTS
File name missing or blank - please enter file name
UNIT 1? LARGE4F1.OUT

WHICH STRESS MAP WOULD YOU LIKE?
ENTER:   1, FOR X-DIRECTION STRESS
         2, FOR Y-DIRECTION STRESS
         3, FOR SHEAR STRESS
         4, FOR FRACTURE FAILURE DIAGRAM
         5, FOR OUTLINE OF DEFORMED MESH ONLY
4
HOW MANY ROWS OF ELEMENTS ARE IN THE
ORIGINAL MESH?
7
HOW MANY COLUMNS OF ELEMENTS ARE IN THE
ORIGINAL MESH?
7
```

Figure 8.38 The interactive session for the program MESHDRAW, used to obtain Figure 8.35.

Smith (1982, p. 131, Figures 5.50 and 5.51) is used. An example published previously is a good test of a finite element analysis program. If a program yields the same results as those published previously, confidence in correct program execution is obtained.

When comparing the output from QUAD to that published in Smith (1982) (Figure 8.39), note that QUAD is modified such that the Gauss points are numbered differently from Smith's. This is done only to facilitate graphical display of results; this change is acknowledged to avoid confusion when making comparisons between QUAD and Smith (1982).

Input data for this analysis (Table 8.4) are used to create a data file (Figure 8.40). When generating an elliptical mesh, note especially the arrays RADIUS and SPOKE. RADIUS contains the NYE + 1 radii, starting at RADIUS(1) = largest radius and moving to RADIUS(NYE + 1) = smallest radius. SPOKE contains factors for describing the elliptical shape of the mesh. In this case, SPOKE(1) = 1.0, where the vertical spoke is associated with the major axis of the ellipse. SPOKE(NXE + 1) is set equal to 0.767, where the ratio of the minor/major axes of the ellipse is 0.767. All other SPOKE values are set equal to the distance from the center of the tunnel to the edge of the ellipse, which

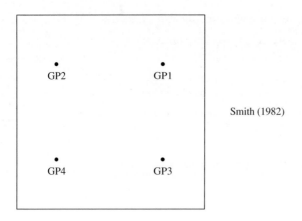

Smith (1982)

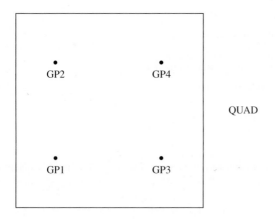

QUAD

Figure 8.39   Gauss point numbering, Smith (1982) versus QUAD.

Table 8.4   Properties for the Mesh Shown in Figure 8.41

NXE = 8	NYE = 12	N = 208
W = 19	NN = 117	RN = 26
NL = 16		
NAIR = 0	NANGLE = 0 (for Section 8.10.3)	
NANGLE = 1 (for Section 8.10.4)		
MESH = 2		

12 values of RADIUS: 12.9, 11., 9., 7., 5., 4., 3., 2.6, 2.25, 2.,
1.75, 1.55, 1.45
8 values of SPOKE: 1.0, 0.977, 0.95, 0.903, 0.861, 0.814,
0.786, 0.772, 0.767

Young's modulus = 500. (units)
Poisson's ratio = 0.20

```
8,12,208,19,117,26,16                                46    99
500.,0.2,0,0,2                                        1     0
12.9,11.,9.,7.,5.,4.,3.,2.6,2.25,2.,1.75,1.55,1.45   0     1
1.,.977,.95,.903,.861,.814,.786,.772,.767            54    100
1                                                     0     1
1                                                     1     0
0                                                     55    108
9                                                     1     0
0                                                     0     1
1                                                     63    109
10                                                    0     1
1                                                     1     0
0                                                     64    117
18                                                    1     0
0                                                     0     1
1                                                     72    193,-.5
19                                                    0     194,-.1951
1                                                     1     195,-.9808
0                                                     73    196,-.3827
27                                                    1     197,-.9239
0                                                     0     198,-.5556
1                                                     81    199,-.8315
28                                                    0     200,-.7071
1                                                     1     201,-.7071
0                                                     82    202,-.8315
36                                                    1     203,-.5556
0                                                     0     204,-.9236
1                                                     90    205,-.3827
37                                                    0     206,-.9808
1                                                     1     207,-.1951
0                                                     91    208,-0.5
45                                                    1
0                                                     0
1
```

Figure 8.40   A listing of the data file for the example in Section 8.10.3.

is this distance divided by the length of the major axis of the ellipse, for each spoke direction.

Suppose a circular tunnel is analyzed. In this case, all NXE + 1 values of the array SPOKE are set equal to 1.0. Further, suppose an elliptical tunnel is to be analyzed whose major axis parallels the horizontal direction. In this case, SPOKE(NXE + 1) is set equal to 1.0, SPOKE(1) is set equal to the ratio of the minor/major axes, and remaining SPOKE values are accordingly scaled. This paragraph simply provides additional information when using the program QUAD.

Results are shown in Figures 8.41 and 8.42 from applying QUAD to the data set in Figure 8.40. A plot of the original mesh (Figure 8.41) verifies correct data file preparation. Figure 8.42 shows the magnitudes of the horizontal stresses around the opening. The interactive session for MESHDRAW is shown in Figure 8.43. Note that a black and white reproduction (Figure 8.42) is presented of the color figure drawn on the screen by the program MESHDRAW. Patterns have been manually added to this figure to indicate the stress regions shown in color by the program MESHDRAW.

8.10.4 Analysis of Vertical Fractures for the Tunnel of Example 8.10.3

Horizontal stress around a tunnel was shown in Figure 8.42. Note that this stress is tensile in the region from 0 to about 45 degrees around the opening.

PLOT OF ORIGINAL MESH

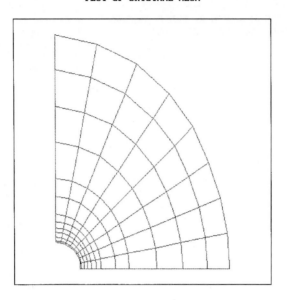

Figure 8.41 The original elliptical mesh, as drawn by the program MESHDRAW.

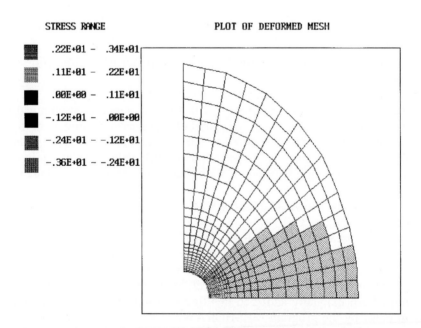

Figure 8.42 The *x*-direction stress for the mesh of Figure 8.41, as drawn by the program MESHDRAW. Each Gaussian point quadrant is outlined in a particular color to show its relative stress, which explains why this figure appears to have more elements than Figure 8.41.

```
IS THE MESH TO BE DRAWN COMPRISED OF BAR ELEMENTS
OR 2D SOLID ELEMENTS?
ENTER 1 FOR BAR ELEMENTS; 2 FOR 2D SOLID ELEMS
2


PLEASE ENTER THE NAME OF THE DATA FILE, UNIT 5,
WHICH STORES THE ORIGINAL NODAL COORDINATES
AND ELEMENT CONNECTIVITY
File name missing or blank - please enter file name
UNIT 5? TUNNELC.OUT


WILL THE ORIGINAL OR THE DEFORMED MESH BE DRAWN?
ENTER 1, FOR THE ORIGINAL, OR 2 FOR THE DEFORMED
2


ENTER THE MAGNIFICATION FACTOR FOR THE PLOTTING
OF THE DEFORMED MESH:   .5, 1.0, 1000., ETC.
1.

ENTER THE NAME OF THE FILE, UNIT 1, WHICH
STORES THE DISPLACEMENTS
File name missing or blank - please enter file name
UNIT 1? TUNNEL.OUT

WHICH STRESS MAP WOULD YOU LIKE?
ENTER:  1, FOR X-DIRECTION STRESS
        2, FOR Y-DIRECTION STRESS
        3, FOR SHEAR STRESS
        4, FOR FRACTURE FAILURE DIAGRAM
        5, FOR OUTLINE OF DEFORMED MESH ONLY
1
```

Figure 8.43   The interactive session used for the program MESHDRAW to obtain Figure 8.42.

PLOT OF DEFORMED MESH

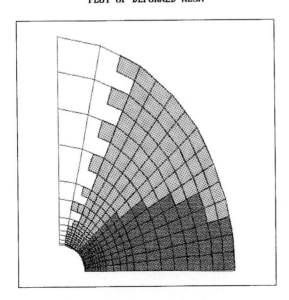

Figure 8.44   The behavior of vertical (90 degree) fractures for the mesh shown in Figure 8.41. Notice that fractures are deemed to be open in the entire region of Figure 8.42 associated with tensile (positive) stress. This is after one load iteration.

PLOT OF DEFORMED MESH

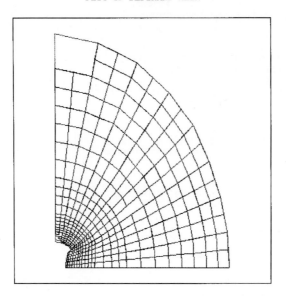

Figure 8.45 By the second load iteration, the entire mesh, except for the upper leftmost element, is associated with fracture opening. Note also the severely deformed tunnel opening after the second load iteration.

In this example, a vertical system of fractures is analyzed to identify the affect the tunnel has on such a fracture system.

Results from two nonlinear, elastic analysis iterations are shown in Figures 8.44 and 8.45. All vertical fractures within the region of tension are shown to be open, as is expected. Outside this region, many of these fractures are in shear failure in the first iteration. By the second iteration, many of the fractures originally in shear failure are shown to be open. Notice the severely deformed shape of the tunnel opening after the second iteration.

8.11 A Note on Boundary Conditions

Boundary conditions are specified in examples presented in this chapter for varying reasons. In some cases, boundary conditions are specified for a particular spatial direction when the finite element mesh provides no stiffness for the direction. For example, the mesh of bar elements in Figure 8.8 offers no y-direction stiffness. Boundary conditions on y are used with this mesh to avoid zero entries on the diagonal of the system stiffness matrix, because such zero entries cause a Gauss elimination equation solver (as implemented in this chapter) to fail.

Boundary conditions are also used when displacements must be absolutely constrained to equal zero for particular spatial directions. Such is the case with the mesh in Figure 8.5. This truss system is a model of a bridge, anchored at each end. Hence, for the finite element model to be realistic, displacements must be constrained to equal zero at the anchors (of course, the x direction is free to displace with the roller anchor used with this mesh).

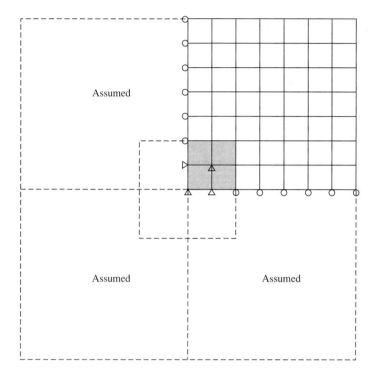

Figure 8.46   The consequence of establishing boundary conditions as shown in Figure 8.26.

A user of finite element analysis should be aware of the consequences of defining boundary conditions. For instance, boundary conditions constrain x displacements to zero at the left margin of the mesh (Figure 8.26); y displacements are constrained to zero along the bottom of the mesh. In this case, imagine each edge of the mesh associated with boundary conditions to be a mirror. Across the boundary, assume that a symmetrical mesh exists, although not explicitly shown.

Indeed, the consequence of establishing boundary conditions is as depicted in Figure 8.46. Therefore, the tunnel opening modeled in Figure 8.26 is only

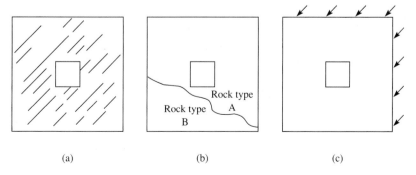

Figure 8.47   Three scenarios that require that the complete tunnel be modeled: (a) a fracture system other than 0 degrees (horizontal) or 90 degrees (vertical); (b) material anisotropy surrounding the tunnel; (c) stress anisotropy surrounding the tunnel.

the upper right quadrant of the actual tunnel, which is perfectly fine as long as there is, in reality, perfect symmetry across a boundary established in a finite element mesh.

What if symmetry does not hold across a boundary? How is such a situation accommodated in finite element analysis? In asymmetrical situations, the entire tunnel must be modeled, as shown in Figure 8.47 for several different scenarios.

8.12 Finite Difference Analysis for Assessing Regional Groundwater Flow

Before leaving this chapter, a discussion is presented on the finite difference method applied to the study of groundwater flow. Up to now this chapter has dealt solely with the finite element method for solving problems. With some problems, the finite difference method is equally useful for obtaining a solution. In particular, the finite difference method has been used much more extensively than the finite element method for solving problems related to groundwater flow (Wang and Anderson, 1982).

8.12.1 Darcy's Law

In late 1855 and early 1856, Henry Darcy conducted experiments to measure the discharge rate Q for water emitted by a column of sand. He concluded that discharge rate was directly proportional to the cross-sectional area A of the sand column and to the head drop $h_2 - h_1$, but was inversely proportional to the length of the sand column $l_2 - l_1$. Darcy introduced a proportionality constant K, called hydraulic conductivity. Hence we have Darcy's law:

$$Q = -KA \frac{h_2 - h_1}{l_2 - l_1}$$

where the negative sign indicates that flow is in the direction of head loss.

If Q/A is written as q, the volume rate of flow per unit area, then Darcy's law can be expressed as a derivative:

$$q = -K \frac{dh}{dl}$$

This equation describes one-dimensional flow, but Darcy's law may be expanded for higher dimensions. For two-dimensional flow,

$$q_x = -K \frac{\partial h}{\partial x}, \qquad q_y = -K \frac{\partial h}{\partial y}$$

For three-dimensional flow, an additional term for z is considered:

$$q_z = -K \frac{\partial h}{\partial z}$$

These equations show that head h is a function of all space coordinates.

Assuming steady-state conditions, the amount of water flowing into a volume (for example, an aquifer) must equal the amount flowing out. With respect to two-dimensional flow (for simplicity), steady-state conditions involve

$$\frac{\partial h}{\partial x} + \frac{\partial h}{\partial y} = 0$$

which is known as the continuity equation (Wang and Anderson, 1982, p. 12). Change in volume rate of flow in the x direction is offset by change in volume rate of flow in the y direction.

8.12.2 Laplace's Equation

Steady-state heat flow and steady-state fluid flow are governed by the same continuity equation. With respect to heat T, Laplace's equation for steady-state flow in two-dimensions is (Chapra and Canale, 1985, p. 253)

$$\frac{\partial^2 T}{\partial x^2} + \frac{\partial^2 T}{\partial y^2} = 0$$

With respect to two-dimensional flow of water through a sand column, Darcy's law is substituted into the continuity equation (Wang and Anderston, 1982, p. 13):

$$\frac{\partial}{\partial x}\left(-K\frac{\partial h}{\partial x}\right) + \frac{\partial}{\partial y}\left(-K\frac{\partial h}{\partial y}\right) = 0$$

Assuming that hydraulic conductivity K is independent of x and y (i.e., the aquifer is homogeneous), this equation simplifies to

$$\frac{\partial^2 h}{\partial x^2} + \frac{\partial^2 h}{\partial y^2} = 0$$

This, too, is Laplace's equation. Hence, Laplace's equation governs both steady-state heat flow and steady-state fluid flow.

8.12.3 Finite Difference Approximation of Laplace's Equation for Steady-State Flow

Finite difference approximation is useful for determining the value of a derivative at a point x (Chapter 1). In this particular instance, finite difference approximation is useful for approximating Laplace's equation for the head at a point (x, y) for two-dimensional flow.

Approximating Laplace's equation using finite differences gives

$$\frac{\partial^2 h}{\partial x^2} \approx \frac{h_{i,j-1} - 2h_{i,j} + h_{i,j+1}}{(dx)^2}, \qquad \frac{\partial^2 h}{\partial y^2} \approx \frac{h_{i-1,j} - 2h_{i,j} + h_{i+1,j}}{(dy)^2}$$

which assumes a grid of nodal points (i, j), where i represents rows in the y direction, and j represents columns in the x direction. For $dx = dy$, substituting these approximations into the Laplace equation gives

$$h_{i,j} = \frac{h_{i,j-1} + h_{i,j+1} + h_{i-1,j} + h_{i+1,j}}{4}$$

8.12.4 Example Application

An example is taken from Toth (1962) as implemented in Wang and Anderson (1982, pp. 32–33). A two-dimensional regional groundwater flow system is shown in Figure 8.48 and idealized in Figure 8.49. The right and left boundaries

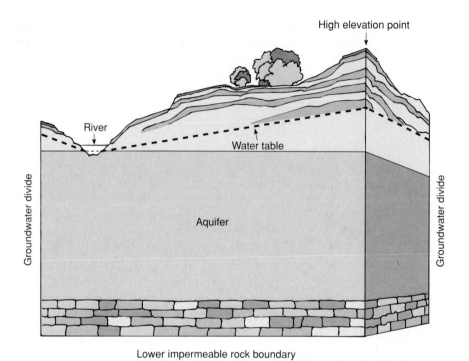

Figure 8.48   A two-dimensional groundwater flow system. The aquifer is underlain by impermeable rock; a groundwater divide (no-flow boundary) is associated with the topographic high and with the valley bottom (river). The aquifer is assumed to be homogeneous, isotropic, and porous. Note the inclined water table. From *Introduction to Groundwater Modeling* by Wang and Anderson. Copyright © 1982 by W.H. Freeman and Company. Reprinted with permission.

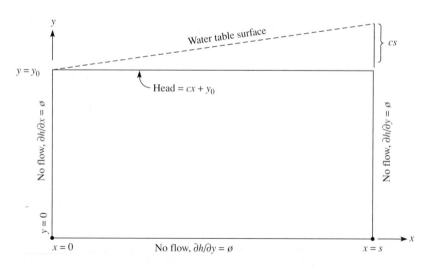

Figure 8.49   An idealization of Figure 8.48 to show boundary conditions and the head h due to the inclined water table. From *Introduction to Groundwater Modeling* by Wang and Anderson. Copyright © 1982 by W.H. Freeman and Company. Reprinted with permission.

are groundwater divides. An impermeable boundary is found at the bottom (the basement rock is assumed to be impermeable). Groundwater flows away from the right boundary (away from the topographic high) and discharges at point A into the river. The objective is to use finite difference approximation to approximate values of head in this system.

A grid of nodal points is developed for this problem (Figure 8.50). Head values are to be approximated at the nodal points. With respect to the idealized diagram (Figure 8.49), let $s = 1000$ m, $y_0 = 500$ m, and $c = 0.03$. The domain (Figure 8.49) is represented by a grid having 6 rows and 11 columns, but 7 rows and 13 columns are shown in Figure 8.50. Columns 1 and 13 and row 7 are added as fictitious rows and columns to implement flow boundary conditions (flow cannot occur across the right and left boundaries; nor can flow occur past the bottom of the grid because of the impermeable lower boundary (Figure 8.48). Boundary conditions for flow are forced using these fictitious columns and rows by setting $h(i, 1) = h(i, 3)$ and $h(i, 13) = h(i, 11)$. In other words, for no flow boundary conditions head must be reflected symmetrically across the boundary (Wang and Anderson, 1982, p. 32), that is, the derivative of head across the no-flow boundary must be zero. Likewise for the lower boundary, $h(7, j) = h(5, j)$ yields a symmetrical reflection across row 6.

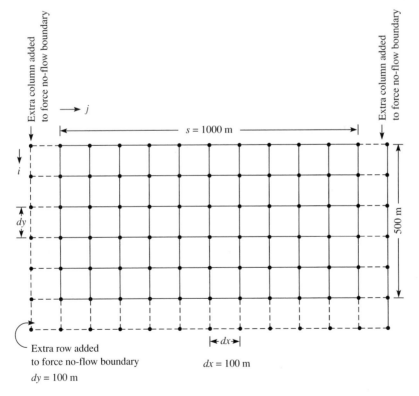

Figure 8.50   The finite difference grid used to analyze the situation (Figures 8.48 and 8.49). Seven rows are used. The seventh (bottom) is used to force a no-flow boundary because the aquifer is underlain by impermeable rock. Thirteen columns are used. The first and thirteenth are used to force no-flow boundaries due to the groundwater divides. From *Introduction to Groundwater Modeling* by Wang and Anderson. Copyright © 1982 by W.H. Freeman and Company. Reprinted with permission.

```
TYPE FDIFF.DAT
7,13
500.
500.
500.
503.0
506.0
509.0
512.0
515.0
518.0
521.0
524.0
527.0
530.0
530.0
```

0;37;44m C:\TEXT\CHAPTER8>

Figure 8.52   A data file used with FDIFF for analyzing the finite difference problem (Figs. 8.48–8.50). This data file was prepared assuming $y_0 = 500$, $c = 0.03$, $s = 1000$. Other parameters are as shown (Figures 8.48–8.50). All grid (head) values are initialized to a value of 500, except for the first (top) row of grid (head) values, which are determined using the equation (Figure 8.49) for h (that for an inclined water table).

The computer program FDIFF (Figure 8.51, page 392) is used to approximate values of head at nodes of the grid in Figure 8.50 using finite difference approximation. This program is a modification of that given in Wang and Anderson (1982, pp. 34–35). A user's guide is given at the beginning of the program listing. Using this program requires that the grid dimensions (M rows and N columns) be specified. A value, X, is required to which head values at

```
ENTER THE NAME OF THE INPUT DATA FILE
ENTER THE NAME OF THE OUTPUT GRID FILE
File name missing or blank - please enter file name
UNIT 1? FDIFF.OUT


    NUMBER OF ITERATIONS =         234

 500.00 503.00 506.00 509.00 512.00 515.00 518.00 521.00 524.00 527.00 530.00

 504.79 505.88 507.79 510.06 512.49 514.99 517.50 519.93 522.20 524.10 525.20

 507.40 507.95 509.23 510.95 512.91 514.99 517.06 519.03 520.75 522.03 522.57

 508.92 509.28 510.23 511.59 513.23 514.98 516.74 518.37 519.74 520.69 521.04

 509.74 510.02 510.80 511.98 513.42 514.98 516.54 517.98 519.16 519.94 520.22

 510.00 510.26 510.99 512.11 513.48 514.98 516.48 517.85 518.97 519.70 519.96
Stop - Program terminated.
```

0;37;44m C:\TEXT\CHAPTER8>

Figure 8.53   Output from FDIFF using the data file (Figure 8.52). Note that 234 iterations are required to converge to a solution. The maximum head value, 530, is that directly beneath the topographic high. The minimum head value, 500, is at the discharge point into the river.

all grid points will be initialized to begin the finite difference approximation; X is set equal to y_0 (Figure 8.49). Finally, the water table head boundary must be established. In this example, the water table is inclined (Figures 8.48 and 8.49). Using the equation for the inclination (Figure 8.49), water table head values are computed for the first row of grid points over the second through twelfth columns. The water table head for $h(1, 1)$ is set equal to $h(1, 2)$; $h(1, 13)$ is set equal to $h(1, 12)$. A data file created using this information is shown in Figure 8.52.

Application of FDIFF to this problem yields the output shown in Figure 8.53. This finite difference program uses Gauss–Seidel iteration to reach a solution. In this method, each iteration uses information from the current iteration and the previous iteration. Calling the current iteration C and the previous $C - 1$, we have

$$h_{i,j}^C = \frac{h_{i,j-1}^C + h_{i-1,j}^C + h_{i,j+1}^{C-1} + h_{i+1,j}^{C-1}}{4}$$

which shows the mixture of information from two iterations. The output (Figure 8.53) shows that 234 iterations were required to converge to a solution such that the maximum absolute value difference between two successive iterations was less than 0.001 (an arbitrary choice).

Contours of equipotential flow lines (equal head) are shown in Figure 8.54. Groundwater flow is perpendicular to the contour lines. Notice the flow is from the topographic high to the river, as expected.

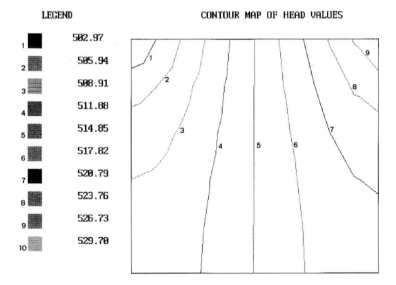

LEGEND   CONTOUR MAP OF HEAD VALUES

1	502.97
2	505.94
3	508.91
4	511.88
5	514.85
6	517.82
7	520.79
8	523.76
9	526.73
10	529.70

Figure 8.54   A contour map produced using the program CTOUR (Chapter 7). The program FDIFF creates an output file that is compatible with the program CTOUR. (*Note:* The FDIFF output file has one value per line, the head at grid point i,j; hence, when using CTOUR, after entering the name of the file for contouring (that which is created by FDIFF), enter the value 1 when asked for the number of values per line, and enter the value 1 again when asked which value is the one to be contoured.) Because this figure is a black and white representation of the color image created by CTOUR, numbers have been added by hand to the output to associate each contour line with a head value. Groundwater flow is perpendicular to the contour lines, and flow is from the high head region (upper right corner of the grid) to the low head region (upper left corner of the grid).

8.13   Finite Element Method in Comparison to the Finite Difference Method

The preceding section reviews the finite difference method in application to the analysis of groundwater flow. In this application, a grid of nodal points was developed. Moreover, no-flow boundary conditions were specified to constrain the solution at locations where the flow conditions were known. Elements were not necessary to obtain a solution.

In the finite difference method a set of simultaneous equations must be solved, yet this set of equations was implicit to the foregoing application. In particular, the Laplace equation must be solved at each nodal point. But the program FDIFF solves this set of equations implicitly using Gauss–Seidel iteration; Gaussian elimination was not used.

In the finite element method, as used in this chapter, materials are discretized into elements using a system of nodal points. Material properties are assigned to each element. Fixed displacement boundary conditions are specified. And a set of simultaneous equations are solved for displacements at the nodal points, given the forces acting on the system.

Similarities exist between the finite element method and the finite difference method (e.g., Cook, 1974). Many engineers prefer finite elements for the flexibility they offer in representing material heterogeneity and complexity of shapes. Many hydrologists, though, rely on the finite difference method because Laplace's differential equation is useful for understanding and modeling groundwater flow. The finite difference method is also ideal for approximating derivatives (Chapter 1).

Chapter Summary

- Hooke's law describes the relationship between force and displacement, likewise the relationship between stress and strain.

- Finite element analysis is based on Hooke's law, adapted for a series of springs or group of elements.

- Finite element analysis involves the discretization of a global system into finite, local pieces (elements) whose individual stiffnesses contribute to a model of the global stiffness. Once this model stiffness is derived, all forces acting on the system are used to assess global deformations, from which are calculated strains and stresses.

- The Rayleigh–Ritz derivation of element stiffness begins with a relationship for potential energy of a system, from which is obtained an integral for the element stiffness. Assumed displacement fields are used to solve for this integral. The Rayleigh–Ritz method is an elegant one for deriving element stiffness matrices. Moreover, as is shown for the bar element, this method is a good model of Hooke's law.

- Nonlinear, elastic analyses are useful for understanding the impact external forces have on fractured rock masses.

- Finite difference analysis is used to approximate the solution to Laplace's equation for steady-state fluid flow, hence it is useful for analyzing regional groundwater flow.

Exercises

(An asterisk indicates an advanced or graduate student problem)

1. In deriving the element stiffness matrix for a bar element, vis-à-vis Figure 8.6 or 8.7, why is the starting point an assumed displacement?

2. Calculate the element stiffness matrix for the following bar element, with an orientation as shown. Assume the following material properties: $E = 1000.0$ MPa, $A = 0.25$ m^2.

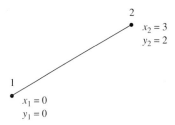

*3. This chapter presents a Rayleigh–Ritz derivation for the element stiffness matrices for bar elements and 2D, solid, isoparametric, quadrilateral elements. A three-node element, a constant strain triangle, may be used in finite element analysis whose element stiffness matrix is intermediate in complexity between the bar element and the 2D, solid, quadrilateral element. Use the Rayleigh–Ritz method to derive the element stiffness matrix for the constant strain triangular element sketched below. Because this is a constant strain element, assume that the element stiffness matrix is obtained as $[k] = [B]^T[E][B]tA$, i.e., without integration. The matrix $[E]$ is that used in this chapter for the quadrilateral element, either for plane stress or plane strain. The element thickness is t, and the area of the element is A. This is a solid element.

In using the Rayleigh–Ritz method to derive this element stiffness matrix, the following assumed displacement fields may be used:

$$u = a_1 + a_2x + a_3y$$
$$v = a_4 + a_5x + a_6y$$

A local coordinate system is not required when deriving the element stiffness matrix for this element.

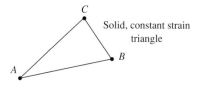

4. Calculate the element stiffness matrix for the following two examples of quadrilateral, isoparametric elements. Assume that Young's modulus is 10 (units) and Poisson's ratio is 0.3. Also assume that the thickness of the element t is equal to 2 for each element.

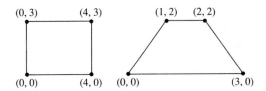

*5. Given the derivation of the element stiffness matrix for a 2D, solid, quadri-lateral, isoparametric element, use the Rayleigh–Ritz method to derive the element stiffness matrix for a 3D, eight-node, isoparametric "brick" element. *Hint:* Use an additional local and an additional global coordinate axis. The modification of the assumed displacement fields is straightforward using the one additional coordinate direction. Take the derivation only up to the point of deriving the matrix [B], because this text does not present the 3D version of the matrix [E].

6. Make a list of all subroutines of the program QUAD and label each according to the function it performs in the program.

7. Use the program TRUSS to analyze the following truss systems (all taken from Hibbeler (1974, p. 187)). Assume that Young's modulus is 1000 lb/ft² and that cross-sectional area A is 0.25 ft² for each truss system. Use

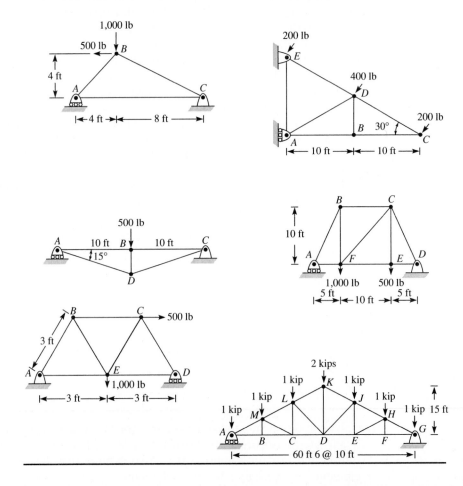

the program MESHDRAW to verify the correctness of each data file (for each mesh) prior to running the program TRUSS. Use MESHDRAW to assess, for each truss system, where tension exists in the mesh and where elements are in compression.

8. Use the program MESHDRAW to display the x-direction and y-direction stresses for the analysis presented in Section 8.10.1.

9. Use QUAD to assess the behavior of vertical (90 degree) fractures, instead of horizontal fractures, for the analysis presented in Section 8.10.2. Use the program MESHDRAW to display the results.

10. Use QUAD to assess the behavior of horizontal fractures, instead of vertical fractures, for the analysis presented in Section 8.10.4.

11. Use the program FDIFF to analyze the groundwater system in Figures 8.48 through 8.50, assuming a sinusoidal water table modeled as

$$h(x, y_0) = y_0 + x \tan(p) + a\{[\sin(bx/\cos(p)]/\cos(p)\}$$

where $p = 1.5°$, $y_0 = 500$ m, $a = 3$ m, and

$$b = \frac{2\pi}{\lambda}, \qquad \lambda = 120 \text{ m}$$

This problem is similar to one given in Wang and Anderson (1982, p. 38). Draw contours by hand of the output from FDIFF. Compare these contours to those obtained in Section 8.12, Figure 8.53.

References and Suggested Readings

CARR, J.R. 1985. A note on the implementation of a continuum finite element approach for stress analysis in jointed rock. *Proc. 26th U.S. Symposium on Rock Mechanics,* Vol. 1. Amsterdam: Balkema, pp. 613–620.

CARR, J.R. 1992. A finite element procedure for calculation of strain from interpolated displacements. *International Journal for Numerical and Analytical Methods in Geomechanics* **16**(3): 211–226.

CHAPRA, S.C., and R.P. CANALE. 1985. *Numerical Methods for Engineers With Personal Computer Applications.* New York: McGraw-Hill.

COOK, R.D. 1974. *Concepts and Applications of Finite Element Analysis.* New York: Wiley.

GOODMAN, R.E. 1976. *Methods of Geological Engineering (in Discontinuous Rock).* St. Paul, MN: West.

GOODMAN, R.E., R.L. TAYLOR, and T.L. BREKKE. 1968. A model for the mechanics of jointed rock. *Journal of the Soil Mechanics and Foundations Division (ASCE)* **94**: 637.

HIBBELER, R.C. 1974. *Engineering Mechanics: Statics.* New York: Macmillan.

POPOV, E.P. 1976. *Mechanics of Materials,* 2d ed. Englewood Cliffs, NJ: Prentice-Hall.

SMITH, I.M. 1982. *Programming the Finite Element Method (with Application to Geomechanics).* New York: Wiley.

TOTH, J. 1962. A theory of groundwater motion in small drainage basins in central Alberta, Canada. *Journal of Geophysical Research* **67**(11): 4375–4387.

WANG, H.F., and M.P. ANDERSON. 1982. *Introduction to Groundwater Modeling: Finite Difference and Finite Element Methods.* New York: Freeman.

ZIENKIEWICZ, O.C., and G.N. PANDE. 1977. Time-dependent multilaminate model of rocks: A numerical study of deformation and failure of rock masses. *International Journal for Numerical and Analytical Methods for Geomechanics* **1**: 219–247.

Figure 8.14 The listing of the program TRUSS.

```
      PROGRAM TRUSS
C
C  SIMPLE FINITE ELEMENT PROGRAM FOR ANALYSIS
C  OF TRUSS (BAR ELEMENT) SYSTEMS
C
C*******************************************************
C                  USER'S GUIDE:   TRUSS
C*******************************************************
C
C  ALL DATA ENTRY IS BY FREE FORMAT
C
C  DATA FILE:
C
C  LINE 1:   READ(5,*) NNODES,NELEMS,NMATLS,NFORCE,MBAND,NBOUND
C  WHERE:
C
C       NNODES = TOTAL NO. OF NODAL POINTS IN MESH
C       NELEMS = TOTAL NO. OF ELEMENTS IN MESH
C       NMATLS = TOTAL NO. OF DIFFERENT MATERIAL TYPES
C                OR MATERIAL CONFIGURATIONS COMPRISING
C                ELEMENTS IN THE MESH.  FOR EXAMPLE,
C                BAR ELEMENTS MAY BE MADE OF STEEL AND
C                ALUMINUM, HENCE THERE ARE TWO MATERIAL
C                TYPES; OR BAR ELEMENTS MAY BE MADE OF
C                STEEL OF TWO DIFFERENT CROSS SECTIONAL
C                SIZES (OR MORE CROSS SECTIONAL SIZES)
C                IN THIS CASE, EACH SIZE IS TREATED AS
C                A DIFFERENT MATERIAL
C       NFORCE = TOTAL NUMBER OF NODAL POINTS ASSOCIATED
C                WITH ACTIVE FORCES
C       MBAND  = MAXIMUM DIFFERENCE IN NODAL POINT NUMBERS
C                FOR ANY ELEMENT
C       NBOUND = TOTAL NUMBER OF NODAL POINTS ASSOC. WITH
C                BOUNDARY CONDITIONS
C       NOTE:  MAXIMUM SIZES FOR THESE PARAMETERS, UNDER
C              CURRENT DIMENSIONING ARE:
C              NNODES = 100
C              NELEMS = 100
C              NMATLS = 10
C              NFORCE = 100
C              MBAND  = 49
C              NBOUND = 100
C  LINE GROUP 2:   NODAL POINT COORDINATES:  NUMBER OF TOTAL
C                  LINES FOR THIS GROUP EQUALS NNODES:
C                  DO I = 1, NNODES
C                    READ(5,*) J, X(J), Y(J)
C                  END DO
C                  WHERE:
C                    J = NODAL POINT NUMBER
C                    X(J) = X-COORDINATE, NODAL POINT J
C                    Y(J) = Y-COORDINATE, NODAL POINT J
C  LINE GROUP 3:   ELEMENT CONNECTIVITY; TOTAL NUMBER OF LINES
C                  FOR THIS GROUP EQUALS NELEMS:
C                  DO I = 1, NELEMS
C                    READ(5,*) J, IL(J), IK(J), MAT(J)
C                  END DO
C                  WHERE:
C                    J = ELEMENT NUMBER
C                    IL(J) = NODAL NUMBER, A
C                    IK(J) = NODAL NUMBER, B
C                    MAT(J) = MATERIAL NUMBER FOR ELEMENT
C                  NOTES:   DEFINE IL AND IK SUCH THAT
C                           IL < IK; FOR MAT, ENTER THE
C                           MATERIAL PROPERTY NUMBER (INTEGER)
C                           DESIGNATING THE MATERIAL PROPERTIES
C                           TO BE DEFINED IN THE NEXT GROUP
C                           OF LINES
C  LINE GROUP 4:   MATERIAL PROPERTIES; TOTAL NUMBER OF LINES
C                  FOR THIS GROUP EQUALS NMATLS:
C                  DO I = 1, NMATL
C                    READ(5,*) A(I), E(I)
C                  END DO
C                  WHERE:
C                    A = CROSS-SECTIONAL AREA FOR BAR
C                    E = YOUNG'S MODULUS FOR MATERIAL
C  LINE GROUP 5:   NODAL ACTIVE FORCE INFORMATION; TOTAL
C                  NUMBER OF LINES = 2 * NFORCE:
```

Figure 8.14 continued

```fortran
C
C
C
C
C
C
      DO I = 1, NFORCE
      READ(5,*) NODE
      READ(5,*) LOADX, LOADY
      END DO
C     WHERE:
C
C           NODE = NODAL POINT NUMBER ASSOC. WITH
C                  AN ACTIVE FORCE OR FORCES;
C           LOADX = X-DIRECTION FORCE; ENTER 0.0 IF
C                   ONLY THE Y-DIRECTION IS LOADED
C           LOADY = Y-DIRECTION FORCE; ENTER 0.0 IF
C                   ONLY THE X-DIRECTION IS LOADED
C
C
C LINE GROUP 6:  BOUNDARY CONDITION INFORMATION; THE NUMBER OF
C     LINES IN THIS GROUP EQUALS NBOUND:
      DO I = 1, NBOUND
      READ(5,*) NODE, IX, IY
      END DO
C     WHERE:
C
C           NODE = NODAL NUMBER ASSOC. WITH A
C                  BOUNDARY CONDITION
C           IX   = 0, IF X-DIRECTION IS FREE TO
C                     MOVE FOR THIS NODAL POINT
C                = 1, IF X-DIRECTION IS RESTRICTED
C           IY   = 0, IF Y-DIRECTION IS FREE TO
C                     MOVE FOR THIS NODAL POINT
C                = 1, IF Y-DIRECTION IS RESTRICTED
C
C*********************************************************
C          END OF USER'S GUIDE AND DISCUSSION
C*********************************************************
C
C     STEP 1:   ESTABLISH ARRAYS AND SIZES
C
      PARAMETER (NNN=100, NEM=100, NMT=10, NBN=100)
      COMMON /OPT/ NNODES, NELEMS, NMATLS, NFORCE
      COMMON /SYS/ STIFF(2*NNN,NBN), LOADS(2*NNN), DISP(2*NNN)
      COMMON /LOCAL/ XSTIF(4,4), DELX, DELY, XA, XE
      COMMON /XDAT/ X(NNN), Y(NNN), IL(NEM), IK(NEM),
     2              A(NMT), E(NMT), MAT(NEM)
      COMMON /BOUND/ NA(NNN,3)
      REAL LOADS
C
C     STEP 2:    OBTAIN MESH INFORMATION
C
C     2A:        OPEN DATA FILE FOR INPUT, UNIT=5
C
      OPEN (5, FILE = ' ')
C
C     2B.        OPEN DATA FILE FOR OUTPUT, UNIT = 1
C
      OPEN(1, FILE = ' ', STATUS = 'NEW')
C
C     2C:   READ INITIAL VALUES
C
      READ(5,*) NNODES, NELEMS, NMATLS, NFORCE, MBAND, NBOUND
      MBAND = (MBAND + 1) * 2
C
C     2D:   READ NODAL COORDINATES
C
      DO 10 I = 1, NNODES
      READ(5,*) J, X(J), Y(J)
   10 CONTINUE
C
C     2E:   READ ELEMENT CONNECTIVITY
C
      DO 20 I = 1, NELEMS
      READ(5,*) J, IL(J), IK(J), MAT(J)
   20 CONTINUE
C
C     2F:   READ MATERIAL INFORMATION
C
      IF (NMATLS .EQ. 1) THEN
      READ(5,*) A(1), E(1)
      ELSE
      DO 30 I = 1, NMATLS
      READ(5,*) J, A(J), E(J)
   30 CONTINUE
      ENDIF
C
```

Figure 8.14 continued

```fortran
C
C     2G:  READ NODAL FORCE (LOAD) INFORMATION
C
      IF (NFORCE .EQ. 1) THEN
        READ (5,*) JL
        KL = (JL - 1) * 2
        READ (5,*) LOADS(KL+1), LOADS(KL+2)
      ELSE
        DO 40 I = 1,NFORCE
        READ (5,*) JL
        KL = (JL - 1) * 2
        READ (5,*) LOADS(KL+1), LOADS(KL+2)
40      CONTINUE
      ENDIF
C
C     2H:  READ BOUNDARY CONDITION INFORMATION
C
      IF (NBOUND .EQ. 1) THEN
        READ(5,*) NA(1,1), NA(1,2), NA(1,3)
      ELSE
        DO 45 I = 1,NBOUND
        READ(5,*) NA(I,1), NA(I,2), NA(I,3)
45      CONTINUE
      ENDIF
C
C     2I:  ECHO THE INPUT INFORMATION
C
      CALL ECHO (NBOUND)
C
C     STEP 3:  FORM SYSTEM STIFFNESS MATRIX, [K]
C
C     3A:  ZERO THE STIFFNESS MATRIX, [K]
C
      DO 50 I = 1,200
      DO 50 J = 1,100
      STIFF(I,J) = 0.0
50    CONTINUE
C
C     3B:  LOOP OVER NELEMS TO POPULATE STIFF
C
      DO 100 I = 1,NELEMS
C


C
C     3C:  FIND NODE NUMBERS, THIS ELEMENT
C
      J = IL(I)
      K = IK(I)
      DELX = X(K)  - X(J)
      DELY = Y(K)  - Y(J)
C
C     3D:  FIND MATERIAL PROPERTIES, THIS ELEMENT
C
      L = MAT(I)
      XA = A(L)
      XE = E(L)
C
C     3E:  FORM THE ELEMENT STIFFNESS MATRIX
C
      CALL MATLOC
C
C     3F:  PUT THE ELEMENT STIFFNESS MATRIX INTO [K]
C
      CALL PUT (J,K)
100   CONTINUE
C
C     STEP 3G:  MODIFY [K] FOR BOUNDARY CONDITIONS
C
      CALL MOD (NBOUND)
C
C     STEP 4:  SOLVE FOR NODAL DISPLACEMENTS
C
      CALL EQSOLV (MBAND)
C
C     STEP 4A:  PRINT DISPLACEMENT RESULTS
C
      WRITE(*,110)
110   FORMAT(1H1, 30X, 'NODAL DISPLACEMENTS', //,
     2 10X,'NODAL POINT',10X,'X-DISP',10X,'Y-DISP',/)
      DO 120 I = 1,NNODES
      K1 = (I-1)*2 + 1
      K2 = (I-1)*2 + 2
      WRITE (*,130) I, DISP(K1), DISP(K2)
      WRITE(1,*) I, DISP(K1), DISP(K2)
120   CONTINUE
130   FORMAT(11X,1I10, 6X, 1E10.3, 6X, 1E10.3)
C
```

Figure 8.14 continued

Figure 8.14 continued

```
C
C     STEP 5:  SOLVE FOR STRESS & STRAIN, EACH ELEMENT
C
      CALL STRAIN
C
C     END EXECUTION
C
      STOP
      END
C
C     SUBROUTINE EQSOLV (MBAND)
C
C     SUBROUTINE TO SOLVE FOR NODAL DISPLACEMENTS
C
      PARAMETER (NNN=100, NEM=100, NMT=10, NBN=100)
      COMMON /OPT/ NNODES, NELEMS, NMATLS, NFORCE
      COMMON /SYS/ S(2*NNN,NBN), R(2*NNN), DISP(2*NNN)
C
C     ESTABLISH ARRAY SIZE, NSIZE
C
      NSIZE = NNODES * 2
C
C     BEGIN FORWARD REDUCTION
C
      DO 100 N = 1,NSIZE
      DO 90 L = 2,MBAND
      IF (S(N,L) .EQ. 0.)  GO TO 90
      I = N + L - 1
      C = S(N,L) / S(N,1)
      J = 0
      DO 80 K = L,MBAND
      J = J + 1
      S(I,J) = S(I,J) - C*S(N,K)
80    S(N,L) = C
90    CONTINUE
100   CONTINUE
C

C
C     FORWARD REDUCTION OF NODAL FORCES
C
      DO 200 N = 1,NSIZE
      DO 150 L = 2,MBAND
      IF (S(N,L) .EQ. 0.)  GO TO 150
      I = N + L - 1
      R(I) = R(I) - S(N,L) * R(N)
150   R(N) = R(N) / S(N,1)
200   CONTINUE
C
C     BACK SUBSTITUTE TO SOLVE FOR UNKNOWNS
C
      DO 300 II = 1,NSIZE
      DISP(II) = R(II)
300
      DO 400 M = 2,NSIZE
      N = NSIZE + 1 - M
      DO 350 L = 2,MBAND
      IF (S(N,L) .EQ. 0.)   GO TO 350
      K = N + L - 1
      DISP(N) = DISP(N) - S(N,L) * DISP(K)
350   CONTINUE
400   CONTINUE
      RETURN
      END
C
C     SUBROUTINE MATLOC
C
C     THIS SUBROUTINE CALCULATES THE ELEMENT STIFFNESS
C     MATRIX FOR EACH ELEMENT
C
      COMMON /LOCAL/ XSTIF(4,4), DELX, DELY, XA, XE
C
C     STEP 1:  CALCULATE ELEMENT LENGTH, COS & SINE TERMS
C
      XL = SQRT(DELX **2 + DELY ** 2)
      XCOS = DELX / XL
      XSIN = DELY / XL
      C2 = XCOS ** 2
      S2 = XSIN ** 2
      CS = XCOS * XSIN
C
```

Figure 8.14 continued

```
C
C      STEP 2:  FORM STIFFNESS MATRIX
C
       XSTIF (1,1) = C2
       XSTIF (1,2) = CS
       XSTIF (1,3) = - C2
       XSTIF (1,4) = - CS
       XSTIF (2,2) = S2
       XSTIF (2,3) = -CS
       XSTIF (2,4) = - S2
       XSTIF (3,3) = C2
       XSTIF (3,4) = CS
       XSTIF (4,4) = S2
C
C      COMPLETE BY SYMMETRY
C
       DO 10 I = 1,4
       DO 10 J = 1,4
       XSTIF(J,I) = XSTIF(I,J)
10     CONTINUE
C
C      MULTIPLY BY MATERIAL CONSTANT
C
       XK = (XA * XE) / XL
       DO 20 I = 1,4
       DO 20 J = 1,4
       XSTIF(I,J) = XSTIF(I,J) * XK
20     CONTINUE
       RETURN
       END
C
C      SUBROUTINE MOD (N)
C
C      THIS SUBROUTINE MODIFIES THE GLOBAL STIFFNESS MATRIX
C      TO ACCOUNT FOR BOUNDARY CONDITIONS.  PROCEDURE:
C      TO RESTRICT DISPLACEMENT, MAKE STIFFNESS INFINITELY
C      LARGE.
C
       PARAMETER (NNN=100, NEM=100, NMT=10, NBN=100)
       COMMON /SYS/ STIFF(2*NNN,NBN), LOADS(2*NNN), DISP(2*NNN)
       COMMON /BOUND/ NA(NNN,3)
       REAL LOADS
       IF (N .EQ. 1) THEN
         K = NA(1,1)
         M = (K - 1) * 2
         IF (NA (1,2) .EQ. 1) THEN
           STIFF (M+1,1) = 10.0 ** 15
         ENDIF
         IF (NA (1,3) .EQ. 1) THEN
           STIFF (M+2,1) = 10.0 ** 15
         ENDIF
       ELSE
         DO 10 I = 1,N
         K = NA (I,1)
         M = (K - 1) * 2
         IF (NA (I,2) .EQ. 1)  THEN
           STIFF (M+1,1) = 10.0 ** 15
         ENDIF
         IF (NA (I,3) .EQ. 1) THEN
           STIFF (M+2,1) = 10.0 ** 15
         ENDIF
10       CONTINUE
       ENDIF
       RETURN
       END
       SUBROUTINE PUT (J,K)
C
C      THIS SUBROUTINE DISASSEMBLES THE ELEMENT STIFFNESS
C      MATRIX AND PLACES IT INTO THE GLOBAL STIFFNESS
C      MATRIX, [K], AND ALSO BANDS THE MATRIX, [K]
```

Figure 8.14 continued

```fortran
c
      PARAMETER (NNN=100, NEM=100, NMT=10, NBN=100)
      COMMON /SYS/ STIFF(2*NNN,NBN), LOADS(2*NNN), DISP(2*NNN)
      COMMON /LOCAL/ XSTIF(4,4), DELX, DELY, XA, XE
      DIMENSION KK(4)
      REAL LOADS
      KK(2) = 2 * J
      KK(4) = 2 * K
      KK(1) = KK(2) - 1
      KK(3) = KK(4) - 1
      DO 100 I = 1,4
      II = KK(I)
      DO 100 M = 1,4
      IF (KK(M) .LT. II) GO TO 100
      JJ = KK(M) - II + 1
      STIFF(II,JJ) = STIFF(II,JJ) + XSTIF(I,M)
  100 CONTINUE
      RETURN
      END
c
c
      SUBROUTINE STRAIN
c
c
c     SUBROUTINE TO COMPUTE STRAIN, THEN STRESS WITHIN
c     EACH BAR ELEMENT COMPONENT OF A TRUSS SYSTEM
c
      PARAMETER (NNN=100, NEM=100, NMT=10, NBN=100)
      COMMON /SYS/ STIFF(2*NNN,NBN), LOADS(2*NNN), DISP(2*NNN)
      COMMON /XDAT/ X(NNN), Y(NNN), IL(NEM), IK(NEM),
     2              A(NMT), E(NMT), MAT(NEM)
      COMMON /OPT/ NNODES, NELEMS, NMATLS, NFORCE
      DIMENSION STRN(NEM), STRESS(NEM)
c
c     STEP 1:  CALCULATE STRAIN IN EACH ELEMENT WHERE
c              STRAIN = CHANGE IN LENGTH/ORIG. LENGTH
c
      DO 100 I = 1,NELEMS
      K = IL(I)
      J = IK(I)
      DELX = X(J) - X(K)
      DELY = Y(J) - Y(K)
      XL = SQRT(DELX ** 2 + DELY ** 2)
      OX1 = X(J)
      OX2 = X(K)
      OY1 = Y(J)
      OY2 = Y(K)
      IKX1 = (K - 1) * 2 + 1
      IKY1 = IKX1 + 1
      IJX1 = (J - 1) * 2 + 1
      IJY1 = IJX1 + 1
      DX1 = OX1 + DISP(IJX1)
      DY1 = OY1 + DISP(IJY1)
      DX2 = OX2 + DISP(IKX1)
      DY2 = OY2 + DISP(IKY1)
      DIFX = DX1 - DX2
      DIFY = DY1 - DY2
      XL1 = SQRT (DIFX ** 2 + DIFY ** 2)
      DIFL = XL - XL1
      STRN(I) = DIFL / XL
  100 CONTINUE
c
c     STEP 2:  CALCULATE STRESS IN EACH ELEMENT;
c              STRESS = STRAIN * ELASTIC MODULUS
c
      DO 200 I = 1,NELEMS
      J = MAT(I)
      EX = E(J)
      STRESS(I) = EX * STRN(I)
  200 CONTINUE
c
c     STEP 3:  PRINT RESULTS
c
      WRITE(*,500)
      DO 400 I = 1,NELEMS
      WRITE(*,510) I, STRESS(I)
      WRITE(1,510) I, STRESS(I)
  400 CONTINUE
  500 FORMAT(1H1, 30X, 'STRESS IN ELEMENTS', //,
     2       13X, 'ELEMENT', 14X, 'STRESS', /,)
  510 FORMAT(10X, 1I10, 10X, 1E15.3)
      RETURN
      END
c
```

Figure 8.14 continued

```fortran
C
      SUBROUTINE ECHO (NBOUND)
C
C     A SUBROUTINE TO PRINT THE INPUT INFORMATION
C
      PARAMETER (NNN=100, NEM=100, NMT=10, NBN=100)
      COMMON /OPT/ NNODES, NELEMS, NMATLS, NFORCE
      COMMON /XDAT/ X(NNN), Y(NNN), IL(NEM), IK(NEM),
     2              A(NMT), E(NMT), MAT(NEM)
      COMMON /SYS/ S(2*NNN,NBN), LOADS(2*NNN), DISP(2*NNN)
      COMMON /BOUND/ NA(NNN,3)
      REAL LOADS
C
C     PRINT INITIAL PARAMETER SETTINGS
C
      WRITE(*,10)
10    FORMAT(1H1)
      WRITE(*,20) NNODES, NELEMS, NMATLS, NFORCE
20    FORMAT(10X,'FINITE ELEMENT ANALYSIS:  TRUSS SYSTEMS',//,
     2       10X,'PARAMETERS:',//,
     3       10X,'NUMBER OF NODES               = ',I5,/,
     4       10X,'NUMBER OF ELEMENTS            = ',I5,/,
     5       10X,'NUMBER OF MATERIALS           = ',I5,/,
     6       10X,'NUMBER OF NODE POINTS W. FORCES  = ',I5,/)
C
C     ECHO NODAL POINT COORDINATES
C
      WRITE(*,30)
30    FORMAT(///,10X,'NODAL POINT COORDINATES',//,
     2       16X,'NODE',19X,'X',19X,'Y',/)
      DO I = 1, NNODES
         WRITE(*,40) I, X(I), Y(I)
      END DO
40    FORMAT(10X,I10,2E20.8)
C
C     ECHO THE ELEMENT CONNECTIVITY
C
      WRITE(*,50)
50    FORMAT(///,10X,'ELEMENT CONNECTIVITY',//,
     2       13X,'ELEMENT',14X,'NODE-A',14X,'NODE-B',
     3       12X,'MATERIAL',/)
      DO I = 1, NELEMS
         WRITE(*,60) I, IL(I), IK(I), MAT(I)
      END DO
60    FORMAT(4I20)
C
C     ECHO THE FORCE INFORMATION
C
      WRITE(*,70)
70    FORMAT(///,10X,'ACTIVE FORCE INFORMATION:',/,
     2       16X,'NODE',13X,'FORCE-X',13X,'FORCE-Y',/)
      DO I = 1,NNODES
         JL = I*2
         JK = JL - 1
         WRITE(*,80) I, LOADS(JK), LOADS(JL)
      END DO
80    FORMAT(I20,2E20.8)
C
C     ECHO THE BOUNDARY CONDITION INFORMATION
C
      WRITE(*,90)
90    FORMAT(///,10X,/'BOUNDARY CONDITIONS:',//,
     2       16X,'NODE',13X,'X-BOUND',13X,'Y-BOUND',/)
      DO I = 1, NBOUND
         WRITE(*,100) NA(I,1), NA(I,2), NA(I,3)
      END DO
100   FORMAT(3I20)
C
C     RETURN TO CALLING PROGRAM
C
      RETURN
      END
```

Figure 8.15 The listing of the program MESHDRAW.

```fortran
C     PROGRAM MESHDRAW
C
C     AN INTERACTIVE, 2D GRAPHICS PROGRAM FOR DRAWING
C     ORIGINAL AND DEFORMED 2D FINITE ELEMENT MESHES.
C
C     ************************************************
C                    USER'S GUIDE
C     ************************************************
C
C     QUESTION 1:  WILL A MESH OF BAR ELEMENTS OR A MESH OF 2D SOLID
C                  ELEMENTS BE PLOTTED?
C                  ENTER 1, FOR BAR ELEMENTS;  2 FOR 2D SOLID ELEM.
C
C     QUESTION 2:  ENTER THE FILE NAME STORING THE ORIGINAL NODAL
C                  POINT COORDINATES AND ELEMENT CONNECTIVITY DATA
C                  THAT IS, THE DATA FILE FOR EITHER TRUSS OR QUAD,
C                  WHICH EVER IS APPLICABLE.
C
C     QUESTION 3:  WILL THE ORIGINAL MESH BE DRAWN, OR THE DEFORMED
C                  MESH?  ENTER 1 FOR THE ORIGINAL MESH;  2 FOR THE
C                  DEFORMED MESH.
C
C     QUESTIONS 3A (ASKED ONLY IF A DEFORMED MESH IS TO BE PLOTTED):
C
C                  WHAT MAGNIFICATION FACTOR WILL BE USED TO PLOT
C                  DEFORMATIONS:  ENTER 0.6, 1.0, 10.0, 1000.0, ETC.
C
C                  WHAT IS THE FILE NAME WHICH STORES THE
C                  DISPLACEMENTS?  (OUTPUT FILE FROM TRUSS OR QUAD)
C                  IF 2D QUAD ELEMENTS ARE USED, ONE ADDITIONAL
C                  QUESTION IS ASKED:
C                      WHICH STRESS COMPONENT IS TO BE MAPPED?:
C                      ENTER 1, FOR STRESS IN X-DIRECTION
C                      ENTER 2, FOR STRESS IN Y-DIRECTION
C                      ENTER 3, FOR SHEAR STRESS
C                      ENTER 4, TO PLOT LOCATIONS OF FAILED FRACTURES
C                          FOR THIS SELECTION, TWO ADDITIONAL
C                          QUESTIONS ARE ASKED:
C                          A.  HOW MANY ROWS OF ELEMENTS ARE
C                              THERE IN THE ORIGINAL MESH
C                          B.  HOW MANY COLUMNS OF ELEMENTS ARE
C                              THERE IN THE ORIGINAL MESH
C                      ENTER 5, FOR A PLOT OF THE DEFORMED MESH
C                              WITHOUT STRESS OR FAILED FRACTURES
C
C     ***********************************************************
C                END OF USER'S GUIDE AND DISCUSSION
C     ***********************************************************
C
      INCLUDE 'FGRAPH.FI'
      INCLUDE 'FGRAPH.FD'
      REAL*8 X(300), Y(300), XX, YY, XMAG, ZDIF, ZMIN, ZMAX
      REAL*8 XMIN, XMAX, YMIN, YMAX, X1(4), Y1(4), XDIF, YDIF
      REAL*8 D(9,2)
      REAL*4 STRESS(900), TEMP(3)
      INTEGER*4 IL(300), IK(300), IP(300), IQ(300), P(900), Q(900)
      INTEGER*4 IFAIL(900,3)
      INTEGER*2 DUMMY2
      RECORD /RCCOORD/ CURPOS
      RECORD /WXYCOORD/ XY
      CHARACTER*50 TITLE
      ISTRES = 0
C
```

Figure 8.15 continued

```fortran
C      ACCESS ALL INPUT DATA INTERACTIVELY
C
       WRITE(*,*) 'IS THE MESH TO BE DRAWN COMPRISED OF BAR ELEMENTS'
       WRITE(*,*) 'OR 2D SOLID ELEMENTS?'
       WRITE(*,*) 'ENTER 1 FOR BAR ELEMENTS; 2 FOR 2D SOLID ELEMS'
       READ(*,*) ITYPE
       WRITE(*,*)
       WRITE(*,*) 'PLEASE ENTER THE NAME OF THE DATA FILE, UNIT 5,'
       WRITE(*,*) 'WHICH STORES THE ORIGINAL NODAL COORDINATES'
       WRITE(*,*) 'AND ELEMENT CONNECTIVITY'
       OPEN(5, FILE = ' ')
       WRITE(*,*)
       WRITE(*,*)
       WRITE(*,*) 'WILL THE ORIGINAL OR THE DEFORMED MESH BE DRAWN?'
       WRITE(*,*) 'ENTER 1, FOR THE ORIGINAL, OR 2 FOR THE DEFORMED'
       READ(*,*) IDRAW
       WRITE(*,*)
       WRITE(*,*)
       IF (IDRAW .EQ. 2) THEN
       WRITE(*,*) 'ENTER THE MAGNIFICATION FACTOR FOR THE PLOTTING'
       WRITE(*,*) 'OF THE DEFORMED MESH:  .5, 1.0, 1000., ETC.'
       READ(*,*) XMAG
       WRITE(*,*)
       WRITE(*,*)
       WRITE(*,*) 'ENTER THE NAME OF THE FILE, UNIT 1, WHICH'
       WRITE(*,*) 'STORES THE DISPLACEMENTS'
       OPEN (1, FILE = ' ')
       IF (ITYPE .EQ. 2) THEN
       WRITE(*,*) 'WHICH STRESS MAP WOULD YOU LIKE?'
       WRITE(*,*) 'ENTER:  1, FOR X-DIRECTION STRESS'
       WRITE(*,*) '        2, FOR Y-DIRECTION STRESS'
       WRITE(*,*) '        3, FOR SHEAR STRESS'
       WRITE(*,*) '        4, FOR FRACTURE FAILURE DIAGRAM'
       WRITE(*,*) '        5, FOR OUTLINE OF DEFORMED MESH ONLY'
       READ(*,*) ISTRES
       IF (ISTRES .EQ. 4) THEN
       WRITE(*,*) 'HOW MANY ROWS OF ELEMENTS ARE IN THE'
       WRITE(*,*) 'ORIGINAL MESH?'
       READ(*,*) NROWS
       WRITE(*,*) 'HOW MANY COLUMNS OF ELEMENTS ARE IN THE'
       WRITE(*,*) 'ORIGINAL MESH?'
       READ(*,*) NCOLS
       END IF
       END IF
       END IF
C
C      READ THE NODAL COORDINATE AND ELEMENT CONNECTIVITY DATA
C
       IF (ITYPE .EQ. 1) THEN
       READ(5,*) NNODES, NELEMS, NMATLS, NFORCE, MBAND, NBOUND
       DO I = 1, NNODES
       READ(5,*) J, X(J), Y(J)
       END DO
       DO I = 1, NELEMS
       READ(5,*) J, IL(J), IK(J), NM
       END DO
       ELSEIF (ITYPE .EQ. 2) THEN
       N = 1
10     CONTINUE
       READ(5,*,END=20) IL(N),IK(N),IP(N),IQ(N)
       READ(5,*) X(IL(N)),Y(IL(N)),
     2           X(IK(N)),Y(IK(N)),
     3           X(IP(N)),Y(IP(N)),
     4           X(IQ(N)),Y(IQ(N))
       N = N + 1
       GO TO 10
20     NELEMS = N - 1
C
C      CALCULATE NNODES
C
       NNODES = 0
       DO I = 1,NELEMS
       IF (IL(I) .GT. NNODES) NNODES = IL(I)
       IF (IK(I) .GT. NNODES) NNODES = IK(I)
       IF (IP(I) .GT. NNODES) NNODES = IP(I)
       IF (IQ(I) .GT. NNODES) NNODES = IQ(I)
       END DO
       END IF
C
```

Figure 8.15 continued

```
C
C     UPDATE NODAL COORDINATES IF THE DEFORMED MESH IS PLOTTED
C
      IF (IDRAW .EQ. 2) THEN
         DO I = 1, NNODES
            READ(1,*) J, XX, YY
            X(J) = X(J) + XX*XMAG
            Y(J) = Y(J) + YY*XMAG
         END DO
         IF (ITYPE .EQ. 2 .AND. ISTRES .EQ. 4) THEN
            NFAIL = 1
30          CONTINUE
            READ(1,*,END=40) XX1,XX2,XX3,P(NFAIL),Q(NFAIL),
     2                       (IFAIL(NFAIL,JK),JK=1,3)
            NFAIL = NFAIL + 1
            GO TO 30
40          NFAIL = NFAIL - 1
         END IF
      DO I = 1, NELEMS
         IF (ITYPE .EQ. 1) THEN
            READ(1,*) J, STRESS(J)
         ELSEIF (ITYPE .EQ. 2 .AND. ISTRES .LE. 3) THEN
            DO J = 1,4
               IZ = (I-1) * 4 + J
               READ(1,*) (TEMP(JK), JK = 1,3)
               STRESS(IZ) = TEMP(ISTRES)
            END DO
         END IF
      END DO
      END IF
      IF (IDRAW .EQ. 2) THEN
C     FIND THE MAXIMUM AND MINIMUM STRESSES
      SMAX = STRESS(1)
      SMIN = STRESS(1)
      IF (ITYPE .EQ. 1) THEN
         DO I = 2, NELEMS
            IF (STRESS(I) .GT. SMAX) THEN
               SMAX = STRESS(I)
            END IF
            IF (STRESS(I) .LT. SMIN) THEN
               SMIN = STRESS(I)
            END IF
         END DO
      ELSEIF (ITYPE .EQ. 2 .AND. ISTRES .LE. 3) THEN
         DO I = 2, NELEMS * 4
            IF (STRESS(I) .GT. SMAX) THEN
               SMAX = STRESS(I)
            END IF
            IF (STRESS(I) .LT. SMIN) THEN
               SMIN = STRESS(I)
            END IF
         END DO
      END IF
C
C     FIND THE MINIMUM AND MAXIMUM X AND Y COORDINATE VALUES
C
      XMIN = X(1)
      XMAX = X(1)
      YMAX = Y(1)
      YMIN = Y(1)
      DO I = 2, NNODES
         IF (X(I) .GT. XMAX) XMAX = X(I)
         IF (Y(I) .GT. YMAX) YMAX = Y(I)
         IF (Y(I) .LT. YMIN) YMIN = Y(I)
         IF (X(I) .LT. XMIN) XMIN = X(I)
      END DO
      XDIF = XMAX - XMIN
      YDIF = YMAX - YMIN
      IF (XDIF .GE. YDIF) THEN
         ZMIN = XMIN
         ZMAX = XMAX
         ZDIF = XMAX - XMIN
      ELSEIF (YDIF .GT. XDIF) THEN
         ZMIN = YMIN
         ZMAX = YMAX
         ZDIF = YMAX - YMIN
      END IF
C
```

Figure 8.15 continued

```fortran
C     DEFINE THE VIDEO MODE
C
      DUMMY2 = SETVIDEOMODE($VRES16COLOR)
C
C     CLEAR THE SCREEN IN PREPARATION FOR PLOTTING
C
      CALL CLEARSCREEN($GCLEARSCREEN)
C
C     DEFINE THE COORDINATE WINDOW
C
      IF (XDIF .EQ. YDIF) THEN
      DUMMY2 = SETWINDOW(.TRUE., (ZMIN-ZDIF/10.), (ZMIN-ZDIF/10.),
     2                          (ZMAX+ZDIF/10.), (ZMAX+ZDIF/10.))
      ELSEIF (XDIF .GT. YDIF) THEN
      DUMMY2 = SETWINDOW(.TRUE., (ZMIN-ZDIF/10.), (ZMIN-ZDIF/10.-
     2         (XDIF-YDIF)/2.), (ZMAX+ZDIF/10.), (ZMAX+ZDIF/10. -
     3         (XDIF-YDIF)/2.))
      ELSEIF (YDIF .GT. XDIF) THEN
      DUMMY2 = SETWINDOW(.TRUE., (ZMIN-ZDIF/10. -(YDIF-XDIF)/2.),
     2         (ZMIN-ZDIF/10.0), (ZMAX+ZDIF/10.0-(YDIF-XDIF)/2.),
     3         (ZMAX+ZDIF/10.))
      ENDIF
C
C     DEFINE THE VIEWING PORT
C
      CALL SETVIEWPORT (201, 26, 600, 425)
C
C     DRAW A BORDER IN BRIGHT WHITE (COLOR 15:   DEFAULT BRT WHITE)
C
      DUMMY2 = SETCOLOR(15)
      IF (XDIF .EQ. YDIF) THEN
      X1(1) = ZMIN - ZDIF/15.
      X1(2) = ZMAX + ZDIF/15.
      Y1(1) = X1(1)
      Y1(2) = X1(2)
      ELSEIF (XDIF .GT. YDIF) THEN
      X1(1) = ZMIN - ZDIF/15.
      Y1(1) = ZMIN - ZDIF/15. - (XDIF-YDIF)/2.
      Y1(2) = ZMAX + ZDIF/15. - (XDIF-YDIF)/2.
      X1(2) = ZMAX + ZDIF/15.
      ELSEIF (YDIF .GT. XDIF) THEN
      X1(1) = ZMIN - ZDIF/15.
      Y1(1) = ZMIN - ZDIF/15.
      Y1(2) = ZMAX + ZDIF/15.
      X1(2) = ZMAX + ZDIF/15. - (YDIF-XDIF)/2.
      END IF
      CALL MOVETO_W(X1(1), Y1(1), XY)
      DUMMY2 = LINETO_W(X1(2), Y1(1))
      DUMMY2 = LINETO_W(X1(2), Y1(2))
      DUMMY2 = LINETO_W(X1(1), Y1(2))
      DUMMY2 = LINETO_W(X1(1), Y1(1))
C
C     DRAW THE MESH:   IN GREEN IF ORIGINAL; OR, IF DEFORMED:
C        RED    = HIGHEST, COMPRESSIONAL STRESS
C        YELLOW = INTERMEDIATE (BETWEEN LOW/HIGH) COMP. STRESS
C        WHITE  = LOWEST COMPRESSIONAL STRESS
C        GREEN  = HIGHEST, TENSILE STRESS
C        MAGENTA = INTERMED. TENSILE STRESS
C        BLUE   = LOWEST, TENSILE STRESS
C
      IF (IDRAW .EQ. 1) KOLOR = 2
      DO I = 1, NELEMS
      L = IL(I)
      K = IK(I)
      IF (ITYPE .EQ. 2) THEN
      LP = IP(I)
      KQ = IQ(I)
      END IF
      IF (IDRAW .EQ. 2) THEN
      IF (ITYPE .EQ. 1) THEN
      IF (STRESS(I) .GT. 0.0) THEN
      KOLOR = INT ((STRESS(I)*3/SMAX) + 1.0)
      IF (KOLOR .LE. 1) THEN
      KOLOR = 7
      ELSEIF (KOLOR .GT. 1 .AND. KOLOR .LE. 3) THEN
      KOLOR = 14
      ELSEIF (KOLOR .GT. 3) THEN
      KOLOR = 4
      END IF
```

Figure 8.15 continued

```fortran
          ELSE
            KOLOR = INT ((STRESS(I)*3/SMIN) + 1.0)
            IF (KOLOR .LE. 1) THEN
              KOLOR = 1
            ELSEIF (KOLOR .GT. 1 .AND. KOLOR .LE. 3) THEN
              KOLOR = 5
            ELSEIF (KOLOR .GT. 3) THEN
              KOLOR = 2
            END IF
          ELSEIF (ITYPE .EQ. 2 .AND. ISTRES .LE. 3) THEN
            KOLOR = 0
            DO J = 1, 4
              IZ = (I-1) * 4 + J
              IF (STRESS(IZ) .GT. 0.0) THEN
                KOLOR = INT ((STRESS(IZ)*3/SMAX)+1.0)
                IF (KOLOR .LE. 1) THEN
                  KOLOR = 7
                ELSEIF (KOLOR .GT. 1 .AND. KOLOR .LE. 3) THEN
                  KOLOR = 14
                ELSEIF (KOLOR .GT. 3) THEN
                  KOLOR = 4
                END IF
              ELSEIF (STRESS(IZ) .LT. 0.0) THEN
                KOLOR = INT ((STRESS(IZ)*3/SMIN) + 1.0)
                IF (KOLOR .LE. 1) THEN
                  KOLOR = 1
                ELSEIF (KOLOR .GT. 1 .AND. KOLOR .LE. 3) THEN
                  KOLOR = 5
                ELSEIF (KOLOR .GT. 3) THEN
                  KOLOR = 2
                END IF
              END IF
              IF (KOLOR .NE. 0) THEN
                DUMMY2 = SETCOLOR(KOLOR)
                D(1,1) = X(L)
                D(1,2) = Y(L)
                D(2,1) = (X(L)+X(KQ))/2.D0
                D(2,2) = (Y(L)+Y(KQ))/2.D0
                D(3,1) = X(KQ)
                D(3,2) = Y(KQ)
                D(4,1) = (X(L)+X(K))/2.D0
                D(4,2) = (Y(L)+Y(K))/2.D0
                D(6,1) = (X(KQ)+X(LP))/2.D0
                D(6,2) = (Y(KQ)+Y(LP))/2.D0
                D(7,1) = X(K)
                D(7,2) = Y(K)
                D(8,1) = (X(K)+X(LP))/2.D0
                D(8,2) = (Y(K)+Y(LP))/2.D0
                D(9,1) = X(LP)
                D(9,2) = Y(LP)
                D(5,1) = (D(8,1)+D(2,1))/2.D0
                D(5,2) = (D(4,2)+D(6,2))/2.D0
                IF (J .EQ. 1) THEN
                  CALL MOVETO_W(D(1,1),D(1,2),XY)
                  DUMMY2 = LINETO_W(D(2,1),D(2,2))
                  DUMMY2 = LINETO_W(D(5,1),D(5,2))
                  DUMMY2 = LINETO_W(D(4,1),D(4,2))
                  DUMMY2 = LINETO_W(D(1,1),D(1,2))
                ELSEIF (J .EQ. 2) THEN
                  CALL MOVETO_W(D(4,1),D(4,2),XY)
                  DUMMY2 = LINETO_W(D(5,1),D(5,2))
                  DUMMY2 = LINETO_W(D(8,1),D(8,2))
                  DUMMY2 = LINETO_W(D(7,1),D(7,2))
                  DUMMY2 = LINETO_W(D(4,1),D(4,2))
                ELSEIF (J .EQ. 3) THEN
                  CALL MOVETO_W(D(2,1),D(2,2),XY)
                  DUMMY2 = LINETO_W(D(3,1),D(3,2))
                  DUMMY2 = LINETO_W(D(6,1),D(6,2))
                  DUMMY2 = LINETO_W(D(5,1),D(5,2))
                  DUMMY2 = LINETO_W(D(2,1),D(2,2))
                ELSEIF (J .EQ. 4) THEN
                  CALL MOVETO_W(D(5,1),D(5,2),XY)
                  DUMMY2 = LINETO_W(D(6,1),D(6,2))
                  DUMMY2 = LINETO_W(D(9,1),D(9,2))
                  DUMMY2 = LINETO_W(D(8,1),D(8,2))
                  DUMMY2 = LINETO_W(D(5,1),D(5,2))
                END IF
              END IF
            END DO
          END IF
        END IF
```

Figure 8.15 continued

```
IF (ITYPE .EQ. 1) THEN
DUMMY2 = SETCOLOR(KOLOR)
CALL MOVETO_W(X(L), Y(L), XY)
DUMMY2 = LINETO_W(X(K), Y(K))
END IF
IF (ITYPE .EQ. 2 ) THEN
IF (IDRAW .EQ. 1 .OR. ISTRES .GT. 3) THEN
DUMMY2 = SETCOLOR(2)
CALL MOVETO_W(X(L),Y(L),XY)
DUMMY2 = LINETO_W(X(K),Y(K))
DUMMY2 = LINETO_W(X(LP),Y(LP))
DUMMY2 = LINETO_W(X(KQ),Y(KQ))
DUMMY2 = LINETO_W(X(L),Y(L))
END IF
IF (ISTRES .EQ. 4) THEN
DO KI = 1, NFAIL
KA = P(KI)
KB = Q(KI)
NUMB = (KA-1)*NROWS + KB
IF (NUMB .EQ. I) THEN
LL = IL(NUMB)
LK = IK(NUMB)
LP = IP(NUMB)
LQ = IQ(NUMB)
D(1,1) = X(LL)
D(1,2) = Y(LL)
D(2,1) = (X(LL)+X(LQ))/2.D0
D(2,2) = (Y(LL)+Y(LQ))/2.D0
D(3,1) = X(LQ)
D(3,2) = Y(LQ)
D(4,1) = (X(LL)+X(LK))/2.D0
D(4,2) = (Y(LL)+Y(LK))/2.D0
D(6,1) = (X(LP)+X(LQ))/2.D0
D(6,2) = (Y(LP)+Y(LQ))/2.D0
D(7,1) = X(LK)
D(7,2) = Y(LK)
D(8,1) = (X(LK)+X(LP))/2.D0
D(8,2) = (Y(LK)+Y(LP))/2.D0
D(9,1) = X(LP)
D(9,2) = Y(LP)
D(5,1) = (D(2,1)+D(8,1))/2.D0
D(5,2) = (D(4,2)+D(6,2))/2.D0
IF (IFAIL(KI,3) .EQ. 1) THEN
DUMMY2 = SETCOLOR(4)
ELSEIF (IFAIL(KI,3) .EQ. 2) THEN
DUMMY2 = SETCOLOR(14)
END IF
IF (IFAIL(KI,1) .EQ. 1 .AND. IFAIL(KI,2) .EQ. 1) THEN
CALL MOVETO_W(D(1,1),D(1,2),XY)
DUMMY2 = LINETO_W(D(2,1),D(2,2))
DUMMY2 = LINETO_W(D(5,1),D(5,2))
DUMMY2 = LINETO_W(D(4,1),D(4,2))
DUMMY2 = LINETO_W(D(1,1),D(1,2))
ELSEIF (IFAIL(KI,1) .EQ. 1 .AND. IFAIL(KI,2) .EQ. 2) THEN
CALL MOVETO_W(D(4,1),D(4,2),XY)
DUMMY2 = LINETO_W(D(5,1),D(5,2))
DUMMY2 = LINETO_W(D(8,1),D(8,2))
DUMMY2 = LINETO_W(D(7,1),D(7,2))
DUMMY2 = LINETO_W(D(4,1),D(4,2))
ELSEIF (IFAIL(KI,1) .EQ. 2 .AND. IFAIL(KI,2) .EQ. 1) THEN
CALL MOVETO_W(D(2,1),D(2,2),XY)
DUMMY2 = LINETO_W(D(3,1),D(3,2))
DUMMY2 = LINETO_W(D(6,1),D(6,2))
DUMMY2 = LINETO_W(D(5,1),D(5,2))
DUMMY2 = LINETO_W(D(2,1),D(2,2))
ELSEIF (IFAIL(KI,1) .EQ. 2 .AND. IFAIL(KI,2) .EQ. 2) THEN
CALL MOVETO_W(D(5,1),D(5,2),XY)
DUMMY2 = LINETO_W(D(6,1),D(6,2))
DUMMY2 = LINETO_W(D(9,1),D(9,2))
DUMMY2 = LINETO_W(D(8,1),D(8,2))
DUMMY2 = LINETO_W(D(5,1),D(5,2))
END IF
END DO
END IF
END IF
END DO
```

c

Figure 8.15 continued

```fortran
C       PLOT A COLOR LEGEND TO SHOW STRESS RANGES
C
        IF (IDRAW .EQ. 2 .AND. ISTRES .LE. 3) THEN
        CALL SETVIEWPORT(1,1,640,480)
        ZK = SMAX / 3.0
        WK = SMIN / 3.0
        IF (SMIN .GT. 0.0) THEN
          WK = 0.0
        END IF
        IF (SMAX .LT. 0.) THEN
          ZK = 0.0
        END IF
        PS = 0.0
        PS1 = ZK
        PS2 = 2*ZK
        PS3 = 3 * ZK
        TS = 0.0
        TS1 = WK
        TS2 = 2 * WK
        TS3 = 3 * WK
        DUMMY2 = SETCOLOR(4)
        DUMMY2 = RECTANGLE($GFILLINTERIOR,10,32,30,52)
        CALL SETTEXTPOSITION(1,1,CURPOS)
        WRITE(*,99)
99      FORMAT( )
        CALL SETTEXTPOSITION (1,7,CURPOS)
        WRITE(*,*) 'STRESS RANGE'
        CALL SETTEXTPOSITION(3,8,CURPOS)
        WRITE(*,99)
        CALL SETTEXTPOSITION(3, 8,CURPOS)
        WRITE(*,100) PS2, PS3
100     FORMAT(E8.2,' - ',E8.2, )
        DUMMY2 = SETCOLOR(14)
        DUMMY2 = RECTANGLE($GFILLINTERIOR,10,64,30,84)
        CALL SETTEXTPOSITION(5, 8, CURPOS)
        WRITE(*,100) PS1, PS2
        DUMMY2 = SETCOLOR(7)
        DUMMY2 = RECTANGLE($GFILLINTERIOR,10,96,30,116)
        CALL SETTEXTPOSITION(7, 8, CURPOS)
        WRITE(*,100) PS, PS1

        DUMMY2 = SETCOLOR(1)
        DUMMY2 = RECTANGLE($GFILLINTERIOR, 10,128, 30, 148)
        CALL SETTEXTPOSITION(9, 8, CURPOS)
        WRITE(*,100) TS1, TS
        DUMMY2 = SETCOLOR(5)
        DUMMY2 = RECTANGLE($GFILLINTERIOR, 10, 160, 30, 180)
        CALL SETTEXTPOSITION(11, 8, CURPOS)
        WRITE(*,100) TS2, TS1
        DUMMY2 = SETCOLOR(2)
        DUMMY2 = RECTANGLE($GFILLINTERIOR, 10, 192, 30, 212)
        CALL SETTEXTPOSITION(13, 8, CURPOS)
        WRITE(*,100) TS3, TS2
        END IF
C
C       PLACE TITLE ABOVE PLOT
C
        IF (IDRAW .EQ. 1) THEN
          TITLE = 'PLOT OF ORIGINAL MESH'
        ELSEIF (IDRAW .EQ. 2) THEN
          TITLE = 'PLOT OF DEFORMED MESH'
        END IF
        KCOL = 27 + (50 - LEN_TRIM(TITLE))/2
        CALL SETTEXTPOSITION(1, KCOL, CURPOS)
        CALL OUTTEXT(TITLE)
C
C       PUSH RETURN KEY TO EXIT PROGRAM
C
        READ(*,*)
        DUMMY2 = SETVIDEOMODE($DEFAULTMODE)
        STOP
        END
```

Figure 8.27 A listing of the program QUAD.

```
C    PROGRAM FOR PLANE STRAIN ANALYSIS OF
C    ELASTIC SOLID USING 4-NODE QUAD ELEMENTS
C
C    THIS PROGRAM IS BASED ON SEVERAL PRESENTED IN:
C    SMITH, I.M. (1982):  PROGRAMMING THE FINITE ELEMENT
C    METHOD, WITH APPLICATIONS TO GEOMECHANICS, WILEY
C    AND SONS.
C
C******************************************************
C               USER'S GUIDE:   QUAD
C******************************************************
C
C    DATA FILE PREPARATION:
C
C    LINE 1:  READ(5,*) NXE,NYE,N,W,NN,RN,NL
C    WHERE:
C
C    NXE =:  1) FOR RECTANGULAR MESH GENERATION, NXE IS
C               THE NUMBER OF COLUMNS OF ELEMENTS IN THE
C               HORIZONTAL DIRECTION;
C            2) FOR TUNNEL MESH GENERATION, NXE IS THE
C               NUMBER OF WEDGES, OR SPOKES, USED TO
C               SUBDIVIDE THE 0-90 DEGREE QUADRANT ADJACENT
C               TO THE TUNNEL OPENING.
C
C    NYE =:  1) FOR RECTANGULAR MESH GENERATION, NYE IS
C               THE NUMBER OF ROWS OF ELEMENTS IN THE
C               VERTICAL DIRECTION;
C            2) FOR TUNNEL MESH GENERATION, NYE IS THE
C               NUMBER OF RADIUSES USED TO SUBDIVIDE THE
C               0-90 DEGREE QUADRANT ADJACENT TO THE TUNNEL
C               OPENING.
C
C    N =:    TOTAL NUMBER OF FREEDOMS IN THE MESH
C            (REGARDLESS OF THE TYPE OF MESH GENERATION
C            USED); N = NNODES*2DOF  - # OF RESTRICTED
C            FREEDOMS;
C
C    W =:    HALF BANDWIDTH = MAXIMUM DIFFERENCE IN

C            FREEDOM NUMBERS FOR ANY GIVEN ELEMENT
C            (W DOES NOT DEPEND ON TYPE OF MESH GENERATION).
C    NN =:   TOTAL NUMBER OF NODAL POINTS IN THE MESH
C            (NN DOES NOT DEPEND ON TYPE OF MESH GENERATION)
C    RN =:   TOTAL # OF NODAL POINTS ASSOC. WITH BOUNDARY
C            CONDITIONS; RN DOES NOT DEPEND ON THE TYPE OF
C            MESH GENERATION USED.
C    NL =:   TOTAL # OF FREEDOMS (FREEDOMS!) ASSOC. WITH
C            ACTIVE FORCES.
C
C    LINE 2:   READ(5,*) E, V, NAIR, NANGLE, MESH
C
C    WHERE:
C
C            E = YOUNG'S MODULUS
C            V = POISSON'S RATIO
C    NAIR = NUMBER OF ELEMENTS COMPRISING A VOID, OR
C           OPENING IN THE MESH; FOR TUNNEL MESH
C           GENERATION, NAIR = 0 (ALWAYS); FOR
C           RECTANGULAR MESH GENERATION, NAIR MAY BE
C           SET TO ZERO, IF NO OPENINGS ARE TO BE
C           MODELED, OR NAIR IS SET EQUAL TO THE NUMBER
C           OF ELEMENTS IN THE MESH REPRESENTING OPENINGS
C    NANGLE = NUMBER OF FRACTURE ORIENTATIONS TO BE
C             ANALYZED FOR A NONLINEAR FINITE ELEMENT
C             ANALYSIS; DO THE FOLLOWING:
C
C             NANGLE = 0; FOR A LINEAR, ELASTIC ANALYSIS;
C                         NO FRACTURES ANALYZED;
C             NANGLE = # OF FRACTURE ORIENTATIONS (UP TO 10)
C                      SETTING NANGLE TO A VALUE OTHER THAN
C                      ZERO AUTOMATICALLY INVOKES A
C                      NONLINEAR, ELASTIC ANALYSIS
C
C               MESH = 1, FOR RECTANGULAR MESH GENERATION
C                    = 2, FOR A TUNNEL MESH GENERATION
C
C    LINE GROUP 3:
C
C    IF MESH = 1, THEN:
C       READ(5,*) AA, BB
C    WHERE:
C
C    AA = HORIZONTAL DIMENSION OF EACH ELEMENT
C    BB = VERTICAL DIMENSION OF EACH ELEMENT
```

Figure 8.27 continued

```
C     ELSEIF MESH = 2, THEN:
C
C       READ(5,*) (RADIUS(J), J = 1, NYE+1)
C       READ(5,*) (SPOKE(J), J = 1, NXE+1)
C     WHERE:
C       RADIUS (J) = JTH DISTANCE FROM THE CENTER OF THE
C                    OF THE TUNNEL; THERE MUST BE NYE+1
C                    OF THESE, THE ADDITIONAL 1 (BEYOND)
C                    NYE IS ADDED FOR THE LAST RADIUS (THAT
C                    CLOSEST TO THE TUNNEL CENTER); NOTE:
C                    J(1) MUST BE THE FARTHEST (LARGEST)
C                    RADIUS FROM THE TUNNEL CENTER; J(2) THE
C                    NEXT LARGEST; ...; J(NYE+1) THE
C                    SMALLEST, OR CLOSEST, RADIUS TO THE
C                    CENTER;
C       SPOKE (J) = THE CHANGE, IF ANY, IN RADIAL LENGTH
C                   FROM THE 90 DEGREE SPOKES (VERTICAL)
C                   TO THE HORIZONAL SPOKES.  IF THE
C                   TUNNEL SHAPE IS PERFECTLY CIRCULAR,
C                   THEN ALL (ALL!) SPOKE VALUES ARE SET
C                   EQUAL TO 1.0: ELSE, IF THE TUNNEL SHAPE
C                   IS ELLIPTICAL, THEN:
C                   SPOKE(1) = 1.0: SPOKE(NXE+1) = RATIO
C                   WHERE RATIO = MINOR/MAJOR AXES LENGTHS
C                   FOR THE ELLIPSE; THEN, ALL OTHER
C                   VALUES OF SPOKE ARE SCALED BETWEEN
C                   1.0 AND RATIO: NOTE THAT SPOKE(1)
C                   , FOR AN ELLIPTICAL TUNNEL, IS ALWAYS
C                   EQUAL TO 1.0; THEN, SPOKE(2) IS THE
C                   NEXT LARGEST VALUE, BUT LESS THAN 1.0;
C                   SPOKE (3) THE NEXT LARGEST AFTER
C                   SPOKE(2); AND SO ON, UP TO SPOKE(NXE+1)
C                   WHICH IS EQUAL TO RATIO.
C
C
C   LINE GROUP 4:
C       IF NAIR = 0; SKIP THIS GROUP
C
C       ELSEIF NAIR > 0, THEN:
C
C         DO I = 1, NAIR
C
C           READ(5,*) IXAIR(I), IYAIR(I)
C         END DO
C       WHERE, FOR MESH = 1 OPTION ONLY:
C         IXAIR = COLUMN POSITION, BETWEEN 1 AND NXE,
C                 FOR THE ELEMENT ASSOC. WITH AN OPENING
C         IYAIR = ROW POSITION, BETWEEN 1 AND NYE, FOR
C                 THE ELEMENT ASSOC. WITH AN OPENING
C       NOTE:  THERE ARE NAIR LINES IN THIS GROUP FOR
C              HOWEVER MANY ELEMENTS THERE ARE IN THE
C              MESH WHICH REPRESENT AN OPENING.
C
C   LINE GROUP 5:  ONLY USED IF NANGLE > 0:
C
C       READ(5,*) (ANGLE(J), J = 1,NANGLE)
C       READ(5,*) (CHESN(J), J = 1,NANGLE)
C       READ(5,*) (FTION(J), J = 1,NANGLE)
C
C       WHERE:
C         ANGLE (J) = JTH FRACTURE ORIENTATION ANGLE,
C                     IN DEGREES (DEGREES!); DEGREES
C                     ARE POSITIVE REPRESENTING
C                     COUNTERCLOCKWISE ORIENTATIONS
C                     FROM 0 = HORIZONTAL, TO 90=
C                     VERTICAL, TO 180=HORIZONTAL.
C         CHESN(J) = COHESION VALUE FOR THE JTH
C                    FRACTURE
C         FTION(J) = THE TANGENT OF THE FRICTION
C                    ANGLE FOR THE JOINT; FOR EXAMPLE,
C                    IF THE FRICTION ANGLE FOR THE
C                    JOINT IS 30 DEGREES, ENTER
C                    FTION(J) AS 0.577.
C
C   LINE GROUP 6:  BOUNDARY CONDTIONS
C
C       THIS GROUP IS ASSOCIATED WITH 3RN TOTAL LINES; FOR
C       EACH OF THE RN RESTRICTED NODES, ENTER THE FOLLOWING,
C       ONE VALUE PER LINE:
```

379

Figure 8.27 continued

```
C
C      DO I = 1, RN:
C
C          READ(5,*) NODAL(I)
C          READ(5,*) X-RESTRICT(I)
C          READ(5,*) Y-RESTRICT(I)
C
C      END DO
C
C   WHERE:  NODAL(I) = ITH RESTRICTED NODE NUMBER
C           X-RESTRICT(I) = : 0, X-FREE TO MOVE
C                             1, X IS RESTRICTED
C           Y-RESTRICT(I) = : 0, Y-FREE TO MOVE
C                             1, Y IS RESTRICTED
C
C LINE GROUP 7:  (LAST GROUP):  ACTIVE FORCE ENTRY
C
C   LOOP OVER TOTAL NUMBER OF FREEDOMS WITH ACTIVE
C   FORCES, NL:
C
C      DO I = 1, NL
C
C          READ(5,*) FREEDOM(I), FORCE(I)
C
C      END DO
C
C   WHERE:
C
C           FREEDOM(I) = ITH FREEDOM NUMBER ASSOC. WITH
C                        A FORCE
C
C           FORCE(I) = FORCE AT FREEDOM NUMBER; NOTE:
C                      ENTER COMPRESSIVE FORCES AS
C                      NEGATIVE NUMBERS; TENSILE FORCES
C                      AS POSITIVE NUMBERS.
C
C **********************************************************
C           END OF USER'S GUIDE AND DISCUSSION
C **********************************************************
```

```
C **********************************************************
C           EXAMPLE:   RECTANGULAR MESH GENERATION
C **********************************************************
C
C   GIVEN THIS MESH:
C
C        1           4           7           10
C   o+----------+-----------+-----------+
C   + 0,1       + 3,4       + 8,9       + 13,14
C   +           +           +           +
C   +           +           +           +
C   + 2         +5          +8          +11
C   +           +-----------+-----------+
C   + 0,2       + 5,6       + 10,11     + 15,16
C   +           +           +           +
C   +   AIR     +           +           +
C   +3          +6          +9          +12
C   o+----------+-----------+-----------+
C   o 0,0       o 7,0       o 12,0      o 17,0
C
C GIVEN:  ONE FRACTURE ORIENTATION IS TO BE ANALYZED, HAVING
C         A 0 DEGREE (HORIZONTAL) ORIENTATION: 1 AIR ELEMENT
C         IS PRESENT; A UNIFORM COMPRESSIVE STRESS OF 500
C         (UNITS) IS PRESENT; YOUNG'S MOD IS 500(UNITS);
C         POISSON'S RATIO IS 0.3; COHESION IS ZERO ALONG THE
C         FRACTURE, WHOSE FRICTION ANGLE IS 30 DEGREES;
C
C         THE SYMBOL:  o, REPRESENTS A BOUNDARY CONDITION IN
C         THIS MESH; NUMBERS SHOWN AT THE NODAL POINTS ARE
C         1) NODAL POINT NUMBER SHOWN ABOVE EACH NODAL PT; &2)
C         FREEDOM NUMBERS; THE FIRST NUMBER BEFORE THE COMMA
C         IS THE NUMBER OF THE X-FREEDOM AT THE NODE; THE
C         SECOND NUMBER, THE ONE AFTER THE COMMA, IS THE
C         Y-FREEDOM NUMBER.
C
C         FOR THIS MESH:   MESH=1 OPTION IS USED TO GENERATE
C         THE MESH; NXE = 3; NYE = 2; N = 17; W = 8;
C         NN = 12; RN = 6; NL = 7; AA=10., BB=10.
C         NAIR = 1; NANGLE = 1; IXAIR=1; IYAIR=2
C
```

Figure 8.27 continued

```
C     THEREFORE, THE DATA FILE FOR THIS MESH LOOKS LIKE THIS:
C
C     3,2,17,8,12,6,7
C     500.0, 0.3, 1,1,1
C     10.0,10.0
C     1,2
C     0.0
C     0.0
C     0.577
C     1
C     1
C     0
C     2
C     1
C     0
C     3
C     1
C     1
C     6
C     0
C     1
C     9
C     0
C     1
C     12
C     0
C     1
C     1,   -2500.0
C     4,   -5000.0
C     9,   -5000.0
C     13,  -2500.0
C     14,  -2500.0
C     15,  -5000.0
C     17,  -2500.0
C
C ****************************************************
C              END OF EXAMPLE
C ****************************************************
C ****************************************************
C
```

```
C
      COMMON /PROP/ ANGLE(10), CHESN(10), FTION(10), NANGLE
      REAL E,V,AA,BB,K1,K2,DET,QUOT
      INTEGER I,J,K,L,P,Q,NXE,NYE,H,N,W,CDMAX,RN,NL,NN,
     2          DOF,NOD,NODOF,R,T,GP,IDER,ICOORD,IJAC,
     3          IJAC1,IDERIV, IDEE,IBEE,IDBEE,IBT,IBTDB,
     4          IKM,INF,ISAMP
      REAL DEE(8,3),SAMP(3,2),COORD(8,2),JAC(8,2),
     2     JAC1(8,2),DER(8,4),DERIV(8,4),BEE(8,8),DBEE(8,8),
     3     BTDB(8,8),KM(8,8),ELD(8),VOL(8),EPS(3),SIGMA(3),
     4     BT(8,3),FUN(4)
      REAL SPOKE(30), RADIUS(30)
      INTEGER G(8), IFAIL(10)
      REAL KB(400,30),LOADS(400),LOADS1(400)
      INTEGER NF(400,2), IXAIR(20), IYAIR(20)
      DATA IDER,ICOORD,IJAC,IJAC1,IDERIV,IDEE/6*8/
      DATA IBEE,IDBEE,IBT,IBTDB,IKM,DOF/6*8/
      DATA ISAMP/3/,NODOF/2/,H/3/,T/2/
C
C
C     DATA INPUT AND INITIALIZATION
C
      OPEN(5, FILE = ' ')
      OPEN(1, FILE = ' ', STATUS = 'NEW')
      OPEN(2, FILE = ' ', STATUS = 'NEW')
      READ(5,*) NXE,NYE,N,W,NN,RN,NL
      READ(5,*) E,V,NAIR,NANGLE,MESH
      IF (NANGLE .NE. 0) THEN
         OPEN(3, FILE = ' ', STATUS = 'NEW')
         OPEN(4, FILE = ' ', STATUS = 'NEW')
      END IF
      IF (MESH .EQ. 1) THEN
         READ(5,*) AA,BB
      ELSEIF (MESH .EQ. 2) THEN
         M = NYE + 1
         READ(5,*) (RADIUS(JK), JK = 1,M)
         M = NXE + 1
         READ(5,*) (SPOKE(JK), JK = 1,M)
      END IF
      E2 = 0.0
      V2 = 0.3
      GP = 2
```

Figure 8.27 continued

```fortran
      IF (NAIR .NE. 0) THEN
        DO I = 1,NAIR
          READ(5,*) IXAIR(I),IYAIR(I)
        END DO
      END IF
      IF (NANGLE .NE. 0) THEN
        READ(5,*) (ANGLE(I), I = 1,NANGLE)
        READ(5,*) (CHESN(I), I = 1,NANGLE)
        READ(5,*) (FTION(I), I = 1,NANGLE)
      END IF
      NOD = 4
      CDMAX = W + 1
      IKB = 400
      IF (NANGLE .EQ. 0) THEN
        NITN = 1
      ELSE
        NITN = 3
      END IF
      INF = 400
      R = N * CDMAX
C     CALL NULVEC(BK,R)
      CALL NULVEC(LOADS,N)
      CALL NULL(DEE,IDEE,H,H)
      CALL NULVEC(LOADS1,N)
      CALL NULL(KB,IKB,N,CDMAX)
      CALL GAUSS(SAMP,ISAMP,GP)
      CALL FORMNF(NF,INF,NN,NODOF,RN)
C
C     NONLINEAR ANALYSIS; STEP 1 =
C     ELEMENT STIFFNESS INTEGRATION AND ASSEMBLY
C
      DO 2000 JJJ = 1,NITN
      IF (JJJ .GT. 1) GO TO 1800
      DO 3 P = 1,NXE
      DO 3 Q = 1,NYE
      IF (MESH .EQ. 1) THEN
        CALL GEOMET(P,Q,NXE,NYE,AA,BB,COORD,ICOORD,G,NF,INF,1)
      ELSEIF (MESH .EQ. 2) THEN
        CALL TUNGEO(P,Q,NXE,NYE,SPOKE,RADIUS,COORD,ICOORD,G,NF,INF,1)
      END IF

      CALL NULL(DEE,IDEE,H,H)
      CALL FORMD(DEE,IDEE,E,V)
      IF (NAIR .NE. 0) THEN
      DO 2 III = 1,NAIR
      IF (P.EQ.IXAIR(III).AND.Q.EQ.IYAIR(III)) THEN
        CALL FORMD(DEE,IDEE,E2,V2)
      ENDIF
2     CONTINUE
      END IF
      CALL NULL(KM,IKM,DOF,DOF)
      DO 4 I = 1,GP
      DO 4 J = 1,GP
      K1 = SAMP(I,2)
      K2 = SAMP(J,2)
      CALL FORMLN(DER,IDER,FUN,SAMP,ISAMP,I,J)
      CALL MATMUL(DER,IDER,COORD,ICOORD,JAC,IJAC,T,NOD,T)
      CALL TWOBY2(JAC,IJAC,JAC1,IJAC1,DET)
      CALL MATMUL(JAC1,IJAC1,DER,IDER,DERIV,IDERIV,T,T,NOD)
      CALL NULL(BEE,IBEE,H,DOF)
      CALL FORMB(BEE,IBEE,DERIV,IDERIV,VOL,NOD)
      CALL MATMUL(DEE,IDEE,BEE,IBEE,DBEE,IDBEE,H,DOF)
      CALL MATRAN(BT,IBT,BEE,IBEE,H,DOF)
      CALL MATMUL(BT,IBT,DBEE,IDBEE,BTDB,IBTDB,DOF,H,DOF)
      QUOT = DET * K1 * K2
      DO 5 K = 1,DOF
      DO 5 L = 1,DOF
5     BTDB(K,L) = BTDB(K,L) * QUOT
      CALL MATADD(KM,IKM,BTDB,IBTDB,DOF,DOF)
4     CONTINUE
      CALL FORMG(KB,IKB,KM,IKM,G,DOF)
3     CONTINUE
1800  CONTINUE
C
C     EQUATION SOLUTION
C
      IF (JJJ .GT. 1) GO TO 900
      DO 6 I = 1,NL
      READ(5,*) K,K1
      LOADS1(K) = K1
6     LOADS(K) = K1
900   CONTINUE
```

Figure 8.27 continued

```fortran
C       CALL PRINTV(LOADS,N)
        CALL BANRED(KB,IKB,N,CDMAX)
        CALL BACK1(KB,IKB,LOADS,N,CDMAX)
        CALL BACK2(KB,IKB,LOADS,N,CDMAX)
        WRITE(*,110)
        DO 100 I = 1,NN
        XX = 0.0
        YY = 0.0
        K = NF(I,1)
        L = NF(I,2)
        IF (K .NE. 0) XX = LOADS(K)
        IF (L .NE. 0) YY = LOADS(L)
        WRITE(*,120) I, XX, YY
        IF (JJJ .EQ. 1) WRITE(2,*) I, XX, YY
        IF (JJJ .EQ. 2) WRITE(3,*) I, XX, YY
        IF (JJJ .EQ. 3) WRITE(4,*) I, XX, YY
100     CONTINUE
110     FORMAT(1H1, 10X, 'NODAL DISPLACEMENTS',//,
     2     10X,'NODE',10X,'X-DISP',10X,'Y-DISP',/)
120     FORMAT(10X,I4,1X,1E15.3,1X,1E15.3)
C
C       CALCULATE STRESS AND STRAIN AT EACH GAUSS POINT
C       OF EVERY ELEMENT
C
        CALL NULL(KB,IKB,N,CDMAX)
        IF (NANGLE .EQ. 0) THEN
        WRITE(*,509)
        ELSE
        WRITE(*,510)
        END IF
509     FORMAT(1H1,10X,'ELEMENT STRAIN AND STRESS ANALYSIS',//,
     2     2X,'STRAIN-X',2X,'STRAIN-Y',5X,'SHEAR',12X,'STRESS-X',
     3     2X,'STRESS-Y',5X,'SHEAR',/)
510     FORMAT(1H1,10X,'JOINT BEHAVIOR SUMMARY',//,
     2     10X,'SIGMA-X',8X,'SIGMA-Y',9X,'TAU-XY',
     3     1X,'COL',1X,'ROW',1X,'GP1',1X,'GP2',
     4     4X,'FAILURE PER ANGLE',///)
        DO 7 P = 1,NXE
        DO 7 Q = 1,NYE
        IF (MESH .EQ. 1) THEN
        CALL GEOMET(P,Q,NXE,NYE,AA,BB,COORD,ICOORD,G,NF,INF,2)

        ELSEIF (MESH .EQ. 2) THEN
        CALL TUNGEO(P,Q,NXE,NYE,SPOKE,RADIUS,COORD,ICOORD,G,NF,INF,2)
        END IF
        CALL NULL(KM,IKM,DOF,DOF)
        KSKIP = 0
        IF (NAIR .NE. 0) THEN
        DO 5000 III = 1,NAIR
        KKK = IXAIR(III)
        LLL = IYAIR(III)
        IF (P.EQ.KKK .AND. Q.EQ.LLL) THEN
        KSKIP = 1
        DO ISIG = 1,3
        SIGMA(ISIG) = 0.0
        END DO
        DO I4 = 1,4
        IF (NANGLE .EQ. 0) WRITE(2,*) (SIGMA(JK), JK = 1,3)
        END DO
        END IF
5000    CONTINUE
        END IF
        IF (KSKIP .NE. 0) GO TO 7
        DO 8 I = 1,GP
        DO 8 J = 1,GP
        CALL FORMLN(DER,IDER,FUN,SAMP,ISAMP,I,J)
        CALL MATMUL(DER,IDER,COORD,ICOORD,JAC,IJAC,T,NOD,T)
        CALL TWOBY2(JAC,IJAC,JAC1,IJAC1,DET)
        CALL MATMUL(JAC1,IJAC1,DER,IDER,DERIV,IDERIV,T,T,NOD)
        CALL NULL(BEE,IBEE,H,DOF)
        CALL FORMB(BEE,IBEE,DERIV,IDERIV,VOL,NOD)
        CALL NULL(DEE,IDEE,H,H)
        CALL FORMD(DEE,IDEE,E,V)
        DO 9 M = 1,DOF
        IF (G(M) .EQ. 0) ELD(M) = 0.0
        IF (G(M) .NE. 0) ELD(M) = LOADS(G(M))
9       CALL MVMULT(BEE,IBEE,ELD,H,DOF,EPS)
        CALL MVMULT(DEE,IDEE,EPS,H,SIGMA)
        IF (NANGLE .EQ. 0) THEN
        WRITE(*,599) (EPS(JK),JK=1,3),(SIGMA(JK),JK=1,3)
599     FORMAT(3E10.3,10X,3E10.3)
        WRITE(2,*) (SIGMA(JK),JK=1,3)
        END IF
```

Figure 8.27 continued

```
         IF (NANGLE .EQ. 0) GO TO 8
         CALL JOINT(SIGMA,H,DEE,IDEE,IFAIL)
         DO 600 IF = 1,NANGLE
         IF (IFAIL(IF) .NE. 0) THEN
            WRITE(*,610)  (SIGMA(JK),  JK=1,3),P,Q,I,J,
     2                    (IFAIL(IW),  IW=1,NANGLE)
         IF (JJJ .EQ. 1) THEN
         WRITE(2,*) (SIGMA(JK),JK=1,3),P,Q,I,J,(IFAIL(IW),IW=1,
     2               NANGLE)
         ELSEIF (JJJ .EQ. 2) THEN
         WRITE(3,*) (SIGMA(JK),JK=1,3),P,Q,I,J,(IFAIL(IW),IW=1,
     2               NANGLE)
         ELSEIF (JJJ .EQ. 3) THEN
         WRITE(4,*) (SIGMA(JK),JK=1,3),P,Q,I,J,(IFAIL(IW),IW=1,
     2               NANGLE)
         END IF
600      CONTINUE
610      FORMAT(2X,3E15.3,4I4,10X,10I3)
         CALL MATMUL(DEE,IDEE,BEE,IBEE,DBEE,IDBEE,H,H,DOF)
         CALL MATRAN(BT,IBT,BEE,IBEE,H,DOF)
         CALL MATMUL(BT,IBT,DBEE,IDBEE,BTDB,IBTDB,DOF,H,DOF)
         K1 = SAMP(I,2)
         K2 = SAMP(J,2)
         QUOT = DET * K1 * K2
         DO 710 K11 = 1,DOF
         DO 710 L11 = 1,DOF
710      BTDB(K11,L11) = BTDB(K11,L11) * QUOT
         CALL MATADD(KM,IKM,BTDB,IBTDB,DOF,DOF)
8        CONTINUE
         CALL FORMG(KB,IKB,KM,IKM,G,DOF)
7        CONTINUE
         DO 2000 KIJ = 1,N
         LOADS(KIJ) = LOADS1(KIJ)
2000     CONTINUE
         STOP
         END
C
C
         SUBROUTINE GEOMET(P,Q,NXE,NYE,AA,BB,COORD,ICOORD,
     2                     G,NF,INF,JRITE)
C
C        SUBROUTINE TO COMPUTE NODAL COORDINATES AND THE
C        STEERING VECTOR FOR A RECTANGULAR MESH OF 4-NODE
C        QUAD PLANE ELEMENTS NUMBERING IN THE Y-DIRECTION.
C
         REAL COORD(ICOORD,1),  AA,BB
         INTEGER G(1),NF(INF,1),P,Q,NXE,NYE,AO,AL,
     2           AM,AN
         AO = (P-1)*(NYE+1)+Q
         AL = AO + 1
         AM = P * (NYE+1) + Q
         AN = AM + 1
         IF (JRITE .EQ. 1) WRITE(1,*) AL, AO, AM, AN
         G(1) = NF(AL,1)
         G(2) = NF(AL,2)
         G(3) = NF(AO,1)
         G(4) = NF(AO,2)
         G(5) = NF(AM,1)
         G(6) = NF(AM,2)
         G(7) = NF(AN,1)
         G(8) = NF(AN,2)
         COORD(1,1) = (P-1) * AA
         COORD(1,2) = (NYE-Q) * BB
         COORD(2,1) = (P-1) * AA
         COORD(2,2) = (NYE-Q+1) * BB
         COORD(3,1) = P * AA
         COORD(3,2) = (NYE-Q+1) * BB
         COORD(4,1) = P * AA
         COORD(4,2) = (NYE-Q) * BB
         IF (JRITE .EQ. 1) THEN
         WRITE(1,10) COORD(1,1),  COORD(1,2),  COORD(2,1),  COORD(2,2),
     2               COORD(3,1),  COORD(3,2),  COORD(4,1),  COORD(4,2)
10       FORMAT(8E10.3)
         END IF
         RETURN
         END
C
```

Figure 8.27 continued

```fortran
C
C     SUBROUTINE NULVEC(VEC,N)
C
C     SUBROUTINE TO INITIALIZE A VECTOR
C
      REAL VEC(1)
      INTEGER I,N
      DO 1 I = 1,N
      VEC(I) = 0.0
    1 CONTINUE
      RETURN
      END
C
C     SUBROUTINE NULL(A,IA,M,N)
C
C     SUBROUTINE TO INITIALIZE A MATRIX
C
      INTEGER IA,M,N,I,J
      REAL A(IA,1)
      DO 1 I = 1,M
      DO 1 J = 1,N
      A(I,J) = 0.0
    1 CONTINUE
      RETURN
      END
C
C     SUBROUTINE FORMD(DEE,IDEE,E,V)
C
C     SUBROUTINE TO FORM ELASTIC STRESS-STRAIN MATRIX
C     FOR A PLANE STRAIN ANALYSIS
C
      INTEGER IDEE,I,J
      REAL DEE(IDEE,1), E,V,V1,VV
      V1 = V/(1.0 - V)
      VV = (1.0 - 2.0*V)*0.5/(1.0 - V)
      DEE(1,1) = 1.0

      DEE(2,2)  = 1.0
      DEE(3,3)  = VV
      DEE(1,2)  = V1
      DEE(2,1)  = V1
      DO 1 I = 1,3
      DO 1 J = 1,3
    1 DEE(I,J) = DEE(I,J)*E/(2.0*(1.0+V)*VV)
      RETURN
      END
C
C     SUBROUTINE GAUSS(SAMP,ISAMP,GP)
C
C     SUBROUTINE TO FORM GAUSS QUADRATURE ABSCISSAE
C     AND WEIGHTS; SHORTENED TO ONLY ALLOW 2 PT RULE
C
      INTEGER ISAMP,GP
      REAL SAMP(ISAMP,1)
      SAMP(1,1) = -1. / SQRT(3.0)
      SAMP(2,1) = - SAMP(1,1)
      SAMP(1,2) = 1.0
      SAMP(2,2) = 1.0
      RETURN
      END
C
C     SUBROUTINE FORMNF(NF,INF,NN,NODOF,RN)
C
C     SUBROUTINE TO FORM NODE FREEDOM ARRAY FOR THE
C     CASE OF MORE THAN ONE FREEDOM PER NODE
C
      INTEGER NF(INF,1),NN,NODOF,RN,I,J,K,L
      DO 1 I = 1,NN
      DO 1 J = 1,NODOF
    1 NF(I,J) = 1
      DO 2 I = 1,RN
      READ(5,*) K
      DO 3 J = 1,NODOF
      READ(5,*) L
    3 IF( L .EQ. 1) NF(K,J) = 0
    2 CONTINUE
```

Figure 8.27 continued

```fortran
      K = 1
      DO 4 I = 1, NN
      DO 4 J = 1, NODOF
      IF(NF(I,J)) 5,4,5
5     K = K + 1
      NF(I,J) = K
4     CONTINUE
      RETURN
      END
C
C
      SUBROUTINE FORMLN(DER,IDER,FUN,SAMP,ISAMP,I,J)
C
C     SUBROUTINE FOR LOCAL COORDINATE SHAPE FUNCTIONS
C     AND THEIR DERIVATIVES FOR 4-NODE QUAD ELEMENTS
C
      INTEGER IDER,ISAMP,I,J
      REAL DER(IDER,1),FUN(1),SAMP(ISAMP,1)
      REAL ETA,XI,ETAM,ETAP,XIM,XIP
      XI = SAMP(I,1)
      ETA = SAMP(J,1)
      ETAM = 0.25 * (1.0-ETA)
      ETAP = 0.25 * (1.0+ETA)
      XIM  = 0.25 * (1.0 - XI)
      XIP  = 0.25 * (1.0 + XI)
      FUN(1) = 4.0 * XIM * ETAM
      FUN(2) = 4.0 * XIM * ETAP
      FUN(3) = 4.0 * XIP * ETAP
      FUN(4) = 4.0 * XIP * ETAM
      DER(1,1) = - ETAM
      DER(1,2) = - ETAP
      DER(1,3) = ETAP
      DER(1,4) = ETAM
      DER(2,1) = - XIM
      DER(2,2) = XIM
      DER(2,3) = XIP
      DER(2,4) = - XIP
      RETURN
      END
C
```

```fortran
C
      SUBROUTINE MATMUL(A,IA,B,IB,C,IC,L,M,N)
C
C     SUBROUTINE FOR MATRIX MULTIPLICATION: A*B=C
C
      INTEGER IA,IB,IC,L,M,N,I,J,K
      REAL A(IA,1), B(IB,1), C(IC,1), X
      DO 1 I = 1,L
      DO 1 J = 1,N
      X = 0.0
      DO 2 K = 1,M
2     X = X + A(I,K) * B(K,J)
      C(I,J) = X
1     CONTINUE
      RETURN
      END
C
C
      SUBROUTINE TWOBY2(JAC,IJAC,JAC1,IJAC1,DET)
C
C     SUBROUTINE TO INVERT THE JACOBIAN MATRIX
C
      INTEGER IJAC,IJAC1,K,L
      REAL JAC(IJAC,1), JAC1(IJAC1,1), DET
      DET = JAC(1,1)*JAC(2,2) - JAC(1,2)*JAC(2,1)
      JAC1(1,1) = JAC(2,2)
      JAC1(1,2) = - JAC(1,2)
      JAC1(2,1) = - JAC(2,1)
      JAC1(2,2) = JAC(1,1)
      DO 1 K = 1,2
      DO 1 L = 1,2
1     JAC1(K,L) = JAC1(K,L) / DET
      RETURN
      END
C
```

Figure 8.27 continued

```fortran
C
      SUBROUTINE FORMB(BEE,IBEE,DERIV,IDERIV,VOL,NOD)
C
C     SUBROUTINE TO FORM STRAIN-DISPLACEMENT MATRIX, B
C
      INTEGER IBEE,IDERIV,NOD,K,L,M
      REAL BEE(IBEE,1),DERIV(IDERIV,1),VOL(1)
      DO 1 M = 1,NOD
      K = 2 * M
      L = K - 1
      VOL(L) = DERIV(1,M)
      BEE(1,L) = VOL(L)
      BEE(3,K) = VOL(L)
      VOL(K) = DERIV(2,M)
      BEE(2,K) = VOL(K)
      BEE(3,L) = VOL(K)
    1 CONTINUE
      RETURN
      END
C
C
      SUBROUTINE MATRAN(A,IA,B,IB,M,N)
C
C     SUBROUTINE TO FORM THE TRANSPOSE OF A MATRIX
C
      INTEGER IA,IB,M,N,I,J
      REAL A(IA,1), B(IB,1)
      DO 1 I = 1,M
      DO 1 J = 1,N
    1 A(J,I) = B(I,J)
      RETURN
      END
C
C
      SUBROUTINE MATADD(A,IA,B,IB,M,N)
C
C     SUBROUTINE TO ADD TWO MATRICES: A=A+B
C
      INTEGER IA,IB,M,N,I,J
      REAL A(IA,1), B(IB,1)
      DO 1 I = 1,M
      DO 1 J = 1,N
    1 A(I,J) = A(I,J) + B(I,J)
      RETURN
      END
C
C
      SUBROUTINE FORMG(KB,IKB,KM,IKM,G,DOF)
C
C     SUBROUTINE TO ASSEMBLE GLOBAL STIFFNESS
C     MATRIX FOR GUASS ELIMINATION
C
      REAL KB(IKB,1), KM(IKM,1)
      INTEGER G(1), DOF, CD
      DO 100 I = 1,DOF
      IF (G(I) .EQ. 0) GO TO 100
      DO 90 J = 1,DOF
      IF (G(J) .EQ. 0) GO TO 90
      CD = G(J) - G(I) + 1
      IF (CD .LT. 1) GO TO 90
      KB(G(I),CD) = KB(G(I),CD) + KM(I,J)
   90 CONTINUE
  100 CONTINUE
      RETURN
      END
C
C
      SUBROUTINE PRINTV(V,L)
C
C     SUBROUTINE TO PRINT A VECTOR
C
      INTEGER L
      REAL V(L)
      WRITE(*,1) V
    1 FORMAT(1H ,6E10.4)
      RETURN
      END
C
```

387

Figure 8.27 continued

```fortran
C       SUBROUTINE MVMULT(M,IM,V,K,L,Y)
C
C       SUBROUTINE TO MULTIPLY A MATRIX AND VECTOR
C
        INTEGER IM,K,L,I,J
        REAL M(IM,1), V(1), Y(1), X
        DO 1 I = 1,K
        X = 0.0
        DO 2 J = 1,L
        X = X + M(I,J) * V(J)
2       Y(I) = X
1       CONTINUE
        RETURN
        END
C
C       SUBROUTINE BANRED(KB,IKB,N,CDMAX)
C
C       SUBROUTINE TO PERFORM GAUSS ELIMINATION
C
C       BANRED, BACK1 AND BACK2 WRITTEN BY J.R. CARR
C
        INTEGER CDMAX
        REAL KB(IKB,1)
        WRITE(*,1000) CDMAX
1000    FORMAT(10X,'CDMAX = ', I10)
        DO 400 I = 1,N
        DO 300 J = 2,CDMAX
        IF (KB(I,J) .EQ. 0.0) GO TO 300
        K        = I + J - 1
        C        = KB(I,J) / KB(I,1)
        L        = 0
        DO 200 M = J,CDMAX
        L        = L + 1
        KB(K,L)  = KB(K,L) - C * KB(I,M)
200     CONTINUE
        KB(I,J)  = C
300     CONTINUE
400     CONTINUE
        RETURN
        END
C
```

```fortran
C       SUBROUTINE BACK1(KB,IKB,LOADS,N,CDMAX)
C
C       FORWARD REDUCTION OF CONSTANTS  (GAUSS ELIMINATION)
C
        INTEGER CDMAX
        REAL KB(IKB,1), LOADS(1)
        DO 400 I = 1,N
        DO 300 J = 2,CDMAX
        IF (KB(I,J) .EQ. 0.0) GO TO 300
        K = I + J - 1
        LOADS(K) = LOADS(K) - KB(I,J) * LOADS(I)
300     CONTINUE
        LOADS(I) = LOADS(I) / KB(I,1)
400     CONTINUE
        RETURN
        END
C
C       SUBROUTINE BACK2(KB,IKB,LOADS,N,CDMAX)
C
C       SOLUTION FOR DISPLACEMENTS BY BACK SUBSTITUTION
C
        INTEGER CDMAX
        REAL KB(IKB,1), LOADS(1)
        DO 400 M = 2,N
        K = N + 1 - M
        DO 300 L = 2,CDMAX
        IF (KB(K,L) .EQ. 0.0) GO TO 300
        J = K + L - 1
        LOADS(K) = LOADS(K) - KB(K,L) * LOADS(J)
300     CONTINUE
400     CONTINUE
        RETURN
        END
C
```

Figure 8.27 continued

```fortran
C
C         SUBROUTINE JOINT (SIGMA, H, DEE, IDEE, IFAIL1)
C
C         SUBROUTINE TO ANALYZE JOINT BEHAVIOR
C
      COMMON /PROP/ ANGLE(10), CHESN(10), FTION(10), NANGLE
      REAL SIGMA(3), DEE(IDEE,1)
      INTEGER NANGLE, H, IFAIL(10)
      DIMENSION SGTRAN(3,3), STTRAN(3,3), SLOCAL(3), SGTT(3,3)
      DIMENSION STTT(3,3), TEMP(3,3), TDE(3,3)
C
C         INITIALIZE MATRIX VALUES
C
      DO 5 I = 1,3
      DO 5 J = 1,3
      TDE(I,J) = DEE(I,J)
    5 CONTINUE
      DO 10 I = 1,NANGLE
      IFAIL(I) = 0
   10 CONTINUE
C
C         CONVERT ANGLE VALUES TO RADIANS AND FORM THE
C         STRESS AND STRAIN TRANSFORMATION MATRICES
C
      DO 100 I = 1,NANGLE
      XANG = ANGLE(I) * 0.01745329
      FCOS = COS(XANG)
      FSIN = SIN(XANG)
      MDEE = 3
      CALL TRANF(SGTRAN, STTRAN, FCOS, FSIN)
C
C         TRANSFORM MATRICES AND VECTORS TO LOCAL COORD.  SYSTEM
C
      CALL MVMLT1(SGTRAN, 3, SIGMA, 3, 3, SLOCAL)
      CALL MATML1(SGTRAN, 3, TDE, MDEE, SGTT, 3, 3, 3)
      CALL MATRN1(STTT, 3, SGTRAN, 3, 3)
      CALL MATML1(SGTT, 3, STTT, 3, TEMP, 3, 3, 3)
C
C         EVALUATE JOINT CLOSURE
C
      IF (SLOCAL(2) .LT. 0.0) GO TO 20
      IFAIL(I) = 1
C

C
C         MODIFY E MATRIX TO ACCOUNT FOR OPENING
C
      DUM5 = TEMP(1,1) - TEMP(1,2) * TEMP(2,1) / TEMP(2,2)
      DO 15 J = 1,3
      DO 15 K = 1,3
      TEMP(J,K) = TEMP(J,K) * 0.01
   15 CONTINUE
      TEMP(1,1) = DUM5
      SLOCAL(2) = 0.0
      SLOCAL(3) = 0.0
      GO TO 90
   20 CONTINUE
C
C         CHECK FOR SHEAR FAILURE
C
      DUM1 = CHESN(I) - (SLOCAL(2) * FTION(I))
      DUM2 = SLOCAL(3)
      DUM3 = ABS(SLOCAL(3))
      IF (DUM3 .LE. DUM1) GO TO 80
      IFAIL(I) = 2
C
C         MODIFY E FOR SHEAR FAILURE
C
      TEMP(3,3) = TEMP(3,3) * 0.01
      IF(DUM2 .LT. 0.0) SLOCAL(3) = - DUM3
      IF (DUM2 .GE. 0.0) SLOCAL(3) = DUM3
   80 CONTINUE
   90 CONTINUE
C
C         TRANSFORM BACK TO GLOBAL SYSTEM
C
      CALL MATRN1(STTT, 3, STTRAN, 3, 3, 3)
      CALL MATML1(STTT, 3, TEMP, 3, SGTT, 3, 3, 3)
      CALL MATML1(SGTT, 3, STTRAN, 3, TDE, MDEE, 3, 3)
      CALL MVMLT1(STTRAN, 3, SLOCAL, 3, 3, SIGMA)
      DO 95 IZ = 1,3
      DO 95 JZ = 1,3
      DEE(IZ,JZ) = TDE(IZ,JZ)
   95 CONTINUE
  100 CONTINUE
      RETURN
      END
C
```

389

Figure 8.27 continued

```fortran
C
        SUBROUTINE MATML1(A,IA,B,IB,C,IC,L,M,N)
        REAL A(3,3), B(3,3), C(3,3)
        DO 10 I = 1,L
        DO 10 J = 1,N
        C(I,J) = 0.0
        DO 10 K = 1,M
10      C(I,J) = C(I,J) + A(I,K) * B(K,J)
        CONTINUE
        RETURN
        END
C
C
        SUBROUTINE MATRN1(A,IA,B,IB,M,N)
        REAL A(3,3) , B(3,3)
        DO 10 I = 1,M
        DO 10 J = 1,N
10      A(J,I) = B(I,J)
        CONTINUE
        RETURN
        END
C
C
        SUBROUTINE MVMLT1(M,IM,V,K,L,Y)
        REAL M(3,3),V(3),Y(3)
        DO 10 I = 1,K
        X = 0.0
        DO 5 J = 1,L
5       X = X + M(I,J) * V(J)
        Y(I) = X
10      CONTINUE
        RETURN
        END
C
C
        SUBROUTINE TRANF(A,B,C,D)
        REAL A(3,3), B(3,3)
        B(1,1) = C * C
        B(1,2) = D * D
        B(1,3) = C * D
        B(2,1) = B(1,2)
        B(2,2) = B(1,1)
        B(2,3) = - B(1,3)
        B(3,1) = - 2.0 * B(1,3)
        B(3,2) = - B(3,1)
        B(3,3) = C * C - D * D
        A(1,1) = B(1,1)
        A(1,2) = B(1,2)
        A(1,3) = B(3,2)
        A(2,1) = B(2,1)
        A(2,2) = B(2,2)
        A(2,3) = B(3,1)
        A(3,1) = B(2,3)
        A(3,2) = B(1,3)
        A(3,3) = B(3,3)
        RETURN
        END
C
C
        SUBROUTINE TUNGEO(P,Q,NXE,NYE,SPOKE,RADIUS,COORD,ICOORD,G,
     2                    NF,INF,IRITE)
C
C       A SUBROUTINE TO FORM A CIRCULAR MESH OF QUADRILATERAL
C       ELEMENTS TO MODEL A TUNNEL
C
        INTEGER P,Q,NXE,NYE,ICOORD,INF,G(1),NF(INF,1),IRITE
        INTEGER AO,AL,AM,AN
        REAL SPOKE(1),RADIUS(1),COORD(ICOORD,1),SL,SR,ANG,PI
        REAL IFACL,IFACR,OFACL,OFACR,CL,CR
C
```

Figure 8.27 continued

```
C
        AO = (Q-1)*(NXE+1)+P
        AM = AO+1
        AL = Q*(NXE+1)+P
        AN = AL+1
        IF (IRITE .EQ. 1) WRITE(1,*) AL,AO,AM,AN
C
C
        G(1) = NF(AL,1)
        G(2) = NF(AL,2)
        G(3) = NF(AO,1)
        G(4) = NF(AO,2)
        G(5) = NF(AM,1)
        G(6) = NF(AM,2)
        G(7) = NF(AN,1)
        G(8) = NF(AN,2)
        PI = 3.141592654
        ANG = PI/(2.*NXE)
        CL = COS(.5*PI-(P-1)*ANG)
        CR = COS(.5*PI-P*ANG)
        SL = SIN(.5*PI-(P-1)*ANG)
        SR = SIN(.5*PI-P*ANG)
C
C
        IFACL = Q*((1.-SPOKE(P))/NYE)+SPOKE(P)
        IFACR = Q*((1.-SPOKE(P+1))/NYE)+SPOKE(P+1)
        OFACL = (Q-1)*((1.-SPOKE(P))/NYE)+SPOKE(P)
        OFACR = (Q-1)*((1.-SPOKE(P+1))/NYE)+SPOKE(P+1)
C
C
        COORD(1,1) = IFACL*RADIUS(Q+1)*CL
        COORD(1,2) = IFACL*RADIUS(Q+1)*SL
        COORD(2,1) = OFACL*RADIUS(Q)*CL
        COORD(2,2) = OFACL*RADIUS(Q)*SL
        COORD(3,1) = OFACR*RADIUS(Q)*CR
        COORD(3,2) = OFACR*RADIUS(Q)*SR
        COORD(4,1) = IFACR*RADIUS(Q+1)*CR
        COORD(4,2) = IFACR*RADIUS(Q+1)*SR
        IF (IRITE .EQ. 1) THEN
          WRITE(1,10) COORD(1,1), COORD(1,2), COORD(2,1), COORD(2,2),
     2                COORD(3,1), COORD(3,2), COORD(4,1), COORD(4,2)
10      FORMAT(8E10.3)
        END IF
C
C
        RETURN
        END
```

391

Figure 8.51 A listing of the computer program FDIFF; a user's guide appears at the beginning of the listing. This program is a modification of one given in Wang and Anderson (1982, Figure 2.10). This program allows variable sized grids, whereas the one given in Wang and Anderson (1982, Figure 2.10) is for a fixed grid size. The original program is from *Introduction to Groundwater Modeling* by Wang and Anderson. Copyright © 1982 by W.H. Freeman and Company. Used by permission.

```
      PROGRAM FDIFF
C
C     THIS PROGRAM IS A SIMPLE, FINITE DIFFERENCE SOLUTION TO
C     LAPLACE'S EQUATION FOR REGIONAL GROUNDWATER FLOW.
C
C     THIS PROGRAM IS MODIFIED FROM ONE GIVEN IN: WANG, H. AND
C     ANDERSON, M., INTRODUCTION TO GROUNDWATER MODELING,
C     NEW YORK: FREEMAN, 1982, PP. 34-35.
C
C*********************************************************
C                   USER'S GUIDE: FDIFF
C*********************************************************
C
C     A DATA FILE IS CREATED AS FOLLOWS:
C
C     LINE 1:  M,N
C          WHERE M = NUMBER OF ROWS (INCLUDING FICTITIOUS ROWS)
C                    COMPOSING THE GRID;
C                N = NUMBER OF COLUMNS (INCLUDING FICTITIOUS COLUMNS)
C                    COMPOSING THE GRID;
C
C     LINE 2:  X
C          WHERE X IS THE VALUE OF HEAD TO WHICH ALL GRID POINTS
C          ARE INITIALIZED TO START
C
C     LINES 3 TO 3+N:  WATER TABLE-J
C                   WHICH IS THE WATER TABLE BOUNDARY HEAD FOR
C                   GRID LOCATION: H(1,J)
C
C*********************************************************
C               END OF USER'S GUIDE: FDIFF
C*********************************************************
C
      DIMENSION H(50,50)
C
C     READ INPUT INFORMATION
C
      WRITE(*,*) 'ENTER THE NAME OF THE INPUT DATA FILE'
      OPEN(5, FILE = ' ')
      WRITE(*,*) 'ENTER THE NAME OF THE OUTPUT GRID FILE'
      OPEN(1, FILE = ' ', STATUS = 'NEW')
      READ(5,*) M,N
      READ(5,*) X
      DO I = 1, N
         READ(5,*) H(1,I)
      END DO
C
C     INITIALIZE ITERATION PROCESS
C
      NUMIT = 0
   35 AMAX  = 0.0
      NUMIT = NUMIT + 1
C
C     FORCE LEFT AND RIGHT NO-FLOW BOUNDARY CONDITIONS
C
C     NO FLOW IS REPRESENTED BY dh/dx OR dh/dy EQUAL TO ZERO;
C     THIS IS FORCED BY SETTING H(i,1) = H(i,3), THEN THE
C     DERIVATIVE OF HEAD ACROSS THE 2ND COLUMN IS ZERO; LIKEWISE,
C     H(i,N-2) = H(i,N) FORCES NO FLOW ACROSS THE NTH COLUMN.
C     MOREOVER, TO ESTABLISH NO-FLOW ACROSS ROW M, ROWS M-2 AND M ARE
C     SET EQUAL.
C
```

Figure 8.51 continued

```
C
C     FOR COLUMNS:

      DO I = 1, M
         H(I,1) = H(I,3)
         H(I,N) = H(I,N-2)
      END DO
C
C     FOR ROWS:

      DO J = 1, N
         H(M,J) = H(M-2,J)
      END DO
C
C     START FINITE DIFFERENCE APPROXIMATION
C
      DO I = 2, M-1
         DO J = 2, N-1
            OLDVAL = H(I,J)
            H(I,J) = (H(I,J-1)+H(I,J+1)+H(I-1,J)+H(I+1,J)) / 4.0
            ERROR = ABS(H(I,J) - OLDVAL)
            IF (ERROR .GT. AMAX) THEN
               AMAX = ERROR
            END IF
         END DO
      END DO
C
C
C     IF AMAX > 0.001, DO ANOTHER ITERATION

      IF (AMAX .GT. 0.001) NUMIT
      WRITE(*,100) NUMIT
      DO I = 1, M-1
         WRITE(*,200) (H(I,JK), JK = 2, N-1)
      END DO
100   FORMAT(//,5X,'NUMBER OF ITERATIONS = ', I10,//)
200   FORMAT(2X,11F7.2,/)
      DO I = 1,M-1
         DO J = 2,N-1
            WRITE(1,*) H(I,J)
         END DO
      END DO
      STOP
      END
```

393

9

Fourier Analysis

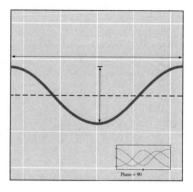

Phase = 90

Is the earth's climate changing such that average atmospheric temperature is increasing as a result of the "greenhouse effect"? Or does average atmospheric temperature oscillate up and down in natural cycles such that at one instant it appears that the earth is warming and at another instant it appears that the earth is cooling? Do periods of increased sunspot activity occur with regular frequency? Is the behavior of the price of gold over time periodic?

The answers to these increasingly important questions involve the analysis of periodic behavior of data. One approach to this kind of analysis is called Fourier analysis, after Jean Baptiste Fourier (1768–1830), who first realized that many types of time series data (or one-dimensional data in general) may be represented as a linear combination of sinusoids. Another name for Fourier analysis is harmonic analysis.

This chapter is an introduction to harmonic (Fourier) analysis. The specific objectives of this chapter are (1) to introduce the notion of sinusoidal waveforms, (2) to use sinusoidal waveforms to represent time series data, and (3) to explain why the representation of time series data as sinusoids is useful. Visual displays are used liberally throughout this chapter to aid understanding.

9.1 Sinusoidal Waveforms

This chapter discusses the analysis of data used primarily to identify periodicities. Identification of periodicities may have important implications: For instance, it may reveal how often droughts or floods occur, whether atmospheric temperature is periodic, and, if so, whether these periods correspond, for instance, to periods of increased sunspot activity.

Trigonometric functions such as sine and cosine are periodic functions. Because these functions are periodic and because periodicities in data, if present, are being analyzed, if the data under study can be accurately modeled using linear combinations of sine and cosine functions, then the periods of these trigonometric functions accurately reflect periods in the data; hence, the periodic nature of the data is identified.

The remainder of this chapter involves the description of data using trigonometric functions. First, examples of sinusoidal functions are used to familiarize the reader with the terms used in this discussion and the influence these terms have on the trigonometric functions.

9.1.1 Cosine Function

Let a function y be written in terms of an independent variable t, such that (after Davis, 1986, p. 253)

$$y(t) = A \left[\cos \left(\frac{2\pi ct}{T} \right) \right]$$

where A is the amplitude, c is the frequency of the waveform, and T is the interval over which y is defined. Another aspect to consider vis-à-vis the cosine waveform is its phase, or phase angle, which results in the offset (Figure 9.1). The phase angle ϕ is incorporated into the equation for y such that

$$y(t) = A \left[\cos \left(\frac{2\pi ct}{T} - \phi \right) \right]$$

As an aid to understanding the terminology used in this chapter, an interactive computer graphics program, CSSINE, is presented in Figure 9.2, page 420). When using this program, interaction is required between the user and the program to define the following terms: (1) amplitude A; (2) frequency c, which describes how many wave cycles are desired; and (3) phase angle ϕ. These terms are illustrated in Figure 9.1. Using the program CSSINE to create a cosine waveform and displaying this waveform using the program GRAPH2D (Chapter 7) shows the effect of each of these parameters. Four examples are displayed in Figure 9.3. The amplitude, frequency, and phase angle used to develop each of these four plots are described in the caption to this figure. These four cosine waveforms are used in the next section to demonstrate the notion of the periodogram.

9.2 The Periodogram

A discussion is presented regarding the transformation of time series data from the spatial to the frequency domain, which will eventually lead to what is called

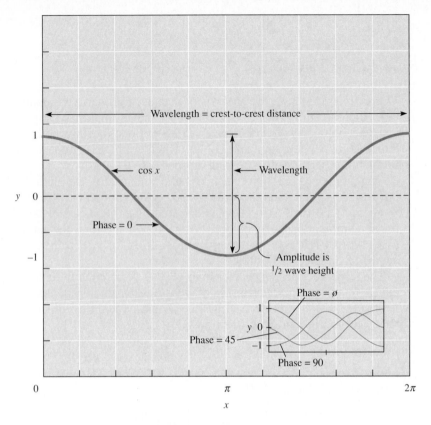

Figure 9.1 A cosine waveform. Amplitude, phase, and wavelength are shown. Frequency, the input parameter required by the program CSSINE, is equal to the inverse of wavelength.

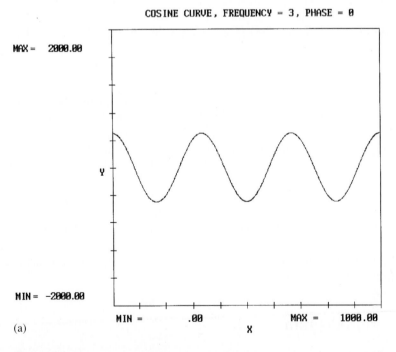

(a)

Figure 9.3 Four cosine waveforms. Each of these curves is composed of 1000 samples. The x axis for each represents the sample interval, and the y axis represents amplitude. (a) A waveform having a frequency of 3 cycles.

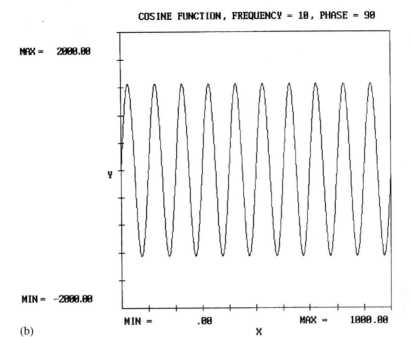

(b)

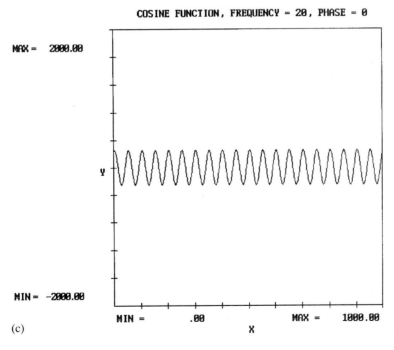

(c)

Figure 9.3 continued   (b) A waveform having a frequency of 10 cycles. (c) A waveform having a frequency of 20 cycles.

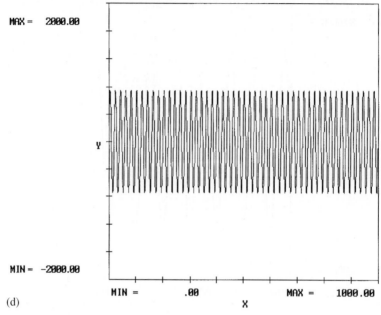

(d)

Figure 9.3 continued   (d) A waveform having a frequency of 50 cycles.

the Fourier transform. Our discussion begins with a trigonometric identity:

$$\cos(A - B) = \cos A \cos B + \sin A \sin B$$

which describes the relationship for the difference between two angles, A and B. This gives a unique way to describe the function y. Given that

$$y(t) = A\left[\cos\left(\frac{2\pi ct}{T} - \phi\right)\right]$$

and letting

$$\theta = \frac{2\pi t}{T}$$

leads to

$$y_c(t) = A_c \cos\phi_c \cos(c\theta) + A_c \sin\phi_c \sin(c\theta)$$

where c is the frequency of wave cycles per interval T. Notice that phase and amplitude are written as functions of c; c is known as a harmonic number (Davis, 1986, p. 253).

Because our major objective in this chapter is to investigate whether a set of time series data is representable as a linear combination of sinusoidal functions, throughout the remainder of this chapter, experiments are conducted to see if a time series function y can be written as follows:

$$y(t) = \sum_{c=0}^{\infty} y_c(t)$$

In other words, the experiment is to see if at any time t (or position t, in general), the total value of function y is equal to the sum of contributions from all harmonic numbers (or frequencies) at position (or time) t.

This expression is known as the Fourier relationship. That this is a linear combination of sinusoidal waveforms is seen more easily if the following is written:

$$y_c(t) = \alpha_c \cos(c\,\theta) + \beta_c \sin(c\,\theta)$$

for which

$$\alpha_c = A_c \cos \phi_c, \qquad \beta_c = A_c \sin \phi_c$$

where cosine and sine terms are equivalent to power terms, and α and β are coefficients solvable using least-squares regression. This is shown in Bloomfield (1976, p. 10):

$$S_c = \sum_{t=1}^{N} [y_t - \alpha_c \cos(c\,\theta) - \beta_c \sin(c\,\theta)]^2$$

where N is the number of samples y_t comprising the time series. Of course, the objective is to minimize the value of S_c.

Using this least-squares procedure, equations for α and β are derived (from Bloomfield, 1976, p. 43):

$$\alpha_c = \frac{2}{N} \sum_{j=0}^{N-1} y_j \cos\left(\frac{2\pi jc}{N}\right)$$

and

$$\beta_c = \frac{2}{N} \sum_{j=0}^{N-1} y_j \sin\left(\frac{2\pi jc}{N}\right)$$

where N is the total number of samples y comprising the time series. With respect to these two coefficients, Davis (1986, p. 255), acknowledges that

$$\beta_0 = 0, \qquad \alpha_0 = \frac{1}{N} \sum_{j=0}^{N-1} y_j = \text{mean of } y$$

Further, the amplitude A_c is a function of α and β:

$$A_c = \sqrt{\alpha_c^2 + \beta_c^2}$$

as is the phase angle:

$$\phi_c = \arctan\left(\frac{\beta_c}{\alpha_c}\right)$$

Finally, a notion of variance of y_c is written as a function of amplitude of the waveform of frequency c such that

$$\text{var}(y_c) = \frac{A_c^2}{2} = \frac{\alpha_c^2 + \beta_c^2}{2}$$

Note that with Fourier analysis a synonym of variance is power. Let power be labeled s^2; then power for a harmonic or frequency c is found as

$$s_c^2 = \text{var}(y_c) = \frac{A_c^2}{2} = \frac{\alpha_c^2 + \beta_c^2}{2}$$

A plot known as the periodogram is developed by plotting s^2 versus the frequency c. This is a simple two-dimensional graph of power versus harmonic number. The notion of the periodogram is explained in the following three examples.

Example 9.1

Recall the graphs of four cosine waveforms in Figure 9.3, each having different frequencies and phase angles (let these four functions be called $f_i(x)$, $i = 1, 2, 3, 4$). The four functions can be combined into one function g as follows:

$$g(x) = \frac{f_1(x) + f_2(x) + f_3(x) + f_4(x)}{4}$$

This expression is plotted in Figure 9.4. As formed, the function g is a linear combination of waveforms f.

The program PERIOD (Figure 9.5, page 421) computes the power s^2 for each frequency (harmonic) c for a particular time series. The output from PERIOD can be displayed using the program GRAPH2D (Chapter 7) by selecting the first data input option for displaying a general x-y file.

Before applying the program PERIOD to the function g, the expected outcome is evaluated. The function g is associated with four distinct frequencies (recall Figure 9.3). Hence the expected outcome of the periodogram analysis of g is the identification of four unique frequencies: 3, 10, 20, and 50 cycles.

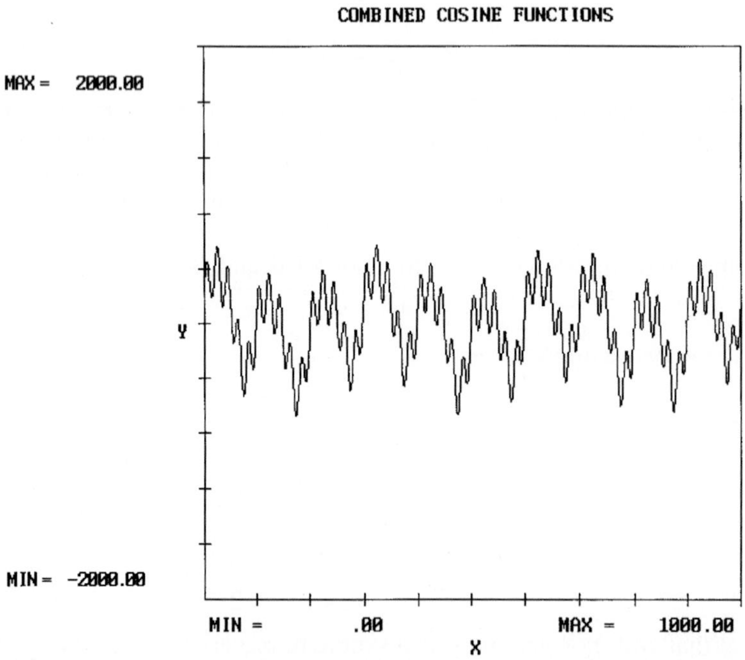

COMBINED COSINE FUNCTIONS

MAX = 2000.00

Y

MIN = -2000.00

MIN =   .00   MAX =   1000.00

X

Figure 9.4 A combination waveform created by using the program ADD (Figure 9.26) to sum the four waveforms displayed in Figure 9.3. The x axis represents sample interval, and the y axis represents amplitude.

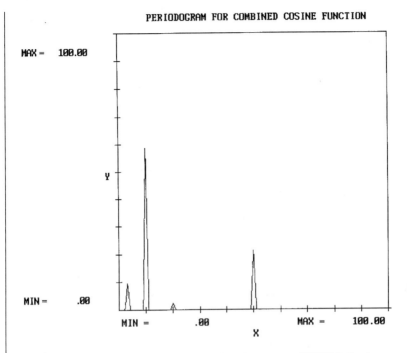

PERIODOGRAM FOR COMBINED COSINE FUNCTION

Figure 9.6 A periodogram calculated using the program PERIOD for the curve plotted in
Figure 9.4. The x axis represents frequency or harmonic number, and the y axis represents
power.

The periodogram of Figure 9.6 indicates that frequencies of 3, 10, 20, and 50
cycles are associated with relatively large power. All other frequencies are associated
with minimal power. True, g is an artificial creation. Yet, this function serves as a
good test of the program PERIOD. Furthermore, because g is developed in a controlled, artificial sense, the identification of all periodicities composing a time series,
which is the major objective of this chapter, is more obvious.

Example 9.2

Given the data listed in Appendix A (from Haan, 1977), an experiment is conducted
to use periodogram analysis to identify the presence, if any, of periodicities. These
data are monthly stream runoff data for Cave Creek, Kentucky. The objective of this
analysis is to examine the regularity at which flow in this stream increases and
decreases. A first step involves simply plotting the data (Figure 9.7). This plot alone
suggests that one prominent waveform, having a frequency of 18, exists in this data
set. The periodogram is useful for showing this and for examining whether other, more
subtle, periodicities are present in these data.

Application of the program PERIOD to the Cave Creek data yields the periodogram
of Figure 9.8. Notice that one harmonic number, $c = 18$, is associated with a large
power, whereas all other harmonic numbers are associated with negligible power. This
underscores the fact that these data are associated with a single prominent waveform
having a frequency of 18. Because 18 cycles exist in these data, which are fairly regularly spaced, and because these data represent an 18-year time span, the wavelength

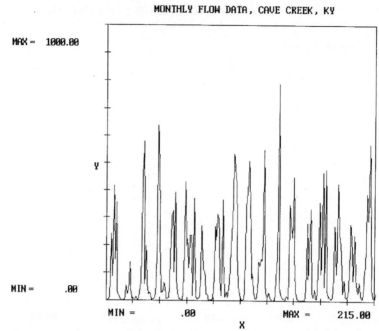

Figure 9.7   A plot of the runoff data for Cave Creek, Kentucky. Note the 18 cycles of peak runoff. These data are taken from Haan (1977). The x axis represents time (215 months), and the y axis represents discharge.

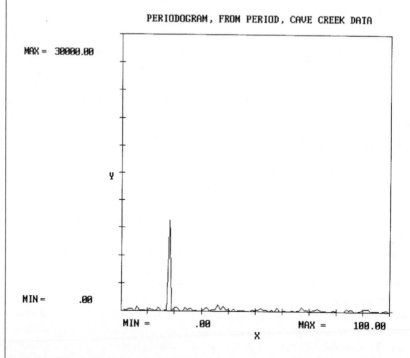

Figure 9.8   A periodogram calculated using the program PERIOD for the Cave Creek, Kentucky, runoff data. The x axis represents frequency or harmonic number, and the y axis represents power.

of each of the 18 cycles is one year. Therefore, an inference might be made that flow in Cave Creek is seasonal, with peak flows occurring from midwinter to early spring.

Figure 9.7 does not look exactly like a perfect cosine waveform having 18 cycles, but it looks approximately like this. An additional, seemingly random component is present in these data which causes the noisiness. This component is considered to be random because no harmonic number other than $c = 18$ is associated with a large power, hence no other periodicities are recognized. The variation (deviation) of the data from the perfect cosine waveform is therefore considered to be random. ∎

Example 9.3

Some data sets are associated with more than one period. For example, a time series for the magnitude of a variable star at midnight for 600 successive nights (Whittaker and Robinson, 1924) appears to be periodic, but the periods appear to involve more than one frequency (Figure 9.9).

A periodogram of these data (Figure 9.10) identifies two harmonic numbers, $c = 21$ cycles and $c = 25$ cycles, associated with large powers, whereas all other harmonics are associated with small powers. This indicates that the variable star data are largely a composite of two waveforms, one having a wavelength of 600/21, or 28.6 days, and the other having a wavelength of 600/26, or 23 days. If the objective of the analysis is to begin to understand how many factors may be influencing the brightness of this particular variable star, then the periodogram for this star indicates that two influences may control the brightness. Astronomers think that the brightness of variable stars is controlled by processes internal to the star, but for some of these stars, brightness is

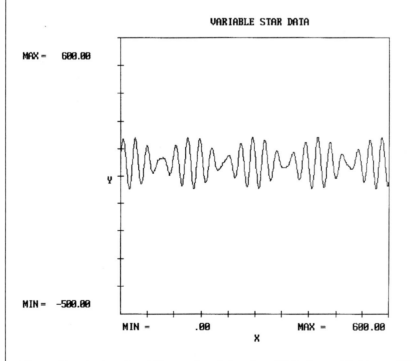

Figure 9.9 A plot of variable star data (taken from Whittaker and Robinson, 1924). The x axis represents sample interval, and the y axis represents brightness.

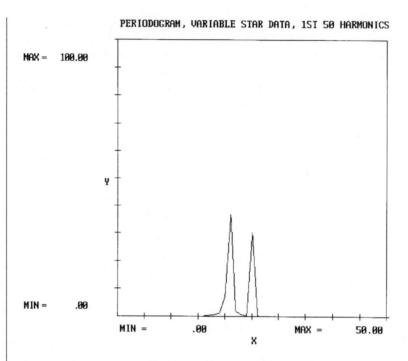

Figure 9.10   A periodogram for the variable star data calculated using the program
   PERIOD. The x axis represents frequency, and the y axis represents power.

also controlled by interference with another astronomical body (an eclipse), such as a
sister star or a planet.

Variable star data are more pertinent to the study of astronomy than to the study of
geology. But these data are used in this text because (1) their periodic nature is
readily apparent (Figure 9.9); (2) they are described in the literature (Whittaker and
Robinson, 1924) and are a classic example of Fourier analysis; and (3) these data are
representative of time series data having more than one period.   ∎

9.3   Fourier Transform

The preceding three examples involve distinctly periodic data. In some
instances, however, a plot of the raw data may not appear to be periodic, even
though subtle periodicities, which may be of great interest in the analysis of
the data, may be present. For such data, a more sophisticated analysis is desired:
the Fourier transform.

In developing the notion of this transform, the Euler relationship is used:

$$\cos(2\pi kt) + i \sin(2\pi kt) = \exp(i2\pi kt)$$

which has a form similar to what is used in Section 9.2:

$$y(t) = \alpha \cos(c\,\theta) + \beta \sin(c\,\theta)$$

This shows that the exponential function is closely related to the cosine and
sine functions. Therefore, the exponential function is additionally useful in
analyzing periodicities in the data.

Euler's relationship is one that involves the imaginary number i. If the Euler relationship is used to represent periodic data y, then they are assumed to be complex numbers. Most often such data are real numbers, but any real number is also a complex number having a zero imaginary component. Complex number arithmetic leads to algebraic simplification, which in turn facilitates the following development of the Fourier transform (Bloomfield, 1976, p. 46).

The formal definition of the Fourier transform involves an integral. For a time series y, its Fourier transform Y is given by

$$Y(c) = \int_{-\infty}^{\infty} y(j)e^{2\pi c j}\, dj$$

where c is the frequency. The inverse of this transform is

$$y(j) = \int_{-\infty}^{\infty} Y(c)e^{-2\pi c j}\, dc$$

which shows that the original time series y is obtainable from its frequency domain counterpart Y.

This integral must be approximated, which leads to the formal presentation of the discrete Fourier transform:

$$Y(c) = \sum_{j=0}^{N-1} y_j e^{2\pi i c j} x$$

where x is the sampling interval for y, j is the particular interval of y, c is the harmonic number, and i is the imaginary number, equal to the square root of -1.

If y is sampled at regular intervals of length x, then

$$Y(c) = x \sum_{j=0}^{N-1} y_j e^{2\pi i j n/N}$$

for which

$$c = \frac{n}{xN}, \qquad n = -\frac{N}{2}, \ldots, \frac{N}{2}$$

Once the discrete Fourier transform of y is obtained, a new form of the periodogram is obtainable. For each frequency c, the power s^2 is obtainable approximately as $2Y(c)^2$. A periodogram for a set of time series data y is therefore obtainable in several ways, one from the Fourier transform and the other from the previous procedure (Section 9.2).

A computer program for calculating the Fourier transform, FORIER (Figure 9.11, page 422), is based on an algorithm given in Yakowitz and Szidarovszky (1989, p. 335), with a few minor modifications: (1) An output file is created that contains the harmonic number c and power $2Y(c)^2$; this file is compatible with the program GRAPH2D (Chapter 7). (2) An output file also is created that stores the actual Fourier transform. This file is useful for data filtering (Section 9.4).

Applying the program FORIER to the Cave Creek data yields the periodogram of Figure 9.12. This figure also demonstrates that the Fourier transform is, indeed, a useful approach to the development of a periodogram, and is analogous to the method of Section 9.2. An additional test of the program FORIER is shown in Figure 9.13 and is an application to the function g (see Example 9.1). The four frequencies identified in Figure 9.13 are the same four frequencies that were identified in Figure 9.6.

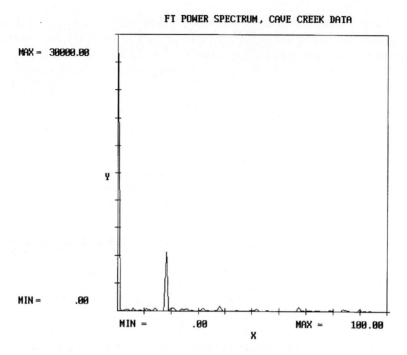

Figure 9.12   A power spectrum (periodogram) calculated using the program FORIER for the Cave Creek, Kentucky, runoff data. Frequency is shown on the *x* axis, and power is shown on the *y* axis.

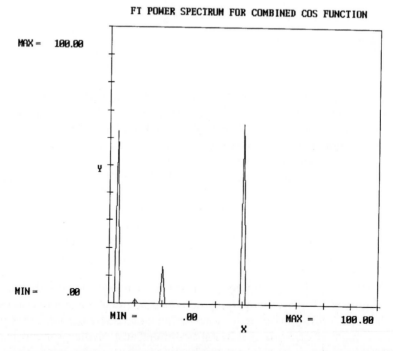

Figure 9.13   A power spectrum (periodogram) calculated using the program FORIER for the curve plotted in Figure 9.4. Frequency is shown on the *x* axis, and power is shown on the *y* axis.

9.4 Justifying the Use of the Fourier Transform

If the Fourier transform leads to the development of a periodogram that is similar to one developed using the program PERIOD, why use the Fourier transform approach, especially because more calculations are involved using complex arithmetic? This is a good question, and the answer is not immediately obvious.

Recall that the Fourier transform of y for the harmonic c is $Y(c)$. Further recall that y can be restored from Y using the inverse Fourier transform, the discrete version of which is

$$y(j) = \sum_{c=0}^{N-1} Y(c)e^{-2\pi ijc/N}$$

This relationship is the primary advantage of the Fourier transform for analyzing periodicity in the data. First, this inverse transform is much easier to implement than modeling y using the method of Section 9.2. Second, the inverse Fourier transform offers an easy and unique way to filter data.

For example, recall the Cave Creek runoff data analyzed previously (Example 9.2). One prominent waveform of frequency $c = 18$ is identified for these data; all other harmonics are inferred to be associated with random noise. Suppose random noise is to be isolated from this waveform. This can be accomplished in three steps: (1) Fourier transform the original data y using the program FORIER; (2) filter the Fourier transform such that the transform $Y(c)$ is set to zero for all harmonics c other than $c = 0$ (which is simply the average value

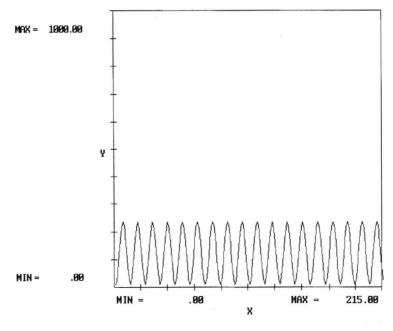

FT FILTERED CAVE CREEK DATA

Figure 9.16 The Cave Creek, Kentucky, runoff data after Fourier filtering all frequencies except those associated with large power. Note the very regular waveform having 18 cycles. This is a model for the actual data displayed in Figure 9.7.

of y), $c = 18$, and, because of symmetry, $c = 197$ (which corresponds to $c = 18$) and $c = 215$ (which corresponds to $c = 0$); and (3) inverse Fourier transform the filtered values Y' to yield the filtered runoff data y'.

A program for computing the inverse Fourier transform, FORINV, (Figure 9.14, page 424) uses as the input file the second output file (unit 2) created by the program FORIER. The Fourier transform for the Cave Creek runoff data is filtered as just described using the program FILTER (Figure 9.15, page 425). After filtering the Fourier transform, an inverse Fourier transform is applied to yield the filtered runoff data (Figure 9.16). Notice that removal of the random component from this data yields a perfect sine waveform, having a frequency of 18. The simplicity by which this filtered result is obtained demonstrates why the Fourier transform is used more often than the procedure described in Section 9.2 for modeling the periodicity in time series data. This demonstration is also a good illustration of the Euler relationship, whereby a waveform is obtainable using the exponential function.

9.5 The Power Spectrum: Spectral Analysis

Often in the literature on Fourier analysis mention is made of a notion called the power spectrum. In fact, this term is used so often that a separate section is devoted to it here. The power spectrum and the periodogram are the same notion. Therefore, in using the program FORIER to Fourier transform time series data, the first output file (unit 1) is the power spectrum for the time series data. The power spectrum is simply the square of the amplitude for each harmonic comprising the data.

9.6 Examples of Fourier Analysis

9.6.1 Sunspot Cycles

Much attention is paid in science today to the concept of "global change." This phrase has broad meaning, from that of the earth's changing climate as a result of the "greenhouse effect" (which is only a hypothesis), to that of earth's changing orbital characteristics, and to the earth changing in all contexts. One hypothesis suggests that the earth's climate is impacted locally in time by sunspot activity. Whether or not this is true has yet to be determined conclusively. However, one step in the testing of this hypothesis is to attempt to understand the periodic nature of sunspot cycles.

Sunspot cycles (Figures 9.17 and 9.18) for the period 1750 through 1980 do appear to be cyclic, but like the variable star data (Example 9.3), several cycles, one of low frequency (long wavelength) and one of higher frequency (shorter wavelength), seem to be present in these data. Two periodograms are computed for the sunspot data, one using the program PERIOD (Figure 9.19) and the other using the program FORIER (Figure 9.20). (The only reason both periodograms are plotted is to verify the correct calculation of the periodogram using two different calculation procedures.) Both periodograms identify a large power for the frequencies in the interval $c = 2$ to $c = 4$ (cycles). This is the low-frequency periodicity suspected to be present in these data. Because these

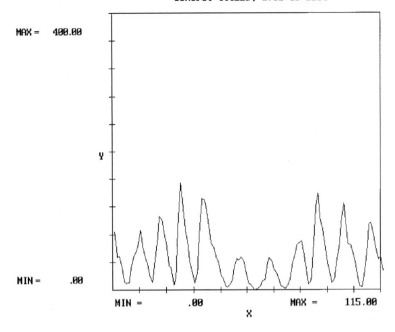

Figure 9.17 Sunspot numbers over the time interval 1750–1864 (from Shove, 1983). The horizontal axis represents the number of years from 1750, and the vertical axis represents the number of sunspots.

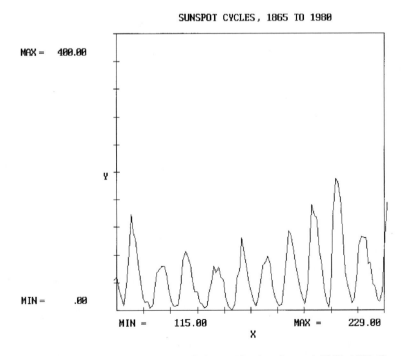

Figure 9.18 Sunspot numbers (cycles) over the time interval 1865–1980 (from Shove, 1983). The horizontal axis represents the number of years from 1865, and the vertical axis represents the number of sunspots.

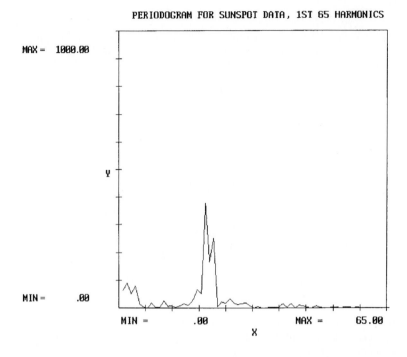

Figure 9.19 Periodogram for the sunspot data, computed using the program PERIOD. Frequency is shown on the x axis, and power is shown on the y axis.

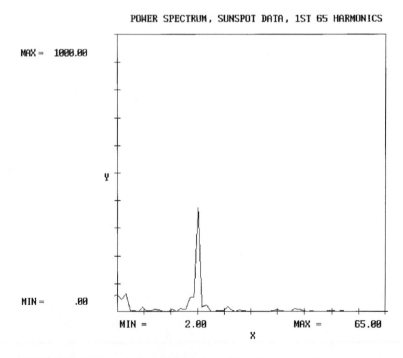

Figure 9.20 Power spectrum (periodogram) for the sunspot data, computed using the program FORIER. Frequency is shown on the x axis, and power is shown on the y axis.

data represent a 230-year time span, 2 to 4 cycles represent a wavelength (or periodicity) of 230/2, or approximately 115 years to 230/4, or about 57.5 years. Both periodograms also identify a higher frequency period for $c = 21$ (cycles), which represents a wavelength (periodicity) of 230/21, or approximately 11 years.

Many researchers feel that increased sunspot activity leads to greater atmospheric temperatures for earth and that droughts are related to these cycles. Researchers often quote the 11-year cycle to see if droughts are similarly cyclic. However, the investigation of this correlation is complicated by the presence of the long wavelength in the sunspot cycles.

Recall that sunspot cycles are largely a combination of two waveforms, one of very low frequency and one of higher frequency. The combination of these two cycles affects the occurrence in time and the severity (number of sunspots) of each sunspot cycle. An end of chapter exercise compares sunspot cycles to those, if any, found in the temperature data (Chapter 4). The intent of this exercise is to see if any correlation between these two phenomena can be inferred.

9.6.2 Analysis of Seismic Data

One of the most obvious applications of Fourier analysis comes from the field of seismology. Earthquakes are natural phenomena associated with a release of energy along faults (or are caused by point sources, such as volcanic and plutonic activity). This energy propagates (radiates) outward in the form of waves, both compressional and shear. Because this energy is a wave phenomenon, ground motion caused by the energy is likewise a wave (that is, cyclic) phenomenon. The problem is that ground motion is composed of waves of many frequencies and phases, making it difficult to understand the nature of this phenomenon without some sort of analysis that can differentiate the various frequency components. Fourier analysis is ideally suited to this application.

Fourier analysis of earthquake ground motion is important for several reasons. (1) The total number of cycles of ground motion occurring during an earthquake is an important measure of its potential for causing damage. Larger earthquakes are associated with a release of greater energy, and consequently more cycles of ground motion occur for these events. (2) If ground motion is measured inside a building, for instance, then understanding how the building interfered, if at all, with the frequencies composing the ground motion is useful for assessing how well the building performed during the earthquake. (3) If ground motion is measured at a "free-field" site, that is, a site not inside a building but instead founded on natural ground, then Fourier analysis of this "free-field" ground motion, in theory, identifies the natural cycles inherent to the ground motion associated with a particular earthquake. If such a free-field power spectrum is compared to one for ground motion measured in a building, cycles present in the spectrum for the building that are not present in the free-field spectrum can be interpreted as being due to the building response. This information is useful for engineering design because in the design of a building, it is useful for the building's motion to have cycles that do not match the natural cycles so that the building will tend to dampen (cancel) the ground motion. (If a building's vibration cycle matches a natural cycle, much more pronounced swaying of the building will occur because the cycles add.)

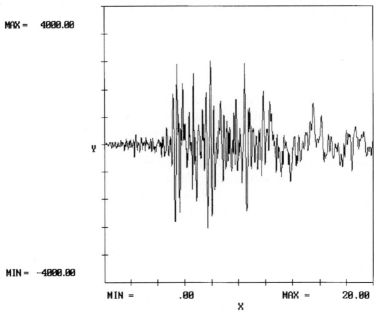

ACCELEROGRAM, 14724 VENTURA BOULEVARD

MAX =   4000.00

Y

MIN = -4000.00

MIN =   .00   MAX =   20.00

X

Figure 9.21   Accelerogram recorded at 14724 Ventura Boulevard, Los Angeles, California (USA), during the 1971 San Fernando earthquake (data from Trifunac et al., 1974). The data represent 1000 samples taken at 0.02-s intervals (x axis); acceleration is shown on the y axis in cm/s². These data are listed in Appendix A: SEISMIC1.DAT.

An example accelerogram, which is a record of horizontal acceleration measured as a percentage of gravitational acceleration (Figure 9.21), shows a 20-second interval, representing 1000 observations at 0.02-second spacings, and was recorded on the first floor of a building located at 14724 Ventura Boulevard, Los Angeles, California (USA) during the 1971 San Fernando, California, earthquake (these data are from Trifunac and others, 1974). Cycles are observed in the figure, yet some hidden cycles may be present that are not readily apparent.

A Fourier transform of these observations using the program FORIER yields the power spectrum of Figure 9.22. Four prominent cycles are identified. The first period, 6 cycles, is a low-frequency component; the other 3 cycles are higher-frequency components representing 33, 51, and 72 cycles. Other than these four most prominent periods, minor periods are found at 95, 113, 138, and 150 cycles. All other frequencies are associated with minimal power.

A hypothesis is forwarded that the low-frequency component of 6 cycles found in the Ventura Boulevard accelerogram is due to the swaying of the building in which the accelerogram was recorded. To test this hypothesis, a free-field accelerogram recorded at Fairmont Reservoir, California, during the 1971 San Fernando earthquake (Figure 9.23) is analyzed for comparison. The periodogram for the Fairmont accelerogram (Figure 9.24) shows that the free-field accelerogram is not associated with a low-frequency component. If this free-field accelerogram is representative of the cycles occurring naturally in the ground motion for the 1971 San Fernando earthquake, a conclusion is

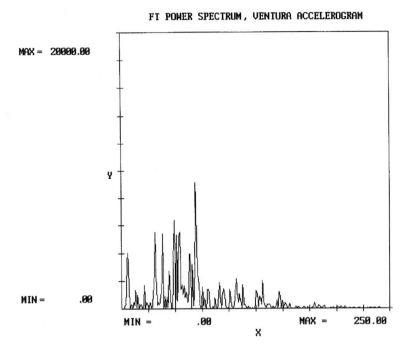

Figure 9.22 Power spectrum (periodogram) for the 14724 Ventura Boulevard accelerogram calculated using the program FORIER. Frequency is shown on the *x* axis, and power is shown on the *y* axis.

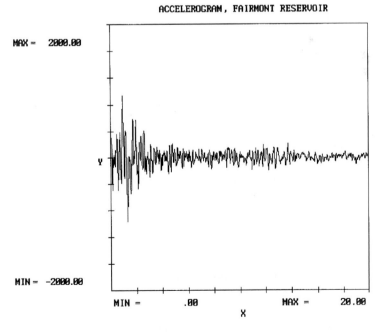

Figure 9.23 Accelerogram recorded at Fairmont Reservoir, California (USA), during the 1971 San Fernando earthquake (data from Trifunac et al., 1974). Data represent 1000 samples taken at 0.02-s intervals (*x* axis); *y* axis shows acceleration in cm/s². These data are listed in Appendix A: SEISMIC2.DAT.

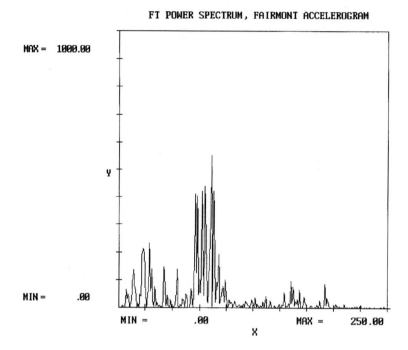

Figure 9.24 Power spectrum (periodogram) for the Fairmont Reservoir accelerogram calculated using the program FORIER. Frequency (*x* axis) is plotted against power (*y* axis).

reached that the low-frequency, 6-cycle component in the Ventura accelerogram is attributable to the building. Of course, analyzing more free-field recordings for this earthquake would help to verify this conclusion.

9.7 Fast Fourier Transform (FFT)

Literature on Fourier transformation often mentions a notion called the fast Fourier transform, or FFT (see, for example, Press et al., 1986, p. 390). The fast Fourier transform yields the same results as the discrete Fourier transform (and inverse transform), implemented herein in the programs FORIER and FORINV. The FFT algorithm, however, avoids complex arithmetic, implicitly accounting for the imaginary number i. Hence the algorithm is faster than that used in this chapter. But the FFT algorithm is more difficult to understand than the direct use of complex arithmetic as programmed in FORIER and FORINV.

The FFT algorithm requires that the number of samples N comprising the time series y be a power of 2. This is not a requirement for the discrete Fourier transform calculated by program FORIER or the inverse transform calculated by program FORINV: The number of samples N should be an even number when using these programs, but N need not be a power of 2. For the sake of simplicity, this text does not implement the FFT algorithm, but instead employs an algorithm that is easier to understand, follow in the program(s), and use.

9.8 A Final Example

An example is presented as a summary and to provide additional clarity, which may be useful if the procedures presented in this chapter are still not clear. The Fourier transform is basically the following notion:

$$FT = data * waveform$$

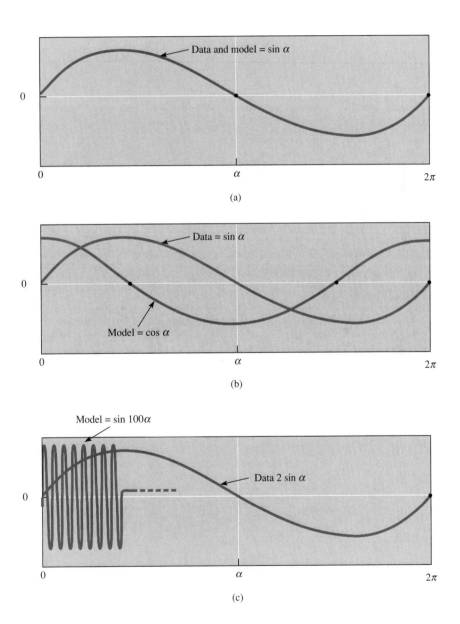

(a)

(b)

(c)

Figure 9.25 The concept of Fourier analysis: (a) if a model waveform matches an actual data waveform, a larger power results; (b and c) if a model waveform imperfectly matches a data waveform, a smaller power is obtained.

Table 9.1   Cumulative Product of One Waveform
with Another

Time Series y	Model Waveform	Summed Product
$\sin(\alpha)$	$\sin(\alpha)$	180.0
$\sin(\alpha)$	$\cos(\alpha)$	0.0
$\sin(\alpha)$	$\sin(100\alpha)$	0.0

Note. Results are for $\alpha = 0$ to 360 degrees and are proportional to total power s^2.

for which the $*$ operator represents a convolution. Compare this expression to that for the discrete Fourier transform:

$$Y(c) = \sum_{j=0}^{N-1} y_j e^{2\pi i c j} x$$

and consider that y_j is the data and $e^{2\pi i c j} x$ is the waveform; Y is the Fourier transform of y.

If the data y are similar in form to the waveform, the Fourier transform has a relatively large value. If, on the other hand, the waveform does not match the data well, they will tend to cancel and the Fourier transform will have a small value.

The importance of this is shown in Figure 9.25, a hypothetical situation of attempting to fit data, in this case represented by $\sin \alpha$, with several waveforms: $\sin \alpha$, a perfect fit; $\cos \alpha$, which is orthogonal to the data; and $\sin(100\alpha)$, an imperfect fit of a waveform having a frequency of 100 to one having a frequency of 1. Calculation of power for each attempted fit of a waveform model to the data is summarized in Table 9.1. Notice that when the waveforms do not coincide, as is the case of attempting to model $\sin \alpha$ using $\cos(\alpha + \phi)$, the power terms are small. But, when the waveforms match exactly, as is the case of representing $\sin \alpha$ with $\sin \alpha$, the power term is large.

The analyses described in this chapter proceed in this same manner, except many more waveforms are used to attempt the fit to the data. One waveform is used for each possible frequency, where the frequencies range from 0 up to the number of samples composing a time series function. Each time one of these waveforms matches a waveform in the data, a large power is obtained. Small power is obtained due to cancellation when a waveform does not match any frequency in the data.

Chapter Summary

- The purpose of harmonic analysis is to identify the presence of periodicities, or waveforms, in time series data. A discrete Fourier transform of the data identifies periodicities, providing insight to what might be controlling the time series.

- If data clearly appear to be periodic in nature, a periodogram of the data may be computed using sine and cosine functions. A periodogram is a function that records the power or variance of the waveform for each unique harmonic number or frequency.

- A Fourier transform approach to developing a periodogram yields similar results to those obtained using sine and cosine functions. The Fourier transform approach, however, offers an easier, more straightforward approach to inverse transformation than using sine and cosine functions. The Fourier transform approach therefore offers an ideal method for filtering unwanted frequencies from data.

- This chapter presents methods for isolating the frequencies of harmonics (waveforms) in data. Please review the figures in this chapter to see how time series data are often the result of composite waveforms. Understanding the periodic nature of time series data has relevance for the study of global change.

Exercises

(An asterisk indicates an advanced or graduate student problem)

1. Use the program CSSINE to form waveforms having the following parameters:

Amplitude A	Frequency c	Phase ϕ
1.0	2.0	0.0
10.0	5.0	45.0
1.0	20.0	30.0
1.0	1.0	90.0

Use the program ADD (Figure 9.26, page 426) to combine these four waveforms. Compute and display the periodogram for this combined waveform using the programs PERIOD and FORIER.

2. Given the following time series data y (taken from Scheid, 1968, p. 303), compute by hand the periodogram using the method of Section 9.2. Use the program PERIOD to verify your calculations.

j:	0	1	2	3	4	5	6	7	8
y_j:	0	4.3	8.5	10.5	16.0	19.0	21.1	24.9	25.9
j:	9	10	11	12	13	14	15	16	17
y_j:	26.3	27.8	30.0	30.4	30.6	26.8	25.7	21.8	18.4
j:	18	19							
y_j:	12.7	7.1							

*3. Consult Press et al. (1986) to find their subroutine FOUR1, a fast Fourier transform. Implement this subroutine in a computer program and compare its results to those obtained using the program FORIER in application to the Cave Creek data.

4. Compute and plot the periodogram for the temperature data given in Chapter 4. Use both the programs PERIOD and FORIER for this purpose to compare results. Are these data associated with periodicities? If so, what is the

wavelength for each significant period? Do any of these wavelengths correspond to those for the sunspot data?

5. Use the program FILTER to filter all harmonics using Thresmn = 0.10 and Thresmx = 10000.0 for the variable star data (listed in Appendix A). Then use the program FORINV to compute the inverse Fourier transform. Use the program GRAPH2D to plot the results. What did you discover?

6. In the analysis of the variable star data, two distinct periods were identified: 21 cycles and 25 cycles. Use the program CSSINE to create two waveforms, one having a frequency of 21 cycles and the other having a frequency of 25 cycles. Use the program ADD to combine these two waveforms, then display this combined result using GRAPH2D. How does this artificially created waveform compare to the variable star plot (Figure 9.9)?

7. Following the procedure described by Figure 9.25 and documented in Table 9.1, compute by hand the product between the following waveforms and time series data for each of the following angles: 0, 20, 40, 60, 80, 100, 120, 140, 160, 180. Then, sum all products to compute a power.

Waveform	Time series data
$\cos \phi$	$\cos \phi$
$\sin \phi$	$\cos \phi$
$\sin \phi$	$\sin(\phi + 10)$

Which of these pairs yields large powers? Which yields small powers? Why is this so?

8. Prepare a data file for the monthly rainfall data for Walnut Gulch near Tombstone, Arizona (see Appendix A for a listing of these data). Compute and then plot the periodogram for these data using both PERIOD and FORIER. What did you discover?

9. Repeat Problem 8, this time using the data set SEISMIC3.DAT (Appendix A). What do the periodograms suggest for these data? Why might this information be useful for engineering design?

10. Find a set of time series data in the literature, such as precipitation data or the change in price of a commodity over time. Compute the periodogram for the data using both PERIOD and FORIER to analyze the periodic nature of the data. Do the periods identified suggest something about the data?

References and Suggested Readings

BLOOMFIELD, P. 1976. *Fourier Analysis of Time Series: An Introduction.* New York: Wiley.

BRACEWELL, R. 1965. *The Fourier Transform and Its Applications.* New York: McGraw-Hill.

DAVIS, J.C. 1986. *Statistics and Data Analysis in Geology, 2d ed.* New York: Wiley.

HAAN, C.T. 1977. *Statistical Methods in Hydrology.* Ames, IA: Iowa State University Press.

PRESS, W.H., B.P. FLANNERY, S.A. TEUKOLSKY, and W.T. VETTERLING. 1986. *Numerical Recipes: The Art of Scientific Programming.* Cambridge, UK: Cambridge University Press.

SCHEID, F. 1968. *Schaum's Outline of Theory and Problems of Numerical Analysis.* New York: McGraw-Hill.

SHOVE, D.J., ed. 1983. *Sunspot Cycles,* Benchmark Papers in Geology/68. Stroudsburg, PA: Hutchinson Ross.

TRIFUNAC, M.D., A.G. BRADY, and D.E. HUDSON. 1974. *Strong Motion Earthquake Accelerograms, Digitized and Plotted Data, Vol. 3, Corrected Accelerograms and Integrated Ground Velocity and Displacement Curves, Parts O and P; Q and R.* Pasadena, CA: California Institute for Technology, prepared for the National Science Foundation, distributed by the National Technical Information Service, United States Department of Commerce.

WHITTAKER, E.T., and G. ROBINSON. 1924. *The Calculus of Observations.* London: Blackie.

YAKOWITZ, S., and F. SZIDAROVSZKY. 1989. *An Introduction to Numerical Computations,* 2d ed. New York: Macmillan.

Figure 9.2 A listing of the program CSSINE. This program was used to create the waveforms plotted in Figure 9.3.

```
      PROGRAM CSSINE
C
C     AN INTERACTIVE PROGRAM FOR EXPERIMENTING
C     WITH COSINE WAVEFORMS
C
C     *****************************************************
C                 USER'S GUIDE:  CSSINE
C     *****************************************************
C
C     A SERIES OF INTERACTIVE QUESTIONS SEEK DATA INPUT
C
C     Q1:  WHAT IS THE DESIRED AMPLITUDE? E.G., .1, 1., 5., 100.,...
C
C     Q2:  WHAT IS THE DESIRED FREQUENCY, C?  C IS EQUAL TO THE
C          THE NUMBER OF DESIRED WAVES, FROM 1 TO N, HOWEVER MANY
C          N IS; FOR EXAMPLE, IF C = 10, THEN 10 WAVES WILL BE DRAWN
C
C     Q3:  WHAT IS THE DESIRED PHASE ANGLE, IN DEGREES?  ENTER A
C          VALUE BETWEEN 0 AND 360.
C
C     THIS PROGRAM CREATES AN OUTPUT FILE COMPATIBLE WITH THE
C     THE PROGRAM, GRAPH2D (CHAPTER 7).  USE THE PROGRAM, GRAPH2D,
C     TO DISPLAY THE RESULT FROM THIS PROGRAM.
C
C     *****************************************************
C            END OF USER'S GUIDE AND DISCUSSION
C     *****************************************************
C
      WRITE(*,*) 'ENTER THE FILENAME FOR OUTPUT, FOR UNIT 1'
      OPEN(1, FILE = ' ', STATUS = 'NEW')
      PI = 4.0 * ATAN(1.0)
      CAPT = 2.0 * PI
      X = CAPT / 1000.
C
C     ENTER INPUT INFORMATION
C
      WRITE(*,*) 'ENTER A VALUE FOR THE AMPLITUDE'
      READ(*,*) A
      WRITE(*,*) 'ENTER A VALUE FOR THE FREQUENCY, C'
      READ(*,*) C
      WRITE(*,*) 'ENTER A VALUE FOR THE PHASE ANGLE'
      READ(*,*) PHI
C
C     COMPUTE THE WAVEFORMS AND SAVE TO OUTPUT FILE
C
      PHI = PHI * PI / 180.0
      DO I = 1, 1000
      T = X * FLOAT(I)
      Y = A * (COS((2*PI*C*T/CAPT) - PHI))
      WRITE(1,*) Y
      END DO
      STOP
      END
```

Figure 9.5 A listing of the program PERIOD.

```fortran
      PROGRAM PERIOD
C
C     A PROGRAM FOR COMPUTING THE COEFFICIENTS:   ALPHA AND BETA,
C     FROM WHICH THE POWER, S2, IS CALCULATED.
C
C***********************************************************************
C                 USER'S GUIDE:   PERIOD
C***********************************************************************
C
C     THE INPUT SIMPLY CONSISTS OF A DATA FILE CONTAINING N
C     VALUES OF Y:
C
C     LINES 1 - N:  Y(J); JTH VALUE OF TIME SERIES, Y
C
C***********************************************************************
C                 END OF USER'S GUIDE
C***********************************************************************
C
      DIMENSION Y(1000)
      OPEN(5, FILE = ' ')
      OPEN(1, FILE = ' ', STATUS = 'NEW')
C
C     ACCESS DATA, Y
C
      N = 1
10    CONTINUE
      READ(5,*, END=20) Y(N)
      N = N + 1
      GO TO 10
20    N = N - 1
C
C     COMPUTE COEFFICIENTS AND POWER
C
      PI = 4.0 * ATAN(1.0)
      DO I = 1,N
        ALPHA = 0.0
        BETA = 0.0
        K = I
        DO L = 1,N
          J = L - 1
          BETA = BETA + Y(L) * SIN(2*PI*J*K/N)
          ALPHA= ALPHA+ Y(L) * COS(2*PI*J*K/N)
        END DO
        ALPHA = 2.0 * ALPHA / N
        BETA  = 2.0 * BETA  / N
        S2    = (BETA*BETA + ALPHA*ALPHA) / 2.0
        WRITE(*,30) K, ALPHA, BETA, S2
30      FORMAT(1X, I5, 3F15.4)
        WRITE(1,*) K, S2
      END DO
      STOP
      END
```

421

Figure 9.11 A listing of the program FORIER.

```fortran
      PROGRAM FORIER

C
C  A FORTRAN-77 PROGRAM TO COMPUTE THE DISCRETE FOURIER
C  TRANSFORM OF A TIME SERIES
C
C  ***********************************************************
C               USER'S GUIDE:  FORIER
C  ***********************************************************
C
C  THE SAME DATA FILE AS IS USED WITH THE PROGRAM, PERIOD,
C  IS USED WITH THIS PROGRAM, FORIER:
C
C  DATA FILE:  CONSISTS OF N LINES:
C
C  LINE 1:  Y(1)
C              TO
C  LINE N:  Y(N)
C
C  DISCUSSION:
C
C  THIS PROGRAM COMPUTES THE DISCRETE FOURIER TRANSFORM USING
C  COMPLEX NUMBER ARITHMETIC.  MOREOVER, THIS PROGRAM CREATES
C  2 OUTPUT FILES, ONE WHICH IS COMPATIBLE WITH THE PROGRAM,
C  GRAPH2D (CHAPTER 7); THIS FILE CONSISTS OF N LINES AS FOLLOWS:
C
C  OUTPUT FILE: (UNIT 1)
C
C  LINE 1:  1, POWER(1)
C              TO
C  LINE N:  N, POWER(N)
C
C  WHERE THE FIRST VALUE IS THE HARMONIC NUMBER, THE SECOND
C  VALUE, POWER, IS TWICE THE SQUARE OF THE FOURIER TRANSFORM OF
C  Y FOR THE HARMONIC.  THEREFORE, UNIT 1 STORES THE PERIODOGRAM
C
C  THE SECOND OUTPUT FILE STORES THE DISCRETE FOURIER TRANS.
C
C  OUTPUT FILE 2:  (UNIT = 2)
C
C  LINE 1:  1, BETA(1)
C              TO
C  LINE N:  N, BETA(N)
C
C  WHERE BETA IS THE DISCRETE FOURIER TRANS FOR THE JTH HARMONIC
C
C  UNIT 2 IS USEFUL FOR INVERSE FOURIER TRANS.  TO FILTER DATA
C
C  ***********************************************************
C             END OF USER'S GUIDE AND DISCUSSION
C  ***********************************************************
C
      IMPLICIT COMPLEX (A-H, O-Z)
      REAL A, B, PI
      DIMENSION Y(1000), BETA(1000)
      REAL Y
      PI = 4.0 * ATAN(1.0)
      EYE = (0., 1.0)
C     EYE = SQUARE ROOT OF -1
      CONS = 2 * PI * EYE
      A = 0.0
C
```

Figure 9.11 continued

```fortran
C     ACCESS THE TIME SERIES DATA, Y
C
      WRITE(*,*) 'ENTER NAME OF INPUT FILE, UNIT=5'
      OPEN (5, FILE = ' ')
      WRITE(*,*) 'ENTER NAME OF OUTPUT FILE, UNIT=1'
      OPEN(1, FILE = ' ', STATUS = 'NEW')
      WRITE(*,*) 'ENTER NAME OF OUTPUT FILE, UNIT=2'
      OPEN(2, FILE = ' ', STATUS = 'NEW')
      N = 1
10    CONTINUE
      READ(5,*,END=20) Y(N)
      N = N + 1
      GO TO 10
20    CONTINUE
      N = N - 1
      CONS = CONS / FLOAT(N)
      CONS2 = FLOAT(N) / FLOAT(N)
C
```

```fortran
C     COMPUTE THE FOURIER TRANSFORM (DISCRETE)
C
      N2 = N/2
      DO J = 1, N
      BETA(J) = (0.,0.)
      DO K = 1, N
      X1 = A + (K-1) * CONS2
      BETA(J) = BETA(J) + Y(K) * CEXP(CONS*(J-1)*X1)/FLOAT(N)
      END DO
      B = 2.0 * REAL(BETA(J))**2
      WRITE(*,*) J, B
      JTEMP = J - 1
      WRITE(1,30) JTEMP, B
      WRITE(2,30) J, REAL(BETA(J))
30    FORMAT(1X, I10, 1E20.6)
      END DO
      STOP
      END
```

Figure 9.14 A listing of the program FORINV.

```fortran
      PROGRAM FORINV
C
C     A FORTRAN-77 PROGRAM TO COMPUTE THE INVERSE FOURIER
C     TRANSFORM OF A TIME SERIES
C
C**********************************************************
C
C                USER'S GUIDE:   FORIER
C
C**********************************************************
C
C     THE INPUT DATA FILE FOR THIS PROGRAM IS THE FOURIER TRANS
C     (UNIT 2) COMPUTED BY THE PROGRAM, FORINV
C
C     WHEN PROMPTED, ENTER THE OUTPUT FILE CREATED BY FORIER
C     AS THE INPUT FILE, UNIT 5
C
C     THIS PROGRAM THEN CREATES AN OUTPUT FILE, UNIT 1, WHICH
C     IS THE INVERSE TRANSFORM; I.E., SHOULD BE THE ORIGINAL
C     DATA (APPROXIMATELY).
C
C**********************************************************
C
C              END OF USER'S GUIDE AND DISCUSSION
C
C**********************************************************
C
      IMPLICIT COMPLEX (A-H, O-Z)
      REAL A, B, PI
      DIMENSION Y(1000), BETA(1000)
      REAL Y,X
      PI = 4.0 * ATAN(1.0)
      EYE = (0., 1.0)
C     EYE = SQUARE ROOT OF -1
      CONS = 2 * PI * EYE
      A = 0.0
C
C     ACCESS THE TIME SERIES DATA, Y
C
      WRITE(*,*) 'ENTER NAME OF INPUT FILE, UNIT=5'
      OPEN (5, FILE = ' ')
      WRITE(*,*) 'ENTER NAME OF OUTPUT FILE, UNIT=1'
      OPEN(1, FILE = ' ', STATUS = 'NEW')
      N = 1
10    CONTINUE
      READ(5,*,END=20) X, Y(N)
      N = N + 1
      GO TO 10
20    CONTINUE
      N = N - 1
      CONS = CONS / FLOAT(N)
      CONS2 = FLOAT(N) / FLOAT(N)
C
C     COMPUTE THE INVERSE TRANSFORM
C
      N2 = N/2
      DO J = 1, N
      BETA(J) = (0.,0.)
      DO K = 1, N
      X1 = A + (K-1) * CONS2
      BETA(J) = BETA(J) + Y(K) * CEXP(-CONS*(J-1)*X1)
      END DO
      WRITE(1,30) J, REAL(BETA(J))
      WRITE(*,*) J, REAL(BETA(J))
30    FORMAT(1X, I10, 1F20.6)
      END DO
      STOP
      END
```

424

Figure 9.15 A listing of the program FILTER.

```fortran
      PROGRAM FILTER
C
C     A PROGRAM TO BAND-PASS FILTER A FOURIER TRANSFORMED
C     TIME SERIES SEQUENCE
C
*******************************************************
C
C          USER'S GUIDE:  FILTER
C
*******************************************************
C
C     INPUT FILE:   THE FOURIER TRANS. AS COMPUTED BY THE
C                   PROGRAM, FORIER (UNIT=2)
C     OUTPUT FILE:  THE FILTERED TRANSFORM, COMPATIBLE WITH THE
C                   PROGRAM, FORINV.
C
C     INTERACTIVE DATA INPUT:
C
C     THE PROGRAM REQUESTS INFORMATION FOR THRESMN AND THRESMAX
C     WHERE:
C          THRESMN = MIN. ABSOLUTE VALUE OF FOUR. TRANS; ALL
C                    FOURIER TRANS. VALUES LESS THAN THRESMN
C                    ARE SET EQUAL TO ZERO;
C          THRESMX = MAX. ABS. VALUE OF FOUR. TRANS; ALL
C                    FOURIER TRANS. VALUES GREATER THAN THRESMX
C                    ARE SET EQUAL TO ZERO
C
C     THRESMN AND THRESMX DEFINE A "BAND" OF FREQUENCY
C     VALUES.  SUPPOSE ALL FREQUENCIES BUT ONE ARE TO BE
C     FILTERED, THEN SET THRESMN SLIGHTLY LESS THAN THE
C     FOUR. TRANS. FOR THAT FREQUENCY, AND SET THRESMX
C     SLIGHTLY GREATER THAN THE FOUR. TRANS. FOR THE
C     FREQ.  AND SO ON.
C
*******************************************************
C
C          END OF USER'S GUIDE AND DISCUSSION
C
*******************************************************
C
      WRITE(*,*) 'ENTER NAME OF INPUT FILE (UNIT 2 FROM FORIER)'
      OPEN (5, FILE = ' ')
      WRITE(*,*) 'ENTER OUTPUT FILE NAME FOR FILTERED RESULT'
      OPEN(1, FILE = ' ', STATUS = 'NEW')
      WRITE(*,*) 'ENTER MINIMUM VALUE, THRESMN, FOR FILTERING'
      READ(*,*) THRESMN
      WRITE(*,*) 'ENTER MAXIMUM VALUE, THRESMX, FOR FILTERING'
      READ(*,*) THRESMX
10    CONTINUE
      READ(5,*,END=20) J, Y
      IF (ABS(Y) .LT. THRESMN) Y = 0.0
      IF (ABS(Y) .GT. THRESMX) Y = 0.0
      WRITE(1,*) J, Y
      GO TO 10
20    STOP
      END
```

Figure 9.26 A listing of the program ADD. This program is used to create Figure 9.4, and is also used in several end-of-chapter exercises.

```
      PROGRAM ADD                              DIMENSION SUM(1000)
C                                              DO I = 1, 1000
C     A PROGRAM TO ADD FILES DEVELOPED USING CSSINE    SUM(I) = 0.0
C                                              END DO
C                                              WRITE(*,*) 'HOW MANY FILES WILL BE COMBINED?'
*****************************************************    READ (*,*) N
C                                              DO I = 1, N
C            USER'S GUIDE:  ADD                    WRITE(*,*) 'ENTER INPUT FILE NAME FOR FILE ', I
C                                                 OPEN(5, FILE = ' ')
*****************************************************    KOUNT = 0
C                                           10    CONTINUE
C     A SERIES OF INTERACTIVE QUESTIONS REQUIRE RESPONSE:    READ(5,*, END = 20) Y
C                                                    KOUNT = KOUNT + 1
C     1.  HOW MANY FILES ARE BEING COMBINED?          SUM(KOUNT) = SUM(KOUNT) + Y
C                                                    GO TO 10
C     2.  ENTER THE FILE NAME FOR EACH, ONE BY ONE 20 CONTINUE
C                                                 CLOSE(5)
C     3.  ENTER THE NAME OF THE OUTPUT FILE (THE COMBINATION)    IF (I .EQ. N) JOUNT = KOUNT
C                                              END DO
*****************************************************    WRITE(*,*) 'ENTER THE NAME OF THE OUTPUT FILE'
C                                              OPEN(1, FILE = ' ', STATUS = 'NEW')
C            END OF USER'S GUIDE:  ADD         DO I = 1, JOUNT
C                                                 SUM(I) = SUM(I) / FLOAT(N)
*****************************************************    WRITE(1,*) SUM(I)
                                               END DO
                                               STOP
                                               END
```

426

10

Numerical Processing of Digital Images

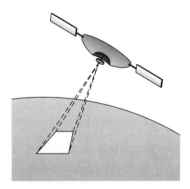

So much geological interpretation is conducted today using multispectral digital imagery that geologists, geological engineers, ecologists, civil engineers, electrical engineers, environmental engineers, and other professionals for which such data are valuable must know how to use and process imagery. In this chapter we review some of the more widely used image-processing techniques to show how digital images are manipulated to enhance interpretation. Numerous examples and software are presented to enable a thorough understanding.

10.1 Digital Imagery

Digital images were introduced earlier in this text. In particular, kriging was used to develop regular grids of estimates (Chapter 6). Each grid is actually a digital image, a two-dimensional array of numbers also called a raster. In the most general of terms, a digital image is a matrix of numbers, usually integers in the range 0 to 255 (i.e., 8-bit integers; Chapter 7 gave an explanation of bit representation of numbers).

Each number comprising a digital image is called a pixel. Pixels were introduced in Chapter 7 wherein they represented the unique dots that comprise a

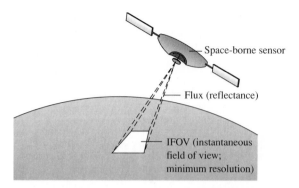

Figure 10.1   The notion of instantaneous field of view (IFOV), which is that portion or area of a surface represented by a pixel. The pixel value represents the average electromagnetic flux reflected by the surface over a particular IFOV.

Figure 10.2   A black and white photograph with a grid arbitrarily superimposed. A digital image may be formed from the photograph by assigning each grid cell a number, say from 0 to 255, representing the intensity of the color white for the cell.

color monitor (or television) screen. With respect to digital imagery, pixels represent average intensity, or brightness, over small subareas (instantaneous fields of view (IFOV) of whatever the digital image represents (Figure 10.1)).

As a further demonstration, a black and white photograph (Figure 10.2) has a grid superimposed which is useful for converting the photograph to a digital image by creating a digital raster (matrix) of pixel values, each pixel value representing the average gray-level intensity for one grid cell. This example shows that pixels comprising digital images represent average brightness over a given area. Furthermore, realizing that a digital image is simply a giant collection of numbers enables the further realization that mathematical manipulations may be used to change the numbers, where the change may lead to better image clarity or to an improved understanding of the information contained in the image.

A digital image is nothing more than a rectangular array of numbers. By using the methods described in Chapter 7, each number comprising an image is converted to a color for display on the monitor screen. By this process, a digital image is transformed into a pseudophotograph.

10.2 Remote Sensing

Satellite scanning of the earth's surface is one example of remote sensing, a process wherein information is acquired about an object without touching the object; in short, acquisition without contact. From this definition, a number of remote sensors come to mind, such as the human eye, ear, and nose. A photograph is an example of remotely acquired data, as are digital images formed by a satellite as it scans the surface of the earth or another planet.

Much of the digital imagery acquired by satellites represent radiation in portions of the electromagnetic spectrum emitted by, scattered by, or reflected from the earth, its atmosphere, or some other astronomical body. A conventional photograph is another example. It is a record of a narrow window of the electromagnetic spectrum, that of visible light.

Before proceeding further in this chapter, the electromagnetic spectrum is introduced and discussed, because no discussion of remote sensing is complete without it. Modern technology for remote sensing uses portions of the electromagnetic spectrum other than the visible portion, and these remote sensing applications are useful for geological interpretation.

10.2.1 The Electromagnetic (EM) Spectrum

The electromagnetic spectrum (Figure 10.3) encompasses visible light, ultraviolet energy, X rays, microwaves, and radio waves. The word wave appears in some of these examples because electromagnetic energy can be considered a wave phenomenon; that is, this energy travels in the form of waves. Hence this energy may be reflected and refracted, like all wave phenomena.

Electromagnetic energy is so named because it consists of an electrical part and a magnetic part. Because it propagates as a wave, electromagnetic energy is associated with varying wavelengths (Figure 10.3). Gamma rays, for example, have wavelengths of approximately 10^{-6} μm (micrometers), whereas radio and television energy has wavelengths of several tens to hundreds of meters. Visible

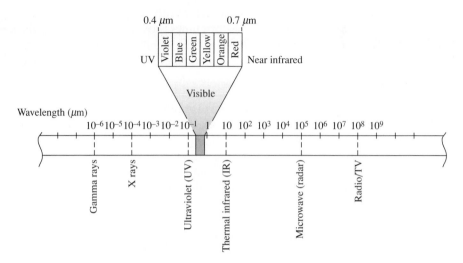

Figure 10.3   The electromagnetic (EM) spectrum. Visible light, x-rays, thermal radiation, radar, and radio waves are all examples of EM energy.

light is composed of wavelengths from 0.4 μm (violet light) to 0.7 μm (red light). Individual wavelengths within the visible spectrum add to give a white light. A prism may be used to refract these waves, the shorter violet wavelengths more than the red, such that a rainbow is differentiated. This simple experiment underscores the wave nature of elecromagnetic energy.

Understanding the electromagnetic spectrum leads to a broader understanding of our natural surroundings. For instance, to understand why the sky is blue, it is necessary to know that visible blue light has a wavelength just a little longer than visible violet light, 0.4 μm. It is further necessary to understand that the molecules of gas composing the earth's atmosphere are larger in comparison to the wavelength of blue light, but are smaller than the wavelengths of visible green light (0.5 μm) or visible red light (0.7 μm). Therefore, blue wavelengths collide with and are scattered by gas molecules of the atmosphere more than are the longer wavelengths. These longer wavelengths essentially pass through the atmosphere unobstructed, whereas blue light bounces around the atmosphere awhile before eventually reaching the earth's surface. Hence, the sky appears blue because it is filled with the scattered blue light. The sky is not violet, because, even though the wavelength of violet light is smaller than that of blue light, it is scattered much more severely and is almost completely filtered (that is, absorbed) by the upper portion of the earth's atmosphere.

Atmospheric scattering and adsorption can adversely affect the remote sensing of the earth's surface using spaceborne (satellite) sensors, which must peer through the earth's atmosphere. Some wavelengths of the electromagnetic spectrum are absorbed by the earth's atmosphere (Figure 10.4). Remote detection of these wavelengths from space is consequently difficult or impossible. If, for instance, a satellite system is designed to detect, wholly or in part, visible and infrared EM radiation reflected or scattered by the earth's surface, the atmosphere can seriously degrade the quality of digital imagery. In this case, image processing is necessary to attempt to remove the atmospheric interference (atmospheric path radiance).

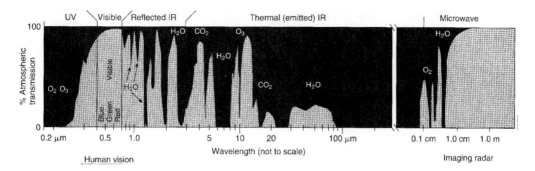

Figure 10.4 Different EM wavelengths absorbed by the earth's atmosphere (Avery and Berlin, 1992). Wavelength portions that are absorbed cannot be imaged from space. Copyright © 1992 by Macmillan Publishing Company. Reprinted with permission.

10.2.2 Satellite Technology

Satellites employ a wide range of different sensors to image varying portions of the electromagnetic spectrum when scanning the earth and other planets. Some early military spy satellites used cameras with film sensitive to visible light or to both visible and near-infrared energy. Other satellites, such as the Landsat satellite, use optics to focus electromagnetic energy onto an array of photoelectric cells. Each cell is sensitive to a particular segment (band) of the electromagnetic spectrum. The intensity of the electromagnetic energy in this band reaching the photoelectric cell governs how much electric current the cell develops in response to the energy. A computer on board the satellite assigns the electric current a number representing the strength of the current. In this manner, a digital image is formed.

Satellites acquire either photographs or images, depending on their design. A diagram of the current Landsat Thematic Mapper satellite is shown in Figure 10.5. This satellite emits no electromagnetic energy of its own. Instead, it

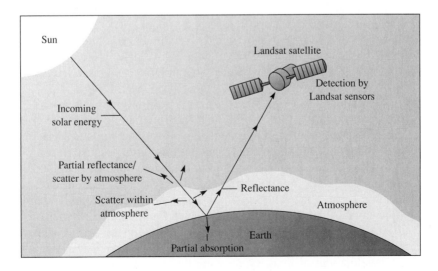

Figure 10.5 A diagram of the Landsat Thematic Mapper satellite, which is a passive remote sensor.

Table 10.1   Spectral Resolution of the Landsat Thematic
Mapper Satellite

Channel	Wavelength Sensitivity Window (microns)	Name of Window
1	0.45–0.52	Blue to cyan
2	0.52–0.60	Green to orange
3	0.63–0.69	Red
4	0.76–0.90	Near infrared
5	1.55–1.75	Mid infrared
6	10.4–12.5	Thermal infrared
7	2.08–2.35	Mid infrared

Note. Data from Lillesand and Kiefer (1987), p. 567.

images reflected solar radiation from the earth's surface. Notice in this figure that before the Landsat satellite can detect the reflected energy, the energy must pass through the earth's atmosphere twice. Recall that the atmosphere scatters visible light; blue is scattered most severely, but all visible wavelengths are scattered to a degree, which affects the quality of the imagery developed by Landsat. In working with this imagery, a digital imagery specialist may desire a contrast enhancement to offset the influence of the atmosphere, or a digital filter may be required to remove unwanted features in the image. Sometimes the affect of the atmosphere may simply be subtracted from the image pixel values (Chavez, 1975). Some of these image-processing techniques are discussed later.

Landsat is designed to detect the reflected flux from the earth's surface using seven photoelectric detectors to image seven different bands of the electromagnetic spectrum (Table 10.1). Landsat's band 4, which is sensitive to the near-infrared portion of the electromagnetic spectrum, is useful for vegetation analysis. Healthy, living vegetation is a strong reflector of near-infrared energy, whereas many other materials absorb this energy. Therefore, healthy vegetation appears bright in images associated with this band. Landsat's band 7 is useful for discriminating clay alteration, although to date its application is best seen in geographic regions, such as the western United States, where vegetation and soil cover are sparse.

The Magellan satellite (Figure 10.6) is presently in orbit around Venus. This satellite first emits, then detects the electromagnetic energy it is designed to image. Because Venus is completely enveloped in clouds, the Magellan satellite emits radar energy, which easily penetrates clouds because its long wavelength (0.1 meter) is not scattered by the molecules composing the clouds. This radar energy is reflected by the surface of Venus back to the satellite, which has a lens that focuses the reflected energy onto photoelectric cells sensitive to radar wavelengths. The intensity of the reflected radar energy depends on the orientation of the Venusian topography and the roughness of the Venusian surface with respect to the position of the Magellan satellite. These variations lead to the development of imagery showing the topography of the Venusian surface. Such images are displayed later in this chapter.

The Landsat satellite is known as a passive remote sensing tool. That is, it

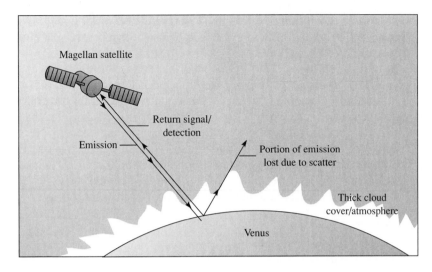

Figure 10.6   A diagram of the Magellan satellite. This is an active remote sensor.

does not emit the electromagnetic energy it eventually detects. Instead, it simply detects the solar energy that is reflected by the earth's surface and atmosphere. In contrast, the Magellan satellite is an active remote sensing tool because it emits the electromagnetic energy it eventually detects after the emission is reflected by an object. In the case of the Magellan satellite, the object is the surface of Venus.

10.2.3 Spatial, Spectral, Temporal, and Radiometric Resolution

Resolution of digital imagery is an important consideration. There are four types of resolution inherent to digital imagery or photography:

1. *Spatial* resolution refers to the area represented by the image or photograph, or portions thereof. In the case of digital images, the total image represents a certain amount of area. Each pixel comprising the digital image represents a subsection of this area as we discussed earlier in this chapter. For example, each pixel comprising a Landsat Thematic Mapper image represents the average radiation reflected from a 30-m × 30-m region on the earth's surface, a size approximately one-third that of a U.S. football field; this is the spatial resolution of Landsat Thematic Mapper imagery.
2. *Spectral* resolution refers to the bandwidth of a portion of the electromagnetic spectrum represented by the imagery. The spectral resolution of the Landsat Thematic Mapper satellite is shown in Table 10.1. Seven distinct bands (images) represent wavelengths from 0.45 μm (visible, blue light) to 10 μm (thermal infrared energy).
3. *Temporal* resolution refers to the time domain: how much time is represented by the imagery or, in the case of a satellite system, how often it repeats its orbital cycle to acquire imagery of a region over many different time periods. Repeat time is referred to as the temporal resolution of a satellite system.
4. *Radiometric* resolution refers to the numerical precision used to record a pixel value. Most pixels are recorded using 8-bit integers, which have a numerical range from 0 to 255; this is 8-bit radiometric resolution.

All of these resolution concepts must be considered when an engineer is designing a satellite system. If a satellite is to be used for repeated scanning, temporal resolution is an important consideration. Spatial resolution governs the smallest object that can be detected by the satellite; this is an important consideration, depending on the needs for the imagery. Spectral resolution is dictated by what the satellite is designed to detect. Radiometric resolution governs how many numerical levels are allowed for the representation of pixel values. The more, the better, but the management of data is an important concern. If, for instance, pixels are represented by 16-bit integers instead of 8-bit integers, twice the number of bits are required to represent the image; therefore, twice the amount of data storage space (magnetic tape, hard disk, floppy disk) is required for storing the image. Data management is an effective constraint on radiometric resolution.

10.3 Image Processing

The primary purpose of this chapter is the display and manipulation of digital imagery. The following types of digital image enhancement are emphasized:

1. *The display of digital images on VGA hardware.* At one time, expensive computer systems were required to display digital images. Now VGA hardware allows single band images to be displayed with acceptable clarity, at least for purposes of education.
2. *The enhancement of contrast in digital imagery.* Often digital images do not display a wide range of contrast (difference in pixel values). By increasing the differences in pixel values, contrast is added to an image, thus improving its clarity.
3. *Filtering digital images.* Sometimes images are contaminated by noise. Filtering this noise improves clarity (Serra (1982) refers to filtering as "picture cleanings"). Filters may also be used to enhance edges (linear features) in imagery. For geological studies, this can be most useful, especially if linear features such as faults and fractures are the focus of the study.
4. *Classification.* This is known also as pattern recognition (cf. Schowengerdt, 1983), and is a technique for dividing an image into different themes. In unsupervised classification, the computer selects ranges of pixel values, called classes or themes, and assigns each pixel to one of the classes depending on its brightness. By displaying each class using a unique color, a colorful map of pixel patterns is created, which may, or may not, accurately reflect actual ground cover themes or features. Alternatively, using supervised classification, the image analyst selects the themes or classes and the computer scans a digital image pixel by pixel to determine the class to which each pixel belongs (if any). This, too, yields a colorful map because each class is again associated with a unique color. Classification is useful for studies of areal ground cover, such as agricultural production, water use patterns, land use planning, and many other applications.

Each of these topics is discussed separately in the following sections. All four or any combination of the four methods may be applied to a digital image.

10.3.1 The Display of Digital Images Using VGA Hardware

Display of graphical images on VGA hardware (Chapter 7) uses a 16-color palette for displaying high-resolution graphical images. Digital images are usefully displayed using a 16-gray-level palette (Table 7.1). Displayed in this manner, the image looks like a black-and-white photograph. The VGA hardware is limited to the display of a single image at a time when using the high-resolution mode; multiple-image composites are not possible without sacrificing spatial resolution. But the VGA hardware is useful for demonstrating many image processing techniques.

In displaying digital images, the only consideration (beyond what is covered in Chapter 7) is the format in which the digital information is stored. Many digital images, such as the Landsat Thematic Mapper images, are written to a file as 8-bit integers. Usually these integers are written to a file in binary form, which means they are written bit by bit. They are also usually written sequentially, row by row. Sometimes, one or more header records are written preceding the image information. A record is simply a line in a data file; a header record is one that precedes other information. With respect to digital images, such records often contain information such as the name of the image, the date it was recorded, the number of pixels per line, total number of lines, and so on. The digital images used in this chapter are written without header records for ease of data access.

Care must be taken when reading 8-bit integers from a file. These integers have the numerical range -127 to 128. The positive integers 0 to 128 represent these same numbers. The actual value of a negative integer, however, is the negative value plus 256; hence -127 corresponds to the number $256 - 127$, or 129; -1 corresponds to the number $256 - 1$, or 255. Therefore, 8-bit integers are associated with the range 0 to 255, for a total of 2^8, or 256 numbers. The conversion of the negative integers to their positive equivalence must be done when reading digital image data stored as 8-bit integers.

The computer program IMAGIN (Figure 10.7, page 469) reads digital image data stored as 8-bit integers and correctly converts negative integers to their positive counterparts. This program can read images as large as 3000 rows by 3000 pixels per row, but will display only the upper left 400 rows by 400 pixels.

When using this program, a few interactive questions need to be answered: (1) The name of the data file that contains the data for the 16-color palette created by the program PALDSGN (Chapter 7) is requested. (2) Following this, the program requests the name of the file containing the digital image. This must be a binary file, with the image written row by row as 8-bit integers, with no header record or records. (3) The program requests information specifying the number of rows and the number of pixels per row composing the image. (4) Finally, the program asks the user to enter a title for the image, up to 50 characters in length.

Following the entry of a title, the program displays the image. Six displays are shown in Figures 10.8 through 10.13. Each of these figures is a black and white conventional photograph of the display created on a VGA or SVGA monitor using IMAGIN. Parameters used with IMAGIN to create each display are listed in the caption for each figure, along with a description of what each image represents.

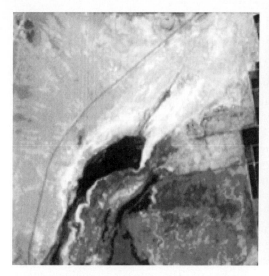

Figure 10.8 A display of an image (HUMB1.IMG) of the sink for the Humboldt River in Nevada (USA). This image has 400 rows and 400 pixels per row.

Figure 10.9 A display of an image (SATURN.IMG) of the planet Saturn and its rings acquired by the Voyager spacecraft. This image was extracted from the three compact disc Joint Education Initiative set (NASA and the United States Geological Survey; see Appendix B for a description of the JEI CD set).

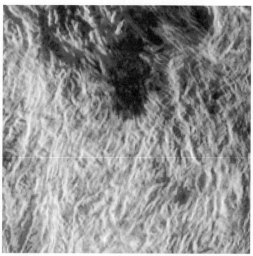

Figure 10.10 A display of the Jovian moon Io. Notice the active volcanic eruption to the upper right side of the moon. This image was also extracted from the JEI CD set (see Appendix B) and was contrast enhanced using CTRAST (Figure 10.28) to enhance the volcanic eruption. This image has 400 rows of pixels, with 400 pixels per row.

Figure 10.11 Unique fractures and folds on Venus. This image was extracted from Magellan CD, F-MIDR 25S009. The fractures and folds are indicative of tectonism on Venus. This image has 400 rows and 400 pixels per row. The data represent a synthetic aperture radar image of the Venusian surface.

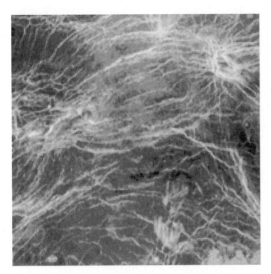

Figure 10.12   More fractures and folds on Venus. The image was extracted from Magellan CD F-MIDR 40N018, and has 400 rows and 400 pixels per row.

Figure 10.13   Pancake volcanic domes extracted from Magellan CD F-MIDR 30S009. The blank rectangular region at the upper edge of the image is attributed to an error caused by a malfunction of the Magellan satellite. The image has 400 rows and 400 pixels per row.

10.3.2 Contrast Enhancement of Digital Images

Often a digital image acquired by a satellite or one acquired by digitizing a photograph lacks sufficient contrast, hence limiting its utility for geologic (or other) interpretation. But because digital images are simply compositions of many hundreds of thousands of numbers (pixels), mathematical manipulation may be applied to change these numbers (pixels) to increase the image contrast.

The idea behind contrast enhancement is to expand the range of a portion of an image's pixels into the entire 0 to 255 8-bit integer range. An image associated with relatively poor contrast has a range smaller than the 0 to 255 maximum range. Therefore, the first step in accomplishing contrast enhancement involves the computation of the histogram of image pixel values to inspect the range and distribution of these values in the image to be enhanced.

The program IMSTAT is presented in Figure 10.14, page 471, for calculating the histogram of image pixels. Although a program HISTO is presented in Chapter 4 for calculating histograms, a digital image often comprises hundreds of thousands, sometimes millions, of digits and HISTO is not an efficient computer algorithm when applied to this much information. The program IMSTAT is designed to take advantage of the fact that image pixels are integers, ranging in value from 0 to 255. Therefore, if an array having 256 addresses is used, as in IMSTAT, then each pixel value in a digital image is also an address for this array and pixel value frequencies are counted accordingly.

For instance, suppose the program encounters the pixel value 112. In this case, the value of the array at the 113th address (a value equal to 112 plus 1 because a pixel value of 0 may exist, but in FORTRAN-77 a zero may not be used as an address; the address of the value 0 is 1) is increased by 1, thus counting the pixel. This is an efficient algorithm for computing image histograms.

When using IMSTAT all that is required for data entry is (1) specifying the name of the file that contains the digital image, the same file displayed by IMAGIN; and (2) specifying the number of rows composing the image and the number of pixels per row. After this information is entered, the histogram is displayed by compressing the 256 bins down to 32 to conveniently fit on one sheet of computer paper. Histograms for the six images (Figures 10.8 through 10.13) are shown in Figures 10.15 through 10.20. We will use these histograms later for contrast-enhancement experiments.

Three types of contrast enhancement are described in this chapter. They are not the only methods available, but they are the most commonly used methods. The first is a simple linear contrast enhancement. Such an enhancement is shown diagrammatically in Figure 10.21. The input pixels are mapped to output pixels by using a linear function. The algorithm is as follows:

1. The minimum pixel value for the pixel range to be enhanced is specified. This need not be the minimum pixel value in the image. In fact, in the simple, linear contrast enhancement, the first step involves determining the range of pixel values to be enhanced (smaller pixel values, midrange pixel values, larger pixel values, or all values). The minimum pixel value is the minimum limit of this range. All image pixels less than this value are set equal to zero; this is called saturation.

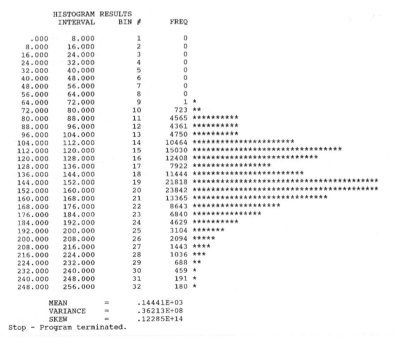

Figure 10.15   A histogram of the Humboldt Sink image shown in Figure 10.8.

```
        HISTOGRAM RESULTS
           INTERVAL      BIN #      FREQ

     .000      8.000       1        506 *
    8.000     16.000       2      25231 ****************************************
   16.000     24.000       3      28388 ****************************************
   24.000     32.000       4      21364 **************************************
   32.000     40.000       5        769 **
   40.000     48.000       6       1129 **
   48.000     56.000       7       2206 ****
   56.000     64.000       8       2884 ******
   64.000     72.000       9       2509 *****
   72.000     80.000      10       1570 ***
   80.000     88.000      11       1004 **
   88.000     96.000      12        906 **
   96.000    104.000      13        668 **
  104.000    112.000      14        650 **
  112.000    120.000      15        916 **
  120.000    128.000      16        980 **
  128.000    136.000      17       1713 ****
  136.000    144.000      18       1898 ****
  144.000    152.000      19       3499 *******
  152.000    160.000      20       3875 *******
  160.000    168.000      21       4239 ********
  168.000    176.000      22       5127 *********
  176.000    184.000      23       5903 ***********
  184.000    192.000      24       6791 ************
  192.000    200.000      25       4823 *********
  200.000    208.000      26       4008 ********
  208.000    216.000      27       3788 *******
  216.000    224.000      28       4005 ********
  224.000    232.000      29       4561 *********
  232.000    240.000      30       4571 *********
  240.000    248.000      31       4251 ********
  248.000    256.000      32       5268 **********

           MEAN      =      .10150E+03
           VARIANCE  =      .13656E+08
           SKEW      =      .31996E+13
Stop - Program terminated.


0;37;44m C:\TEXT>
```

Figure 10.16 A histogram of the Saturn image shown in Figure 10.9. Note that most of the pixels are near zero and are associated with space; a relatively few pixels are higher in value and are associated with Saturn and its rings.

```
        HISTOGRAM RESULTS
           INTERVAL      BIN #      FREQ

     .000      8.000       1      86650 ****************************************
    8.000     16.000       2         51 *
   16.000     24.000       3         38 *
   24.000     32.000       4         52 *
   32.000     40.000       5         15 *
   40.000     48.000       6         57 *
   48.000     56.000       7         11 *
   56.000     64.000       8         39 *
   64.000     72.000       9         26 *
   72.000     80.000      10         43 *
   80.000     88.000      11         28 *
   88.000     96.000      12         48 *
   96.000    104.000      13         27 *
  104.000    112.000      14         68 *
  112.000    120.000      15         26 *
  120.000    128.000      16         96 *
  128.000    136.000      17        331 *
  136.000    144.000      18        178 *
  144.000    152.000      19        531 *
  152.000    160.000      20       1922 **
  160.000    168.000      21       3379 **
  168.000    176.000      22      11852 *******
  176.000    184.000      23       9501 ******
  184.000    192.000      24      12797 ********
  192.000    200.000      25      20275 ************
  200.000    208.000      26       9604 ******
  208.000    216.000      27       2258 **
  216.000    224.000      28         94 *
  224.000    232.000      29          3 *
  232.000    240.000      30          0
  240.000    248.000      31          0
  248.000    256.000      32          0

           MEAN      =      .86042E+02
           VARIANCE  =      .29093E+08
           SKEW      =      .12847E+14
Stop - Program terminated.


0;37;44m C:\TEXT>
```

Figure 10.17 A histogram of Io and its volcanism. As with the Saturn image, many pixels are near zero and are associated with space; those pixels representing Io are higher in value.

```
        HISTOGRAM RESULTS
        INTERVAL      BIN #       FREQ

   .000    8.000        1           0
  8.000   16.000        2           0
 16.000   24.000        3           0
 24.000   32.000        4           0
 32.000   40.000        5           0
 40.000   48.000        6           0
 48.000   56.000        7           0
 56.000   64.000        8           0
 64.000   72.000        9           0
 72.000   80.000       10          26 *
 80.000   88.000       11        1882 ***
 88.000   96.000       12       11542 *******************
 96.000  104.000       13       20015 *********************************
104.000  112.000       14       30378 ***************************************************
112.000  120.000       15       31465 ***************************************************
120.000  128.000       16       25178 ******************************************
128.000  136.000       17       17736 *****************************
136.000  144.000       18       10951 ****************
144.000  152.000       19        6045 **********
152.000  160.000       20        2854 *****
160.000  168.000       21        1269 ***
168.000  176.000       22         453 *
176.000  184.000       23         168 *
184.000  192.000       24          36 *
192.000  200.000       25           2 *
200.000  208.000       26           0
208.000  216.000       27           0
216.000  224.000       28           0
224.000  232.000       29           0
232.000  240.000       30           0
240.000  248.000       31           0
248.000  256.000       32           0

        MEAN      =    .11809E+03
        VARIANCE  =    .38101E+08
        SKEW      =    .14208E+14
Stop - Program terminated.


0;37;44m C:\TEXT>
```

Figure 10.18   A histogram of the Magellan image displayed in Figure 10.11. Notice the relatively poor contrast (small data range); contrast adjustment is warranted for these data (see the end of chapter exercises).

```
        HISTOGRAM RESULTS
        INTERVAL      BIN #       FREQ

   .000    8.000        1           0
  8.000   16.000        2           0
 16.000   24.000        3           0
 24.000   32.000        4           0
 32.000   40.000        5           0
 40.000   48.000        6           0
 48.000   56.000        7           0
 56.000   64.000        8           0
 64.000   72.000        9          24 *
 72.000   80.000       10        1734 ***
 80.000   88.000       11       19983 *************************
 88.000   96.000       12       41667 ******************************************
 96.000  104.000       13       39936 *****************************************
104.000  112.000       14       27292 ***********************************
112.000  120.000       15       16559 *********************
120.000  128.000       16        8007 **********
128.000  136.000       17        3248 ****
136.000  144.000       18        1112 **
144.000  152.000       19         336 *
152.000  160.000       20          87 *
160.000  168.000       21          13 *
168.000  176.000       22           2 *
176.000  184.000       23           0
184.000  192.000       24           0
192.000  200.000       25           0
200.000  208.000       26           0
208.000  216.000       27           0
216.000  224.000       28           0
224.000  232.000       29           0
232.000  240.000       30           0
240.000  248.000       31           0
248.000  256.000       32           0

        MEAN      =    .10143E+03
        VARIANCE  =    .37919E+08
        SKEW      =    .16309E+14
Stop - Program terminated.


0;37;44m C:\TEXT>
```

Figure 10.19   A histogram of the Magellan image displayed in Figure 10.12. Again notice the relatively poor contrast.

```
        HISTOGRAM RESULTS
        INTERVAL        BIN #        FREQ

   .000     8.000        1         1670 **
  8.000    16.000        2           11 *
 16.000    24.000        3            9 *
 24.000    32.000        4           16 *
 32.000    40.000        5           14 *
 40.000    48.000        6           15 *
 48.000    56.000        7           29 *
 56.000    64.000        8          273 *
 64.000    72.000        9         4479 *****
 72.000    80.000       10        32923 ************************************
 80.000    88.000       11        45826 ******************************************
 88.000    96.000       12        38385 *****************************************
 96.000   104.000       13        20528 ************************
104.000   112.000       14         9402 ***********
112.000   120.000       15         3525 ****
120.000   128.000       16         1599 **
128.000   136.000       17          691 *
136.000   144.000       18          319 *
144.000   152.000       19          153 *
152.000   160.000       20           82 *
160.000   168.000       21           29 *
168.000   176.000       22           17 *
176.000   184.000       23            5 *
184.000   192.000       24            0
192.000   200.000       25            0
200.000   208.000       26            0
208.000   216.000       27            0
216.000   224.000       28            0
224.000   232.000       29            0
232.000   240.000       30            0
240.000   248.000       31            0
248.000   256.000       32            0

        MEAN      =      .88378E+02
        VARIANCE  =      .31703E+08
        SKEW      =      .13317E+14
Stop - Program terminated.


0;37;44m C:\TEXT>
```

Figure 10.20   A histogram of the Magellan image displayed in Figure 10.13. Again a poor contrast is noted. The relatively large number of zero pixel values is the result of that portion of the image associated with error.

2. The maximum pixel value for the pixel range to be enhanced is specified. As with step 1, this need not be the maximum pixel value in the image. All pixel values greater than this maximum pixel value are set equal to 255, and this, too, is called saturation.
3. All pixel values are rescaled as follows:
   (a) if $pixel_i$ < minimum pixel value, then $pixel_i = 0$
   (b) if $pixel_i$ > maximum pixel value, then $pixel_i = 255$
   (c) if minimum < $pixel_i$ < maximum, then $pixel_i = 255(pixel_i - minimum)/(maximum - minimum)$

An example of this type of enhancement, applied to the Humboldt Sink image (Figure 10.8), is shown in Figure 10.22. A simple, linear stretch is applied, using a minimum pixel value of 77 and a maximum pixel value of 200. The resultant histogram (Figure 10.23) is computed using IMSTAT. Comparing this histogram to that for the original image (Figure 10.15) reveals the numerical consequence of the contrast enhancement. The visual consequence of the contrast enhancement is obvious (Figure 10.22 compared to Figure 10.8).

Another type of contrast enhancement, which offers more flexibility in comparison to the simple, linear enhancement, is bilinear contrast enhancement. This method (Figure 10.21) uses two linear functions to enhance two portions of a digital image differently. For example, image clarity may be improved by decreasing the contrast for one portion of the image and increasing the contrast for another.

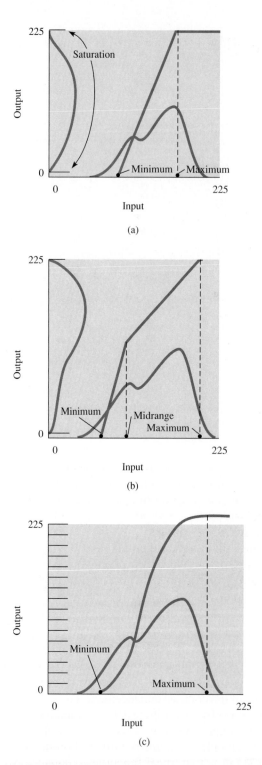

Figure 10.21   Diagrams of three types of contrast enhancement: (a) linear; (b) bilinear; and (c) histogram equalization.

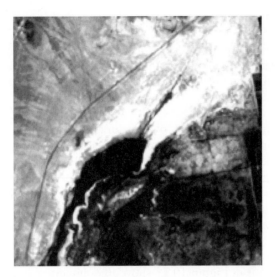

Figure 10.22   The Humboldt Sink image after application of linear contrast enhancement (parameters of the enhancement are given in the text).

```
           HISTOGRAM RESULTS
           INTERVAL      BIN #      FREQ

     .000     8.000        1        743 ***
    8.000    16.000        2       2470 *********
   16.000    24.000        3       9143 ******************************
   24.000    32.000        4       9333 ******************************
   32.000    40.000        5       9249 ******************************
   40.000    48.000        6       5778 ******************
   48.000    56.000        7       6105 *********************
   56.000    64.000        8       6029 ********************
   64.000    72.000        9       9891 **********************************
   72.000    80.000       10      12166 *****************************************
   80.000    88.000       11      11357 **************************************
   88.000    96.000       12      10728 ***********************************
   96.000   104.000       13       9980 *********************************
  104.000   112.000       14       7616 **************************
  112.000   120.000       15       5016 *****************
  120.000   128.000       16       6900 ***********************
  128.000   136.000       17       7900 **************************
  136.000   144.000       18      10800 ************************************
  144.000   152.000       19      13049 ********************************************
  152.000   160.000       20      15093 **************************************************
  160.000   168.000       21      13831 **********************************************
  168.000   176.000       22       7711 **************************
  176.000   184.000       23       8389 ****************************
  184.000   192.000       24       6460 **********************
  192.000   200.000       25       5781 *******************
  200.000   208.000       26       5238 *****************
  208.000   216.000       27       4602 ***************
  216.000   224.000       28       3748 ************
  224.000   232.000       29       2315 ********
  232.000   240.000       30       2556 *********
  240.000   248.000       31       2019 *******
  248.000   256.000       32       8829 ****************************

           MEAN       =     .12340E+03
           VARIANCE   =     .52842E+08
           SKEW       =     .44413E+14
Stop - Program terminated.


0;37;44m C:\TEXT=
```

Figure 10.23   A histogram of the image shown in Figure 10.22; compare this histogram to that shown in Figure 10.15 and notice the effect of the linear contrast adjustment.

The algorithm for bilinear contrast enhancement is as follows:

1. Specify a minimum, midrange, and maximum pixel value (Figure 10.21). Minimum and midrange values define the range of smaller pixel values to be enhanced; midrange and maximum values define the range of larger pixel values to be enhanced.
2. The algorithm is
   (a) if $pixel_i$ < minimum, then $pixel_i$ = 0 (saturation)
   (b) if $pixel_i$ > maximum, then $pixel_i$ = 255 (saturation)
   (c) if minimum < $pixel_i$ < midrange, then $pixel_i$ = 128($pixel_i$ − minimum)/(midrange − minimum)
   (d) if midrange < $pixel_i$ < maximum, then $pixel_i$ = 127 + 128($pixel_i$ − midrange)/(maximum-midrange)
   (e) if $pixel_i$ = midrange, then $pixel_i$ = 128

This type of enhancement is demonstrated in Figure 10.24 for the Humboldt Sink image. The histogram in Figure 10.25 shows the numerical consequence. In this enhancement, minimum is set equal to 77, midrange is set equal to 160, and maximum is set to 200. By examining the original image histogram (Figure 10.15), bilinear contrast enhancement is seen to have resulted in a reduction in contrast for the pixel range 160 to 255 and an increase in contrast for the pixel range 0 to 160. The visual inspection of the enhancement in Figure 10.24 shows that image clarity is improved for smaller pixel values.

The third type of contrast enhancement to be described in this chapter is known as histogram equalization. This method (Figure 10.21) processes pixel values such that all pixel values between 0 and 255 are represented in equal proportions in the image and that no one portion of the pixel value range 0 to 255 dominates the image.

Figure 10.24 The Humboldt Sink image after the application of a bilinear contrast enhancement (parameters of the enhancement are given in the text).

```
          HISTOGRAM RESULTS
              INTERVAL    BIN #      FREQ

      .000       8.000      1       1535 *****
     8.000      16.000      2       8077 **********************
    16.000      24.000      3      12077 ********************************
    24.000      32.000      4      10900 ****************************
    32.000      40.000      5       8452 **********************
    40.000      48.000      6      10102 ***************************
    48.000      56.000      7      13556 ************************************
    56.000      64.000      8      14894 ****************************************
    64.000      72.000      9      15974 ******************************************
    72.000      80.000     10      11429 ******************************
    80.000      88.000     11       8608 **********************
    88.000      96.000     12       8695 **********************
    96.000     104.000     13      10961 ****************************
   104.000     112.000     14      18993 **************************************************
   112.000     120.000     15      18934 **************************************************
   120.000     128.000     16      15153 ******************************************
   128.000     136.000     17       4937 *************
   136.000     144.000     18       6000 ****************
   144.000     152.000     19       3453 **********
   152.000     160.000     20       4454 ************
   160.000     168.000     21       2883 ********
   168.000     176.000     22       4172 ***********
   176.000     184.000     23       2517 *******
   184.000     192.000     24       3518 **********
   192.000     200.000     25       2119 ******
   200.000     208.000     26       2713 ********
   208.000     216.000     27       1604 *****
   216.000     224.000     28       2055 ******
   224.000     232.000     29       1212 ****
   232.000     240.000     30       1571 *****
   240.000     248.000     31        936 ***
   248.000     256.000     32       8341 **********************

            MEAN       =    .97909E+02
            VARIANCE   =    .37426E+08
            SKEW       =    .35915E+14
   Stop - Program terminated.


   0;37;44m C:\TEXT=
```

Figure 10.25   A histogram of the image shown in Figure 10.24; compare this histogram to those shown in Figures 10.8 and 10.23.

The algorithm for this type of enhancement is as follows:

1. As with the simple, linear enhancement, specify a minimum and maximum pixel value defining the range of pixel values to be enhanced. These need not coincide with actual image minimum and maximum values; therefore, some saturation may occur.
2. Determine the histogram for pixel values between minimum and maximum using the program IMSTAT.
3. Process pixel values, such that
   (a) if $pixel_i$ < minimum, then $pixel_i$ = 0 (saturation)
   (b) if $pixel_i$ > maximum, then $pixel_i$ = 255 (saturation)
   (c) if minimum < $pixel_i$ < maximum, then $pixel_i$ = 255(cumulative relative frequency for $pixel_i$)

This type of contrast enhancement is demonstrated in Figure 10.26 for the Humboldt Sink image. When all pixel values occur with equal frequency in an image, a substantial benefit in contrast is achieved. The histogram for this enhanced image (Figure 10.27) shows the consequence of this type of contrast enhancement.

10.3.3 Using Look-up Tables for Improved Computation Efficiency

The foregoing discussion of contrast enhancement involves pixel specific operations. To achieve greater contrast in an image, individual pixel values are rescaled, regardless of the values of neighboring pixels. Effecting contrast

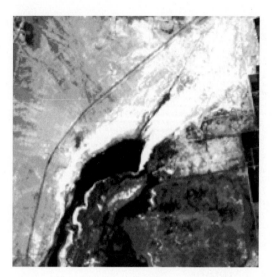

Figure 10.26   The Humboldt Sink image after the application of a histogram equalization contrast enhancement. The same minimum and maximum values as were used for the linear enhancement were used for the histogram equalization. Ten bins were used for the histogram, and the frequency for each bin was set equal to 0.1. Hence, the cumulative frequency for each bin was set equal to 0.1 (bin number).

```
        HISTOGRAM RESULTS
        INTERVAL        BIN #    FREQ

   .000    8.000         1        0
  8.000   16.000         2        0
 16.000   24.000         3        0
 24.000   32.000         4     1396 **
 32.000   40.000         5        0
 40.000   48.000         6        0
 48.000   56.000         7     7744 ***********
 56.000   64.000         8        0
 64.000   72.000         9        0
 72.000   80.000        10     9422 *************
 80.000   88.000        11        0
 88.000   96.000        12        0
 96.000  104.000        13    23049 ******************************
104.000  112.000        14        0
112.000  120.000        15        0
120.000  128.000        16    16583 *********************
128.000  136.000        17        0
136.000  144.000        18        0
144.000  152.000        19        0
152.000  160.000        20    20869 ****************************
160.000  168.000        21        0
168.000  176.000        22        0
176.000  184.000        23    38265 *************************************************
184.000  192.000        24        0
192.000  200.000        25        0
200.000  208.000        26    18882 ************************
208.000  216.000        27        0
216.000  224.000        28        0
224.000  232.000        29    11319 ***************
232.000  240.000        30        0
240.000  248.000        31        0
248.000  256.000        32    12471 ****************

        MEAN       =    .15869E+03
        VARIANCE   =    .62515E+09
        SKEW       =    .30379E+16
Stop - Program terminated.

0;37;44m C:\TEXT>
```

Figure 10.27   Histogram for the image shown in Figure 10.26. Compare this histogram to those shown in Figures 10.8, 10.23, and 10.25.

enhancement of a 400-row by 400-pixel image, though, requires up to 160,000 calculations, depending on the amount of saturation desired. As Schowengerdt (1983) discusses, this number can be reduced to just 256 by using a look-up table.

The idea of a look-up table is a simple one: Because 8-bit integers have a numerical range of 0 to 255, for a total of 256 possible values, instead of recalculating the value of each image pixel, a vector is formed, called LOOK for instance, having 256 addresses. Each address from 1 to 256 is associated with an image pixel value. For example, each value of LOOK at a particular address is set equal to the contrast-enhanced equivalent of the image pixel used for the address. This is somewhat confusing to describe with words, therefore a simple calculation example is included.

Example 10.1

For simple, linear contrast enhancement, we use a minimum pixel value of 50 and a maximum pixel of 200:

$$\text{LOOK}(I) = 0, \qquad I = 1, 2, \ldots, 50$$
$$\text{LOOK}(I) = 255, \qquad I = 201, 202, \ldots, 256$$
$$\text{LOOK}(I) = 255(I - 50)/(200 - 50), \quad I = 51, \ldots, 200$$

Once the array LOOK is formed, an image is enhanced as follows:

$$\text{new pixel value} = \text{LOOK}(\text{old pixel value})$$

where this is a simple replacement, rather than a calculation. Every pixel in the original image, referred to as "old" pixel values, is changed to a "new" pixel value by finding the value of the vector LOOK at the address of the old pixel value. Suppose the original value is 175; then, using the vector LOOK as defined in this example, the new pixel value is equal to LOOK(175); in this case, this new pixel value is equal to 212. The benefit of the look-up table is obvious. Only 256 calculations are required to fill the vector LOOK. Thereafter, the whole image is enhanced using replacement by values from the array LOOK rather than through calculation. ∎

10.3.4 CTRAST: A Computer Algorithm for Contrast Enhancement

All previous examples of contrast enhancement were developed using the program CTRAST (Figure 10.28, page 473) to create an enhanced image, and using the program IMAGIN to display the enhanced image. CTRAST is based on the efficient algorithm for contrast enhancement using a look-up table called LOOK. This algorithm is fast, even when implemented on microcomputers lacking math coprocessors.

When using CTRAST, the following information is requested interactively by the program:

1. The name of the input image file to be enhanced.
2. The name of the output image file to be created. This file contains the enhanced image and is written in a format compatible with the program IMAGIN.
3. The number of rows and pixels per row comprising the image.
4. The type of contrast enhancement desired: simple, linear; bilinear; histogram equalization; or user specified (an option enabling a user to transcend the current chapter to implement contrast enhancement methods not described herein).

(a) If a simple, linear contrast enhancement is selected, values defining minimum and maximum must be entered to establish the range of pixel values to be enhanced.

(b) If a bilinear contrast enhancement is chosen, values defining minimum, midrange, and maximum must be entered where these establish two image pixel intervals to be enhanced.

(c) If histogram equalization contrast enhancement is chosen, values defining minimum and maximum must be entered to specify the pixel range to be enhanced. The number of bins used to represent the histogram between minimum and maximum must be entered, and the cumulative relative frequency for each bin must be specified.

(d) If the fourth option is chosen, the user must first create the vector LOOK and store it in a file; 256 values must be stored. This option allows any type of contrast enhancement to be used. For example, suppose an experiment is conducted to examine the type of contrast enhancement afforded by a nonlinear procedure of the form

$$LOOK(I) = 0, \qquad I = 1 \text{ to } 50$$
$$LOOK(I) = 2.55(I - 50) - 255[(I - 50)^3/(2(150^3)], \qquad I = 51 \text{ to } 200$$
$$LOOK(I) = 255, \qquad I = 201 \text{ to } 256$$

This is simply one example of a nonlinear function. In using such an enhancement, a data file is created containing 256 values, LOOK(1) through LOOK(256), where each value is defined by the function just shown. Then, the fourth, or user-defined, option in CTRAST is chosen to implement the contrast enhancement.

Once this information is entered (steps 1 through 4), the output image is created. This image is displayed using the program IMAGIN to see the visual consequence of the contrast enhancement.

10.3.5 Filtering Digital Images

The major goal of this chapter is the application of numerical methods to digital images to enhance their interpretation. If images are contaminated by considerable noise (in this instance, noise is any extraneous information in an image detracting attention from a feature or features of interest), the ability to interpret the images is impaired. A numerical method for removing noise is desirable.

Perhaps the goal of interpreting images is the identification of linear features, which may be faults or fractures. Such analyses are useful for active fault studies or groundwater flow studies. In this case, it would be advantageous to enhance edges and lines in an image and to remove other information, such as tone, which may impair our ability to recognize lines. Tone, in this instance, is considered to be noise.

Removing information from an image is called filtering. One method for filtering images is spatial convolution filtering using a box filter algorithm. Spatial convolution filtering involves the convolution between a point spread function (PSF) and an image f, such that (Schowengerdt, 1983)

$$g(x, y) = PSF(x, y) * f(x, y)$$

where the asterisk represents convolution, and g represents the filtered image.

Convolution requires integration. In this case,

$$g(x, y) = \int_{-\infty}^{\infty} \int_{-\infty}^{\infty} f(x', y')\text{PSF}(x - x', y - y') \, dx' \, dy'$$

where x' and y' show shifts around the pixel at location x, y. A point spread function, in an optical sense, is the two-dimensional image of a dot (from Schowengerdt, 1983, p. 18). In a numerical sense, a PSF can be designed using weights. In fact, a PSF is exactly analogous to weights used in kriging (Chapter 6; Carr, 1990).

Spatial convolution filtering is approximated as a double summation (Schowengerdt, 1983, p. 31):

$$g(i, j) = \sum_{m=-(M-1)/2}^{(M-1)/2} \sum_{n=-(N-1)/2}^{(N-1)/2} f(i - m, j - n)\text{PSF}(m, n)$$

where M, N are odd numbers describing the rectangular size of the PSF filter, i ranges from 1 to the total number of rows composing the image being filtered, and j ranges from 1 to the total number of pixels per row composing the image to be filtered (this equation begs the question of boundary conditions; actually, g is slightly smaller than f because the filter PSF cannot be centered over pixels on the boundary of f).

Looking at this filtering process in more practical terms, at least in the context of this chapter, PSF is designed to be a two-dimensional square matrix of weights, where the dimension (size) of PSF is an odd number: 3, 5, 7, and so on. For example, suppose noise from an image is to be filtered. Noise can be a variety of things in an image, but, in general, noise is considered to be a contrast between adjacent pixel values. An edge is one example of noise. Therefore, noise may be a high-frequency phenomenon. To remove this noise, a low-pass filter is used. Such a filter does not filter (i.e., "passes") low frequencies in an image, but does attenuate (filter) high frequencies, such as edges. An example of a low-frequency phenomenon in an image is tone, such as the dark tone of a lake, or the light or dark tone of a type of rock or soil.

One way to design a low-pass filter is to use a PSF as follows:

$$[\text{PSF}] = \begin{vmatrix} \frac{1}{9} & \frac{1}{9} & \frac{1}{9} \\ \frac{1}{9} & \frac{1}{9} & \frac{1}{9} \\ \frac{1}{9} & \frac{1}{9} & \frac{1}{9} \end{vmatrix}$$

which is a 3×3 matrix of weights, each equal to the reciprocal of the total number of entries in PSF. This PSF is applied to an image as follows:

$$\begin{aligned} g(i, j) = {} & \text{PSF}(1, 1)f(i - 1, j - 1) + \text{PSF}(1, 2)f(i - 1, j) \\ & + \text{PSF}(1, 3)f(i - 1, j + 1) + \text{PSF}(2, 1)f(i, j - 1) \\ & + \text{PSF}(2, 2)f(i, j) + \text{PSF}(2, 3)f(i, j + 1) \\ & + \text{PSF}(3, 1)f(i + 1, j - 1) + \text{PSF}(3, 2)f(i + 1, j) \\ & + \text{PSF}(3, 3)f(i + 1, j + 1) \end{aligned}$$

which shows a shift in f around the central pixel (i, j). This shift is written specifically for a PSF of size 3×3. This example emphasizes the notion of the shift shown before in the integral definition of the convolution.

In summary, a common way to effect a low-pass filter is to design a PSF having weights such that a pixel averaging is obtained. The visual consequence of low-pass filtering is a blurring of the original image. For instance, using a

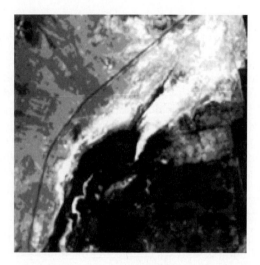

Figure 10.29   The Humboldt Sink image after the application of a 3 × 3 low-pass filter. Notice how the image has been blurred by this filter process.

3 × 3 low-pass filter yields an image (Figure 10.29) that appears blurry in comparison to the original image (Figure 10.8). Tones are emphasized by this low-pass filter, which aids rock-type discrimination, for instance, but image sharpness, an effect rendered by edges, is sacrificed.

The filtered image (Figure 10.29) is developed using the program CONVOL (Figure 10.30, page 476). This program requires the following information:

1. The name of the file containing the input image to be filtered.
2. The name of the output image file, written in binary but using 32-bit integers.
3. The name of the file containing the PSF and the size of the PSF (Figure 10.31).
4. The number of rows and pixels per row comprising the image to be filtered.

Once the filtering process has been completed using CONVOL, a second program, IMAG41 (Figure 10.32, page 478), is run to convert the output 32-bit integer image to an 8-bit integer image compatible with the program IMAGIN. The program IMAG41 is simple to use:

1. Specify the name of the input 32-bit integer image created by program CONVOL.
2. Specify the number of rows and pixels per row comprising the input filtered image. Note that the number of rows of the filtered image is equal to that for the original image minus the row size of the PSF filter. Thus in Figure 10.29, the row size is 397, equal to 400 (original rows) minus 3 (PSF row

```
3
0.11111   0.11111    0.11111
0.11111   0.11111    0.11111
0.11111   0.11111    0.11111

0;37;44m C:\TEXT>
```

Figure 10.31   A listing of the point spread function (PSF) data file used to develop Figure 10.29.

size). Likewise, the number of columns of the filtered image is equal to 397, the number of columns for the original image minus the column size of the PSF.

3. Specify the name of the output 8-bit integer image file to be created.
4. Specify the number of rows and pixels per row comprising the input image (the 32-bit integer image from CONVOL).

In addition to low-pass filtering, spatial convolution filtering may be used also to implement a high-pass filter. In such a filter, low frequencies, such as tones, are attenuated, and high frequencies, such as edges, are accentuated. One way to implement a high-pass filter is to design the following PSF:

$$[PSF] = \begin{vmatrix} -\frac{1}{25} & -\frac{1}{25} & -\frac{1}{25} & -\frac{1}{25} & -\frac{1}{25} \\ -\frac{1}{25} & -\frac{1}{25} & -\frac{1}{25} & -\frac{1}{25} & -\frac{1}{25} \\ -\frac{1}{25} & -\frac{1}{25} & \frac{24}{25} & -\frac{1}{25} & -\frac{1}{25} \\ -\frac{1}{25} & -\frac{1}{25} & -\frac{1}{25} & -\frac{1}{25} & -\frac{1}{25} \\ -\frac{1}{25} & -\frac{1}{25} & -\frac{1}{25} & -\frac{1}{25} & -\frac{1}{25} \end{vmatrix}$$

Notice that, whereas the weights associated with a low-pass PSF sum to 1, the weights for a high-pass PSF sum to zero (in the PSF just shown, notice that 24 entries, each equal to $-\frac{1}{25}$, and one entry at the center of the matrix, equal to $\frac{24}{25}$, sum to zero).

Application of this PSF to the image of Saturn is shown in Figure 10.33. Tones, which are low-frequency phenomena, are completely filtered, but edges, such as the rings of Saturn, are greatly enhanced. This demonstrates the advantage of high-pass filtering if the interpretation of edges is important. With respect to the image of Saturn, edge detection is useful for discriminating the different rings and especially for highlighting rings having a subtle (less obvious) presence in the image.

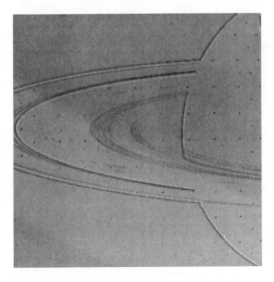

Figure 10.33   An omnidirectional high-pass filter of the Saturn image. Note the enhancement of the rings.

Edge-detection or high-pass filters may be additionally designed to enhance particular spatial directions. For instance, notice in the image of the Humboldt Sink that a highway, U.S. Interstate 80, extends from the upper right portion to the lower left portion of this image. This highway has a general northeast–southwest trend. Other naturally occurring lineaments in this image have a similar trend.

To enhance these edges best, a directional high-pass filter is designed to be orthogonal to these linear features. In this case, a high-pass filter which operates in the northwest–southeast direction is sought. Such a PSF is designed as follows:

$$[\text{PSF}] = \begin{vmatrix} -0.2 & 0 & 0 & 0 & 0 \\ 0 & -0.2 & 0 & 0 & 0 \\ 0 & 0 & 0.8 & 0 & 0 \\ 0 & 0 & 0 & -0.2 & 0 \\ 0 & 0 & 0 & 0 & -0.2 \end{vmatrix}$$

This [PSF] is a small realization of space. Nonzero weights in this matrix describe a northwest–southeast direction; all other weights are zero. The sum of these weights is zero.

Applying this PSF to the image of the Humboldt Sink yields the image shown in Figure 10.34. Tones are filtered, but the highway and other linear features orthogonal to the PSF are greatly enhanced. A directional high-pass filter applied in the northeast–southwest direction (Figure 10.35) virtually obscures the highway because it is a low-frequency feature in this direction.

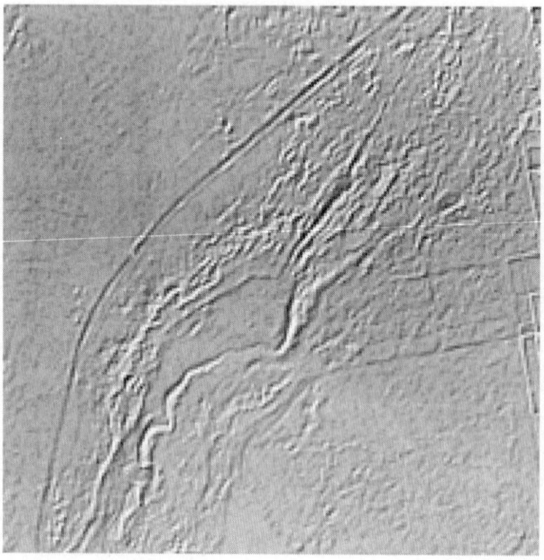

Figure 10.34 A northwest–southeast directional high-pass filter of the Humboldt Sink image. Note the enhancement of all features having a northeast–southwest strike, including the highway.

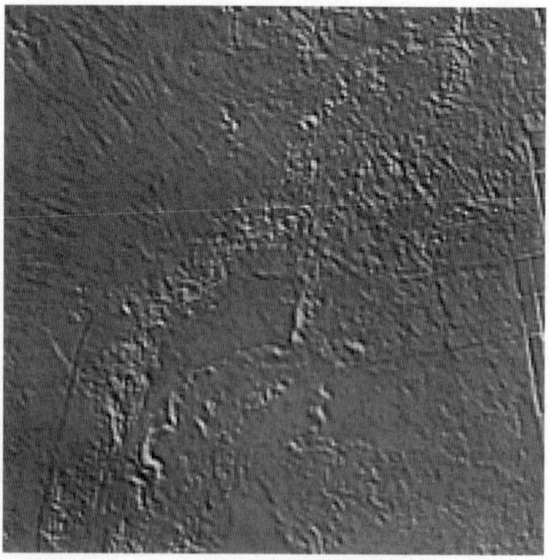

Figure 10.35 A northeast–southwest directional high-pass filter of the Humboldt Sink image. Note that all features having a northeast–southwest strike are now filtered.

The PSF used to implement the northeast–southwest high-pass filter is

$$[PSF] = \begin{vmatrix} 0 & 0 & 0 & 0 & -0.2 \\ 0 & 0 & 0 & -0.2 & 0 \\ 0 & 0 & 0.8 & 0 & 0 \\ 0 & -0.2 & 0 & 0 & 0 \\ -0.2 & 0 & 0 & 0 & 0 \end{vmatrix}$$

10.3.6 Fourier (Frequency) Filtering of Digital Images

The preceding section discusses filtering images in the spatial domain. In some cases, filtering is more conveniently applied in the frequency domain, especially if a periodic noise contaminates a digital image and this noise is to be removed. Both high-pass and low-pass filtering can be accomplished, often more creatively or selectively, using frequency domain filtering. However, Fourier transformation, at least in the manner implemented in computer programs in this text, is less efficient than spatial convolution. This statement is made relative to low-pass and high-pass filtering; identifying and removing a periodic noise from an image requires Fourier transformation.

Suppose a digital image is contaminated by periodic noise. Such noise is simulated (Figure 10.36b) using a two-dimensional sine function:

$$G(x, y) = 255[\sin(30x + 10y)]$$

where G is the resultant simulated image. The periodic nature of the noisy image G is obvious (Figure 10.36b).

If the image of G is superimposed on that for the Humboldt Sink (Figure 10.36a), a "noisy" image is created (Figure 10.36c) by simply adding the image of G to that of the Humboldt Sink and rescaling pixel values of this combined image to the range 0 to 255. Such a noisy image can result in practice if the sensor (satellite, in this case) that acquired the data was influenced by, for instance, a periodic vibration or an electrical noise. If a digital image appears to be contaminated by a periodic noise, such as is artificially shown in Figure 10.36c, then Fourier analysis is the logical tool to use to identify the periodic nature of the noise and to filter it from the image data.

The program FORIER2D (Figure 10.37, page 479) is an extension of the program FORIER (Chapter 9) to two-dimensional analyses. The input to this program is a file storing the digital image (up to 400 rows by 400 pixels per row) to be analyzed.

Two-dimensional Fourier transformation can be accomplished as follows (Gonzalez and Wintz, 1977, pp. 41–50): For a two-dimensional input function $f(x, y)$, one example of which is a digital image, its Fourier transform $F(u, v)$, where u and v are frequency terms, is computed as

$$F(u, v) = \frac{1}{MN} \sum_{x=0}^{M-1} \sum_{y=0}^{N-1} f(x, y) \exp\left[-i2\pi\left(\frac{ux}{M} + vy/N\right)\right]$$

wherein i is an imaginary number (the square root of -1). For computational purposes, the two-dimensional Fourier transform is separated more conveniently

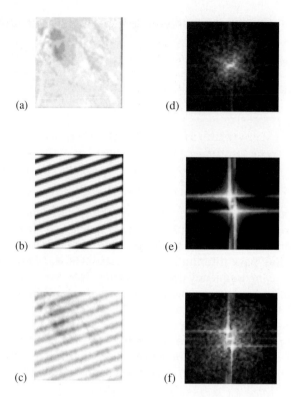

Figure 10.36   A composite image: (a) the upper 100 × 100 region of the Humboldt Sink image; (b) a 100 × 100 image of artificially created periodic noise; (c) the superimposition of the periodic noise onto the 100 × 100 subimage of the Humboldt Sink image; (d) the Fourier transform image of (a); (e) the Fourier transform image of (b); and (f) the Fourier transform image of (c).

as follows:

$$F(u, v) = \frac{1}{N} \sum_{x=0}^{N-1} \exp\left(\frac{-i2\pi ux}{N}\right) \sum_{y=0}^{N-1} f(x, y) \exp\left(\frac{-i2\pi vy}{N}\right)$$

which now assumes that M equals N (i.e., $f(x, y)$ is square); this equation is used in FORIER2D. Further, for each of these two equations, $u, v = 0, 1, \ldots, N - 1$.

Application of FORIER2D to the images in Figure 10.36 yields the two-dimensional Fourier transforms, which are also displayed. Each two-dimensional Fourier transform is converted to a digital image using the programs READRI (Figure 10.38, page 480) and IMAG41 (i.e., the output from READRI is in a form that is compatible with IMAG41). Actually, what is displayed as the Fourier transform is $\log\{1 + abs[F(u, v)]\}$ because this yields a more interesting display. The value 1 is added to the absolute value of F to avoid an error when taking the logarithm should F be equal to zero.

Notice where each Fourier transform image is bright, and also where it is dark. Bright areas identify periods, or wavelengths, composing the original image; dark areas are associated with frequencies that do not occur in the original image. The Fourier transforms in Figure 10.36 have been enhanced to change their original contrast. After computing the two-dimensional Fourier transform using FORIER2D and creating a digital image of the transform using the program combination READRI and IMAG41, the image is often too bright. In all of the cases in Figure 10.36, the program CTRAST was used to decrease the contrast so as to either improve the display or emphasize features of importance, such as the presence of noise.

Because Figure 10.36b is used to artificially create Figure 10.36c and because Figure 10.36c is an example image from which the periodic noise is to be removed, a Fourier transform is computed of Figure 10.36c using FORIER2D and is displayed in Figure 10.36f. Recall that Figure 10.36b is an image of the periodic noise alone and that Figure 10.36e is the Fourier transform of this noise. It is useful, therefore, to compare Figure 10.36f to Figure 10.36e to understand what portion of the two-dimensional Fourier transform shown in Figure 10.36f is associated with the periodic noise. Most of the original image data are associated with frequencies that plot at or near the center of the Fourier transform image; the noise plots in the same position as it does in Figure 10.36e. The objective is to remove the bright portion associated with noise from Fourier transform image shown in Figure 10.36f, then inverse Fourier transform the image back to the spatial domain. If conducted properly, the filtering should yield the original Humboldt Sink image.

The program FORFILT (Figure 10.39, page 481) is used to filter frequencies from the Fourier transform image. Input to this program is the Fourier transform image created by FORIER2D. Interactive input is also used to specify the pixel region in the Fourier transform image that is to be set equal to zero (hence filtering frequencies in the bounded interval). This pixel region is defined by the minimum and maximum row and pixel numbers of the bounded region. Once this region is defined, the program creates the output, filtered Fourier domain image.

Once the filtering is complete, the program FORINV2 (Figure 10.40, page 482) is executed to inverse Fourier transform the filtered image. This two-dimensional inverse is (e.g., Gonzalez and Wintz, 1977, p. 50)

$$f(x, y) = \frac{1}{N} \sum_{u=0}^{N-1} \exp\left(\frac{i2\pi ux}{N}\right) \sum_{v=0}^{N-1} F(u, v) \exp\left(\frac{i2\pi vy}{N}\right)$$

for which $x, y = 0, 1, \ldots, N - 1$. Once FORINV2 is completed, the program INTFAC (Figure 10.41, page 483) is run to create a digital image from the output file that is created by FORINV2.

Application of FORFILT to the Fourier domain image of the Humboldt Sink yields the image in Figure 10.42b. This figure is an image of the Fourier transform after filtering the periodic noise component, which appears in the Fourier transform image as the star-shaped bright regions. Figure 10.42b shows that only the central, brightest regions of this noise are filtered (these regions are associated with the frequencies of noise having the greatest power). The program FORINV2 is used to compute the inverse Fourier transform (Figure 10.42c).

Notice in Figure 10.42c that the periodic noise is successfully removed from the center of the image, but not from the edges. As Schowengerdt (1983,

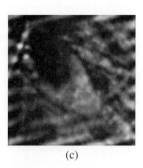

(a)   (b)   (c)

Figure 10.42   A composite image: (a) the image shown in Figure 10.36c; (b) the Fourier transform shown in Figure 10.36e, but the two bright areas associated with noise are filtered; the filtering was done using FORFILT applied to the RI file created by FORIER2D; (c) an image after applying the program FORINV2 to the filtered Fourier transform. Note that the periodic noise has been removed from the center of the image, but not from the edges.

p. 97) notes, if noise has a high enough frequency such that it plots far away (in the image of the Fourier transform) from the spectrum for the image, noise is easily filtered. In this case (Figure 10.42b), noise has a high enough frequency to plot somewhat away from the image spectrum (Figure 10.36f), but is still close enough that its total removal is difficult. Hence (Figure 10.42c) a partial success is achieved in removing the periodic noise from the image data.

Fourier domain filtering also may be used to achieve high-pass and low-pass filtering. Of course, such filtering was accomplished previously using the spatial domain, but more creative filtering can be achieved in the frequency domain. For example, four images are shown in Figure 10.43. The upper left image (Figure 10.43a) is the original image. Filtering only the zero-frequency component from the Fourier transform yields a high-pass filter (Figure 10.43b). A more pronounced high-pass filter is implemented by filtering the first 10 frequency components from the Fourier transform (Figure 10.43c); these are low-frequency components. Finally, a low-pass filter is implemented by filtering all frequency components except the first 10 (Figure 10.43d).

Details are given regarding the calculation of the filtered images (Figure 10.43). First, the program FORIER2D was applied to the original image, followed by program READRI to create an output file. A file editor (such as EDIT, an MS-DOS editor) was used to filter the first line from the file created by READRI; the first line is the zero-frequency component. After editing, FORINV2 and INTFAC were used to create the filtered image (Figure 10.43b). The program FORFILT was applied to the output file from READRI to create Figures 10.43c and 10.43d.

Anytime a digital image is contaminated by what visually appears to be periodic noise, experiments should be conducted with Fourier transformation to see if the periodic noise can be isolated from the image information, thereby allowing the noise to be successfully removed through simple filtering and inverse Fourier transformation. Furthermore, if a more precise frequency filtering is desired (in comparison to what is offered using spatial convolution filtering), Fourier transformation should be considered.

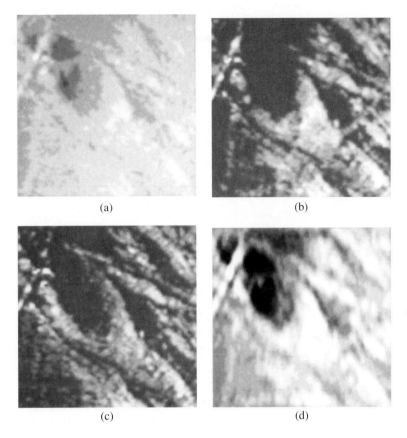

Figure 10.43   A composite image: (a) the image shown in Figure 10.36a; (b) the image after Fourier transformation, filtering the zero frequency component, then inverse Fourier transformation; (c) the image after Fourier transformation, removing the first 10 lowest frequencies, then inverse Fourier transformation; (d) the image after Fourier transformation, removing all frequencies except the first 10 lowest frequencies, then inverse Fourier transformation.

10.3.7 Classification of Digital Images

Feature extraction is an important aspect of digital image analysis. Contrast enhancement is used to accentuate differences between pixels to render certain image features more visible. Filters are used to remove some image information, thus making certain features more recognizable, such as linear features or tones. Digital imagery may be used for pattern recognition, that is, as an aid to identifying portions of an image that are statistically similar. Such statistical signatures are found using image classification.

In this chapter, single band images are classified by selecting training sites. A training site is a small subregion of a digital image representing, as much as is possible, a homogeneous portion of the image associated with a particular feature on the ground, such as corn (for crop identification), alluvium, granite (for geological applications), and water. A training site is used to identify statistical attributes of a ground feature in terms of its pixel representation.

Pixel values are compared to the statistical attributes of each training class to determine the class to which the pixel most likely belongs. This is known as supervised classification because the user must select training sites. An alternative method is unsupervised classification, in which the computer algorithm automatically selects ground classes based on groupings of statistical characteristics. The advantage to unsupervised classification, in comparison to supervised classification, lies in the ease of application. The disadvantage of unsupervised classification to the user is that the computer algorithm may choose ground classes that are not meaningful to the user.

In supervised classification, one method that may be used to determine to which class a pixel belongs is Bayes' rule (from Schowengerdt, 1983, p. 38): the conditional probability $p(i \mid x)$, the probability of having a class i given the pixel value x, is given by

$$p(i \mid x) = \frac{p(x \mid i)p(i)}{p(x)}$$

where $p(x)$ is given by

$$p(x) = \sum_{i=1}^{N} p(x \mid i)p(i)$$

where N is the number of different classes to which a pixel x may be assigned.

The probability $p(x)$, that is, the probability of having a pixel value x, is the same for all N classes. Therefore, Bayes' rule simplifies to

$$p(i|x) = p(x|i)p(i)$$

In this equation, $p(i)$, the probability of having a class i, is simply $1/N$, where N is the number of different classes. Training data are used to determine $p(x|i)$, the probability of having a pixel value x in a class i. Recall that training sites are small subregions of a digital image. Once chosen, a histogram of pixels composing each training site is developed to define $p(x|i)$.

For instance, suppose pixel values for a training site range from 40 to 75 in integer value. Obviously, given a pixel value 20, its probability of belonging to this training class is zero. But, a pixel value of 50, which is inside the interval 40 to 75 has a probability of belonging to the class equal to the relative frequency for the pixel value 50 within the training class.

Therefore, the Bayesian classification algorithm is as follows:

1. Choose from 1 to N training sites, where each describes a class.
2. For each class, determine the histogram of pixel values within each training site.
3. Form a look-up table, called LOOK.
   (a) For each of the N classes i, calculate $p(i|x)$, and then LOOK$(X) = I$, where X is a pixel value from 0 to 255, and I is the class of the N total associated with the largest probability $p(i|x)$. This is called the maximum likelihood supervised classification algorithm. If $p(i|x)$ is zero for all N classes, then LOOK$(X) = 0$.
4. Classify all image pixels using the look-up table such that class = LOOK(X), where the pixel value X is an address for the array LOOK.

MXLIKE, a computer algorithm for maximum-likelihood Bayesian classification, is presented in Figure 10.44, page 484, and a complete user's guide is shown at the beginning of the program. An image to be classified is displayed by MXLIKE using a 16-color palette specified by the user. Once the image is displayed, the user is asked to specify training sites interactively:

1. First, the user is asked how many training classes N are desired; the program allows up to 10.
2. For each of N training classes, the program asks the user to specify the row, then pixel coordinates for the upper left and lower right corners of each training site.
3. The user is asked to specify a threshold probability value for all training classes. A pixel will not be assigned to any class unless its probability of belonging to any one class is equal to or greater than this threshold probability. Thresholding is used to minimize classification errors (assigning pixels to a class when it does not belong to the class).

An example is presented using the Humboldt Sink image. Two training classes are chosen, one of alluvium and the other of vegetation. Pixel coordinates for these two classes (Table 10.2) are entered and the program MXLIKE draws the training sites on the displayed image (Figure 10.45). This program, in effect, produces both an unsupervised and supervised classification. In preparation for classification, MXLIKE remaps palettes after the initial display of the image, regardless of the palette chosen by the user. Once the 16 colors are remapped, a brilliant pseudocolor display of the image is achieved. The program pauses at this point waiting for the user to push the enter key should a photograph be desired of the screen. This pseudocolor display is a form of unsupervised classification (Figure 10.46).

Once the 16 colors are remapped, the program performs supervised classification (Figure 10.47) using the maximum-likelihood Bayesian algorithm. In this example, alluvium is displayed in red and vegetation in green. Notice how classification has successfully extracted two features, alluvium and vegetation, from the image, thus enhancing them. Unsupervised and supervised classifications are useful for discriminating ground type, and may provide a useful platform for initial geologic mapping in a region.

As an alternative to Bayesian supervised classification, which is, admittedly, somewhat cumbersome to use, a relatively simple minimum-distance classifier

Table 10.2 Pixel Coordinates for the Aluvium and Vegetation Training Sites, Humboldt Sink Image

Alluvium Training Site

Row number, upper left corner	200
Pixel (column) number, upper left corner	1
Row number, lower right corner	210
Pixel (column) number, lower right corner	10

Vegetation Training Site

Row number, upper left corner	95
Pixel (column) number, upper left corner	395
Row number, lower right corner	100
Pixel (column) number, lower right corner	400

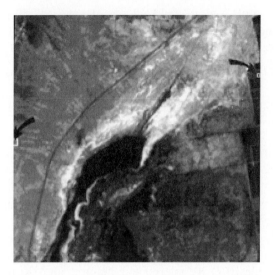

Figure 10.45 Two training sites as described in Table 10.2 are shown, drawn by MXLIKE; arrows are added for easy recognition.

may be used. Such an option is allowed in the program MXLIKE. A minimum-distance classifier is implemented using the following algorithm:

1. Compute the mean value of the pixels composing each training site.
2. For a pixel in the image to be classified, compute the Euclidean distance between the numerical value of the pixel and the mean value of each class.

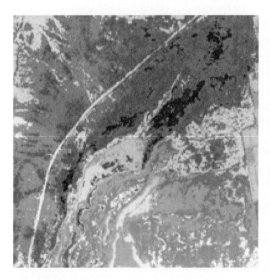

Figure 10.46 The pseudocolor, pseudo-unsupervised classification of the Humboldt Sink image. This is an interesting way to discriminate different ranges of pixel values in the image to note their spatial correlation. It is also an interesting way to define or determine ground classes (and patterns).

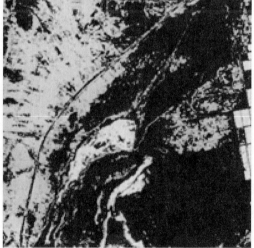

Figure 10.47 A maximum-likelihood classification of the Humboldt Sink image to classify alluvium (red) and vegetation (green). Compare this classification to that shown in Figure 10.46 to see how the unsupervised classification treated the alluvium and vegetation.

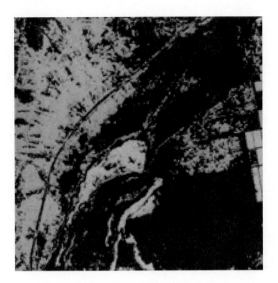

Figure 10.48 A minimum-distance-to-mean classification (supervised) of the Humboldt
Sink image to classify alluvium (red) and vegetation (green). This outcome is virtually
identical to that shown in Figure 10.47.

3. Classify the pixel in the class to which it is the closest (i.e., in the class to
   which the pixel value is closest to the mean value).
4. Apply some sort of threshold scheme to minimize the number of misclassified
   pixels.

A minimum-distance classifier is a special form of the maximum-likelihood
classifier (Schowengerdt, 1983, pp. 49–54). By experimenting with the threshold values, a supervised classification can be obtained using the minimum-
distance algorithm that is similar to one obtained using the more general
maximum-likelihood (Bayesian) algorithm (Figure 10.48).

10.4 Magellan Radar Images of Venus

A special section is devoted to the digital images of reflected radar imagery
acquired by the Magellan satellite, which was placed in orbit around the planet
Venus in 1992. This section is included because (1) these images are newly
acquired and recently described (see the special issue of the *Journal of Geophysical Research* **97**(E8), 25 August 1992); (2) these data are remarkably well
displayed using VGA hardware; and (3) these images capture volcanic, tectonic,
and geomorphic features on Venus, which are, to say the least, interesting and
may help us to better understand similar features on earth.

Much of the Magellan information is available on compact disks (CDs) (see
Appendix B for a description). These compact disks are compatible with a
variety of operating systems, including MS-DOS. Several images are extracted
from these disks (Figures 10.49 through 10.54). In particular, notice pancake
volcanic domes (Figure 10.49), fracture systems (Figure 10.50), wind deposit
features (Figure 10.51), volcano and related flows (Figure 10.52), a double-
impact crater system (Figure 10.53), and what appears to be a corona (Figure

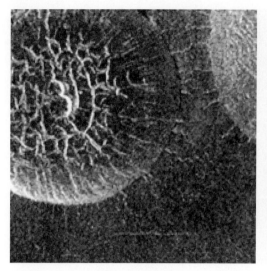

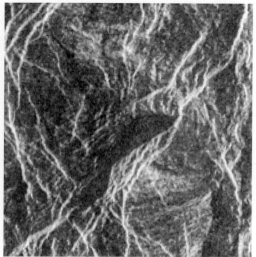

Figure 10.49   A closeup image of two of the Venus pancake volcanic domes shown in Figure 10.13. This image has 400 rows with 400 pixels per row, and was extracted from image FF31.IMG of the subdirectory F30S009 on the Magellan CD F-MIDR 30S009.

Figure 10.50   Fracture systems observed on Venus, extracted from image FF53.IMG from subdirectory F05N065 from Magellan CD F-MIDR 05N065. This image has 400 rows with 400 pixels per row.

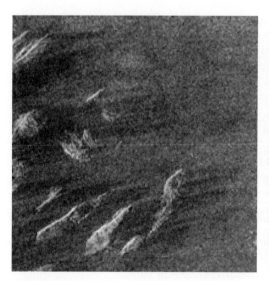

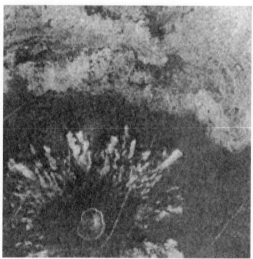

Figure 10.51   A series of (presumably sand) dunes on Venus. Notice the wind shadows to the right of each dune; therefore, the wind direction is from left to right. This image was extracted from image FF38.IMG from subdirectory F00N070 from Magellan CD F-MIDR 00N070.

Figure 10.52   A volcano (note the crater at the bottom center of the image, which is ringed by a series of flows). The upper portion of this image is a large volcanic flow. This image has 400 rows and 400 pixels per row and was extracted from image FF22.IMG from subdirectory F10S200 from Magellan CD F-MIDR 10S200.

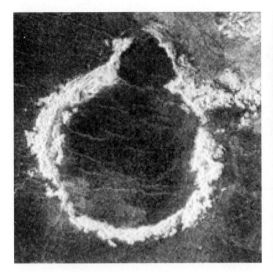 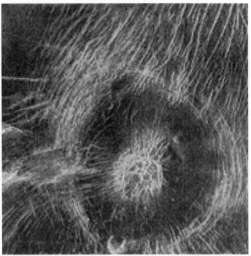

Figure 10.53   The double (possibly triple) crater, which has been named Heloise (see Schaber et al., 1992). This image has 400 rows with 400 pixels per row and is a merger of extracted portions of images FF27.IMG and FF28.IMG from subdirectory F40N053 of Magellan CD F-MIDR 40N053.

Figure 10.54   A possible corona, a tectonic feature. Notice both the radial and concentric circular fractures. See Soloman et al. (1992) for a description of a corona. The one in this figure was found in image C1F01.IMG from subdirectory C145S159 from Magellan CD C1-MIDR 45S159. The image has 400 rows with 400 pixels per row.

10.54). Exercises at the end of this chapter stimulate the enhancement of these images to better enable their interpretation.

10.5   Examples of Airborne-Sensor-Acquired Digital Imagery

In addition to satellite-acquired imagery, an increasing number of airborne sensors are being developed that are flown on aircraft, such as NASA's U2 plane. One example is the Thermal Infrared Multispectral Scanner (TIMS). Data are acquired in six bands of the electromagnetic spectrum in the wavelength range 8.2 to 12.2 μm. An example TIMS image (Figure 10.55) is particularly useful for geologic mapping because the spectral excitance of silicate minerals is distinctive over these EM wavelengths (Lillesand and Kiefer, 1987, p. 463).

Another type of airborne-acquired data is the radar imagery acquired by the AIRSAR sensor. An example of an AIRSAR image is shown in Figure 10.56. AIRSAR is a synthetic aperture radar (SAR) system that uses a short physical antenna, but data processing allows this short antenna to emulate a much longer one. This provides a very narrow radar beam width. Radar wavelengths can penetrate clouds and are most useful for imaging topographic relief. This is most evident in the Venus data.

Aircraft-acquired imagery provide better spatial resolution than satellite-acquired imagery. For instance, TIMS pixel size is 2.5 m compared to the

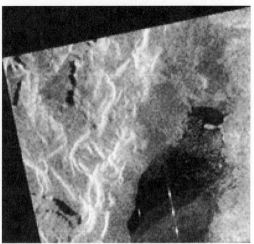

Figure 10.55 A TIMS image of Lunar Lake, California. This image has 400 rows with 400 pixels per row and was extracted from NASA's nine CD set for the Geologic Remote Sensing Field Experiment.

Figure 10.56 An AIRSAR image of Lunar Lake, California. The image is registered to that displayed in Figure 10.55. It has 400 rows with 400 pixels per row and was extracted from NASA's nine CD set for the Geologic Remote Sensing Field Experiment.

Landsat Thematic Mapper pixel size of 30 m. Moreover, atmospheric path interference less severely influences aircraft-acquired data because the sensor is closer to the ground in comparison to satellite systems, hence emittance and reflectance have shorter path lengths through the atmosphere.

Chapter Summary

- A digital image is simply a two-dimensional array of numbers. Each number (digit) is called a pixel. Digital images may be formed in a variety of ways: by digitizing a conventional photograph, by a satellite or aircraft system that detects reflected electromagnetic energy and represents the intensity of this reflected energy as a number, and so on.

- Much geologic knowledge of the earth and all geologic knowledge of neighboring planets has been obtained through remote sensing.

- Satellite and airborne sensors are designed to image bands of the electromagnetic spectrum. In this chapter, though, only single band images are displayed because the high-resolution VGA graphics mode is used; hence spectral resolution is sacrificed in favor of spatial resolution.

- Image processing consists of two basic types of operations: (1) pixel-specific operations, such as contrast adjustment; and (2) neighborhood processing (involving more than one pixel at a time), such as spatial convolution filtering.

- Classification, both unsupervised and supervised, is useful for pattern recognition and feature extraction.

Exercises

(An asterisk indicates an advanced or graduate student problem)

1. If the sky is blue at noon, why are sunrises and sunsets red?
2. Why are clouds white?
3. The following digital image is given:

$$
\begin{array}{ccccc}
20 & 55 & 65 & 32 & 46 \\
20 & 43 & 31 & 27 & 1 \\
0 & 5 & 11 & 45 & 59 \\
10 & 73 & 21 & 89 & 99 \\
0 & 1 & 10 & 56 & 78
\end{array}
$$

Use a look-up table procedure to enhance the contrast of these pixels (i.e., create a new, enhanced 5×5 image) using the following:
   (a) simple, linear algorithm, minimum = 15, maximum = 82
   (b) bilinear algorithm, minimum = 10, midrange = 50, maximum = 90
   (c) histogram equalization, minimum = 0, maximum = 99.

*4. How can the program CTRAST be used to create a binary (black and white) contrast enhancement of a digital image? (*Hint:* Consider the fourth option in CTRAST for the type of contrast enhancement.) Experiment using the Humboldt Sink image.

5. Given the following 10×10 digital image:
   (a) Compute by hand a low-pass filtered image using a 3×3 PSF.
   (b) Compute by hand a high-pass filtered image using a 3×3 PSF.

$$
\begin{array}{cccccccccc}
134 & 141 & 140 & 141 & 142 & 141 & 135 & 138 & 143 & 144 \\
131 & 134 & 134 & 138 & 140 & 142 & 145 & 146 & 147 & 148 \\
131 & 135 & 136 & 140 & 144 & 147 & 146 & 143 & 143 & 143 \\
138 & 141 & 141 & 148 & 154 & 152 & 148 & 137 & 137 & 147 \\
154 & 148 & 146 & 151 & 155 & 148 & 143 & 135 & 143 & 151 \\
158 & 157 & 149 & 145 & 143 & 139 & 135 & 136 & 148 & 151 \\
155 & 154 & 148 & 146 & 138 & 127 & 131 & 148 & 139 & 136 \\
154 & 149 & 145 & 139 & 132 & 123 & 130 & 139 & 139 & 138 \\
151 & 148 & 141 & 136 & 134 & 128 & 133 & 137 & 138 & 140
\end{array}
$$

6. Form a digital image from Figure 10.2 by using the grid superimposed over the photograph. Create this as a binary image: that is, if a grid cell is more dark than bright, make the pixel value for the cell equal to zero; if the cell is more bright (light toned) than dark, make the pixel value equal to 255. Create an output file of your results, for which each line in the data file is

   row number, pixel number, pixel value

   Do this for each cell shown in Figure 10.2. When completed, use the program INTFAC to form a file compatible with the program IMAGIN. Once INTFAC is completed, display your result using IMAGIN. How does the display compare to Figure 10.2?

*7. Compute a high-boost filtered image for
   (a) the Humboldt Sink image
   (b) the image of Saturn and its rings

Hint: A high boost filter yields an image "crispening." Implement such a filter by using the 5×5 high-pass filter applied to the image of Saturn, but change the central weight from 24/25 to 49/25 or greater. A high-boost filter does not filter tones, but does enhance edges; weights sum to 1 or greater.

8. To understand high-pass and directional high-pass filtering better, the following calculation experiments are suggested.

(a) Consider the following two images:

1.
$$\begin{bmatrix} 1 & 1 & 1 \\ 1 & 1 & 1 \\ 1 & 1 & 1 \end{bmatrix}$$
2.
$$\begin{bmatrix} 1 & 1 & 1 \\ 1 & 9 & 1 \\ 1 & 1 & 1 \end{bmatrix}$$

and a 3×3 high-pass filter having weights

$$\begin{bmatrix} -\frac{1}{9} & -\frac{1}{9} & -\frac{1}{9} \\ -\frac{1}{9} & \frac{8}{9} & -\frac{1}{9} \\ -\frac{1}{9} & -\frac{1}{9} & -\frac{1}{9} \end{bmatrix}$$

In this case, the PSF and images are the same size, hence only one pixel, the central one, is changed. Apply this PSF to images 1 and 2 and observe the result (in this case, the result will be one filtered pixel value per image because the PSF and images are the same size). Image 1 is associated with pixels of equivalent value, hence there are no high frequencies (edges) in this image. Image 2 is composed of eight identical pixel values and one pixel value that is different (an edge), hence high-frequency information is present in this image. How do the results of filtering relate to the high-frequency content of the image?

(b) Consider the image

```
0 0 0 0 0 0 0 0 0 1
0 0 0 0 0 0 0 0 1 1
0 0 0 0 0 0 0 1 1 1
0 0 0 0 0 0 1 1 1 1
0 0 0 0 0 1 1 1 1 1
0 0 0 0 1 1 1 1 1 1
0 0 0 1 1 1 1 1 1 1
0 0 1 1 1 1 1 1 1 1
0 1 1 1 1 1 1 1 1 1
1 1 1 1 1 1 1 1 1 1
```

which contains an edge having an orientation N 45 E. Apply the following directional high-pass filters to obtain an output image:

1.
$$\begin{bmatrix} 0 & 0 & -\frac{1}{3} \\ 0 & \frac{2}{3} & 0 \\ -\frac{1}{3} & 0 & 0 \end{bmatrix}$$
2.
$$\begin{bmatrix} -\frac{1}{3} & 0 & 0 \\ 0 & \frac{2}{3} & 0 \\ 0 & 0 & -\frac{1}{3} \end{bmatrix}$$

Which filter enhanced the edge better? Why?

9. Using the histogram of the Saturn image, use your own judgment to experiment with several bilinear contrast adjustments.

10. Enhance the contrast for all Venus images in Figures 10.11 through 10.13 and Figures 10.49 through 10.54. It is evident by their histograms that the original images have a relatively low contrast. See if changing the contrast aids the interpretation of these images.

11. Compute a horizontal high-pass filter for the Saturn image. Try a vertical high-pass filter as well.

12. Experiment with image classification using the Venus images. Using Figure 10.52, for instance, select training sites representing volcanic flows. How successful was the classification?

13. Two training sites have been chosen for an image. Class A is associated with the following histogram:

Pixel value:	10	11	12	13	14	15
Relative frequency:	.05	.2	.45	.45	.2	.05

Class B is associated with the following histogram:

Pixel value:	13	14	15	16	17	18	19	20
Relative frequency:	.01	.07	.09	.32	.5	.4	.1	.04

Given the following pixel values, use Bayes' rule to determine to which class each pixel most likely belongs; assume a threshold probability of 0:

Pixel value: 10 13 14 15 5 19 67

14. Compute unsupervised and supervised classifications of the TIMS image (TIMSM001.IMG). How is TIMS imagery useful for geologic interpretation? Do these classification maps aid the interpretation of geology? Compare the classifications to geologic maps for Lunar Lake, California.

15. Experiment with unsupervised and supervised classification of the Venus imagery. Discuss problems associated with the radiometric classification of radar imagery (*Hint:* Does the ''salt-and-pepper'' texture (a noise), known as speckle, interfere with classification?)

References and Suggested Readings

AVERY, T.E., and G.L. BERLIN. 1992. *Fundamentals of Remote Sensing and Airphoto Interpretation,* 5th ed. New York: Macmillan.

CARR, J.R. 1990. Application of spatial filter theory to kriging. *Mathematical Geology* **22**(8): 1063–1080.

CHAVEZ, P., JR. 1975. Atmospheric, Solar, and MTF corrections for ERTS digital imagery. *Proc. American Society of Photogrammetry Fall Convention.*

GONZALEZ, R.C., and P. WINTZ. 1977. *Digital Image Processing.* Reading, MA: Addison-Wesley.

LILLESAND, T.M., and R.W. KIEFER. 1987. *Remote Sensing and Image Interpretation,* 2d ed. New York: Wiley.

LINDLEY, C.A. 1991. *Practical Image Processing in C.* New York: Wiley.

SCHALKOFF, R.J. 1989. *Digital Image Processing and Computer Vision.* New York: Wiley.

SCHOWENGERDT, R.A. 1983. *Techniques for Image Processing and Classification in Remote Sensing.* New York: Academic.

SERRA, J. 1982. *Image Analysis and Mathematical Morphology.* London: Academic.

Selected Articles Describing Magellan Images

BAKER, V.R., G. KOMATSU, T.J. PARKER, V.C. GULICK, J.S. KARGEL, and J.S. LEWIS. 1992. Channels and valleys on Venus: preliminary analysis of Magellan data. *Journal of Geophysical Research* **97**(E8): 13421–13444.

GREELEY, R., R.E. ARVIDSON, C. ELACHI, M.A. GERINGER, J.J. PLAUT, R.S. SAUNDERS, G. SCHUBERT, E.R. STOFAN, E.J.P. THOUVENOT, S.D. WALL, and C.M. WEITZ. 1992. Aeolian features on Venus: preliminary Magellan results. *Journal of Geophysical Research* **97**(E8): 13319–13345.

HEAD, J.W., L.S. CRUMPLER, J.C. AUBELE, J.E. GUEST, and R.S. SAUNDERS. 1992. Venus volcanism: classification of volcanic features and structures, associations, and global distribution from Magellan data. *Journal of Geophysical Research* **97**(E8): 13153–13197.

SCHABER, G.G., R.G. STROM, H.J. MOORE, L.A. SODERBLOM, R.L. KIRK, D.J. CHADWICK, D.D. DAWSON, L.R. GADDIS, J.M. BOYCE, and J. RUSSEL. 1992. Geology and distribution of impact craters on Venus: What are they telling us? *Journal of Geophysical Research* **97**(E8): 13257–13301.

SOLOMAN, S.C., S.E. SMREKAR, D.L. BINDSCHADLER, R.E. GRIMM, W.M. KAULA, G.E. McGILL, R.J. PHILLIPS, R.S. SAUNDERS, G. SCHUBERT, S.W. SQUYRES, and E.R. STOFAN. 1992. Venus tectonics: an overview of Magellan observations. *Journal of Geophysical Research* **97**(E8): 13199–13255.

Figure 10.7 A listing of the image display program IMAGIN.

```fortran
      PROGRAM IMAGIN
C
C
C     THIS PROGRAM DISPLAYS AN ENTIRE IMAGE IF THE SIZE IS
C     EQUAL TO OR LESS THAN 400 ROWS BY 400 PIXELS PER ROW.
C     OR, IF LARGER, THIS PROGRAM DISPLAYS THE UPPER, LEFT
C     400 X 400 REGION OF THE IMAGE.  IMAGE DATA ARE ASSUMED
C     TO BE STORED IN A DATA FILE AS FOLLOWS:
C
C        1.  NO HEADER RECORD(S) PRECEED IMAGE INFORMATION
C        2.  DATA ARE WRITTEN ROW BY ROW
C        3.  DATA ARE WRITTEN IN BINARY AS 8-BIT INTEGERS
C
C     IMAGIN DISPLAYS THE IMAGE USING A 16 COLOR PALETTE
C     DESIGNED USING THE PROGRAM, PALDSGN (CHAPTER 7).  A COMMON
C     CHOICE FOR DISPLAYING IMAGES IS A 16 GREY LEVEL DISPLAY
C     HENCE, USE THE PALETTE AS IS SHOWN IN TABLE 7.1.
C
C******************************************************************
C                         USER'S GUIDE
C******************************************************************
C
C     THIS PROGRAM REQUIRES A PALETTE OF 16 COLORS BE DESIGNED
C     USING THE PROGRAM, PALDSGN (CHAPTER 7), STORED TO A FILE
C     BY THAT PROGRAM, AND ENTERED INTO THIS PROGRAM (IMAGIN).
C
C     THERE ARE ONLY A FEW INTERACTIVE QUESTIONS TO ANSWER WHEN
C     USING THIS PROGRAM:
C
C     QUESTION 1.  WHAT IS THE NAME OF THE INPUT FILE CONTAINING
C                  THE PALETTE (UNIT = 1)?
C
C     QUESTION 2.  WHAT IS THE NAME OF THE INPUT FILE CONTAINING
C                  THE DIGITAL IMAGE (UNIT = 5)?
C
C     QUESTION 3.  HOW MANY ROWS OF PIXELS COMPRISE THE IMAGE?
C     QUESTION 4.  HOW MANY PIXELS PER ROW COMPRISE THE IMAGE?
C     QUESTION 5.  WHAT IS THE TITLE FOR THE IMAGE DISPLAY, UP
C                  TO 50 CHARACTERS IN LENGTH?
C
C     AFTER THE 5TH QUESTION, THE IMAGE IS DISPLAYED.
C
C******************************************************************
C              END OF USER'S GUIDE AND DISCUSSION
C******************************************************************
C
      INCLUDE 'FGRAPH.FI'
      INCLUDE 'FGRAPH.FD'
      INTEGER*1 ITEMP(3000)
      INTEGER*2 DUMMY2, ITP
      INTEGER*4 DUMMY4, IR, IB, IG, IBUF(400), RGB, TMP
      CHARACTER*50 TITLE
      RECORD /RCCOORD/ CURPOS
      RECORD /XYCOORD/ XY
C
C     ENTER THE FILE NAMES FOR THE PALETTE AND IMAGE
C
      WRITE(*,*) 'ENTER FILENAME FOR PALETTE, UNIT 1'
      OPEN(1, FILE = ' ')
      WRITE(*,*)
      WRITE(*,*) 'ENTER FILENAME FOR IMAGE, UNIT 5'
      OPEN(5, FILE = ' ', FORM = 'BINARY')
C
C     ACQUIRE ADDITIONAL INPUT
C
      WRITE(*,*)
      WRITE(*,*) 'HOW MANY ROWS COMPRISE THE IMAGE?'
      READ(*,*) NROW
      WRITE(*,*)
      WRITE(*,*) 'HOW MANY PIXELS ARE THERE PER ROW?'
      READ(*,*) NCOL
      WRITE(*,*)
      WRITE(*,*) 'ENTER THE TITLE FOR THIS IMAGE, 50 CHAR. MAX.'
      READ(*,10) TITLE
   10 FORMAT(1A50)
C
```

469

Figure 10.7 continued

```fortran
C
C     READ THE INFORMATION FOR THE PALETTE AND DESIGN THE COLORS
C
      DUMMY2 = SETVIDEOMODE($VRES16COLOR)
      DO I = 1,16
        READ(1,*) IR, IG, IB
        TMP = RGB(IR, IG, IB)
        DUMMY4 = REMAPPALETTE (I-1, TMP)
      END DO
C
C     ATTENTION:   COLOR 15 IS USED TO DRAW A BORDER AROUND THE
C                  IMAGE, REGARDLESS OF THE PALETTE
C
C     DRAW A BORDER AROUND THE IMAGE
C
      CALL CLEARSCREEN($GCLEARSCREEN)
      CALL SETVIEWPORT(1, 1, 640, 480)
      DUMMY2 = SETCOLOR(15)
      CALL MOVETO(190, 30, XY)
      DUMMY2 = LINETO(610,30)
      DUMMY2 = LINETO(610,450)
      DUMMY2 = LINETO(190,450)
      DUMMY2 = LINETO(190,30)
C
C     READ THE IMAGE DATA AND DISPLAY, ROW BY ROW, IN ONE LOOP
C
      IF (NROW .GT. 400) NROW = 400
      DO I = 1, NROW
        DO K = 1, 3000
          ITEMP(K) = 0
        END DO
        READ(5) (ITEMP(JK), JK = 1,NCOL)
        DO J = 1, 400
          ITP = ITEMP(J)
          IF (ITP .GE. 0) THEN
            IBUF(J) = ITP
          ELSE
            IBUF(J) = 256 + ITP
          END IF
        END DO

C
C     DISPLAY
      DO J = 1, 400
        K = (IBUF(J) / 16) - 1
        IF (K .LT. 0) K = 0
        IF (K .GT. 15) K = 15
        DUMMY2 = SETCOLOR(K)
        DUMMY2 = SETPIXEL(J+200, I+40)
      END DO
      END DO
C
C     PRINT THE TITLE
C
      KCOL = 27 + (50 - LEN TRIM(TITLE))/ 2
      CALL SETTEXTPOSITION (1, KCOL, CURPOS)
      CALL OUTTEXT(TITLE)
C
C     PUSH THE ENTER KEY TO TERMINATE THE PROGRAM
C
      READ(*,*)
      DUMMY2 = SETVIDEOMODE($DEFAULTMODE)
      STOP
      END
C
C     RGB FUNCTION
C
      INTEGER*4 FUNCTION RGB(R, G, B)
      INTEGER*4 R, G, B
      RGB = ISHL(ISHL(B,8) .OR. G,8) .OR. R
      RETURN
      END
```

Figure 10.14 A listing of the program IMSTAT for calculating histograms of 8-bit, binary digital images.

```
      PROGRAM IMSTAT
C
C**********************************************
C            USER'S GUIDE FOR IMSTAT
C**********************************************
C
C     IMSTAT IS A PROGRAM TO PERFORM A STATISTICAL ANALYSIS
C     ON DIGITAL IMAGES
C
C     YOU WILL BE PROMPTED TO ENTER THE NAME OF THE FILE
C     WHICH CONTAINS THE DIGITAL IMAGE
C
C     THE PROGRAM THEN ASKS YOU SEVERAL QUESTIONS:
C
C     1.  HOW MANY ROWS COMPRISE THE IMAGE?
C     2.  HOW MANY PIXELS PER ROW?
C
C**********************************************
C            END OF INPUT GUIDE
C**********************************************
C
      COMMON /BIN/ NBIN, IBIN(32), XMIN, XMAX, ASTAT(4)
      INTEGER*1 Z(3000)
      INTEGER IBIN, JBIN(256)
      NBIN = 32
      XMIN = 0.0
      XMAX = 256.0
C
C     ACCESS INPUT DATA
C
      OPEN(5, FILE = ' ', FORM = 'BINARY')
      WRITE(*,*) 'HOW MANY ROWS COMPRISE THE IMAGE?'
      READ(*,*) NROW
      WRITE(*,*) 'HOW MANY PIXELS ARE THERE PER ROW?'
      READ(*,*) NCOL
C
C
C     CALCULATE THE HISTOGRAM
C
      NDAT = NROW * NCOL
      DO I = 1, 256
        JBIN(I) = 0
      END DO
      DO I = 1, NROW
        READ(5) (Z(JK), JK = 1,NCOL)
        DO J = 1, NCOL
          ITP = Z(J)
          IF (ITP .LT. 0) ITP = 256 + ITP
          JBIN(ITP+1) = JBIN(ITP+1) + 1
        END DO
      END DO
C
C     COMPRESS JBIN, WITH 256 BINS, DOWN TO IBIN, WITH 32 BINS
C
      DO I = 1,32
        IBIN(I) = 0
      END DO
      DO I = 1,32
        DO J = 1, 8
          K = (I-1) * 8 + J
          IBIN(I) = IBIN(I) + JBIN(K)
        END DO
      END DO
      SUM = 0.0
      SUM2 = 0.0
      SUM3 = 0.0
      SUM4 = 0.0
      DO I = 1, 256
        SUM = SUM + FLOAT(I) * FLOAT(JBIN(I))
        SUM2=SUM2 + (FLOAT(I) * FLOAT(JBIN(I)))**2
        SUM3=SUM3 + (FLOAT(I) * FLOAT(JBIN(I)))**3
        SUM4=SUM4 + (FLOAT(I) * FLOAT(JBIN(I)))**4
      END DO
```

Figure 10.14 continued

```fortran
      ASTAT(1) = SUM / FLOAT(NDAT)
      ASTAT(2) = (FLOAT(NDAT)*SUM2-SUM*SUM)/(FLOAT(NDAT)*
     2           FLOAT(NDAT - 1))
      ASTAT(3) = (SUM3/FLOAT(NDAT))-3.*ASTAT(1)*SUM2/FLOAT(NDAT)
     2           + 3.0 * (ASTAT(1) * ASTAT(1)) * SUM/FLOAT(NDAT) -
     3           ASTAT(1) * ASTAT(1) * ASTAT(1)
      ASTAT(4) = (SUM4 / FLOAT(NDAT)) - 4.0 * ASTAT(1) * SUM3
     2           /FLOAT(NDAT)+6.0*ASTAT(1) * ASTAT(1) * SUM2/FLOAT(NDAT)
     3           -4.0 * ASTAT(1) * ASTAT(1) * ASTAT(1) * SUM/FLOAT(NDAT) +
     4           ASTAT(1) * ASTAT(1) * ASTAT(1) * ASTAT(1)
C
C     PLOT HISTOGRAM
C
      CALL PLOT
C
C     TERMINATE PROGRAM
C
      STOP
      END
C
C
      SUBROUTINE PLOT
      COMMON /BIN/ NBIN, IBIN(32), XMIN, XMAX, ASTAT(4)
      INTEGER IBIN
      CHARACTER*1 ASTER
      ASTER = '*'
      WRITE(*,10)
 10   FORMAT(1H1, 10X, 'HISTOGRAM RESULTS',/,
     2 12X,'INTERVAL',5X,'BIN #',6X,'FREQ',/)
      AINT = (XMAX - XMIN) / FLOAT(NBIN)
      IBINMAX = IBIN(1)
      DO 15 I = 2, NBIN
      IF (IBIN(I) .GT. IBINMAX) IBINMAX = IBIN(I)
 15   CONTINUE
      DO 20 I = 1,NBIN
      W = XMIN + AINT * FLOAT(I-1)
      V = W + AINT
      IF (IBIN(I) .EQ. 0) THEN
         WRITE(*,30) W,V,I,IBIN(I)
      ELSE
         KBIN = INT(50 * IBIN(I) / IBINMAX) + 1
         WRITE(*,40) W,V,I,IBIN(I),(ASTER,J=1,KBIN)
      ENDIF
 20   CONTINUE
 30   FORMAT(2F10.3,I10,I10)
 40   FORMAT(2F10.3,I10,I10,1X,92A1)
      WRITE(*,50) (ASTAT(I), I = 1,3)
 50   FORMAT(1H0, 9X,'MEAN        = ',1E15.5,/,
     2           10X,'VARIANCE    = ',1E15.5,/,
     3           10X,'SKEW        = ',1E15.5)
      RETURN
      END
```

472

Figure 10.28 A listing of the program CTRAST.

```
      PROGRAM CTRAST
C
C     A PROGRAM TO ENHANCE THE CONTRAST OF A DIGITAL IMAGE
C     USING A LOOKUP TABLE TO SAVE CALCULATIONS
C
C     ***************************************************
C                     USER'S GUIDE
C     ***************************************************
C
C     THIS PROGRAM READS AN INPUT IMAGE, ENHANCES THE CONTRAST,
C     THEN CREATES AN OUTPUT FILE CONTAINING THE ENHANCED IMAGE
C     FOR DISPLAY BY THE PROGRAM, IMAGIN.
C
C     INTERACTIVE DATA ENTRY IS REQUIRED AS FOLLOWS:
C
C     1.  ENTER THE FILENAME WHICH STORES THE IMAGE TO BE ENHANCED
C         AND THE NAME OF THE OUTPUT FILE STORING THE ENHANCED IMAGE
C         ALSO, ENTER THE NUMBER OF ROWS, AND TOTAL PIXELS PER ROW,
C         COMPRISING THE IMAGE.
C
C     2.  SELECT THE TYPE OF CONTRAST ENHANCEMENT:
C
C         A.  SIMPLE, LINEAR; REQUIRES THE SPECIFICATION OF
C             MIN AND MAX WHERE ALL IMAGE PIXEL VALUES LESS THAN
C             MIN ARE SET TO ZERO, ALL PIXEL VALUES GREATER THAN
C             MAX ARE SET TO 255, AND ALL PIXEL VALUES BETWEEN
C             MIN AND MAX ARE MAPPED TO THE INTERVAL:  0,255.
C
C         B.  BI-LINEAR; REQUIRES THE SPECIFICATION OF MIN, MID,
C             AND MAX; WHERE ALL IMAGE PIXEL VALUES LESS THAN
C             MIN ARE SET TO ZERO, ALL IMAGE PIXEL VALUES GREATER
C             THAN MAX ARE SET TO 255, PIXELS IN THE RANGE,
C             MIN TO MID ARE MAPPED TO THE INTERVAL:  0,127;
C             PIXELS IN THE RANGE, MID TO MAX, ARE MAPPED TO THE
C             INTERVAL:  127,255.
C
C         C.  HISTOGRAM EQUILIZATION:  REQUIRES THE SPECIFICATION
C             OF THE NUMBER OF BINS USED TO COMPUTE THE HISTOGRAM
C             FOR THE IMAGE TO BE ENHANCED (USING THE PROGRAM,
C             IMSTAT) AND THE CUMULATIVE RELATIVE FREQENCY FOR
C             EACH BIN, ENTERED AS A NUMBER IN THE RANGE:  0,1.
C             THIS OPTION ALSO REQUIRES THE SPECIFICATION OF THE
C             MINIMUM (MIN) AND MAXIMUM (MAX) PIXEL VALUES FOR
C             THE STRETCH
C
C         D.  OR, IN LIEU OF THE OPTIONS ABOVE, A THROUGH C, YOU
C             MAY SELECT THIS OPTION, D, TO ENTER YOUR OWN LOOK-UP
C             TABLE; IN SELECTING THIS OPTION, YOU MUST FIRST
C             CREATE AN INPUT FILE (CALLED UNIT 3 HEREIN) WHICH
C             CONTAINS THE 256 INTEGER VALUES OF THE LOOK-UP TABLE
C             CREATE THIS FILE SUCH THAT ONE INTEGER VALUE OCCURS ON
C             EACH LINE OF THE FILE, HENCE THE FILE WILL BE 256
C             LINES LONG; ENTER THESE 256 VALUES USING FREE-FORMAT
C
C     TO ENABLE THE DETERMINATION OF INPUT REQUIRED FOR
C     2A - 2C, THE PROGRAM IMSTAT IS RUN TO OBTAIN A HISTOGRAM
C     OF THE IMAGE TO BE ENHANCED.
C
C     ***************************************************
C                END OF USER'S GUIDE AND DISCUSSION
C     ***************************************************
C
      DIMENSION BIN(100)
      INTEGER*1 IBUF(3000), JBUF(3000)
      INTEGER*4 ITP, LOOK(256), MIN, MID, MAX
C
```

Figure 10.28 continued

```fortran
C     ACCESS IMAGE DATA
C

      WRITE(*,*) 'ENTER THE NAME OF THE INPUT IMAGE FILE, UNIT 5'
      OPEN(5, FILE = ' ', FORM = 'BINARY')
      WRITE(*,*)
      WRITE(*,*) 'ENTER THE NAME OF THE OUTPUT IMAGE FILE, UNIT 1'
      OPEN(1, FILE = ' ', FORM = 'BINARY', STATUS = 'NEW')
      WRITE(*,*)
      WRITE(*,*) 'HOW MANY ROWS COMPRISE THE IMAGE TO BE ENHANCED?'
      READ(*,*) NROW
      WRITE(*,*)
      WRITE(*,*) 'HOW MANY PIXELS ARE THERE PER ROW IN THE IMAGE?'
      READ(*,*) NCOL
      WRITE(*,*)
      WRITE(*,*) 'WHAT TYPE OF CONTRAST ENHANCEMENT DO YOU DESIRE?'
      WRITE(*,*)
      WRITE(*,*) 'ENTER:          1, FOR SIMPLE, LINEAR'
      WRITE(*,*) '                2, FOR BI-LINEAR'
      WRITE(*,*) '                3, FOR HISTOGRAM EQUILIZATION'
      WRITE(*,*) '                4, FOR USER-DEFINED LOOK-UP TABLE'
      WRITE(*,*) 'PLEASE ENTER YOUR SELECTION NOW:'
      READ(*,*) ISELCT
      IF (ISELCT .EQ. 1) THEN
      WRITE(*,*)
      WRITE(*,*) 'YOU SELECTED A SIMPLE, LINEAR ENHANCEMENT.'
      WRITE(*,*) 'THERFORE, PLEASE ENTER TWO VALUES:  MIN, MAX:'
      READ(*,*) MIN, MAX
      ELSEIF (ISELCT .EQ. 2) THEN
      WRITE(*,*)
      WRITE(*,*) 'YOU HAVE SELECTED A BI-LINEAR ENHANCEMENT.'
      WRITE(*,*) 'THERFORE, PLEASE ENTER 3 VALUES:  MIN,MID,MAX:'
      READ(*,*) MIN, MID, MAX
      ELSEIF (ISELCT .EQ. 3) THEN
      WRITE(*,*)
      WRITE(*,*) 'YOU HAVE SELECTED A HISTOGRAM EQUILIZATION'
      WRITE(*,*) 'THERFORE, PLEASE ENTER NO. OF BINS IN YOUR'
      WRITE(*,*) 'HISTOGRAM, THEN ENTER CUM. REL. FREQ. FOR'
      WRITE(*,*) 'EACH BIN, ONE AT A TIME:'
      WRITE(*,*)
      WRITE(*,*) 'ENTER NO. OF BINS NOW:'
      READ(*,*) NBIN
      DO I = 1, NBIN
      WRITE(*,*) 'ENTER CUM. REL. FREQ FOR BIN ',I
      READ(*,*) BIN(I)
      END DO
      WRITE(*,*) 'ENTER THE MINIMUM (MIN) AND MAXIMUM'
      WRITE(*,*) 'PIXEL RANGE FOR THE HISTO EQ'
      WRITE(*,*) 'ENTER MIN,MAX NOW:'
      READ(*,*) MIN, MAX
      ELSEIF (ISELCT .EQ. 4) THEN
      WRITE(*,*) 'ENTER THE FILENAME STORING THE LOOK-UP TABLE'
      OPEN(3, FILE = ' ')
      DO I = 1, 256
      READ(3,*) LOOK(I)
      END DO
      END IF
C
```

Figure 10.28 continued

```fortran
C      COMPUTE THE 256 LOOKUP TABLES NOW
C
       IF (ISELCT .EQ. 1) THEN
          LOOK(1) = 0
          DO I = 2, MIN+1
             LOOK(I) = 0
          END DO
          DO I = MAX, 256
             LOOK(I) = 255
          END DO
          IDIF = MAX - MIN
          DO I = MIN+1, MAX-1
             LOOK(I) = INT(255*(I-MIN)/IDIF)
             IF (LOOK(I) .GT. 255) LOOK(I) = 255
          END DO
       ELSEIF (ISELCT .EQ. 2) THEN
          IDIF1 = MID - MIN
          IDIF2 = MAX - MID
          LOOK(1) = 0
          DO I = 2, MIN+1
             LOOK(I) = 0
          END DO
          DO I = MAX, 256
             LOOK(I) = 255
          END DO
          LOOK(MID) = 127
          DO I = MIN+1, MID-1
             LOOK(I) = INT(127 * (I - MIN) / IDIF1)
          END DO
          DO I = MID+1, MAX-1
             LOOK(I) = INT(128*(I-MID)/IDIF2) + 127
             IF (LOOK(I) .GT. 255) LOOK(I) = 255
          END DO
       ELSEIF (ISELCT .EQ. 3) THEN
          LOOK(1) = 0
          DO I = 2, MIN+1
             LOOK(I) = 0
          END DO
          DO I = MAX, 256
             LOOK(I) = 255
          END DO
          IDIF = MAX - MIN
          DO I = MIN+1, MAX-1
             J = INT(NBIN*(I-MIN)/IDIF) + 1
             IF (J .LT. 1) J = 1
             IF (J .GT. NBIN) J = NBIN
             LOOK(I) = INT(255 * BIN(J))
          END DO
       END IF
C
C      ENHANCE THE IMAGE; CREATE THE OUTPUT FILE
C
       DO I = 1, NROW
          READ(5) (IBUF(JK), JK=1,NCOL)
          DO J = 1, NCOL
             ITP = IBUF(J)
             IF (ITP .LT. 0) ITP = 256 + ITP
             JBUF(J) = INT1(LOOK(ITP+1))
          END DO
          WRITE(1) (JBUF(JK), JK = 1,NCOL)
       END DO
       STOP
       END
```

Figure 10.30 A listing of the program CONVOL.

```fortran
      PROGRAM CONVOL

C
C     A PROGRAM TO PERFORM SPATIAL CONVOLUTION FILTERING
C     USING SQAURE BOX FILTERS OF ODD NUMBER SIZES:  3, 5, 7 ...
C
C     ********************************************************
C                          USER'S GUIDE
C     ********************************************************
C
C     THIS PROGRAM REQUIRES TWO INPUT FILES, ONE WHICH CONTAINS
C     THE SIZE OF THE FILTER, THE BOX SIZE, FOLLOWED BY THE WEIGHTS,
C     AND THE OTHER CONTAINS THE DIGITAL IMAGE.
C
C     USER'S GUIDE FOR PREPARING THE FILE CONTAINING THE FILTER:
C
C     LINE 1:  READ(5,*) NSIZE
C              WHERE:
C                     NSIZE IS THE FILTER SIZE; ENTER 3, IF 3X3;
C                     5, IF 5X5, 7 IF 7X7; AND SO ON.
C
C     LINES 2 TO NSIZE+1:  THE FILTER WEIGHTS:
C              DO I = 1, NSIZE
C                 READ(5,*) BOX(I,JK), JK = 1, NSIZE
C              END DO
C
C     USER'S GUIDE FOR IMAGE FILE:  THIS IS THE SAME IMAGE FILE
C     AS IS DISPLAYED BY THE PROGRAM, IMAGIN, THEREFORE, NO
C     SPECIAL USER INFORMATION IS REQUIRED FOR THIS
C
C     USER'S GUIDE FOR CONVOL:
C
C     STEP 1:  ENTER THE NAME OF THE IMAGE FILE
C     STEP 2:  ENTER THE NAME OF THE OUTPUT FILTERED FILE
C     STEP 3:  ENTER THE NAME OF THE FILE CONTAINING THE FILTER
C     STEP 4:  ENTER THE NUMBER OF ROWS, AND PIXELS PER ROW, FOR
C              THE IMAGE TO BE FILTERED
C
C     *************************************************************
C                    END OF USER'S GUIDE AND DISCUSSION
C     *************************************************************
C
      DIMENSION BOX(10,10),  SUMCOL(10)
      INTEGER*1 ITEMP(3000)
      INTEGER KTEMP(3000)
      INTEGER*4 IBUF(10,3000)
      WRITE(*,*) 'ENTER THE NAME OF THE INPUT IMAGE FILE'
      OPEN(5, FILE = ' ', FORM = 'BINARY')
      WRITE(*,*) 'ENTER THE NAME OF THE OUTPUT, FILTERED IMAGE FILE'
      OPEN(1, FILE = ' ', FORM = 'BINARY', STATUS = 'NEW')
      WRITE(*,*) 'ENTER NAME OF FILE WHICH CONTAINS THE FILTER'
      OPEN(2, FILE = ' ')
      READ(2,*) NSIZE
      DO I = 1, NSIZE
         READ(2,*) (BOX(I,JK), JK = 1,NSIZE)
      END DO
      WRITE(*,*) 'HOW MANY ROWS COMPRISE THE INPUT IMAGE?'
      READ(*,*) NROW
      WRITE(*,*) 'HOW MANY PIXELS PER ROW COMPRISE THE IMAGE?'
      READ(*,*) NCOL
C
```

Figure 10.30 continued

```
C       READ NSIZE LINES INITIALLY INTO IBUF
C
        DO I = 1, NSIZE
          READ(5) (ITEMP(JK), JK=1,NCOL)
          DO J = 1, NCOL
            ITP = ITEMP(J)
            IF (ITP .LT. 0) ITP = 256 + ITP
            IBUF(I,J) = ITP
          END DO
        END DO
C
C       FILTER THE DATA
C
        LN = (NSIZE-1) / 2
        DO I = 1, NROW-NSIZE
          JSWIT = 0
          DO K = LN, NCOL-LN
            KN = K - LN + 1
            SUM = 0.0
            IF (JSWIT .EQ. 0) THEN
              DO LINIT = 1, NSIZE
                SUMCOL(LINIT) = 0.0
              END DO
              DO L = 1,NSIZE
                DO J = 1,5
                  SUMCOL(L) = SUMCOL(L) + IBUF(J, KN+L-1) * BOX(J,L)
                END DO
              END DO
            ELSE
              SUMCOL(NSIZE) = 0.0
              DO J = 1, NSIZE
                SUMCOL(NSIZE) = SUMCOL(NSIZE) + IBUF(J, KN+NSIZE-1) *
     2                          BOX(J,NSIZE)
              END DO
            END IF

            DO L = 1, NSIZE
              SUM = SUM + SUMCOL(L)
            END DO
            KTEMP(K) = INT(SUM)
            DO L = 2, NSIZE
              SUMCOL(L-1) = SUMCOL(L)
            END DO
          END DO
          WRITE(1) (KTEMP(JK), JK = 1,NCOL-NSIZE)
          DO J = 2,NSIZE
            DO K = 1,NCOL
              IBUF(J-1,K) = IBUF(J,K)
            END DO
          END DO
          READ(5) (ITEMP(JK), JK = 1,NCOL)
          DO LK = 1, NCOL
            ITP = ITEMP(LK)
            IF (ITP .LT. 0) ITP = 256 + ITP
            IBUF(NSIZE,LK) = ITP
          END DO
        END DO
        STOP
        END
```

Figure 10.32 A listing of the program IMAG41.

```fortran
      PROGRAM IMAG41
C
C     A PROGRAM TO BE USED TO CREATE 8-BIT DIGITAL IMAGES.
C     THIS PROGRAM READS AN OUTPUT, 32 BIT INTEGER IMAGE FILE
C     AND REWRITES/RESCALES THE IMAGE TO 8-BIT INTEGER DATA.
C
C*******************************************************************
C                    USER'S GUIDE:     IMAG41
C*******************************************************************
C
C     1.  SPECIFY THE INPUT IMAGE FILENAME; THIS IS A 32-BIT
C         INTEGER FILE
C
C     2.  SPECIFY THE OUTPUT IMAGE FILENAME; THIS IS THE ONE TO
C         BE DISPLAYED BY IMAGIN
C
C     3.  SPECIFY THE NUMBER OF ROWS AND COLUMNS COMPOSING THE
C         INPUT IMAGE; THE OUTPUT IMAGE WILL BE OF THE SAME SIZE
C
C*******************************************************************
C                    END OF USER'S GUIDE
C*******************************************************************
C
      INTEGER JBUF(3000)
      INTEGER*1 IBUF(3000)
C
C     ACCESS IMAGE DATA; SCAN COMPLETE IMAGE TO FIND
C     MAXIMA AND MINIMA
C
      WRITE(*,*) 'ENTER NAME OF INPUT IMAGE'
      OPEN(5, FILE = ' ', FORM = 'BINARY')
      WRITE(*,*) 'HOW MANY ROWS COMPRISE THIS IMAGE?'
      READ(*,*) NROW
      WRITE(*,*) 'HOW MANY PIXELS PER ROW ARE THERE?'
      READ(*,*) NCOL
      MINPIX = 1000000
      MAXPIX = -1000000
      DO I = 1, NROW
      READ(5) (JBUF(JK), JK = 1, NCOL)
      DO J = 1, NCOL
      IF (JBUF(J) .LT. MINPIX) THEN
          MINPIX = JBUF(J)
      END IF
      IF (JBUF(J) .GT. MAXPIX) THEN
          MAXPIX = JBUF(J)
      END IF
      END DO
      END DO
      WRITE(*,*) MINPIX, MAXPIX
C
C
      REWIND UNIT 5; REWRITE TO UNIT 1 AS 8 BIT INTEGERS
C
      WRITE(*,*) 'ENTER NAME OF OUTPUT FILE TO BE CREATED'
      OPEN(1, FILE = ' ', FORM = 'BINARY', STATUS = 'NEW')
      REWIND 5
      DO I = 1, NROW
      READ(5) (JBUF(JK), JK = 1, NCOL)
      DO J = 1, NCOL
      XTEMP = FLOAT(JBUF(J)) - FLOAT(MINPIX)
      YTEMP = FLOAT(MAXPIX) - FLOAT(MINPIX)
      KTEMP = INT4(255. * XTEMP / YTEMP)
      IBUF(J) = INT1(KTEMP)
      END DO
      WRITE(1) (IBUF(JK), JK = 1, NCOL)
      END DO
      STOP
      END
```

Figure 10.37 A listing of the program FORIER2D.

```fortran
      PROGRAM FOR2D

C     A FORTRAN-77 PROGRAM TO COMPUTE THE FOURIER
C     TRANSFORM OF A TWO DIMENSIONAL DIGITAL IMAGE
C
C
*********************************************************
C              USER'S GUIDE:    FORIER2D
C
*********************************************************
C
C     THE INPUT DATA FILE FOR THIS PROGRAM IS A 2D DIGITAL IMAGE
C     STORED AS 8-BIT INTEGERS; THE IMAGE MUST BE NO LARGER THAN
C     400 ROWS BY 400 PIXELS PER ROW.
C
C     WHEN PROMPTED, ENTER THE FILE NAME WHICH STORES THE IMAGE
C     THIS INPUT FILE IS DESIGNATED AS UNIT 5.
C
C     THIS PROGRAM CREATES ONE OUTPUT FILE (CALLED AN "RI" FILE)
C     ON UNIT 1, THE REAL AND IMAGINARY PARTS OF THE FOURIER
C     TRANSFORM ARE STORED TO ENABLE FOURIER FILTERING;
C
*********************************************************
C          END OF USER'S GUIDE AND DISCUSSION
C
*********************************************************
C
      IMPLICIT COMPLEX (A-H, O-Z)
      INTEGER*1 BUF1(400,400)
      INTEGER*4 ITB
      REAL A, PI, XBETA, YBETA
      DIMENSION  BETA(400)
      PI = 4.0 * ATAN(1.0)
      EYE = (0., 1.0)
C     EYE = SQUARE ROOT OF -1
      CONS = 2 * PI * EYE
      A = 0.0
C
C     ACCESS THE TIME SERIES DATA, Y
C
      WRITE(*,*) 'ENTER NAME OF INPUT FILE, UNIT=5'
      OPEN (5, FILE = ' ', FORM = 'BINARY')
      WRITE(*,*) 'ENTER NAME OF OUTPUT FILE, UNIT=1'
      OPEN (1, FILE = ' ', STATUS = 'NEW')
      WRITE(*,*) 'HOW MANY ROWS COMPOSE THE INPUT IMAGE?'
      WRITE(*,*) 'THE NUMBER OF ROWS MUST EQUAL NO. OF COLUMNS'
      WRITE(*,*) 'ENTER NUMBER OF ROWS NOW:'
      READ (*,*) N
      DO I = 1,N
         READ(5) (BUF1(I,JK), JK = 1,N)
      END DO
      N = 100
      CONS = CONS / FLOAT(N)
      CONS2 = FLOAT(N) / FLOAT(N)
C
C     COMPUTE THE INVERSE TRANSFORM
C
      N2 = N
      DO I = 1,N
         WRITE (*,*) 'ROW = ', I
         DO J = 1,N
            WRITE(*,*) 'COLUMN = ', J
            BETA( J) = (0.,0.)
            DO K = 1,N
               X1 = (K-1) * CONS2
               SUML = (0., 0.)
               DO L = 1,N
                  X2 = A + (L-1) * CONS2
                  ITB = INT4(BUF1(K,L))
                  IF (ITB .LT. 0) ITB = 256 + ITB
                  XBETA = FLOAT(ITB)
                  YBETA = 0.0
                  XTEMP = CMPLX(XBETA,YBETA)
                  SUML = SUML + XTEMP  * CEXP( CONS* (J-1)*X2)
               END DO
               BETA( J) = BETA( J) + SUML*CEXP( CONS*(I-1)*X1)
            END DO
         END DO
         WRITE(1,*) (REAL(BETA(JK)),AIMAG(BETA(JK)),JK=1,N)
      END DO
      STOP
      END
```

479

Figure 10.38 A listing of the program READRI.

```fortran
      PROGRAM RDRI
C
C     PROGRAM TO READ OUTPUT (UNIT 1) FILES FROM FORIER2D
C
C     *******************************************************
C                    USER'S GUIDE:  READRI
C     *******************************************************
C
C     PROGRAM READRI IS DESIGNED FOR USE WITH FORIER2D; THE OUTPUT
C     FROM FORIER2D IS AN ASCII FILE STORING REAL AND IMAGINARY
C     PARTS OF THE 2D FOURIER TRANSFORM OF A 2D IMAGE; EACH LINE
C     IN THE RI FILE STORES ONE REAL AND ONE IMAGINARY VALUE (TWO
C     TOTAL VALUES) FOR EACH PIXEL IN THE INPUT IMAGE.
C
C     PURPOSE OF READRI:  TO CREATE A 32 BIT INTEGER DIGITAL IMAGE
C     OF THE FOURIER TRANSFORM RESULTS; EACH PIXEL IN THE NEWLY
C     CREATED IMAGE IS ACTUALLY EQUAL TO 1 + LOG(FT), WHERE FT
C     IS THE FOURIER TRANSFORM VALUE.
C
C     NOTICE:  THE PROGRAM, IMAG41, MUST BE USED AFTER READRI TO
C              CREATE AN 8-BIT INTEGER IMAGE, COMPATIBLE WITH
C              IMAGIN, FROM THE 32-BIT PIXEL IMAGE.
C
C     INPUT TO READRI:  THE OUTPUT FILE FROM FORIER2D
C                       AFTER THE INPUT FILE NAME IS ENTERED
C                       THE PROGRAM REQUESTS THE NAME OF THE
C                       OUTPUT FILE (A NEW FILE NAME IS REQUIRED);
C                       THEN, THE PROGRAM REQUESTS THE SIZE OF THE
C                       IMAGE, N; E.G., IF THE IMAGE IS 100 X 100,
C                       THEN ENTER ONLY THE ONE NUMBER, 100.
C                       NOTE:  PROGRAM ASSUMES SQUARE IMAGES
C
C     OUTPUT FROM READRI:  A FILE STORING THE 32-BIT IMAGE
C
C     *******************************************************
C                 END OF USER'S GUIDE:  READRI
C     *******************************************************
C
      DIMENSION XR(3000), XI(3000)
      INTEGER*4 BUF(100,100)
      WRITE(*,*) 'ENTER NAME OF INPUT FILE (I.E., OUTPUT FILE'
      WRITE(*,*) 'CREATED BY THE PROGRAM:  FORIER2D'
      WRITE(*,*)
      OPEN(5, FILE = ' ')
      WRITE(*,*) 'ENTER NAME OF OUTPUT FILE:  MUST BE A NEW NAME'
      WRITE(*,*)
      OPEN(1, FILE = ' ', FORM = 'BINARY', STATUS = 'NEW')
      WRITE(*,*) 'ENTER NUMBER OF ROWS,N ALSO = NO COLS'
      READ(*,*) N
      K = N/2
      DO I = 1, N
      READ(5,*) (XR(JK), XI(JK), JK = 1, N)
      DO J = 1, N
      XTEMP = ALOG10(1.0 + ABS(XR(J)))
      IF (I.LE. K .AND. J.LE. K) THEN
      BUF( K    -I+1,  K  -J+1) = INT4(XTEMP*25.)
      ELSEIF (I.LE. K .AND. J.GT. K ) THEN
      BUF( K    -I+1, N -J + K + 1) = INT4(XTEMP*25.)
      ELSEIF (I.GT. K .AND. J.LE. K ) THEN
      BUF(N  - I + K + 1,  K   -J+1) = INT4(XTEMP*25.)
      ELSEIF (I.GT. K .AND. J.GT. K ) THEN
      BUF(N-I+K+1, N-J+K+1) = INT4(XTEMP*25.)
      END IF
      END DO
      END DO
      DO I = 1, N
      WRITE(1) (BUF(I,JK), JK = 1,N)
      END DO
      STOP
      END
```

480

Figure 10.39 A listing of the program FORFILT.

```
      PROGRAM FILTFOR
C
C     A PROGRAM TO FILTER RIR FILES
C
C     *********************************************************
C              USER'S GUIDE:   FILTFOR
C     *********************************************************
C
C     PROGRAM FILTFOR IS DESIGNED TO MODIFY THE RI FILES CREATED BY
C     FORIER2D; THE OUTPUT FILE IS WRITTEN IN A FORMAT WHICH IS
C     COMPATIBLE WITH THE PROGRAM, FORINV2.
C
C     INPUT:
C
C       1.   INPUT FILE NAME (THE RI FILE CREATED BY FORIER2D)
C       2.   OUTPUT FILE NAME (MUST BE A NEW NAME)
C       3.   SIZE OF THE ORIGINAL IMAGE WHICH WAS TRANSFORMED BY
C            FORIER2D.
C       4.   ENTER THE NUMBER OF REGIONS IN THE IMAGE OF THE
C            FOURIER TRANSFORM WHICH YOU WISH TO FILTER (SET TO
C            ZERO); THE NOISE YOU WISH TO FILTER WILL APPEAR
C            AS A BRIGHT REGION IN THE IMAGE OF THE FT; IT IS
C            THESE AREAS TO FILTER:
C       5.   FOR EACH REGION SELECTED IN NO. 4, SPECIFY THE
C            MINIMUM AND MAXIMUM ROW NUMBER AND MINIMUM AND
C            MAXIMUM COLUMN NUMBER OF THE REGION TO BE FILTERED
C
C     OUTPUT FILE:   A FILE WHOSE FORMAT IS COMPATIBLE WITH FORINV2
C
C     *********************************************************
C              END OF USER'S GUIDE:   FILTFOR
C     *********************************************************
C
      DIMENSION X(100,100), Y(100,100)
      WRITE(*,*) 'ENTER THE NAME OF THE INPUT FILE'
      WRITE(*,*)
      OPEN(5, FILE = ' ')
      WRITE(*,*) 'ENTER THE NAME OF THE OUTPUT FILE'
      WRITE(*,*)
      OPEN(1, FILE = ' ', STATUS = 'NEW')
      WRITE(*,*) 'ENTER THE SIZE, N, OF THE ORIGINAL IMAGE'
      READ (*,*) N
      DO I = 1, N
         READ(5,*) (X(I,J), Y(I,J), J = 1, N)
      END DO
      WRITE(*,*) 'HOW MANY REGIONS DO YOU WISH TO FILTER?'
      READ (*,*) N
      DO I = 1, N
         WRITE(*,*) 'ENTER MIN, MAX ROW NUMBERS FOR REGION', I
         READ (*,*) MINROW, MAXROW
         WRITE(*,*) 'ENTER MIN, MAX COL NUMBERS FOR REGION', I
         READ (*,*) MINCOL, MAXCOL
         DO J = MINROW, MAXROW
            DO K = MINCOL, MAXCOL
               X(J,K) = 0.0
               Y(J,K) = 0.0
            END DO
         END DO
      END DO
      DO I = 1, N
         DO J = 1, N
            WRITE(1,*) X(I,J), Y(I,J)
         END DO
      END DO
      STOP
      END
```

481

Figure 10.40 A listing of the program FORINV2.

```fortran
      PROGRAM FORINV2
C
C   A FORTRAN-77 PROGRAM TO COMPUTE THE INVERSE FOURIER
C   TRANSFORM OF A TWO DIMENSIONAL FUNCTION
C
C
************************************************************
C
C              USER'S GUIDE:   FORINV2
C
************************************************************
C
C   THE INPUT DATA FILE FOR THIS PROGRAM IS THE FOURIER TRANS
C   (UNIT 2) COMPUTED BY THE PROGRAM, FORIER2, OR SPFM2D
C
C   WHEN PROMPTED, ENTER THE OUTPUT FILE CREATED BY FORIER2
C   OR SPFM2D AS THE INPUT FILE, UNIT 5
C
C   THIS PROGRAM THEN CREATES AN OUTPUT FILE, UNIT 1, WHICH
C   IS THE INVERSE TRANSFORM; I.E., SHOULD BE THE ORIGINAL
C   DATA (APPROXIMATELY).
C
C
************************************************************
C
C              END OF USER'S GUIDE AND DISCUSSION
C
************************************************************
C
      IMPLICIT COMPLEX (A-H, O-Z)
      REAL A, PI
      DIMENSION Y(128,128), BETA(128,128)
      REAL YR(128), YI(128), XBETA
      PI = 4.0 * ATAN(1.0)
      EYE = (0., 1.0)
C     EYE = SQUARE ROOT OF -1
      CONS = 2 * PI * EYE
      A = 0.0
C
C   ACCESS THE TIME SERIES DATA, Y
C
      WRITE(*,*) 'ENTER NAME OF INPUT FILE, UNIT=5'
      OPEN (5, FILE = ' ')
      WRITE(*,*) 'ENTER NAME OF OUTPUT FILE, UNIT=1'
      OPEN(1, FILE = ' ', STATUS = 'NEW')
      READ(5,*) N

      DO I = 1,N
        READ(5,*) (YR(JK), YI(JK), JK = 1,N)
        DO J = 1,N
          Y(I,J) = CMPLX(YR(J), YI(J))
        END DO
      END DO
      CONS = CONS / FLOAT(N)
      CONS2 = FLOAT(N) / FLOAT(N)
C
C   COMPUTE THE INVERSE TRANSFORM
C
      N2 = N
      DO I = 1,N
      WRITE (*,*) 'ROW = ', I
      DO J = 1,N
      WRITE(*,*) 'COLUMN = ', J
        BETA(I,J) = (0.,0.)
      DO K = 1,N2
      X1 = (K-1) * CONS2
      SUML = (0., 0.)
      DO L = 1,N2
      X2 = A + (L-1) * CONS2
      SUML = SUML + Y(K,L) * CEXP(-CONS*(J-1)*X2)
      END DO
      BETA(I,J) = BETA(I,J) + SUML*CEXP(-CONS*(I-1)*X1)
      END DO
      END DO
      DO I = 1,N
      DO J = 1, N
      XBETA = REAL(BETA(I,J))
      IF (XBETA .LT. 0.0) THEN
        XBETA = 0.0
      END IF
      WRITE(1,*) FLOAT(I), FLOAT(J), XBETA
      END DO
      END DO
      STOP
      END
```

Figure 10.41 A listing of the program INTFAC.

```fortran
      PROGRAM INTFAC
C
C     A PROGRAM TO CONVERT THE OUTPUT FROM FORINV2 TO
C     DIGITAL IMAGES COMPATIBLE WITH THE DISPLAY PROGRAM,
C     IMAGIN (CHAPTER 10).
C
***************************************************
C                 USER'S GUIDE:   INTFAC
C
***************************************************
C
C     THE INPUT TO THIS PROGRAM IS PERFORMED INTERACTIVELY.
C
C     1.  SPECIFY THE SIZE, N, OF THE DIGITAL IMAGE TO BE CREATED
C
C     2.  SPECIFY THE NAME OF THE INPUT FILE; THIS FILE IS THAT
C         CREATED BY FORINV2.
C
C     3.  SPECIFY THE NAME OF THE OUTPUT FILE; THIS WILL CONTAIN
C         THE DIGITAL IMAGE, A FILE WHICH IS COMPATIBLE WITH
C         IMAGIN (CHAPTER 10).
C
C
***************************************************
C            END OF USER'S GUIDE:   INTFAC
C
***************************************************
C
      INTEGER*1 BUF (512)
C
C     INTERACTION TO DEFINE OPTIONS AND DATA FILES
C
      WRITE(*,*) 'ENTER THE SIZE, N, OF THE SIMULATION'
      READ (*,*) N
      WRITE(*,*)
      WRITE(*,*) ' ENTER THE NAME OF THE INPUT FILE'
      OPEN (5, FILE = ' ')
      WRITE(*,*) 'ENTER THE NAME OF THE OUTPUT FILE'
      OPEN(1, FILE = ' ', STATUS = 'NEW', FORM = 'BINARY')
C
C     SCAN INPUT FILE TO DETERMINE MIN AND MAX DATA VALUES
C
      READ (5,*) XI, XJ, XZ
      YMIN = XZ
      YMAX = XZ
10    CONTINUE
      READ (5,*,END=20) XI, XJ, XZ
      IF (XZ .LT. YMIN) YMIN = XZ
      IF (XZ .GT. YMAX) YMAX = XZ
      GO TO 10
20    CONTINUE
      REWIND 5
C
C     CONVERT SIMULATION VALUES TO 8-BIT INTEGERS AND CREATE IMAGE
C
      DO I = 1, N
      DO J = 1, N
      READ(5,*) XI, XJ, XZ
      XZN = 255.0 * ((XZ - YMIN) / (YMAX - YMIN))
      BUF(J) = INT1(XZN)
      END DO
      WRITE(1) (BUF(JK), JK = 1, N)
      END DO
C
C     TERMINATE PROGRAM
C
      STOP
      END
```

483

Figure 10.44 A listing of the program MXLIKE.

```fortran
      PROGRAM MXLIKE
C
C     THIS PROGRAM USES APPROXIMATELY THE ENTIRE PROGRAM, IMAGIN,
C     FOR IMAGE DISPLAY, THEN, THROUGH INTERACTION WITH THE USER,
C     ACCESSES PIXEL VALUES WITHIN TRAINING SITES TO PERFORM
C     A MAXIMUM LIKLIHOOD, SUPERVISED CLASSIFICATION OF A DIGITAL
C     IMAGE
C
C ******************************************************************
C                      USER'S GUIDE: MXLIKE
C ******************************************************************
C
C     THIS PROGRAM REQUIRES A PALETTE OF 16 COLORS BE DESIGNED
C     USING THE PROGRAM, PALDSGN (CHAPTER 7), STORED TO A FILE
C     BY THAT PROGRAM, AND ENTERED INTO THIS PROGRAM (IMAGIN).
C
C     THERE ARE ONLY A FEW INTERACTIVE QUESTIONS TO ANSWER WHEN
C     USING THIS PROGRAM:
C
C     QUESTION 1.  WHAT IS THE NAME OF THE INPUT FILE CONTAINING
C                  THE PALETTE (UNIT = 1)?
C
C     QUESTION 2.  WHAT IS THE NAME OF THE INPUT FILE CONTAINING
C                  THE DIGITAL IMAGE (UNIT = 5)?
C
C     QUESTION 3.  HOW MANY ROWS OF PIXELS COMPRISE THE IMAGE?
C     QUESTION 4.  HOW MANY PIXELS PER ROW COMPRISE THE IMAGE?
C     QUESTION 5.  WHAT IS THE TITLE FOR THE IMAGE DISPLAY, UP
C                  TO 50 CHARACTERS IN LENGTH?
C
C     NOTICE THAT THESE FIVE QUESTIONS ARE THOSE USED WITH
C     IMAGIN; NOW, SOME ADDITIONAL QUESTIONS MUST BE ANSWERED FOR
C     THE SUPERVISED CLASSIFICATION:
C
C     QUESTION 6:  HOW MANY TRAINING CLASSES WILL BE SELECTED?
C                  THERE ARE UP TO 10 ALLOWABLE BY THE PROGRAM
C
C     FOR EACH TRAINING CLASS:
C
C     QUESTION 7:  WHAT ARE THE ROW AND PIXEL NUMBERS OF THE
C                  UPPER, LEFT CORNER OF THE TRAINING SITE FOR
C                  THIS CLASS?  ENTER LIKE THIS: ROW, PIXEL
C                  FOR EXAMPLE, 100, 244
C
C     QUESTION 8:  WHAT ARE THE ROW AND PIXEL NUMBERS OF THE
C                  LOWER, RIGHT CORNER OF THE TRAINING SITE FOR
C                  THIS CLASS?  ENTER LIKE THIS: ROW, PIXEL
C
C     AT THIS POINT, THE PROGRAM DRAWS THIS TRAINING SITE ON
C     THE IMAGE FOR VERIFICATION
C
C     QUESTION 9:  WHAT THRESHOLD PROBABILITY DO YOU
C                  DESIRE FOR ALL CLASSES; IF PROB.
C                  THAT A PIXEL BELONGS TO A CLASS IS
C                  LESS THAN THRESH, THEN PIXEL IS NOT
C                  ASSIGNED TO THE CLASS
C
C     QUESTION 10: WHAT TYPE OF CLASSIFICATION DO YOU WANT:
C                  A.  MINIMUM DISTANCE TO MEAN
C                  B.  MAXIMUM LIKELIHOOD
C
C ******************************************************************
C               END OF USER'S GUIDE AND DISCUSSION
C ******************************************************************
C
      INCLUDE 'FGRAPH.FI'
      INCLUDE 'FGRAPH.FD'
      REAL HIST(10,256), XMEAN(10)
      INTEGER TRAIN(10,4), LOOK(256)
      INTEGER*1 ITEMP(3000)
      INTEGER*2 DUMMY2, ITP
      INTEGER*4 DUMMY4, IR, IB, IG, IBUF(400), RGB, TMP
      CHARACTER*50 TITLE
      RECORD /RCCORD/ CURPOS
      RECORD /XYCOORD/ XY
C
```

Figure 10.44 continued

```fortran
C
C      ENTER THE FILE NAMES FOR THE PALETTE AND IMAGE

       WRITE(*,*) 'ENTER FILENAME FOR PALETTE, UNIT 1'
       OPEN(1, FILE = ' ')
       WRITE(*,*)
       WRITE(*,*) 'ENTER FILENAME FOR IMAGE, UNIT 5'
       OPEN(5, FILE = ' ', FORM = 'BINARY')
C
C      ACQUIRE ADDITIONAL INPUT
C
       WRITE(*,*)
       WRITE(*,*) 'HOW MANY ROWS COMPRISE THE IMAGE?'
       READ(*,*) NROW
       WRITE(*,*)
       WRITE(*,*) 'HOW MANY PIXELS ARE THERE PER ROW?'
       READ(*,*) NCOL
       WRITE(*,*)
       WRITE(*,*) 'ENTER THE TITLE FOR THIS IMAGE, 50 CHAR. MAX.'
       READ(*,10) TITLE
10     FORMAT(1A50)
C
C      READ THE INFORMATION FOR THE PALETTE AND DESIGN THE COLORS
C
       DUMMY2 = SETVIDEOMODE($VRES16COLOR)
       DO I = 1,16
         READ(1,*) IR, IG, IB
         TMP = RGB(IR, IG, IB)
         DUMMY4 = REMAPPALETTE (I-1, TMP)
       END DO
C
C      ATTENTION:   COLOR 15 IS USED TO DRAW A BORDER AROUND THE
C                   IMAGE, REGARDLESS OF THE PALETTE
C
C      DRAW A BORDER AROUND THE IMAGE
C
       CALL CLEARSCREEN ($GCLEARSCREEN)
       CALL SETVIEWPORT(1, 1, 640, 480)
       DUMMY2 = SETCOLOR(15)
       CALL MOVETO(190, 30, XY)
       DUMMY2 = LINETO(610,30)
       DUMMY2 = LINETO(610,450)
       DUMMY2 = LINETO(190,450)
       DUMMY2 = LINETO(190,30)
C
C      DRAW TICS ON BORDER TO AID IN TRAINING SITE SELECTION
C
       DO I = 1,9
         JK = I * 42
         CALL MOVETO(190+JK,30,XY)
         DUMMY2 = LINETO(190+JK, 25)
         CALL MOVETO(190,30+JK,XY)
         DUMMY2 = LINETO(185,30+JK)
         CALL MOVETO(610,30+JK,XY)
         DUMMY2 = LINETO(615,30+JK)
         CALL MOVETO(190+JK,450,XY)
         DUMMY2 = LINETO(190+JK,455)
       END DO
C
C      READ THE IMAGE DATA AND DISPLAY, ROW BY ROW, IN ONE LOOP
C
       IF (NROW .GT. 400) NROW = 400
       DO I = 1, NROW
         DO K = 1, 3000
           ITEMP(K) = 0
         END DO
         READ(5) (ITEMP(JK), JK = 1,NCOL)
         DO J = 1, 400
           ITP = ITEMP(J)
           IF (ITP .GE. 0) THEN
             IBUF(J) = ITP
           ELSE
             IBUF(J) = 256 + ITP
           END IF
         END DO
C
C        DISPLAY
         DO J = 1, 400
           K = (IBUF(J) / 16) - 1
           IF (K .LT. 0) K = 0
           IF (K .GT. 15) K = 15
           DUMMY2 = SETCOLOR(K)
           DUMMY2 = SETPIXEL(J+200, I+40)
         END DO
       END DO
C
```

485

Figure 10.44 continued

```fortran
C       PRINT THE TITLE
C
        KCOL = 27 + (50 - LEN_TRIM(TITLE))/ 2
        CALL SETTEXTPOSITION (1, KCOL, CURPOS)
        CALL OUTTEXT(TITLE)
C
C       SELECT TRAINING SITES THROUGH INTERACTIVE Q/A
C
        CALL SETTEXTPOSITION(1,1,CURPOS)
        WRITE(*,20)
20      FORMAT( )
        WRITE(*,*) 'SELECT TRAINING SITES'
        CALL SETTEXTPOSITION(2,1,CURPOS)
        WRITE(*,*) 'NUMBER OF SITES=?'
        READ(*,*) NCLASS
        DO I = 1, NCLASS
          CALL SETTEXTPOSITION(4,1,CURPOS)
          WRITE(*,20)
          WRITE(*,*) 'FOR CLASS', I
          CALL SETTEXTPOSITION(5,1,CURPOS)
          WRITE(*,20)
          WRITE(*,*) 'UP. LEFT ROW #=?'
          READ(*,*) TRAIN(I,1)
          WRITE(*,*) 'UP. LEFT PIXEL #=?'
          READ(*,*) TRAIN(I,2)
          WRITE(*,*) 'LOW. RT. ROW #=?'
          READ(*,*) TRAIN(I,3)
          WRITE(*,*) 'LOW. RT. PIXEL #=?'
          READ(*,*) TRAIN(I,4)
          DO ICLER = 1,12
            CALL SETTEXTPOSITION(ICLER,1,CURPOS)
            WRITE(*,20)
            WRITE(*,*) ' '
          END DO
        END DO

        CALL SETTEXTPOSITION(14,1,CURPOS)
        WRITE(*,20)
        WRITE(*,*) 'ENTER THRESHOLD NOW'
        READ(*,*) THRESH
        CALL SETTEXTPOSITION (16,1,CURPOS)
        WRITE(*,20)
        WRITE(*,*) 'ENTER TYPE OF'
        WRITE(*,*) ' CLASSIFICATION:'
        CALL SETTEXTPOSITION(18,1,CURPOS)
        WRITE(*,20)
        WRITE(*,*) '1=MINDIST; 2=MAXLIKE'
        READ(*,*) ITCLAS
C
C       DRAW TRAINING SITES
C
        CALL SETTEXTPOSITION(20,1,CURPOS)
        WRITE(*,20)
        WRITE(*,*) 'DRAWING TRAIN. SITES'
        DO I = 1, NCLASS
          DUMMY2 = SETCOLOR(15)
          DUMMY2 = RECTANGLE($GBORDER, TRAIN(I,2)+199,TRAIN(I,1)+39,
2                            TRAIN(I,4)+201, TRAIN(I,3)+41)
        END DO
        CALL SETTEXTPOSITION(22,1,CURPOS)
        WRITE(*,*) 'HIT ENTER TO PROCEED'
        READ(*,*)
C
C       COMPUTE THE TRAINING STATISTICS (HISTOGRAM)
C
        DO I = 1,10
          XMEAN(I) = 0.0
          DO J = 1, 256
            HIST(I,J) = 0
          END DO
        END DO
```

Figure 10.44 continued

```
C      GET PIXEL VALUES, EACH TRAINING SITE
       DO I = 1, NCLASS
         IT = TRAIN(I,1)
         IU = TRAIN(I,3)
         JT = TRAIN(I,2)
         JU = TRAIN(I,4)
         JCOUNT = 0
         XMEAN(I) = 0.0
         DO J = IT, IU
           DO K = JT, JU
             JCOUNT = JCOUNT + 1
             DUMMY2 = GETPIXEL(K+200, J+40)
             ITP = (DUMMY2+1) * 16
             IF (ITP .GT. 255) ITP = 255
             XMEAN(I) = XMEAN(I) + FLOAT(ITP)
             HIST(I,ITP+1) = HIST(I,ITP+1) + 1./((IU-IT+1)*(JU-JT+1))
           END DO
         END DO
         XMEAN(I) = XMEAN(I) / FLOAT(JCOUNT)
       END DO
C
C      CALCULATE THE LOOKUP TABLE USING BAYE'S THEOREM OR MIN. DIST.
C
C      IF (ITCLAS .EQ. 1) THEN
C      MINIMUM DISTANCE ALGORITHM (I.E., MIN. DIST. TO MEAN)
         DO I = 1, 256
           K = 0
           DISTMN = 1000.
           DO J = 1, NCLASS
C            DISTANCE TO EACH MEAN VALUE
             DIST = ABS(XMEAN(J) - FLOAT(I))
             IF (DIST .LT. DISTMN) THEN
               DISTMN = DIST
               K      = J
             END IF
           END DO
           IF (DISTMN .LE. THRESH) THEN
             LOOK(I) = K
           ELSE
             LOOK(I) = 0
           END IF
         END DO
       ELSEIF (ITCLAS .EQ. 2) THEN
         PI = FLOAT(1./ NCLASS)
         DO I = 1,256
           PMAX = 0.0
           K = 0
           DO J = 1, NCLASS
             PXI = HIST(J,I)
             PIX = PXI * PI
             IF (PIX .GT. PMAX) THEN
               PMAX = PIX
               K = J
             END IF
           END DO
           IF (PMAX .GE. THRESH) THEN
             LOOK(I) = K
           ELSE
             LOOK(I) = 0
           END IF
         END DO
       END IF
C
```

487

Figure 10.44 continued

```
C     DRAW CLASSIFICATION MAP
C
      TMP = RGB(63,0,0)
      DUMMY4 = REMAPPALETTE(1,TMP)
      TMP = RGB(0,63,0)
      DUMMY4 = REMAPPALETTE(2,TMP)
      TMP = RGB(0,0,63)
      DUMMY4 = REMAPPALETTE(3,TMP)
      TMP = RGB(63,63,0)
      DUMMY4 = REMAPPALETTE(4,TMP)
      TMP = RGB(0,63,63)
      DUMMY4 = REMAPPALETTE(5,TMP)
      TMP = RGB(63,0,63)
      DUMMY4 = REMAPPALETTE(6,TMP)
      TMP = RGB(63,63,63)
      DUMMY4 = REMAPPALETTE(7,TMP)
      TMP = RGB(63,32,0)
      DUMMY4 = REMAPPALETTE(8,TMP)
      TMP = RGB(32,0,63)
      DUMMY4 = REMAPPALETTE(9,TMP)
      TMP = RGB(0,32,63)
      DUMMY4 = REMAPPALETTE(10,TMP)
      TMP = RGB(0,32,0)
      DUMMY4 = REMAPPALETTE(11,TMP)
      TMP = RGB(32,0,0)
      DUMMY4 = REMAPPALETTE(12,TMP)
      TMP = RGB(0,0,32)
      DUMMY4 = REMAPPALETTE(13,TMP)
      TMP = RGB(32,32,0)
      DUMMY4 = REMAPPALETTE(14,TMP)
      TMP = RGB(0,32,32)
      DUMMY4 = REMAPPALETTE(15,TMP)
      TMP = RGB(0,0,0)
      DUMMY4 = REMAPPALETTE(0,TMP)

      CALL SETTEXTPOSITION(20,1,CURPOS)
      WRITE(*,*) 'HIT ENTER TO PROCEED'
      READ(*,*)
      REWIND 5
      DO I = 1, NROW
      READ(5)(ITEMP(JK), JK = 1,NCOL)
      DO J = 1, 400
      ITP = ITEMP(J)
      IF (ITP .LT. 0) ITP = 256 + ITP
      ITP = (ITP/16) - 1
      IF (ITP .LT. 0) ITP = 0
      IF (ITP .GT. 15) ITP = 15
      ITP = (ITP+1) * 16
      IF (ITP .GT. 255) ITP = 255
      L = LOOK(ITP+1)
      DUMMY2 = SETCOLOR(L)
      DUMMY2 = SETPIXEL(J+200, I+40)
      END DO
      END DO

C
C
      PUSH THE ENTER KEY TO TERMINATE THE PROGRAM

      READ(*,*)
      DUMMY2 = SETVIDEOMODE($DEFAULTMODE)
      STOP
      END

C
C
      RGB FUNCTION

      INTEGER*4 FUNCTION RGB(R, G, B)
      INTEGER*4 R, G, B
      RGB = ISHL(ISHL(B,8) .OR. G,8) .OR. R
      RETURN
      END
```

11

Fractals

Euclid, a Greek mathematician who lived circa 300 B.C., contributed a most important work to mathematics: the *Elements*. This treatise comprises thirteen volumes on various subjects, including plane geometry. The first several of these volumes encompass what is essentially taught today in high school geometry courses. This shows the influence Euclid's writings have after more than two millennia.

Euclid's notion of geometry is an ideal one: that shapes recognizable in nature can be described as perfect planes, triangles, circles, lines, spheres, cubes, and so on. But nature is rarely associated with ideal shapes, thus Euclid's idealism is inadequate. As one example, attempt the measure of the length of a coastline or river using several different ruler sizes and observe the outcome. Further, observe a snowflake and attempt a description of it using Euclidean idealism. It has taken more than 2000 years for us to understand, through observation, that there is more to the universe than what can possibly be described using an ideal notion of geometry.

In this chapter several experiments are presented as a means of introducing the notion of *fractals,* objects that are, by one definition, infinitely complex. More important, in this chapter you are challenged to enhance your observational skills. Dr. Benoit Mandelbrot, on page 461 of his book, *The Fractal Geometry of Nature* (New York: Freeman, 1983) addresses this issue bluntly. Paraphrasing his remarks: Science cannot advance if modeling and theorizing are favored over, and thereby usurp, observation. The point is taken that observation must come first, on which is founded hypothesis.

Observation tells us that Euclidean geometry is wholly inadequate when applied to the description of certain natural phenomena, such as the shape of a coastline, a geomorphologic notion. So, how can these natural phenomena be adequately described? The answer to this question leads to the essence of the fractal, a bizarre, yet wondrous notion. Fractals offer a unique way to observe and describe natural phenomena.

A few practical applications to the study of earth sciences are touched upon to show why the study of fractals may become an important part of the study of geological sciences in the future. First, a historical review is presented to show the origin of the concept of fractals. This review is necessary for a thorough understanding, but it is presented for interest as well.

11.1 Nineteenth-Century Discoveries

One of Mandelbrot's major contributions to science and mathematics is the seemingly insignificant coining in 1975 of the term "fractal." Many properties of fractals, and some objects that are called fractals, were first recognized a little over one hundred years ago. The history of fractals predating 1975 is therefore reviewed.

11.1.1 Poincaré

Jules Henri Poincaré was one of the foremost French physicists and mathematicians of the nineteenth century. He was born in Nancy and educated at École Polytechnique and the École Superieur des Mines in Paris. Poincaré anticipated chaos theory (Gleick, 1987), and the notion of fractals, in particular, tessellations (checkered mosaic constructions) for simulating what are known today as fractals. Some of these constructions have been used by the artist M.C. Escher for his artistic experiments on "infinite smallness."

Poincaré's anticipation of chaos is obvious in a passage from his book *Science and Method* (as reproduced in Gleick (1987, p. 321)):

> A very small cause which escapes our notice determines a considerable effect that we cannot fail to see, and then we say that the effect is due to chance. If we knew exactly the laws of nature and the situation of the universe at the initial moment, we could predict exactly the situation of that same universe at a succeeding moment. But even if it were the case that the natural laws had no longer any secret for us, we could still know the situation approximately. If that enabled us to predict the succeeding situation with the same approximation, that is all we require, and we should say that the phenomenon had been predicted, that it is governed by the laws. But it is not always so; it may happen that small differences in the initial conditions produce very great ones in the final phenomena. A small error in the former will produce an enormous error in the latter. Prediction becomes impossible.

In the twentieth century, Edward Lorenz presented a paper regarding atmospheric dynamics entitled "Predictability: Does the flap of a butterfly's wings in Brazil set off a tornado in Texas?" (Lorenz, 1979). Notice that this title implies what Poincaré recognized, that very small changes at one time or place may bring about great changes at a later time or another place.

Turbulence is an example of a physical phenomenon that exhibits chaotic behavior. Turbulence is also fractal in that it comprises many different scales of swirls and eddys. The turbulence caused by a butterfly is similar, but smaller in scale, to that caused by a bird or an airplane.

The notions of chaos and fractal geometry are interrelated. The remainder of this chapter is focused on fractal geometry, but one example related to chaos is presented later. Chaos is introduced in this discussion of Poincaré to demonstrate that the foundation for fractal geometry and related mathematical notions was laid in the nineteenth century.

11.1.2 Fricke and Klein

Poincaré's algorithms for tessellations were used by Robert Fricke and Felix Klein to produce illustrations in their book *Vorlesungen uber die Theorie der automorphen Funktionen*, published in 1897. One of these illustrations (that involving a tessellation using recursive circles) is compared to a computer construction in Mandelbrot's *Fractal Geometry of Nature* (1983, p. 179). Therefore, tessellations that are used to generate artificial fractals were well known by the dawn of the twentieth century. In fact, Fricke's "hyperbolic tessellation" is also a foundation for some of the artwork by Escher.

11.1.3 Weierstrass and duBois Reymond

An example of a continuous, but everywhere nondifferentiable function was given by K. Weierstrass and is a complex (i.e., real and imaginary) function, the sum of the following series:

$$W_0(t) = \frac{1}{\sqrt{1 - w^2}} \sum_{n=0}^{\infty} w^n e^{2\pi i b^n t}$$

for which b is a real number greater than 1, and $w = b^H$, with $0 < H < 1$, or $w = b^{D-2}$, with $1 < D < 2$; this function is nowhere differentiable.

Weierstrass developed this function in 1870, but did not formally publish it until 1895. But Weierstrass' effort was noted in duBois Reymond (1875), marking the date of the formal notice of the nondifferentiable functions to the mathematical community. In his 1875 paper, duBois Reymond acknowledges that the function hides many puzzles, and these may lead to "the limit of our intellect."

Recall from Chapter 1 that for a function f to be differentiable, it must be a continuous function. The Weierstrass function is continuous, but is not differentiable; this function is fractal. That this is so may be imagined as follows. A fractal is (by one definition) a mathematical function that is self-similar; that is, it appears the same regardless of the scale at which it is viewed. Such a function is also infinitely complex. A derivative is defined as the slope of a function over an infinitely small region (Chapter 1). But a fractal function is infinitely complex, even over an infinitely small region, hence no unique derivative (unique slope) can be determined.

11.1.4 Other Nineteenth-Century Discoveries

Mathematical monsters were discovered at the dawn of the twentieth century which many mathematicians at the time chose to ignore (out of fright, if for no other reason). Cantor sets (named for Georg Cantor, a nineteenth-century German mathematician who developed the theory of sets), Weierstrass functions, Peano curves (named for G. Peano), curves without derivatives, and lines that can fill space are a few of these monsters. All are described in Mandelbrot (1983). At the turn of the twentieth century, mathematicians assumed that such monsters had no natural relevancy. However, Mandelbrot has shown otherwise. A coastline, for instance, is one such monster, a line that is to a degree space filling. Computer graphical images help us to understand that these ''mathematical monsters'' represent the geometrical language of nature.

11.2 Fractals in the Twentieth Century

11.2.1 Fatou and Julia

In the early part of the twentieth century, mathematical monsters continued to be unearthed. In 1918, prefractal notions are found in the work by Fatou and Julia (Mandelbrot (1983) describes ''julia sets,'' for instance). A fundamental, precursory research effort to the notion of fractals was undertaken by Louis Richardson just subsequent to the work by Fatou and Julia. How fundamental this research effort was is shown in the next section.

11.2.2 Richardson

Lewis F. Richardson spent much of his life studying the causes of war. Much of this work was published subsequent to his death in an article, *Statistics of Deadly Quarrels* (Richardson, 1960). An important appendix to this article was published separately: *The Problem of Contiguity* (Richardson, 1961). This latter article had overwhelming implications for fractal geometry.

Part of Richardson's study of the root causes of war focused on the likelihood of war occurring between nations sharing a common boundary. Of course, war cannot be attributed to a common boundary alone, for citing civil war is sufficient to reject this hypothesis (unless the internal strife is caused by some sort of boundary, such as a cultural one). The fact that a common border exists may be a contributing factor to some wars.

This argument notwithstanding, Richardson's study of boundaries between nations led to some remarkable discoveries. Some of these boundaries are formed by rivers and Richardson found that as he attempted to measure the length of such boundaries, the length seemingly increased without bound as he used smaller and smaller step sizes (ruler sizes) to trace the length. This discovery by itself violates Euclidean principles, for in ideal Euclidean space, length is a constant, regardless of the scale at which it is measured. Richardson went further to investigate the lengths of coastlines and discovered that, as for rivers, length seemingly increases without bound as the ruler size used to measure the length decreases. Three experiments of Richardson's show this (*The Problem of Contiguity*, 1961, pp. 168 and 169):

The West Coast of Great Britain (from Land's End to Duncansby Head)

Ruler (step) size, km:	971	490	200	100	30	10
Number of steps:	1	2	5.9	15.4	69.1	293.1
Total length, km:	971	980	1180	1540	2073	2931

The Coast of the Australian Mainland

Ruler (step) size, km:	2000	1000	500	250	100
Number of steps:	4.7	11.04	23.9	52.8	144.2
Total length, km:	9400	11040	11950	13200	14420

Coast of South Africa

Ruler (step) size, km:	1000	500	215.5	100
Number of steps:	4.12	8.31	19.78	43.34
Total length, km:	4120	4155	4263	4334

Notice that in each of these three cases, the total length does seem to increase without bound as the rule (step) size decreases. Once again, this behavior between length and ruler size is incompatible with the notion of ideal plane geometry vis-à-vis Euclid.

Richardson (1961, p. 170) presents some conclusions on inscribing polygons of equal sides to frontiers. (Richardson equated the ruler (step) sizes to sides of polygons; his use of the word frontiers refers to boundaries between countries, coastlines, and so on.) These conclusions are understatements and not all are reviewed here.

First, Richardson noted that it is difficult to select a ruler size that can be "walked" between two points on a coastline in a whole number of steps; usually, the final step is a fraction of the ruler size. This explains why in the three tables above the number of ruler steps is often a whole number plus a fractional part. Further, Richardson doubted whether the total length of coastlines, rivers, etc. tends to any limit as the ruler size tends to zero. Finally, Richardson makes a gross understatement by noting the following proportionality:

$$\text{total length} \propto (\text{ruler (step) size})^{-\alpha}$$

that is, the total length is proportional to the ruler size (used to measure the length) raised to some power. The fact that this power is negative suggests an inverse proportionality between total length and ruler size. For the three coastlines presented, the power term α is as follows: 0.25 for the west coast of Great Britain, 0.13 for the coastline of the Australian mainland, and 0.02 for the coastline of South Africa (all from Richardson (1961)).

The proportional relationship between total length and ruler size is an integral part of fractal geometry. Before proceeding to the formal presentation of fractals, an experiment is presented for measuring the length of the west coast of Vancouver Island, British Columbia, Canada. This example is included to show exactly what a ruler (step) size is and how to count the number of steps when tracing total length.

Example 11.1   An Experiment to Measure the Length of the West Coast of Vancouver Island from Victoria to Cape Scott

An outline of the coast of Vancouver Island in British Columbia, Canada (Figure 11.1), reveals one of the most intricate coastlines in the world. An experiment is attempted to measure the length of the western coast, starting at Victoria, the capital

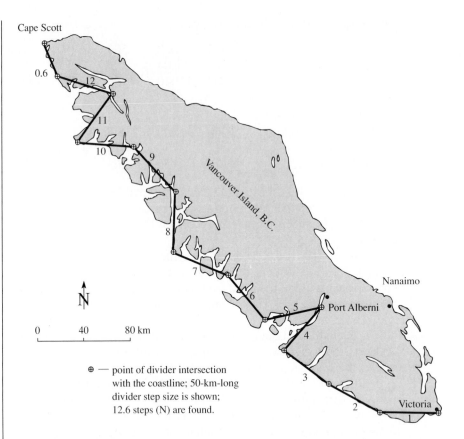

Cape Scott

N

0   40   80 km

⊕ — point of divider intersection
with the coastline; 50-km-long
divider step size is shown;
12.6 steps (N) are found.

Figure 11.1 Outline of the coast of Vancouver Island, British Columbia, Canada. This figure also shows the experiment to measure the coastal distance from Victoria to Cape Scott using a 50-km ruler.

of British Columbia, and proceeding to the western most point of Vancouver Island, Cape Scott.

To begin this experiment, a ruler is selected that is the straight line distance from Victoria to Cape Scott. This distance is found to be equal to 464 km using the scale shown in Figure 11.1.

Other ruler sizes are arbitrarily chosen as follows: 200, 100, 50, and 10 km. For each of these rulers, a pair of dividers is opened to a distance equal to these scaled distances. Beginning at Victoria, the pair of dividers is "walked" between Victoria and Cape Scott, counting the number of steps along the way; this is shown explicitly in Figure 11.1 for the 50-km ruler size. The results for this experiment are summarized:

West Coast of Vancouver Island

Ruler (step) size, km:	464	200	100	50	10
Number of steps:	1	2.36	4.91	12.6	105.5
Total length, km:	464	472	491	630	1055

This experiment dramatically illustrates what Richardson acknowledges in *The Problem of Contiguity,* in reference to Henri Poincaré's *Science and Method* (as quoted earlier in this chapter): the notion that insignificant causes produce very noticeable effects. Richardson acknowledges that an experiment to measure the length of a coastline, such as this one using Vancouver Island, is a good example of Poincaré's notion because one ruler size may just miss a peninsula or bay, whereas a ruler just a little bit smaller may catch the peninsula or bay. Notice in the experiment to measure the west coast of Vancouver Island that rulers equal or larger to 100 km miss most of the bays and peninsulas on the western coast. But the ruler size of 50 km catches many of these bays and peninsulas, yielding a dramatically larger total length than the next larger ruler, 100 km. Therefore, Vancouver Island is one of the most fascinating coastlines for conducting an experiment such as those documented by Richardson.   ∎

11.2.3 The Sketches of M.C. Escher: A Notion of Infinite Smallness

Maurits Cornelis Escher was born June 17, 1898. His beautiful prints are known worldwide for their repetitive patterns, teasing plays on perspective, and patterns that transcend into other patterns. Many of Escher's sketches are predicated on some mathematical algorithm, such as the Poincaré or Fricke tessellations.

Many who are familiar with fractals feel that Escher's sketches are fractal recreations, but have difficulty explaining why. Perhaps Escher's own words are of help. In an essay entitled *Oneindigheidsbenaderingen* (Escher, 1959, as reproduced by Doris Schattschneider, *Visions of Symmetry,* 1990, p. 241). Escher wrote:

> Anyone who plunges into infinity, in both time and space, farther and farther without stopping, needs fixed points, mileposts as he flashes by, for otherwise his movement is indistinguishable from standing still. There must be stars past which he shoots, beacons by which he can measure the path he has traveled. He must mark off his universe into units of a certain length, into compartments which repeat one another in endless succession. Each time he crosses the border from one compartment to another, his clock ticks.

This statement is suggestive of the notion of self-similarity, a fundamental property of fractals. A most elegant, visual translation of this thesis is found in Escher's print *Smaller and Smaller* (Figure 11.2). This print is Escher's notion of "infinite smallness." Notice how the pattern of lizards repeats, becoming smaller and smaller toward the center of the print.

Escher illustrated the tessellation on which this print is based with a drawing (Figure 11.3). This algorithm is described as follows: Beginning with a right triangle, smaller triangles are produced adjacent to the right triangle by reflecting the triangle across its hypotenuse, splitting the reflection into two congruent right triangles. The newly created triangles are scaled by 1/SQRT(2). Each newly created right triangle becomes the beginning right triangle and the halving is repeated. This is done ad infinitum. This algorithm is like the creation of a von Koch artificial coastline (as is described in Mandelbrot's *The Fractal Geometry of Nature* and described later in this chapter). Both Figures 11.2 and 11.3 are illustrations of self-similarity (of shape) and of infinite complexity, or infinite repetition of shape with changing scale (Figure 11.2).

11.2.4 Hausdorff and Besicovitch

Suppose the area of a surface S is to be approximated. One way to approach this approximation is to cover the surface with balls of radius p. For two-dimensional surfaces, the total surface area is proportional to p^2. For a d-dimensional generalization, the area is proportional to p^d.

In 1919 F. Hausdorff allowed d to be fractional (before Hausdorff's work, d was accepted to be a whole number). Hausdorff further assumed that d is positive and finite. In 1935, A. Besicovitch extended Hausdorff's work to

Figure 11.2   M.C. Escher's print, *Smaller and Smaller,* completed in October 1956. Notice
how the pattern of lizards repeats and becomes infinitely small toward the center of the
print. (This print is reproduced by permission from Cordon Art, B.V., Nieuwstraat 6, 3743
BL Baarn, The Netherlands.)

include two situations: where d is zero and where d is infinitely large. The
dimension d is called the Hausdorff–Besicovitch dimension, has been called
a "fractional dimension" in the past, and is loosely and sometimes incorrectly
referred to as the "fractal" dimension.

11.2.5 Mandelbrot

Born in Poland in 1924, Benoit B. Mandelbrot was educated in France and
the United States. He has taught at Harvard and Yale universities and since
1958 has been affiliated with IBM's Thomas B. Watson Research Center in
New York.

   Mandelbrot was intrigued by Richardson's experiments to measure the length
of coastlines. Six years after the publication of Richardson's *The Problem of
Contiguity,* Mandelbrot published the landmark paper "How long is the coast
of Great Britain? Statistical self-similarity and fractional dimension" in the
journal *Science* (Vol. 155, 1967). This publication marked the birth of the
mathematical notion, and a separate branch of mathematics, known as fractals
(so named by Mandelbrot in 1975).

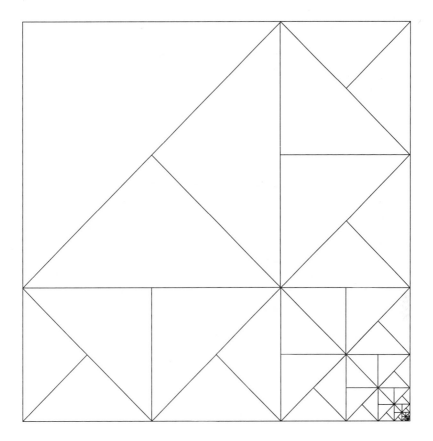

Figure 11.3 Algorithm (tessellation) used by Escher to create *Smaller and Smaller* (Figure 11.2).

Recall Richardson's proportionality: that total length is proportional to the ruler size, raised to a power equal to the negative of α. The power α intrigued Mandelbrot, for the essence of a fractal object is that its dimension is crucial to its definition. Euclid allowed us to think only in terms of one, two, or three dimensions. A line, for example, is the epitome of one-dimensional space, a plane epitomizes two-dimensional space, and a cube epitomizes three-dimensional space. But with the power term α, Mandelbrot suddenly recognized that a limitless number of dimensions are intermediate to the ideal Euclidean dimensions. A coastline, for instance, may be much more complex than an ideal line, as is the case with the coasts of Great Britain and Vancouver Island. But these coastlines are not so complex that they completely fill a plane. Therefore, their dimensions lie somewhere between these two Euclidean ideal dimensions.

Mandelbrot expanded upon Richardson's proportionality and obtained the following relationship:

$$L = Ny^D$$

where L is a factor of proportionality; N, the number of steps of length y needed to walk along the curve, is proportional to $1/y^D$; and D is a power term (compare it to the discussion of the Hausdorff–Besicovitch dimension, Section 11.2.4).

For an ideal Euclidean line, $L = NY$ is the proper relationship, which may also be written as $L = Ny^1$. Notice that in the case of an ideal Euclidean line, the power D is equal to 1 (this is also known as the topological dimension of the line). For fractal lines, D is in the range $]1, 2]$ (i.e., 1 is excluded but 2 is included).

If Mandelbrot's relationship is written as $Ly^{-D} = N$ and if the property of logarithms is used to write $\log L - D \log y = \log N$, then an equation of the form $b - mx = y$, the standard equation for a line, is obtained. Using the linear regression procedure outlined in Chapter 2, the data are used from an experiment to measure the length of a coastline to solve for the power D:

$$D = -\frac{n\Sigma(\log(y)\log(N)) - \Sigma\log(y)\Sigma\log(N)}{n\Sigma\log(y)^2 - (\Sigma\log(y))^2}$$

where n is the number of different ruler sizes used in the experiment, and N is the number of ruler steps found for each ruler size y. Applying this equation to the Richardson data for the coasts of Great Britain, Australia, and South Africa, as well as to the data for the west coast of Vancouver Island, the power D is found to be 1.25 for the west coast of Great Britain, 1.13 for the coastline of the Australian mainland, 1.02 for the coastline of South Africa, and 1.23 for the west coast of Vancouver Island.

In reviewing these results, notice that the power term D is equal to Richardson's power term α plus one. Therefore, Richardson's power term and Mandelbrot's power term are the same notion; that is,

$$D = 1 + \alpha$$

Mandelbrot, though, called his power term a dimension. This is, in fact, what the power term D represents, since an irregular coastline has a dimension between two Euclidean ideals: a line and a plane. Notice that the power term D for the west coast of Great Britain is, in fact, 1.25, a value intermediate between the ideal Euclidean dimensions. Therefore, D is the dimension of the feature under study (and is also, by this definition, the Hausdorff–Besicovitch dimension).

11.3 Fractals: A Summary of Definitions

Poincaré, Fricke, Weierstrass, Richardson, and others discovered the notion and properties of the fractal, Escher provided unique visual translations of fractal tessellations, and Mandelbrot formalized the theory of fractals and named them. Mandelbrot invented the word fractal from the Latin *fractus,* an adjective having a corresponding Latin verb, *frangere,* which means to break or create irregular fragments.

Recall from Hadamard (Chapter 1) that the creation of a word or a notation for a class of ideas is often a scientific fact of very great importance, because it means connecting these ideas together in our subsequent thought. This notion underscores the tremendously important contribution made by Mandelbrot just in coining the term fractal. At the very least, this has given birth to a new discipline, both scientific and mathematical. Moreover, after two millennia, the ideal geometry of Euclid has been supplemented by a natural geometry, that of fractals.

The term fractal has been used up to now without the presentation of a formal definition. No single, superb definition for a fractal yet exists, but some conceptualizations are available. For instance, a fractal is an object or feature whose dimension D must be used in the description of its size. A fractal is also a feature whose appearance (shape and so on) is the same regardless of the scale at which it is viewed, or is a process or feature whose statistical properties are the same regardless of the scale over which they are measured. Mandelbrot (1989) forwards perhaps the best definition of a fractal: something that exhibits invariance under contraction or dilation.

An example of a fractal, an artificial one, is the von Koch "coastline" (Figure 11.4), a good illustration of invariance under contraction or dilation. This coastline, or "snowflake," as this geometric construction is sometimes called, is obtained by using equilateral triangles of smaller and smaller size. Notice in this construction that the original starting shape (construction a) is repeated ad infinitum, providing the invariance of shape with contraction. This

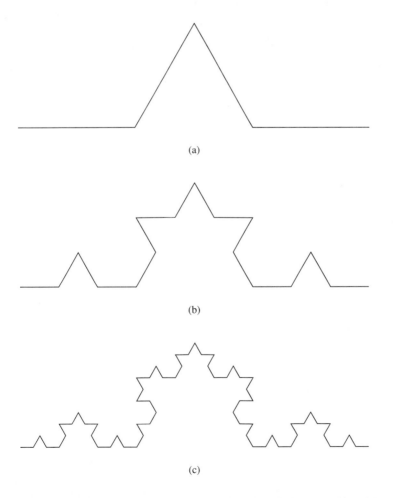

(a)

(b)

(c)

Figure 11.4 The von Koch curve (or coastline) tessellation. Drawing (a) is the algorithm, repeated to yield drawings (b) and (c).

construction is not unlike the tessellation described by Escher in his creation of the print *Smaller and Smaller* (Figures 11.2 and 11.3).

In review, a fractal has the following characteristics:

1. Its "dimension" (i.e., Hausdorff–Besicovitch dimension) is intermediate between Euclidean ideals. A fractal line has a dimension $1 < D < 2$; a fractal surface has a dimension $2 < D < 3$; a fractal volume has a dimension $3 < D < 4$; and so on.
2. a. Its appearance is the same regardless of the scale at which it is viewed; this is called self-similarity.
   b. It is associated with the property of statistical self-similarity; that is, the statistical attributes of a phenomenon, such as variance, are the same, regardless of the scale over which they are measured. This notion is discussed later.

Many natural phenomena have been shown to be fractal. Geomorphic features such as coastlines and rivers have already been discussed, but topography is also fractal (Mark and Aronson (1984) provide one example). Clouds and galaxy clusters are also fractal, as are stones used for concrete aggregate (Carr et al., 1990, 1992). Many natural features are fractal, which explains why this new discipline is associated with an intense degree of interest inside and outside the earth sciences.

11.4 A Spectral Notion for Fractal Dimension

Three methods are available for estimating the fractal dimension for an object, function, process, and so on. One of these methods is the divider, or compass, technique that was used originally to demonstrate the fractal concept (Mandelbrot, 1967). The other two calculation methods resolve the stochastic character of what are known as random fractals. A power spectrum method is used for estimating fractal dimension of random fractals and is based on the fractal character of brown noise. A brown noise concept also forms the basis for a fractal dimension estimation method based on the semivariogram.

11.4.1 The Random Fractal

Fractals created using some sort of recursive algorithm, such as the tessellations used by Escher or the construction used to create the von Koch coastline, are deterministic fractals. Another type of fractal is the random fractal. In addition to some sort of recursive, deterministic rule, a degree of randomness is added to simulate such fractals.

Recent research, in part based on Sayles and Thomas (1978) but applied specifically for fractal dimension estimation (e.g., Berry and Lewis, 1980; Mandelbrot et al., 1984), resolves the stochastic nature of an object, function, etc., using a method based on the power spectrum or periodogram. If, for example, the power spectrum $P(f)$ is calculated for brown noise (e.g., BROWN.DAT, Appendix A), then a plot of $\log[P(f)]$ versus $\log f$ is found to be linear; a least-squares, linear regression of this plot has a slope of -2. A fractal dimension is estimated using this slope as (Berry and Lewis, 1980)

$$D = \frac{5 - |\beta|}{2}$$

where β is the slope of the least-squares, linear regression line fitted to the $\log[P(f)]$ versus $\log f$ plot. Using this equation, the fractal dimension D for brown noise is $(5 - 2)/2$, or 1.5. Detailed development of this spectral method for fractal analysis is presented next.

11.4.2 Background

Three statistical properties hold for brown noise. First, let Y represent a brown noise function. The increment

$$Y(x_2) - Y(x_1)$$

has a normal distribution. Moreover,

$$E\{[Y(x_2) - Y(x_1)]^2\} \propto |x_2 - x_1|$$

The log-semivariogram versus log–lag distance plot (Figure 11.5) shows that spatial variance for brown noise varies linearly with lag and the plot has a slope of 1; hence this proportionality holds. Finally, increments of Y are statistically self-similar; that is, $Y(x)$ and $Y(ax)$ are statistically indistinguishable.

Berry and Lewis (1980) present the relationship between fractal dimension D and the slope of the periodogram β with limited proof. A detailed and clearly presented proof is found in Peitgen and Saupe (1988, pp. 91–92). For example, let the spectral density (periodogram) of fractal Brownian motion (brown noise) be called $P(f)$, which has the following proportionality:

$$P(f) \propto \frac{1}{f^{\beta}}$$

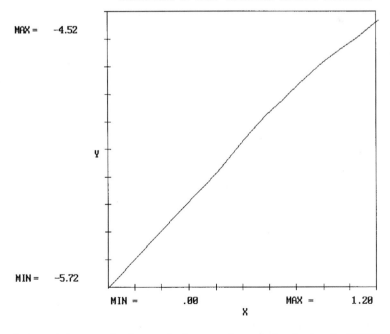

LOGARITHMS OF SEMI-VARIOGRAM VALUES, BROWN.DAT

MAX = −4.52

Y

MIN = −5.72

MIN =   .00   MAX =   1.20

X

Figure 11.5 A semivariogram for brown noise, calculated using the BROWN.DAT data set (Appendix A). This plot is actually one of the logarithms of the semivariogram results. Note that the slope at the origin is 1, which is equal to $2H$.

Moreover, let the following be written:

$$E\{[Y(x_2) - Y(x_1)]^2\} \propto |x_2 - x_1|^{2H}, \qquad 0 \leq H \leq 1$$

If $H = \frac{1}{2}$, the relationship as given for brown noise is obtained. If $H = 0$, no matter how Y is squeezed or stretched, it looks the same. That is, if $H = 0$, then $[Y(x_2) - Y(x_1)]^2$ is proportional to 1 as is $[bY(x_2) - bY(x_1)]^2$ with the stretching factor b. Furthermore, if $H = 0$, then $E\{[y(x_2) - Y(x_1)]^2\}$ is independent of x_1 and x_2. This suggests that Y is plane filling when $H = 0$; therefore the fractal dimension of Y is 2 when $H = 0$.

White noise is a good example of a function for which $H = 0$. Let $W(x)$ represent a white noise function having a mean of zero and finite variance. For such a function, $E[W(x_i)] = 0$; further, $E\{[W(x_i) - W(x_j)]^2\} = 2\mathrm{var}(W) =$ constant for any i and j. This requires $H = 0$ with a proportionality factor of $2\mathrm{var}(W)$. If W is rescaled by a factor b, the variance would still be constant for any $bW(x_i)$ and $bW(x_j)$; therefore H will still be 0.

At the other extreme, $H = 1$, stretching Y in the x direction must be compensated for by rescaling $Y(x)$. As an example, consider an ideal Euclidean line inclined at an angle α. The fractal dimension for such a line is one. Let $Y(x)$ represent the elevation of the line at a point x. Moreover, let $Y(x_1) = 0$ and $x_1 = 0$ (for simplicity):

$$Y(x_i) = [|x_i - x_1|] * \tan \alpha$$

From this expression is inferred

$$E\{[Y(x_i) - Y(x_1)]^2\} \propto |x_i - x_1|^2$$

with a proportionality factor of $\tan^2 \alpha$; therefore $H = 1$.

Based on these inferences regarding H, the relationship between H and fractal dimension D is inferred to be

$$D = 2 - H$$

Furthermore, returning to the relationship between power and frequency for a Brownian motion function,

$$P(f) \propto \frac{1}{f^\beta}$$

Peitgen and Saupe (1988) show that

$$\beta = 2H + 1 = 5 - 2D$$

This is an important relationship for the spectral notion of fractal dimension and is demonstrated empirically. Given a function Y, which is a fractal Brownian motion function, and given that the power spectrum P_Y of Y conforms to

$$P_Y(f) \propto \frac{1}{f^\beta}$$

then

$$T(j) = \frac{Y(rj)}{r^H}$$

is a proper rescaling of Y for $r > 0$ (this follows from the statistical self-similarity property of Y stated previously), provided that $H = 2 - D$ and D is the fractal dimension of Y. In this case, T and Y have the same statistical properties. This notion is now tested empirically.

The program SPFM1D (Figure 11.6, page 531) implements the algorithm (Peitgen and Saupe, 1988, p. 94). This program generates the frequency domain analog of fractal Brownian motion having a fractal dimension $2 - H$. Once this program is completed, the program CFRINV (Figure 11.7, page 533) is used to inverse Fourier transform the frequency domain simulation to the spatial domain. CFRINV is a slightly modified version of the program FORINV (Chapter 9) and treats the vector Y as a complex vector (it is a real-valued vector in FORINV). The modification is necessary to maintain compatibility with SPFM1D.

Three simulations are shown in Figures 11.8 through 11.10 for $H = 0.5$ ($D = 1.5$), $H = 0.2$ ($D = 1.8$), and $H = 0.9$ ($D = 1.1$). For each of these profiles Y, the same statistical properties should be obtained for T as are obtained for Y only when Y is rescaled by r^H, where H is the same value as is used to obtain each simulation. That is, given a function Y, which has a fractal dimension $2 - H$, a function T is statistically self-similar to Y only when H is exactly equal to $2 - D$, where D is the fractal dimension. For any other value of H, T is not statistically self-similar to Y. Three experiments demonstrate this.

EXPERIMENT 1

Rescale $Y_{H=0.5}$ (Figure 11.8) using the program HSCALE (Figure 11.11, page 534) to yield T such that

$$T(j) = \frac{Y(2j)}{\sqrt{2}}$$

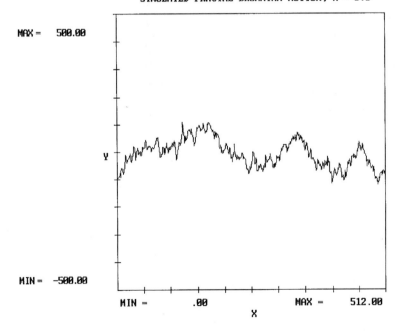

Figure 11.8 A simulation using SPFM1D/CFRINV with $H = 0.5$. This figure was created using GRAPH2D (Chapter 7) by choosing option 6 (for displaying one-dimensional functions) and rescaling y by 100.

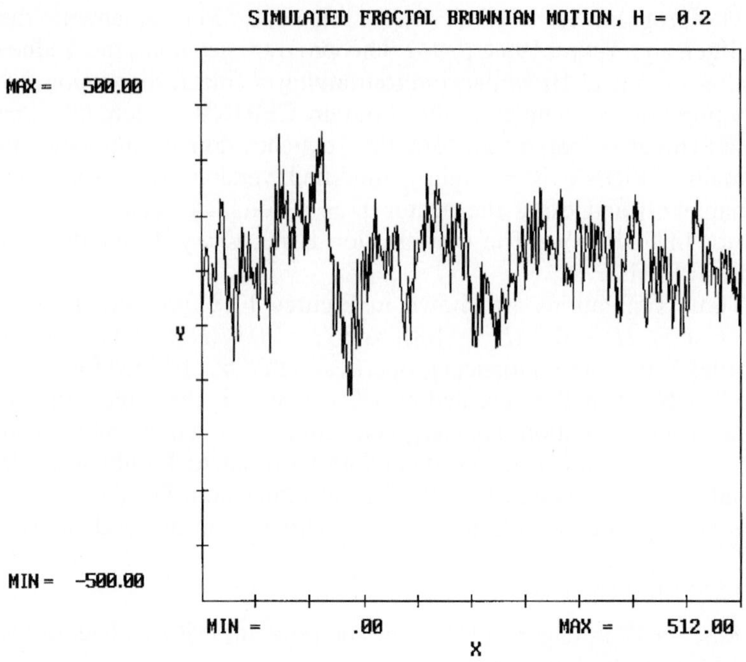

SIMULATED FRACTAL BROWNIAN MOTION, H = 0.2
MAX = 500.00
Y
MIN = -500.00
MIN = .00
MAX = 512.00
X

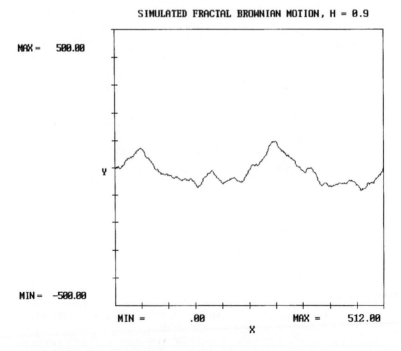

Figure 11.9 A simulation using SPFM1D/CFRINV with $H = 0.2$. This figure was displayed using GRAPH2D (Chapter 7), in the same manner as is described for Figure 11.8.

SIMULATED FRACTAL BROWNIAN MOTION, H = 0.9
MAX = 500.00
Y
MIN = -500.00
MIN = .00
MAX = 512.00
X

Figure 11.10 A simulation using SPFM1D/CFRINV with $H = 0.9$. This figure was displayed using GRAPH2D (Chapter 7), in the same manner as is described for Figure 11.8.

where r is arbitrarily chosen to be equal to 2, and H is implicitly included and is equal to 0.5. Because the fractal dimension for Y is 1.5 (i.e., $H = 0.5$), Y and T should have the same statistical properties. One such statistical property is the semivariogram (Chapter 6). A table of semivariogram values reveals that T and Y are statistically self-similar:

Lag	Semivariogram Values for Y	Semivariogram Values for T
2	0.01497	0.01456
4	0.02799	0.02251
6	0.03597	0.02991
8	0.04395	0.03443
10	0.05215	0.04373
12	0.05896	0.05419
14	0.06390	0.06449
16	0.06914	0.07296
18	0.07770	0.08498

When statistical variability is considered, this is a rather close match and seems to empirically show a statistical similarity between functions Y and T, which is formed by resampling Y.

If Y is rescaled such that

$$T(j) = \frac{Y(2j)}{2^{0.90}}$$

then Y and T are not statistically self-similar because a value of $H = 0.90$ is used, which is not consistent with the fractal dimension of Y ($D = 1.5$). The following table of semivariogram values shows what happens:

Lag	Semivariogram Values for Y	Semivariogram Values for T
2	0.01497	0.00443
4	0.02799	0.00836
6	0.03597	0.01718
8	0.04395	0.01978
10	0.05215	0.02511
12	0.05896	0.03113
14	0.06390	0.03704
16	0.06914	0.04191
18	0.07770	0.04881

This empirically shows that $T(j) = Y(rj)/r^H$ is a proper rescaling only when H is equal to $2 - D$, where D is the fractal dimension of Y. When Y is rescaled using $H = 0.5$ to yield T, semivariogram values for T are similar to those for Y. But when Y is rescaled using $H = 0.9$, semivariogram values for T are not similar to those for Y.

EXPERIMENT 2

Rescale $Y_{H=0.2}$ (Figure 11.9) using HSCALE to yield T such that

$$T(j) = \frac{Y(2j)}{2^{0.20}}$$

where r is arbitrarily selected to be equal to 2, and H is equal to 0.20. In this case, Y and T are statistically indistinguishable, as is shown by semivariogram values:

Lag	Semivariogram Values for Y	Semivariogram Values for T
2	0.20516	0.17787
4	0.25371	0.27725
6	0.33926	0.34870
8	0.39038	0.41755
10	0.45689	0.45483
12	0.47977	0.47142
14	0.54399	0.51466
16	0.56258	0.52138
18	0.57159	0.52680

Recall that for $H = 0.2$, fractal dimension is 1.8. Therefore Y is approaching white (purely random) noise, hence more variability is expected. The outcome of this experiment is good considering this aspect of Y.

Consider using $H = 0.9$ for the rescaling, such that $T(j) = Y(2j)/2^{0.90}$. Now Y and T are not statistically self-similar:

Lag	Semivariogram Values for Y	Semivariogram Values for T
2	0.20516	0.05997
4	0.25371	0.10506
6	0.33926	0.13213
8	0.39038	0.15822
10	0.45690	0.17235
12	0.47977	0.17864
14	0.54399	0.19502
16	0.56258	0.19757
18	0.57159	0.19962

This experiment demonstrates, as does Experiment 1, that T is a proper rescaling of Y only when $H = 2 - D$.

EXPERIMENT 3

Rescale $Y_{H=0.9}$ (Figure 11.10) using HSCALE to yield T such that

$$T(j) = \frac{Y(2j)}{2^{0.90}}$$

where r is again arbitrarily selected to equal 2, and H is equal to 0.90. In this case, Y and T are statistically indistinguishable as shown by values of their semivariograms:

Lag	Semivariogram Values for Y	Semivariogram Values for T
2	0.001100	0.001036
4	0.003610	0.003322
6	0.007128	0.006520
8	0.011534	0.010022
10	0.016840	0.013721

12	0.022617	0.017192
14	0.028606	0.020698
16	0.034816	0.024259
18	0.041281	0.027985

This is not as good an outcome as is obtained with experiments 1 and 2. But it is a better outcome than that obtained using $H = 0.2$ for the rescaling, as is demonstrated by the following semivariogram values:

Lag	Semivariogram Values for Y	Semivariogram Values for T
2	0.001100	0.002734
4	0.003610	0.008766
6	0.007128	0.017205
8	0.011534	0.026448
10	0.016840	0.036209
12	0.022617	0.045369
14	0.028606	0.054621
16	0.034816	0.064020
18	0.041281	0.073704

These experiments serve to demonstrate that T is a proper rescaling of Y, provided $H = 2 - D$, for a range of fractal dimensions: $D = 1.1$, $D = 1.5$, and $D = 1.8$. Moreover, these experiments show that Y and T are statistically self-similar, herein demonstrated by their semivariograms. Although not explicitly shown, the outcomes of the three experiments imply that the power spectra for Y and T are also identical. Semivariogram results are presented for the three experiments because they show the statistical self-similarity more clearly than power spectrum results, which tend to be more variable. This statement is made with respect to the discrete Fourier transform, which is computed in Chapter 9.

With respect to power spectra, Peitgen and Saupe (1988, p. 92) show that

$$F_T(f) = \int_0^X T(x)e^{-2\pi i f x}\,dx = \frac{1}{r^H}\int_0^{rX} Y(s)e^{-2\pi i (f/r)s}\,\frac{ds}{r}$$

where F_T is the continuous Fourier transform of T, X is the resampled length of T, rX is the total length for Y, and H is related to the fractal dimension of Y as $H = 2 - D$. If, as Peitgen and Saupe (1988, p. 92) show, s/r is substituted for x and ds/r is substituted for dx in the second integral, then

$$F_T(f) = \frac{1}{r^{H+1}} F_Y\left(\frac{f}{r}\right)$$

and consequently for power spectra,

$$P_T(f) = \frac{1}{r^{2H+1}} P_Y\left(\frac{f}{r}\right)$$

Because T and Y are statistically indistinguishable, $P_T = P_Y$, and

$$P_Y(f) = \frac{1}{r^{2H+1}} P_Y\left(\frac{f}{r}\right)$$

Now if f is formally set equal to 1 and $1/r$ is replaced by f, then

$$P_Y(f) \propto \frac{1}{f^{2H+1}} = \frac{1}{f^\beta}$$

from which is realized

$$\beta = 2H + 1$$

Intuition showed that $H = 2 - D$; therefore

$$\beta = 2H + 1 = 2(2 - D) + 1; \qquad D = \frac{5 - \beta}{2}$$

Whereas a formal, rigorous proof of this relationship has not been given, the three experiments showing that T and Y are statistically indistinguishable provide an empirical path to the relationship.

Example 11.2   Calculation of Fractal Dimension Using the Periodogram (Power Spectrum)

Several one-dimensional profiles are measured across rock surfaces at Yucca Mountain, Nevada (Carr, 1989). The data for one such profile (profile A) are listed in Appendix A.

A periodogram is calculated using the program FORIER (Chapter 9). The plot of the base-ten logarithm of power versus the base-ten logarithm of frequency is shown in Figure 11.12. Notice that no one frequency predominates and a linear decay in power is observed (this linear decay in power is a typical statistical signature of ran-

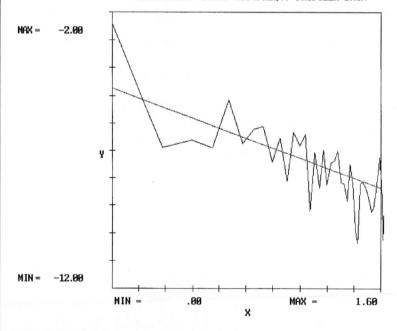

Figure 11.12   A plot using the logarithms of the power spectrum for the rock surface, Profile A (Appendix A).

dom fractals). The linear and decreasing behavior demonstrates the proportionality presented earlier:

$$P(f) \propto \frac{1}{f^\beta}$$

A least-squares, best linear fit to these data is shown in Figure 11.12. The absolute value of the slope of this line is equal to the parameter β shown in the foregoing proportionality. The slope is equal to -2.28, hence the fractal dimension for the profile is $(5 - 2.28)/2$, or 1.36.

This example visually demonstrates that this rock profile is statistically similar to fractal Brownian motion having a fractal dimension equal to 1.36. Random fractals are associated with power spectra which, if plotted as logarithms, display a linear decay in power with no one frequency predominating. In contrast, nonfractal functions, such as the variable star data (Chapter 9), are associated with spectra wherein one frequency or a few frequencies do predominate, whereas all other frequencies are associated with negligible power.   ∎

11.5 Use of the Semivariogram for Fractal Studies

Recall the following proportionality for brown noise:

$$E\{[Y(x_i) - Y(x_j)]^2\} \propto |x_i - x_j|^{2H}$$

Letting $x_i - x_j$ be written as h_{ij} allows the following:

$$E\{[Y(x_i) - Y(x_j)]^2\} \propto |h_{ij}|^{2H}$$

Consider the semivariogram defined as follows (e.g., Journel and Huijbregts, 1978):

$$\gamma(h_{ij}) = \tfrac{1}{2}E\{[Y(x_i) - Y(x_j)]^2\}$$

Then

$$\gamma(h_{ij}) \propto 2|h_{ij}|^{2H}$$

If

$$\log[\gamma(h_{ij})]$$

is plotted versus $\log(h_{ij})$, the slope of this plot is $2H$. Therefore, the semivariogram may be used as an alternative to the power spectrum for determining random fractal dimension.

Example 11.3   Estimating Random Fractal Dimension Using the Semivariogram

Rock profile A (Appendix A; Example 11.2) is again analyzed to determine its fractal dimension, this time by using the semivariogram (Figure 11.13). A plot of the base-ten logarithms of semivariance versus base-ten logarithms of lag distance (Figure 11.14) has a slope over the first eight increments of lag distance approximately equal to 0.90. Therefore, H is 0.90/2, or 0.45. In this case, the resultant random fractal dimension is $2 - H$, or 1.55, which differs from the fractal dimension estimated using the power spectrum (Example 11.2). But, if the first five cycles from the power spectrum are not used, the slope of the remaining cycles is 1.9, and the fractal dimension is equal to 1.55, exactly what is obtained using the semivariogram.

SEMI-VARIOGRAM FOR PROFILEA DATA

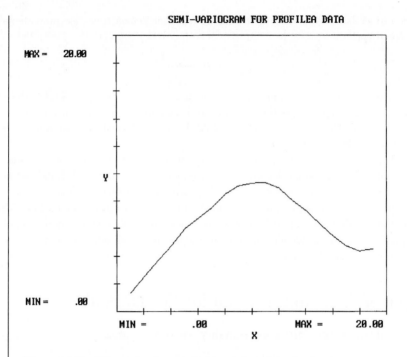

Figure 11.13   The semivariogram for the rock surface, Profile A.

LOGARITHMS OF SEMI-VARIOGRAM VALUES, PROFILEA.DAT

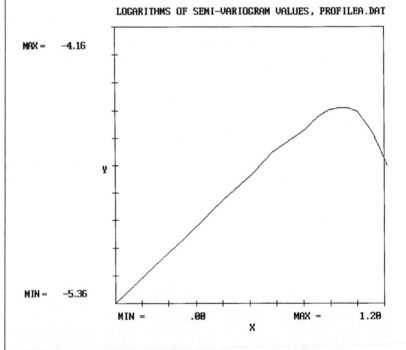

Figure 11.14   A plot of the logarithms of the semivariogram results for the rock surface, Profile A. This figure was plotted using GRAPH2D (Chapter 7) by choosing option 5 for the display of semivariogram results from FGAM (Chapter 6) and choosing the option for converting the data to logarithms; no axis rescaling was chosen.

The semivariogram and power spectrum analysis of random fractal dimension should in theory yield the same result. In practice, however, they may yield different results, due to differences in calculation procedures. Therefore, both tools really should be used for comparison when estimating random fractal dimension. Furthermore, the fractal characteristics of a random fractal may not be constant over all frequencies analyzed by a power spectrum. This is true of the rock surface profile, profile A. A rule has not been established for determining which frequencies to use when estimating the random fractal dimension from a power spectrum. This underscores the necessity to use the semivariogram in conjunction with the periodogram for comparison.

In this example, the slope of the plot based on logarithms from the semivariogram calculation is determined using only the first eight lag increments. This is appropriate because this method is based on that part of the semivariogram which shows increasing behavior away from the origin. Notice that the semivariogram attains a sill, then decreases (Figure 11.22). Using the first eight lag increments avoids that part of the semivariogram where the slope begins to decrease as the sill value is approached. ∎

11.6 Estimating the Fractal Dimension in Practice

A fundamental question arises regarding fractal dimension: Is the fractal dimension estimated using the divider, or compass, method the same as that resolved using the power spectrum or semivariogram? The answer to this question appears to be no. Hough (1989), for example, acknowledges that estimation of fractal dimension based on the spectral notion is not necessarily consistent with estimation based on the divider method.

In application to coastlines, fractal dimensions estimated using dividers differ substantially from those estimated using semivariograms, as seen in Table 11.1 (from Carr and Benzer, 1991). This table lists coastlines in descending order with respect to divider dimensions. A poor correlation in terms of numerical magnitude is found between these dimensions and those estimated using semivariograms. The second largest random fractal dimension, 1.77, is obtained for

Table 11.1 Estimates of Fractal Dimension for Selected Coastlines

Coastline	Divider	Variogram
West Coast, Great Britain	1.27	1.67
von Koch (artificial)	1.21	1.67
North Coast, Australia	1.19	1.50
West Shore, Puget Sound, USA	1.19	1.86
East Shore, Puget Sound, USA	1.15	1.70
South Coast, Australia	1.13	1.30
East Shore, Hood Canal, Puget Sound, USA	1.08	1.77
West Shore, Gulf of California, Mexico	1.03	1.33
East Shore, Gulf of California, Mexico	1.02	1.22

Fractal Dimension Estimates (spanning header over Divider and Variogram)

Source: From Carr and Benzer (1991).

the eastern shore of Hood Canal (Puget Sound, Washington State), which is associated with the third smallest divider dimension, and this is just one example. Intuitively, there is no reason to expect any correlation between the two fractal dimension concepts.

Recall that the semivariogram and power spectrum methods resolve the stochastic nature of random fractals, features associated with a random component. Fractal dimension estimated by the semivariogram or power spectrum indicates the similarity of this random component to noise, white noise being the most extreme. Fractal dimension estimated using dividers, on the other hand, in application to lines, resolves the change in length as the scale of measure changes. This change is related to the complexity of the shape of the line. For instance, if one considers rock surfaces (Table 11.2; from Carr and Benzer, 1991), specifically planar surfaces in rock, profiles across these surfaces appear visually similar to straight lines, with subtle aberrations. Divider fractal dimensions are nearly equal to one. Random fractal dimensions, though, may be much greater.

Topography, including the texture of rock surfaces, is said to be self-affine, rather than self-similar (Mandelbrot, 1985). An affine transformation (in two dimensions, for instance) is one that treats the two dimensions differently. In the three experiments (Section 11.4) showing the proper rescaling of Y, the y direction was scaled by r^H whereas the x direction was scaled by r; hence the three realizations of Y presented earlier are self-affine. Because the y direction and the x direction are associated with different scaling properties, the divider method (which is predicated on the assumption that the feature under study is self-similar) may yield ambiguous results for self-affine fractals (Peitgen and Saupe, 1988, pp. 63–64). (Notice that the word ambiguous is used, not the word incorrect.)

Given a function T, for example, suppose its fractal dimension is to be determined using the divider method based on "rulers" y. In this case, y is a function of T and x such that

$$y = \sqrt{[\Delta T(x)]^2 + (\Delta x)^2}$$

If $T(x) = x^H$ (Peitgen and Saupe, 1988, p. 63; this assumes that $T(x)$ is a self-affine process), then

$$y = \sqrt{\Delta T(x)^2 + (\Delta x)^2} = \sqrt{(\Delta x^H)^2 + (\Delta x)^2}$$

which simplifies to

$$y = \Delta x \sqrt{1 + \frac{1}{\Delta x^{2-2H}}}$$

Table 11.2   Estimates of Fractal Dimensions for Rock Surfaces

	Divider	Variogram	Spectral
Surface 1	1.034	1.559	1.420
Surface 2	1.003	1.219	1.340
Surface 3	1.002	1.593	1.470

Source: Carr and Benzer (1991).

If

$$\Delta x \ll 1$$

then the second term of the square root becomes large, and

$$y \approx \frac{\Delta x}{\Delta x^{1-H}} \approx \Delta x^H$$

Suppose that the total length of T is 1 and is regularly sampled at N equal intervals:

$$\Delta x = \frac{1}{N}$$

In this case, if

$$y \approx \Delta x^H$$

then with the divider method N is proportional to y^{-D}, and D must equal $1/H$ because

$$N \propto \frac{1}{y^D}, \qquad \wedge : y \approx \Delta x^H$$

$$\therefore N \propto \frac{1}{(\Delta x^H)^D}, \qquad \because N = \frac{1}{\Delta x}$$

$$\text{then } D \approx \frac{1}{H}$$

But if

$$\Delta x \gg 1$$

then the second term of the square root becomes insignificant. In this case,

$$y \approx \Delta x$$

Therefore, D is approximately 1, hence the ambiguity in the value of the fractal dimension depending on the size of the rulers y chosen for the divider method (that is, the size of the increment in the x direction represented by the ruler length y).

Notice in this development that if y is much less than a characteristic length (1 in this example), then D is approximately equal to $1/H$; if y is much greater than this characteristic length, then D is approximately equal to 1. Recall the discussion of rock surfaces, cf. Table 11.2. Rulers used in the divider method to obtain fractal dimensions listed in this table were relatively large, and self-similar fractal dimensions (those based on the divider method) are approximately equal to 1.

As Peitgen and Saupe (1988, p. 63) note, the association of a self-similar fractal dimension D with a self-affine fractal is implicitly imposing a scaling between the y direction and the x direction. The resultant dimension D depends strongly on the scale, large or small (relative to the "characteristic" length), at which the self-affine fractal is examined. Herein is shown that D is approximately equal to $1/H$ if a small-scale examination is used, but is closer to 1 if a large-scale examination is used.

More is said about this discrepancy in Section 11.8.1. An awareness of self-similar versus self-affine fractals is important as is recognizing how estimates of fractal dimension will vary depending on the method used for the estimation.

The method for estimating fractal dimension depends on the type of fractal to which the fractal dimension estimation method is being applied. As shown in Section 11.8.1, the method chosen to estimate fractal dimension also depends on the intended use of the fractal dimension.

11.7 Simulation of Fractal Topography and Clouds

Now that the concept of a random fractal has been introduced, simulations of random fractals are presented to suggest some applications of the study of random fractals to the earth sciences. Peitgen and Saupe (1988, pp. 95–112) describe algorithms for simulating surface topography, clouds (actually simulated topography and simulated clouds are the same thing, depending on what label one wishes to assign to the simulation), and ocean waves. Herein an example is presented showing how to simulate topography/clouds.

The computer program SPFM2D (Figure 11.15, page 535) is based on the algorithm by Peitgen and Saupe (1988, p. 108). This algorithm generates a two-dimensional simulation of fractal Brownian motion (such a fractal is discussed in Section 11.4) having a fractal dimension equal to $3 - H$, where H is a parameter entered into the program and has a range $0 < H < 1$ (H is also explained in Section 11.4).

Once the program SPFM2D has been successfully completed, the program FORINV2 (Chapter 10) is used to inverse Fourier transform the frequency domain image to the spatial domain. Once this task is complete, the simulation

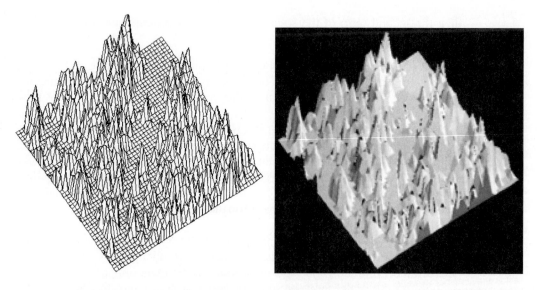

Figure 11.16 Simulated topography using SPFM2D with a beginning random number seed of 993399 and $H = 0.2$. The simulation was inverse Fourier transformed using FORINV2 (Chapter 10). The display on the left is from GRID3D (Chapter 7) and shows only the grid outlines. A tip angle of -20 and a spin angle of 145 were used to obtain this display. The display on the right is also from GRID3D and is the shaded relief map using a sun angle defined by DIX $= -100$, DIY $= -0.1$, and DIZ $= +50$.

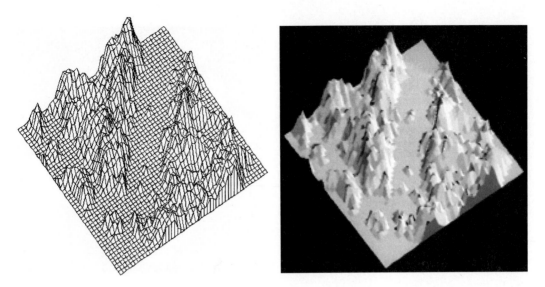

Figure 11.17   Simulated topography using $H = 0.5$, created and displayed as described for Figure 11.16.

is complete and can be considered a model of topography or clouds, depending on the application.

Several simulations (Figures 11.16 through 11.19) represent topography. In the first simulation (Figure 11.16), a 50 × 50 grid is developed using SPFM2D. The fractal dimension parameter H is set equal to 0.2 for this simulation and the fractal dimension is equal to $3 - H$, or 2.8 in this case, representing rough topography.

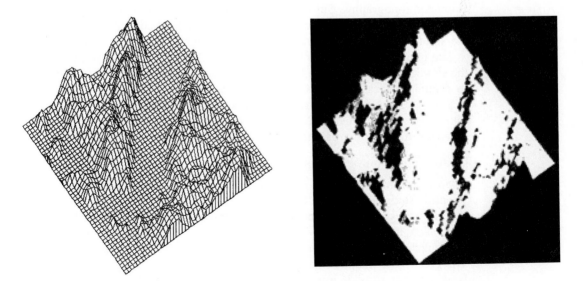

Figure 11.18   Simulated topography using $H = 0.7$, created and displayed as described for Figure 11.16.

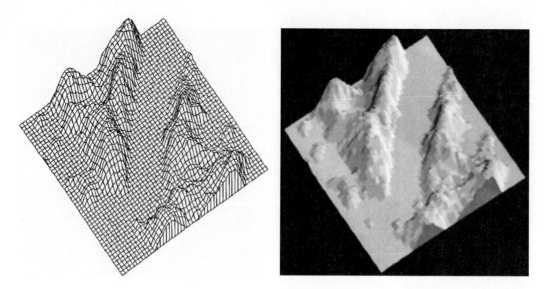

Figure 11.19 Simulated topography using $H = 0.9$, created and displayed as described for Figure 11.16.

Adjusting the fractal dimension parameter H adjusts the "smoothness" of the topography. For example, 50×50 simulations developed using H equal to 0.5 (Figure 11.17), 0.7 (Figure 11.18), and 0.9 (Figure 11.19) show how the "roughness" of the topography can be decreased by increasing the value of parameter H.

11.7.1 Clouds

Previous displays show how simulations from SPFM2D can be considered simulated topography. But by "slicing" horizontally through the simulations, clouds can be created. Such creations can be accomplished simply by displaying the simulations as digital images using the program IMAGIN (Chapter 10).

The interface program INTFC4 (Figure 11.20, page 537) is used to convert the simulations after inverse Fourier transformation by FORINV2 (Chapter 10) to digital images. An example of such a "cloud," produced by slicing a 128×128 simulation (Figure 11.21) was created by converting the simulation into a digital image using INTFC4, then displaying the image using IMAGIN (Chapter 10) with the palette GREY16.PAL (Chapter 10).

11.7.2 Why Are These Simulations Useful?

Showing simulations alone is a shallow exercise if the purpose of this text is scientific study. Fractals are extremely useful for classification. In other words, the irregularity of a coastline is, in a sense, classified by its associated fractal dimension; the larger the fractal dimension is, the more irregular is the coastline.

Similarly, fractal dimension of surface topography is useful for classification. The greater the fractal dimension of the surface is, the "rougher" the surface is, which may provide important clues regarding age of the surface, lithology, extent of erosion, and so on. Simulated topographic models (Figures 11.16 through 11.19) help to show the appearance of topography as a function of fractal dimension of the surface. Computing fractal dimension for an actual

SIMULATED CLOUDS

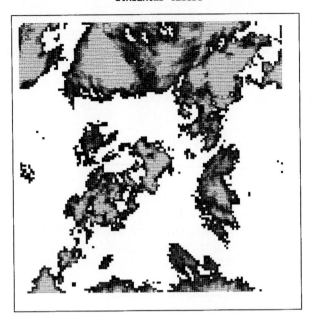

Figure 11.21   Simulated clouds. The simulation was created using SPFM2D/FORINV2 with $H = 0.5$. The digital image was then created using INTFAC and displayed using IMAGIN (Chapter 10) with the palette GREY16.PAL (Chapter 10).

surface may be an arduous task (one approach to this is to compute the fractal dimension of a profile across the surface and add 1 to this value to obtain the fractal dimension of the surface). Instead, the simulations can be compared to the actual surface to approximately determine its fractal dimension. Once the fractal dimension for the surface is known or approximated, its numerical size can be evaluated against the geomorphologic and lithologic information available for the actual surface to determine whether fractal dimension is useful for classifying landscapes.

For instance, a display of actual surface topography was presented in Chapter 7 to demonstrate the use of a three-dimensional grid display program, GRID3D. Its appearance shows some similarity to simulated topography (Figure 11.17), a simulation having a fractal dimension of 2.5 (i.e., $3 - H$). But some of the actual topography bears a resemblance to another simulation (Figure 11.19), a smoother topographic simulation developed using $H = 0.9$. In this case, actual topography is largely smooth, but the smoothness is interrupted occasionally by sudden abrupt changes in slope: mountain ranges change abruptly to flat valley floors. This digital topography is for a portion of the Great Basin, near Caliente, Nevada (USA), a tectonically active basin-and-range province. The fact that this region is tectonically active may explain why the actual topography bears a resemblance to simulated topography for two different fractal dimensions ($D = 2.5$ and $D = 2.1$).

Just as simulations of topography are useful for classifying actual topographic surfaces, simulations of clouds are useful for comparison to actual weather patterns. Perhaps different weather fronts will be found to have unique fractal signatures. For example, cloud formations associated with isolated thunderstorms may be different from those associated with a cold front. If so, perhaps the fractal character of clouds will be useful for weather prediction.

Of course, the simulated clouds may be considered to be other features of more geologic significance. For example, they may represent the spatial pattern of an ore deposit. In this sense, the simulation to yield ''clouds'' is similar to nonconditional simulation, a geostatistical method (Journel, 1974; a computer program is given in Carr and Myers, 1985).

11.8 Selected Applications of the Fractal Dimension

11.8.1 Analysis of Rock Surfaces

Fractal description of rock surfaces (Table 11.2) from self-similar dimension correlates very well with a notion called the joint roughness coefficient, JRC (Barton and Choubey, 1977). This coefficient describes the similarity of the shape of a rock surface to one of ten standard profiles (Figure 11.22). The peak shear strength for a rock surface is related to the joint roughness coefficient

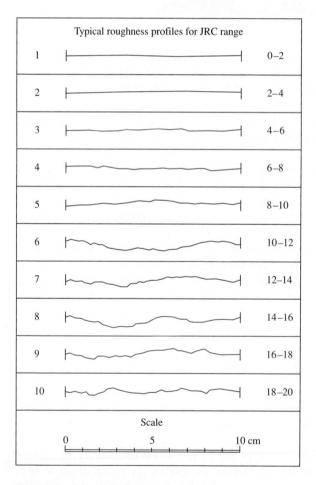

Figure 11.22   The 10 standard rock profiles describing joint roughness coefficient (JRC) (from Barton and Choubey, 1977).

by Barton (1974) as follows:

$$\tau = \sigma \tan\left[\phi + \text{JRC}\left(\log\frac{\sigma_j}{\sigma}\right)\right]$$

where τ is the peak shear strength, σ is the normal stress on the rock surface, ϕ is the friction angle for the rock surface, and σ_j is the uniaxial compressive strength of the rock.

The joint roughness coefficient JRC and the self-similar fractal dimension are related as (Carr and Warriner, 1989)

$$\text{JRC} = -1022.55 + (1023.92)D$$

or approximately as

$$\text{JRC} = 1000(D - 1)$$

In further work, Lee and others (1990) found this relationship to be nonlinear of the form

$$\text{JRC} = -0.87804 + 37.7844\left(\frac{D-1}{0.015}\right) - 16.9304\left(\frac{D-1}{0.015}\right)^2$$

Both the Carr and Warriner (1989) and Lee and others (1990) equations are applied in the range $1 < D < 1.015$ (for greatest accuracy). Each of these equations yields similar results for small values of self-similar fractal dimension ($1 < D < 1.006$).

Is this a valid example? Recall from Section 11.6 that the self-similar fractal dimension may be ambiguous when measured for a self-affine fractal. In this particular application, the objective is to measure the shear resistance of rock surfaces. The magnitude y of the surface undulations (topography) relative to the horizontal direction x is of paramount importance to this application (see, for example, the ten standard profiles in Figure 11.22). In this case, the y direction cannot be assumed to scale differently than (i.e., be independent of) the x direction. If y is very small relative to x, as is the case with profiles 1–6 (Figure 11.22), the surface has little resistance to shear. On the other hand, if y is larger, as with the remaining profiles, the surface topography provides increased shear resistance. In this case, because the size of the y-direction asperities relative to the x direction is of utmost importance, an affine transformation to scale the y-direction measures differently than the x-direction measures is definitely not desirable. Treating the profile as a self-similar profile where y and x are scaled the same is exactly what is desirable for an engineering application to estimate JRC. Therefore, this is a valid application.

Because JRC is a measure of the relative size of the surface undulations y relative to the horizontal direction x, the self-similar dimension is the more appropriate descriptor of the rock surface (if the objective is to estimate JRC). Moreover, the self-similar dimension correlates well with JRC (Carr and Warriner, 1989; Lee et al., 1990). In contrast, the fractal dimension assessed using the power spectrum does not seem to correlate well with JRC (Carr and Benzer, 1991; Carr, 1989).

11.8.2 The Shape and Texture of Concrete Aggregate

Peleg and Normand (1985) investigated the shape and texture of instant coffee particles. The intent of this research was to correlate surface characteristics with properties of the coffee particles, such as rate of dissolvability. Nevertheless, this

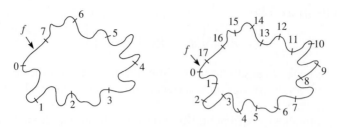

Figure 11.23   The outline of a particle showing several experiments to measure
circumferential length. This is an application of the divider method for computing fractal
dimension.

research was a landmark application of fractals for characterizing the shape of
particles. In this case, the notion of the self-similar fractal is used. Two experiments to measure the circumference of the particle are shown in Figure 11.23.
Depending on the complexity of the shape of the particle, the circumferential
length increases as the ruler (step) size used to measure it decreases.

   Research by Peleg and Normand (1985) prompted other research devoted
to the characterization of particle shape. In particular, an effort was made by
Carr et al. (1990) to characterize the shape of concrete aggregate pieces. In
the design of concrete, including both asphalt and portland cement, particle
shape governs the quality of the bond between aggregate and binder (angular,
rough pieces bond better than do smooth, rounded pieces), workability of the
mix (flat, elongated pieces hinder workability; rounded pieces aid workability),
air voids, and amount of binder required (the more surface area the aggregate
has, the more binder is required for the mix (Carr and others, 1992)).

   Photographs (Figures 11.24 and 11.25) of silhouettes of several aggregate
pieces are made by illuminating particles from behind; fractal dimensions of
these silhouettes are shown in the figure captions. In general, the fractal dimension adequately characterizes particle shape, being greater for rough, irregular
particles, and smaller for smooth, rounded or smooth, flat pieces. This dimension

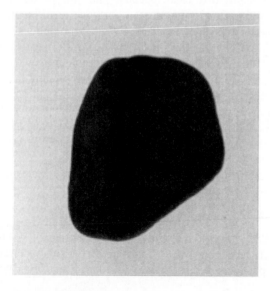

Figure 11.24   A smooth concrete
aggregate particle silhouette that has
a fractal dimension of 1.04.

Figure 11.25   A particle of concrete aggregate that is rougher than that shown in Figure 11.24, having a fractal dimension of 1.08.

cannot distinguish, though, between a smooth sphere and a smooth cube. Notwithstanding this problem, the fractal dimension seems to yield an accurate measure of surface area for large particles, something that is difficult to measure (Carr and others, 1992).

This example suggests an application of the fractal dimension for the study of particles comprising clastic sedimentary rocks, particularly conglomerates and breccias. Whether a clastic sedimentary rock is composed of rough, angular grains or smooth, rounded grains says something about its origin. Therefore, the fractal dimension might prove to be a useful descriptor for such rocks as an aid to determining origin of the sediments.

11.8.3 Chaos? Characteristics of Rainfall for a Storm

Recall that Poincaré was the first (at least in writing) to contemplate the notion of chaos. Although no single definition has yet been chosen for chaos, Poincaré's original speculation is most widely accepted: that [chaos] describes a seemingly small disturbance which may become a very large disturbance elsewhere (a spatial notion) or at a later time (a temporal notion).

An example of chaos is the physical phenomenon known as turbulence. This phenomenon is well known to geomorphologists studying stream flow, civil engineers designing water delivery systems, and atmospheric physicists studying the behavior of clouds and wind. Chaos offers a deterministic approach to the modeling of turbulence, a process once considered to be completely random (Berge et al., 1984).

Turbulence also displays a fractal geometry in that it is composed of endless scales of swirls and eddys. Storms are atmospheric phenomena which display turbulent behavior. Herein is examined the rainfall amounts recorded over the duration of a single storm which happened in Boston, Massachusetts, on 25 October 1980 (Rodriguez-Iturbe et al., 1989). If clouds and storms display turbulent behavior, it is interesting to ponder the influence this turbulence has on rainfall amounts and whether these amounts are likewise chaotic.

Rainfall amounts for the Boston storm (Figure 11.26; Appendix A) may be fractal in time. Because these data represent a time series and because the *y*-direction scale for rainfall amount differs from the *x*-direction scale for time, these data display an affine geometry. Because these data represent a time series amenable to Fourier analysis and because these data represent affine geometry, a power spectrum approach is chosen to estimate fractal dimension.

A log-power versus log-frequency plot (Figure 11.27) based on the power spectrum for the Boston rainfall data shows a linear decay in power with frequency, a typical spectral signature for fractals. These data are therefore considered to be fractal, having an estimated random fractal dimension of 1.77. That these rainfall data are fractal leads to the interesting conclusion that the rainfall amounts are also chaotic, that small disturbances in the storm may have a large impact on rainfall amounts later on.

This example is for one storm. As Rodriguez-Iturbe et al. (1989) acknowledge, this is a preliminary finding and much more rainfall data is needed to fully understand the meaning of this one outcome. For example, do different types of storms (thunderstorms, cold fronts, snow versus rain, etc.) have different fractal signatures? If so, these signatures could be extremely valuable for classifying storms, hence predicting their behavior. Moreover, how storms are related spatially as well as temporally is still unclear, so considerable data are still needed to enable an adequate study of this relationship. Therefore, this example shows an exciting application of fractals to the study of hydrology and meteorology and also shows the need for considerable future research.

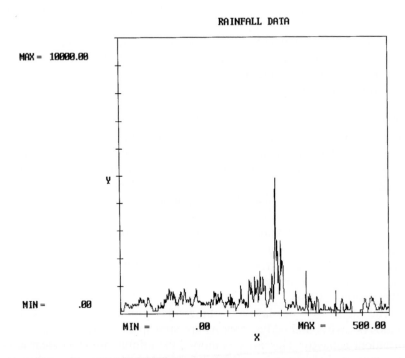

Figure 11.26   Rainfall data from a storm that occurred at Boston, Massachusetts (USA) on 25 October 1980. The data were measured directly from a figure presented in Rodriguez-Iturbe et al. (1989). These data were displayed using GRAPH2D (Chapter 7) by choosing option 6 for displaying one-dimensional functions and rescaling *y* by 100.

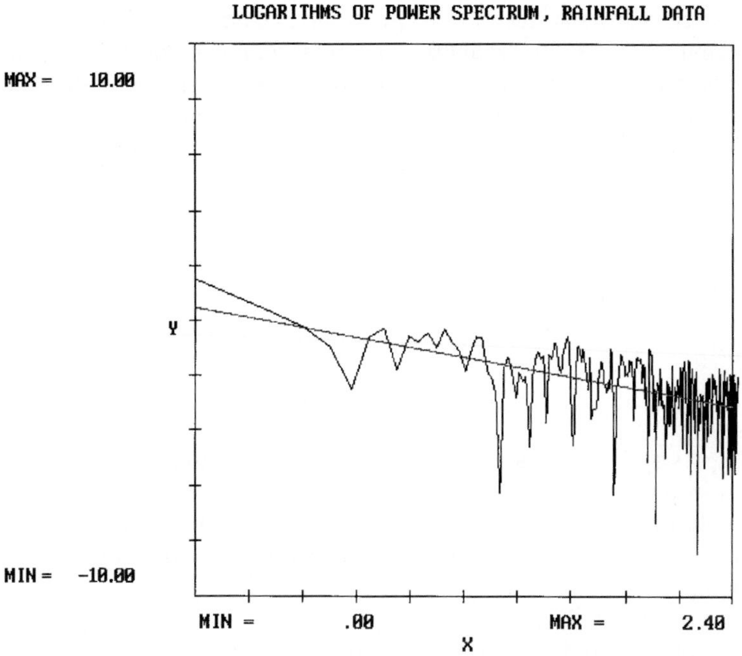

LOGARITHMS OF POWER SPECTRUM, RAINFALL DATA

MAX =   10.00

Y

MIN =   -10.00

MIN =   .00   MAX =   2.40

X

Figure 11.27   A plot of the logarithms of the power spectrum for the rainfall data. This plot was created using GRAPH2D (Chapter 7) to display the power spectrum computed by FORINV. Option 1 for displaying general files was chosen along with the option to convert the data to logarithms.

11.9   Visualizing Periodicities and Fractal Behavior in One-Dimensional Time Series: The Wavelet Transform

A final aspect of this chapter deals with the display of digital images created from one-dimensional time series; this is a visual way to approach the analyses described in Chapter 9 and in this chapter. A transform based on the use of wavelets is used to form two-dimensional images from one-dimensional time series. The resultant image is also a way in which the fractal aspects of the time-series profile may be assessed.

In brief, a wavelet transform is a convolution, exactly the type of filtering process that was described in Section 10.3.5. The exception is that rather than convolving a two-dimensional image with a two-dimensional PSF, a one-dimensional function is convolved with a one-dimensional filter (wavelet). Yet, how does this yield a two-dimensional image?

In answering this question, acknowledgment is given to the fact that the size of the wavelet is a function of the row number of the image. Therefore, to develop a two-dimensional image of a one-dimensional time series, the time series is convolved with a wavelet N times, where the size of the filter is equal to i, $i = 1, 2, \ldots, N$ (that is, the filter size becomes larger and larger). These N filtered profiles are assembled as rows of a digital image, which forms the two-dimensional image.

Wavelets are described in varying detail in Combes et al., (1989). An excellent description of wavelet transforms is given by Piech and Piech (1990), but they do not refer to their transform as a wavelet transform; instead they use the term "fractal fingerprint."

A simple wavelet transform is given by

$$g(i,j) = \sum_{k=j-(i-1)/2}^{j+(i-1)/2} \text{PSF}_i(L_k)f(x_k), \qquad L_k = 1, \dots, i$$

for which j ranges from 1 to N, the size of the time series, and i, the filter size, ranges from 1 to N. Moreover,

$$\text{PSF}_i(L) = \frac{1}{i}$$

Papers by Combes et al. (1989) describe the computation of wavelet convolutions in the frequency domain. Schowengerdt (1983), though, shows that the convolution between a function and a PSF is applicable in both the spatial and frequency domains; a wavelet transform is more easily computed in the spatial domain.

The program WAVELET (Figure 11.28, page 538) is applied to the time series for the variable star data (Chapter 9; Appendix A). The resultant transform (Figure 11.29) displayed using the palette shown in Table 11.3 shows regular periodicities. In a wavelet transform, the "hills" and "valleys" of a time-series profile are smeared out in the vertical direction. The more dominant the hill or valley, the farther it extends, from top to bottom in the image. In some instances, hills and valleys blend together; this, too, is a measure of the

Table 11.3 Intensity Levels for the Palette BINARY.PAL

	Intensity		
Red	Blue	Green	Color
0	0	0	Black
63	63	63	White
0	0	0	Black
63	63	63	White
0	0	0	Black
63	63	63	White
0	0	0	Black
63	63	63	White
0	0	0	Black
63	63	63	White
0	0	0	Black
63	63	63	White
0	0	0	Black
63	63	63	White
0	0	0	Black
63	63	63	White

WAVELET TRANSFORM, VARIABLE STAR DATA

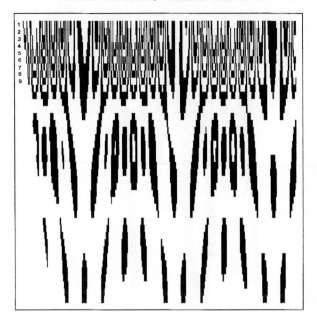

Figure 11.29   A wavelet transform of the variable star data (Chapter 9). This display was
created as follows: (1) the wavelet transform was created by WAVELET, then converted
to 8-bit image data using IMAG41 (Chapter 10); (2) a palette, BINARY.PAL, was created
(see Table 11.3); (3) the wavelet transform was displayed using IMAGIN (Chapter 10)
using the palette BINARY.PAL.

dominance of these features in a time series. The wavelet transform is simply
a different approach, in comparison to Fourier transforms, for identifying periodicities in data.

As a last example, a one-dimensional fractal profile is developed using the
first row in the image of the Humboldt Sink (Chapter 10). That this profile is
fractal is demonstrated momentarily. The wavelet transform for this profile
(Figure 11.30) shows that for fractal profiles the periodicities are not as easy
to follow; in fact, they are very complex. Notice in Figure 11.30 that hills and
valleys merge in bizarre ways.

Aside from a visual display of fractal behavior, a wavelet transform can be
used to quantitatively assess fractal characteristics of the function under study
(this discussion is developed from Piech and Piech (1990)). Recall the Mandelbrot relationship: $L = Ny^D$. This relationship is used to establish an analogy:
$G = NK^D$, where N is the number of features (the hills/valleys, i.e., the black
features in Figures 11.29 and 11.30) that are resolved by each filter size K,
and D is a power term. Once N is determined for several filter sizes K, these
data are transformed to logarithms and plotted. If the slope of the plot is linear
and different from zero, a power law is observed between N and K and the
feature under study possesses scaling (hence is fractal).

A regular wavelet transform is seen in Figure 11.29, developed using the
variable star data (Chapter 9), data that are associated with only two principal
periodicities. The first 6 filter steps are highlighted in this figure for clarity.
For each of these filter steps, the number of features (black) are simply counted.

WAVELET TRANSFORM, 1ST ROW OF HUMBOLDT IMAGE

Figure 11.30   A wavelet transform using the first row of the digital image of Humboldt Sink, Nevada (Chapter 10). This transform was created and displayed in the same manner as that described for Figure 11.29.

This yields the following data:

Filter Size	Number of Features
1	90
2	95
3	92
4	98
5	97
6	91

Therefore, for the variable star data, D is approximately zero; hence the data are not fractal.

In contrast, the Humboldt Sink image data are found to be fractal. The first 6 filter steps are highlighted for clarity in Figure 11.30. The number of features (black) for each filter step are as follows:

Filter Size	Number of Features
1	43
2	28
3	22
4	17
5	15
6	14

The slope of the log–log plot of these data is −0.63, hence D is 0.63 and a power scaling is observed. *Note:* The power term D for these count data is a fractional dimension, but is not associated with the same numerical range as is observed for the coastline relationship. In this case, D is a similar notion to the power term for the Sierpinski carpet (Mandelbrot, 1983), which has a value of $\log(3)/\log(4)$, or 0.79. What matters in this case is whether the log–log plot of the count data is linear and that the slope is significantly different than zero. A zero or near-zero slope is indicative of nonfractal behavior, such as is observed for the variable star data.

11.9.1 Specific Information for the Program WAVELET

A detailed user's guide for the program WAVELET is shown in Figure 11.28. In using this program, an input file is required for the one-dimensional time series. The format for this file is identical to that used in Chapter 9. Once the name of this input file is specified, WAVELET requests a name be entered for the output, 32-bit binary integer file to contain the digital image of the wavelet transform. This is the total information required for input to WAVELET.

Once WAVELET has successfully created the 32-bit integer output file, the program IMAG41 (Chapter 10) is used to convert this image file to an 8-bit binary integer file compatible with the image display program IMAGIN (Chapter 10). When using IMAG41, a user must know how many rows and pixels per row comprise the wavelet transform image. These values are determined as follows. Given that the one-dimensional time series is composed of N values, then

$$\text{number of rows} = \frac{10(N-4)}{4}$$

$$\text{number of pixels per row} = N$$

The equation for the number of rows comes from the following information:

1. i, the row increment and filter size for the wavelet transform, increments from 3 to $N/2$, by increments of 2 such that i is always an odd number.
2. Each row of the wavelet transform is actually written 10 times to rescale the vertical direction in proportion to the horizointal direction. This is done simply to make the digital image larger for a better display by IMAGIN.

Chapter Summary

- The mathematical foundation for the notion of fractals was laid at the end of the nineteenth century by several mathematicians.
- Lewis F. Richardson encountered the problem of measuring the lengths of rivers and coastlines when the lengths of these features seemingly increase without bound as ruler sizes used to measure them decrease. He introduced a power relation between total length and ruler size to explain his observations. This power turns out to be the fractal dimension.
- Benoit B. Mandelbrot coined the term fractal, thereby grouping all concepts, such as Richardson's power term, into a single discipline. Mandelbrot's development of the theory of fractals requires a rigorous understanding of set theory. However, as this chapter shows, numerous practical applications exist for fractals that do not require an understanding of set theory.

- Mandelbrot's development of the notion of fractals is one of the most important developments in mathematics since the time of Carl Gauss.

- Two different fractal notions exist: the self-similar fractal whose dimension may be estimated using, for example, the divider method, and the random fractal (which may be self-similar or self-affine).

- Fractal dimension is useful for characterizing the shape of rock surfaces and concrete aggregate and shapes inherent to the earth sciences in general.

- Topographic simulations are an interesting application of fractals and have been used extensively for computer graphics simulations.

- Chaos is a notion first envisioned by Henri Poincaré and is related to fractal geometry. Chaos is a function of the random interaction of physical influences on a process. Rainfall amounts in time over the duration of a storm are fractal, hence rainfall appears to be chaotic. This knowledge may lead to important discoveries in hydrology and meteorology.

- Wavelet transforms offer a visual way to interpret the fractal nature of one-dimensional functions. A fractional dimension D may be interpreted from the wavelet transform.

Exercises

(An asterisk indicates an advanced or graduate student problem)

1. Repeat Example 11.1 using Figure 11.1 and the same ruler sizes.

2. Select a map of a coastline (one not used in this chapter) *and* a river. Compute their fractal dimensions using the divider method. What did you learn from this exercise?

3. Continue with Problem 2: Find two rivers, each flowing over a different rock type (or tectonic province). How do their fractal dimensions compare? Do you see any promise for using the fractal dimension to classify stream drainage patterns on different lithologies or different tectonic provinces?

4. Compute the random fractal dimension of PROFILEB (Appendix A), using the power spectrum and semivariogram methods. What did you discover? How does this profile compare to brown noise? How do the results from the semivariogram compare to those from the power spectrum?

*5. Explore the use of the fractal dimension for the classification of conglomerates and breccias. Use the application by Carr et al. (1990) for the characterization of the shape and roughness of concrete aggregate as a guide.

6. Are the time series data of runoff values for Cave Creek, Kentucky (Chapter 9), fractal? Discuss. What about the sunspot or variable star data (Chapter 9)? Discuss.

7. Select a topographic map showing some variable terrain. Arbitrarily choose two points on this map and draw a line between them. Then measure the elevations along this line. What is the random fractal dimension for this profile? Use the semivariogram method to make this determination.

8. After completing Problem 7, use the program SPFM2D to simulate topography using $H = 2 - D$, where D is the fractal dimension you found for

your profile in Problem 7. Use the program FORINV2 to inverse Fourier transform your simulation, then display the simulation using GRID3D (Chapter 7). Do you think the simulation has an appearance similar to the actual topography (as best as you can judge from the topographic map)?

*9. Comment on the similarity between the simulations yielded by SPFM2D and nonconditional simulation (Carr and Myers (1985) give a computer program for nonconditional simulation). Experiment with the two methods, comparing the semivariograms for simulations.

10. Use the programs SPFM1D and CFRINV to simulate several one-dimensional fractal profiles. Then use the program WAVELET to produce a wavelet transform image, as described in Section 11.9. Compute the fractal dimension of each profile using the wavelet transform, following the procedure outlined in Section 11.9. How do these dimensions compare to the fractal dimensions of the simulations based on the parameters H that you used to create each simulation?

11. Create some cloud simulations on your own using SPFM2D, converting them to digitial images using INTFAC, then displaying them using IMAGIN (Chapter 10) with the palette GREY16.PAL (Chapter 10).

12. Assess the fractal dimension of the rainfall data using the semivariogram method. How does this fractal dimension compare to the value of 1.77 determined using the power spectrum (see Section 11.8.3)?

References and Suggested Readings

BARTON, N.R. 1974. Estimating the shear strength of rock joints. *Advances in Rock Mechanics* **IIA:** 219–220.

BARTON, N.R., and V. CHOUBEY. 1977. The shear strength of rock joints in theory and practice. *Rock Mechanics* **10:** 1–54.

BERGE, P., Y. POMEAU, and C. VIDAL. 1984. *Order Within Chaos: Towards a Deterministic Approach to Turbulence.* New York: Wiley-Interscience.

BERRY, M.V., and Z.V. LEWIS. 1980. On the Weierstrass–Mandelbrot fractal function. *Proceedings of the Royal Society, London, Series A* **370:** 459–484.

CARR, J.R. 1989. Fractal characterization of joint surface roughness in welded tuff at Yucca Mountain, Nevada. In A.W. Khair (ed.), *Rock Mechanics as a Guide for Efficient Utilization of Natural Resources: Proc. 30th U.S. Symposium,* Rotterdam: Balkema, pp. 193–200.

CARR, J.R., and W.B. BENZER. 1991. On the practice of estimating fractal dimension. *Mathematical Geology* **23**(7): 945–958.

CARR, J.R., and D.E. MYERS. 1985. COSIM: A FORTRAN IV program for coconditional simulation. *Computers and Geosciences* **11**(6): 675–706.

CARR, J.R., and J.B. WARRINER. 1989. Relationship between the fractal dimension and joint roughness coefficient. *Bulletin of the Association of Engineering Geologists* **26**(2): 253–263.

CARR, J.R., G.M. NORRIS, and D.E. NEWCOMB. 1990. Characterization of aggregate shape using fractal dimension. *Transportation Research Record, 1278: Soils, Geology and Foundations,* pp. 43–50.

CARR, J.R., M. MISRA, and J. LITCHFIELD. 1992. Estimating surface area for aggregate in the size range: one millimeter or larger. *Transportation Research Record. 1362: Soils, Geology, and Foundations,* pp. 20–27.

COOMBS, J.M., A. GROSSMANN, and P. TCHAMITCHIAN, eds. 1989. *Wavelets: Time–Frequency Methods and Phase Space.* Berlin: Springer.

DUBOIS REYMOND, P. 1875. Versuch einer Classification der Willkurlichen Functionen reeller

Argument nach ihren Anderungen in den kleinsten Intervallen. *J. für die reine und angewandte Mathematik (Crelle)* **79**: 21–37.

GLEICK, J. 1987. *Chaos: Making a New Science.* New York: Viking.

HOUGH, S.E. 1989. On the use of spectral methods for the determination of fractal dimension. *Geophysical Research Letters* **16**(7): 673–676.

JOURNEL, A.G. 1974. Geostatistics for conditional simulation of ore bodies. *Economic Geology* **69**(5): 673–687.

JOURNEL, A.G., and C.J. HUIJBREGTS. 1978. *Mining Geostatistics.* London: Academic.

LEE, Y.H., J.R. CARR, D.J. BARR, and C.J. HAAS. 1990. The fractal dimension as a measure of the roughness of rock discontinuity profiles. *International Journal of Rock Mechanics and Geomechanics Abstracts* **27**(6): 453–464.

LORENZ, E. 1979. Predictability: Does the flap of a butterfly's wings in Brazil set off a tornado in Texas? Address at the annual meeting of the American Association for the Advancement of Science, Washington, DC, 29 December 1979.

MANDELBROT, B.B. 1967. How long is the coast of Great Britain? Statistical self-similarity and the fractional dimension. *Science* **156**: 636–638.

MANDELBROT, B.B. 1983. *The Fractal Geometry of Nature.* New York: Freeman.

MANDELBROT, B.B. 1985. Self-affine fractals and the fractal dimension. *Physica Scripta* **32**: 257–260.

MANDELBROT, B.B. 1989. Fractal geometry: What is it, and what does it do? In M. Fleischmann, D.J. Tildesley, and R.C. Ball (eds.), *Fractals in the Natural Sciences.* Princeton, NJ: Princeton University Press, pp. 3–16.

MANDELBROT, B.B., D.E. PASSOJA, and A.J. PAULLAY. 1984. Fractal character of fracture surfaces of metals. *Nature* **308**(19): 721–722.

MARK, D.M., and P.B. ARONSON. 1984. Scale-dependent fractal dimensions of topographic surfaces: an empirical investigation, with applications in geomorphology and computer mapping. *Journal of the International Association for Mathematical Geology* **16**(7): 671–684.

PEITGEN, H.O., and D. SAUPE, eds. 1988. *The Science of Fractal Images.* New York: Springer.

PELEG, M., and M.D. NORMAND. 1985. Characterization of the ruggedness of instant coffee particle shape by natural fractals. *Journal of Food Science* **50.** pp. 829–831.

PIECH, M.A., and K.R. PIECH. 1990. Fingerprints and fractal terrain. *Mathematical Geology* **22**(4): 457–486.

RICHARDSON, L.F. 1960. In Q. Wright and C.C. Lineau (eds.), *Statistics of Deadly Quarrels.* Pittsburgh: Boxwood Press.

RICHARDSON, L.F. 1961. The problem of contiguity: an appendix to *Statistics of Deadly Quarrels.* In A. Rapoport, L. von Bertalanfly, and R.L. Meier (eds.), *General Systems: Yearbook of the Society for General Systems Research,* Vol. 6, pp. 139–187.

RODRIGUEZ-ITURBE, I., B. FEBRES DE POWER, M.B. SHARIFI, and K.P. GEORGAKAKOS. Chaos in rainfall. *Water Resources Research* **25**(7): 1667–1675.

SAYLES, R.S., and T.R. THOMAS. 1978. Surface topography as a nonstationary random process. *Nature* **271**: 431–434.

SCHATTSCHNEIDER, D. 1990. *Visions of Symmetry: Notebooks, Periodic Drawings, and Related Work of M.C. Escher.* New York: Freeman.

SCHOWENGERDT, R.A. 1983. *Techniques for Image Processing and Classification in Remote Sensing.* New York: Academic.

Figure 11.6 A listing of the program SPFM1D, a FORTRAN-77 implementation of the pseudo-computer algorithm SpectralSynthesisFM1D (Peitgen and Saupe, 1988, p. 94).

```
      PROGRAM SPFM1D
C
C     A FORTRAN-77 VERSION OF THE COMPUTER ALGORITHM:
C
C               SpectralSynthesisFM1D
C
C     GIVEN BY DIETMAR SAUPE IN:  PEITGEN AND SAUPE,
C     1988, THE SCIENCE OF FRACTAL IMAGES, CHAPTER 2,
C     P. 94.
C
C     THIS PROGRAM SYNTHESIZES A 1/(F**(BETA)) FUNCTION,
C     WHERE BETA=2H+1, WHERE FRACTAL DIMENSION IS 2-H.
C     THE SIMULATED FUNCTION HAS 512 REALIZATIONS (VALUES).
C     THE FUNCTION, X = 1/(F**(BETA)), IS WRITTEN TO A
C     FILE; THE PROGRAM, CFRINV, MUST BE USED
C     TO INVERSE FOURIER TRANSFORM X.
C
C*********************************************************
C
C                 USER'S GUIDE:  SPFM1D
C
C*********************************************************
C
C     THIS PROGRAM REQUIRES INTERACTIVE DATA INPUT AS FOLLOWS:
C
C     1.  THE PROGRAM PROMPTS THE USER TO SPECIFY A BEGINNING
C         RANDOM NUMBER SEED FOR THE SIMULATION.  USE ANY
C         ODD NUMBER FROM 1 TO 2147483647.  ENTER THIS AS A
C         DOUBLE PRECISION NUMBER, E.G.:  3333.D0 OR 123.D0
C
C     2.  ONCE THE BEGINNING RANDOM NUMBER SEED IS ENTERED,
C         THE PROGRAM REQUESTS THAT THE VALUE FOR H BE SPECIFIED.
C         THIS VALUE DEFINES THE FRACTAL DIMENSION FOR THE
C         SIMULATION, SUCH THAT THE FRACTAL DIMENSION, D,
C         EQUALS 2 - H.  ENTER A VALUE FOR H THAT IS BETWEEN
C         0 AND 1.
C
C     3.  AFTER H IS SPECIFIED, THE SIMULATION IS COMPUTED.  THEN,
C         THE PROGRAM REQUESTS THAT A FILE NAME FOR UNIT 1 BE
C         SPECIFIED; THIS IS THE OUTPUT FILE WHICH STORES THE
C         SIMULATION.  MOREOVER, THIS OUTPUT FILE IS THE INPUT
C         FILE FOR CFRINV.
C
C*********************************************************
C                 END OF USER'S GUIDE:   SPFM1D
C*********************************************************
```

531

Figure 11.6 continued

```fortran
C
      REAL X(512), A(512), B(512)
      DOUBLE PRECISION BSEED
      NRAND = 4
      ARAND = 1.0
      GSADD = SQRT(3 * NRAND)
      GSFAC = 2*GSADD/NRAND
      WRITE(*,*) 'ENTER BEGINNING RANDOM NUMBER SEED; MUST'
      WRITE(*,*) 'BE AN ODD NUMBER UP TO 8 DIGITS IN LENGTH'
      WRITE(*,*) 'WRITTEN LIKE THIS:  111119.D0 OR 33367.D0'
      WRITE(*,*) 'ETC, THE .D0 ADDED BECAUSE THIS NUMBER IS'
      WRITE(*,*) 'DOUBLE PRECISION.'
      READ(*,*) BSEED
      N     = 1024
      WRITE(*,*) 'ENTER FRACTAL DIMENSION PARAMETER, H'
      WRITE(*,*) 'WHERE H IS IN THE INTERVAL, 0 AND 1'
      WRITE(*,*) 'NOT INCLUDING 0 AND NOT INCLUDING 1'
      READ (*,*) H
      BETA = 2.0 * H + 1
      DO I = 1,N/2
         RAD = FLOAT(I) ** (-BETA/2)
         RAD = RAD * GAUSS(NRAND,BSEED,GSADD,GSFAC)
         PHASE = 2 * 3.141592 * DRAND(BSEED)
         A(I) = RAD * COS(PHASE)
         B(I) = RAD * SIN(PHASE)
      END DO
C
C     CREATE THE OUTPUT FILE, STORING A AND B
C
      OPEN(1, FILE = ' ', STATUS = 'NEW')
      DO I = 1,N/2
         WRITE(1,*) A(I), B(I)
      END DO
      STOP
      END
C

C
      FUNCTION GAUSS(NRAND, BSEED, GSADD, GSFAC)
      DOUBLE PRECISION BSEED
      SUM = 0.0
      DO I = 1,NRAND
         SUM = SUM + DRAND(BSEED)
      END DO
      GAUSS = GSFAC * SUM - GSADD
      RETURN
      END
C
C
      FUNCTION DRAND(IX)
C
C     THIS FUNCTION IS FROM SCHRAGE, L., "A MORE PORTABLE
C     RANDOM NUMBER GENERATOR," ACM TRANSACTIONS ON
C     MATHEMATICAL SOFTWARE, VOL. 5, NO. 2, JUNE, 1979.
C
      DOUBLE PRECISION A, P, IX, B15, B16, XHI, XALO, LEFTLO,
      DOUBLE PRECISION FHI, K
      A   = 16807.D0
      B15 = 32768.D0
      B16 = 65536.D0
      P   = 2147483647.D0
      XHI = IX / B16
      XHI = XHI - DMOD(XHI,1.D0)
      XALO = (IX-XHI*B16)*A
      LEFTLO = XALO/B16
      LEFTLO = LEFTLO - DMOD(LEFTLO,1.D0)
      FHI = XHI * A + LEFTLO
      K = FHI / B15
      K = K - DMOD(K,1.D0)
      IX = (((XALO-LEFTLO*B16)-P)+(FHI-K*B15)*B16) + K
      IF (IX .LT. 0.D0) IX = IX + P
      DRAND = IX * 4.656612875D-10
      RETURN
      END
```

Figure 11.7 A listing of the program CFRINV, a program to inverse Fourier transform the simulation produced by SPFM1D.

```fortran
      PROGRAM CFRINV
C
C     A FORTRAN-77 PROGRAM TO COMPUTE THE INVERSE FOURIER
C     TRANSFORM OF THE SIMULATION PRODUCED BY SPFM1D
C
C     **********************************************************
C
C                   USER'S GUIDE:    CFRINV
C
C     **********************************************************
C
C     THE INPUT DATA FILE FOR CFRINV IS THE SIMULATION OUTPUT FILE
C     CREATED BY SPFM1D
C
C     WHEN PROMPTED, ENTER THE OUTPUT FILE CREATED BY SPFM1D
C     AS THE INPUT FILE, UNIT 5
C
C     THIS PROGRAM THEN CREATES AN OUTPUT FILE, UNIT 1, WHICH
C     IS THE INVERSE TRANSFORM.  YOU ARE PROMPTED TO ENTER THE
C     NAME OF THIS FILE.
C
C     **********************************************************
C
C                END OF USER'S GUIDE AND DISCUSSION
C
C     **********************************************************
C
      IMPLICIT COMPLEX (A-H, O-Z)
      REAL A, B, PI
      DIMENSION Y(1000), BETA(1000)
      REAL DUM1, DUM2
      PI = 4.0 * ATAN(1.0)
      EYE = (0., 1.0)
C     EYE = SQUARE ROOT OF -1
      CONS = 2 * PI * EYE
      A = 0.0
C
C     ACCESS THE TIME SERIES DATA, Y
C
      WRITE (*,*) 'ENTER NAME OF INPUT FILE, UNIT=5'
      OPEN (5, FILE = ' ')
      WRITE (*,*) 'ENTER NAME OF OUTPUT FILE, UNIT=1'
      OPEN (1, FILE = ' ', STATUS = 'NEW')
      N = 1
10    CONTINUE
      READ (5,*,END=20) DUM1, DUM2
         Y(N) = CMPLX(DUM1, DUM2)
      N = N + 1
      GO TO 10
20    CONTINUE
      N = N - 1
      CONS = CONS / FLOAT(N)
      CONS2 = FLOAT(N) / FLOAT(N)
C
C     COMPUTE THE INVERSE TRANSFORM
C
      N2 = N/2
      DO J = 1,N
         BETA(J) = (0.,0.)
         DO K = 1,N
            X1 = A + (K-1) * CONS2
            BETA(J) = BETA(J) + Y(K) * CEXP(-CONS*(J-1)*X1)
         END DO
         WRITE (1,*)   REAL(BETA(J))
         WRITE (*,*) J, REAL(BETA(J))
      END DO
      STOP
      END
```

533

Figure 11.11 A listing of the program HSCALE. This program rescales simulations from SPFM1D/CFRINV using a resampling interval r. The program also requires the parameter H be entered.

```
      PROGRAM HSCALE
C
C     PROGRAM TO RESCALE BROWN NOISE BY 1/R^H.
C
C******************************************************
C
C            USER'S GUIDE:  HSCALE
C
C******************************************************
C
C     THIS PROGRAM ASKS FOR THE INPUT FILE NAME (E.G.,
C     A SIMULATION FORMED BY SPFM1D, OR SOME OTHER TIME SERIES).
C
C     ONCE THIS FILE NAME IS SPECIFIED, THE PROGRAM ASKS YOU TO
C     ENTER THE RESAMPLING INTERVAL (CALLED IA FOR THIS PROGRAM);
C     NOTE, IA IS AN INTEGER, SO ENTER A WHOLE NUMBER WITHOUT A
C     DECIMAL POINT; E.G. 2, 3, 4 ....
C
C     ONCE IA IS SPECIFIED, THE PROGRAM ASKS YOU TO ENTER THE
C     PARAMETER, H, WHICH IS RELATED TO THE FRACTAL DIMENSION, D,
C     AS:  H = 2 - D.
C
C     THE PROGRAM WILL THEN REQUEST THE NAME FOR AN OUTPUT FILE
C     WHICH WILL STORE THE RESCALED PROFILE.
C
C******************************************************
C
C            END OF USER'S GUIDE FOR HSCALE
C
C******************************************************

C
      DIMENSION B(5000), Y(5000)
      OPEN (5, FILE = ' ')
      N = 1
10    READ(5,*,END=20)  B(N)
      N = N + 1
      GO TO 10
20    N = N - 1
      WRITE(*,*) 'ENTER TIMESCALE VALUE, IA:    INTEGER VALUE'
      READ(*,*) IA
      WRITE(*,*) 'ENTER SCALING FACTOR, H'
      READ(*,*) H
C
C     CREATE RESCALED TIME SERIES, Y
C
      XSCAL = 1. / (FLOAT(IA) ** H)
      DO I = 2,N
      Y(I) = B((I-1)*IA) * XSCAL
      END DO
      OPEN (1, FILE = ' ', STATUS = 'NEW')
      DO I = 1,N
      WRITE(1,*) Y(I)
      END DO
      STOP
      END
```

Figure 11.15 A listing of the program SPFM2D, a FORTRAN-77 implementation of the pseudo-computer algorithm SpectralSynthesisFM2D (Peitgen and Saupe, 1988, p. 108).

```
      PROGRAM SPFM2D
C
C     A FORTRAN-77 VERSION OF THE COMPUTER ALGORITHM:
C
C              SpectralSynthesisFM2D
C
C     GIVEN BY DIETMAR SAUPE IN: PEITGEN AND SAUPE,
C     1988, THE SCIENCE OF FRACTAL IMAGES, CHAPTER 2,
C     P. 108.
C
C     ***********************************************
C              USER'S GUIDE:  SPFM2D
C     ***********************************************
C
C     THIS PROGRAM IS RELATIVELY SIMPLE TO USE; ITS INPUT GUIDE
C     IS SIMILAR TO THAT FOR SPFM1D
C
C     SEVERAL INTERACTIVE QUESTIONS/ANSWERS ARE CALLED FOR:
C
C     1. ENTER THE BEGINNING RANDOM NUMBER SEED: ENTER ANY ODD
C        ATTENTION: ODD, NUMBER BETWEEN 1 AND 2147483647.
C        ENTER THIS NUMBER IN DOUBLE PRECISION FORMAT:
C        E.G.:  3333.D0, 7777773.D0, ETC.
C
C     2. ENTER THE SIZE OF THE SIMULATION, N, WHERE THE SIMULATION
C        IS AN N X N SQUARE GRID. N CAN BE NO LARGER THAN 128
C        UNDER CURRENT DIMENSIONING
C
C     3. ONCE THE BEGINNING RANDOM NUMBER SEED IS ENTERED,
C        ENTER A VALUE FOR THE PARAMETER, H; THIS PARAMETER
C        DEFINES THE FRACTAL DIMENSION, D, SUCH THAT D = 2-H.
C
C     4. AFTER H IS SPECIFIED, THE SIMULATION IS COMPUTED; THE
C        PROGRAM WILL REQUEST THAT A NAME BE SPECIFIED FOR A
C        FILE TO STORE THE SIMULATION; THIS FILE IS THE INPUT
C        REQUIRED BY FORINV2 (CHAPTER 10). ONCE THIS FILE NAME
C        IS SPECIFIED, THE PROGRAM IS FINISHED.
C
C     ***********************************************
C            END OF USER'S GUIDE:   SPFM2D
C     ***********************************************
C
      COMPLEX  A(128,128)
      DOUBLE PRECISION BSEED
      NRAND = 4
      ARAND = 1.0
      GSADD = SQRT(3 * NRAND)
      GSFAC = 2*GSADD/NRAND
      WRITE(*,*) 'ENTER RANDOM NUMBER SEED, AN ODD NUMBER UP TO'
      WRITE(*,*) '8 DIGITS IN LENGTH; THIS NUMBER SHOULD BE '
      WRITE(*,*) 'ENTERED LIKE THIS:   EXAMPLE:   4444449.D0'
      READ(*,*) BSEED
      WRITE(*,*) 'ENTER THE SIZE OF THE IMAGE, N'
      READ(*,*) N
      WRITE(*,*) 'ENTER FRACTAL DIMENSION PARAMETER, H'
      WRITE(*,*) 'WHERE H IS IN THE INTERVAL, 0 AND 1'
      WRITE(*,*) 'NOT INCLUDING 0 AND NOT INCLUDING 1'
      READ (*,*) H
      BETA = -(H+1.)/2.
      DO I = 1, (N/2)+1
      DO J = 1, (N/2)+1
      PHASE = 2.*3.141592*DRAND(BSEED)
      IF ((I-1) .NE. 0 .OR. (J-1) .NE. 0) THEN
      RAD = (FLOAT((I-1)**2 + (J-1)**2)) ** BETA
      RAD = RAD * GAUSS(NRAND, BSEED, GSADD, GSFAC)
      ELSE
      RAD = 0.0
      END IF
      A(I,J) = CMPLX(RAD*COS(PHASE),RAD*SIN(PHASE))
      IF (I .EQ. 1) THEN
      I0 = 1
      ELSE
      I0 = N - I + 1
      END IF
      IF (J .EQ. 1) THEN
      J0 = 1
      ELSE
      J0 = N - J + 1
      END IF
      A(I0,J0) = CMPLX(RAD*COS(PHASE),-RAD*SIN(PHASE))
      END DO
      END DO
```

Figure 11.15 continued

```fortran
        DUMR = REAL(A((N/2)+1,1))
        DUMI = AIMAG(A((N/2)+1,1))
        DUMI = 0.0
        A((N/2)+1,1) = CMPLX(DUMR,DUMI)
        DUMR = REAL(A(1,(N/2)+1))
        DUMI = AIMAG(A(1,(N/2)+1))
        A(1,(N/2)+1) = CMPLX(DUMR,DUMI)
        DUMR = REAL(A((N/2)+1,(N/2)+1))
        DUMI = AIMAG(A((N/2)+1,(N/2)+1))
        A((N/2)+1,(N/2)+1) = CMPLX(DUMR,DUMI)
        DO J = 2, N/2
        DO I = 2, N/2
          PHASE = 2*3.141592*DRAND(BSEED)
          RAD = (FLOAT(I*I + J*J)) ** BETA
          RAD = RAD * GAUSS(NRAND, BSEED, GSADD, GSFAC)
          A(I,N-J) = CMPLX(RAD*COS(PHASE), RAD*SIN(PHASE))
          A(N-I,J) = CMPLX(RAD*COS(PHASE), -RAD*SIN(PHASE))
        END DO
      END DO
      OPEN(2, FILE = '        ', STATUS = 'NEW')
      WRITE(2,*) N
      DO I = 1,N
        WRITE(2,*) (REAL(A(I,JK)),AIMAG(A(I,JK)),JK=1,N)
      END DO
      STOP
      END
C
      FUNCTION GAUSS(NRAND, BSEED, GSADD, GSFAC)
      DOUBLE PRECISION BSEED
      SUM = 0.0
      DO I = 1, NRAND
        SUM = SUM + DRAND(BSEED)
      END DO
      GAUSS = GSFAC * SUM - GSADD
      RETURN
      END
C
C
      FUNCTION DRAND(IX)
C
C     THIS FUNCTION IS FROM SCHRAGE, L., "A MORE PORTABLE
C     RANDOM NUMBER GENERATOR," ACM TRANSACTIONS ON
C     MATHEMATICAL SOFTWARE, VOL. 5, NO. 2, JUNE, 1979.
C
      DOUBLE PRECISION A, P, IX, B15, B16, XHI, XALO, LEFTLO
      DOUBLE PRECISION FHI, K
      A   = 16807.D0
      B15 = 32768.D0
      B16 = 65536.D0
      P   = 2147483647.D0
      XHI = IX / B16
      XHI = XHI - DMOD(XHI,1.D0)
      XALO = (IX-XHI*B16)*A
      LEFTLO = XALO/B16
      LEFTLO = LEFTLO - DMOD(LEFTLO,1.D0)
      FHI = XHI * A + LEFTLO
      K = FHI / B15
      K = K - DMOD(K,1.D0)
      IX = (((XALO-LEFTLO*B16)-P)+(FHI-K*B15)*B16) + K
      IF (IX .LT. 0.D0) IX = IX + P
      DRAND = IX * 4.656612875D-10
      RETURN
      END
```

Figure 11.20 A listing of the program INTFC4, a program to convert a simulation from SPFM2D/FORINV2 into a digital image compatible with the image display program IMAGIN (Chapter 10). This program is almost identical to program INTFAC (Chapter 10), except INTFC4 yields images four times the size of the original grid. In other words, if a 50 × 50 grid is entered into INTFC4, the resultant image is 200 × 200.

```
      PROGRAM INTFC4
C
C   A PROGRAM TO CONVERT THE OUTPUT FROM FORINV2 TO
C   DIGITAL IMAGES COMPATIBLE WITH THE DISPLAY PROGRAM,
C   IMAGIN (CHAPTER 10).
C
C   *******************************************************
C                  USER'S GUIDE:  INTFC4
C   *******************************************************
C
C   THE INPUT TO THIS PROGRAM IS PERFORMED INTERACTIVELY.
C
C   1.  SPECIFY THE SIZE, N, OF THE SIMULATION (THAT YIELDED
C       BY SPFM2D/FORINV2).
C
C   2.  SPECIFY THE NAME OF THE INPUT FILE; THIS FILE IS THAT
C       CREATED BY FORINV2.
C
C   3.  SPECIFY THE NAME OF THE OUTPUT FILE; THIS WILL CONTAIN
C       THE DIGITAL IMAGE, A FILE WHICH IS COMPATIBLE WITH
C       IMAGIN (CHAPTER 10).  OUTPUT IMAGE SIZE = 4N X 4N.
C
C   *******************************************************
C                END OF USER'S GUIDE:  INTFC4
C   *******************************************************
C
      INTEGER*1 BUF(512)
C
C   INTERACTION TO DEFINE OPTIONS AND DATA FILES
C
      WRITE(*,*) 'ENTER THE SIZE, N, OF THE SIMULATION'
      READ (*,*) N
      WRITE(*,*)
      WRITE(*,*) ' ENTER THE NAME OF THE INPUT FILE'
      OPEN (5, FILE = ' ')
      WRITE(*,*) 'ENTER THE NAME OF THE OUTPUT FILE'
      OPEN(1, FILE = ' ', STATUS = 'NEW', FORM = 'BINARY')
C


C   SCAN INPUT FILE TO DETERMINE MIN AND MAX DATA VALUES
C
      READ(5,*) XI, XJ, XZ
      YMIN = XZ
      YMAX = XZ
10    CONTINUE
      READ(5,*,END=20) XI, XJ, XZ
      IF (XZ .LT. YMIN) YMIN = XZ
      IF (XZ .GT. YMAX) YMAX = XZ
      GO TO 10
20    CONTINUE
      REWIND 5
C
C   CONVERT SIMULATION VALUES TO 8-BIT INTEGERS AND CREATE IMAGE
C
      DO I = 1, N
      DO J = 1, N
      JK = (J-1) * 4
      READ(5,*) XI, XJ, XZ
      XZN = 255.0 * ((XZ - YMIN) / (YMAX - YMIN))
      DO K = 1, 4
      JJK = JK + K
      BUF(JJK) = INT1(XZN)
      END DO
      END DO
      DO K = 1, 4
      WRITE(1) (BUF(JK), JK = 1, 4*N)
      END DO
      END DO
C
C   TERMINATE PROGRAM
C
      STOP
      END
C
```

537

Figure 11.28 A listing of the program WAVELET for creating images of wavelet transforms.

```fortran
      PROGRAM WAVELT
C
C     A PROGRAM TO CONVOLV A TIME SERIES FUNCTION WITH A
C     WAVELET OVER SUCCESSIVE TIME STEPS; THE
C     RESULT IS A DIGITAL IMAGE
C
C***********************************************************
C***********************************************************
C
C                     USER'S GUIDE
C
C***********************************************************
C***********************************************************
C
C     THIS PROGRAM REQUIRES ONLY TWO ITEMS FOR INPUT:
C
C     1.  THE NAME OF THE INPUT FILE CONTAINING THE ONE-DIMENSIONAL
C         TIME SERIES; THE FORMAT FOR THIS FILE IS:
C
C         LINE 1:  Y(1)
C                   .
C                   .
C         LINE N:  Y(N)
C
C         THIS IS THE SAME FORMAT USED IN CHAPTER 9
C
C     2.  THE NAME OF THE OUTPUT FILE TO STORE THE 32 BIT,
C         BINARY INTEGER DIGITAL IMAGE FILE FOR THE WAVELET
C         TRANSFORM; THIS FILE IS COMPATIBLE WITH THE PROGRAM,
C         IMAG41.
C
C         THIS FILE MUST NOT ALREADY EXIST
C
C***********************************************************
C                 END OF USER'S GUIDE
C***********************************************************
C
      DIMENSION Y(2000), G(2000)
      INTEGER*4 KTEMP(2000)
      WRITE(*,*) 'ENTER NAME OF INPUT DATA FILE'
      OPEN(5, FILE = ' ')
      WRITE(*,*) 'ENTER NAME OF OUTPUT FILE TO STORE DIGITAL IMAGE'
      OPEN(1, FILE = ' ', STATUS='NEW', FORM='BINARY')
      N = 1
10    CONTINUE
      READ(5,*,END=20) Y(N)
      N = N + 1
      GO TO 10
20    N = N - 1
      K = 1
      DO I = 3,N/2,2
      K = K + 1
      DO J = 1,I
      G(J) = 1./FLOAT(I)
      END DO
      DO L = K, N-K
      SUM = 0.0
      KK = (I-1)/2
      MM = L - KK
      MK = 0
      DO MN = MM, L+KK
      MK = MK + 1
      SUM = SUM + G(MK)*Y(MN)
      END DO
      KTEMP(L) = INT(SUM)
      END DO
      DO IRITE = 1,10
      WRITE(1) (KTEMP(JK), JK = 1,N)
      END DO
      DO NB = 1, 2000
      KTEMP(NB) = 0
      END DO
      END DO
      STOP
      END
```

Appendix A

Listings are presented of data sets that are used throughout this text. Data sets are listed in the order shown subsequently; chapters in which these data sets are used are shown in parentheses. The source for each data set is shown by superscript number; a list of these sources follows.

Data Sets

1. NVSIM.DAT (Chapter 6)[1]
2. NVSIMLOG.DAT (Chapter 6)[1]
3. LONGB.DAT (Chapter 6)[2]
4. SANFER.DAT (Chapter 6)[2]
5. AREA1.DAT (Chapter 6)[3]
6. AREA2.DAT (Chapter 6)[3]
7. CALIENTE.DAT (Chapters 7 and 11)[4]
8. CAVECREK.DAT (Chapter 9)[5]
9. WALGULCH.DAT (Chapter 9)[5]
10. VARSTAR.DAT (Chapter 9)[6]
11. SUNSPOT.DAT (Chapter 9)[7]
12. SEISMIC1.DAT (Chapter 9)[8]
13. SEISMIC2.DAT (Chapter 9)[8]
14. SEISMIC3.DAT (Chapter 9)[8]
15. BROWN.DAT (Chapter 11)[9]
16. PROFILEA.DAT (Chapter 11)[10]
17. PROFILEB.DAT (Chapter 11)[10]
18. RAINFALL.DAT (Chapter 11)[11]

Sources for Data

1. These data were simulated using the conditional simulation technique known as turning bands. The computer algorithm used to obtain the simulation is that given in J.R. Carr and D.E. Myers, COSIM: A FORTRAN IV program for coconditional simulation. *Computers and Geosciences* **11**(6): 675–706 (1985). The NVSIMLOG.DAT data set was developed simply by converting the data values in the NVSIM.DAT data set to natural logarithms.

2. Both the LONGB.DAT (Mercalli intensity data for the southern California region affected by the 1933 Long Beach, California, earthquake, magnitude 6 to 6.3, northwest trending strike-slip fault mechanism) and SANFER.DAT (Mecalli intensity data for the southern California region affected by the 1971 San Fernando, California, earthquake, magnitude 6.5, focus on a previously unknown thrust fault) data sets are taken from a United States government publication, *United States Earthquakes,* published annually by the United States Department of Commerce, Coast and Geodetic Survey from 1928 to 1968, the NOAA National Ocean Survey in 1969, the NOAA Environmental Data Service from 1970 to 1972, and jointly by the NOAA Environmental Data and Information Service and the United States Geological Survey thereafter.

3. The author of this text was one of the 12 participants in the United States Environmental Protection Agency sponsored experiment described by E. Englund, A variance of geostatisticians. *Mathematical Geology* **22**(4): 417–456 (1990). The AREA1.DAT and AREA2.DAT data sets were distributed to each participant in this study.

4. The CALIENTE.DAT data set is a 50 × 50 subimage of an original 1200 × 1200 image of a digital elevation model (DEM) for a region of southern Nevada including Caliente (Lincoln County). Source of DEM data: the United States Geological Survey.

5. The CAVECREK.DAT (runoff for Cave Creek, Kentucky) and WALGULCH.DAT (monthly rainfall data for Walnut Gulch near Tombstone, Arizona) data sets were taken from C.T. Haan, *Statistical Methods in Hydrology.* Ames, IA: Iowa State University Press, 1977. The CAVECREK.DAT data come from a U.S.G.S. Water Supply Paper. The WALGULCH.DAT data are from Dr. Leonard Lane, U.S.D.A., A.R.S., Southwest Watershed Research Center, Tucson, Arizona.

6. These variable star data were taken from E.T. Whittaker and G. Robinson, *The Calculus of Observations.* London: Blackie, 1924.

7. These sunspot data were estimated from Figure 5.13 given in P. Bloomfield, *Fourier Analysis of Time Series: An Introduction.* New York: Wiley, 1976. More accurate data are found in D.J. Schove, ed., *Sunspot Cycles* (Benchmark Papers in Geology/68). Stroudsburg, PA: Hutchinson Ross, 1983, Table 1, p. 10.

8. The three data sets SEISMIC1.DAT, SEISMIC2.DAT, and SEISMIC3.DAT are acceleration data taken from M.D. Trifunac, A.G. Brady, and D.E. Hudson, *Strong Motion Earthquake Accelerograms, Digitized and Plotted Data,* Vol. 2, *Corrected Accelerograms and Integrated Ground Velocity and Displacement Curves, Parts O and P; Q and R.* Pasadena, CA: California Institute of Technology; prepared for the National Science Foundation, distributed by the National Technical Information Service, United States Department of Commerce, 1974. These acceleration data were recorded during the 1971 San Fernando, California, earthquake.

9. These brown noise data were simulated using the algorithm given in H.-O. Peitgen and D. Saupe, eds., *The Science of Fractal Images.* New York: Springer, 1988.

10. Data sets PROFILEA.DAT and PROFILEB.DAT are stringline elevations across two joint surfaces, measured by the author of this textbook at Yucca Mountain, Nevada Test Site, Nye County, Nevada, in May 1988.

11. These rainfall data are for one storm. Amounts were measured atop a building on the campus of the Massachusetts Institute of Technology. These data were estimated from I. Rodriguez-Iturbe, B. Febres de Power, M.B. Charifi, and K.P. Georgakakos, Chaos in rainfall. *Water Resources Research* **25**(7): 1667–1675 (1989).

NVSIM.DAT:

DATA VALUE	Y	X
14.588290	490.000000	50.000000
13.040420	490.000000	140.000000
15.407950	490.000000	190.000000
16.072600	490.000000	380.000000
26.464770	490.000000	480.000000
30.897050	480.000000	300.000000
17.112880	480.000000	350.000000
24.340530	480.000000	450.000000
38.464350	470.000000	0.000000E+00
13.624570	470.000000	50.000000
20.199100	470.000000	460.000000
30.228660	460.000000	10.000000
-4.188071	460.000000	190.000000
41.297290	460.000000	280.000000
19.568400	460.000000	300.000000
40.821910	460.000000	490.000000
29.945730	450.000000	20.000000
31.311920	450.000000	60.000000
7.515244	450.000000	140.000000
33.275210	450.000000	250.000000
23.258970	450.000000	260.000000
18.244340	450.000000	450.000000
43.733750	440.000000	10.000000
14.383250	440.000000	210.000000
12.088290	440.000000	230.000000
16.345070	440.000000	330.000000
15.460940	440.000000	380.000000
54.138230	440.000000	480.000000
33.578250	430.000000	10.000000
19.740000	430.000000	320.000000
21.937400	430.000000	420.000000
6.053269	420.000000	310.000000
31.648230	420.000000	410.000000
21.685130	410.000000	0.000000E+00
25.916410	410.000000	10.000000
30.522300	410.000000	190.000000
22.907420	410.000000	260.000000
9.016590	410.000000	290.000000
33.614800	400.000000	60.000000
22.530340	400.000000	80.000000
23.061490	400.000000	130.000000
4.340898	400.000000	280.000000
25.249410	400.000000	350.000000
33.669570	400.000000	420.000000
26.449070	400.000000	450.000000
48.083870	400.000000	490.000000
-3.132410	390.000000	300.000000
23.405660	380.000000	80.000000
14.273520	380.000000	150.000000
19.912590	380.000000	160.000000
24.428670	380.000000	190.000000
13.407360	380.000000	260.000000
11.673490	380.000000	290.000000
24.412500	380.000000	360.000000
27.165390	380.000000	380.000000
32.834420	380.000000	430.000000
14.903040	370.000000	30.000000
20.212080	370.000000	80.000000
26.673560	370.000000	90.000000
33.469250	370.000000	200.000000
24.623690	370.000000	220.000000
24.245940	370.000000	390.000000
31.755230	370.000000	470.000000
32.988140	360.000000	100.000000
14.771860	360.000000	260.000000
10.040490	360.000000	320.000000
16.037430	350.000000	250.000000
26.641630	350.000000	260.000000

DATA VALUE	Y	X
25.579370	350.000000	300.000000
12.981440	350.000000	390.000000
11.112410	340.000000	30.000000
23.822160	340.000000	40.000000
35.985910	340.000000	50.000000
16.191340	340.000000	200.000000
33.398190	340.000000	360.000000
32.217920	340.000000	370.000000
22.604930	330.000000	100.000000
28.306800	330.000000	110.000000
18.039830	330.000000	150.000000
11.090060	330.000000	170.000000
31.925580	330.000000	250.000000
36.098050	330.000000	360.000000
15.936400	330.000000	440.000000
7.831378	330.000000	450.000000
32.941430	330.000000	480.000000
19.103430	330.000000	490.000000
13.224470	320.000000	10.000000
8.201035	320.000000	20.000000
21.474270	320.000000	70.000000
15.325230	320.000000	150.000000
10.310100	320.000000	170.000000
30.525730	320.000000	330.000000
20.372390	320.000000	340.000000
18.927800	320.000000	410.000000
18.626160	320.000000	470.000000
30.598730	320.000000	490.000000
10.810120	310.000000	190.000000
18.997780	300.000000	30.000000
18.265000	300.000000	50.000000
15.584230	300.000000	100.000000
48.734540	300.000000	280.000000
22.158550	300.000000	340.000000
9.724548	290.000000	80.000000
3.491447	290.000000	110.000000
18.604950	290.000000	140.000000
51.449930	290.000000	230.000000
42.071300	290.000000	280.000000
4.922839	280.000000	130.000000
44.411510	280.000000	290.000000
5.811162	280.000000	420.000000
43.283430	270.000000	270.000000
15.177070	270.000000	430.000000
10.522340	270.000000	470.000000
23.187890	260.000000	160.000000
46.123890	260.000000	190.000000
41.024540	260.000000	230.000000
42.225940	260.000000	300.000000
29.443330	260.000000	410.000000
6.293068	250.000000	70.000000
38.774110	250.000000	210.000000
29.186480	250.000000	340.000000
23.231670	250.000000	450.000000
34.929870	240.000000	220.000000
34.469520	240.000000	360.000000
40.994670	240.000000	400.000000
19.038110	230.000000	30.000000
29.252440	230.000000	120.000000
33.488130	230.000000	140.000000
11.133550	230.000000	190.000000
46.366280	230.000000	380.000000
39.268320	230.000000	400.000000
40.486870	230.000000	420.000000
20.714650	230.000000	450.000000
34.918410	220.000000	30.000000
20.353010	220.000000	80.000000
32.477310	220.000000	240.000000
41.827570	220.000000	360.000000
45.087770	210.000000	20.000000

DATA VALUE	Y	X	DATA VALUE	Y	X
23.869210	210.000000	170.000000	27.036300	60.000000	0.000000E+00
21.854900	210.000000	220.000000	28.855130	60.000000	10.000000
22.028640	210.000000	290.000000	15.687590	60.000000	110.000000
32.313190	200.000000	110.000000	25.961510	60.000000	190.000000
22.711230	200.000000	140.000000	33.167910	60.000000	210.000000
27.902730	200.000000	170.000000	20.134470	60.000000	290.000000
23.680440	200.000000	490.000000	26.412820	60.000000	360.000000
38.843430	190.000000	50.000000	14.540280	60.000000	460.000000
33.794610	190.000000	110.000000	33.272840	60.000000	490.000000
1.551135	190.000000	210.000000	27.054810	50.000000	20.000000
4.003802	190.000000	240.000000	22.723480	50.000000	90.000000
33.759270	190.000000	360.000000	18.239000	50.000000	160.000000
27.198390	180.000000	10.000000	25.774860	50.000000	220.000000
38.044370	180.000000	40.000000	29.819790	50.000000	230.000000
22.163910	180.000000	130.000000	11.629660	50.000000	320.000000
12.372140	180.000000	260.000000	12.854900	40.000000	30.000000
12.202270	180.000000	280.000000	6.823383	40.000000	140.000000
26.536700	180.000000	290.000000	29.673670	30.000000	100.000000
38.054370	170.000000	70.000000	12.550250	30.000000	180.000000
15.630470	170.000000	190.000000	29.221740	30.000000	310.000000
5.447649	170.000000	240.000000	33.501160	30.000000	330.000000
29.025120	170.000000	490.000000	25.728130	30.000000	370.000000
20.926060	160.000000	150.000000	32.938560	30.000000	400.000000
27.393150	160.000000	170.000000	23.023370	30.000000	420.000000
9.462253	160.000000	240.000000	26.153090	30.000000	470.000000
29.877280	160.000000	270.000000	42.175740	20.000000	20.000000
29.132990	150.000000	180.000000	18.169260	20.000000	150.000000
15.573120	150.000000	230.000000	15.195660	20.000000	170.000000
24.094360	150.000000	260.000000	33.977890	10.000000	110.000000
29.040450	140.000000	30.000000	19.070130	10.000000	160.000000
31.867460	140.000000	160.000000	16.542390	10.000000	280.000000
30.593260	140.000000	250.000000	29.337790	10.000000	290.000000
25.328140	140.000000	310.000000	51.345600	10.000000	350.000000
25.087100	140.000000	430.000000	33.084640	10.000000	490.000000
31.935950	130.000000	10.000000	20.062710	0.000000E+00	290.000000
37.012120	130.000000	90.000000	33.958640	0.000000E+00	320.000000
21.284980	130.000000	240.000000	35.435110	0.000000E+00	340.000000
20.225980	130.000000	250.000000	31.555150	0.000000E+00	370.000000
19.269580	130.000000	340.000000	26.842080	0.000000E+00	410.000000
26.118450	130.000000	350.000000	45.446010	0.000000E+00	440.000000
20.953470	130.000000	360.000000			
26.852630	130.000000	460.000000			
37.478770	130.000000	470.000000			
39.320630	120.000000	120.000000			
32.547480	120.000000	200.000000			
18.948390	120.000000	250.000000			
32.759210	120.000000	440.000000			
23.088030	110.000000	70.000000			
22.779650	110.000000	90.000000			
27.850800	110.000000	100.000000			
42.963850	110.000000	170.000000			
23.715220	110.000000	370.000000			
29.119170	110.000000	470.000000			
42.858610	110.000000	490.000000			
26.557770	100.000000	30.000000			
27.782690	100.000000	310.000000			
26.249260	100.000000	440.000000			
26.794990	90.000000	190.000000			
30.241580	90.000000	210.000000			
21.494640	90.000000	300.000000			
11.462920	90.000000	430.000000			
54.736730	90.000000	480.000000			
19.971510	80.000000	60.000000			
23.177950	80.000000	130.000000			
26.929500	80.000000	190.000000			
30.997530	80.000000	250.000000			
15.674490	70.000000	20.000000			
27.928080	70.000000	210.000000			
13.095410	70.000000	330.000000			
31.197420	70.000000	480.000000			

NVSIMLOG.DAT

VALUE	Y	X	VALUE	Y	X
			12.909160	350.000000	300.000000
			3.662493	350.000000	390.000000
4.300920	490.000000	50.000000	3.038126	340.000000	30.000000
3.684158	490.000000	140.000000	10.828870	340.000000	40.000000
4.668300	490.000000	190.000000	36.546700	340.000000	50.000000
4.989122	490.000000	380.000000	5.048716	340.000000	200.000000
14.104260	490.000000	480.000000	28.214020	340.000000	360.000000
21.970600	480.000000	300.000000	25.073010	340.000000	370.000000
5.536088	480.000000	350.000000	9.587815	330.000000	100.000000
11.405010	480.000000	450.000000	16.956990	330.000000	110.000000
46.825840	470.000000	0.000000E+00	6.073791	330.000000	150.000000
3.905778	470.000000	50.000000	3.031344	330.000000	170.000000
7.537647	470.000000	460.000000	24.350630	330.000000	250.000000
20.550110	460.000000	10.000000	36.958840	330.000000	360.000000
0.000000E+00	460.000000	190.000000	4.921631	330.000000	440.000000
62.161080	460.000000	280.000000	2.188328	330.000000	450.000000
7.076928	460.000000	300.000000	26.954300	330.000000	480.000000
59.275200	460.000000	490.000000	6.755405	330.000000	490.000000
19.976830	450.000000	20.000000	3.752593	320.000000	10.000000
22.901260	450.000000	60.000000	2.270735	320.000000	20.000000
2.120230	450.000000	140.000000	8.562798	320.000000	70.000000
27.869170	450.000000	250.000000	4.629843	320.000000	150.000000
10.235860	450.000000	260.000000	2.803896	320.000000	170.000000
6.199286	450.000000	450.000000	21.169740	320.000000	330.000000
79.310850	440.000000	10.000000	7.669405	320.000000	340.000000
4.213632	440.000000	210.000000	6.637795	320.000000	410.000000
3.349560	440.000000	230.000000	6.440563	320.000000	470.000000
5.126930	440.000000	330.000000	21.324850	320.000000	490.000000
4.693103	440.000000	380.000000	2.947661	310.000000	190.000000
224.488200	440.000000	480.000000	6.684410	300.000000	30.000000
28.726640	430.000000	10.000000	6.212106	300.000000	50.000000
7.199417	430.000000	320.000000	4.751323	300.000000	100.000000
8.968694	430.000000	420.000000	130.771800	300.000000	280.000000
1.831851	420.000000	310.000000	9.169246	300.000000	340.000000
23.684550	420.000000	410.000000	2.644428	290.000000	80.000000
8.745270	410.000000	0.000000E+00	1.417854	290.000000	110.000000
13.351660	410.000000	10.000000	6.426917	290.000000	140.000000
21.162490	410.000000	190.000000	171.570300	290.000000	230.000000
9.882267	410.000000	260.000000	67.163510	290.000000	280.000000
2.463687	410.000000	290.000000	1.636049	280.000000	130.000000
28.831830	400.000000	60.000000	84.872580	280.000000	290.000000
9.516565	400.000000	80.000000	1.788033	280.000000	420.000000
10.035700	400.000000	130.000000	75.818530	270.000000	270.000000
1.543558	400.000000	280.000000	4.561753	270.000000	430.000000
12.490160	400.000000	350.000000	2.864042	270.000000	470.000000
28.990180	400.000000	420.000000	10.163360	260.000000	160.000000
14.082140	400.000000	450.000000	100.724500	260.000000	190.000000
122.533800	400.000000	490.000000	60.488540	260.000000	230.000000
0.000000E+00	390.000000	300.000000	68.210200	260.000000	300.000000
10.387110	380.000000	80.000000	18.997990	260.000000	410.000000
4.167649	380.000000	150.000000	1.876310	250.000000	70.000000
7.324750	380.000000	160.000000	48.299000	250.000000	210.000000
11.505980	380.000000	190.000000	18.516240	250.000000	340.000000
3.821855	380.000000	260.000000	10.207590	250.000000	450.000000
3.213462	380.000000	290.000000	32.884030	240.000000	220.000000
11.487390	380.000000	360.000000	31.404530	240.000000	360.000000
15.127870	380.000000	380.000000	60.308140	240.000000	400.000000
26.667400	380.000000	430.000000	6.711423	230.000000	30.000000
4.438445	370.000000	30.000000	18.638770	230.000000	120.000000
7.547437	370.000000	80.000000	28.468920	230.000000	140.000000
14.401840	370.000000	90.000000	3.044556	230.000000	190.000000
28.415220	370.000000	200.000000	103.195800	230.000000	380.000000
11.732570	370.000000	220.000000	50.745950	230.000000	400.000000
11.297640	370.000000	390.000000	57.322140	230.000000	420.000000
23.939340	370.000000	470.000000	7.936441	230.000000	450.000000
27.080500	360.000000	100.000000	32.846370	220.000000	30.000000
4.380601	360.000000	260.000000	7.654556	220.000000	80.000000
2.729311	360.000000	320.000000	25.731890	220.000000	240.000000
4.971606	350.000000	250.000000	65.546330	220.000000	360.000000
14.355930	350.000000	260.000000	90.810680	210.000000	20.000000

VALUE	Y	X	VALUE	Y	X
10.879940	210.000000	170.000000	14.933840	60.000000	0.000000E+00
8.895006	210.000000	220.000000	17.912750	60.000000	10.000000
9.050899	210.000000	290.000000	4.800687	60.000000	110.000000
25.313020	200.000000	110.000000	13.412010	60.000000	190.000000
9.690276	200.000000	140.000000	27.571740	60.000000	210.000000
16.285470	200.000000	170.000000	7.489088	60.000000	290.000000
10.676490	200.000000	490.000000	14.031180	60.000000	360.000000
48.634980	190.000000	50.000000	4.280321	60.000000	460.000000
29.354940	190.000000	110.000000	27.862560	60.000000	490.000000
1.167791	190.000000	210.000000	14.961510	50.000000	20.000000
1.492392	190.000000	240.000000	9.702155	50.000000	90.000000
29.251390	190.000000	360.000000	6.195976	50.000000	160.000000
15.177880	180.000000	10.000000	13.164000	50.000000	220.000000
44.899960	180.000000	40.000000	19.726820	50.000000	230.000000
9.174161	180.000000	130.000000	3.199409	50.000000	320.000000
3.445999	180.000000	260.000000	3.616440	40.000000	30.000000
3.387957	180.000000	280.000000	1.978499	40.000000	140.000000
14.206080	180.000000	290.000000	19.440660	30.000000	100.000000
44.944890	170.000000	70.000000	3.507926	30.000000	180.000000
4.773344	170.000000	190.000000	18.581640	30.000000	310.000000
1.724203	170.000000	240.000000	28.506040	30.000000	330.000000
18.219860	170.000000	490.000000	13.102630	30.000000	370.000000
8.106011	160.000000	150.000000	26.946570	30.000000	400.000000
15.476380	160.000000	170.000000	9.997520	30.000000	420.000000
2.575968	160.000000	240.000000	13.671440	30.000000	470.000000
19.840550	160.000000	270.000000	67.868630	20.000000	20.000000
18.417460	150.000000	180.000000	6.152915	20.000000	150.000000
4.746047	150.000000	230.000000	4.570241	20.000000	170.000000
11.127680	150.000000	260.000000	29.897920	10.000000	110.000000
18.247810	140.000000	30.000000	6.732947	10.000000	160.000000
24.209520	140.000000	160.000000	5.229099	10.000000	280.000000
21.313190	140.000000	250.000000	18.798530	10.000000	290.000000
12.588880	140.000000	310.000000	169.789600	10.000000	350.000000
12.289070	140.000000	430.000000	27.343100	10.000000	490.000000
24.375900	130.000000	10.000000	7.435538	0.000000E+00	290.000000
40.496350	130.000000	90.000000	29.840430	0.000000E+00	320.000000
8.402238	130.000000	240.000000	34.588140	0.000000E+00	340.000000
7.557936	130.000000	250.000000	23.465120	0.000000E+00	370.000000
6.868585	130.000000	340.000000	14.646600	0.000000E+00	410.000000
13.624160	130.000000	350.000000	94.122870	0.000000E+00	440.000000
8.128262	130.000000	360.000000			
14.662060	130.000000	460.000000			
42.430910	130.000000	470.000000			
51.012100	120.000000	120.000000			
25.913090	120.000000	200.000000			
6.651477	120.000000	250.000000			
26.467590	120.000000	440.000000			
10.062370	110.000000	70.000000			
9.756804	110.000000	90.000000			
16.201110	110.000000	100.000000			
73.433840	110.000000	170.000000			
10.713690	110.000000	370.000000			
18.392020	110.000000	470.000000			
72.665070	110.000000	490.000000			
14.236040	100.000000	30.000000			
16.091140	100.000000	310.000000			
13.803550	100.000000	440.000000			
14.577790	90.000000	190.000000			
20.576670	90.000000	210.000000			
8.580258	90.000000	300.000000			
3.146504	90.000000	430.000000			
238.334000	90.000000	480.000000			
7.368034	80.000000	60.000000			
10.153260	80.000000	130.000000			
14.775200	80.000000	190.000000			
22.192470	80.000000	250.000000			
4.794402	70.000000	20.000000			
16.326800	70.000000	210.000000			
3.704473	70.000000	330.000000			
22.640540	70.000000	480.000000			

LONGB.DAT

DATA VALUE	Y	X		DATA VALUE	Y	X
				7.000	61.150	129.790
				9.000	61.150	139.060
4.000	137.860	136.650		7.000	91.170	132.360
6.000	94.500	141.590		7.000	76.710	137.090
4.000	107.840	143.340		7.000	74.490	116.730
7.000	86.720	118.500		4.000	200.120	138.040
7.000	66.710	163.110		4.000	210.130	120.490
7.000	71.150	148.250		5.000	102.280	153.560
4.000	67.820	171.440		6.000	85.610	144.430
5.000	98.950	163.760		4.000	93.390	148.080
8.000	83.380	138.900		4.000	106.730	135.030
8.000	74.490	143.590		8.000	43.360	162.380
6.000	93.390	116.610		5.000	103.400	119.320
4.000	76.710	166.730		7.000	74.490	148.220
7.000	72.270	155.650		5.000	93.390	186.020
4.000	120.070	205.170		6.000	63.370	169.630
4.000	110.070	60.100		6.000	107.840	46.240
4.000	96.720	169.330		4.000	150.090	143.950
4.000	112.290	98.930		6.000	102.280	141.530
4.000	85.610	183.320		4.000	131.190	83.130
4.000	72.270	195.490		6.000	71.150	164.930
7.000	42.250	167.960		6.000	91.170	180.490
8.000	46.690	162.350		3.000	73.380	189.000
6.000	87.830	116.640		7.000	61.150	116.810
4.000	108.950	210.840		4.000	182.330	142.780
7.000	67.820	136.230		4.000	96.720	171.180
7.000	77.820	143.570		5.000	116.740	113.700
5.000	100.060	136.920		4.000	96.720	144.350
4.000	157.870	191.850		7.000	55.590	129.820
5.000	76.710	116.710		8.000	55.590	169.700
4.000	100.060	199.840		6.000	87.830	110.160
3.000	166.770	118.010		5.000	126.740	56.360
6.000	132.300	70.190		8.000	57.810	148.350
4.000	96.720	204.500		5.000	104.510	148.920
7.000	71.150	163.070		8.000	63.370	139.970
6.000	74.490	129.700		8.000	80.050	135.210
6.000	102.280	127.660		5.000	98.950	139.710
7.000	77.820	123.190		5.000	95.610	99.940
4.000	92.280	194.360		7.000	66.710	126.040
6.000	63.370	128.840		4.000	102.280	77.700
7.000	71.150	116.750		4.000	112.290	133.140
8.000	46.690	154.000		6.000	57.810	172.460
7.000	83.380	132.410		4.000	96.720	185.990
4.000	97.840	161.920		7.000	82.270	117.600
4.000	110.070	136.850		6.000	116.740	35.130
5.000	107.840	136.870		8.000	77.820	132.450
6.000	96.720	176.740		3.000	196.780	67.190
4.000	163.430	142.920		6.000	83.380	152.780
4.000	140.080	170.800		8.000	76.710	134.310

SANFER.DAT

VALUE	Y	X	VALUE	Y	X	VALUE	Y	X
			6.000	76.710	160.250	6.000	96.720	148.980
			6.000	73.380	153.790	6.000	84.500	208.330
5.000	137.860	136.650	6.000	88.940	162.000	5.000	96.720	171.180
6.000	102.280	80.480	5.000	96.720	176.740	6.000	96.720	144.350
6.000	118.960	134.020	6.000	160.100	114.360	6.000	98.950	144.330
7.000	94.500	141.590	6.000	66.710	143.650	5.000	55.590	129.820
5.000	101.170	192.420	6.000	163.430	142.920	5.000	166.770	90.350
7.000	107.840	143.340	5.000	72.270	121.370	6.000	55.590	169.700
6.000	66.710	163.110	5.000	177.880	74.630	6.000	77.820	148.200
7.000	102.280	131.360	6.000	143.420	156.920	6.000	87.830	110.160
5.000	71.150	148.250	5.000	140.080	170.800	5.000	126.740	56.360
5.000	98.950	163.760	6.000	61.150	129.790	6.000	115.620	90.590
6.000	94.500	161.950	6.000	61.150	139.060	8.000	131.190	105.300
5.000	83.380	138.900	5.000	65.590	149.220	5.000	57.810	148.350
5.000	82.270	141.680	7.000	91.170	132.360	7.000	102.280	127.660
5.000	74.490	143.590	5.000	81.160	182.440	7.000	102.280	115.630
7.000	93.390	116.610	5.000	76.710	137.090	6.000	104.510	148.920
5.000	76.710	166.730	6.000	114.510	201.530	5.000	63.370	139.970
5.000	72.270	155.650	6.000	88.940	92.570	5.000	58.920	190.990
6.000	110.070	60.100	6.000	74.490	116.730	6.000	80.050	135.210
7.000	105.620	100.810	6.000	38.910	130.860	7.000	98.950	139.710
6.000	64.480	129.760	6.000	83.380	137.040	5.000	91.170	143.470
6.000	141.200	97.860	5.000	118.960	47.140	5.000	65.590	155.710
6.000	31.130	151.340	5.000	57.810	156.700	5.000	102.280	118.410
5.000	96.720	169.330	5.000	87.830	199.040	5.000	101.170	212.770
6.000	112.290	98.930	8.000	103.400	130.430	7.000	98.950	129.530
5.000	85.610	183.320	5.000	200.120	138.040	7.000	114.510	126.650
5.000	96.720	181.360	5.000	210.130	120.490	5.000	56.700	149.290
6.000	68.930	132.510	6.000	102.280	153.560	5.000	55.590	148.370
5.000	72.270	195.490	5.000	91.170	181.420	7.000	104.510	106.370
5.000	42.250	167.960	5.000	85.610	144.430	5.000	211.240	113.130
6.000	46.690	162.350	6.000	93.390	148.080	5.000	96.720	148.050
6.000	94.500	164.730	7.000	106.730	135.030	6.000	104.510	78.620
6.000	91.170	194.370	6.000	115.620	72.100	6.000	95.610	99.940
6.000	87.830	116.640	6.000	112.290	186.760	6.000	66.710	126.040
5.000	83.380	170.380	5.000	111.180	149.790	7.000	112.290	133.140
6.000	77.820	143.570	6.000	105.620	72.140	6.000	57.810	172.460
5.000	102.280	160.030	8.000	127.850	104.390	6.000	96.720	185.990
5.000	187.890	163.920	5.000	43.360	162.380	8.000	112.290	124.820
6.000	91.170	152.720	5.000	77.820	195.440	6.000	88.940	195.320
5.000	76.710	116.710	7.000	103.400	119.320	6.000	127.850	171.830
5.000	48.920	180.890	7.000	111.180	106.330	7.000	105.620	113.760
6.000	157.870	122.670	5.000	74.490	148.220	6.000	82.270	117.600
7.000	103.400	110.080	5.000	130.080	36.950	5.000	116.740	35.130
5.000	100.060	199.840	5.000	137.860	41.550	7.000	110.070	134.080
6.000	133.410	131.140	6.000	93.390	186.020	7.000	85.610	134.250
7.000	105.620	139.660	5.000	63.370	169.630	6.000	118.960	147.880
6.000	82.270	147.240	6.000	107.840	46.240	6.000	86.720	168.500
5.000	96.720	204.500	5.000	90.050	106.450	6.000	94.500	162.880
5.000	54.480	158.580	7.000	115.620	116.480	5.000	57.810	155.770
6.000	71.150	163.070	6.000	150.090	143.950	5.000	196.780	67.190
5.000	61.150	162.230	5.000	63.370	120.500	6.000	83.380	152.780
6.000	74.490	129.700	6.000	57.810	125.170	5.000	61.150	130.710
6.000	102.280	167.430	7.000	111.180	115.580	7.000	102.280	101.760
5.000	173.440	76.490	5.000	74.490	139.890	5.000	127.850	189.380
8.000	115.620	107.230	7.000	102.280	141.530	5.000	72.270	172.330
6.000	82.270	158.350	5.000	141.200	164.320	11.000	124.178	129.870
5.000	63.370	128.840	5.000	82.270	203.720			
6.000	76.710	123.200	5.000	126.740	193.090			
5.000	71.150	116.750	5.000	81.160	148.170			
7.000	96.720	124.920	5.000	133.410	188.400			
8.000	132.300	110.830	6.000	71.150	164.930			
6.000	46.690	154.000	6.000	93.390	56.450			
6.000	83.380	132.410	5.000	91.170	180.490			
5.000	85.610	164.800	6.000	98.950	44.410			
5.000	76.710	121.340	6.000	128.970	143.180			
5.000	50.030	172.530	6.000	61.150	116.810			
5.000	97.840	161.920	7.000	106.730	102.660			
7.000	110.070	136.850	5.000	60.040	124.230			
7.000	107.840	136.870	5.000	182.330	142.780			

AREA1.DAT:

DATA VALUE	Y	X		DATA VALUE	Y	X
				13.830000	495.000000	365.000000
				3.840000	495.000000	515.000000
558.530000	25.000000	205.000000		23.050000	495.000000	665.000000
5.070000	65.000000	15.000000		5.390000	495.000000	815.000000
28.500000	75.000000	55.000000		992.200000	495.000000	965.000000
48.220000	105.000000	415.000000		52.530000	495.000000	1115.000000
12.600000	105.000000	865.000000		62.140000	495.000000	1265.000000
185.570000	135.000000	215.000000		40.120000	495.000000	1415.000000
345.840000	145.000000	165.000000		47.550000	495.000000	1565.000000
53.100000	165.000000	745.000000		201.140000	495.000000	1715.000000
382.240000	185.000000	505.000000		0.000000E+00	635.000000	65.000000
99.930000	235.000000	815.000000		254.530000	635.000000	215.000000
317.910000	255.000000	895.000000		190.210000	635.000000	365.000000
103.470000	275.000000	805.000000		8.430000	635.000000	515.000000
13.360000	285.000000	735.000000		45.760000	635.000000	665.000000
12.150000	295.000000	675.000000		28.170000	635.000000	815.000000
43.620000	345.000000	365.000000		76.330000	635.000000	965.000000
191.300000	395.000000	725.000000		431.100000	635.000000	1115.000000
2.860000	435.000000	585.000000		43.570000	635.000000	1265.000000
17.080000	465.000000	135.000000		882.120000	635.000000	1415.000000
146.090000	535.000000	475.000000		179.510000	635.000000	1565.000000
321.110000	575.000000	465.000000		4.080000	635.000000	1715.000000
184.110000	615.000000	345.000000		0.000000E+00	775.000000	65.000000
83.940000	625.000000	555.000000		28.340000	775.000000	215.000000
39.860000	635.000000	175.000000		108.660000	775.000000	365.000000
29.950000	655.000000	735.000000		203.450000	775.000000	515.000000
103.280000	665.000000	455.000000		493.900000	775.000000	665.000000
464.760000	675.000000	475.000000		12.690000	775.000000	815.000000
4.190000	685.000000	875.000000		137.750000	775.000000	965.000000
11.270000	695.000000	125.000000		28.820000	775.000000	1115.000000
59.760000	715.000000	335.000000		5.790000	775.000000	1265.000000
215.770000	775.000000	605.000000		481.250000	775.000000	1415.000000
4.020000	75.000000	65.000000		183.780000	775.000000	1565.000000
204.520000	75.000000	215.000000		56.460000	775.000000	1715.000000
397.180000	75.000000	365.000000		0.000000E+00	915.000000	65.000000
37.770000	75.000000	515.000000		1.000000E-02	915.000000	215.000000
123.420000	75.000000	665.000000		15.280000	915.000000	365.000000
42.930000	75.000000	815.000000		49.250000	915.000000	515.000000
2.600000E-01	75.000000	965.000000		244.390000	915.000000	665.000000
1.100000E-01	75.000000	1115.000000		915.920000	915.000000	815.000000
4.800000E-01	75.000000	1265.000000		335.900000	915.000000	965.000000
1.100000E-01	75.000000	1415.000000		11.610000	915.000000	1115.000000
1.630000	75.000000	1565.000000		526.800000	915.000000	1265.000000
12.720000	75.000000	1715.000000		1.020000	915.000000	1415.000000
10.710000	215.000000	65.000000		14.980000	915.000000	1565.000000
177.360000	215.000000	215.000000		66.020000	915.000000	1715.000000
16.610000	215.000000	365.000000		3.880000	1055.000000	65.000000
2.570000	215.000000	515.000000		0.000000E+00	1055.000000	215.000000
33.280000	215.000000	665.000000		3.460000	1055.000000	365.000000
90.960000	215.000000	815.000000		3.700000E-01	1055.000000	515.000000
18.000000	215.000000	965.000000		6.400000E-01	1055.000000	665.000000
1.340000	215.000000	1115.000000		525.450000	1055.000000	815.000000
60.070000	215.000000	1265.000000		15.230000	1055.000000	965.000000
5.400000E-01	215.000000	1415.000000		232.950000	1055.000000	1115.000000
2.100000E-01	215.000000	1565.000000		9.190000	1055.000000	1265.000000
7.100000E-01	215.000000	1715.000000		5.760000	1055.000000	1415.000000
4.420000	355.000000	65.000000		877.540000	1055.000000	1565.000000
20.250000	355.000000	215.000000		58.480000	1055.000000	1715.000000
33.600000	355.000000	365.000000				
7.820000	355.000000	515.000000				
9.840000	355.000000	665.000000				
84.140000	355.000000	815.000000				
34.210000	355.000000	965.000000				
7.870000	355.000000	1115.000000				
857.650000	355.000000	1265.000000				
6.040000	355.000000	1415.000000				
1.120000	355.000000	1565.000000				
1.000000E-02	355.000000	1715.000000				
0.000000E+00	495.000000	65.000000				
134.680000	495.000000	215.000000				

AREA2.DAT:

DATA VALUE	Y	X
0.000000E+00	22.000000	0.000000E+00
0.000000E+00	23.000000	217.000000
1.260000	29.000000	283.000000
6.270000	92.000000	434.000000
4.340000	4.000000	568.000000
3.910000	73.000000	715.000000
2.320000	10.000000	944.000000
3.040000	3.000000	1003.000000
32.180000	47.000000	1181.000000
23.620000	46.000000	1271.000000
0.000000E+00	145.000000	2.000000
11.830000	167.000000	149.000000
16.920000	114.000000	367.000000
7.420000	191.000000	493.000000
3.980000	108.000000	635.000000
10.140000	167.000000	730.000000
3.670000	124.000000	875.000000
1.170000	136.000000	1020.000000
0.000000E+00	181.000000	1131.000000
19.140000	127.000000	1277.000000
1.360000	240.000000	42.000000
5.330000	283.000000	276.000000
14.430000	233.000000	299.000000
20.680000	299.000000	526.000000
7.890000	209.000000	636.000000
8.910000	235.000000	706.000000
6.070000	225.000000	886.000000
7.550000	245.000000	1025.000000
0.000000E+00	215.000000	1144.000000
7.050000	244.000000	1380.000000
6.040000	395.000000	55.000000
13.800000	341.000000	242.000000
2.700000	357.000000	380.000000
16.800000	313.000000	481.000000
1.270000	367.000000	593.000000
15.460000	331.000000	802.000000
20.830000	334.000000	972.000000
13.630000	309.000000	987.000000
0.000000E+00	308.000000	1128.000000
13.100000	324.000000	1322.000000
4.400000	424.000000	102.000000
16.160000	499.000000	153.000000
5.750000	408.000000	379.000000
5.170000	486.000000	466.000000
12.230000	423.000000	676.000000
23.900000	433.000000	820.000000
22.110000	476.000000	935.000000
14.480000	456.000000	1085.000000
2.200000	412.000000	1159.000000
3.650000	434.000000	1323.000000
3.440000	520.000000	66.000000
4.920000	524.000000	167.000000
16.290000	550.000000	328.000000
20.900000	561.000000	500.000000
2.680000	501.000000	638.000000
6.500000	520.000000	794.000000
2.180000	593.000000	843.000000
17.670000	529.000000	1061.000000
6.120000	529.000000	1160.000000
7.330000	587.000000	1319.000000
0.000000E+00	614.000000	29.000000
10.390000	699.000000	275.000000
9.780000	685.000000	368.000000
11.480000	601.000000	429.000000
3.230000	693.000000	569.000000
1.730000	631.000000	801.000000
6.100000E-01	671.000000	876.000000
28.860000	646.000000	1099.000000

DATA VALUE	Y	X
6.380000	636.000000	1178.000000
1.200000E-01	643.000000	1328.000000
0.000000E+00	769.000000	59.000000
3.700000	756.000000	213.000000
14.550000	721.000000	370.000000
9.030000	733.000000	493.000000
5.650000	781.000000	691.000000
9.350000	751.000000	834.000000
9.600000E-01	712.000000	904.000000
20.920000	800.000000	1047.000000
3.480000	705.000000	1214.000000
4.430000	779.000000	1336.000000
2.130000	893.000000	78.000000
22.900000	824.000000	276.000000
24.290000	844.000000	295.000000
12.030000	859.000000	547.000000
8.470000	871.000000	580.000000
8.890000	829.000000	716.000000
24.200000	807.000000	922.000000
21.020000	827.000000	1094.000000
6.570000	837.000000	1152.000000
1.000000	815.000000	1315.000000
3.700000E-01	919.000000	28.000000
4.220000	907.000000	234.000000
4.180000	909.000000	305.000000
19.350000	976.000000	441.000000
21.650000	934.000000	601.000000
19.640000	941.000000	739.000000
35.210000	944.000000	942.000000
6.790000	938.000000	1068.000000
24.930000	911.000000	1174.000000
0.000000E+00	973.000000	1264.000000
4.100000	1041.000000	23.000000
4.800000	1037.000000	262.000000
44.810000	1087.000000	316.000000
12.630000	1013.000000	492.000000
9.260000	1090.000000	588.000000
21.560000	1038.000000	718.000000
17.690000	1025.000000	875.000000
13.350000	1072.000000	985.000000
11.580000	1080.000000	1124.000000
0.000000E+00	1013.000000	1273.000000
26.860000	1152.000000	58.000000
22.240000	1193.000000	242.000000
22.870000	1195.000000	281.000000
10.630000	1194.000000	465.000000
1.750000	1124.000000	568.000000
3.380000	1170.000000	833.000000
13.000000	1168.000000	925.000000
8.490000	1159.000000	1059.000000
25.330000	1117.000000	1146.000000
11.300000	1121.000000	1309.000000
24.070000	1215.000000	135.000000
14.360000	1293.000000	151.000000
28.130000	1290.000000	310.000000
0.000000E+00	1231.000000	520.000000
4.000000E-01	1285.000000	614.000000
5.960000	1228.000000	727.000000
17.970000	1247.000000	874.000000
3.610000	1209.000000	1044.000000
20.630000	1246.000000	1131.000000
7.160000	1249.000000	1384.000000
44.610000	1302.000000	79.000000
4.670000	1324.000000	189.000000
42.720000	1309.000000	314.000000
0.000000E+00	1355.000000	511.000000
4.000000E-01	1381.000000	563.000000
29.800000	1316.000000	790.000000
24.030000	1342.000000	860.000000
4.840000	1318.000000	1006.000000

DATA VALUE	Y	X
15.990000	1362.000000	1172.000000
8.430000	1357.000000	1345.000000
0.000000E+00	1463.000000	37.000000
9.820000	1462.000000	221.000000
16.400000	1413.000000	328.000000
8.890000	1480.000000	532.000000
11.180000	1478.000000	689.000000
4.130000	1463.000000	826.000000
3.520000	1455.000000	858.000000
2.080000	1463.000000	1118.000000
7.900000E-01	1474.000000	1180.000000
35.030000	1412.000000	1319.000000
7.520000	1569.000000	69.000000
0.000000E+00	1533.000000	268.000000
3.350000	1503.000000	296.000000
4.950000	1533.000000	542.000000
20.780000	1570.000000	651.000000
3.430000	1589.000000	810.000000
3.200000E-01	1574.000000	953.000000
19.620000	1566.000000	1095.000000
4.500000E-01	1590.000000	1248.000000
11.580000	1513.000000	1371.000000
8.010000	1674.000000	8.000000
6.080000	1644.000000	171.000000
1.870000	1666.000000	342.000000
21.550000	1679.000000	525.000000
11.380000	1630.000000	603.000000
1.400000E-01	1684.000000	712.000000
18.320000	1644.000000	862.000000
21.390000	1649.000000	1087.000000
4.010000	1612.000000	1211.000000
7.990000	1651.000000	1356.000000
1.900000E-01	1707.000000	90.000000
8.760000	1788.000000	276.000000
4.420000	1727.000000	331.000000
3.780000	1779.000000	544.000000
13.140000	1780.000000	591.000000
4.790000	1730.000000	701.000000
20.780000	1741.000000	978.000000
12.460000	1728.000000	1073.000000
35.790000	1737.000000	1134.000000
6.300000E-01	1782.000000	1384.000000
1.720000	1880.000000	26.000000
13.300000	1840.000000	143.000000
4.620000	1863.000000	323.000000
0.000000E+00	1840.000000	502.000000
6.330000	1892.000000	588.000000
5.900000E-01	1830.000000	809.000000
4.630000	1809.000000	881.000000
26.960000	1893.000000	1064.000000
54.830000	1816.000000	1150.000000
0.000000E+00	1830.000000	1394.000000

CALIENTE.DAT

Y	X	ELEVATION
1.000000	1.000000	0.000000E+00
1.000000	2.000000	0.000000E+00
1.000000	3.000000	0.000000E+00
1.000000	4.000000	0.000000E+00
1.000000	5.000000	0.000000E+00
1.000000	6.000000	0.000000E+00
1.000000	7.000000	0.000000E+00
1.000000	8.000000	0.000000E+00
1.000000	9.000000	0.000000E+00
1.000000	10.000000	0.000000E+00
1.000000	11.000000	0.000000E+00
1.000000	12.000000	0.000000E+00
1.000000	13.000000	0.000000E+00
1.000000	14.000000	0.000000E+00
1.000000	15.000000	0.000000E+00
1.000000	16.000000	0.000000E+00
1.000000	17.000000	0.000000E+00
1.000000	18.000000	0.000000E+00
1.000000	19.000000	3.000000
1.000000	20.000000	13.000000
1.000000	21.000000	8.000000
1.000000	22.000000	8.000000
1.000000	23.000000	4.000000
1.000000	24.000000	2.000000
1.000000	25.000000	2.000000
1.000000	26.000000	2.000000
1.000000	27.000000	3.000000
1.000000	28.000000	5.000000
1.000000	29.000000	5.000000
1.000000	30.000000	9.000000
1.000000	31.000000	6.000000
1.000000	32.000000	5.000000
1.000000	33.000000	6.000000
1.000000	34.000000	6.000000
1.000000	35.000000	9.000000
1.000000	36.000000	12.000000
1.000000	37.000000	19.000000
1.000000	38.000000	27.000000
1.000000	39.000000	25.000000
1.000000	40.000000	10.000000
1.000000	41.000000	8.000000
1.000000	42.000000	7.000000
1.000000	43.000000	8.000000
1.000000	44.000000	6.000000
1.000000	45.000000	6.000000
1.000000	46.000000	8.000000
1.000000	47.000000	6.000000
1.000000	48.000000	7.000000
1.000000	49.000000	9.000000
1.000000	50.000000	6.000000
2.000000	1.000000	0.000000E+00
2.000000	2.000000	0.000000E+00
2.000000	3.000000	0.000000E+00
2.000000	4.000000	0.000000E+00
2.000000	5.000000	0.000000E+00
2.000000	6.000000	0.000000E+00
2.000000	7.000000	0.000000E+00
2.000000	8.000000	0.000000E+00
2.000000	9.000000	0.000000E+00
2.000000	10.000000	0.000000E+00
2.000000	11.000000	0.000000E+00
2.000000	12.000000	0.000000E+00
2.000000	13.000000	0.000000E+00
2.000000	14.000000	0.000000E+00
2.000000	15.000000	0.000000E+00
2.000000	16.000000	0.000000E+00
2.000000	17.000000	10.000000
2.000000	18.000000	4.000000
2.000000	19.000000	6.000000
2.000000	20.000000	14.000000

Y	X	ELEVATION
2.000000	21.000000	10.000000
2.000000	22.000000	7.000000
2.000000	23.000000	6.000000
2.000000	24.000000	6.000000
2.000000	25.000000	6.000000
2.000000	26.000000	5.000000
2.000000	27.000000	6.000000
2.000000	28.000000	8.000000
2.000000	29.000000	13.000000
2.000000	30.000000	10.000000
2.000000	31.000000	7.000000
2.000000	32.000000	7.000000
2.000000	33.000000	7.000000
2.000000	34.000000	7.000000
2.000000	35.000000	7.000000
2.000000	36.000000	7.000000
2.000000	37.000000	11.000000
2.000000	38.000000	15.000000
2.000000	39.000000	22.000000
2.000000	40.000000	22.000000
2.000000	41.000000	12.000000
2.000000	42.000000	12.000000
2.000000	43.000000	14.000000
2.000000	44.000000	8.000000
2.000000	45.000000	7.000000
2.000000	46.000000	7.000000
2.000000	47.000000	7.000000
2.000000	48.000000	7.000000
2.000000	49.000000	7.000000
2.000000	50.000000	7.000000
3.000000	1.000000	0.000000E+00
3.000000	2.000000	0.000000E+00
3.000000	3.000000	0.000000E+00
3.000000	4.000000	0.000000E+00
3.000000	5.000000	0.000000E+00
3.000000	6.000000	0.000000E+00
3.000000	7.000000	0.000000E+00
3.000000	8.000000	0.000000E+00
3.000000	9.000000	0.000000E+00
3.000000	10.000000	0.000000E+00
3.000000	11.000000	0.000000E+00
3.000000	12.000000	0.000000E+00
3.000000	13.000000	0.000000E+00
3.000000	14.000000	0.000000E+00
3.000000	15.000000	0.000000E+00
3.000000	16.000000	4.000000
3.000000	17.000000	22.000000
3.000000	18.000000	7.000000
3.000000	19.000000	7.000000
3.000000	20.000000	8.000000
3.000000	21.000000	13.000000
3.000000	22.000000	14.000000
3.000000	23.000000	13.000000
3.000000	24.000000	9.000000
3.000000	25.000000	9.000000
3.000000	26.000000	8.000000
3.000000	27.000000	13.000000
3.000000	28.000000	14.000000
3.000000	29.000000	22.000000
3.000000	30.000000	14.000000
3.000000	31.000000	6.000000
3.000000	32.000000	5.000000
3.000000	33.000000	5.000000
3.000000	34.000000	5.000000
3.000000	35.000000	6.000000
3.000000	36.000000	7.000000
3.000000	37.000000	7.000000
3.000000	38.000000	7.000000
3.000000	39.000000	12.000000
3.000000	40.000000	10.000000

Y	X	ELEVATION	Y	X	ELEVATION
3.000000	41.000000	9.000000	5.000000	11.000000	7.000000
3.000000	42.000000	6.000000	5.000000	12.000000	0.000000E+00
3.000000	43.000000	6.000000	5.000000	13.000000	2.000000
3.000000	44.000000	6.000000	5.000000	14.000000	7.000000
3.000000	45.000000	6.000000	5.000000	15.000000	29.000000
3.000000	46.000000	6.000000	5.000000	16.000000	23.000000
3.000000	47.000000	6.000000	5.000000	17.000000	37.000000
3.000000	48.000000	6.000000	5.000000	18.000000	15.000000
3.000000	49.000000	6.000000	5.000000	19.000000	12.000000
3.000000	50.000000	6.000000	5.000000	20.000000	14.000000
4.000000	1.000000	0.000000E+00	5.000000	21.000000	19.000000
4.000000	2.000000	0.000000E+00	5.000000	22.000000	19.000000
4.000000	3.000000	0.000000E+00	5.000000	23.000000	25.000000
4.000000	4.000000	0.000000E+00	5.000000	24.000000	22.000000
4.000000	5.000000	0.000000E+00	5.000000	25.000000	29.000000
4.000000	6.000000	0.000000E+00	5.000000	26.000000	20.000000
4.000000	7.000000	0.000000E+00	5.000000	27.000000	17.000000
4.000000	8.000000	0.000000E+00	5.000000	28.000000	22.000000
4.000000	9.000000	0.000000E+00	5.000000	29.000000	14.000000
4.000000	10.000000	0.000000E+00	5.000000	30.000000	9.000000
4.000000	11.000000	0.000000E+00	5.000000	31.000000	5.000000
4.000000	12.000000	0.000000E+00	5.000000	32.000000	3.000000
4.000000	13.000000	1.000000	5.000000	33.000000	1.000000
4.000000	14.000000	0.000000E+00	5.000000	34.000000	2.000000
4.000000	15.000000	7.000000	5.000000	35.000000	3.000000
4.000000	16.000000	7.000000	5.000000	36.000000	1.000000
4.000000	17.000000	28.000000	5.000000	37.000000	0.000000E+00
4.000000	18.000000	12.000000	5.000000	38.000000	0.000000E+00
4.000000	19.000000	7.000000	5.000000	39.000000	1.000000
4.000000	20.000000	12.000000	5.000000	40.000000	1.000000
4.000000	21.000000	16.000000	5.000000	41.000000	3.000000
4.000000	22.000000	14.000000	5.000000	42.000000	3.000000
4.000000	23.000000	15.000000	5.000000	43.000000	4.000000
4.000000	24.000000	14.000000	5.000000	44.000000	4.000000
4.000000	25.000000	14.000000	5.000000	45.000000	3.000000
4.000000	26.000000	14.000000	5.000000	46.000000	3.000000
4.000000	27.000000	14.000000	5.000000	47.000000	3.000000
4.000000	28.000000	15.000000	5.000000	48.000000	3.000000
4.000000	29.000000	15.000000	5.000000	49.000000	4.000000
4.000000	30.000000	11.000000	5.000000	50.000000	4.000000
4.000000	31.000000	6.000000	6.000000	1.000000	0.000000E+00
4.000000	32.000000	4.000000	6.000000	2.000000	0.000000E+00
4.000000	33.000000	3.000000	6.000000	3.000000	0.000000E+00
4.000000	34.000000	4.000000	6.000000	4.000000	0.000000E+00
4.000000	35.000000	6.000000	6.000000	5.000000	7.000000
4.000000	36.000000	7.000000	6.000000	6.000000	5.000000
4.000000	37.000000	6.000000	6.000000	7.000000	0.000000E+00
4.000000	38.000000	6.000000	6.000000	8.000000	0.000000E+00
4.000000	39.000000	5.000000	6.000000	9.000000	14.000000
4.000000	40.000000	5.000000	6.000000	10.000000	7.000000
4.000000	41.000000	5.000000	6.000000	11.000000	0.000000E+00
4.000000	42.000000	5.000000	6.000000	12.000000	1.000000
4.000000	43.000000	5.000000	6.000000	13.000000	13.000000
4.000000	44.000000	5.000000	6.000000	14.000000	16.000000
4.000000	45.000000	5.000000	6.000000	15.000000	27.000000
4.000000	46.000000	5.000000	6.000000	16.000000	23.000000
4.000000	47.000000	5.000000	6.000000	17.000000	15.000000
4.000000	48.000000	5.000000	6.000000	18.000000	14.000000
4.000000	49.000000	5.000000	6.000000	19.000000	14.000000
4.000000	50.000000	5.000000	6.000000	20.000000	15.000000
5.000000	1.000000	0.000000E+00	6.000000	21.000000	22.000000
5.000000	2.000000	0.000000E+00	6.000000	22.000000	28.000000
5.000000	3.000000	0.000000E+00	6.000000	23.000000	36.000000
5.000000	4.000000	0.000000E+00	6.000000	24.000000	36.000000
5.000000	5.000000	0.000000E+00	6.000000	25.000000	33.000000
5.000000	6.000000	0.000000E+00	6.000000	26.000000	19.000000
5.000000	7.000000	0.000000E+00	6.000000	27.000000	14.000000
5.000000	8.000000	0.000000E+00	6.000000	28.000000	14.000000
5.000000	9.000000	0.000000E+00	6.000000	29.000000	15.000000
5.000000	10.000000	10.000000	6.000000	30.000000	7.000000

Y	X	ELEVATION	Y	X	ELEVATION
6.000000	31.000000	5.000000	8.000000	1.000000	0.000000E+00
6.000000	32.000000	3.000000	8.000000	2.000000	0.000000E+00
6.000000	33.000000	0.000000E+00	8.000000	3.000000	0.000000E+00
6.000000	34.000000	0.000000E+00	8.000000	4.000000	0.000000E+00
6.000000	35.000000	0.000000E+00	8.000000	5.000000	0.000000E+00
6.000000	36.000000	0.000000E+00	8.000000	6.000000	0.000000E+00
6.000000	37.000000	0.000000E+00	8.000000	7.000000	0.000000E+00
6.000000	38.000000	0.000000E+00	8.000000	8.000000	5.000000
6.000000	39.000000	0.000000E+00	8.000000	9.000000	6.000000
6.000000	40.000000	0.000000E+00	8.000000	10.000000	5.000000
6.000000	41.000000	1.000000	8.000000	11.000000	6.000000
6.000000	42.000000	2.000000	8.000000	12.000000	7.000000
6.000000	43.000000	2.000000	8.000000	13.000000	13.000000
6.000000	44.000000	3.000000	8.000000	14.000000	19.000000
6.000000	45.000000	2.000000	8.000000	15.000000	22.000000
6.000000	46.000000	2.000000	8.000000	16.000000	22.000000
6.000000	47.000000	3.000000	8.000000	17.000000	22.000000
6.000000	48.000000	3.000000	8.000000	18.000000	18.000000
6.000000	49.000000	3.000000	8.000000	19.000000	19.000000
6.000000	50.000000	3.000000	8.000000	20.000000	22.000000
7.000000	1.000000	0.000000E+00	8.000000	21.000000	25.000000
7.000000	2.000000	0.000000E+00	8.000000	22.000000	29.000000
7.000000	3.000000	0.000000E+00	8.000000	23.000000	34.000000
7.000000	4.000000	0.000000E+00	8.000000	24.000000	35.000000
7.000000	5.000000	0.000000E+00	8.000000	25.000000	22.000000
7.000000	6.000000	0.000000E+00	8.000000	26.000000	21.000000
7.000000	7.000000	3.000000	8.000000	27.000000	15.000000
7.000000	8.000000	7.000000	8.000000	28.000000	13.000000
7.000000	9.000000	5.000000	8.000000	29.000000	9.000000
7.000000	10.000000	0.000000E+00	8.000000	30.000000	7.000000
7.000000	11.000000	4.000000	8.000000	31.000000	4.000000
7.000000	12.000000	7.000000	8.000000	32.000000	2.000000
7.000000	13.000000	9.000000	8.000000	33.000000	0.000000E+00
7.000000	14.000000	14.000000	8.000000	34.000000	0.000000E+00
7.000000	15.000000	22.000000	8.000000	35.000000	0.000000E+00
7.000000	16.000000	22.000000	8.000000	36.000000	0.000000E+00
7.000000	17.000000	19.000000	8.000000	37.000000	0.000000E+00
7.000000	18.000000	14.000000	8.000000	38.000000	0.000000E+00
7.000000	19.000000	21.000000	8.000000	39.000000	0.000000E+00
7.000000	20.000000	22.000000	8.000000	40.000000	0.000000E+00
7.000000	21.000000	24.000000	8.000000	41.000000	0.000000E+00
7.000000	22.000000	29.000000	8.000000	42.000000	0.000000E+00
7.000000	23.000000	29.000000	8.000000	43.000000	0.000000E+00
7.000000	24.000000	29.000000	8.000000	44.000000	1.000000
7.000000	25.000000	22.000000	8.000000	45.000000	0.000000E+00
7.000000	26.000000	18.000000	8.000000	46.000000	0.000000E+00
7.000000	27.000000	14.000000	8.000000	47.000000	0.000000E+00
7.000000	28.000000	13.000000	8.000000	48.000000	1.000000
7.000000	29.000000	9.000000	8.000000	49.000000	1.000000
7.000000	30.000000	7.000000	8.000000	50.000000	2.000000
7.000000	31.000000	5.000000	9.000000	1.000000	0.000000E+00
7.000000	32.000000	2.000000	9.000000	2.000000	0.000000E+00
7.000000	33.000000	0.000000E+00	9.000000	3.000000	0.000000E+00
7.000000	34.000000	0.000000E+00	9.000000	4.000000	0.000000E+00
7.000000	35.000000	0.000000E+00	9.000000	5.000000	0.000000E+00
7.000000	36.000000	0.000000E+00	9.000000	6.000000	0.000000E+00
7.000000	37.000000	0.000000E+00	9.000000	7.000000	0.000000E+00
7.000000	38.000000	0.000000E+00	9.000000	8.000000	1.000000
7.000000	39.000000	0.000000E+00	9.000000	9.000000	5.000000
7.000000	40.000000	0.000000E+00	9.000000	10.000000	7.000000
7.000000	41.000000	0.000000E+00	9.000000	11.000000	10.000000
7.000000	42.000000	1.000000	9.000000	12.000000	14.000000
7.000000	43.000000	1.000000	9.000000	13.000000	17.000000
7.000000	44.000000	2.000000	9.000000	14.000000	24.000000
7.000000	45.000000	1.000000	9.000000	15.000000	22.000000
7.000000	46.000000	1.000000	9.000000	16.000000	22.000000
7.000000	47.000000	1.000000	9.000000	17.000000	22.000000
7.000000	48.000000	2.000000	9.000000	18.000000	21.000000
7.000000	49.000000	2.000000	9.000000	19.000000	21.000000
7.000000	50.000000	3.000000	9.000000	20.000000	25.000000

Y	X	ELEVATION	Y	X	ELEVATION
9.000000	21.000000	28.000000	10.000000	41.000000	0.000000E+00
9.000000	22.000000	35.000000	10.000000	42.000000	0.000000E+00
9.000000	23.000000	44.000000	10.000000	43.000000	0.000000E+00
9.000000	24.000000	36.000000	10.000000	44.000000	0.000000E+00
9.000000	25.000000	25.000000	10.000000	45.000000	0.000000E+00
9.000000	26.000000	22.000000	10.000000	46.000000	0.000000E+00
9.000000	27.000000	18.000000	10.000000	47.000000	0.000000E+00
9.000000	28.000000	14.000000	10.000000	48.000000	0.000000E+00
9.000000	29.000000	10.000000	10.000000	49.000000	0.000000E+00
9.000000	30.000000	6.000000	10.000000	50.000000	1.000000
9.000000	31.000000	4.000000	11.000000	1.000000	0.000000E+00
9.000000	32.000000	2.000000	11.000000	2.000000	0.000000E+00
9.000000	33.000000	1.000000	11.000000	3.000000	0.000000E+00
9.000000	34.000000	0.000000E+00	11.000000	4.000000	0.000000E+00
9.000000	35.000000	0.000000E+00	11.000000	5.000000	0.000000E+00
9.000000	36.000000	0.000000E+00	11.000000	6.000000	0.000000E+00
9.000000	37.000000	0.000000E+00	11.000000	7.000000	0.000000E+00
9.000000	38.000000	0.000000E+00	11.000000	8.000000	6.000000
9.000000	39.000000	0.000000E+00	11.000000	9.000000	7.000000
9.000000	40.000000	0.000000E+00	11.000000	10.000000	10.000000
9.000000	41.000000	0.000000E+00	11.000000	11.000000	14.000000
9.000000	42.000000	0.000000E+00	11.000000	12.000000	18.000000
9.000000	43.000000	0.000000E+00	11.000000	13.000000	19.000000
9.000000	44.000000	0.000000E+00	11.000000	14.000000	21.000000
9.000000	45.000000	0.000000E+00	11.000000	15.000000	22.000000
9.000000	46.000000	0.000000E+00	11.000000	16.000000	22.000000
9.000000	47.000000	0.000000E+00	11.000000	17.000000	27.000000
9.000000	48.000000	0.000000E+00	11.000000	18.000000	29.000000
9.000000	49.000000	1.000000	11.000000	19.000000	30.000000
9.000000	50.000000	1.000000	11.000000	20.000000	39.000000
10.000000	1.000000	0.000000E+00	11.000000	21.000000	43.000000
10.000000	2.000000	0.000000E+00	11.000000	22.000000	54.000000
10.000000	3.000000	0.000000E+00	11.000000	23.000000	47.000000
10.000000	4.000000	0.000000E+00	11.000000	24.000000	34.000000
10.000000	5.000000	0.000000E+00	11.000000	25.000000	29.000000
10.000000	6.000000	0.000000E+00	11.000000	26.000000	24.000000
10.000000	7.000000	0.000000E+00	11.000000	27.000000	24.000000
10.000000	8.000000	3.000000	11.000000	28.000000	12.000000
10.000000	9.000000	6.000000	11.000000	29.000000	6.000000
10.000000	10.000000	7.000000	11.000000	30.000000	3.000000
10.000000	11.000000	13.000000	11.000000	31.000000	1.000000
10.000000	12.000000	22.000000	11.000000	32.000000	0.000000E+00
10.000000	13.000000	23.000000	11.000000	33.000000	0.000000E+00
10.000000	14.000000	27.000000	11.000000	34.000000	0.000000E+00
10.000000	15.000000	22.000000	11.000000	35.000000	0.000000E+00
10.000000	16.000000	22.000000	11.000000	36.000000	0.000000E+00
10.000000	17.000000	27.000000	11.000000	37.000000	0.000000E+00
10.000000	18.000000	23.000000	11.000000	38.000000	0.000000E+00
10.000000	19.000000	28.000000	11.000000	39.000000	0.000000E+00
10.000000	20.000000	29.000000	11.000000	40.000000	0.000000E+00
10.000000	21.000000	34.000000	11.000000	41.000000	0.000000E+00
10.000000	22.000000	46.000000	11.000000	42.000000	0.000000E+00
10.000000	23.000000	52.000000	11.000000	43.000000	0.000000E+00
10.000000	24.000000	44.000000	11.000000	44.000000	0.000000E+00
10.000000	25.000000	25.000000	11.000000	45.000000	0.000000E+00
10.000000	26.000000	22.000000	11.000000	46.000000	0.000000E+00
10.000000	27.000000	33.000000	11.000000	47.000000	0.000000E+00
10.000000	28.000000	18.000000	11.000000	48.000000	0.000000E+00
10.000000	29.000000	7.000000	11.000000	49.000000	0.000000E+00
10.000000	30.000000	4.000000	11.000000	50.000000	0.000000E+00
10.000000	31.000000	3.000000	12.000000	1.000000	0.000000E+00
10.000000	32.000000	1.000000	12.000000	2.000000	0.000000E+00
10.000000	33.000000	0.000000E+00	12.000000	3.000000	0.000000E+00
10.000000	34.000000	0.000000E+00	12.000000	4.000000	0.000000E+00
10.000000	35.000000	0.000000E+00	12.000000	5.000000	0.000000E+00
10.000000	36.000000	0.000000E+00	12.000000	6.000000	0.000000E+00
10.000000	37.000000	0.000000E+00	12.000000	7.000000	0.000000E+00
10.000000	38.000000	0.000000E+00	12.000000	8.000000	7.000000
10.000000	39.000000	0.000000E+00	12.000000	9.000000	7.000000
10.000000	40.000000	0.000000E+00	12.000000	10.000000	7.000000

Y	X	ELEVATION	Y	X	ELEVATION
12.000000	11.000000	12.000000	13.000000	31.000000	0.000000E+00
12.000000	12.000000	13.000000	13.000000	32.000000	0.000000E+00
12.000000	13.000000	14.000000	13.000000	33.000000	0.000000E+00
12.000000	14.000000	17.000000	13.000000	34.000000	0.000000E+00
12.000000	15.000000	20.000000	13.000000	35.000000	0.000000E+00
12.000000	16.000000	22.000000	13.000000	36.000000	0.000000E+00
12.000000	17.000000	28.000000	13.000000	37.000000	0.000000E+00
12.000000	18.000000	34.000000	13.000000	38.000000	0.000000E+00
12.000000	19.000000	37.000000	13.000000	39.000000	0.000000E+00
12.000000	20.000000	45.000000	13.000000	40.000000	0.000000E+00
12.000000	21.000000	43.000000	13.000000	41.000000	0.000000E+00
12.000000	22.000000	43.000000	13.000000	42.000000	0.000000E+00
12.000000	23.000000	37.000000	13.000000	43.000000	0.000000E+00
12.000000	24.000000	29.000000	13.000000	44.000000	0.000000E+00
12.000000	25.000000	22.000000	13.000000	45.000000	0.000000E+00
12.000000	26.000000	22.000000	13.000000	46.000000	0.000000E+00
12.000000	27.000000	14.000000	13.000000	47.000000	0.000000E+00
12.000000	28.000000	7.000000	13.000000	48.000000	0.000000E+00
12.000000	29.000000	4.000000	13.000000	49.000000	0.000000E+00
12.000000	30.000000	1.000000	13.000000	50.000000	0.000000E+00
12.000000	31.000000	0.000000E+00	14.000000	1.000000	0.000000E+00
12.000000	32.000000	0.000000E+00	14.000000	2.000000	0.000000E+00
12.000000	33.000000	0.000000E+00	14.000000	3.000000	0.000000E+00
12.000000	34.000000	0.000000E+00	14.000000	4.000000	0.000000E+00
12.000000	35.000000	0.000000E+00	14.000000	5.000000	0.000000E+00
12.000000	36.000000	0.000000E+00	14.000000	6.000000	2.000000
12.000000	37.000000	0.000000E+00	14.000000	7.000000	0.000000E+00
12.000000	38.000000	0.000000E+00	14.000000	8.000000	0.000000E+00
12.000000	39.000000	0.000000E+00	14.000000	9.000000	6.000000
12.000000	40.000000	0.000000E+00	14.000000	10.000000	7.000000
12.000000	41.000000	0.000000E+00	14.000000	11.000000	7.000000
12.000000	42.000000	0.000000E+00	14.000000	12.000000	14.000000
12.000000	43.000000	0.000000E+00	14.000000	13.000000	14.000000
12.000000	44.000000	0.000000E+00	14.000000	14.000000	15.000000
12.000000	45.000000	0.000000E+00	14.000000	15.000000	16.000000
12.000000	46.000000	0.000000E+00	14.000000	16.000000	21.000000
12.000000	47.000000	0.000000E+00	14.000000	17.000000	29.000000
12.000000	48.000000	0.000000E+00	14.000000	18.000000	22.000000
12.000000	49.000000	0.000000E+00	14.000000	19.000000	25.000000
12.000000	50.000000	0.000000E+00	14.000000	20.000000	22.000000
13.000000	1.000000	0.000000E+00	14.000000	21.000000	26.000000
13.000000	2.000000	0.000000E+00	14.000000	22.000000	21.000000
13.000000	3.000000	0.000000E+00	14.000000	23.000000	19.000000
13.000000	4.000000	0.000000E+00	14.000000	24.000000	18.000000
13.000000	5.000000	0.000000E+00	14.000000	25.000000	15.000000
13.000000	6.000000	0.000000E+00	14.000000	26.000000	14.000000
13.000000	7.000000	0.000000E+00	14.000000	27.000000	7.000000
13.000000	8.000000	4.000000	14.000000	28.000000	2.000000
13.000000	9.000000	6.000000	14.000000	29.000000	0.000000E+00
13.000000	10.000000	7.000000	14.000000	30.000000	0.000000E+00
13.000000	11.000000	8.000000	14.000000	31.000000	0.000000E+00
13.000000	12.000000	12.000000	14.000000	32.000000	0.000000E+00
13.000000	13.000000	14.000000	14.000000	33.000000	0.000000E+00
13.000000	14.000000	16.000000	14.000000	34.000000	0.000000E+00
13.000000	15.000000	18.000000	14.000000	35.000000	0.000000E+00
13.000000	16.000000	21.000000	14.000000	36.000000	0.000000E+00
13.000000	17.000000	29.000000	14.000000	37.000000	0.000000E+00
13.000000	18.000000	37.000000	14.000000	38.000000	0.000000E+00
13.000000	19.000000	47.000000	14.000000	39.000000	0.000000E+00
13.000000	20.000000	35.000000	14.000000	40.000000	0.000000E+00
13.000000	21.000000	30.000000	14.000000	41.000000	0.000000E+00
13.000000	22.000000	33.000000	14.000000	42.000000	0.000000E+00
13.000000	23.000000	28.000000	14.000000	43.000000	0.000000E+00
13.000000	24.000000	22.000000	14.000000	44.000000	0.000000E+00
13.000000	25.000000	22.000000	14.000000	45.000000	0.000000E+00
13.000000	26.000000	15.000000	14.000000	46.000000	0.000000E+00
13.000000	27.000000	11.000000	14.000000	47.000000	0.000000E+00
13.000000	28.000000	5.000000	14.000000	48.000000	0.000000E+00
13.000000	29.000000	2.000000	14.000000	49.000000	0.000000E+00
13.000000	30.000000	0.000000E+00	14.000000	50.000000	0.000000E+00

Y	X	ELEVATION	Y	X	ELEVATION
15.000000	1.000000	0.000000E+00	16.000000	21.000000	10.000000
15.000000	2.000000	0.000000E+00	16.000000	22.000000	9.000000
15.000000	3.000000	0.000000E+00	16.000000	23.000000	8.000000
15.000000	4.000000	7.000000	16.000000	24.000000	7.000000
15.000000	5.000000	7.000000	16.000000	25.000000	5.000000
15.000000	6.000000	7.000000	16.000000	26.000000	3.000000
15.000000	7.000000	4.000000	16.000000	27.000000	0.000000E+00
15.000000	8.000000	4.000000	16.000000	28.000000	0.000000E+00
15.000000	9.000000	7.000000	16.000000	29.000000	0.000000E+00
15.000000	10.000000	7.000000	16.000000	30.000000	0.000000E+00
15.000000	11.000000	8.000000	16.000000	31.000000	0.000000E+00
15.000000	12.000000	14.000000	16.000000	32.000000	0.000000E+00
15.000000	13.000000	14.000000	16.000000	33.000000	0.000000E+00
15.000000	14.000000	14.000000	16.000000	34.000000	0.000000E+00
15.000000	15.000000	14.000000	16.000000	35.000000	0.000000E+00
15.000000	16.000000	17.000000	16.000000	36.000000	0.000000E+00
15.000000	17.000000	18.000000	16.000000	37.000000	0.000000E+00
15.000000	18.000000	17.000000	16.000000	38.000000	0.000000E+00
15.000000	19.000000	17.000000	16.000000	39.000000	0.000000E+00
15.000000	20.000000	21.000000	16.000000	40.000000	0.000000E+00
15.000000	21.000000	15.000000	16.000000	41.000000	0.000000E+00
15.000000	22.000000	14.000000	16.000000	42.000000	0.000000E+00
15.000000	23.000000	14.000000	16.000000	43.000000	0.000000E+00
15.000000	24.000000	13.000000	16.000000	44.000000	0.000000E+00
15.000000	25.000000	9.000000	16.000000	45.000000	0.000000E+00
15.000000	26.000000	7.000000	16.000000	46.000000	0.000000E+00
15.000000	27.000000	3.000000	16.000000	47.000000	0.000000E+00
15.000000	28.000000	0.000000E+00	16.000000	48.000000	0.000000E+00
15.000000	29.000000	0.000000E+00	16.000000	49.000000	0.000000E+00
15.000000	30.000000	0.000000E+00	16.000000	50.000000	0.000000E+00
15.000000	31.000000	0.000000E+00	17.000000	1.000000	3.000000
15.000000	32.000000	0.000000E+00	17.000000	2.000000	7.000000
15.000000	33.000000	0.000000E+00	17.000000	3.000000	7.000000
15.000000	34.000000	0.000000E+00	17.000000	4.000000	7.000000
15.000000	35.000000	0.000000E+00	17.000000	5.000000	7.000000
15.000000	36.000000	0.000000E+00	17.000000	6.000000	7.000000
15.000000	37.000000	0.000000E+00	17.000000	7.000000	7.000000
15.000000	38.000000	0.000000E+00	17.000000	8.000000	7.000000
15.000000	39.000000	0.000000E+00	17.000000	9.000000	7.000000
15.000000	40.000000	0.000000E+00	17.000000	10.000000	7.000000
15.000000	41.000000	0.000000E+00	17.000000	11.000000	10.000000
15.000000	42.000000	0.000000E+00	17.000000	12.000000	14.000000
15.000000	43.000000	0.000000E+00	17.000000	13.000000	7.000000
15.000000	44.000000	0.000000E+00	17.000000	14.000000	7.000000
15.000000	45.000000	0.000000E+00	17.000000	15.000000	7.000000
15.000000	46.000000	0.000000E+00	17.000000	16.000000	7.000000
15.000000	47.000000	0.000000E+00	17.000000	17.000000	7.000000
15.000000	48.000000	0.000000E+00	17.000000	18.000000	7.000000
15.000000	49.000000	0.000000E+00	17.000000	19.000000	6.000000
15.000000	50.000000	0.000000E+00	17.000000	20.000000	7.000000
16.000000	1.000000	0.000000E+00	17.000000	21.000000	6.000000
16.000000	2.000000	0.000000E+00	17.000000	22.000000	5.000000
16.000000	3.000000	0.000000E+00	17.000000	23.000000	4.000000
16.000000	4.000000	6.000000	17.000000	24.000000	3.000000
16.000000	5.000000	7.000000	17.000000	25.000000	2.000000
16.000000	6.000000	7.000000	17.000000	26.000000	0.000000E+00
16.000000	7.000000	7.000000	17.000000	27.000000	0.000000E+00
16.000000	8.000000	6.000000	17.000000	28.000000	0.000000E+00
16.000000	9.000000	7.000000	17.000000	29.000000	0.000000E+00
16.000000	10.000000	7.000000	17.000000	30.000000	0.000000E+00
16.000000	11.000000	8.000000	17.000000	31.000000	0.000000E+00
16.000000	12.000000	13.000000	17.000000	32.000000	0.000000E+00
16.000000	13.000000	10.000000	17.000000	33.000000	0.000000E+00
16.000000	14.000000	10.000000	17.000000	34.000000	0.000000E+00
16.000000	15.000000	12.000000	17.000000	35.000000	0.000000E+00
16.000000	16.000000	14.000000	17.000000	36.000000	0.000000E+00
16.000000	17.000000	12.000000	17.000000	37.000000	0.000000E+00
16.000000	18.000000	11.000000	17.000000	38.000000	0.000000E+00
16.000000	19.000000	11.000000	17.000000	39.000000	0.000000E+00
16.000000	20.000000	11.000000	17.000000	40.000000	0.000000E+00

Y	X	ELEVATION	Y	X	ELEVATION
17.000000	41.000000	0.000000E+00	19.000000	11.000000	5.000000
17.000000	42.000000	0.000000E+00	19.000000	12.000000	4.000000
17.000000	43.000000	0.000000E+00	19.000000	13.000000	3.000000
17.000000	44.000000	0.000000E+00	19.000000	14.000000	3.000000
17.000000	45.000000	0.000000E+00	19.000000	15.000000	3.000000
17.000000	46.000000	0.000000E+00	19.000000	16.000000	3.000000
17.000000	47.000000	0.000000E+00	19.000000	17.000000	2.000000
17.000000	48.000000	0.000000E+00	19.000000	18.000000	2.000000
17.000000	49.000000	0.000000E+00	19.000000	19.000000	2.000000
17.000000	50.000000	0.000000E+00	19.000000	20.000000	1.000000
18.000000	1.000000	7.000000	19.000000	21.000000	0.000000E+00
18.000000	2.000000	14.000000	19.000000	22.000000	0.000000E+00
18.000000	3.000000	8.000000	19.000000	23.000000	3.000000
18.000000	4.000000	13.000000	19.000000	24.000000	3.000000
18.000000	5.000000	7.000000	19.000000	25.000000	4.000000
18.000000	6.000000	7.000000	19.000000	26.000000	4.000000
18.000000	7.000000	7.000000	19.000000	27.000000	3.000000
18.000000	8.000000	7.000000	19.000000	28.000000	2.000000
18.000000	9.000000	7.000000	19.000000	29.000000	0.000000E+00
18.000000	10.000000	7.000000	19.000000	30.000000	0.000000E+00
18.000000	11.000000	8.000000	19.000000	31.000000	0.000000E+00
18.000000	12.000000	6.000000	19.000000	32.000000	0.000000E+00
18.000000	13.000000	5.000000	19.000000	33.000000	0.000000E+00
18.000000	14.000000	5.000000	19.000000	34.000000	0.000000E+00
18.000000	15.000000	5.000000	19.000000	35.000000	0.000000E+00
18.000000	16.000000	5.000000	19.000000	36.000000	0.000000E+00
18.000000	17.000000	4.000000	19.000000	37.000000	0.000000E+00
18.000000	18.000000	4.000000	19.000000	38.000000	0.000000E+00
18.000000	19.000000	4.000000	19.000000	39.000000	0.000000E+00
18.000000	20.000000	4.000000	19.000000	40.000000	0.000000E+00
18.000000	21.000000	3.000000	19.000000	41.000000	0.000000E+00
18.000000	22.000000	2.000000	19.000000	42.000000	0.000000E+00
18.000000	23.000000	2.000000	19.000000	43.000000	0.000000E+00
18.000000	24.000000	0.000000E+00	19.000000	44.000000	0.000000E+00
18.000000	25.000000	0.000000E+00	19.000000	45.000000	0.000000E+00
18.000000	26.000000	0.000000E+00	19.000000	46.000000	0.000000E+00
18.000000	27.000000	0.000000E+00	19.000000	47.000000	0.000000E+00
18.000000	28.000000	0.000000E+00	19.000000	48.000000	0.000000E+00
18.000000	29.000000	0.000000E+00	19.000000	49.000000	0.000000E+00
18.000000	30.000000	0.000000E+00	19.000000	50.000000	0.000000E+00
18.000000	31.000000	0.000000E+00	20.000000	1.000000	11.000000
18.000000	32.000000	0.000000E+00	20.000000	2.000000	12.000000
18.000000	33.000000	0.000000E+00	20.000000	3.000000	14.000000
18.000000	34.000000	0.000000E+00	20.000000	4.000000	10.000000
18.000000	35.000000	0.000000E+00	20.000000	5.000000	7.000000
18.000000	36.000000	0.000000E+00	20.000000	6.000000	7.000000
18.000000	37.000000	0.000000E+00	20.000000	7.000000	4.000000
18.000000	38.000000	0.000000E+00	20.000000	8.000000	3.000000
18.000000	39.000000	0.000000E+00	20.000000	9.000000	3.000000
18.000000	40.000000	0.000000E+00	20.000000	10.000000	1.000000
18.000000	41.000000	0.000000E+00	20.000000	11.000000	2.000000
18.000000	42.000000	0.000000E+00	20.000000	12.000000	2.000000
18.000000	43.000000	0.000000E+00	20.000000	13.000000	2.000000
18.000000	44.000000	0.000000E+00	20.000000	14.000000	1.000000
18.000000	45.000000	0.000000E+00	20.000000	15.000000	1.000000
18.000000	46.000000	0.000000E+00	20.000000	16.000000	1.000000
18.000000	47.000000	0.000000E+00	20.000000	17.000000	0.000000E+00
18.000000	48.000000	0.000000E+00	20.000000	18.000000	0.000000E+00
18.000000	49.000000	0.000000E+00	20.000000	19.000000	0.000000E+00
18.000000	50.000000	0.000000E+00	20.000000	20.000000	0.000000E+00
19.000000	1.000000	15.000000	20.000000	21.000000	0.000000E+00
19.000000	2.000000	15.000000	20.000000	22.000000	0.000000E+00
19.000000	3.000000	15.000000	20.000000	23.000000	5.000000
19.000000	4.000000	19.000000	20.000000	24.000000	8.000000
19.000000	5.000000	14.000000	20.000000	25.000000	9.000000
19.000000	6.000000	7.000000	20.000000	26.000000	11.000000
19.000000	7.000000	6.000000	20.000000	27.000000	9.000000
19.000000	8.000000	6.000000	20.000000	28.000000	7.000000
19.000000	9.000000	6.000000	20.000000	29.000000	6.000000
19.000000	10.000000	5.000000	20.000000	30.000000	4.000000

Y	X	ELEVATION	Y	X	ELEVATION
20.000000	31.000000	2.000000	22.000000	1.000000	0.000000E+00
20.000000	32.000000	0.000000E+00	22.000000	2.000000	0.000000E+00
20.000000	33.000000	0.000000E+00	22.000000	3.000000	6.000000
20.000000	34.000000	0.000000E+00	22.000000	4.000000	7.000000
20.000000	35.000000	0.000000E+00	22.000000	5.000000	5.000000
20.000000	36.000000	0.000000E+00	22.000000	6.000000	2.000000
20.000000	37.000000	0.000000E+00	22.000000	7.000000	0.000000E+00
20.000000	38.000000	0.000000E+00	22.000000	8.000000	0.000000E+00
20.000000	39.000000	0.000000E+00	22.000000	9.000000	0.000000E+00
20.000000	40.000000	0.000000E+00	22.000000	10.000000	0.000000E+00
20.000000	41.000000	0.000000E+00	22.000000	11.000000	0.000000E+00
20.000000	42.000000	0.000000E+00	22.000000	12.000000	0.000000E+00
20.000000	43.000000	0.000000E+00	22.000000	13.000000	0.000000E+00
20.000000	44.000000	0.000000E+00	22.000000	14.000000	0.000000E+00
20.000000	45.000000	0.000000E+00	22.000000	15.000000	0.000000E+00
20.000000	46.000000	0.000000E+00	22.000000	16.000000	0.000000E+00
20.000000	47.000000	0.000000E+00	22.000000	17.000000	0.000000E+00
20.000000	48.000000	0.000000E+00	22.000000	18.000000	0.000000E+00
20.000000	49.000000	0.000000E+00	22.000000	19.000000	0.000000E+00
20.000000	50.000000	0.000000E+00	22.000000	20.000000	0.000000E+00
21.000000	1.000000	7.000000	22.000000	21.000000	2.000000
21.000000	2.000000	6.000000	22.000000	22.000000	8.000000
21.000000	3.000000	10.000000	22.000000	23.000000	14.000000
21.000000	4.000000	12.000000	22.000000	24.000000	21.000000
21.000000	5.000000	6.000000	22.000000	25.000000	27.000000
21.000000	6.000000	4.000000	22.000000	26.000000	30.000000
21.000000	7.000000	1.000000	22.000000	27.000000	35.000000
21.000000	8.000000	0.000000E+00	22.000000	28.000000	21.000000
21.000000	9.000000	0.000000E+00	22.000000	29.000000	21.000000
21.000000	10.000000	0.000000E+00	22.000000	30.000000	21.000000
21.000000	11.000000	0.000000E+00	22.000000	31.000000	20.000000
21.000000	12.000000	0.000000E+00	22.000000	32.000000	12.000000
21.000000	13.000000	0.000000E+00	22.000000	33.000000	14.000000
21.000000	14.000000	0.000000E+00	22.000000	34.000000	10.000000
21.000000	15.000000	0.000000E+00	22.000000	35.000000	10.000000
21.000000	16.000000	0.000000E+00	22.000000	36.000000	0.000000E+00
21.000000	17.000000	0.000000E+00	22.000000	37.000000	0.000000E+00
21.000000	18.000000	0.000000E+00	22.000000	38.000000	0.000000E+00
21.000000	19.000000	0.000000E+00	22.000000	39.000000	0.000000E+00
21.000000	20.000000	0.000000E+00	22.000000	40.000000	0.000000E+00
21.000000	21.000000	0.000000E+00	22.000000	41.000000	0.000000E+00
21.000000	22.000000	4.000000	22.000000	42.000000	0.000000E+00
21.000000	23.000000	9.000000	22.000000	43.000000	0.000000E+00
21.000000	24.000000	14.000000	22.000000	44.000000	0.000000E+00
21.000000	25.000000	18.000000	22.000000	45.000000	0.000000E+00
21.000000	26.000000	18.000000	22.000000	46.000000	0.000000E+00
21.000000	27.000000	17.000000	22.000000	47.000000	0.000000E+00
21.000000	28.000000	13.000000	22.000000	48.000000	0.000000E+00
21.000000	29.000000	11.000000	22.000000	49.000000	0.000000E+00
21.000000	30.000000	9.000000	22.000000	50.000000	0.000000E+00
21.000000	31.000000	10.000000	23.000000	1.000000	0.000000E+00
21.000000	32.000000	6.000000	23.000000	2.000000	0.000000E+00
21.000000	33.000000	6.000000	23.000000	3.000000	1.000000
21.000000	34.000000	3.000000	23.000000	4.000000	3.000000
21.000000	35.000000	0.000000E+00	23.000000	5.000000	2.000000
21.000000	36.000000	0.000000E+00	23.000000	6.000000	0.000000E+00
21.000000	37.000000	0.000000E+00	23.000000	7.000000	0.000000E+00
21.000000	38.000000	0.000000E+00	23.000000	8.000000	0.000000E+00
21.000000	39.000000	0.000000E+00	23.000000	9.000000	0.000000E+00
21.000000	40.000000	0.000000E+00	23.000000	10.000000	0.000000E+00
21.000000	41.000000	0.000000E+00	23.000000	11.000000	0.000000E+00
21.000000	42.000000	0.000000E+00	23.000000	12.000000	0.000000E+00
21.000000	43.000000	0.000000E+00	23.000000	13.000000	0.000000E+00
21.000000	44.000000	0.000000E+00	23.000000	14.000000	0.000000E+00
21.000000	45.000000	0.000000E+00	23.000000	15.000000	0.000000E+00
21.000000	46.000000	0.000000E+00	23.000000	16.000000	0.000000E+00
21.000000	47.000000	0.000000E+00	23.000000	17.000000	0.000000E+00
21.000000	48.000000	0.000000E+00	23.000000	18.000000	0.000000E+00
21.000000	49.000000	0.000000E+00	23.000000	19.000000	0.000000E+00
21.000000	50.000000	0.000000E+00	23.000000	20.000000	1.000000

Y	X	ELEVATION	Y	X	ELEVATION
23.000000	21.000000	4.000000	24.000000	41.000000	0.000000E+00
23.000000	22.000000	11.000000	24.000000	42.000000	0.000000E+00
23.000000	23.000000	25.000000	24.000000	43.000000	0.000000E+00
23.000000	24.000000	35.000000	24.000000	44.000000	0.000000E+00
23.000000	25.000000	46.000000	24.000000	45.000000	0.000000E+00
23.000000	26.000000	55.000000	24.000000	46.000000	0.000000E+00
23.000000	27.000000	44.000000	24.000000	47.000000	0.000000E+00
23.000000	28.000000	33.000000	24.000000	48.000000	0.000000E+00
23.000000	29.000000	32.000000	24.000000	49.000000	0.000000E+00
23.000000	30.000000	40.000000	24.000000	50.000000	0.000000E+00
23.000000	31.000000	36.000000	25.000000	1.000000	3.000000
23.000000	32.000000	22.000000	25.000000	2.000000	3.000000
23.000000	33.000000	37.000000	25.000000	3.000000	2.000000
23.000000	34.000000	18.000000	25.000000	4.000000	2.000000
23.000000	35.000000	5.000000	25.000000	5.000000	2.000000
23.000000	36.000000	0.000000E+00	25.000000	6.000000	1.000000
23.000000	37.000000	0.000000E+00	25.000000	7.000000	1.000000
23.000000	38.000000	0.000000E+00	25.000000	8.000000	0.000000E+00
23.000000	39.000000	0.000000E+00	25.000000	9.000000	0.000000E+00
23.000000	40.000000	0.000000E+00	25.000000	10.000000	0.000000E+00
23.000000	41.000000	0.000000E+00	25.000000	11.000000	0.000000E+00
23.000000	42.000000	0.000000E+00	25.000000	12.000000	0.000000E+00
23.000000	43.000000	0.000000E+00	25.000000	13.000000	0.000000E+00
23.000000	44.000000	0.000000E+00	25.000000	14.000000	0.000000E+00
23.000000	45.000000	0.000000E+00	25.000000	15.000000	0.000000E+00
23.000000	46.000000	0.000000E+00	25.000000	16.000000	0.000000E+00
23.000000	47.000000	0.000000E+00	25.000000	17.000000	2.000000
23.000000	48.000000	0.000000E+00	25.000000	18.000000	3.000000
23.000000	49.000000	0.000000E+00	25.000000	19.000000	3.000000
23.000000	50.000000	0.000000E+00	25.000000	20.000000	5.000000
24.000000	1.000000	0.000000E+00	25.000000	21.000000	7.000000
24.000000	2.000000	0.000000E+00	25.000000	22.000000	13.000000
24.000000	3.000000	0.000000E+00	25.000000	23.000000	19.000000
24.000000	4.000000	0.000000E+00	25.000000	24.000000	26.000000
24.000000	5.000000	0.000000E+00	25.000000	25.000000	36.000000
24.000000	6.000000	0.000000E+00	25.000000	26.000000	34.000000
24.000000	7.000000	0.000000E+00	25.000000	27.000000	30.000000
24.000000	8.000000	0.000000E+00	25.000000	28.000000	31.000000
24.000000	9.000000	0.000000E+00	25.000000	29.000000	28.000000
24.000000	10.000000	0.000000E+00	25.000000	30.000000	23.000000
24.000000	11.000000	0.000000E+00	25.000000	31.000000	24.000000
24.000000	12.000000	0.000000E+00	25.000000	32.000000	18.000000
24.000000	13.000000	0.000000E+00	25.000000	33.000000	15.000000
24.000000	14.000000	0.000000E+00	25.000000	34.000000	16.000000
24.000000	15.000000	0.000000E+00	25.000000	35.000000	8.000000
24.000000	16.000000	0.000000E+00	25.000000	36.000000	7.000000
24.000000	17.000000	0.000000E+00	25.000000	37.000000	0.000000E+00
24.000000	18.000000	0.000000E+00	25.000000	38.000000	0.000000E+00
24.000000	19.000000	0.000000E+00	25.000000	39.000000	0.000000E+00
24.000000	20.000000	3.000000	25.000000	40.000000	0.000000E+00
24.000000	21.000000	6.000000	25.000000	41.000000	0.000000E+00
24.000000	22.000000	13.000000	25.000000	42.000000	0.000000E+00
24.000000	23.000000	21.000000	25.000000	43.000000	0.000000E+00
24.000000	24.000000	32.000000	25.000000	44.000000	0.000000E+00
24.000000	25.000000	44.000000	25.000000	45.000000	0.000000E+00
24.000000	26.000000	48.000000	25.000000	46.000000	0.000000E+00
24.000000	27.000000	50.000000	25.000000	47.000000	0.000000E+00
24.000000	28.000000	52.000000	25.000000	48.000000	0.000000E+00
24.000000	29.000000	49.000000	25.000000	49.000000	0.000000E+00
24.000000	30.000000	49.000000	25.000000	50.000000	0.000000E+00
24.000000	31.000000	53.000000	26.000000	1.000000	6.000000
24.000000	32.000000	36.000000	26.000000	2.000000	6.000000
24.000000	33.000000	27.000000	26.000000	3.000000	5.000000
24.000000	34.000000	17.000000	26.000000	4.000000	4.000000
24.000000	35.000000	7.000000	26.000000	5.000000	4.000000
24.000000	36.000000	2.000000	26.000000	6.000000	4.000000
24.000000	37.000000	0.000000E+00	26.000000	7.000000	2.000000
24.000000	38.000000	0.000000E+00	26.000000	8.000000	2.000000
24.000000	39.000000	0.000000E+00	26.000000	9.000000	1.000000
24.000000	40.000000	0.000000E+00	26.000000	10.000000	1.000000

Y	X	ELEVATION	Y	X	ELEVATION
26.000000	11.000000	0.000000E+00	27.000000	31.000000	2.000000
26.000000	12.000000	0.000000E+00	27.000000	32.000000	2.000000
26.000000	13.000000	0.000000E+00	27.000000	33.000000	1.000000
26.000000	14.000000	0.000000E+00	27.000000	34.000000	0.000000E+00
26.000000	15.000000	0.000000E+00	27.000000	35.000000	0.000000E+00
26.000000	16.000000	0.000000E+00	27.000000	36.000000	0.000000E+00
26.000000	17.000000	5.000000	27.000000	37.000000	0.000000E+00
26.000000	18.000000	13.000000	27.000000	38.000000	0.000000E+00
26.000000	19.000000	6.000000	27.000000	39.000000	0.000000E+00
26.000000	20.000000	6.000000	27.000000	40.000000	0.000000E+00
26.000000	21.000000	9.000000	27.000000	41.000000	0.000000E+00
26.000000	22.000000	13.000000	27.000000	42.000000	0.000000E+00
26.000000	23.000000	19.000000	27.000000	43.000000	0.000000E+00
26.000000	24.000000	22.000000	27.000000	44.000000	0.000000E+00
26.000000	25.000000	33.000000	27.000000	45.000000	0.000000E+00
26.000000	26.000000	44.000000	27.000000	46.000000	0.000000E+00
26.000000	27.000000	20.000000	27.000000	47.000000	0.000000E+00
26.000000	28.000000	16.000000	27.000000	48.000000	0.000000E+00
26.000000	29.000000	12.000000	27.000000	49.000000	3.000000
26.000000	30.000000	10.000000	27.000000	50.000000	6.000000
26.000000	31.000000	9.000000	28.000000	1.000000	14.000000
26.000000	32.000000	7.000000	28.000000	2.000000	14.000000
26.000000	33.000000	7.000000	28.000000	3.000000	19.000000
26.000000	34.000000	6.000000	28.000000	4.000000	13.000000
26.000000	35.000000	0.000000E+00	28.000000	5.000000	14.000000
26.000000	36.000000	0.000000E+00	28.000000	6.000000	7.000000
26.000000	37.000000	0.000000E+00	28.000000	7.000000	6.000000
26.000000	38.000000	0.000000E+00	28.000000	8.000000	3.000000
26.000000	39.000000	0.000000E+00	28.000000	9.000000	3.000000
26.000000	40.000000	0.000000E+00	28.000000	10.000000	2.000000
26.000000	41.000000	0.000000E+00	28.000000	11.000000	1.000000
26.000000	42.000000	0.000000E+00	28.000000	12.000000	1.000000
26.000000	43.000000	0.000000E+00	28.000000	13.000000	0.000000E+00
26.000000	44.000000	0.000000E+00	28.000000	14.000000	0.000000E+00
26.000000	45.000000	0.000000E+00	28.000000	15.000000	1.000000
26.000000	46.000000	0.000000E+00	28.000000	16.000000	5.000000
26.000000	47.000000	0.000000E+00	28.000000	17.000000	11.000000
26.000000	48.000000	0.000000E+00	28.000000	18.000000	12.000000
26.000000	49.000000	0.000000E+00	28.000000	19.000000	7.000000
26.000000	50.000000	1.000000	28.000000	20.000000	11.000000
27.000000	1.000000	9.000000	28.000000	21.000000	14.000000
27.000000	2.000000	9.000000	28.000000	22.000000	22.000000
27.000000	3.000000	7.000000	28.000000	23.000000	23.000000
27.000000	4.000000	7.000000	28.000000	24.000000	20.000000
27.000000	5.000000	7.000000	28.000000	25.000000	15.000000
27.000000	6.000000	7.000000	28.000000	26.000000	13.000000
27.000000	7.000000	4.000000	28.000000	27.000000	9.000000
27.000000	8.000000	3.000000	28.000000	28.000000	6.000000
27.000000	9.000000	2.000000	28.000000	29.000000	2.000000
27.000000	10.000000	1.000000	28.000000	30.000000	0.000000E+00
27.000000	11.000000	1.000000	28.000000	31.000000	0.000000E+00
27.000000	12.000000	0.000000E+00	28.000000	32.000000	0.000000E+00
27.000000	13.000000	0.000000E+00	28.000000	33.000000	0.000000E+00
27.000000	14.000000	0.000000E+00	28.000000	34.000000	0.000000E+00
27.000000	15.000000	0.000000E+00	28.000000	35.000000	0.000000E+00
27.000000	16.000000	4.000000	28.000000	36.000000	0.000000E+00
27.000000	17.000000	7.000000	28.000000	37.000000	0.000000E+00
27.000000	18.000000	8.000000	28.000000	38.000000	0.000000E+00
27.000000	19.000000	7.000000	28.000000	39.000000	0.000000E+00
27.000000	20.000000	7.000000	28.000000	40.000000	0.000000E+00
27.000000	21.000000	12.000000	28.000000	41.000000	0.000000E+00
27.000000	22.000000	18.000000	28.000000	42.000000	0.000000E+00
27.000000	23.000000	29.000000	28.000000	43.000000	0.000000E+00
27.000000	24.000000	22.000000	28.000000	44.000000	0.000000E+00
27.000000	25.000000	22.000000	28.000000	45.000000	0.000000E+00
27.000000	26.000000	28.000000	28.000000	46.000000	0.000000E+00
27.000000	27.000000	14.000000	28.000000	47.000000	7.000000
27.000000	28.000000	14.000000	28.000000	48.000000	7.000000
27.000000	29.000000	6.000000	28.000000	49.000000	8.000000
27.000000	30.000000	3.000000	28.000000	50.000000	13.000000

Y	X	ELEVATION	Y	X	ELEVATION
29.000000	1.000000	25.000000	30.000000	21.000000	16.000000
29.000000	2.000000	26.000000	30.000000	22.000000	18.000000
29.000000	3.000000	28.000000	30.000000	23.000000	14.000000
29.000000	4.000000	22.000000	30.000000	24.000000	10.000000
29.000000	5.000000	14.000000	30.000000	25.000000	6.000000
29.000000	6.000000	9.000000	30.000000	26.000000	3.000000
29.000000	7.000000	7.000000	30.000000	27.000000	1.000000
29.000000	8.000000	5.000000	30.000000	28.000000	0.000000E+00
29.000000	9.000000	4.000000	30.000000	29.000000	0.000000E+00
29.000000	10.000000	3.000000	30.000000	30.000000	0.000000E+00
29.000000	11.000000	2.000000	30.000000	31.000000	0.000000E+00
29.000000	12.000000	2.000000	30.000000	32.000000	0.000000E+00
29.000000	13.000000	3.000000	30.000000	33.000000	0.000000E+00
29.000000	14.000000	4.000000	30.000000	34.000000	0.000000E+00
29.000000	15.000000	5.000000	30.000000	35.000000	0.000000E+00
29.000000	16.000000	6.000000	30.000000	36.000000	0.000000E+00
29.000000	17.000000	7.000000	30.000000	37.000000	0.000000E+00
29.000000	18.000000	11.000000	30.000000	38.000000	0.000000E+00
29.000000	19.000000	12.000000	30.000000	39.000000	0.000000E+00
29.000000	20.000000	15.000000	30.000000	40.000000	0.000000E+00
29.000000	21.000000	19.000000	30.000000	41.000000	0.000000E+00
29.000000	22.000000	22.000000	30.000000	42.000000	0.000000E+00
29.000000	23.000000	15.000000	30.000000	43.000000	0.000000E+00
29.000000	24.000000	15.000000	30.000000	44.000000	0.000000E+00
29.000000	25.000000	14.000000	30.000000	45.000000	0.000000E+00
29.000000	26.000000	6.000000	30.000000	46.000000	0.000000E+00
29.000000	27.000000	4.000000	30.000000	47.000000	0.000000E+00
29.000000	28.000000	2.000000	30.000000	48.000000	8.000000
29.000000	29.000000	0.000000E+00	30.000000	49.000000	18.000000
29.000000	30.000000	0.000000E+00	30.000000	50.000000	28.000000
29.000000	31.000000	0.000000E+00	31.000000	1.000000	22.000000
29.000000	32.000000	0.000000E+00	31.000000	2.000000	20.000000
29.000000	33.000000	0.000000E+00	31.000000	3.000000	16.000000
29.000000	34.000000	0.000000E+00	31.000000	4.000000	14.000000
29.000000	35.000000	0.000000E+00	31.000000	5.000000	12.000000
29.000000	36.000000	0.000000E+00	31.000000	6.000000	10.000000
29.000000	37.000000	0.000000E+00	31.000000	7.000000	8.000000
29.000000	38.000000	0.000000E+00	31.000000	8.000000	6.000000
29.000000	39.000000	0.000000E+00	31.000000	9.000000	6.000000
29.000000	40.000000	0.000000E+00	31.000000	10.000000	5.000000
29.000000	41.000000	0.000000E+00	31.000000	11.000000	4.000000
29.000000	42.000000	0.000000E+00	31.000000	12.000000	4.000000
29.000000	43.000000	0.000000E+00	31.000000	13.000000	4.000000
29.000000	44.000000	0.000000E+00	31.000000	14.000000	4.000000
29.000000	45.000000	0.000000E+00	31.000000	15.000000	5.000000
29.000000	46.000000	0.000000E+00	31.000000	16.000000	6.000000
29.000000	47.000000	0.000000E+00	31.000000	17.000000	7.000000
29.000000	48.000000	7.000000	31.000000	18.000000	7.000000
29.000000	49.000000	19.000000	31.000000	19.000000	8.000000
29.000000	50.000000	22.000000	31.000000	20.000000	10.000000
30.000000	1.000000	24.000000	31.000000	21.000000	10.000000
30.000000	2.000000	21.000000	31.000000	22.000000	11.000000
30.000000	3.000000	19.000000	31.000000	23.000000	18.000000
30.000000	4.000000	15.000000	31.000000	24.000000	18.000000
30.000000	5.000000	12.000000	31.000000	25.000000	6.000000
30.000000	6.000000	10.000000	31.000000	26.000000	2.000000
30.000000	7.000000	8.000000	31.000000	27.000000	0.000000E+00
30.000000	8.000000	6.000000	31.000000	28.000000	0.000000E+00
30.000000	9.000000	5.000000	31.000000	29.000000	0.000000E+00
30.000000	10.000000	4.000000	31.000000	30.000000	0.000000E+00
30.000000	11.000000	3.000000	31.000000	31.000000	0.000000E+00
30.000000	12.000000	3.000000	31.000000	32.000000	0.000000E+00
30.000000	13.000000	3.000000	31.000000	33.000000	0.000000E+00
30.000000	14.000000	4.000000	31.000000	34.000000	0.000000E+00
30.000000	15.000000	5.000000	31.000000	35.000000	0.000000E+00
30.000000	16.000000	6.000000	31.000000	36.000000	0.000000E+00
30.000000	17.000000	7.000000	31.000000	37.000000	0.000000E+00
30.000000	18.000000	9.000000	31.000000	38.000000	0.000000E+00
30.000000	19.000000	15.000000	31.000000	39.000000	0.000000E+00
30.000000	20.000000	19.000000	31.000000	40.000000	0.000000E+00

Y	X	ELEVATION	Y	X	ELEVATION
31.000000	41.000000	0.000000E+00	33.000000	11.000000	6.000000
31.000000	42.000000	0.000000E+00	33.000000	12.000000	5.000000
31.000000	43.000000	0.000000E+00	33.000000	13.000000	5.000000
31.000000	44.000000	0.000000E+00	33.000000	14.000000	5.000000
31.000000	45.000000	0.000000E+00	33.000000	15.000000	6.000000
31.000000	46.000000	0.000000E+00	33.000000	16.000000	6.000000
31.000000	47.000000	0.000000E+00	33.000000	17.000000	7.000000
31.000000	48.000000	6.000000	33.000000	18.000000	7.000000
31.000000	49.000000	21.000000	33.000000	19.000000	7.000000
31.000000	50.000000	30.000000	33.000000	20.000000	7.000000
32.000000	1.000000	22.000000	33.000000	21.000000	7.000000
32.000000	2.000000	18.000000	33.000000	22.000000	6.000000
32.000000	3.000000	15.000000	33.000000	23.000000	5.000000
32.000000	4.000000	14.000000	33.000000	24.000000	4.000000
32.000000	5.000000	12.000000	33.000000	25.000000	1.000000
32.000000	6.000000	11.000000	33.000000	26.000000	0.000000E+00
32.000000	7.000000	9.000000	33.000000	27.000000	0.000000E+00
32.000000	8.000000	8.000000	33.000000	28.000000	0.000000E+00
32.000000	9.000000	7.000000	33.000000	29.000000	0.000000E+00
32.000000	10.000000	6.000000	33.000000	30.000000	0.000000E+00
32.000000	11.000000	5.000000	33.000000	31.000000	0.000000E+00
32.000000	12.000000	5.000000	33.000000	32.000000	0.000000E+00
32.000000	13.000000	4.000000	33.000000	33.000000	0.000000E+00
32.000000	14.000000	5.000000	33.000000	34.000000	0.000000E+00
32.000000	15.000000	5.000000	33.000000	35.000000	0.000000E+00
32.000000	16.000000	6.000000	33.000000	36.000000	0.000000E+00
32.000000	17.000000	7.000000	33.000000	37.000000	0.000000E+00
32.000000	18.000000	7.000000	33.000000	38.000000	0.000000E+00
32.000000	19.000000	7.000000	33.000000	39.000000	0.000000E+00
32.000000	20.000000	7.000000	33.000000	40.000000	0.000000E+00
32.000000	21.000000	7.000000	33.000000	41.000000	0.000000E+00
32.000000	22.000000	7.000000	33.000000	42.000000	0.000000E+00
32.000000	23.000000	12.000000	33.000000	43.000000	0.000000E+00
32.000000	24.000000	14.000000	33.000000	44.000000	0.000000E+00
32.000000	25.000000	4.000000	33.000000	45.000000	0.000000E+00
32.000000	26.000000	0.000000E+00	33.000000	46.000000	0.000000E+00
32.000000	27.000000	0.000000E+00	33.000000	47.000000	0.000000E+00
32.000000	28.000000	0.000000E+00	33.000000	48.000000	6.000000
32.000000	29.000000	0.000000E+00	33.000000	49.000000	10.000000
32.000000	30.000000	0.000000E+00	33.000000	50.000000	16.000000
32.000000	31.000000	0.000000E+00	34.000000	1.000000	18.000000
32.000000	32.000000	0.000000E+00	34.000000	2.000000	15.000000
32.000000	33.000000	0.000000E+00	34.000000	3.000000	14.000000
32.000000	34.000000	0.000000E+00	34.000000	4.000000	14.000000
32.000000	35.000000	0.000000E+00	34.000000	5.000000	14.000000
32.000000	36.000000	0.000000E+00	34.000000	6.000000	13.000000
32.000000	37.000000	0.000000E+00	34.000000	7.000000	12.000000
32.000000	38.000000	0.000000E+00	34.000000	8.000000	11.000000
32.000000	39.000000	0.000000E+00	34.000000	9.000000	10.000000
32.000000	40.000000	0.000000E+00	34.000000	10.000000	9.000000
32.000000	41.000000	0.000000E+00	34.000000	11.000000	7.000000
32.000000	42.000000	0.000000E+00	34.000000	12.000000	6.000000
32.000000	43.000000	0.000000E+00	34.000000	13.000000	6.000000
32.000000	44.000000	0.000000E+00	34.000000	14.000000	6.000000
32.000000	45.000000	0.000000E+00	34.000000	15.000000	6.000000
32.000000	46.000000	0.000000E+00	34.000000	16.000000	6.000000
32.000000	47.000000	0.000000E+00	34.000000	17.000000	7.000000
32.000000	48.000000	5.000000	34.000000	18.000000	7.000000
32.000000	49.000000	15.000000	34.000000	19.000000	7.000000
32.000000	50.000000	24.000000	34.000000	20.000000	7.000000
33.000000	1.000000	21.000000	34.000000	21.000000	7.000000
33.000000	2.000000	17.000000	34.000000	22.000000	5.000000
33.000000	3.000000	15.000000	34.000000	23.000000	3.000000
33.000000	4.000000	14.000000	34.000000	24.000000	0.000000E+00
33.000000	5.000000	13.000000	34.000000	25.000000	0.000000E+00
33.000000	6.000000	12.000000	34.000000	26.000000	0.000000E+00
33.000000	7.000000	11.000000	34.000000	27.000000	0.000000E+00
33.000000	8.000000	9.000000	34.000000	28.000000	0.000000E+00
33.000000	9.000000	8.000000	34.000000	29.000000	0.000000E+00
33.000000	10.000000	7.000000	34.000000	30.000000	0.000000E+00

Y	X	ELEVATION	Y	X	ELEVATION
34.000000	31.000000	0.000000E+00	36.000000	1.000000	17.000000
34.000000	32.000000	0.000000E+00	36.000000	2.000000	16.000000
34.000000	33.000000	0.000000E+00	36.000000	3.000000	16.000000
34.000000	34.000000	0.000000E+00	36.000000	4.000000	16.000000
34.000000	35.000000	0.000000E+00	36.000000	5.000000	16.000000
34.000000	36.000000	0.000000E+00	36.000000	6.000000	15.000000
34.000000	37.000000	0.000000E+00	36.000000	7.000000	15.000000
34.000000	38.000000	0.000000E+00	36.000000	8.000000	14.000000
34.000000	39.000000	0.000000E+00	36.000000	9.000000	14.000000
34.000000	40.000000	0.000000E+00	36.000000	10.000000	13.000000
34.000000	41.000000	0.000000E+00	36.000000	11.000000	11.000000
34.000000	42.000000	0.000000E+00	36.000000	12.000000	9.000000
34.000000	43.000000	0.000000E+00	36.000000	13.000000	8.000000
34.000000	44.000000	0.000000E+00	36.000000	14.000000	7.000000
34.000000	45.000000	0.000000E+00	36.000000	15.000000	7.000000
34.000000	46.000000	0.000000E+00	36.000000	16.000000	6.000000
34.000000	47.000000	1.000000	36.000000	17.000000	7.000000
34.000000	48.000000	7.000000	36.000000	18.000000	7.000000
34.000000	49.000000	12.000000	36.000000	19.000000	7.000000
34.000000	50.000000	19.000000	36.000000	20.000000	7.000000
35.000000	1.000000	16.000000	36.000000	21.000000	7.000000
35.000000	2.000000	14.000000	36.000000	22.000000	6.000000
35.000000	3.000000	15.000000	36.000000	23.000000	4.000000
35.000000	4.000000	14.000000	36.000000	24.000000	3.000000
35.000000	5.000000	14.000000	36.000000	25.000000	1.000000
35.000000	6.000000	14.000000	36.000000	26.000000	0.000000E+00
35.000000	7.000000	14.000000	36.000000	27.000000	0.000000E+00
35.000000	8.000000	13.000000	36.000000	28.000000	0.000000E+00
35.000000	9.000000	12.000000	36.000000	29.000000	0.000000E+00
35.000000	10.000000	10.000000	36.000000	30.000000	0.000000E+00
35.000000	11.000000	9.000000	36.000000	31.000000	0.000000E+00
35.000000	12.000000	7.000000	36.000000	32.000000	0.000000E+00
35.000000	13.000000	6.000000	36.000000	33.000000	0.000000E+00
35.000000	14.000000	6.000000	36.000000	34.000000	0.000000E+00
35.000000	15.000000	6.000000	36.000000	35.000000	0.000000E+00
35.000000	16.000000	6.000000	36.000000	36.000000	0.000000E+00
35.000000	17.000000	7.000000	36.000000	37.000000	0.000000E+00
35.000000	18.000000	7.000000	36.000000	38.000000	0.000000E+00
35.000000	19.000000	7.000000	36.000000	39.000000	0.000000E+00
35.000000	20.000000	7.000000	36.000000	40.000000	0.000000E+00
35.000000	21.000000	7.000000	36.000000	41.000000	0.000000E+00
35.000000	22.000000	5.000000	36.000000	42.000000	0.000000E+00
35.000000	23.000000	2.000000	36.000000	43.000000	0.000000E+00
35.000000	24.000000	0.000000E+00	36.000000	44.000000	0.000000E+00
35.000000	25.000000	0.000000E+00	36.000000	45.000000	0.000000E+00
35.000000	26.000000	0.000000E+00	36.000000	46.000000	0.000000E+00
35.000000	27.000000	0.000000E+00	36.000000	47.000000	6.000000
35.000000	28.000000	0.000000E+00	36.000000	48.000000	10.000000
35.000000	29.000000	0.000000E+00	36.000000	49.000000	17.000000
35.000000	30.000000	0.000000E+00	36.000000	50.000000	21.000000
35.000000	31.000000	0.000000E+00	37.000000	1.000000	19.000000
35.000000	32.000000	0.000000E+00	37.000000	2.000000	19.000000
35.000000	33.000000	0.000000E+00	37.000000	3.000000	19.000000
35.000000	34.000000	0.000000E+00	37.000000	4.000000	18.000000
35.000000	35.000000	0.000000E+00	37.000000	5.000000	19.000000
35.000000	36.000000	0.000000E+00	37.000000	6.000000	18.000000
35.000000	37.000000	0.000000E+00	37.000000	7.000000	17.000000
35.000000	38.000000	0.000000E+00	37.000000	8.000000	16.000000
35.000000	39.000000	0.000000E+00	37.000000	9.000000	15.000000
35.000000	40.000000	0.000000E+00	37.000000	10.000000	14.000000
35.000000	41.000000	0.000000E+00	37.000000	11.000000	14.000000
35.000000	42.000000	0.000000E+00	37.000000	12.000000	12.000000
35.000000	43.000000	0.000000E+00	37.000000	13.000000	10.000000
35.000000	44.000000	0.000000E+00	37.000000	14.000000	9.000000
35.000000	45.000000	0.000000E+00	37.000000	15.000000	7.000000
35.000000	46.000000	0.000000E+00	37.000000	16.000000	7.000000
35.000000	47.000000	2.000000	37.000000	17.000000	7.000000
35.000000	48.000000	8.000000	37.000000	18.000000	7.000000
35.000000	49.000000	13.000000	37.000000	19.000000	7.000000
35.000000	50.000000	21.000000	37.000000	20.000000	7.000000

Y	X	ELEVATION	Y	X	ELEVATION
37.000000	21.000000	7.000000	38.000000	41.000000	0.000000E+00
37.000000	22.000000	5.000000	38.000000	42.000000	0.000000E+00
37.000000	23.000000	4.000000	38.000000	43.000000	0.000000E+00
37.000000	24.000000	2.000000	38.000000	44.000000	0.000000E+00
37.000000	25.000000	1.000000	38.000000	45.000000	0.000000E+00
37.000000	26.000000	0.000000E+00	38.000000	46.000000	0.000000E+00
37.000000	27.000000	0.000000E+00	38.000000	47.000000	7.000000
37.000000	28.000000	0.000000E+00	38.000000	48.000000	15.000000
37.000000	29.000000	0.000000E+00	38.000000	49.000000	22.000000
37.000000	30.000000	0.000000E+00	38.000000	50.000000	29.000000
37.000000	31.000000	0.000000E+00	39.000000	1.000000	25.000000
37.000000	32.000000	0.000000E+00	39.000000	2.000000	25.000000
37.000000	33.000000	0.000000E+00	39.000000	3.000000	25.000000
37.000000	34.000000	0.000000E+00	39.000000	4.000000	24.000000
37.000000	35.000000	0.000000E+00	39.000000	5.000000	23.000000
37.000000	36.000000	0.000000E+00	39.000000	6.000000	23.000000
37.000000	37.000000	0.000000E+00	39.000000	7.000000	22.000000
37.000000	38.000000	0.000000E+00	39.000000	8.000000	21.000000
37.000000	39.000000	0.000000E+00	39.000000	9.000000	21.000000
37.000000	40.000000	0.000000E+00	39.000000	10.000000	24.000000
37.000000	41.000000	0.000000E+00	39.000000	11.000000	22.000000
37.000000	42.000000	0.000000E+00	39.000000	12.000000	19.000000
37.000000	43.000000	0.000000E+00	39.000000	13.000000	15.000000
37.000000	44.000000	0.000000E+00	39.000000	14.000000	13.000000
37.000000	45.000000	0.000000E+00	39.000000	15.000000	11.000000
37.000000	46.000000	0.000000E+00	39.000000	16.000000	8.000000
37.000000	47.000000	13.000000	39.000000	17.000000	7.000000
37.000000	48.000000	18.000000	39.000000	18.000000	7.000000
37.000000	49.000000	22.000000	39.000000	19.000000	7.000000
37.000000	50.000000	37.000000	39.000000	20.000000	7.000000
38.000000	1.000000	22.000000	39.000000	21.000000	7.000000
38.000000	2.000000	22.000000	39.000000	22.000000	7.000000
38.000000	3.000000	22.000000	39.000000	23.000000	7.000000
38.000000	4.000000	21.000000	39.000000	24.000000	7.000000
38.000000	5.000000	21.000000	39.000000	25.000000	7.000000
38.000000	6.000000	20.000000	39.000000	26.000000	3.000000
38.000000	7.000000	20.000000	39.000000	27.000000	0.000000E+00
38.000000	8.000000	19.000000	39.000000	28.000000	0.000000E+00
38.000000	9.000000	18.000000	39.000000	29.000000	0.000000E+00
38.000000	10.000000	16.000000	39.000000	30.000000	0.000000E+00
38.000000	11.000000	17.000000	39.000000	31.000000	0.000000E+00
38.000000	12.000000	15.000000	39.000000	32.000000	0.000000E+00
38.000000	13.000000	12.000000	39.000000	33.000000	0.000000E+00
38.000000	14.000000	11.000000	39.000000	34.000000	0.000000E+00
38.000000	15.000000	9.000000	39.000000	35.000000	0.000000E+00
38.000000	16.000000	7.000000	39.000000	36.000000	0.000000E+00
38.000000	17.000000	7.000000	39.000000	37.000000	0.000000E+00
38.000000	18.000000	7.000000	39.000000	38.000000	0.000000E+00
38.000000	19.000000	7.000000	39.000000	39.000000	0.000000E+00
38.000000	20.000000	7.000000	39.000000	40.000000	0.000000E+00
38.000000	21.000000	7.000000	39.000000	41.000000	0.000000E+00
38.000000	22.000000	7.000000	39.000000	42.000000	0.000000E+00
38.000000	23.000000	6.000000	39.000000	43.000000	0.000000E+00
38.000000	24.000000	5.000000	39.000000	44.000000	0.000000E+00
38.000000	25.000000	5.000000	39.000000	45.000000	0.000000E+00
38.000000	26.000000	1.000000	39.000000	46.000000	1.000000
38.000000	27.000000	0.000000E+00	39.000000	47.000000	6.000000
38.000000	28.000000	0.000000E+00	39.000000	48.000000	11.000000
38.000000	29.000000	0.000000E+00	39.000000	49.000000	20.000000
38.000000	30.000000	0.000000E+00	39.000000	50.000000	29.000000
38.000000	31.000000	0.000000E+00	40.000000	1.000000	28.000000
38.000000	32.000000	0.000000E+00	40.000000	2.000000	29.000000
38.000000	33.000000	0.000000E+00	40.000000	3.000000	29.000000
38.000000	34.000000	0.000000E+00	40.000000	4.000000	29.000000
38.000000	35.000000	0.000000E+00	40.000000	5.000000	29.000000
38.000000	36.000000	0.000000E+00	40.000000	6.000000	29.000000
38.000000	37.000000	0.000000E+00	40.000000	7.000000	32.000000
38.000000	38.000000	0.000000E+00	40.000000	8.000000	28.000000
38.000000	39.000000	0.000000E+00	40.000000	9.000000	29.000000
38.000000	40.000000	0.000000E+00	40.000000	10.000000	46.000000

Y	X	ELEVATION	Y	X	ELEVATION
40.000000	11.000000	34.000000	41.000000	31.000000	0.000000E+00
40.000000	12.000000	26.000000	41.000000	32.000000	0.000000E+00
40.000000	13.000000	19.000000	41.000000	33.000000	0.000000E+00
40.000000	14.000000	16.000000	41.000000	34.000000	0.000000E+00
40.000000	15.000000	13.000000	41.000000	35.000000	0.000000E+00
40.000000	16.000000	11.000000	41.000000	36.000000	0.000000E+00
40.000000	17.000000	9.000000	41.000000	37.000000	0.000000E+00
40.000000	18.000000	7.000000	41.000000	38.000000	0.000000E+00
40.000000	19.000000	7.000000	41.000000	39.000000	0.000000E+00
40.000000	20.000000	7.000000	41.000000	40.000000	0.000000E+00
40.000000	21.000000	7.000000	41.000000	41.000000	0.000000E+00
40.000000	22.000000	7.000000	41.000000	42.000000	0.000000E+00
40.000000	23.000000	7.000000	41.000000	43.000000	0.000000E+00
40.000000	24.000000	7.000000	41.000000	44.000000	0.000000E+00
40.000000	25.000000	7.000000	41.000000	45.000000	1.000000
40.000000	26.000000	5.000000	41.000000	46.000000	4.000000
40.000000	27.000000	3.000000	41.000000	47.000000	8.000000
40.000000	28.000000	1.000000	41.000000	48.000000	14.000000
40.000000	29.000000	0.000000E+00	41.000000	49.000000	22.000000
40.000000	30.000000	0.000000E+00	41.000000	50.000000	29.000000
40.000000	31.000000	0.000000E+00	42.000000	1.000000	37.000000
40.000000	32.000000	0.000000E+00	42.000000	2.000000	37.000000
40.000000	33.000000	0.000000E+00	42.000000	3.000000	37.000000
40.000000	34.000000	0.000000E+00	42.000000	4.000000	35.000000
40.000000	35.000000	0.000000E+00	42.000000	5.000000	34.000000
40.000000	36.000000	0.000000E+00	42.000000	6.000000	32.000000
40.000000	37.000000	0.000000E+00	42.000000	7.000000	30.000000
40.000000	38.000000	0.000000E+00	42.000000	8.000000	29.000000
40.000000	39.000000	0.000000E+00	42.000000	9.000000	33.000000
40.000000	40.000000	0.000000E+00	42.000000	10.000000	40.000000
40.000000	41.000000	0.000000E+00	42.000000	11.000000	33.000000
40.000000	42.000000	0.000000E+00	42.000000	12.000000	28.000000
40.000000	43.000000	0.000000E+00	42.000000	13.000000	22.000000
40.000000	44.000000	0.000000E+00	42.000000	14.000000	22.000000
40.000000	45.000000	0.000000E+00	42.000000	15.000000	22.000000
40.000000	46.000000	2.000000	42.000000	16.000000	20.000000
40.000000	47.000000	7.000000	42.000000	17.000000	12.000000
40.000000	48.000000	12.000000	42.000000	18.000000	10.000000
40.000000	49.000000	19.000000	42.000000	19.000000	8.000000
40.000000	50.000000	38.000000	42.000000	20.000000	7.000000
41.000000	1.000000	32.000000	42.000000	21.000000	7.000000
41.000000	2.000000	35.000000	42.000000	22.000000	7.000000
41.000000	3.000000	33.000000	42.000000	23.000000	7.000000
41.000000	4.000000	30.000000	42.000000	24.000000	12.000000
41.000000	5.000000	29.000000	42.000000	25.000000	9.000000
41.000000	6.000000	36.000000	42.000000	26.000000	7.000000
41.000000	7.000000	37.000000	42.000000	27.000000	6.000000
41.000000	8.000000	37.000000	42.000000	28.000000	3.000000
41.000000	9.000000	36.000000	42.000000	29.000000	1.000000
41.000000	10.000000	45.000000	42.000000	30.000000	1.000000
41.000000	11.000000	39.000000	42.000000	31.000000	1.000000
41.000000	12.000000	29.000000	42.000000	32.000000	1.000000
41.000000	13.000000	22.000000	42.000000	33.000000	1.000000
41.000000	14.000000	21.000000	42.000000	34.000000	1.000000
41.000000	15.000000	19.000000	42.000000	35.000000	1.000000
41.000000	16.000000	13.000000	42.000000	36.000000	1.000000
41.000000	17.000000	11.000000	42.000000	37.000000	0.000000E+00
41.000000	18.000000	9.000000	42.000000	38.000000	0.000000E+00
41.000000	19.000000	7.000000	42.000000	39.000000	0.000000E+00
41.000000	20.000000	7.000000	42.000000	40.000000	0.000000E+00
41.000000	21.000000	7.000000	42.000000	41.000000	0.000000E+00
41.000000	22.000000	7.000000	42.000000	42.000000	0.000000E+00
41.000000	23.000000	7.000000	42.000000	43.000000	0.000000E+00
41.000000	24.000000	7.000000	42.000000	44.000000	1.000000
41.000000	25.000000	7.000000	42.000000	45.000000	4.000000
41.000000	26.000000	7.000000	42.000000	46.000000	6.000000
41.000000	27.000000	5.000000	42.000000	47.000000	9.000000
41.000000	28.000000	2.000000	42.000000	48.000000	13.000000
41.000000	29.000000	0.000000E+00	42.000000	49.000000	29.000000
41.000000	30.000000	0.000000E+00	42.000000	50.000000	29.000000

Y	X	ELEVATION	Y	X	ELEVATION
43.000000	1.000000	44.000000	44.000000	21.000000	7.000000
43.000000	2.000000	44.000000	44.000000	22.000000	7.000000
43.000000	3.000000	40.000000	44.000000	23.000000	7.000000
43.000000	4.000000	37.000000	44.000000	24.000000	7.000000
43.000000	5.000000	37.000000	44.000000	25.000000	7.000000
43.000000	6.000000	35.000000	44.000000	26.000000	7.000000
43.000000	7.000000	32.000000	44.000000	27.000000	6.000000
43.000000	8.000000	30.000000	44.000000	28.000000	4.000000
43.000000	9.000000	33.000000	44.000000	29.000000	3.000000
43.000000	10.000000	37.000000	44.000000	30.000000	3.000000
43.000000	11.000000	36.000000	44.000000	31.000000	3.000000
43.000000	12.000000	29.000000	44.000000	32.000000	3.000000
43.000000	13.000000	24.000000	44.000000	33.000000	3.000000
43.000000	14.000000	21.000000	44.000000	34.000000	4.000000
43.000000	15.000000	18.000000	44.000000	35.000000	4.000000
43.000000	16.000000	15.000000	44.000000	36.000000	5.000000
43.000000	17.000000	13.000000	44.000000	37.000000	3.000000
43.000000	18.000000	11.000000	44.000000	38.000000	2.000000
43.000000	19.000000	10.000000	44.000000	39.000000	2.000000
43.000000	20.000000	8.000000	44.000000	40.000000	1.000000
43.000000	21.000000	7.000000	44.000000	41.000000	1.000000
43.000000	22.000000	7.000000	44.000000	42.000000	1.000000
43.000000	23.000000	7.000000	44.000000	43.000000	3.000000
43.000000	24.000000	14.000000	44.000000	44.000000	5.000000
43.000000	25.000000	12.000000	44.000000	45.000000	6.000000
43.000000	26.000000	7.000000	44.000000	46.000000	7.000000
43.000000	27.000000	6.000000	44.000000	47.000000	11.000000
43.000000	28.000000	4.000000	44.000000	48.000000	16.000000
43.000000	29.000000	2.000000	44.000000	49.000000	22.000000
43.000000	30.000000	2.000000	44.000000	50.000000	27.000000
43.000000	31.000000	2.000000	45.000000	1.000000	34.000000
43.000000	32.000000	2.000000	45.000000	2.000000	37.000000
43.000000	33.000000	2.000000	45.000000	3.000000	42.000000
43.000000	34.000000	2.000000	45.000000	4.000000	44.000000
43.000000	35.000000	2.000000	45.000000	5.000000	43.000000
43.000000	36.000000	3.000000	45.000000	6.000000	43.000000
43.000000	37.000000	1.000000	45.000000	7.000000	41.000000
43.000000	38.000000	1.000000	45.000000	8.000000	36.000000
43.000000	39.000000	0.000000E+00	45.000000	9.000000	37.000000
43.000000	40.000000	0.000000E+00	45.000000	10.000000	37.000000
43.000000	41.000000	0.000000E+00	45.000000	11.000000	37.000000
43.000000	42.000000	0.000000E+00	45.000000	12.000000	30.000000
43.000000	43.000000	1.000000	45.000000	13.000000	25.000000
43.000000	44.000000	3.000000	45.000000	14.000000	21.000000
43.000000	45.000000	5.000000	45.000000	15.000000	18.000000
43.000000	46.000000	7.000000	45.000000	16.000000	16.000000
43.000000	47.000000	11.000000	45.000000	17.000000	14.000000
43.000000	48.000000	15.000000	45.000000	18.000000	12.000000
43.000000	49.000000	22.000000	45.000000	19.000000	11.000000
43.000000	50.000000	29.000000	45.000000	20.000000	9.000000
44.000000	1.000000	44.000000	45.000000	21.000000	8.000000
44.000000	2.000000	46.000000	45.000000	22.000000	7.000000
44.000000	3.000000	44.000000	45.000000	23.000000	7.000000
44.000000	4.000000	42.000000	45.000000	24.000000	7.000000
44.000000	5.000000	38.000000	45.000000	25.000000	7.000000
44.000000	6.000000	38.000000	45.000000	26.000000	7.000000
44.000000	7.000000	35.000000	45.000000	27.000000	7.000000
44.000000	8.000000	33.000000	45.000000	28.000000	5.000000
44.000000	9.000000	35.000000	45.000000	29.000000	4.000000
44.000000	10.000000	37.000000	45.000000	30.000000	4.000000
44.000000	11.000000	33.000000	45.000000	31.000000	4.000000
44.000000	12.000000	29.000000	45.000000	32.000000	4.000000
44.000000	13.000000	25.000000	45.000000	33.000000	4.000000
44.000000	14.000000	21.000000	45.000000	34.000000	5.000000
44.000000	15.000000	17.000000	45.000000	35.000000	6.000000
44.000000	16.000000	14.000000	45.000000	36.000000	7.000000
44.000000	17.000000	13.000000	45.000000	37.000000	4.000000
44.000000	18.000000	12.000000	45.000000	38.000000	3.000000
44.000000	19.000000	10.000000	45.000000	39.000000	3.000000
44.000000	20.000000	9.000000	45.000000	40.000000	3.000000

Y	X	ELEVATION	Y	X	ELEVATION
45.000000	41.000000	3.000000	47.000000	11.000000	39.000000
45.000000	42.000000	3.000000	47.000000	12.000000	31.000000
45.000000	43.000000	4.000000	47.000000	13.000000	26.000000
45.000000	44.000000	6.000000	47.000000	14.000000	22.000000
45.000000	45.000000	7.000000	47.000000	15.000000	20.000000
45.000000	46.000000	7.000000	47.000000	16.000000	16.000000
45.000000	47.000000	12.000000	47.000000	17.000000	14.000000
45.000000	48.000000	16.000000	47.000000	18.000000	13.000000
45.000000	49.000000	22.000000	47.000000	19.000000	12.000000
45.000000	50.000000	29.000000	47.000000	20.000000	11.000000
46.000000	1.000000	29.000000	47.000000	21.000000	9.000000
46.000000	2.000000	29.000000	47.000000	22.000000	8.000000
46.000000	3.000000	32.000000	47.000000	23.000000	7.000000
46.000000	4.000000	36.000000	47.000000	24.000000	7.000000
46.000000	5.000000	40.000000	47.000000	25.000000	7.000000
46.000000	6.000000	44.000000	47.000000	26.000000	7.000000
46.000000	7.000000	44.000000	47.000000	27.000000	7.000000
46.000000	8.000000	44.000000	47.000000	28.000000	7.000000
46.000000	9.000000	40.000000	47.000000	29.000000	7.000000
46.000000	10.000000	42.000000	47.000000	30.000000	6.000000
46.000000	11.000000	39.000000	47.000000	31.000000	7.000000
46.000000	12.000000	33.000000	47.000000	32.000000	7.000000
46.000000	13.000000	27.000000	47.000000	33.000000	7.000000
46.000000	14.000000	22.000000	47.000000	34.000000	7.000000
46.000000	15.000000	19.000000	47.000000	35.000000	7.000000
46.000000	16.000000	16.000000	47.000000	36.000000	7.000000
46.000000	17.000000	14.000000	47.000000	37.000000	7.000000
46.000000	18.000000	13.000000	47.000000	38.000000	6.000000
46.000000	19.000000	11.000000	47.000000	39.000000	6.000000
46.000000	20.000000	10.000000	47.000000	40.000000	6.000000
46.000000	21.000000	9.000000	47.000000	41.000000	6.000000
46.000000	22.000000	7.000000	47.000000	42.000000	7.000000
46.000000	23.000000	7.000000	47.000000	43.000000	8.000000
46.000000	24.000000	7.000000	47.000000	44.000000	10.000000
46.000000	25.000000	7.000000	47.000000	45.000000	10.000000
46.000000	26.000000	7.000000	47.000000	46.000000	12.000000
46.000000	27.000000	7.000000	47.000000	47.000000	14.000000
46.000000	28.000000	6.000000	47.000000	48.000000	14.000000
46.000000	29.000000	5.000000	47.000000	49.000000	21.000000
46.000000	30.000000	5.000000	47.000000	50.000000	22.000000
46.000000	31.000000	6.000000	48.000000	1.000000	14.000000
46.000000	32.000000	6.000000	48.000000	2.000000	19.000000
46.000000	33.000000	6.000000	48.000000	3.000000	23.000000
46.000000	34.000000	7.000000	48.000000	4.000000	31.000000
46.000000	35.000000	7.000000	48.000000	5.000000	37.000000
46.000000	36.000000	7.000000	48.000000	6.000000	42.000000
46.000000	37.000000	5.000000	48.000000	7.000000	44.000000
46.000000	38.000000	5.000000	48.000000	8.000000	39.000000
46.000000	39.000000	4.000000	48.000000	9.000000	37.000000
46.000000	40.000000	4.000000	48.000000	10.000000	40.000000
46.000000	41.000000	4.000000	48.000000	11.000000	35.000000
46.000000	42.000000	5.000000	48.000000	12.000000	28.000000
46.000000	43.000000	6.000000	48.000000	13.000000	25.000000
46.000000	44.000000	7.000000	48.000000	14.000000	22.000000
46.000000	45.000000	7.000000	48.000000	15.000000	21.000000
46.000000	46.000000	9.000000	48.000000	16.000000	19.000000
46.000000	47.000000	11.000000	48.000000	17.000000	17.000000
46.000000	48.000000	15.000000	48.000000	18.000000	14.000000
46.000000	49.000000	19.000000	48.000000	19.000000	13.000000
46.000000	50.000000	22.000000	48.000000	20.000000	12.000000
47.000000	1.000000	22.000000	48.000000	21.000000	10.000000
47.000000	2.000000	29.000000	48.000000	22.000000	8.000000
47.000000	3.000000	29.000000	48.000000	23.000000	7.000000
47.000000	4.000000	31.000000	48.000000	24.000000	7.000000
47.000000	5.000000	36.000000	48.000000	25.000000	7.000000
47.000000	6.000000	44.000000	48.000000	26.000000	7.000000
47.000000	7.000000	44.000000	48.000000	27.000000	7.000000
47.000000	8.000000	42.000000	48.000000	28.000000	7.000000
47.000000	9.000000	40.000000	48.000000	29.000000	7.000000
47.000000	10.000000	44.000000	48.000000	30.000000	7.000000

Y	X	ELEVATION	Y	X	ELEVATION
48.000000	31.000000	8.000000	50.000000	1.000000	10.000000
48.000000	32.000000	8.000000	50.000000	2.000000	16.000000
48.000000	33.000000	9.000000	50.000000	3.000000	22.000000
48.000000	34.000000	9.000000	50.000000	4.000000	30.000000
48.000000	35.000000	8.000000	50.000000	5.000000	36.000000
48.000000	36.000000	8.000000	50.000000	6.000000	43.000000
48.000000	37.000000	8.000000	50.000000	7.000000	44.000000
48.000000	38.000000	7.000000	50.000000	8.000000	37.000000
48.000000	39.000000	7.000000	50.000000	9.000000	35.000000
48.000000	40.000000	8.000000	50.000000	10.000000	29.000000
48.000000	41.000000	8.000000	50.000000	11.000000	29.000000
48.000000	42.000000	8.000000	50.000000	12.000000	29.000000
48.000000	43.000000	11.000000	50.000000	13.000000	27.000000
48.000000	44.000000	13.000000	50.000000	14.000000	26.000000
48.000000	45.000000	13.000000	50.000000	15.000000	26.000000
48.000000	46.000000	14.000000	50.000000	16.000000	29.000000
48.000000	47.000000	17.000000	50.000000	17.000000	22.000000
48.000000	48.000000	16.000000	50.000000	18.000000	28.000000
48.000000	49.000000	20.000000	50.000000	19.000000	20.000000
48.000000	50.000000	26.000000	50.000000	20.000000	14.000000
49.000000	1.000000	12.000000	50.000000	21.000000	11.000000
49.000000	2.000000	20.000000	50.000000	22.000000	9.000000
49.000000	3.000000	26.000000	50.000000	23.000000	7.000000
49.000000	4.000000	31.000000	50.000000	24.000000	7.000000
49.000000	5.000000	40.000000	50.000000	25.000000	9.000000
49.000000	6.000000	44.000000	50.000000	26.000000	10.000000
49.000000	7.000000	42.000000	50.000000	27.000000	11.000000
49.000000	8.000000	37.000000	50.000000	28.000000	12.000000
49.000000	9.000000	36.000000	50.000000	29.000000	12.000000
49.000000	10.000000	33.000000	50.000000	30.000000	12.000000
49.000000	11.000000	29.000000	50.000000	31.000000	13.000000
49.000000	12.000000	28.000000	50.000000	32.000000	13.000000
49.000000	13.000000	25.000000	50.000000	33.000000	13.000000
49.000000	14.000000	23.000000	50.000000	34.000000	13.000000
49.000000	15.000000	22.000000	50.000000	35.000000	13.000000
49.000000	16.000000	22.000000	50.000000	36.000000	11.000000
49.000000	17.000000	22.000000	50.000000	37.000000	11.000000
49.000000	18.000000	19.000000	50.000000	38.000000	11.000000
49.000000	19.000000	14.000000	50.000000	39.000000	13.000000
49.000000	20.000000	13.000000	50.000000	40.000000	16.000000
49.000000	21.000000	10.000000	50.000000	41.000000	20.000000
49.000000	22.000000	8.000000	50.000000	42.000000	20.000000
49.000000	23.000000	7.000000	50.000000	43.000000	21.000000
49.000000	24.000000	7.000000	50.000000	44.000000	24.000000
49.000000	25.000000	7.000000	50.000000	45.000000	21.000000
49.000000	26.000000	8.000000	50.000000	46.000000	24.000000
49.000000	27.000000	8.000000	50.000000	47.000000	26.000000
49.000000	28.000000	8.000000	50.000000	48.000000	22.000000
49.000000	29.000000	8.000000	50.000000	49.000000	21.000000
49.000000	30.000000	7.000000	50.000000	50.000000	21.000000
49.000000	31.000000	10.000000			
49.000000	32.000000	11.000000			
49.000000	33.000000	11.000000			
49.000000	34.000000	11.000000			
49.000000	35.000000	11.000000			
49.000000	36.000000	10.000000			
49.000000	37.000000	9.000000			
49.000000	38.000000	10.000000			
49.000000	39.000000	10.000000			
49.000000	40.000000	12.000000			
49.000000	41.000000	13.000000			
49.000000	42.000000	13.000000			
49.000000	43.000000	14.000000			
49.000000	44.000000	17.000000			
49.000000	45.000000	17.000000			
49.000000	46.000000	18.000000			
49.000000	47.000000	22.000000			
49.000000	48.000000	20.000000			
49.000000	49.000000	20.000000			
49.000000	50.000000	21.000000			

CAVECREK.DAT: only y (runoff) values are shown; assume these data are regularly sampled in the x direction at a spacing of 1 (units). Y values are shown in order by column.

Col 1	Col 2	Col 3	Col 4
2	236	788	130
5	19	45	389
19	369	21	291
240	170	11	568
86	12	8	206
416	6	2	38
147	9	16	14
354	17	15	6
31	270	7	27
18	195	347	
7	112	276	
1	102	230	
0	24	449	
2	24	146	
4	5	31	
54	4	8	
22	2	5	
40	3	1	
139	36	2	
35	269	4	
8	219	2	
7	313	2	
6	291	48	
14	68	281	
2	19	79	
4	364	202	
30	138	332	
73	14	25	
463	30	14	
579	12	41	
59	52	11	
197	79	7	
55	204	119	
24	295	357	
28	532	97	
3	476	161	
4	414	466	
6	159	50	
13	48	476	
59	18	33	
637	4	14	
469	2	15	
192	6	7	
28	76	9	
32	346	38	
64	401	271	
38	508	135	
8	330	98	
7	79	425	
10	96	238	
172	30	199	
308	8	91	
325	7	29	
103	39	75	
392	141	16	
68	124	14	
24	150	22	
6	146	112	
5	548	278	
2	52	216	
3	25	73	
106	14	237	
432	29	74	
200	11	40	
221	3	27	
117	1	66	
235	4	17	
	3	7	
	87	25	
	173	91	

WALNUT GULCH data; only y (rainfall amounts) is shown; assume sampling is regular in the x direction with a sampling interval of 1 (units). Y values are shown in order by column.

Col 2	Col 3	Col 4
1.00	.05	.00
1.55	4.39	.04
.69	4.89	.89
.00	2.41	1.81
.38	.00	1.30
.44	.20	.48
.10	.14	2.27
.00	.00	1.07
.00	.09	.06
.00	.02	
.24	.14	
2.07	.16	
4.11	.19	
.83	1.75	
1.32	1.31	
.37	2.15	
.54	.37	
.74	.04	
.07	3.44	
.00	.55	
.00	.69	
.00	.76	
.08	.11	
2.78	.00	
.09	.00	
.93	2.13	
.36	5.64	
.58	.43	
.91	.00	
.17	.28	
.33	.35	
.01	.10	
.05	.60	
.00	.16	
.00	.00	
2.83	.20	
3.02	.00	
.85	4.26	
.62	3.78	
1.19	2.49	
.19	.00	
.32	.26	
.01	.51	
.32	.00	
.28	.28	
.00	1.16	
.00	.04	
3.03	.00	
1.56	.18	
2.47	1.48	
.42	4.06	
.96	.83	
.09	.16	
.28	.00	
.08	.42	
.18	.05	
.00	.20	
.00	.00	
.04	.45	
3.06	.00	
.77	.07	
.78	1.67	
.00	3.11	
.12	2.16	
2.91	1.66	
.48	.07	
.90	1.21	
.00	.00	
.31	.00	
.00	.00	

Col 1

Col 1
.62
.20
.11
.00
.10
.50
7.57
4.08
.17
.18
.00
.12
.61
.10
.00
.00
.00
.38
2.08
1.31
.00
.05
.05
.06
.87
.04
.97
.00
.04
.66
2.75
1.75
.00
.83
.00
.21
.00
.84
1.33
.62
.18
1.06
2.22
3.74
2.33
1.11
.29
.00
.00
.42
.00
.08
.00
.51
4.74
3.83
.41
1.80
.64
.96
1.01
.57
.22
.00
.00
.11
1.04

VARSTAR.DAT (Variable Star Data); only intensity, y, is given; assume
x increment is regular and equal to 1.0 (units). Y values
are arranged in order by column.

Col 1	Col 2	Col 3	Col 4	Col 5	Col 6	Col 7	Col 8	Col 9
18	9	2	13	5	17	13	15	21
20	8	3	12	3	17	10	16	19
22	7	6	11	1	16	7	16	17
23	7	8	11	1	15	4	16	15
23	6	11	10	2	14	3	16	13
22	7	13	9	3	13	1	16	11
19	8	15	9	5	12	0	15	9
17	8	18	8	7	11	1	14	7
15	9	21	8	9	11	2	14	6
13	10	22	8	11	10	3	13	4
10	10	23	9	14	10	4	13	2
8	11	24	9	16	9	5	13	0
6	12	23	9	18	9	7	12	0
4	12	21	10	21	10	8	12	1
3	13	20	10	23	10	10	12	3
2	13	19	11	24	10	11	11	5
0	13	17	11	24	10	14	11	8
0	13	16	11	23	10	16	11	10
0	13	13	11	21	10	19	10	13
1	14	10	11	20	10	21	10	15
2	14	8	12	19	10	24	10	18
5	14	6	12	17	10	24	10	20
8	14	2	12	14	10	23	10	22
10	14	0	13	11	11	20	9	23
13	14	0	13	9	11	18	9	24
14	14	0	13	7	11	16	9	24
16	14	1	14	5	12	14	9	23
18	15	2	14	3	12	11	9	21
20	14	3	14	1	13	9	9	19
22	14	5	15	0	13	7	9	17
23	14	6	15	0	14	5	10	16
23	13	8	15	1	14	3	10	13
22	13	11	15	2	15	2	10	12
21	12	13	15	4	15	1	11	9
19	11	15	15	5	16	1	12	7
17	9	16	14	7	16	2	13	5
14	8	18	14	10	17	3	14	3
12	7	21	13	12	17	5	15	2
9	6	23	13	15	17	8	16	2
6	6	23	12	17	16	11	16	2
5	7	22	11	19	15	13	17	3
3	7	20	10	21	15	15	18	3
2	8	17	8	22	14	16	18	4
2	8	15	6	22	13	18	17	
2	9	13	5	21	11	20	16	
3	9	10	5	20	9	21	15	
4	10	9	6	19	8	22	14	
5	10	7	6	17	7	22	13	
7	11	5	7	15	6	21	11	
9	13	3	8	13	5	19	10	
10	15	3	9	11	4	18	9	
12	16	4	12	9	4	17	7	
14	17	4	13	7	4	16	5	
16	18	5	14	5	5	15	4	
18	19	5	16	4	6	13	4	
19	20	6	17	3	8	12	3	
20	20	8	19	4	10	10	4	
21	19	9	20	4	12	8	5	
20	18	10	21	5	14	7	6	
19	16	12	21	6	15	6	8	
18	15	14	20	8	17	5	10	
17	14	16	19	9	19	5	11	
16	12	17	18	10	22	5	13	
15	10	18	16	12	22	6	15	
13	9	19	15	13	22	8	17	
12	8	19	12	15	20	10	19	
10	5	18	10	16	19	11	21	
	3	17	9	16	17	13	22	
	3	16	7	17	16	14	22	
	2	14		17	14		22	

SUNSPOT.DAT: only y values (sun spot numbers) are shown; these are recorded yearly, hence assume x sampling interval is regular and equal to 1 (year). Y values are shown in order by column; the first Y value corresponds to the year, 1750; the last to the year, 1980.

Col 1	Col 2	Col 3	Col 4
83.4	13.9	63.5	4.4
47.4	35.4	52.2	38.0
47.8	45.8	25.4	141.7
30.7	41.1	13.1	190.2
12.2	30.1	6.8	184.8
9.6	23.9	6.3	159.0
10.2	15.6	7.1	112.3
32.4	6.6	35.6	53.9
47.6	4.0	73.0	37.5
54.0	1.8	85.1	27.9
62.9	8.5	78.0	10.2
85.9	16.6	64.0	15.1
61.2	36.3	41.8	47.0
45.1	49.6	26.2	93.8
36.4	64.2	26.7	105.9
20.9	67.0	12.1	105.5
11.4	70.9	9.5	104.5
37.8	47.8	2.7	66.6
69.8	27.5	5.0	68.9
106.1	8.5	24.4	38.0
100.8	13.2	42.0	34.5
81.6	56.9	63.5	15.5
66.5	121.5	53.8	12.6
34.8	138.3	62.0	27.5
30.6	103.2	48.5	92.5
7.0	85.7	43.9	155.4
19.8	64.6	18.6	154.6
92.5	36.7	5.7	
154.4	24.2	3.6	
125.9	10.7	1.4	
84.8	15.0	9.6	
68.1	40.1	47.4	
38.5	61.5	57.1	
22.8	98.5	103.9	
10.2	124.7	80.6	
24.1	96.3	63.6	
82.9	66.6	37.6	
132.0	64.5	26.1	
130.9	54.1	14.2	
118.1	39.0	5.8	
89.9	20.6	16.7	
66.6	6.7	44.3	
60.0	4.3	63.9	
46.9	22.7	69.0	
41.0	54.8	77.8	
21.3	93.8	64.9	
16.0	95.8	35.7	
6.4	77.2	21.2	
4.1	59.1	11.1	
6.8	44.0	5.7	
14.5	47.0	8.7	
34.0	30.5	36.1	
45.0	16.3	79.7	
43.1	7.3	114.4	
47.5	37.6	109.6	
42.2	74.0	88.8	
28.1	139.0	67.8	
10.1	111.2	47.5	
8.1	101.6	30.6	
2.5	66.2	16.3	
0.0	44.7	9.6	
1.4	17.0	33.2	
5.0	111.3	92.6	
12.2	12.4	151.6	
	3.4	136.3	
	6.0	134.7	
	32.3	83.9	
	54.3	69.4	
	59.7	31.5	
	63.7	13.9	

SEISMIC1.DAT: only y (acceleration) values are shown; assume
x-increment is regular and equal to 1 (units).
Y values arranged in order by column.

Col 1	Col 2	Col 3	Col 4	Col 5	Col 6	Col 7	Col 8
25	47	213	-71	835	34	201	-406
8	64	207	12	870	232	-49	20
4	50	178	64	-126	394	-412	759
10	44	104	15	-1418	558	-378	1308
23	15	-117	-133	-1952	579	-330	1310
35	-68	-133	-291	-1974	431	-83	908
33	-185	-71	-364	-1520	315	318	174
7	-163	51	-3	-646	294	644	-496
1	-108	40	229	194	152	691	-759
1	-20	-11	-58	469	-129	523	-757
1	71	48	-109	305	-313	304	-539
-1	45	11	25	159	-338	306	-288
-18	-69	40	135	-12	-221	660	38
-36	19	78	219	419	-405	964	124
-16	159	-73	217	942	-361	1155	-135
4	114	-192	38	1467	81	1266	-317
20	68	-80	-227	1781	323	1062	54
31	57	107	-341	1058	435	1035	464
62	42	196	-275	377	1015	671	366
60	29	144	-317	128	848	57	89
60	-58	100	-230	237	117	-139	-36
54	-140	38	-204	589	108	-191	-150
24	-139	39	-239	390	584	-351	-153
-22	-32	58	-38	140	346	-711	-307
-69	-2	18	410	572	-280	-1082	-800
-93	64	-6	482	593	-764	-954	-1130
-70	101	43	367	206	-571	-38	-916
11	30	-72	378	-137	91	850	-387
55	-81	-127	278	-47	508	796	-226
60	-44	-84	175	-52	864	297	-210
34	104	-65	16	-282	1325	-148	-177
7	140	-113	-219	-472	1504	-457	144
-38	82	-93	-310	-681	1138	-469	805
-63	-17	-30	-264	-656	617	-508	1526
-25	-59	41	-321	-272	-446	-714	1916
-4	-28	157	-202	-377	-1355	-496	2344
29	-48	132	49	-449	-2080	119	1798
63	-82	-15	121	-230	-2433	425	542
85	-23	-102	-40	193	-1958	668	-156
3	-15	-209	81	793	-1335	476	-1053
-9	36	-143	221	936	-914	34	-1689
-39	64	71	41	470	-503	-234	-1717
-85	124	192	-21	218	-117	-486	-1855
-94	139	140	-119	255	115	-441	-1929
-44	-74	94	-551	-13	480	6	-1675
-51	-367	33	-636	62	928	389	-1379
-54	-347	-8	-448	25	1544	198	-951
-52	-191	-116	-25	-181	2323	-257	343
25	3	-62	906	56	2421	-288	1018
70	224	-121	1472	-155	1616	204	762
13	287	-253	1465	-812	348	486	457
-33	284	-133	1393	-1306	-1076	245	365
-32	262	-62	1248	-1191	-1862	-24	693
-34	128	-88	713	-1	-2091	-366	1060
-34	18	-91	-328	1600	-1839	-569	1036
-48	-85	-200	-1127	2058	-1193	-636	556
-41	-157	-210	-1556	1402	-471	-779	569
-24	-150	-44	-1886	589	209	-745	477
-60	-43	-40	-2235	-215	984	-568	-68
35	66	27	-2247	-573	1125	-391	-623
120	71	60	-1884	403	506	112	-717
93	48	63	-1161	-367	-77	147	-462
38	-17	42	-11	-465	-674	-227	-41
-37	-13	36	1554	-346	-1215	-265	409
-44	-63	139	2316	-614	-1331	482	359
3	-233	97	2146	-960	-908	910	364
56	-208	93	1513	-1335	-576	587	460
	-84	122	635	-1643	-494	223	370
	-27	89	295	-1296	-275	-298	64
	175	-58	543	-664	6	-596	-63

Col 9	Col 10	Col 11	Col 12	Col 13	Col 14	Col 15
-154	274	-301	449	-115	298	-306
-527	282	-112	399	21	299	-228
-498	22	-181	447	137	256	-93
-214	-44	-168	553	208	234	-72
-43	74	-93	747	233	221	-122
79	-43	18	978	167	240	-183
189	-57	164	1158	125	276	-214
239	-74	192	1206	81	262	-211
173	-291	190	1123	42	135	-201
65	-696	232	1001	-102	-128	-152
-293	-814	216	942	-315	-301	-84
-617	-583	-1	865	-369	-458	-85
-730	-199	-294	649	-380	-616	-85
-664	161	-483	333	-395	-716	-120
-564	292	-604	198	-403	-781	-243
-116	57	-624	157	-405	-667	-317
336	-1	-456	91	-305	-365	-253
398	-157	-332	4	-137	-120	-117
83	-402	-367	-85	9	78	-53
179	-376	-505	-80	30	194	9
430	-139	-556	43	26	280	58
458	-80	-271	115	-57	315	183
-1	-114	31	142	-98	364	306
-321	-187	151	172	-120	405	
-418	-277	267	128	-126	433	
-613	-410	244	45	-80	459	
-917	-531	139	50	-140	506	
-655	-593	-150	91	-217	502	
79	-770	-385	119	-160	444	
659	-943	-512	118	-101	390	
848	-911	-466	133	-20	298	
1072	-753	-465	210	45	212	
1371	-480	-461	326	40	206	
1544	-209	-417	368	-51	213	
1316	-195	-301	426	-174	201	
1058	-479	-192	617	-295	154	
700	-584	-57	772	-411	-47	
321	-522	23	814	-515	-145	
-63	-199	75	834	-585	-131	
-396	211	105	722	-621	-100	
-585	266	73	478	-543	8	
-356	62	-20	228	-384	103	
-45	-98	-103	157	-289	133	
427	-121	-146	31	-288	137	
361	-63	-102	-94	-373	115	
324	72	-55	-295	-454	108	
397	103	8	-457	-539	108	
452	-4	145	-509	-593	159	
428	-59	272	-363	-662	235	
296	-74	340	-241	-637	264	
321	-152	306	-254	-447	295	
66	-172	274	-277	-207	289	
-181	-173	301	-353	-9	230	
-93	-176	311	-398	26	190	
383	-176	339	-355	-93	210	
971	-322	297	-325	-307	265	
1092	-584	229	-296	-376	348	
1175	-588	2	-310	-266	358	
1237	-392	-188	-309	-115	312	
936	-258	-126	-271	68	264	
672	-408	-40	-164	170	163	
674	-768	33	-67	285	116	
760	-994	75	75	413	95	
884	-1039	130	255	364	7	
899	-1088	355	283	221	-85	
453	-1055	628	181	64	-229	
164	-910	668	87	-4	-286	
215	-782	610	-27	3	-337	
125	-722	617	-146	93	-321	
142	-622	575	-230	269	-312	

SEISMIC2.DAT: only y (acceleration) values are shown; assume
              x-increment is regular and equal to 1 (units).
              Y values arranged in order by column.

Col 1	Col 2	Col 3	Col 4	Col 5	Col 6	Col 7	Col 8
238	-302	-182	89	-62	-64	-85	35
211	-160	-47	23	48	-77	-91	-72
301	16	22	-55	134	-42	-53	-109
290	9	6	-29	100	16	28	-115
-12	-49	27	28	-48	75	94	-144
-362	-28	85	23	-134	120	80	-149
-493	-187	19	-21	-71	85	27	-128
-295	-468	-71	-36	50	48	9	-104
-4	-540	-96	-49	91	30	54	-76
-48	-514	-37	17	28	18	93	-47
-139	-427	-24	103	6	-9	84	28
-18	-181	-163	64	19	-37	64	79
41	-91	-291	-53	10	-48	37	48
37	-112	-257	-73	5	-18	-39	-5
-54	-11	-49	-90	57	65	-55	-12
-77	215	146	-218	87	74	6	17
-130	457	204	-240	25	65	20	37
-70	580	106	-144	-24	68	25	25
109	528	-13	-54	0	24	34	1
95	143	-41	75	-24	-67	36	5
82	-100	-30	183	-47	-67	0	14
242	-27	-44	187	47	-20	-27	-1
355	16	-67	64	55	-12	-24	5
338	-14	-138	35	-64	-19	35	24
154	-11	-235	-4	-34	-18	81	1
-175	71	-168	-122	91	3	103	-43
-403	223	-70	-215	111	48	110	-70
-489	440	45	-166	6	34	75	-78
-416	576	146	-19	-46	-53	39	-73
-76	486	146	119	-24	-81	-26	-57
60	254	123	196	16	-101	-123	-29
94	-106	97	214	14	-138	-150	-11
113	-255	54	205	-40	-76	-60	-25
245	-206	-2	99	-120	69	78	-43
386	-111	-78	-14	-153	108	130	-85
204	-157	-99	-68	-139	82	156	-118
-138	-359	-81	-68	-112	73	146	-108
-272	-474	-60	2	-78	61	120	-27
-319	-419	-16	62	12	39	116	7
-388	-310	69	118	86	-4	140	-12
-205	43	167	100	108	-66	97	24
97	238	135	1	16	-97	50	47
319	140	77	-40	-80	-158	3	68
678	95	52	-24	-100	-174	-68	62
934	81	15	-53	-90	-110	-92	13
925	-24	-10	-123	-57	-7	-60	9
599	-37	58	-113	-44	28	10	34
327	172	68	-92	-54	-5	48	74
219	331	-41	-99	-52	-39	78	103
51	328	-87	-16	-28	-16	114	97
-126	262	46	105	31	48	104	52
-279	123	116	158	79	86	21	-5
-304	-29	70	136	64	86	-10	-26
2	-110	49	48	31	10	15	-27
320	-128	2	-45	23	-34	11	-20
411	-171	-59	-13	21	-37	-60	-7
412	-103	-101	49	-7	-14	-113	-35
354	176	-190	24	-15	34	-122	-148
209	367	-160	-12	8	53	-41	-186
37	410	-7	-14	10	4	56	-73
-102	405	42	49	-9	-28	109	63
-473	273	-25	91	-37	-41	131	95
-812	-31	-126	92	-34	-6	108	28
-971	-201	-94	42	56	56	32	-2
-813	-105	36	7	118	74	-68	-27
-495	20	71	-12	42	39	-67	-75
-320	19	27	-85	-45	-13	-4	-125
	-91	31	-113	-26	-33	20	-100
	-199	82	-144	14	-30	40	-70
	-243	106	-160	-28	-57	63	-45

Col 9	Col 10	Col 11	Col 12	Col 13	Col 14	Col 15
-1	50	34	9	15	51	35
71	89	30	5	-11	18	41
142	73	18	-12	-44	-13	49
133	57	10	-42	-57	-30	46
79	100	-2	-47	-59	-43	37
-28	154	-24	-22	-56	-44	16
-97	143	-54	-4	-41	-20	-12
-86	94	-60	-1	-23	4	-25
-66	4	1	-6	-7	5	-21
-39	-116	63	-8	9	19	-4
3	-173	99	-14	14	37	20
66	-147	106	-32	5	23	51
105	-130	82	-34	-6	-6	59
99	-116	14	-24	-11	-1	40
7	-72	-69	-37	-8	16	23
-102	-13	-41	-50	2	21	14
-135	66	9	-44	24	32	10
-104	138	19	-39	55	49	13
-80	145	13	-52	59	43	28
-46	146	-16	-75	21	13	39
-15	132	-35	-45	-9	4	40
-6	88	7	35	-13	36	45
39	53	38	70	8	52	55
75	26	48	54	26	31	
50	-1	46	41	30	-15	
21	-30	49	37	26	-51	
40	-56	42	25	12	-58	
77	-69	30	9	6	-34	
15	-73	41	-6	-1	8	
-88	-103	27	-22	-26	35	
-143	-90	-15	-66	-52	27	
-131	-48	-32	-106	-59	7	
-23	-33	-35	-90	-28	-5	
126	-55	3	-54	32	-11	
191	-55	2	-56	53	-22	
148	-7	-15	-75	25	-39	
45	-2	1	-66	-8	-51	
-18	-14	31	-17	-16	-36	
-57	-27	53	19	-16	-12	
-34	-26	58	19	-6	-10	
-24	-7	58	9	21	-41	
-52	27	59	14	35	-79	
-42	64	63	19	24	-106	
35	77	74	-6	13	-100	
108	80	65	-67	25	-49	
127	66	36	-95	29	-9	
123	9	41	-83	1	-20	
122	-57	50	-52	4	-40	
131	-105	44	-21	27	-36	
115	-126	46	4	22	-23	
51	-124	39	35	13	1	
-28	-69	26	37	25	21	
-121	-29	9	6	32	25	
-194	-46	-1	-19	14	20	
-198	-60	-1	-27	-9	-9	
-127	3	4	-46	-20	-52	
-20	136	-3	-67	-19	-55	
41	211	-46	-71	1	-30	
46	177	-63	-36	23	-9	
44	84	-40	-9	24	-6	
27	43	-17	-7	12	4	
-26	41	-20	-9	15	8	
-45	23	-29	-11	39	-3	
-29	-14	-24	-10	54	9	
-23	-64	-3	-14	44	40	
-54	-113	29	-26	38	36	
-83	-58	69	-44	54	14	
-91	49	81	-12	72	-3	
-104	78	67	28	83	11	
-53	47	36	28	77	28	

SEISMIC3.DAT only y (acceleration) values are shown; assume x increment is regular and equal to 1.0 (units). Y values are arranged in order by column.

Col 1	Col 2	Col 3	Col 4
0	260	0	0
0	0	50	0
0	-150	40	15
0	100	0	5
10	260	-50	5
0	0	0	-5
0	-175	30	0
0	0	0	0
10	175	-30	5
0	0	20	10
10	-100	10	0
-10	80	10	-5
0	0	-10	0
0	-100	40	10
0	10	0	0
10	-10	-20	-5
-10	100	-10	-5
0	0	0	-5
0	-100	0	0
-10	0	30	-10
10	50	0	-5
0	-30	-10	0
0	60	40	10
5	-90	30	7
0	0	30	7
10	50	-15	0
-20	-30	0	5
20	0	-50	-5
-5	-20	-25	5
15	80	0	10
-2	0	-20	
10	-50	0	
-15	0	60	
0	30	20	
5	-20	0	
0	50	-10	
2	0	0	
-20	-70	0	
5	0	-50	
0	50	-25	
30	0	0	
20	-80	30	
0	0	50	
-10	120	30	
-10	0	0	
-100	-80	0	
-150	0	-10	
10	90	0	
150	0	0	
268	-90	20	
150	0	0	
0	50	-10	
-20	-30	-10	
-20	0	-20	
-50	-130	0	
0	-20	15	
10	0	30	
-120	100	15	
30	190	-30	
210	100	-50	
100	0	-30	
-100	-30	0	
-130	-95	0	
100	-90	15	
-100	-100	0	
-150	0	-15	
0	50	0	
	60	30	
	0	30	
	-80	0	

BROWN.DAT (brownian noise): only y values are shown; assume regular sampling
in the x direction at a spacing of 1 (units). Y values are shown
in order by column.

Col 1	Col 2	Col 3	Col 4
0.000000E+00	-3.923032E-03	-1.845082E-02	-1.456679E-02
1.301304E-03	-9.862649E-05	-2.319654E-02	-1.208190E-02
-2.028677E-03	-1.608113E-05	-2.155735E-02	-1.351789E-02
-7.516351E-04	1.730308E-03	-2.355491E-02	-1.174164E-02
1.979042E-03	-1.221162E-04	-2.126773E-02	-1.091615E-02
3.099980E-03	-2.772357E-03	-2.256932E-02	-1.103491E-02
4.818126E-04	-4.738064E-03	-2.514453E-02	-1.173493E-02
2.221277E-05	-2.021770E-03	-2.465866E-02	-1.004953E-02
-1.537204E-03	-6.049884E-04	-2.538723E-02	-1.241491E-02
-2.411068E-03	-3.030275E-03	-2.528605E-02	-1.393544E-02
-4.708492E-03	-4.769617E-03	-2.442699E-02	-1.627056E-02
-9.112715E-03	-7.314606E-03	-2.383362E-02	-1.498133E-02
-1.310688E-02	-8.120785E-03	-2.411104E-02	-1.200931E-02
-1.219735E-02	-1.023679E-02	-2.304187E-02	-9.619028E-03
-1.122111E-02	-6.232040E-03	-2.521294E-02	-1.029859E-02
-1.413746E-02	-8.988733E-03	-2.930518E-02	-9.489239E-03
-1.426600E-02	-1.080562E-02	-2.894016E-02	-9.249258E-03
-1.560065E-02	-1.026992E-02	-2.667741E-02	-9.776304E-03
-1.375246E-02	-9.928189E-03	-2.347597E-02	-9.714281E-03
-1.365455E-02	-1.360420E-02	-2.479110E-02	-6.780630E-03
-1.032220E-02	-1.511998E-02	-2.408650E-02	-8.764992E-03
-8.639626E-03	-1.726114E-02	-2.436055E-02	-8.268514E-03
-7.088254E-03	-1.856627E-02	-2.354888E-02	-5.957316E-03
-2.161114E-03	-1.848161E-02	-2.179933E-02	-8.379582E-03
-1.647037E-03	-2.286713E-02	-2.260440E-02	-8.087924E-03
2.605190E-03	-2.451802E-02	-1.881029E-02	-6.809943E-03
7.186642E-05	-2.188788E-02	-1.922456E-02	-5.675588E-03
-4.128597E-03	-2.292609E-02	-1.917456E-02	-9.006862E-03
-3.263127E-03	-2.135297E-02	-2.043449E-02	-8.488228E-03
-2.721325E-03	-2.093215E-02	-2.209965E-02	-9.129416E-03
-3.064618E-03	-2.437212E-02	-2.005049E-02	-7.364722E-03
-1.336150E-03	-2.652268E-02	-2.102426E-02	-9.005941E-03
-4.180605E-03	-2.597382E-02	-2.161746E-02	-7.396690E-03
-3.571960E-03	-2.390735E-02	-2.251263E-02	-6.823562E-03
-3.965040E-03	-2.382700E-02	-2.204356E-02	-8.478159E-03
-4.881975E-03	-2.283915E-02	-2.144825E-02	-6.763028E-03
-4.998179E-03	-2.367324E-02	-2.399369E-02	-1.106161E-02
-3.642043E-03	-2.657093E-02	-2.267022E-02	-1.319309E-02
-2.123334E-03	-2.799019E-02	-2.123942E-02	-8.784901E-03
-7.218836E-03	-2.481046E-02	-2.019827E-02	-1.097457E-02
-6.645920E-03	-2.441436E-02	-1.904371E-02	-8.297058E-03
-8.879788E-03	-2.264080E-02	-1.602733E-02	-7.661715E-03
-1.028754E-02	-2.359474E-02	-1.476046E-02	-8.771754E-03
-1.138575E-02	-2.592207E-02	-1.275814E-02	-6.420104E-03
-9.502468E-03	-2.742948E-02	-1.315964E-02	-4.361963E-03
-8.786577E-03	-2.773080E-02	-1.150090E-02	-3.891848E-03
-6.921098E-03	-2.607132E-02	-1.554226E-02	-3.599314E-03
-7.630855E-03	-2.407920E-02	-1.513648E-02	-4.639288E-03
-5.594240E-03	-2.437471E-02	-1.762913E-02	-4.357625E-03
-5.110017E-03	-2.466814E-02	-1.934552E-02	-6.750476E-03
-1.940723E-03	-2.638189E-02	-1.982864E-02	-5.892825E-03
-4.104814E-03	-2.835704E-02	-1.678291E-02	-5.145686E-03
-4.943149E-03	-2.655039E-02	-1.458178E-02	-7.268922E-03
-5.719834E-03	-2.425007E-02	-1.265614E-02	-7.627223E-03
-6.549878E-03	-2.445487E-02	-1.229166E-02	-8.867638E-03
-8.961871E-03	-2.454338E-02	-1.385078E-02	-1.008884E-02
-1.139421E-02	-2.212556E-02	-1.283901E-02	-1.361484E-02
-1.206537E-02	-2.227588E-02	-1.255316E-02	-1.164935E-02
-9.907226E-03	-2.470115E-02	-1.187003E-02	-1.195662E-02
-1.124220E-02	-2.440148E-02	-9.524214E-03	-1.195771E-02
-7.545363E-03	-2.247234E-02	-1.056775E-02	-1.270181E-02
-7.670116E-03	-2.168964E-02	-1.268004E-02	-1.458769E-02
-9.152333E-03	-2.170830E-02	-1.283803E-02	-1.382769E-02
-7.890402E-03	-1.788250E-02	-1.575661E-02	-1.252956E-02
-4.751983E-03	-1.757067E-02	-1.419603E-02	-1.144960E-02
	-1.896046E-02	-1.237393E-02	-1.092683E-02
	-1.875285E-02	-1.585163E-02	-8.121088E-03
	-1.937105E-02	-1.699019E-02	-6.028024E-03
	-2.256405E-02	-1.872079E-02	-3.716608E-03
	-2.003155E-02	-1.884798E-02	-2.198257E-03

Col 5	Col 6	Col 7	Col 8
1.898800E-03	5.035824E-03	-1.124343E-02	-3.199494E-02
-3.293361E-03	9.770486E-03	-8.138428E-03	-3.424527E-02
-3.409667E-03	8.861019E-03	-8.154464E-03	-3.118093E-02
-3.801445E-03	1.015709E-02	-9.100311E-03	-3.170424E-02
-3.276489E-03	1.336889E-02	-1.212496E-02	-3.001328E-02
-4.432378E-04	1.423217E-02	-1.053150E-02	-2.839643E-02
-4.607353E-03	1.117736E-02	-1.328286E-02	-2.747274E-02
-2.159097E-03	1.128837E-02	-1.128121E-02	-2.746360E-02
1.977827E-03	1.299509E-02	-1.261863E-02	-2.741818E-02
-5.329911E-04	1.102990E-02	-1.307044E-02	-2.965927E-02
-1.053435E-03	1.274853E-02	-1.432064E-02	-2.796829E-02
2.718557E-03	1.626023E-02	-1.791814E-02	-2.780904E-02
-2.512701E-05	1.637192E-02	-1.845879E-02	-2.948280E-02
-1.679401E-03	1.732427E-02	-1.995964E-02	-3.237467E-02
-2.646691E-03	1.966322E-02	-1.960885E-02	-2.946267E-02
-1.412945E-03	1.575609E-02	-2.120066E-02	-3.162767E-02
-2.994181E-03	1.649857E-02	-2.134092E-02	-3.209224E-02
-4.943427E-03	1.608991E-02	-2.094935E-02	-3.498922E-02
-2.995268E-03	1.467419E-02	-2.093749E-02	-3.558895E-02
1.401203E-03	1.682198E-02	-2.000128E-02	-3.134372E-02
2.513709E-03	1.748594E-02	-1.789198E-02	-3.102462E-02
4.669006E-03	1.834039E-02	-1.666026E-02	-2.776334E-02
7.419820E-03	2.077655E-02	-1.963812E-02	-2.591738E-02
7.831821E-03	2.160600E-02	-2.001821E-02	-2.426796E-02
9.052078E-03	1.811054E-02	-2.051074E-02	-2.539200E-02
8.885730E-03	1.875366E-02	-2.135593E-02	-2.577376E-02
1.063115E-02	2.229582E-02	-2.036181E-02	
1.094851E-02	2.040479E-02	-2.134815E-02	
1.060718E-02	1.829763E-02	-2.309526E-02	
1.092868E-02	1.875568E-02	-2.106837E-02	
1.047303E-02	1.617078E-02	-2.153659E-02	
1.230077E-02	1.333503E-02	-2.085959E-02	
1.219267E-02	1.305540E-02	-2.286549E-02	
1.287718E-02	1.274084E-02	-2.594936E-02	
1.379274E-02	1.001127E-02	-2.867975E-02	
1.287902E-02	8.546386E-03	-2.640998E-02	
1.081049E-02	1.004912E-02	-2.442520E-02	
8.767114E-03	9.862284E-03	-2.065626E-02	
5.955959E-03	1.005391E-02	-1.971469E-02	
7.450716E-03	1.276841E-02	-1.633793E-02	
5.158220E-03	1.126954E-02	-1.726745E-02	
7.145986E-03	1.408372E-02	-1.766361E-02	
5.868100E-03	1.282824E-02	-1.983094E-02	
4.969914E-03	9.285911E-03	-2.099735E-02	
7.633627E-03	8.871122E-03	-2.279199E-02	
6.691378E-03	9.590931E-03	-2.487652E-02	
6.025186E-03	8.566418E-03	-2.758007E-02	
1.372597E-03	6.287867E-03	-2.768771E-02	
7.414677E-04	7.528653E-03	-2.468118E-02	
1.276511E-03	6.528670E-03	-2.452746E-02	
5.985897E-04	7.067994E-03	-2.200538E-02	
3.557625E-03	7.284072E-03	-2.119285E-02	
7.419475E-03	9.977655E-03	-2.262941E-02	
6.250981E-03	9.238459E-03	-2.000788E-02	
7.106575E-03	6.217464E-03	-2.006477E-02	
8.484484E-03	2.724712E-03	-1.945749E-02	
9.631275E-03	2.924115E-03	-1.788953E-02	
8.981088E-03	5.841082E-04	-1.998375E-02	
8.662858E-03	-7.106699E-04	-2.188511E-02	
8.177643E-03	-1.527796E-03	-2.320756E-02	
1.079500E-02	1.520792E-04	-2.500004E-02	
1.031531E-02	-4.159751E-03	-2.653839E-02	
9.934975E-03	-3.526846E-03	-2.933487E-02	
8.390347E-03	-5.513001E-03	-2.932584E-02	
5.900468E-03	-5.944207E-03	-3.132925E-02	
7.670368E-03	-5.888996E-03	-3.252823E-02	
4.481207E-03	-7.169272E-03	-3.318581E-02	
3.378386E-03	-1.025779E-02	-3.028029E-02	
3.771228E-03	-9.712312E-03	-3.214114E-02	
3.738323E-03	-1.148955E-02	-3.241629E-02	

PROFILEA.DAT: only y (elevation) values are shown; assume x-increment
is regular and equal to 1.0 (units).

ELEVATION

4.477941E-02	.0364853	3.519118E-02	3.889706E-02
3.955882E-02	3.626471E-02	3.397059E-02	3.867647E-02
3.933824E-02	4.004412E-02	.05075	3.845588E-02
3.911765E-02	4.382354E-02	4.852942E-02	3.823529E-02
3.889706E-02	4.360295E-02	4.830883E-02	3.801471E-02
3.867647E-02	3.838236E-02	4.808824E-02	4.279412E-02
3.845588E-02	4.316177E-02	4.786765E-02	4.257353E-02
3.823529E-02	4.294118E-02	4.764706E-02	4.235295E-02
3.801471E-02	4.472059E-02	4.742647E-02	4.213236E-02
4.279412E-02	.0455	4.720589E-02	4.691177E-02
4.257353E-02	4.827942E-02	.0469853	4.669118E-02
4.235295E-02	5.205883E-02	4.676471E-02	4.947059E-02
4.213236E-02	4.683824E-02	4.654412E-02	.05125
4.691177E-02	4.761765E-02	4.232354E-02	5.102942E-02
4.669118E-02	4.339706E-02	4.210295E-02	4.580883E-02
4.947059E-02	4.117647E-02	4.588235E-02	4.558824E-02
.05125	4.095589E-02	4.566177E-02	4.536765E-02
5.102942E-02	.0407353	4.544118E-02	4.514706E-02
4.580883E-02	4.051471E-02	4.722059E-02	3.992647E-02
4.558824E-02	4.029412E-02	.045	3.970588E-02
4.536765E-02	4.007353E-02	4.477941E-02	.0364853
4.514706E-02	3.685294E-02	3.955882E-02	3.626471E-02
3.992647E-02	3.463236E-02	3.933824E-02	4.004412E-02
3.970588E-02	3.441177E-02	3.911765E-02	4.382354E-02

PROFILEB.DAT: only y (elevation) values are shown; assume a regular sampling
in the x direction at a spacing of 1 (units). Y values are
shown in order by column.

Col 1	Col 2	Col 3	Col 4
.55	.535	.085	.102
.555	.53	.08	.108
.57	.485	.062	.118
.575	.455	.03	.11
.55	.392	.03	.098
.52	.352	.023	.1
.51	.304	.035	.11
.515	.262	.022	.122
.52	.24	.01	.118
.52	.225	.008	.112
.54	.215	.02	.115
.54	.205	.022	.114
.542	.18	.03	.12
.56	.13	.048	.115
.552	.12	.07	
.545	.105	.102	

RAINFALL.DAT: only y (rainfall) amounts are shown; assume
x-increment is regular and equal to 1.0 (units)
Y values arranged in order by column.

Col 1	Col 2	Col 3	Col 4	Col 5	Col 6	Col 7	Col 8
2	1	5	6	8	1	1	1
1	1	6	4	9	5	3	1
1	1	7	2	8	6	1	1
1	2	6	4	5	4	3	3
1	3	9	3	14	7	1	2
1	2	7	4	11	4	2	3
2	2	4.5	2	8	5	1	2
3	2	4	1	4	1	1	1
4	4	4.5	1.5	9	4	1	2
4	2	4	2	15	1	1	1.5
3	3	4.5	2.5	44	1	3	2
3.5	2	4	3	49	2	4	2
3	2	4	3	29	3	3	
3.5	3	3	3	19	4	0	
3	2	4	5	26	1	0	
2.5	5	3	4	11	0	0	
2	4	4	8	17	1	0	
2.5	5	4	10	12	6	0	
3	4	3	5.5	7	5	0	
2	5	3.5	3.5	12	4	0	
2.5	7	3	4	19	3	0	
2	5	3	3.5	26	0	0	
3.5	6	3	5	15	1	0	
3	7	4	4	19	1.5	0	
3	9	3.5	3	18	1	0	
3	7	3	4	14	1.5	0	
3.5	5	4	2	10	1	0	
3	7	4	4	5	1	0	
3.5	8	4	3	4	1	1	
3	7	3	2	3	1	1	
3	5	2	3	2	3	1	
3.5	7	4	7	2	3	1	
3	8	5	12	2	1	1	
4	5	4	10	1	1	2	
5	7	3	7	3	1	3	
4	4	3	11	2	1	3	
6	6	4	7	3	1	5	
5	3	8	9	3	1	5	
5	4	6	5	2	2	4	
4	3.5	7.5	4	3	2	4	
4	3	7	4.5	2	1	2	
5	4	3	4	3	1	2	
4	5	5	13	4	2	0	
5	6	3.5	7	3	0	1	
4	7	7	9	4	1	2	
3	5	4	7	3	1	3	
2.5	8	6	8.5	2	1	5	
3	7	4	12	2	1	5	
3.5	6	6	11	4	1	5	
3	3.5	5	4	8	0	6	
6	4	8	7	3	3	4	
5	6	5	8	2	2	5	
6	9	4	15	1	8	5	
5	8	4	7	1	1	4	
4	7	3.5	8	1	2	3	
3	5	3	7	2.5	1	3	
3.5	4	2.5	8	2	1	2	
2.5	5	2	13	1	1	2	
3	5	2.5	10	1	1	1	
2	4	4	8	1	0	1	
1	5	3	9	2	2	1	
1	6	4	10	1	3	1	
1	4	6	9	1	4	2	
1	3	4	5	1	5	3	
2	3.5	4	3	4	4	4	
1	2.5	5	2	15	2	5	
	3	3	3	3	1	1	
	3	3	4	2	0	2	
	4	7	4		1		
		2	4				

Appendix B

This appendix provides information regarding the Joint Education Initiative (originally the JEDI program, now referred to as the JEI program) CD-ROM discs. Information is also provided regarding the source of the NASA (National Aeronautics and Space Administration) CD-ROM discs of the Magellan radar data for Venus.

Joint Education Initiative

JEI is an informal group of government agencies, companies, educators, and individuals formed in 1990 "to broaden educational horizons and promote the advancement of science." This group has now produced a series of four CD-ROMs containing data and software in use in academia and three United States government agencies: USGS (United States Geological Survey), NASA (National Aeronautics and Space Administration), and NOAA (National Oceanic and Atmospheric Administration). The goal was to provide children, teachers, and the general public with science data and software programs in use by scientists and researchers in the aforementioned government agencies.

The original set of three CD-ROMs (from which the images of Saturn and Io used in this textbook were obtained) are available for a fee. (As of September 1993, the fee was $31 plus $3.50 shipping and handling, but the fee is likely to change over time.) Contact the following address for the CD-ROMs and fee information:

Office of Technology Liaison
University of Maryland
4312 Knox Road
College Park, MD 20742 (USA)

A fourth CD-ROM, Introduction to Remote Sensing, is now available for a fee. (As of September 1993, the fee was $24.95, including shipping and handling, but the fee is likely to change over time.) Contact the following address for this CD-ROM and fee information:

Remote Sensing Disc
Joint Education Initiative
3433 A.V. Williams Bldg.
University of Maryland
College Park, MD 20742-3281 (USA)

Viking (Mars), Voyager (Outer Planets: Jupiter, Saturn, Uranus, and Neptune), and Magellan (Venus) Data

NASA has published data from these satellites on CD-ROMs. For information concerning the availability and acquisition of these CD-ROMs, contact

Goddard Space Flight Center
Request Coordination Office, Code 633
Greenbelt, MD 20771 (USA)
Phone: 301-286-6695

Index

Accelerograms (earthquake), 411–414, 573–577
Additive color (in computer graphics), 230–231
Airborne acquired digital imagery, 463–464
  thermal infrared multispectral scanner (TIMS) imagery,
    463–464
AIRSAR radar imagery, 463–464
Antiderivative, 3, 6–9, 10
Area
  approximate, 1, 2–3, 11, 13, 19–27, 188–191
  of a circle, 11
  by double integration, 11, 13
  exact, 2, 6, 157–159
  of a rectangle, 13
Artificial (cosine) data, 400–401
Atmospheric absorption of electromagnetic radiation,
    429–431
Australia, fractal dimension of coastline, 492–493, 498
Autocorrelation, 77–78, 79–81, 155, 160–184
  spatial, see Semivariogram

Bayes, Thomas (Reverend), 68
Bayes' theorem (rule), 68, 83–84, 458
Besicovitch, A., 495–496
Bilinear contrast adjustment of digital images, 441–444
Bit, 229–230, 427
  defined, 229
  used for representing numbers, 229–230, 427
Block kriging, 188–191
  Gauss quadrature used for, 188–191
  see also Kriging
Boundary condition
  in finite difference analysis, 357
  in finite element analysis, 299, 303–306, 312, 339–354
Byte, 229–230

Calculus, 1–16
  antiderivative of a function, 3, 6–9, 10
  applications, 9–14
  circle, area of, 11
  cone, volume of, 9, 12
  derivative of a function, 3–6
    partial derivative, concept of, 5–6
  double integral, 11, 13

Fundamental Theorem of Integral Calculus, 6–9
  integral, 6
  limit, 3
  numerical approximation of derivatives, 4–5, 354–359
  numerical approximation of integrals, 12, 14–15,
    19–27, 153, 188–191, 327–334, 405
  rectangular area and volume, 13–14
  triple integrals, 12, 14
  in the twentieth century, 14–15
Cave Creek, Kentucky, stream runoff data, 401–403, 569
Chaos, 490–491, 521–523
  and fractals, 522
  and rainfall data, 521–523
Chi-square statistic, 104–106
Circle, area, of 11
Class size (in semivariogram calculations), 160–179
  see also Lag
Classification of digital imagery, 457–461
  supervised
    maximum likelihood (Bayesian), 458–459
      MXLIKE computer program for, 484–488
    minimum distance to mean, 459–461
      MXLIKE computer program for, 484–488
  unsupervised, 458
Clouds, fractal simulation of, 516–518
Coastlines, fractal dimension of, 498, 511
Coefficient of skewness, 76
Color cube, 230–231
Color density slicing, 255–258
  COLORDEN computer program for 286–288
Color palettes, 229–232
  PALDSGN computer program for design of, 264–265
Computer graphics, 228–293
  additive colors in, 230–231
  color cube, 230–231
  color density slicing, 255–258
    COLORDEN computer program for, 286–288
  color palettes, 229–232
    PALDSGN computer program for design of,
      264–265
  contour mapping, 195, 201, 202, 204, 256–261
    CTOUR computer program for, 289–293
  hardware considerations, 229

Computer Graphics (*continued*)
  pel (picture element), see Pixel
  three-dimensional coordinate rotations, 248–250
  three-dimensional graphical images, 247–255
    coordinate rotations used in, 248–250
    forgetting transform for display, 246
    GRID3D computer program, 276–285
    hidden line removal, 250–255
    plotting three dimensional grids, 250–255
    shading algorithms, 250–255
    vanishing points for, 250
  two-dimensional coordinate rotation, 245–247
  two-dimensional graphical images, 237–245
    GRAPH2D computer program, 269–275
  vector generation, 235–237
    BOXDRAW computer program for demonstration, 266–268
  VGA graphics environment, 229–232
  viewing and windowing, 232–235
    BOXDRAW computer program for demonstration, 266–268
Concrete aggregate, fractal dimension of, 519–521
Conditional probability, 67–68
Continuous random variables, 68–70
Contour mapping, 195, 201, 202, 204, 256–261
  algorithm for, 258–259
  CTOUR computer program for, 289–293
Contrast adjustment of digital images, 437–448
  bilinear, 441–444
  CTRAST computer program for, 473–475
  histogram equalization, 444–445
  linear, 438–441
  user defined, 447–448
Coordinate rotation
  in finite element analysis, 310–311, 337
  for three dimensional graphics, 248–250
  for two dimensional graphics, 245–247
Correlation, 77
  and autocorrelation, 77–78, 155, 160–184
Correspondence analysis, 102–125
  applications of, 115–125
  CORSPOND program for, 138–145
  example hand calculations, 109–111
  factors in, 109
  graphical display of results, 239–240
  handling missing data in, 111–114, 125
Cosine function, 395
  ADD computer program for combinations, 426
  CSSINE computer program for, 420
Covariance, 77, 155, 160–184
  see also Autocorrelation; Correlation
Cross-validation used in kriging, 193
  graphical display of results from JCBLOK, 241–245
Cumulative distribution functions, 70–74
  Gaussian (normal), 70
  lognormal, 70
  Poisson, 71
Cumulative relative frequency, 72

Darcy, Henry, 354
Darcy's law, 354–355
Density function, 69
Derivative, 3–6
  defined, 4
  example, 4
  Fermat's notion of, 3–4
  finite difference approximation of, 4–5, 354–359
  partial derivative, 5–6
Digital image, 427–429
Digital image processing, see Image processing
Digital imagery, 427–429, 435–437, 461–463
  airborne acquired, 463–464
  spaceborne acquired, 431–433, 461–463
Discriminant analysis, 125–130
  applications, 128–130
  DISCRIM computer program for, 146–149
Discriminant function, 125
Displaying digital images, 435–437
  IMAGIN computer program for, 469–470
Distribution
  cumulative, 70–74
  of a population, 71–72
  and probability, 69–76
Double integral 11, 13

Earthquake data, 176–178, 205, 411–414, 546–547, 573–577
  Mercalli intensity, 1933 Long Beach and 1971 San Fernando, 176–178, 205, 546–547
  accelerograms, 1971 San Fernando earthquake, 411–414, 573–577
Eckart–Young theorem, 106–108
Eigendecomposition, 50–57, 96–98, 100–102, 106–111
  EIGEND computer program for, 61–63
  eigenvalue, defined, 50
  eigenvalue, calculation of, 50–57
  eigenvector, defined, 50
  eigenvector, calculation of, 52–57
  iterative solution for, 56
Electromagnetic spectrum, 429–431
Encke, J.F., 31
Equation solution
  Gauss elimination, 31–35
  Gauss–Seidel iteration, 359–360
  LU decomposition, 34
Error trapping (in finite element analysis), 314
  MESHDRAW computer program for, 371–377
Escher, M.C., 495
Estimation variance, 152–160, 186–187, 187–188
  see also Kriging variance
Euclid, 3, 489–490
Euler's relationship (in Fourier analysis), 404–405
Expected value, 69–70, 152–162, 501–509
  defined, 69
  and estimation variance in kriging, 152–160
  example calculation, 70
  and properties of brown noise, 501

and representation of semivariogram, 161–162

Exponential semivariogram model, 182

Factor analysis, 99–102
example, 100–102
R-mode, 100
Q-mode, 100–102
Fast Fourier transform (FFT), 414
Fatou, P., 492
Fermat, Pierre, 3
Filtering digital images, 448–457
spatial convolution filtering, 448–453
Fourier domain filtering, 453–457
Filtering in Fourier analysis, 407–408
FILTER computer program for, 425
Finite (divided) difference approximation, 4–5, 354–359
application to groundwater flow, 354–359
backward difference, 5
boundary conditions, in 357
centered difference, 5
comparison to finite element analysis, 360
of derivatives, 4–5
example calculations, 5
forward difference, 5
groundwater flow, Laplace's equation in, 355
Finite element analysis, 294–393
applications
fractured rock analysis, 335–338, 343–346, 349–352
stress beneath a spread footing, 339–343
truss systems, 315–319
tunnels, 346–352
bar (two dimensional) elements, 298–319
TRUSS computer program for, 364–370
boundary conditions in, 299, 303–306, 312, 339–354
comparison to finite difference analysis, 360
coordinate rotation for, 310–311, 337
element connectivity, 311–314, 338–352
error trapping in, 314
MESHDRAW computer program for, 371–377
Hooke's law in, 295–297, 299, 329
isoparametric elements used in, 319–352
Jacobian matrix in, 321, 323–328, 329–334
linear elastic analysis, 335
material constitutive matrix in, 328–329
plane-strain formulation, 329
plane-stress formulation, 328
mesh, 298
nodal point, 298
coordinates, use of, 298, 303, 315–318, 324, 330–331
nonlinear elastic analysis in, 335–338, 343–346, 349–352
Poisson's ratio in, 297–298, 328–329
quadrilateral elements, 319–352
Gauss quadrature for stiffness calculation, 327–334
QUAD computer program for, 378–391
Rayleigh–Ritz derivation of stiffness matrices
bar elements, 306–311

quadrilateral elements, 321–334
triangular elements, 361
stiffness matrices in
bar elements, 300–303, 306–311
example calculations, 303–306, 329–334
quadrilateral elements, 321–334
system stiffness matrix, 303–306
triangular elements, 361
strain–displacement matrix, 307–311, 321–334
strain and stress, 295–297, 314, 318, 328–329, 334
system stiffness matrix, 303–306
truss systems (see bar elements), 298–319
TRUSS computer program for, 364–370
Young's modulus in, 296–297, 328–329
Forgetting transform (for three-dimensional graphics), 246
Fourier, Jean Baptiste, 394
Fourier analysis, 394–426, 453–457, 500–514
cosine function for use in, 395
ADD computer program for combinations, 426
CSSINE computer program for, 420
data applied to
artificial (cosine) data, 400–401
Cave Creek, Kentucky data, 401–403, 569
seismic data, 411–414, 546–547, 573–577
sunspot data, 408–411, 572
variable star data, 403–404, 571
of digital images, 453–457
two-dimensional Fourier transform, 453–455
FORIER2D computer program for, 479
two-dimensional inverse Fourier transform, 455–457
FORINV2 computer program for, 482
filtering, 407–408
FILTER computer program, for 425
Fourier transform, fast, 414
Fourier transform, slow, 404–407
FORIER computer program for, 422–423
inverse Fourier transform, 407, 455–457
FORINV computer program for, 424
FORINV2 computer program for, 482
periodogram, 395–404
PERIOD computer program, for 421
and similarity to power spectrum, 408
power spectrum, 408
FORIER computer program for, 422–423
and fractal dimension, 500–514
and similarity to periodogram, 408
Fourier transform, 404–407, 414, 453–457
Fractals, 6, 489–538
and chaos, 490–491
clouds as fractals, 516–518
and coastline of Australia, 492–493, 498
and coastline of Great Britain, 492–493, 498
and coastline of South Africa, 492–493, 498
and coastline of Vancouver Island, 493–494, 498
coastlines as fractals, 498
definitions for, 498–499
dimension, 498–514
self-affine notion, 500–514
self-similar notion, 498–500

Fractals (*continued*)
divider technique applied to self-affine fractals, 511–514
divider technique used for analysis, 492–493, 498, 511–514
Escher, M.C., sketches and, 495
Euclid and, 489–490
Fatou and, 492
Fricke and, 491
Hausdorff and Besicovitch and, 495
Hausdorff dimension, 495
Julia and, 492
Klein and, 491
Mandelbrot and, 496–498
Poincaré and, 490–491
power spectrum used for analysis, 500–514
random fractals, 500–514
Reymond and, 491
Richardson and, 492–495
self-affine notion of, 500–514
self-similar notion of, 498–500
semivariogram used for analysis, 500–514
simulation of
one-dimensional profiles, 503–507
SPFM1D/CFRINV computer programs used for, 531–533
two-dimensional surfaces and clouds, 514–518
SPFM2D computer program used for, 535–536
spectral notion for, 500–514
topography as fractals, 514–516
wavelets and fractals, 523–527
Weierstrass and, 491
Fractal dimension
of coastlines, 492–493, 498
of concrete aggregate, 519–521
divider calculation method, 492–493, 498, 511–514
power spectrum calculation method, 500–514
of rainfall data, 521–523
of rock surfaces, 518–519
and self-similar fractals, 498–500, 511–514
and self-affine fractals, 500–514
semivariogram calculation method, 500–514
wavelet transform calculation method, 523–527
Fractured rock masses, 335–338, 343–346, 349–352, 518–519
Frequency, 72
cumulative, relative, 70–74
relative, 72
Fricke, Robert (and fractals), 491

Gauss, Carl Friedrich, 17–43
early years, 17–18
and linear, least squares regression, 27–30
LINEAR computer program for, 38
and normal (Gaussian) distribution, 19
and numerical approximation of integrals, 19–27
and solving simultaneous equations, 31–35
GAUSS computer program for, 39–40
GAUSSIT computer program for, 41–43

and symmetry of arithmetic progressions, 18
and theory of errors, 18
Gauss elimination, 31–35
example, 33
GAUSS computer program for, 39–40
iterative improvement for, 34–35
GAUSSIT computer program, 41–43
precision in, 34–35
Gauss quadrature, 19–27
general form for, 24
table of weights and locations, 26
three point approximation example, 22–24
two point approximation example, 20–22
used for block kriging, 188–191
used in finite element analysis, 327–334
Gauss–Seidel iterative equation solution, 359–360
Gaussian (normal) distribution, 19, 70
and geostatistics, 150–204
Gaussian semivariogram model, 182
Geostatistics, 150–204
kriging, concept of, 152–204
semivariogram, concept of, 161–204
variogram, concept of, 152, 160–161
Graphics, *see* Computer graphics
Great Britain, fractal dimension of coastline, 492–493, 498
Gridding (and kriging), 193–204

Hadamard, Jacques Solomon, 6
Hausdorff, F., 495
Hidden line removal (for three dimensional graphics), 250–255
High boost filtering of digital images, 465–466
CONVOL computer program for, 476–477
weights used to implement, 466
High pass filtering of digital images, 451–453, 456–457
directional high pass filtering, 452–453
Fourier transform approach, 456–457
spatial convolution approach, 451–453
CONVOL computer program for, 476–477
weights used to implement, 451–453
Histogram, 69, 71–72, 437–445
defined, 71–72
of digital images, 437–445
IMSTAT computer program for, 471–472
example, 72, 81–83
HISTO computer program for, 88–90
role in geostatistics, 150–204
Histogram equalization of digital images, 444–445
Hooke, Robert, 295
Hooke's Law, 295–297, 299, 329

Image, *see* Digital image
Image processing, 434–488
classification, 457–461
supervised
maximum likelihood (Bayesian), 458–459
MXLIKE computer program for, 484–488

minimum distance to mean, 459–461
  MXLIKE computer program for, 484–488
  unsupervised, 458
contrast enhancement, 437–448
  bilinear, 441–444
  CTRAST computer program for, 473–475
  histogram equalization, 444–445
  linear, 438–441
  user-defined, 447–448
displaying images using VGA graphics domain,
    435–437
  IMAGIN computer program for, 469–470
filtering, 448–457
  Fourier domain filtering, 453–457
    FORFILT computer program for, 481
  spatial convolution filtering, 448–453
    CONVOL computer program for, 476–477
look-up tables used in, 445–447
statistical analysis, 437–445
  IMSTAT computer program for, 471–472
Indicator function, 76, 179–180, 192–193, 201–204
Indicator transform, 76, 179–180, 192–193, 201–204
Intrinsic hypothesis, 154
  and kriging, 154–161
  and semivariogram, 161–162
Inverse Fourier transform, 407, 455
  FORINV computer program for, 424
  FORINV2 computer program for, 482
Inverse of a matrix, 48, 49, 58
Isoparametric finite elements, 319–352

Jacobian matrix (in finite element analysis), 321,
    323–328, 329–334
Joint probability density function, 77
Julia, G., 492
Julia sets, 492

Klein, Felix, 491
Kriging, 6, 152–204
  block kriging, 188–191
  constraint for unbiased estimation, 157–161
  data transforms
    for indicator transform, 179–180, 192–193, 201–204
    for lognormal transform, 179–180, 192–193,
        194–201
  derivation of equation for estimation weights, 154–161
  displaying cross-validation results from JCBLOK,
    241–245
  displaying results
    as three-dimensional grid, 254–255
    using color density slicing, 255–258
    using contour mapping, 195, 201, 202, 204, 258–260
  error variance in, 152–160, 186–188
  and estimation weights, 154–161
  exact interpolation characteristics, 187
  example kriging calculation, 184–187
  example JCBLOK calculations
    NVSIM.DAT data set, 193–194
    NVSIMLOG.DAT data set, 194–201
  expected value representation for error variance,
    152–160
  Gauss quadrature used in block kriging, 188–191
  and Gauss' theory of errors, 152–161
  and intrinsic hypothesis, 154–161
  integral definition for kriging, 153
  JCBLOK computer program, 212–227
  Lagrangian function for unbiased constraint, 157–161
  numerical approximation of integral for kriging, 153
  punctual kriging, 152–188
  sample support in, 188
  semivariogram used in, 161–204
Kriging variance, 152–160, 186–188

Lag, 78
  and autocorrelation, 78
  concept of, 78
  and variogram, semivariogram, 160–179
Lagrangian function (and kriging), 157–161
Landsat satellite, 431–432
Laplace, Pierre Simon, Marquis de, 19, 28, 355
Laplace's equation, 355
  finite difference approximation of, 354–359
Legendre, Adrien Marie, 18, 19
Leibniz, Gottfried Wilhelm, 6
Limit, definition of, 3
Linear, elastic finite element analysis, 294–334
Linear, least squares regression, 18–19, 27–30, 47–50
  example, 29
  Gauss and, 27–30
  Laplace and, 28
  Legendre and, 18
  LINEAR computer program for, 38
  matrix solution for, 47–50
Linear contrast enhancement of digital images, 438–441
Linear semivariogram model, 183
Lognormal distribution, 70, 179, 194–201
Long Beach, California, Mercalli intensity data for 1933
    earthquake, 176–179, 546
Look-up table (for image processing), 445–447
Lorenz, Edward, 490
Low-pass filtering of digital images, 449–451, 456–457
  Fourier transform approach, 456–457
  spatial convolution approach, 449–451
    CONVOL computer program for, 476–477
  weights used to implement, 449
LU decomposition, 34–35

Magellan satellite, 432–433
  radar imagery of Venus from, 435–437, 461–463
Mandelbrot, Benoit B., 6, 496–498
Material constitutive matrix in finite element analysis,
    328–329
Matheron, Georges, 6, 151–152
Matrix, defined, 45
Matrix algebra, 44–63
  determinants of 3 × 3 matrices, 58
  determinants of 2 × 2 matrices, 49

Matrix algebra (*continued*)
  eigendecomposition
    EIGEND computer program for, 61–63
    iterative solution for, 56
    of 3 × 3 matrices, 54–55
    of 2 × 2 matrices, 50–54
  eigenvalues, 50–57
  eigenvectors, 50–57
  identity matrix, 48
  inverse, calculation of, 48, 49, 58
  multiplication rule for, 45
    MULMAT computer program, 60
  trace, 55
  transpose, 48
Maximum likelihood classification of digital images,
    457–461
Mean
  defined, 65
  examples, 65, 78
Median
  defined, 66
  examples, 66, 78
Mesh (used in finite element analysis), 298
Minimum distance to mean classification of digital
    images, 459–461
Multivariate data analysis, 91–149
  correspondence analysis, 102–125
  discriminant analysis, 125–130
  factor analysis, 99–102
  principal components analysis (PCA), 96–99

Negative semidefiniteness and semivariogram models,
    183
Nested semivariogram models, 183–184
Newton, Sir Isaac, 3
Nodal point (in finite element analysis), 298
  coordinates, use of, 298, 303, 315–318, 324, 330–331
Nonlinear, elastic finite element analysis, 335–338,
    343–346, 349–352
Normal (Gaussian) distribution, 19, 70, 150–204
  lognormal distribution, 70, 179, 194–201
Nugget value (of a semivariogram)
  defined, 167
  illustrated, 162, 184
  modeled, 180–183, 193–204
  used in example calculation, 184–187

Palette, *see* Color palettes
Partial derivative, 5–6
Pel (picture element), *see* Pixel
Periodogram, 395–404
  example calculations, 400–404
  PERIOD computer program for, 421
  and similarity to power spectrum, 408
Pixel, 229, 427–429
Poincaré, Jules Henri, 490–491
Poisson, S.D., 297
Poisson distribution, 71
Poisson's ratio, 297–298, 328–329

Power spectrum, 408
  and calculation of fractal dimension, 500–514
  FORIER computer program for, 422–423
  and similarity to periodogram, 408
Precision in Gauss elimination, 34–35
Principal components analysis (PCA), 96–99
  example calculations, 96–99
  graphical representation, 98–99
Probability, 67–74
  Bayes' theorem, 68, 458
  conditional probability, 67–68
  defined, 67
  density functions, 69
  and expected value, 69
  Gaussian (normal) distribution, 70
  lognormal distribution, 70
  normal (Gaussian) distribution, 70
  Poisson distribution, 71
Punctual kriging, *see* Kriging

Quadrature, *see* Gauss quadrature
Quadrilateral finite element, 319–352
  Gauss quadrature used for, 327–334
  isoparametric formulation for, 319–352
  QUAD computer program for, 378–391
  Rayleigh–Ritz derivation of stiffness, 321–334

Radar imagery, 435–437, 461–464
  AIRSAR airborne acquired imagery, 463–464
  Magellan spaceborne acquired imagery of Venus,
    435–437, 461–463
Radiometric resolution in digital imagery, 433–434
Rainfall data, fractal dimension of, 521–523
Random fractals, 500–514
Random variable, *see* Continuous random variables
Range (of a semivariogram)
  defined, 166
  illustrated, 162, 184
  modeled, 180–183, 193–204
  used in example calculation, 184–187
Rayleigh–Ritz derivation of finite element stiffness
  assumed displacement fields used in, 308–311
  for quadrilateral elements, 321–334
  for triangular elements, 361
  for truss (bar) elements, 306–311
Rectangular area and volume, 13–14
Regression, *see* Linear, least squares regression
Relative frequency, 72
Remote sensing (concept of), 429
Reymond, P. Dubois, 491
Richardson, Lewis F., 492–495
Rock surface roughness, *see* Surface roughness
Roots of algebraic equations, 51, 96, 100, 106
  and relationship to eigendecomposition, 50–57

Sample support, 188
  block, 188
  and geostatistics, 188
  punctual, 188

San Fernando, California
  accelerograms recorded during 1971 earthquake, 411–414, 573–577
  Mercalli intensity data for 1971 earthquake, 205, 547
Satellite technology, 431–433
  Landsat satellite, 431–432, 436
  Magellan satellite, 432–433, 461–463
  Voyager satellite program, 436
Seismic data, *see* Earthquake data
Self-affinity and fractals, 500–514
Self-similarity and fractals, 498–500
Semivariogram, 161–204
  calculation of, 162–167
  class size for calculation of, 160–179
  data transform
    for indicator, 179–180
    for lognormal, 179
  for determining fractal dimension, 509
  directional calculation of, 169–172
  example calculations, 162–167, 168–179, 193–204
  expected value representation of, 161–162, 509
  FGAM computer program for, 207–211
  graphical display of FGAM results, 240–241
  and intrinsic hypothesis, 161–162
  models for
    exponential, 182
    Gaussian, 182
    linear, 183
    spherical, 181–182
  and negative semidefinite requirements for, 183
  nested models for, 183–184
  nugget value, 162, 167, 180–184, 193–204
  omnidirectional calculation of, 172–174
  range, 162, 166, 180–184, 193–204
  sill, 162, 166, 180–184, 193–204
Shading algorithm for three-dimensional graphics, 250–255
Sill value (of a semivariogram)
  defined, 166
  illustrated, 162, 184
  modeled, 180–183
  used in example calculation, 184–187
Simulation of fractals
  clouds, 516–518
  profiles, 503–507
  topography, 514–516
Sinusoidal waveform (and Fourier analysis), 395, 415–416
Skewness, 74–76
  coefficient, of 76
  defined, 76
South Africa, fractal dimension of coastline, 492–493, 498
Spaceborne acquired digital imagery
  Landsat satellite data, 431–432, 436
  Magellan satellite data, 432–433, 461–463
  Voyager satellite data, 436
Spatial analysis
  autocovariance, 155, 160–184

semivariogram, 161–204
variogram, 152, 160–161
see also Kriging
Spatial convolution filtering of digital images
  CONVOL computer program for, 476–477
  high boost filtering, 465–466
  high pass filtering, 451–453
  low pass filtering, 449–451
Spatial resolution of digital images, 433
Spectral resolution of digital images, 433
Spherical semivariogram model, 181–182
Standard deviation, 67
Statistics, 64–90
  autocorrelation, 77–78, 79–81
  continuous random variables, 68–70
  correlation, 77
  covariance, 77
  histogram, 69, 71–72
  mean, 65, 78
  median, 66, 78
  nonparametric statistics, 76
  and probability, 67–74
  standard deviation, 67
  variance, 66, 78–79
Stillwater Lakes, Nevada, 92, 115–125
  correspondence analysis of, 115–125
  and environmental problem, 92
  multivariate data for, 94–95
Sunspot data, 408–411, 572
Supervised classification of digital images, 458–461
Surface roughness, fractal dimension of, 518–519
System stiffness matrix (in finite element analysis), 303–306

Temperature data, 80
Temporal resolution of digital imagery, 433
Thermal infrared airborne acquired imagery (TIMS imagery), 463–464
Three-dimensional computer graphics, 247–255
  coordinate rotation in, 248–250
  forgetting transform and, 246
  GRID3D computer program for, 276–285
  hidden line removal, 250–255
  plotting three dimensional grids, 250–255
  shading algorithm, 250–255
Three-dimensional coordinate rotations, 248–250
TIMS imagery, 463–464
Topographic data (of Caliente, Nevada), 551–568
Topography and fractals, 514–516
Trace of a matrix, 55
Triple integral, 12, 14
Truss systems, 298–319
  TRUSS computer program for, 364–370
Two-dimensional coordinate rotations, 245–247, 310–311, 337
Two-dimensional drawing of graphs, 237–245
  applications
    displaying correspondence analysis results, 239–240
    displaying semivariogram results, 240–241

Two-dimensional drawing of graphs (*continued*)
    displaying cross-validation results from JCBLOK,
        241–245
    drawing simple graphs, 239
    GRAPH2D computer program for, 269–275

Unsupervised classification of digital images, 458

Vancouver Island, British Columbia, Canada, fractal
    dimension of coastline, 493–494, 498
Vanishing point (used in three dimensional graphics), 250
Variable star data, 403–404, 571
Variables
  continuous random, 68–70
  regionalized, 150–204
    defined, 152
    spatial, 150–204
Variance, 66
  example calculations, 66, 78–79
  methods for calculation, 66
Variogram, 152
  see also Semivariogram
Vector generation (for computer graphics), 235–237

BOXDRAW computer program for demonstration,
    266–268
Venus, Magellan satellite radar data of, 435–437,
    461–463
VGA graphics environment, 229–232, 435–437
  color palettes, 229–232
    PALDSGN computer program for, 264–265
  resolution, 229–232
Viewing and windowing (in computer graphics), 232–235
  BOXDRAW computer program for demonstration,
    266–268
Volume
  of cone, 9, 12
  of rectangular solid, 14
  by triple integration, 12, 14

Wavelet transforms, 523–527
  for calculating fractal dimension, 525–527
  WAVELET computer program for, 538
Wavelets, 523–527
Weierstrass, K., 491

Young, Thomas, 296
Young's modulus, 296–297, 328–329